Low-Temperature Physics

Christian Enss   Siegfried Hunklinger

Low-Temperature Physics

With 421 Figures

Prof. Dr. Christian Enss
Prof. Dr. Siegfried Hunklinger
Universität Heidelberg
Kirchhoff-Institut für Physik
Im Neuenheimer Feld 227
69120 Heidelberg, Germany

christian.enss@physik.uni-heidelberg.de
siegfried.hunklinger@physik.uni-heidelberg.de

ISBN-13 978-3-642-06216-2 e-ISBN-13 978-3-540-26619-8

Springer is a part of Springer Science+Business Media
springeronline.com

Printed in Germany

Cover design: Erich Kirchner

Preface

Science is often a journey to the limits of the feasible and ascertainable. In low-temperature physics this journey strives towards absolute zero. When *Louis Cailletet* on December 2nd, 1877, realized a major step in terms of the production of low temperatures, namely the first liquefaction of oxygen, he could hardly imagine the wealth of exciting physical phenomena that would be discovered in this field. Despite the anticipation from everyday experience, which generally equates cold with discomfort and stiffening, condensed matter at low temperatures reveals a wide array of fascinating properties. As the most prominent examples let us mention superfluidity and superconductivity, whose attraction is undiminished since their discovery. With every step towards lower temperatures numerous new insights have resulted, which make the traditional subject of low-temperature physics an attractive and modern research topic.

The present book is based on material from lectures that both authors have given several times at the universities of Heidelberg, Bayreuth and Konstanz. It is focused on the discussion of physical phenomena that become most apparent at low temperatures. The book is mainly aimed at students, and provides a compact and comprehensible introduction to various topics of low-temperature physics. Selection and emphasis of the material is subjective and certainly reflects our personal preferences. However, we have tried to give room for as wide a spectrum of topics as possible. The contents are organized in three parts, entitled quantum fluids, solids at low temperatures and principles of refrigeration and thermometry. Quantum fluids, with their diverse and exotic properties, are discussed in the first five chapters of the book. Here, many aspects of the extraordinary liquid, superfluid ^{3}He, could only be touched upon since a thorough discussion of this topic is beyond the scope of the book. Chapters six to ten cover aspects of solids at low temperatures. Naturally, superconductivity has been given the largest space here. Atomic tunneling systems, a topic of our own research, has also been discussed in some detail, since these degrees of freedom considerably influence the properties of many solids at low temperatures. The last two chapters of the book are devoted to the common physical principles and methods of low-temperature production and thermometry. In this section, we have intentionally omitted many technical details – which are admittedly often important for everyday work in the laboratory, but have little to do with the understanding of the underlying physics.

The citations in the text are intended to provide references for the reader to selected important articles and reviews, and to some historically interesting articles. For further studies, problems related to the material discussed are given at the end of each chapter. In addition, some historic anecdotes have been included in the text to introduce some variety.

This book would never have appeared without the help of many colleagues and coworkers. S. Bandler and G. Seidel have read major parts of the manuscript and have given invaluable advice. We are deeply thankful for the enormous time they have committed to identify errors and shortcomings and to make this book much more readable. In addition, selected topics have been read by D. Einzel, A. Fleischmann, R. Kühn, H.v. Löhneysen, D. Vollhardt, M.v. Schickfus, G. Thummes, and V. Mitrovic. Their suggestions have certainly improved the quality of the book and we are most grateful for their help. Special thanks go to R. Weis who skillfully produced all the figures.

Heidelberg, *C. Enss*
February 2005 *S. Hunklinger*

Contents

Part I Quantum Fluids

Part II Solids at Low Temperatures

Part III Principles of Refrigeration and Thermometry

Part I

Quantum Fluids

1 Helium – General Properties

Evidence for the existence of the rare noble gas helium was first obtained by the French astronomer *Janssen* in the visible spectrum of solar protuberances during a total eclipse in 1868 in India [1]. Using a spectrometer, he noticed a hitherto unknown yellow line. Shortly afterwards, his observation was confirmed by *Lockyer*, an English astronomer [2]. He related the new spectral line to a chemical element that had not been seen on Earth before and suggested the name helium for it after helios, the Greek name for sun. On Earth, the helium was discovered in 1895 by *Ramsay* [3] and independently by *Cleve* and *Langlet* [4] in celveite, a rocksand mineral that contains uranium and thorium. Later in the same year, helium was also found in gas evolving from a spring in Bad Wildbad in the Black Forest in Germany by *Kayser* [5].

Only a few years after the discovery of helium on Earth, a race took place between several competing low-temperature laboratories with the aim of being the first to liquefy this last of the so-called permanent gases. One major difficulty was to obtain sufficient quantities of helium gas. With the help of his brother, who was the director of the Department of Foreign Trade Relations in Amsterdam, *Kamerlingh Onnes*, the head of the laboratory in Leiden, managed to get a ship load of monazite sand from North Carolina from which he and his coworkers extracted 360 liters of helium gas. He reached the goal of liquefaction of helium on 10 July 1908 [6]. With the achievement of this milestone he established modern low-temperature physics. For many years, the laboratory in Leiden was the leader in this new field.

1.1 Basic Facts

At room temperature, helium is a light inert gas. It is odorless, colorless, and tasteless, and, after hydrogen, the second most abundant element in the universe. Despite its very simple structure, helium exhibits numerous exotic phenomena in condensed form whose theoretical descriptions are rather complex in many cases. We will take a look at some of the unusual characteristics of quantum fluids and quantum solids in the following chapters. To begin, however, let us introduce some general properties of helium in the remaining part of this chapter.

1.1.1 Terrestrial Occurrence

Helium exists in two stable isotopes. ^{4}He makes up about 5.2 ppm of the Earth's atmosphere. This trace amount of helium is not gravitationally bound to the Earth and is constantly lost into space. The Earth's atmospheric helium is replaced by the decay of radioactive elements in the crust of the Earth. Whereas in the early days of helium research, ^{4}He gas was mainly obtained from minerals, today it is recovered exclusively from natural-gas deposits. The largest sources of helium-rich natural gas are located in the United States. The chemical composition and the helium content of natural gas differs widely, depending on the geological strata. In some natural-gas wells a helium content up to 7% has been found.

The lighter isotope ^{3}He was discovered and identified in 1933 by *Oliphant, Kinsey* and *Rutherford* [7]. At first, however, it was believed that ^{3}He should not be stable. This misconception was disproved in 1939 by *Alvarez* and *Cornog* in cyclotron accelerator experiments [8]. To obtain ^{3}He in quantities for use in low-temperature physics experiments, it has to be produced artificially via nuclear reactions, because the concentration of ^{3}He in helium from natural-gas sources is only 0.14 ppm. Therefore, the production of ^{3}He relies on waste from nuclear reactors or the waste from various constituents of hydrogen bombs. The nuclear reaction is

$$
\begin{aligned}
{}^6\mathrm{Li} + \mathrm{n} \longrightarrow \quad & {}^3\mathrm{H} + {}^4\mathrm{He} \\
& \hookrightarrow {}^3\mathrm{He} + \mathrm{e}^- + \overline{\nu}_\mathrm{e} \,.
\end{aligned}
\tag{1.1}
$$

In the first step of this reaction, tritium is produced that has a half-life of 12.5 years. It undergoes beta decay with ^{3}He as a product. Significant amounts of ^{3}He were not available before the end of the 1950s. Nevertheless, *Sydoriak, Grilly* and *Hammel* managed to liquefy ^{3}He as early as 1949 and investigated its properties [9].

1.1.2 Basic Atomic and Nuclear Properties

The electronic structure of helium is the simplest many-body system of all atoms. With two electrons completely filling the K shell (1s^2) it has a spherical shape and thus no permanent electric dipole moment. Helium has the smallest known atomic polarizability of $\alpha = 0.1232\,\mathrm{cm^3\,mol^{-1}}$ and only a very weak diamagnetic susceptibility $\chi_\mathrm{d} = -1.9 \times 10^{-6}\,\mathrm{cm^3\,mol^{-1}}$. With an atomic radius of only 31 pm it is the smallest atom. It has the highest ionization energy of about 24.6 eV.

The liquid of both isotopes is colorless and since their index of refraction is very close to unity, they are very difficult to see. Because of their nuclear spin $I = 1/2$, ^{3}He atoms are fermions and obey Fermi–Dirac statistics, whereas ^{4}He atoms are bosons with nuclear spin $I = 0$. Besides the two stable isotopes ^{3}He and ^{4}He, there exist two unstable helium isotopes with relatively long half-lives: ^{6}He ($T_{1/2} = 0.82$ s), and ^{8}He ($T_{1/2} = 0.12$ s).

1.1.3 Van der Waals Bond

Van der Waals forces act between all atoms. They are based upon the electrical dipole–dipole interaction. At first glance the appearance of this type of force between helium atoms is surprising, because, as mentioned above, they have perfect spherical symmetry and thus no permanent electrical dipole moment. However, one has to take into account that there are always zero-point fluctuations present in the charge distribution, which result in fluctuating dipole moments. The associated electrical fields induce fluctuating dipole moments at neighboring atoms as well and thus lead to a force between the atoms. This force can be described using the Van der Waals potential $\phi(r) \propto 1/r^6$, where r denotes the interatomic distance. In the framework of a quantum-mechanical treatment of the Van der Waals force one can show that this type of interaction always leads to an energy reduction and thus to an attractive force. The strength of this force is given by the polarizability of the interacting atoms. Since the polarizability is small in the case of helium atoms, the binding force is very weak.

By adding a repulsive potential of the form $\phi(r) \propto 1/r^{12}$ one obtains the well-known *Lennard–Jones potential* [10]

$$\phi(r) = 4\varepsilon \left[\left(\frac{\sigma}{r}\right)^{12} - \left(\frac{\sigma}{r}\right)^{6} \right] , \tag{1.2}$$

with the characteristic parameters ε and σ. For both helium isotopes we have $\varepsilon/k_\mathrm{B} = 10.2\,\mathrm{K}$ and $\sigma = 2.56\,\text{Å}$. The potential energy is obtained by integrating (1.2)

$$E_\mathrm{pot} = \frac{1}{2} \int\limits_0^\infty \phi(r)\, 4\pi r^2 n(r)\, \mathrm{d}r , \tag{1.3}$$

taking into account the radial density function $n(r)$. The difference in the potential energy of liquid and solid helium originates from the difference in $n(r)$ of the two phases.

In addition to the very weak binding forces between helium atoms, there is a reduction of the binding by the zero-point motion. Although a detailed calculation of the zero-point energy of helium in the liquid phase is rather complex, we can estimate the size of this effect in a simple model in which we assign each atom a certain 'cage' volume V. The ground-state energy (zero-point energy) of a particle with mass m in such a cage is given by

$$E_0 = \frac{3\hbar^2\pi^2}{2mV^{2/3}} . \tag{1.4}$$

From this result, we can directly see that the zero-point energy for atoms with a small mass like helium is large and that it increases with decreasing molar volume V_m. In Fig. 1.1a the zero-point energy of ^{4}He is plotted as a function of the molar volume along with the curves for the potential energy of liquid and solid ^{4}He.

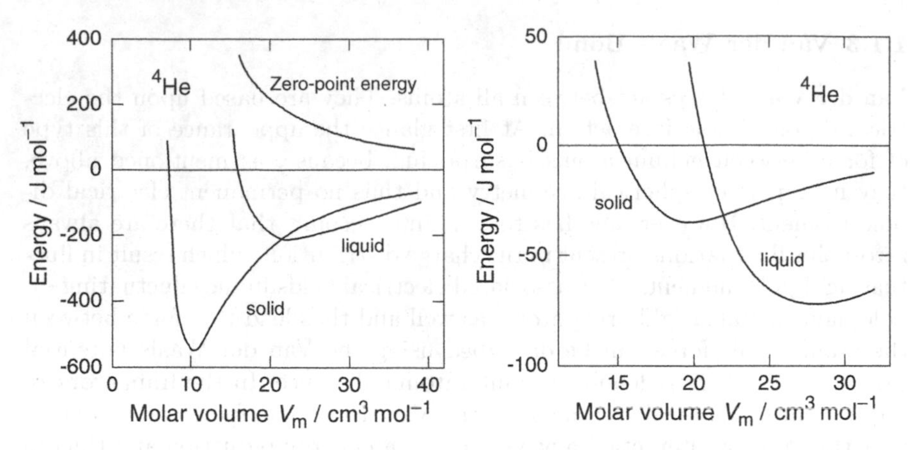

Fig. 1.1. (**a**) Zero-point energy and potential energy of liquid and solid ^{4}He at $T = 0$ as function of the molar volume [11]. (**b**) Total energy of ^{4}He at $T = 0$ as function of the molar volume

Neglecting the zero-point motion, the solid phase of ^{4}He should be stable at a molar volume of approximately $10\,\mathrm{cm}^3\,\mathrm{mol}^{-1}$. However, if we consider the total energy, which is plotted in Fig. 1.1b, we find that the liquid phase is energetically more favorable. In the liquid phase, helium has a molar volume of $28\,\mathrm{cm}^3\,\mathrm{mol}^{-1}$ at $T = 0$. According to (1.4) the zero-point motion of ^{3}He leads to an even stronger reduction of the binding energy than for ^{4}He. Although the sizes of ^{4}He and ^{3}He atoms are essentially identical, the molar volume of liquid ^{3}He is about $40\,\mathrm{cm}^3\,\mathrm{mol}^{-1}$, and thus much larger than that of liquid ^{4}He. We conclude that the interplay of the weak binding force and the large zero-point energy is responsible for the fact that helium is a permanent liquid, which means that it remains liquid under saturated vapor pressure even for $T \to 0$. In addition, this is also why both helium isotopes have the lowest boiling temperatures and critical points known for any substance in nature. Some important parameters of liquid ^{3}He and ^{4}He are listed in Table 1.1.

Table 1.1. Some important material parameters of ^{3}He and ^{4}He. After [12,13]

	^{3}He	^{4}He
boiling temperature at normal pressure T_b (K)	3.19	4.21
critical temperature T_c (K)	3.32	5.19
critical pressure p_c (bar)	1.16	2.29
density for $T \to 0$ ϱ_0 ($\mathrm{g\,cm}^{-3}$)	0.076	0.145
density at boiling point ϱ_b ($\mathrm{g\,cm}^{-3}$)	0.055	0.125

1.2 Thermodynamic Properties

In the following sections we will briefly consider some basic thermodynamic properties of the two helium isotopes in their liquid form, including a brief discussion of the specific heat. A more detailed analysis of the specific heat of ^{4}He is presented in Chap. 2, and for ^{3}He in Chaps. 3 and 4.

1.2.1 Density

During the first liquefaction of helium in 1908, Kamerlingh Onnes already realized that the density of liquid helium is exceptionally small. The values for the densities of ^{3}He and ^{4}He at their boiling points are given in Table 1.1. In addition, in 1911 Kamerlingh Onnes made the surprising observation that the density of ^{4}He exhibits a maximum at about 2 K [14]. Later investigations showed that there is a sharp kink in the temperature dependence of the density at 2.17 K, and that ^{4}He expands again below that temperature [15].

In Fig. 1.2 the density of liquid ^{4}He and ^{3}He is shown as a function of temperature. Measurements of the density of liquid ^{3}He were carried out in 1949 when ^{3}He was liquefied for the first time [16]. The density of ^{3}He did not show a maximum and, as expected, it was much smaller than that of ^{4}He.

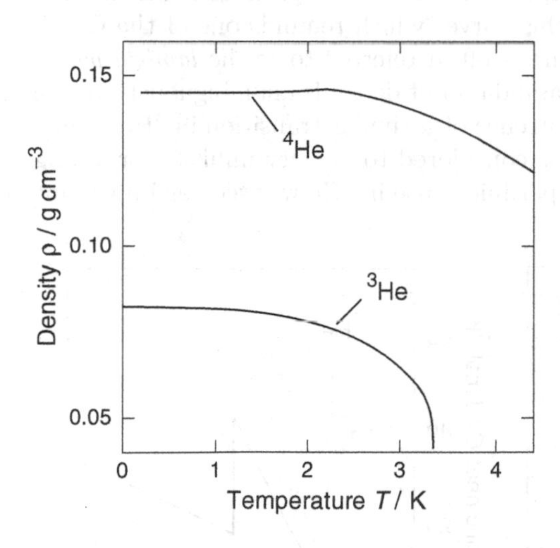

Fig. 1.2. Temperature dependence of the density of liquid ^{3}He and ^{4}He [17]

1.2.2 Specific Heat

The first measurements of the specific heat of liquid ^{4}He were performed by *Dana* and *Kamerlingh Onnes* in 1923. They found an abnormal rise of the specific heat around 2 K. In the publication of their results in 1926 they decided to leave out these data points, because they feared that this anomaly might have been caused by experimental problems [18]. In 1932 *Keesom* and *Clusius* investigated the specific heat of liquid ^{4}He again and observed a pronounced maximum at about 2.17 K, which they attributed to a phase transition [19].

Since the true nature of the phase transition was unclear for a long time, the two phases were distinguished by naming them helium I and helium II, where helium I denotes the liquid phase above the transition. It was at first believed that helium II represented a crystalline phase under normal pressure. Within this description the fact that it still looked like a fluid was explained in terms of a liquid crystal with flexible planes. This misconception was disproved in 1938 when X-ray diffraction measurements showed undoubtedly that helium II is, in fact, a liquid phase. Surprisingly, it took more than 30 years from the initial observation to the successful explanation of this phase transition. As we will discuss in Sect. 2.3, the nature of the phase transition at 2.17 K can be understood as Bose–Einstein condensation. One of the most intriguing features of helium II is certainly its ability to flow through narrow capillaries without any friction at all. Following the naming of the frictionless transport of electrons in metals as the superconducting state one often refers to helium II as superfluid helium.

Figure 1.3a shows more recent data of the specific heat of liquid ^{4}He as a function of temperature. At a temperature of 2.17 K a pronounced maximum occurs. Because of the shape of this curve, which reminds one of the Greek letter λ, the transition temperature is often referred to as the *lambda point*. Since the phase transition at the lambda point depends unambiguously on the bosonic character of ^{4}He, the occurrence of a similar transition in ^{3}He, which carries a nuclear spin $I = 1/2$, was considered to be very unlikely for a long time. Instead, the absence of a superfluid state in ^{3}He was seen as important

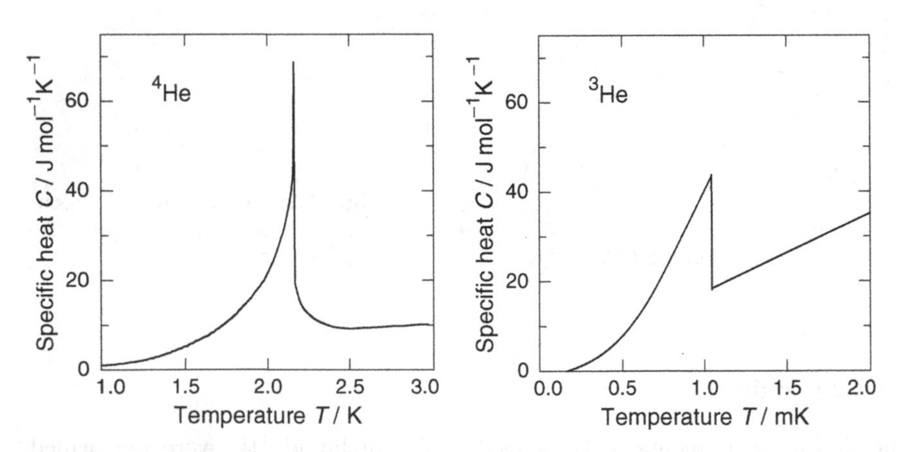

Fig. 1.3. Specific heat of (**a**) ^{4}He [20] and (**b**) ^{3}He [21] in the temperature range where the transition from the normal to the superfluid phase occurs, as a function of temperature

evidence for the validity of the interpretation of the phase transition in ^{4}He as a Bose–Einstein condensation. However, the explanation of the microscopic nature of superconductivity in the framework of BCS theory (see Sect. 10.3) in 1957 changed that viewpoint and intensified the search for a superfluid phase of ^{3}He. Finally in 1972, superfluid phases of ^{3}He were discovered by *Osheroff*, *Richardson* and *Lee* in nuclear magnetic resonance (NMR) measurements [22]. In contrast to ^{4}He, which exhibits just one superfluid phase, ^{3}He has three different superfluid phases, depending on temperature, magnetic field, and pressure. Figure 1.3b displays the temperature dependence of the specific heat of liquid ^{3}He in the vicinity of one such transition at normal pressure and zero magnetic field. Compared to ^{4}He the phase transition in ^{3}He occurs at much lower temperatures, namely in the low millikelvin range.

1.2.3 Latent Heat

Dana and Kamerlingh Onnes were also the first to measure the latent heat L of vaporization of ^{4}He in 1924 [18]. They observed a minimum of L at the lambda point. The latent heat of ^{3}He is much smaller and, as expected, no corresponding minimum occurs around 2 K. Figure 1.4 shows the latent heat of vaporization of ^{4}He and ^{3}He as a function of temperature.

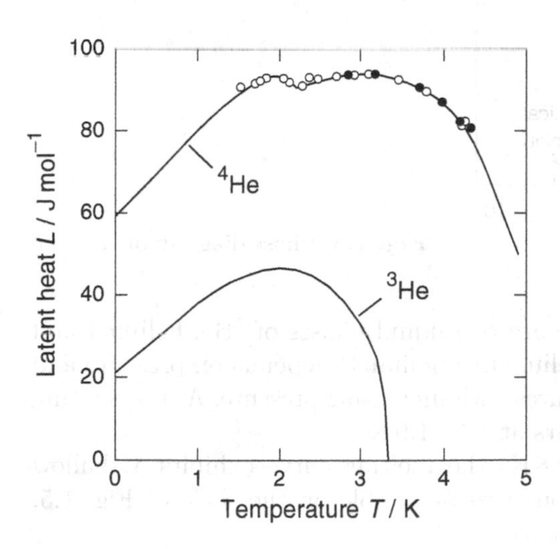

Fig. 1.4. Latent heat of vaporization of ^{3}He and ^{4}He as a function of temperature. The data points for ^{4}He are taken from [18, 23]. The *solid line* has been deduced from the vapor-pressure measurements using the Clausius–Clapeyron equation [24]. In the case of ^{3}He the *solid line* represents the measured temperature dependence of L according to [17]

1.3 Phase Diagrams

The phase diagrams of ^{4}He and ^{3}He are, in several ways, remarkably different from these of all other known substances. In the following, we will take a brief look at the phase diagrams of the two helium isotopes and point out their distinctive features.

1.3.1 ^{4}He

The p–T phase diagram of ^{4}He is shown in Fig. 1.5. It is most remarkable that helium has no triple point where gas, liquid and solid phase intersect. It remains liquid under normal pressure even at $T = 0$ as discussed before. Solid helium can only be produced at pressures above 25 bar. Depending on temperature and pressure one finds three different crystalline modifications. In the whole temperature range – but not at all pressures – solid helium with hcp structure exists. In a small pressure range at temperatures between 1.45 K and 1.75 K, ^{4}He first solidifies into a bcc structure. For pressures exceeding 1 kbar and temperatures above 15 K, ^{4}He shows a fcc phase (not shown in Fig. 1.5).

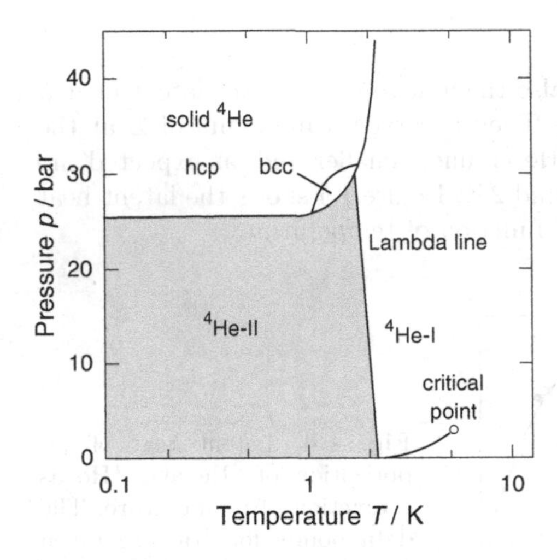

Fig. 1.5. Phase diagram of ^{4}He [25]

As we have seen before, there are two liquid phases of ^{4}He: helium I and helium II. The transition from helium I to helium II depends on pressure and is shifted towards lower temperatures with increasing pressure. At the melting curve the lambda transition occurs at $T = 1.9$ K.

At low temperatures ($T \approx 0.8$ K) the melting curve exhibits a shallow minimum, which is not deep enough to be visible on the scale of Fig. 1.5. Using the *Clausius–Clapeyron equation*

$$\left.\frac{\partial p}{\partial T}\right|_{\text{meltingcurve}} = \frac{S_\ell - S_\text{s}}{V_\ell - V_\text{s}} , \tag{1.5}$$

we can draw conclusions about the entropies of liquid and solid ^{4}He from the melting curve. Here, S and V represent entropy and volume per mole, respectively. The indices ℓ and s denote liquid and solid state, respectively. Since the molar volume of liquid ^{4}He is always larger than that of solid ^{4}He, i.e., $V_\ell - V_\text{s} > 0$, we can conclude from $\partial p/\partial T < 0$ the surprising fact that below $T \approx 0.8\,\mathrm{K}$ the entropy of the solid phase is larger than the entropy of the liquid phase. In other words, the disorder in the solid is larger than in the liquid. The entropy of solid and liquid ^{4}He is determined by thermal excitations in this temperature range. It turns out that solid ^{4}He has a slightly higher phonon heat capacity than liquid ^{4}He, because of the low transverse sound velocity in solid ^{4}He. Therefore, at low temperatures the entropy of solid ^{4}He is larger than that of liquid ^{4}He.

1.3.2 ^{3}He

The phase diagram of ^{3}He (Fig. 1.6) looks qualitatively very similar to that of ^{4}He, except for the much more pronounced bcc phase. The hcp phase exists in ^{3}He only at pressures above 100 bar. As mentioned before, ^{3}He also exhibits a transition from a normal fluid to a superfluid phase, or more precisely into three different superfluid phases. We will return to this point in Chap. 4. The transition temperatures are between 1 mK and 3 mK and thus three orders of magnitude lower than in ^{4}He.

For $T \to 0$, ^{3}He solidifies at pressures above 33 bar. Between 30 and 100 bar one finds a bcc structure and above about 100 bar an hcp lattice. The bcc phase in solid ^{3}He is much more extended than in ^{4}He. The reason is the higher zero-point energy that favors a smaller packing density. For temperatures above 18 K and for pressures exceeding 1.3 kbar, ^{3}He crystallizes into a fcc structure. As in ^{4}He, the boundary line between liquid and solid ^{3}He – the melting curve – shows an anomaly at low temperatures, which we will discuss in the following section.

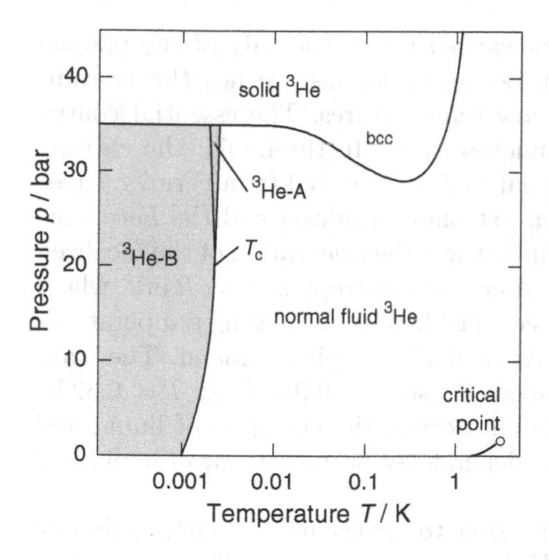

Fig. 1.6. Phase diagram of ^{3}He [26]

Melting Curve

Figure 1.7a shows again the temperature dependence of the melting curve of ^{3}He, which exhibits a pronounced minimum at a temperature of about $T = 320$ mK. Using the same reasoning as we did for ^{4}He we can conclude that the entropy of solid ^{3}He is larger than that of liquid ^{3}He at temperatures below the minimum.

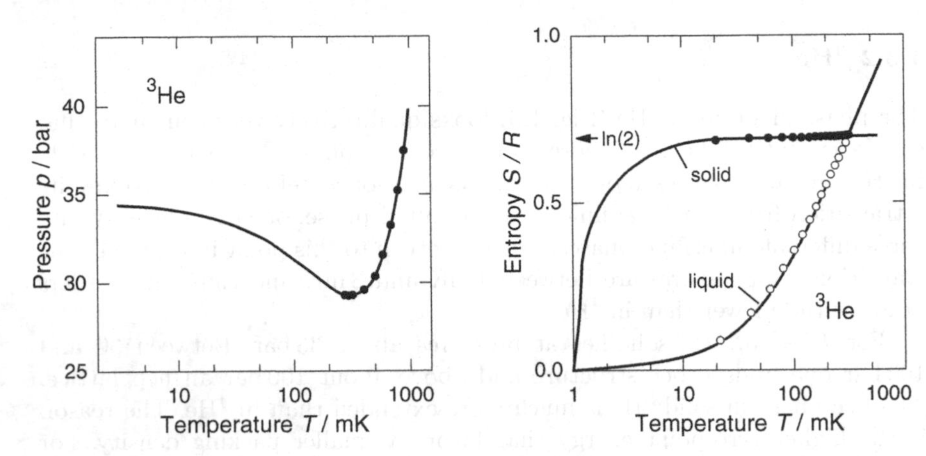

Fig. 1.7. (**a**) Melting curve of ^{3}He. The *solid line* and the points are from different measurements [27–29]. (**b**) Reduced entropy S/R of liquid and solid ^{3}He as a function of temperature. The *solid lines* represent the expected curves for the two phases. *Open* and *closed circles* mark experimental data [30]

This amazing phenomenon is not caused by an anomaly of the phonon spectrum, as was the case for ^{4}He, because phonons are not the relevant degrees of freedom in ^{3}He at such low temperatures. The essential contribution to the entropy comes from nuclear spins. In the liquid, the entropy varies at low temperatures proportional to T as expected for a Fermi gas (see Sect. 3.1). In solid ^{3}He the atoms are strongly localized and the Fermi-gas model is not applicable. At high temperatures the orientation of the localized spins is statistical and their contribution to the entropy is $S_s = R \ln 2$, where R is the universal gas constant (see Sect. 11.5). With decreasing temperatures a transition to an antiparallel arrangement of the spins is found. The transition temperature to the antiferromagnetic state is 0.9 mK. At $T = 0.32$ K, the temperature at which the minimum occurs, the entropies of liquid and solid ^{3}He are equal. The temperature dependence of the entropy of liquid and solid ^{3}He is shown in Fig. 1.7b

As we will see in Sect. 11.5, it is possible to use the melting curve anomaly as a cooling mechanism below 0.32 K, by applying pressure. This mechanism is called *Pomeranchuk cooling*. With this technique, temperatures down to about 1 mK can be obtained.

Exercises

1.1 Calculate the zero-point energy of a particle in a cube of side length L. Determine the zero-point temperature of H_2, ^{3}He, ^{4}He, Ne and Ar under the assumption that the diameter of the atoms is equal to L.

1.2 Quantum effects are expected to become important if the wavelength of neighboring atoms noticeably overlap, i.e., if the thermal de Broglie wavelength becomes equal to the interatomic distance. The latter is given approximately by the diameter of the atoms. Calculate the corresponding temperatures at which the above condition is met for the systems mentioned in 1.1.

1.3 The latent heat of liquid helium at $T = 1\,\mathrm{K}$ is $L = 2.156 \times 10^6\,\mathrm{J\,m^{-3}}$. Calculate the binding energy of a helium atom on the liquid surface.

1.4 Estimate the temperature of the melting-curve minimum for ^{3}He.

2 Superfluid ^{4}He – Helium II

One of the most striking properties of helium II is its ability to flow through very small capillaries or narrow channels without experiencing any friction at all. This phenomenon was discovered in 1938 independently by *Kapitza* [31] and by *Allen* and his coworker *Misener* [32]. Kapitza named it superfluidity in analogy to superconductivity, which denotes the lossless transport of electrons in superconductors. To explain this discovery, *F. London* suggested in 1938 that superfluidity is related to the occurrence of an ordered state in momentum space, as would be expected for a Bose–Einstein condensate [33]. Adopting this viewpoint, *Tisza* postulated in the same year the phenomenological two-fluid model, which nicely describes many properties of helium II [34]. A great success of this model was the prediction of second sound, which was experimentally observed in superfluid helium a few years later. Between 1941 and 1947 *Landau* published three landmark papers on two-fluid hydrodynamics, in which he explained the phenomenon of superfluidity as a consequence of the excitation spectrum of helium II [35]. Finally, *Feynman* showed that the excitation spectrum postulated by Landau can be derived within a quantum-mechanical description [36].

In this chapter, we shall discuss the extraordinary properties of superfluid ^{4}He starting with some basic experimental observations, which in particular demonstrate the peculiar behavior of this fascinating liquid. Following this we introduce the two-fluid model and discuss whether Bose–Einstein condensation and the collective excitations proposed by Landau can be used to understand the phenomenon of superfluidity, and whether they provide a microscopic foundation for the two-fluid model.

2.1 Experimental Observations

At high temperatures, liquid ^{4}He behaves as a dense classical gas, but at the lambda point at $T_\lambda = 2.17\,\mathrm{K}$, almost all properties of liquid ^{4}He change. Even with the naked eye one can see this dramatic transition. In the normal-fluid state ^{4}He boils like an ordinary liquid with bubbles rising constantly in the liquid. Below the lambda transition, however, it suddenly becomes deadly quiet and evaporation only takes place at the free surface of the liquid. In the following we present some additional experimental observations of helium II.

2.1.1 Viscosity and Superfluidity

The first indications for the occurrence of superfluidity came from flow measurements through very thin capillaries and narrow slits [31, 32]. Using the Hagen–Poiseuille law

$$\dot{V} = \frac{\pi r^4}{8} \frac{1}{\eta} \frac{\Delta p}{L}, \tag{2.1}$$

one can conclude from measurements of the flow velocity in narrow capillaries that the viscosity of helium II is several orders of magnitude lower than that of helium I. The quantity L denotes the length of the capillary, r the radius, Δp the pressure drop along the capillary and $\dot{V}$ the volume rate of helium transported through it. Some measurements that demonstrate the typical variation of flow velocity $v = \dot{V}/(\pi r^2)$ with pressure are shown in Fig. 2.1a. Besides the extremely low viscosity, two other very remarkable observations can be made, namely that the flow velocity is nearly independent of the pressure gradient along the capillary, and that the flow velocity *increases* with *decreasing* diameter of the capillary. The temperature dependence of the viscosity deduced from flow measurements through narrow capillaries is shown in Fig. 2.1b. Above the lambda point, the viscosity is nearly temperature independent, but it falls to an undetectably low value for $T < T_\lambda$.

An important question in this context is whether the viscosity becomes extremely small but finite or whether it actually becomes zero below the lambda transition. To answer this question *persistent-mass flows* have been generated and monitored [37,38], analogous to persistent-current experiments with superconductors (see Chap. 10). A torus, containing compressed fine powder is filled with liquid helium and set into rotation above the lambda

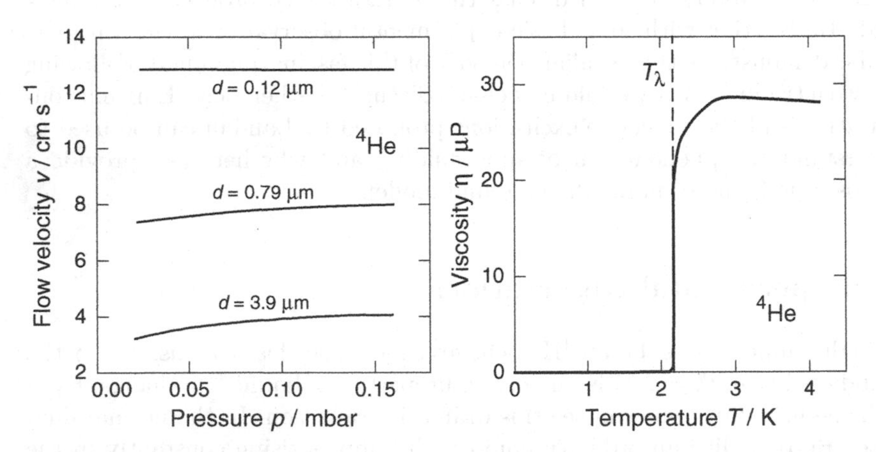

Fig. 2.1. (**a**) Flow velocity of helium II through capillaries with different diameter as a function of the applied pressure [39, 40]. (**b**) Temperature dependence of the viscosity of liquid helium as determined from flow experiments with thin capillaries

point. Because of its viscosity, the helium is dragged along with the torus under these conditions. The rotating torus is cooled below T_λ and is gently brought to rest. Subsequently, the evolution of the angular velocity of the helium with time is determined. In several experiments this has been achieved by implementing the torus as part of a superfluid gyroscope. From the observation of a constant angular velocity for many hours one can conclude that the viscosity drops at the lambda point by at least eleven orders of magnitude. Within the accuracy of the experiment, this means that helium II is truly flowing without dissipation.

It has been observed, however, that the results of viscosity measurements on helium II, but not on helium I, depend on the measuring method employed. As we will see later, this very peculiar phenomenon can be explained in the framework of the two-fluid model. Before introducing this model in Sect. 2.2, we will take a brief look at the results obtained with two standard techniques for measuring viscosity: the rotary viscosimeter (Fig. 2.2a) and the oscillating-disc method (Fig. 2.2b). In both experiments, the viscosity does not drop instantaneously to zero at the lambda point but remains finite well below the phase transition. For the rotary viscosimeter, the measured viscosity even increases again on cooling below about 1.8 K and substantially exceeds the viscosity of helium I below 1 K. In contrast, η drops steadily below T_λ with decreasing temperature if measured with an oscillating disc. In addition, in these experiments with oscillating discs, the damping has quite a strong affect on the amplitude of the oscillation, indicating a nonlinear behavior.

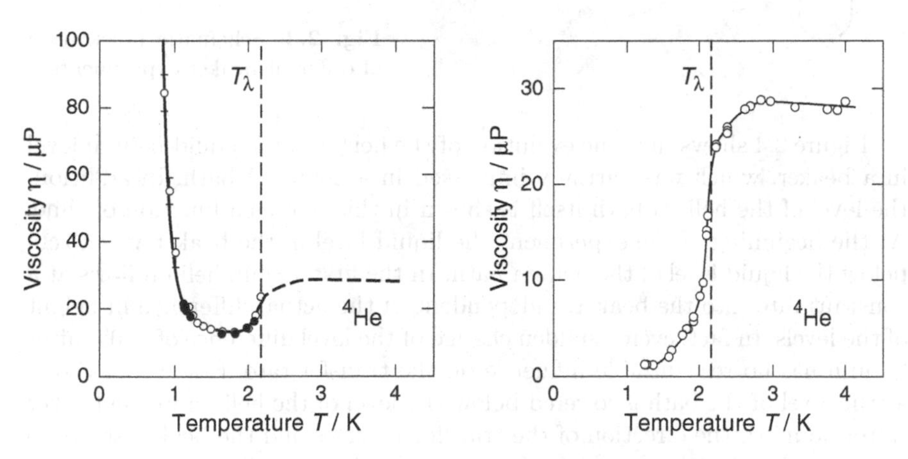

Fig. 2.2. Viscosity of liquid helium as a function of temperature measured (**a**) with a rotary viscosimeter [41, 42] and (**b**) with an oscillating disc [43]

2.1.2 Beaker Experiments

Further extraordinary observations were made in connection with so-called *beaker experiments* [44,45]. Three basic configurations of this experiment are shown schematically in Fig. 2.3. Superfluid helium will flow over the rim into an empty beaker dipped into a bath of helium II until the levels of helium inside and outside the beaker are equal (*left*). Raising the beaker afterwards causes the helium II to flow back into the bath until equalization of the levels has taken place again (*middle*). If the beaker is lifted completely out of the bath, helium II will flow over the rim of the beaker and will drop back into the bath until the beaker is empty (*right*).

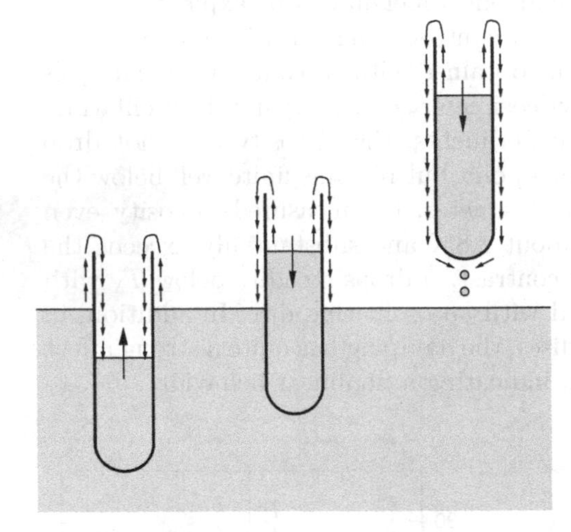

Fig. 2.3. Schematic illustration of different beaker experiments

Figure 2.4 shows the time evolution of the height of the liquid-helium level in a beaker, which was partially immersed in a helium II bath. In addition, the level of the helium bath itself is shown in this plot as a function of time. At the beginning of the experiment the liquid level in the beaker was 13 cm below the liquid level of the helium bath. In the first 45 min helium flows at a constant rate into the beaker, independent of the actual difference in height of the levels. In fact, even a sudden change of the level difference of 100% after 30 min has no recognizable influence on the transfer rate. However, as soon as the level of the bath is lowered below the level of the helium in the beaker (after 45 min), the direction of the transfer reverses and the beaker starts to empty again with the same transfer rate as during the filling procedure.

Figure 2.5a shows a schematic illustration of an experimental setup that was used for a detailed investigation of the equalization of levels. The liquid level in a long thin tube open at the top is monitored with time. In this way, level changes caused by very small transfer rates can be determined with high accuracy. The result of such a measurement is shown in Fig. 2.5b. At the beginning of the experiment the difference in level height decreases linearly with

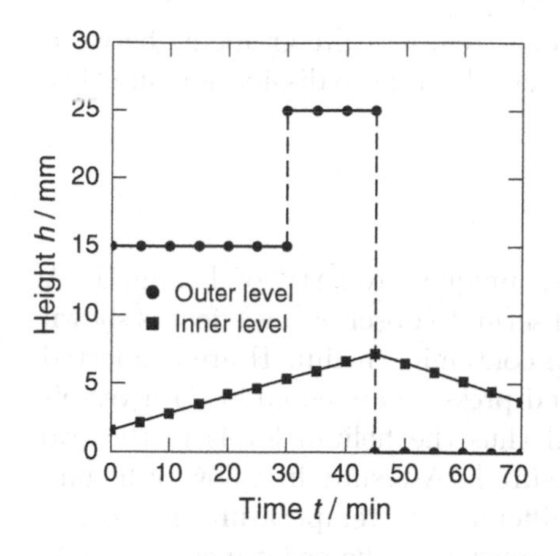

Fig. 2.4. Level height of a beaker as a function of time. The level was 13 cm below the surface of the surrounding helium bath at the beginning of the experiment [44]

time, which means that the transfer rate is independent of the level difference, as in the experiments discussed previously. But surprisingly, at the time when the level difference has almost vanished, an undamped oscillation of the helium level in the beaker is observed. In the particular experiment discussed here, the amplitude of this oscillation was about 0.35 mm. The origin of this oscillation is an overshoot of the flow because of the inertia of the flowing helium film every time the levels are equal, leading to a periodic reversal of the flow. Because of superfluidity the oscillation persists nearly undamped over several minutes in the experiment discussed here. This does not always

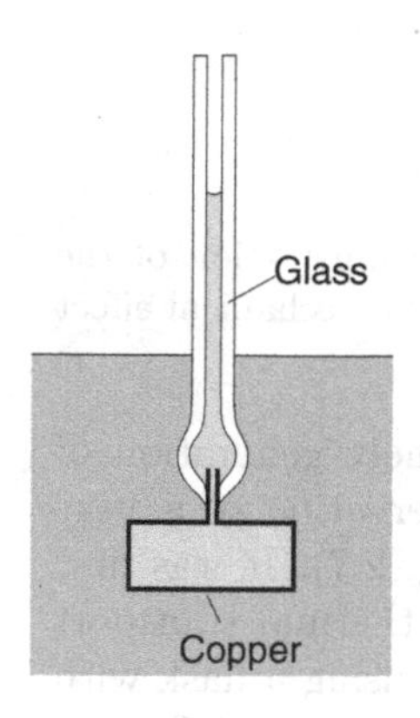

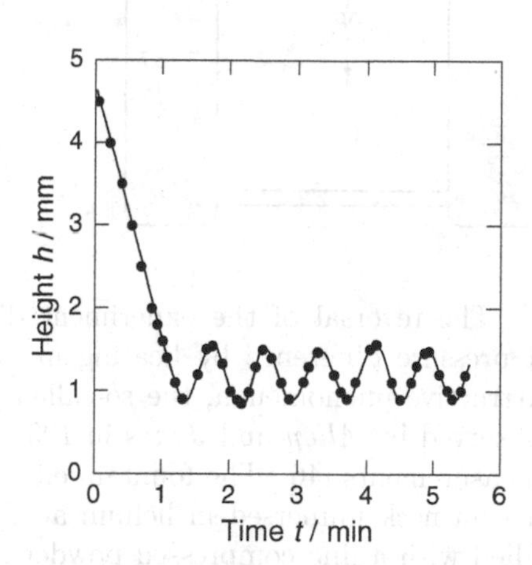

Fig. 2.5. (**a**) Schematic illustration of a beaker experiment that allows a detailed determination of the helium transfer. The inner diameter of the tube was only 0.58 mm. The copper base of the beaker was incorporated to provide good thermal contact. For clarity, the helium in the beaker is drawn in light grey, although there is no difference between the inner and outer helium. (**b**) Time evolution of the helium height in the beaker [45]

happen, because, unless special care is taken, temperature gradients between the inside and the outside of the beaker occur, leading to dissipation and thus to a rapid damping of the oscillation.

2.1.3 Thermomechanical Effect

The *thermomechanical effect* is another unique property of helium II. A schematic illustration of an experimental setup to observe this effect is shown in Fig. 2.6. Two vessels (A and B), both containing helium II are connected via a very thin capillary. Temperature and pressure are equal in both vessels at the beginning of the experiment and thus the helium levels in the two vessels are the same. Increasing the pressure in A results in a flow of helium towards B. Surprisingly, this causes a difference in temperature in the two vessels. The temperature in B decreases somewhat, whereas it increases in A. Equalizing the pressure difference again brings the system back to its starting condition indicating that this is a reversible process. This experiment clearly shows that there is mass flow in helium II associated with the heat flow. However, the fact that the direction of heat flow is actually opposite to the flow of mass is very peculiar.

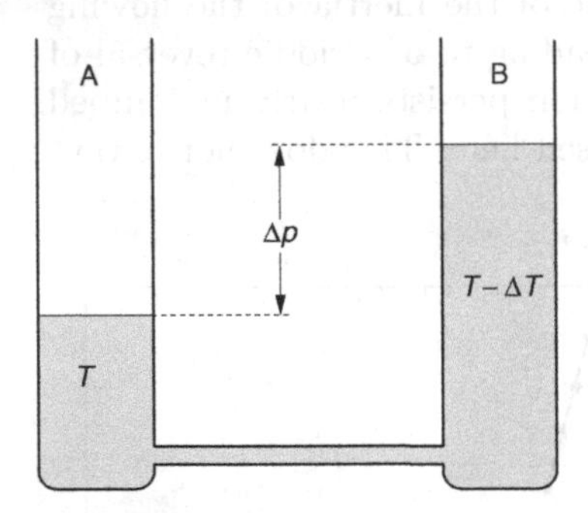

Fig. 2.6. Schematic illustration of the principle of the thermomechanical effect

The reversal of the experiment discussed above, namely generation of a pressure difference by heating makes possible the observation of a very attractive phenomenon, the so-called *fountain effect* (Fig. 2.7). It was first observed by *Allen* and *Jones* in 1938 in connection with thermal transport measurements [46]. The fountain effect can be realized by using a flask with a thin neck immersed in helium at $T < T_\lambda$. The lower part of the flask is filled with a fine compressed powder and is open at the bottom. Above the powder tablet an electrical heater is located in the flask. Without heating, the flask fills up with helium until the level of the bath is reached. Heating the helium in the flask results in a fountain of helium ejected from the top of the flask due to the thermomechanical effect. Stationary fountains with heights up to 30 cm have been achieved in this way. Usually, such fountains show turbulent flow. However, under certain conditions (low heater power,

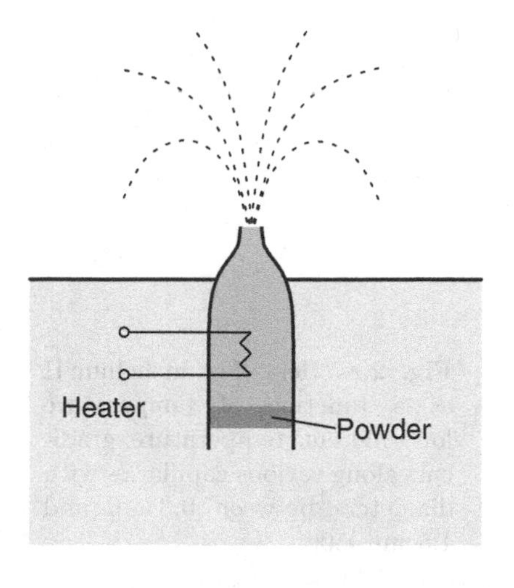

Fig. 2.7. (**a**) Schematic sketch of an experimental setup used to demonstrate the fountain effect. The helium inside and outside the flask has been drawn in a slightly different shade for clarity. (**b**) Photo of a fountain generated in helium II [47]

low temperatures, etc.) fountains can be produced exhibiting pure potential flow, like the one shown in Fig. 2.7b.

2.1.4 Heat Transport

Early experiments on heat transport in superfluid ^{4}He indicated that the thermal conductivity of helium II is more than five orders of magnitude larger than that of helium I [48,49]. This extremely high thermal conductivity of the superfluid immediately explains the remarkable observation that the boiling of liquid helium stops suddenly when passing the lambda transition. The temperature distribution becomes homogeneous within the liquid and thus evaporation takes place only at the free surface.

Not only is the heat transport of helium II very high, it also has a number of other unusual properties. Figure 2.8 shows that under certain circumstances a pronounced maximum of the heat current density is observed at about 1.8 K. Using capillaries with large diameters one finds, in addition, that the heat-current density $\dot{q}$ rises proportional to $|\mathrm{grad}\, T|^{1/3}$. This means, that the thermal transport cannot be described by the usual expression $\dot{\boldsymbol{q}} = -\Lambda \,\mathrm{grad}\, T$, because the thermal conductivity Λ would not be constant but would diverge for small temperature gradients as $\Lambda \propto |\mathrm{grad}\, T|^{-2/3}$.

Detailed investigations of the heat flux $\dot{Q}$ of helium II through very thin capillaries have shown that for small temperature differences the heat flux

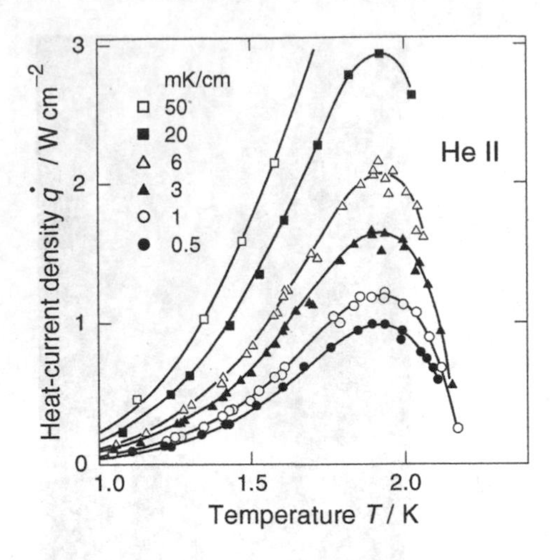

Fig. 2.8. Heat flow in helium II as a function of temperature for different temperature gradients along various capillaries with diameters between 0.3 mm and 1.5 mm [50]

is indeed proportional to grad T as expected. In fact, one finds in helium II a linear relation between heat flow and temperature gradient in a variety of different experiments in which the heat flux does not exceed a certain critical value. The conditions for which this is true is called the *linear regime*. Results of experiments that exhibit this behavior at low heat flux are shown in Fig. 2.9a. Clearly, the heat flow in these experiments depends linearly on the temperature gradient for small values and not too high temperatures.

It is important to note that the presence of a linear regime is *not* only due to the fact that the nonlinear effects seen at high heat fluxes are small. This becomes clear in Fig. 2.9b in which the thermal resistance $\Delta T/\dot{Q}$ of helium II

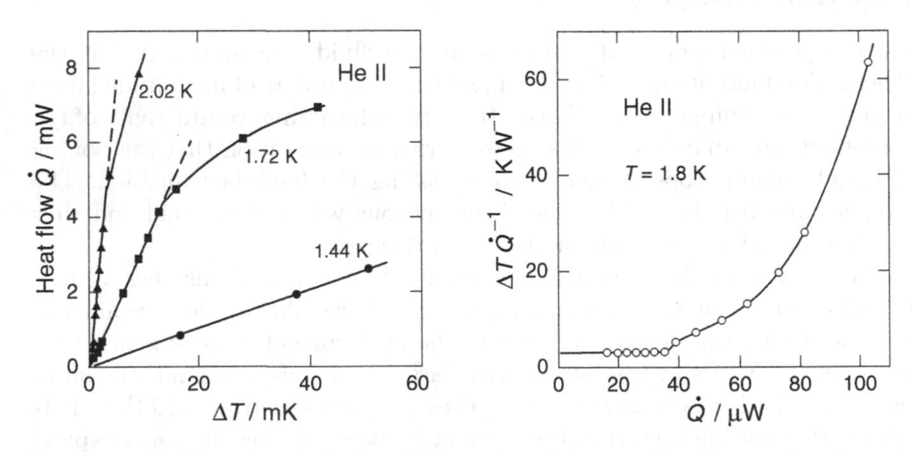

Fig. 2.9. **(a)** Heat flow in helium II through a 2.4 μm wide slit as a function of the temperature difference ΔT along the slit for three different temperatures [51]. **(b)** Thermal resistance $\Delta T/\dot{Q}$ of helium II in a thin capillary (diameter 107 μm, length 10 cm) as a function of the heat current [52]

in a thin capillary is plotted as a function of the heat current. At small heat currents the thermal resistance is constant, but changes suddenly at a certain value of the heat flux. As we shall discuss later in more detail, the heat transport in helium II is associated with a mass flow for which turbulences in the liquid arise at a certain *critical velocity*. This, in turn, leads to a sudden increase of the thermal resistance.

2.1.5 Second Sound

Temperature waves, which propagate with a characteristic velocity, are another very remarkable feature of helium II. Since the propagation of such waves is similar to that of ordinary sound this phenomenon was named *second sound*. The first experimental observation of second sound was made by *Peshkov* in 1944. In his early experiments he traced the temperature variation associated with propagating second-sound waves. Later, he improved the accuracy of his measurements by generating standing temperature waves. A sketch of his experimental setup for the investigation of standing second-sound waves is shown in Fig. 2.10a.

Using an electrical heater, periodic temperature waves are generated in a resonator with variable length L containing helium II. The temperature distribution in the resonator is monitored with a thermometer that can be moved with respect to the heater position. At resonance, periodic variations of the temperature in the liquid are observed. The result of measurements at two different frequencies is shown in Fig. 2.10b, indicating the presence of standing waves. In this experiment, the velocity of second sound can be

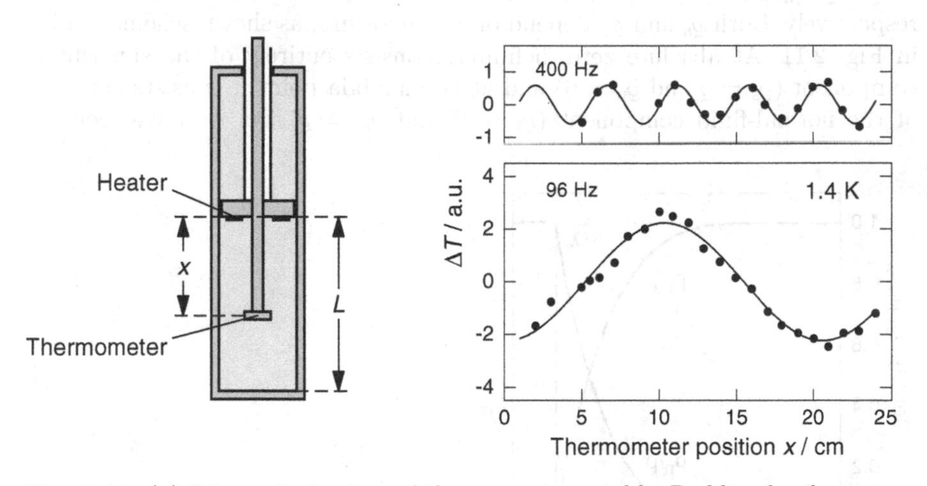

Fig. 2.10. (**a**) Schematic drawing of the apparatus used by Peshkov for the generation and detection of standing temperature waves in helium II. (**b**) Temperature of superfluid helium as a function of the thermometer position obtained at two different resonator modes [53]

determined by the simple relation $v_2 = 2L\nu/n$ for longitudinal resonances. Here, ν denotes the heater frequency and n the number of half-waves in the resonator. With this setup, it is possible to generate temperature waves with frequencies up to 100 kHz. It is remarkable that the velocity of second sound has been found to be independent of the frequency of the heat pulses up to this experimental limit.

2.2 Two-Fluid Model

In this section, we will see that the anomalous properties of helium II can be described phenomenologically with the so-called *two-fluid model*. The basic idea of this concept was first suggested in 1938 by *Tisza*, in order to describe transport phenomena of helium II. According to this model, helium II behaves *as if* it were a mixture of two completely interpenetrating fluids with different properties, although in reality this is not the case. To avoid any misunderstanding, it must be clearly stated at the outset that the two fluids cannot be physically separated; it is not permissible even to regard some atoms as belonging to the normal fluid and the remainder to the superfluid component, since all ^{4}He atoms are identical. But accepting these limits of the physical interpretation, many of the phenomena just described can be relatively clearly understood by formally expressing the density of helium II as the sum of a normal-fluid and a superfluid component:

$$\varrho = \varrho_n + \varrho_s \,, \tag{2.2}$$

where ϱ, ϱ_n and ϱ_s denote the total, normal-fluid and superfluid densities, respectively. Both ϱ_s and ϱ_n depend on temperature, as shown schematically in Fig. 2.11. At absolute zero, helium II consists entirely of the superfluid component ($\varrho_s = \varrho$ and $\varrho_n = 0$) and at the lambda point it consists entirely of the normal-fluid component ($\varrho_s = 0$ and $\varrho_n = \varrho$). As we have seen in

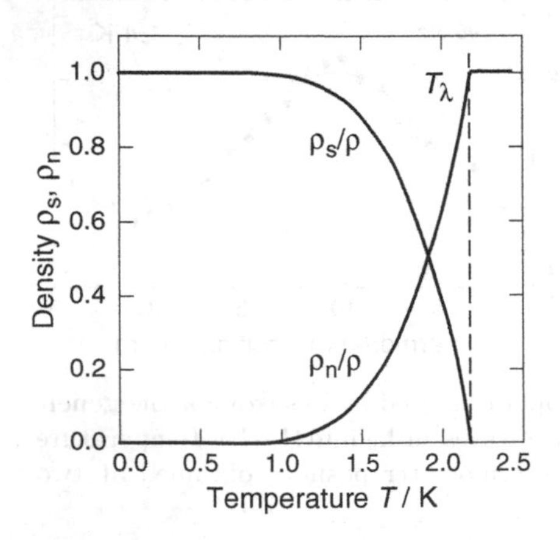

Fig. 2.11. Density of the superfluid and normal-fluid component in helium II as a function of temperature

Sect. 1.2, the total density ϱ is also slightly temperature dependent (see Fig. 1.2). However, in the following description this weak dependence will be neglected. Furthermore, it is assumed that the superfluid component carries no entropy, exhibits no viscous friction and shows no turbulence. The normal-fluid component, in contrast, is assumed to carry the total entropy of the fluid and to exhibit a finite viscosity. The basic assumptions of the two-fluid model are summarized in Table 2.1.

Table 2.1. Basic assumptions of the two-fluid model

	density	viscosity	entropy
normal-fluid component	ϱ_n	$\eta_n = \eta$	$S_n = S$
superfluid component	ϱ_s	$\eta_s = 0$	$S_s = 0$

We shall see that these simple assumptions lead to a satisfying phenomenological description of many different transport properties of helium II. After introducing the hydrodynamic equations we shall discuss the experimental observations presented in Sect. 2.1 in terms of the two-fluid model.

2.2.1 Two-Fluid Hydrodynamics

In this section, we will look at the basic hydrodynamic equations of the two component fluids. First, we introduce the momentum density $\boldsymbol{j}$ of mass flow per unit volume

$$\boldsymbol{j} = \varrho_n \boldsymbol{v}_n + \varrho_s \boldsymbol{v}_s \,. \tag{2.3}$$

Here, $\boldsymbol{v}_n$ and $\boldsymbol{v}_s$ denote the velocity of the normal and superfluid component, respectively. Mass conservation is expressed by the continuity equation

$$\frac{\partial \varrho}{\partial t} = -\operatorname{div} \boldsymbol{j} \,. \tag{2.4}$$

Since the viscosity of the normal-fluid component is very low – several orders of magnitude lower than that of water at 300 K – and its influence in most experiments is only a higher-order effect, we shall neglect the normal-fluid viscosity to a first approximation. In this case, helium II is considered as an ideal fluid, which can be described by the *Euler equation*, the equivalent of Newton's second law of motion for continua

$$\frac{\partial \boldsymbol{j}}{\partial t} + \underbrace{\varrho \boldsymbol{v} \cdot \operatorname{grad} \boldsymbol{v}}_{\approx 0} = -\operatorname{grad} p \,, \tag{2.5}$$

where p denotes the pressure. If the velocities of the two fluids are not too high, to a good approximation we can neglect terms quadratic in the velocities as indicated in (2.5).

Since viscous friction is neglected, no dissipative processes need to be considered and the motion of the fluids is reversible. Therefore, the entropy of helium II is conserved in this approximation, which can be expressed by the equation

$$\frac{\partial(\varrho S)}{\partial t} = -\mathrm{div}(\varrho S \boldsymbol{v}_\mathrm{n}), \tag{2.6}$$

where S represents the entropy per unit mass and ϱS the entropy density.

Finally, we need to set up the equations of motion for the two phases. A formal derivation of these equations, such as the one given by *Dingle* in [54], is very involved and will therefore not be presented here. Following *Landau*, we will motivate the equations of motion for the two components using a simple *thought experiment*. We imagine that a portion of superfluid phase is added to a helium II bath at constant volume. The resulting change of the internal energy is given by $\mathrm{d}U = T\,\mathrm{d}S - p\,\mathrm{d}V + G\,\mathrm{d}m$. Here, G denotes the Gibbs free energy per unit mass, which is here identical with the chemical potential μ. In the given situation we have $\mathrm{d}V = 0$ and $\mathrm{d}S = 0$ and the change of the internal energy reduces to $\mathrm{d}U = G\,\mathrm{d}m$. Consequently, the procedure described causes just a change of mass of the superfluid component. Therefore, we can interpret the Gibbs free energy G as the potential energy of this component and $-\mathrm{grad}\,G$ as the corresponding force. As a result, we find

$$\frac{\mathrm{d}\boldsymbol{v}_\mathrm{s}}{\mathrm{d}t} = -\mathrm{grad}\,\mu\,. \tag{2.7}$$

Using the thermodynamic relation

$$\mathrm{d}\mu = -S\,\mathrm{d}T + \frac{1}{\varrho}\,\mathrm{d}p\,, \tag{2.8}$$

we can substitute $\mathrm{grad}\,\mu$, and we obtain the equation of motion for the superfluid component

$$\frac{\partial \boldsymbol{v}_\mathrm{s}}{\partial t} = S\,\mathrm{grad}\,T - \frac{1}{\varrho}\,\mathrm{grad}\,p\,. \tag{2.9}$$

In writing (2.9) we have also performed the transition from the total to the partial differential $\partial \boldsymbol{v}_\mathrm{s}$. This is possible since the two differentials differ only by the term $\boldsymbol{v}_\mathrm{s}\cdot\mathrm{grad}\,\boldsymbol{v}_\mathrm{s}$ containing only quadratic terms in the velocity, which we have neglected in the approximation discussed here.

The equation of motion for the normal-fluid component can easily be calculated by substituting $\partial \boldsymbol{v}_\mathrm{s}/\partial t$ in (2.9) using (2.5). We obtain

$$\frac{\partial \boldsymbol{v}_\mathrm{n}}{\partial t} = -\frac{\varrho_\mathrm{s}}{\varrho_\mathrm{n}}\,S\,\mathrm{grad}\,T - \frac{1}{\varrho}\,\mathrm{grad}\,p\,. \tag{2.10}$$

We emphasize again that the equations obtained are only valid for the *linear regime*. Of course, more general considerations are possible and have been carried out, but at the expense of a significantly increased complexity.

2.2.2 Viscosity Measurements

In this section, we shall revisit the viscosity measurements, which have already been presented in Sect. 2.1. Here, we discuss the surprisingly different results in the context of the two-fluid model.

Flow Through Thin Capillaries Due to viscous damping the normal-fluid component is almost completely blocked in thin capillaries ($\boldsymbol{v}_\mathrm{n} \approx 0$). Only the superfluid component is mobile and is observed in such experiments. Since its motion is frictionless, the measured viscosity is zero below the lambda transition.

Rotary Viscosimeter A rotary viscosimeter consists of two hollow cylinders of different size, one rotating inside the other. The viscosity of the liquid between the cylinders is determined via the torque $M_\mathrm{r} = \pi\eta\omega d_\mathrm{r}^2 d_\mathrm{s}^2/(d_\mathrm{s}^2 - d_\mathrm{r}^2)$ transferred from the rotating inner cylinder with diameter d_r to the outer stationary cylinder with diameter d_s. Here, ω denotes the angular velocity of the rotation. Since the viscosity of the superfluid component is zero, it applies no torque onto the stationary cylinder. Therefore, only the viscosity of the normal-fluid component $M_\mathrm{r} \propto \eta = \eta_\mathrm{n}$ is measured in such experiments, which is nonzero even below T_λ.

The temperature dependence of η_n is mainly given by the mean free path ℓ_n of the thermal excitations in helium. The increase of the mean free path with decreasing temperature below about 1.8 K can be explained according to the theory of *Landau* and *Khalatnikov*, by the reduction of the scattering of thermal excitations. In their model, they assumed a dilute gas of excitations. This assumption is not valid above 1.8 K and therefore the temperature dependence of the normal-fluid viscosity in this temperature range cannot be explained with this theory.

Oscillating-Disc Viscosimeter The viscosity measurements made with this technique are based on the torque $M_\mathrm{d} = \pi\sqrt{\varrho\eta}\,\omega^{3/2} r^4\, \Theta(\omega)$ acting on an oscillating disc with radius r in the liquid. Here, $\Theta(\omega) = \Theta_0 \cos(\omega t - \pi/4)$ denotes the angle of deflection and ω the angular frequency of the oscillation. The crucial point is that, in this experiment, it is not just the viscosity that is measured, but the product $\varrho\eta$ of density and viscosity.

Since the superfluid component does not contribute below T_λ ($\eta_\mathrm{s} = 0$), the apparent viscosity is given by $\varrho_\mathrm{n}\eta_\mathrm{n}$. Therefore, the temperature dependence shown in Fig. 2.5 is understandable because the density of the normal-fluid component decreases rapidly below the lambda point. Using the result for η_n from the measurement with rotary viscosimeters, it is possible to draw conclusions about the temperature dependence of ϱ_n. The first direct measurement of this quantity is discussed in the next section.

2.2.3 Determination of ϱ_n/ϱ

A schematic drawing of the specially designed torsion oscillator that *Andronikashvili* used in 1948 to determine the normal-fluid density ϱ_n is shown in Fig. 2.12 [55]. The torsion pendulum consisted of 50 equally spaced aluminum discs with a diameter of 3.5 cm and a thickness of only 13 μm. The discs were mounted on a rod separated from each other by 210 μm. The rod with the discs was suspended by a torsion fiber forming an oscillator. The lower part of the oscillator was immersed in liquid helium. In the actual experiment, slow oscillations with a typical period of 30 s were excited and the resulting deflection amplitude was detected optically via a mirror that was rigidly attached to the axis of the pendulum.

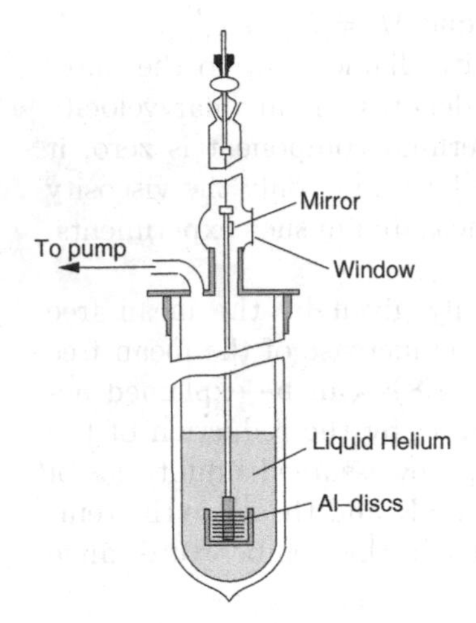

Fig. 2.12. Schematic drawing of the apparatus used by Andronikashvili to determine the normal-fluid density ϱ_n of helium II

The total mass of the pendulum bob determines its moment of inertia. The oscillator was driven at the resonant frequency of its torsional mode. This frequency depends on the ratio of the torsion constant and the moment of inertia of the pendulum. The discs of the viscosimeter were constructed from very light material to achieve as large a change as possible in the moment of inertia for a given change of the entrained mass. The spacing d between the discs and the period of oscillation were chosen so that the viscous penetration depth $\delta = \sqrt{2\eta_n/\varrho_n\omega}$ was larger than the distance between the aluminum discs throughout the entire experiment. Therefore, the complete normal-fluid component ϱ_n, but not the superfluid component ϱ_s, was dragged with the discs above and below the lambda point.

In this way, the temperature dependence of ϱ_n can be measured directly. The results of the original measurements by Andronikashvili are shown in Fig. 2.13. Above T_λ, all the fluid is dragged with the discs and the data reflect the temperature dependence of the density of helium I. Below T_λ,

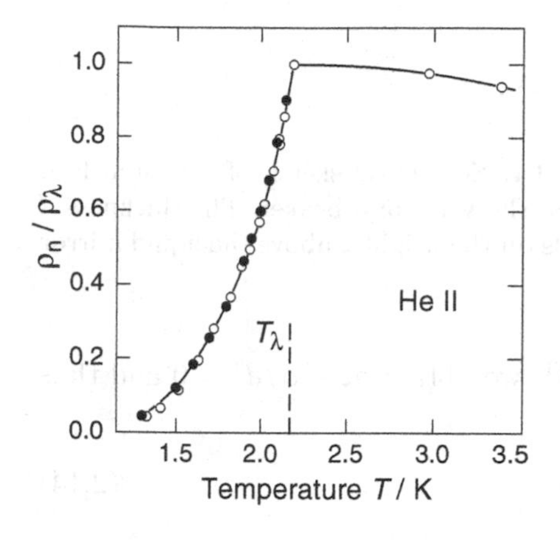

Fig. 2.13. Temperature dependence of the normal-fluid density ϱ_n normalized to the density ϱ_λ at T_λ. The data were obtained with two different methods: ∘ Andronikashvili viscosimeter, and • second-sound measurements (after [56])

however, the period of oscillation decreases sharply, indicating that the fluid in the spaces was not completely entrained by the discs. This result confirms the prediction that the superfluid component has no effect on the torsion pendulum. The temperature dependence of ϱ_n below the lambda point can be described by the empirical formula

$$\varrho_n = \varrho_\lambda \left(\frac{T}{T_\lambda}\right)^{5.6}, \tag{2.11}$$

where ϱ_λ denotes the density at the lambda point. For comparison, the normal-fluid density ϱ_n as determined by second-sound measurements is also shown in Figure 2.13. As can be seen, the results of the two methods are in very good agreement. In more recent experiments, the ratio ϱ_s/ϱ has been determined over a wider temperature range (see Sect. 2.5).

2.2.4 Beaker Experiments

The explanation of the peculiar behavior of helium II in the beaker experiments is straightforward. The geometrical conditions on which we base our considerations are shown in Fig. 2.14. A helium film forms on surfaces above a bath of helium II. The Van der Waals force between helium atoms and the walls plays a crucial role here. The properties of such helium films can be discussed in a simple way in terms of the chemical potential μ. For films in saturated vapor and thermodynamic equilibrium we have

$$\mu_f = \mu_g = \mu_\ell \,. \tag{2.12}$$

Here, the indices f, g and ℓ refer to film, gas and liquid.

The fact that the film is located on the beaker walls above the liquid surface indicates that the influence of gravity is compensated by Van der Waals forces. Therefore, we can write

$$\mu_f = \mu_\ell \underbrace{+\mu_{grav} + \mu_{vdW}}_{=0} = \mu_\ell \,. \tag{2.13}$$

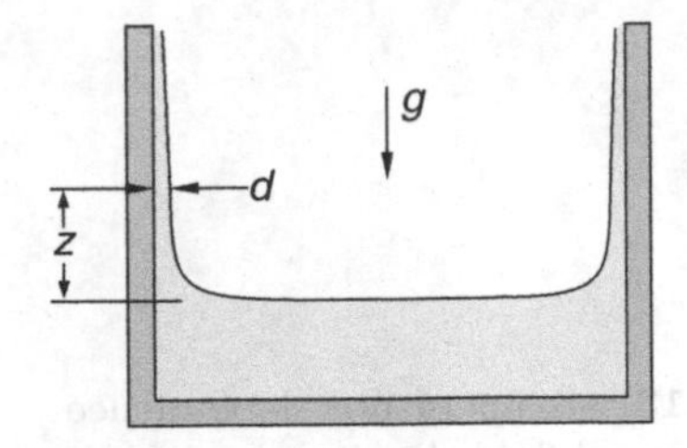

Fig. 2.14. Schematic sketch of a helium film covering the walls of a beaker. The thickness d depends on the height z above the liquid mirror surface

Inserting $\mu_{\rm grav} = gz$ and $\mu_{\rm vdW} = -\alpha/d^3$, we obtain $gz - \alpha/d^3 = 0$ and thus for the thickness[1]

$$d = \sqrt[3]{\frac{\alpha}{gz}} . \tag{2.14}$$

Here, α represents the *Hamaker constant*, the magnitude of which is determined by the dielectric properties of the wall and the helium atoms. Essentially, the atomic polarizability enters here. In saturated vapor, we find at $z = 10\,\mathrm{cm}$ a typical film thickness of 200 Å. The superfluid properties are *unimportant* for the film thickness, but play a crucial role in the film flow. The superfluid component can flow without friction so as to minimize the chemical potential in the entire system. An interesting point is the behavior of the entropy, because the rest of the fluid in the beaker should warm up proportional to $\varrho_{\rm n}/\varrho_{\rm s}$ if the superfluid component is flowing out. However, this is not observed in experiments because the surrounding helium gas leads to thermalization.

Films in an unsaturated vapor can be described in a similar way. According to the kinetic theory of gases, the chemical potential of an unsaturated vapor is given by the expression

$$\mu_{\rm g}(p) = \mu_{\rm g}(p_0) + \frac{k_{\rm B}T}{m_4} \ln\left(\frac{p}{p_0}\right) . \tag{2.15}$$

Here, p_0 denotes the saturated-vapor pressure of the liquid, R the universal gas constant and m_4 the mass of a helium atom. As we have seen before, the chemical potential in saturated vapor is identical with that of the liquid $\mu_{\rm g}(p_0) = \mu_\ell$. In very thin films, the contribution of gravity to the chemical potential of the film is negligible. In this case, we obtain from $\mu_{\rm g}(p) = \mu_{\rm f} = \mu_\ell + \mu_{\rm vdW}$ the relation

$$\frac{\alpha}{d^3} = \frac{k_{\rm B}T}{m_4} \ln\left(\frac{p_0}{p}\right) , \tag{2.16}$$

and thus the film thickness is given by

[1] Note that $\mu_{\rm vdW} = -\alpha/d^3$ is only valid for thin films ($d < 30\,\mathrm{nm}$) because in thicker films the retardation of the Van der Waals forces due to the finite light velocity has to be taken into account. For very thick films ($d > 80\,\mathrm{nm}$) the chemical potential is better described by $\mu = -\alpha/d^4$, resulting in $z \propto d^4$.

$$d = \sqrt[3]{\frac{\alpha\, m_4}{k_{\mathrm{B}}T(\ln p_0 - \ln p)}}\,. \tag{2.17}$$

This expression corresponds to a Van der Waals *adsorption isotherm* and reflects the fact that the thickness of the film can be reduced by lowering the pressure. Even submonolayers can be produced in this way. In experiments with very thin films it has been demonstrated that the onset of superfluidity occurs at about 2.1 monolayers, and that the first layer is solid [57].

A plot of (2.17) is shown in Fig. 2.15. Neglecting gravity, arbitrarily thick films can, in principle, be produced for pressures $p \to p_0$.

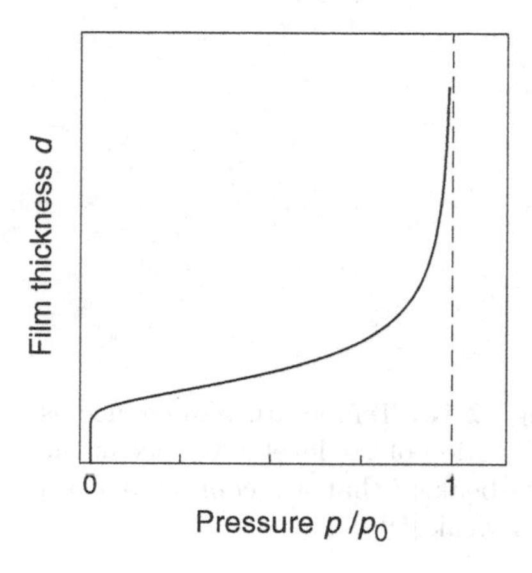

Fig. 2.15. Thickness of helium films as a function of pressure according to (2.17)

2.2.5 Thermomechanical Effect

The nature of the experiments showing the thermomechanical effect have been discussed in Sect. 2.1. Two beakers (A and B) filled with helium II are connected at the bottom via a thin capillary, which acts as a so-called *superleak*, which blocks the normal-fluid component and allows the superfluid component to pass through. Increasing the pressure in A results in a flow of superfluid component through the superleak to B. Because of this, the ratio of ϱ_s/ϱ in both beakers changes and thus a temperature gradient between A and B rises. In equilibrium, we can describe the situation by using (2.9)

$$\frac{\partial \boldsymbol{v}_{\mathrm{s}}}{\partial t} = S\,\mathrm{grad}\,T - \frac{1}{\varrho}\,\mathrm{grad}\,p = 0 \tag{2.18}$$

and obtain

$$\frac{\Delta p}{\Delta T} = \varrho S\,. \tag{2.19}$$

The latter expression was first derived by *H. London* and thus is often called the *London equation* [58]. Figure 2.16 shows experimental values of the temperature difference between the two beakers as a function of the level difference, which corresponds to a pressure difference.

The data demonstrate nicely the linear relation between Δp and ΔT predicted by (2.19). As expected, the thermomechanical effect weakens with increasing temperature. At $\Delta h = 2\,\mathrm{cm}$ and $T = 1.5\,\mathrm{K}$ a temperature difference of about 1 mK is found. In principle, one could use the thermomechanical effect for cooling, but this method is very inefficient.

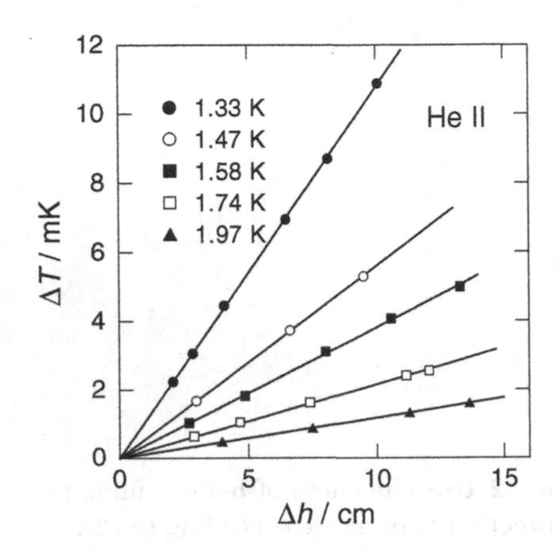

Fig. 2.16. Temperature difference as a function of the level difference in the two beakers that are connected via a superleak [59]

2.2.6 Heat Transport

So far we have discussed the mass transport of helium II in thin capillaries under the assumption that the normal-fluid component is completely blocked. However, for capillaries with finite width this is only approximately true. In fact, even in equilibrium ($\Delta p = \varrho S \Delta T$), there is always a flow of normal-fluid component ϱ_n from the warm to the cold end. The superfluid component ϱ_s moves in the opposite direction. Because of the difference in entropy of the two components, this counterflow is associated with entropy transport and thus with the transport of heat. The heat flow is only limited by the viscosity of the normal-fluid component. We describe the flow of ϱ_n within classical hydrodynamics by

$$\dot{V}_\mathrm{n} = \frac{\beta}{\eta_\mathrm{n}} \frac{\Delta p}{L} . \tag{2.20}$$

Here, L denotes the length of the flow channel and β is a constant that is determined by the geometry of the flow channel. For capillaries the *Hagen–Poiseuille law* is valid and thus $\beta \propto r^4$. For heat transport through small slits with width d, we have $\beta \propto d^3$. The entropy transported is given by $\dot{V}_n \varrho S$. Under the assumption that the flow of the two components is reversible we find

$$\dot{Q} = T \dot{V}_n \varrho S \,. \tag{2.21}$$

After inserting this result in (2.20) and using the London equation (2.19) we eventually obtain:

$$\dot{Q} = \frac{\beta T (\varrho S)^2}{\eta_n L} \Delta T \,. \tag{2.22}$$

As expected for the linear regime, the heat flow is proportional to ΔT. Equation (2.22) also agrees quantitatively with the data very well. An example of a measurement of the heat transport through slits with different widths is shown in Fig. 2.17. At all three temperatures one finds $\dot{Q} \propto \beta \propto d^3$, in agreement with (2.22).

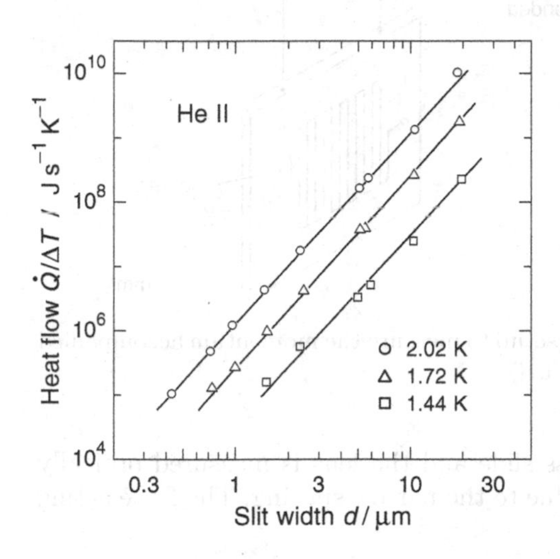

Fig. 2.17. Heat flux $\dot{Q}/\Delta T$ of helium II at three different temperatures as a function of the slit width d. The *solid lines* correspond to a variation of the heat transport proportional to d^3 [51]

2.2.7 Momentum of the Heat Flow

A further remarkable feature of the heat flow in helium II that we have not discussed yet is the fact that a momentum transport is associated with a heat current in helium II. This phenomenon was first discovered by *Kapitza* in 1941 [60].

The momentum current per unit volume is given by $\varrho \boldsymbol{v} \cdot \boldsymbol{v}$. The resulting pressure acting on a heat source is therefore

$$p = \varrho_n v_n^2 + \varrho_s v_s^2 \,. \tag{2.23}$$

We write $\varrho_n v_n + \varrho_s v_s = 0$ since no mass transport takes place. If we take into account that the heat flux per unit area is given by $W = \varrho S T v_n$ we find that the pressure associated with a unidirectional heat current is given by

$$p = \frac{\varrho_n W^2}{\varrho_s \varrho T^2 S^2} = \frac{W^2}{v_2^2 \varrho C T} . \tag{2.24}$$

An experimental setup for the determination of the momentum of the heat flux is sketched in Fig. 2.18. An electrical heater evaporated on a thin glass slide (drawn enlarged and somewhat rotated on the right side of the figure) is suspended freely in front of a convex lens. The whole device is immersed in helium II. Because of the poor thermal conductance of the glass lens, the heat flux generated by the heater flows through the tube. Thus, the heat flow causes a deflection of the glass slide towards the lens.

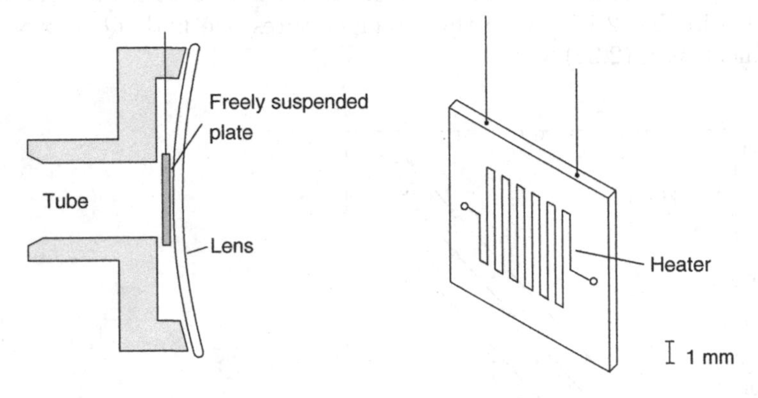

Fig. 2.18. Sketch of an experimental setup to measure the momentum accompanied by a heat current in helium II (after [61])

The distance between the glass slide and the lens is measured optically via Newton rings that are visible due to the narrow spacing. The force acting on the glass slide is given by

$$F = \frac{\beta' \dot{Q}^2}{A} \left(\frac{1}{v_2^2 \varrho C_p T} \right) . \tag{2.25}$$

The quantity β' is a geometry-dependent constant, which is mainly determined by the size of the heater, and A stands for the cross-sectional area of the tube. The results of measurements with sightly different experimental setups ($\beta' = 0.22 \ldots 0.75$) are shown in Fig. 2.19. The force on the glass slide increases near T_λ and at low temperatures, since the condition for massless heat transport $\varrho_n v_n + \varrho_s v_s = 0$ has to be fulfilled. The two-fluid model (solid line) provides a very good description over the whole temperature range. We should emphasize that there is no adjustable parameter used in the calculation of the theoretical curve.

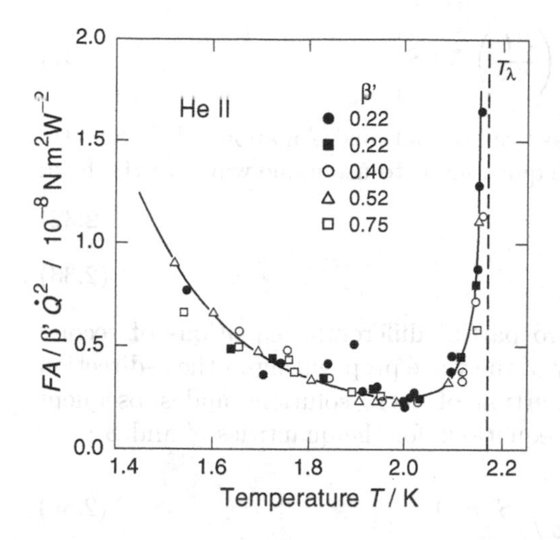

Fig. 2.19. Normalized force acting on a glass slide caused by a unidirectional heat flux in helium II as a function of temperature. The *solid line* represents a theoretical calculation according to the two-fluid model [61]

2.2.8 Sound Propagation

In this section, we use the hydrodynamic equations to discuss the propagation of sound in helium II. First, we differentiate (2.4) with respect to time and insert the result into (2.5). One obtains

$$\frac{\partial^2 \varrho}{\partial t^2} = \nabla^2 p \,. \tag{2.26}$$

In the next step we eliminate all terms containing $\boldsymbol{v}_\mathrm{s}$ and $\boldsymbol{v}_\mathrm{n}$ in (2.6) and (2.9), since these quantities cannot be observed experimentally. After some simple algebra, and disregarding terms of higher order, one finally obtains

$$\frac{\partial^2 S}{\partial t^2} = \frac{\varrho_\mathrm{s}\, S^2}{\varrho_\mathrm{n}} \nabla^2 T \,. \tag{2.27}$$

Using these two equations the sound propagation in helium II can thoroughly be discussed in the framework of the approximations made above. We have in total four variables (ϱ, S, p and T), only two of which are independent. In the following, we chose the density ϱ and the entropy S as independent quantities and express their variation with pressure and temperature by

$$\delta p = \left(\frac{\partial p}{\partial \varrho}\right)_S \delta\varrho + \left(\frac{\partial p}{\partial S}\right)_\varrho \delta S \,, \tag{2.28}$$

$$\delta T = \left(\frac{\partial T}{\partial \varrho}\right)_S \delta\varrho + \left(\frac{\partial T}{\partial S}\right)_\varrho \delta S \,. \tag{2.29}$$

Insertion of these two expressions into (2.26) and (2.27) leads to:

$$\frac{\partial^2 \varrho}{\partial t^2} = \left(\frac{\partial p}{\partial \varrho}\right)_S \nabla^2 \varrho + \left(\frac{\partial p}{\partial S}\right)_\varrho \nabla^2 S \,, \tag{2.30}$$

$$\frac{\partial^2 S}{\partial t^2} = \frac{\varrho_s}{\varrho_n} S^2 \left[\left(\frac{\partial T}{\partial \varrho}\right)_S \nabla^2 \varrho + \left(\frac{\partial T}{\partial S}\right)_\varrho \nabla^2 S \right] . \tag{2.31}$$

Equations of this form are known as 'wave equations'. An approach frequently employed in physics to solve such equations is to use plane waves of the form

$$\varrho = \varrho_0 + \varrho' \, e^{i\omega(t-x/v)} , \tag{2.32}$$

$$S = S_0 + S' \, e^{i\omega(t-x/v)} , \tag{2.33}$$

as an ansatz for solving these two partial differential equations of second order. Here, v denotes the velocity of the wave propagating in the x-direction and ω the angular frequency. Insertion of these solutions and subsequent differentiation leads to two linear equations for the quantities ϱ' and S':

$$\left[\left(\frac{v}{v_1}\right)^2 - 1 \right] \varrho' + \left(\frac{\partial p}{\partial S}\right)_\varrho \left(\frac{\partial \varrho}{\partial p}\right)_S S' = 0 , \tag{2.34}$$

$$\left(\frac{\partial T}{\partial \varrho}\right)_S \left(\frac{\partial S}{\partial T}\right)_\varrho \varrho' + \left[\left(\frac{v}{v_2}\right)^2 - 1 \right] S' = 0 . \tag{2.35}$$

Here, the abbreviations

$$v_1^2 = \left(\frac{\partial p}{\partial \varrho}\right)_S \quad \text{and} \quad v_2^2 = \frac{\varrho_s}{\varrho_n} S^2 \left(\frac{\partial T}{\partial S}\right)_\varrho \tag{2.36}$$

were used. Thus, the problem is reduced to solving a system of linear equations. The constraints equation for the coefficients is given by

$$\left[\left(\frac{v}{v_1}\right)^2 - 1 \right] \left[\left(\frac{v}{v_2}\right)^2 - 1 \right] = \left(\frac{\partial p}{\partial S}\right)_\varrho \left(\frac{\partial \varrho}{\partial p}\right)_S \left(\frac{\partial T}{\partial \varrho}\right)_S \left(\frac{\partial S}{\partial T}\right)_\varrho . \tag{2.37}$$

Using thermodynamic identities, this equation can be transformed into

$$\left[\left(\frac{v}{v_1}\right)^2 - 1 \right] \left[\left(\frac{v}{v_2}\right)^2 - 1 \right] = \frac{C_p - C_V}{C_p} . \tag{2.38}$$

Each of the two expressions in the square brackets is a dispersion relation for a certain type of wave. Both waves are weakly coupled through $(C_p - C_V)/C_p$. Since the specific heats at constant pressure C_p and at constant volume C_V are almost identical for helium II we may approximate (2.38) by

$$\left[\left(\frac{v}{v_1}\right)^2 - 1 \right] \left[\left(\frac{v}{v_2}\right)^2 - 1 \right] \approx 0 . \tag{2.39}$$

In the following sections we will discuss further details regarding the propagation of sound waves. In these considerations we will always neglect the coupling between the two types of waves.

First Sound

The propagation of first sound with velocity v_1 is given by (2.34) in the case where $\varrho' \neq 0$ and $S' = 0$. Under these conditions the temperature gradient vanishes, i.e. $\operatorname{grad} T \approx 0$, just as it does for an ordinary sound wave. We can use this approximation together with (2.5) and (2.9) to obtain the important relation

$$\varrho_\mathrm{n} \frac{\partial}{\partial t} (\boldsymbol{v}_\mathrm{n} - \boldsymbol{v}_\mathrm{s}) = \varrho S \operatorname{grad} T = 0 \,. \tag{2.40}$$

Therefore, we finally get

$$\boldsymbol{v}_\mathrm{n} = \boldsymbol{v}_\mathrm{s} \,. \tag{2.41}$$

From this we can conclude that the two components of helium II move in phase, resulting in an adiabatic density variation. In this case, helium II behaves like an ordinary liquid in which sound waves propagate with velocity $v = v_1$. For the low-temperature limit $T \to 0$, we find $v_1 \approx 238\,\mathrm{m\,s^{-1}}$. This result is strictly valid only if the two types of waves are completely decoupled. The actual corrections are important only near the lambda point.

The temperature dependence of the velocity of first sound in liquid helium is shown in Fig. 2.20. The data points are from measurements at 1 MHz and 14 MHz. The velocity of first sound increases from about $180\,\mathrm{m\,s^{-1}}$ at the boiling point of helium to $238\,\mathrm{m\,s^{-1}}$ at low temperatures. As expected, an anomaly in the elastic properties of liquid helium is visible in the vicinity of the lambda transition.

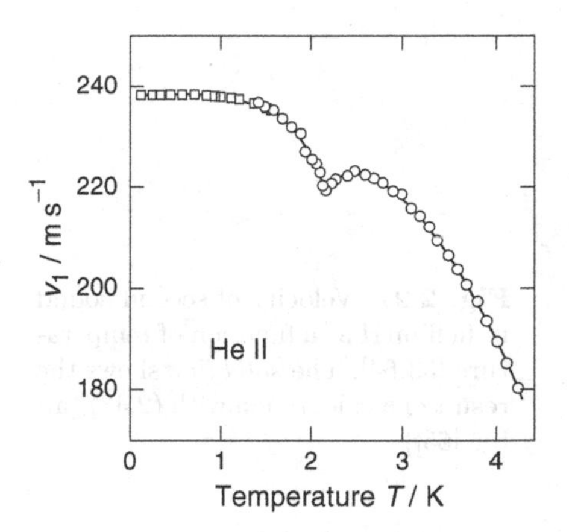

Fig. 2.20. Velocity of first sound in liquid helium as a function of temperature [62, 63]

Second Sound

In the case where $S' \neq 0$ and $\varrho' = 0$, we find temperature waves described by (2.35), which propagate with the velocity $v = v_2$. From (2.26) it follows that $\operatorname{grad} p = 0$, and together with the Euler equation (2.5) we obtain

$$\frac{\partial(\varrho_n \boldsymbol{v}_n)}{\partial t} + \frac{\partial(\varrho_s \boldsymbol{v}_s)}{\partial t} = 0, \tag{2.42}$$

which means that the momentum density $\boldsymbol{j} = \varrho_n \boldsymbol{v}_n + \varrho_s \boldsymbol{v}_s$ is either constant or zero. Since no constant mass flow can occur in a closed container, it follows that $\varrho_n \boldsymbol{v}_n + \varrho_s \boldsymbol{v}_s = 0$. This means that the motion of the two components takes place in phase opposition. The velocity of the temperature waves that propagate through helium II is given by

$$v_2 = \sqrt{\frac{\varrho_s}{\varrho_n} S^2 \left(\frac{\partial T}{\partial S}\right)_\varrho} = \sqrt{\frac{\varrho_s}{\varrho_n} \frac{T S^2}{C_p}}. \tag{2.43}$$

As we will see, according to the Landau model of helium II, the only excitations at very low temperatures are phonons. In the framework of this description we expect for the velocity of second sound $v_2 \to v_1/\sqrt{3} \approx 137\,\mathrm{m\,s^{-1}}$ for $T \to 0$. This limiting value is obtained by insertion of $\varrho_s \approx \varrho$, $S = AT^3$, $C_p = 3AT^3$ and $\varrho_n = A\varrho T^4/v_1^2$, with $A = 2\pi^2 k_B^4/(45\hbar^3 v_1^3 \varrho)$. Here, we have used in advance the expressions (2.87) and (2.89) for C_p and ϱ_n, respectively.

The temperature dependence of the sound velocity of second sound is shown in Fig. 2.21. Over a wide temperature range the experimental data agrees very well with the calculated curve. Below 0.5 K the generation and detection of second sound becomes very difficult. Thus, the limiting value for $T \to 0$ of approximately $v_1/\sqrt{3}$ cannot be verified experimentally.

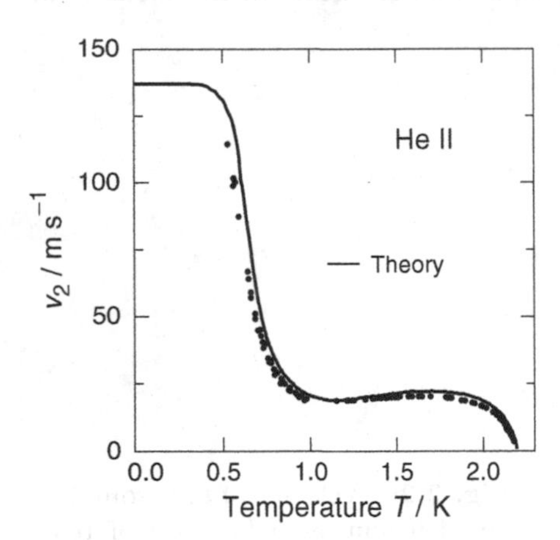

Fig. 2.21. Velocity of second sound in helium II as a function of temperature [53,64]. The *solid line* shows the result of a calculation with (2.43) (after [65])

It is interesting to note that in the first attempts to look for second sound in helium II it was not attempted to generate directly temperature waves via periodical heating, but rather one tried to detect the (weak) temperature wave that accompanies the propagation of ordinary sound in helium II. Such a temperature variation occurs because the density variation due to the first sound leads to slight local variations of the ratio ϱ_s/ϱ_n, which is equivalent to a temperature change (see (2.37)).

Third Sound

Propagating surface waves on thin helium films are called *third sound*. Such waves can be produced by local periodic heating of helium II films in the audio-frequency range. The heating causes oscillations of the film thickness, which then propagate as a surface wave on the film.

The profile of the film and thus the amplitude of the thickness oscillations can be measured using optical methods.[2] A third-sound experiment is illustrated in Fig. 2.22. With such a setup it is possible to determine the velocity and the damping of third sound. In the case where the liquid levels in both beakers are different, third sound in moving films can be investigated. We will return to this interesting option at the end of this section, but for now we will assume stationary films.

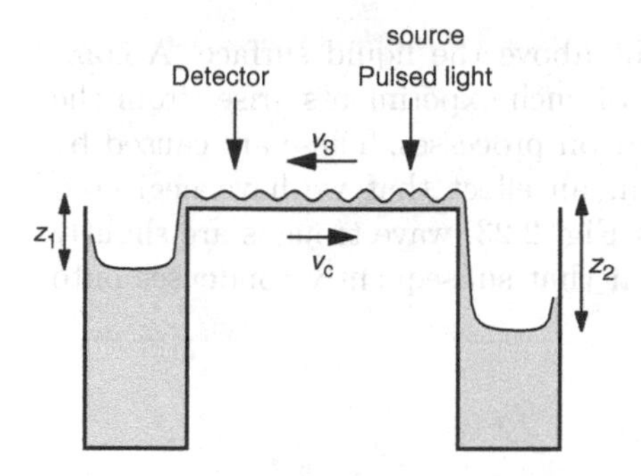

Fig. 2.22. Schematic illustration of an experimental setup to measure the velocity of third sound. The quantity v_c denotes the flow velocity, which under most experimental conditions is identical with the critical velocity, except for very small level differences

Under most experimental conditions, the wavelength is much larger than the film thickness ($\lambda \gg d$) and thus helium can only move parallel to the substrate ($v_y = v_z = 0$). The motion of the normal-fluid component is strongly damped due to its viscous friction. We therefore can consider ϱ_n, to a good approximation, as completely immobile ($\boldsymbol{v}_n = 0$). In contrast, the superfluid component is mobile and flows to the heated region due to the fountain effect. There, it causes a local increase $\xi(x,t)$ of the film thickness d. The resulting thickness variation propagates like a surface wave. Under these conditions the continuity equation (2.4) can be written in the form

$$\varrho \frac{\partial \xi}{\partial t} = -\varrho_s d \frac{\partial v_{sx}}{\partial x} \,. \tag{2.44}$$

Using (2.7) and assuming for a rough approximation that there are no temperature variations in the film ($\operatorname{grad} T \approx 0$), we obtain $\dot{v}_{sx} = -\varrho^{-1}(\partial p/\partial x)$. Neglecting gravity, the pressure can be expressed in terms of the Van der

[2] We should note that in most experiments nowadays third sound is not generated and detected optically, but is investigated in specially designed resonators in which standing waves of third sound can be produced (see, for example, [66]).

Waals force $f = 3\alpha/d^4$, and is given by $p(x) = \varrho f\,[d + \xi(x)]$. Inserting this relation leads to the equation

$$\frac{\partial v_{\mathrm{sx}}}{\partial t} = f\frac{\partial \xi}{\partial x}\,. \tag{2.45}$$

By differentiating (2.44) with respect to time and (2.45) with respect to the x-coordinate, and combining the results we finally obtain the wave equation

$$\frac{\partial^2 \xi}{\partial t^2} = fd\frac{\varrho_{\mathrm{s}}}{\varrho}\frac{\partial^2 \xi}{\partial x^2}\,, \tag{2.46}$$

which we solve assuming a plane wave of the form $\xi = \xi_0 \exp[\mathrm{i}\omega(t - x/v)]$. Inserting (2.14) for the thickness d of the film, we finally find for the velocity of third sound in saturated films

$$v_3^2 = \frac{\varrho_{\mathrm{s}}}{\varrho}\,3gz\,, \tag{2.47}$$

where $z = z_1 = z_2$ represents the height above the liquid surface. A complication in the quantitative description of such experiments arises from the accompanying evaporation and condensation processes. These are caused by local temperature differences in the film, an effect that we have neglected in our treatment above. As visualized in Fig. 2.23, wave troughs are slightly warmer, leading to evaporation of helium that subsequently condenses onto the colder crests.

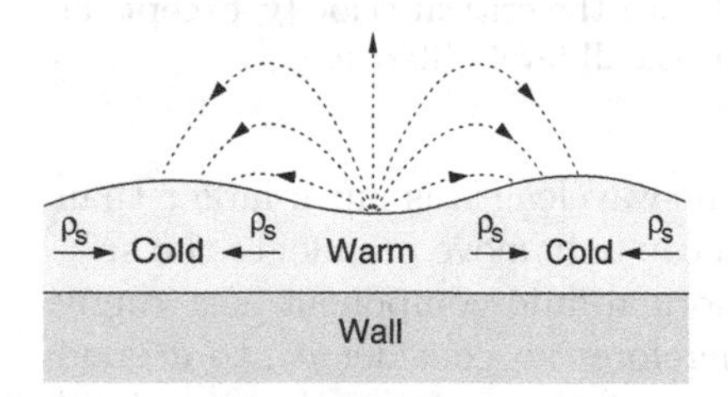

Fig. 2.23. Schematic illustration of the evaporation and condensation effect accompanying the propagation of third sound in superfluid helium films

This mechanism not only enhances the amplitude of the third-sound wave, but also influences its velocity. In a more elaborate model, one can find the following expression [67] for the velocity of third sound:

$$v_3^2 \approx \frac{\varrho_{\mathrm{s}}}{\varrho}\,3gz\left(1 + \frac{TS}{L}\right)\,. \tag{2.48}$$

Here, L denotes the latent heat of vaporization. The factor TS/L is about 0.15 at T_λ and decreases with decreasing temperature to a value of only 0.01 at $T = 1\,\mathrm{K}$. Figure 2.24 shows the velocity of third sound in helium II films on different substrates (• roughened steel, ◦ polished steel, ■ nickel) as a function of the height z above the liquid level. According to (2.14) each height corresponds to a certain film thickness. The *solid line* represents the dependence of v_3 on z as expected from (2.47). It fits the data reasonably

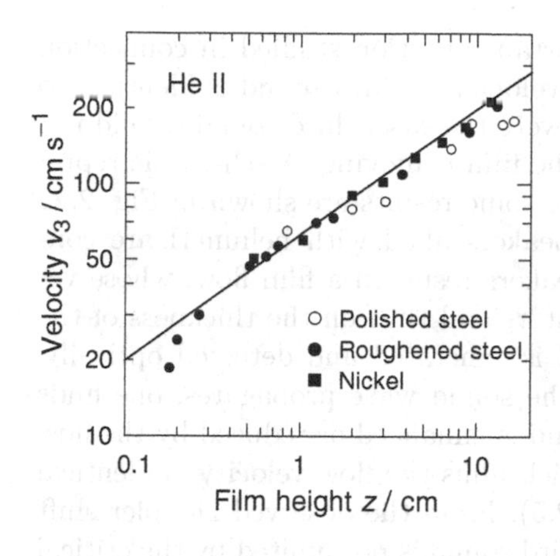

Fig. 2.24. Velocity of third sound as a function of the height z above the liquid level [68]. The *solid line* represents the variation according to (2.47)

well over a wide range of film heights. According to the data in Fig. 2.24, the velocity of third sound seems to be almost independent of the substrate.

In Fig. 2.25 the temperature dependence of the velocity of third sound is shown. With decreasing temperature, v_3 increases, and for $T \to 0$ reaches a value of $150\,\mathrm{cm\,s^{-1}}$ for films 9 cm above the liquid level. The data shown in Fig. 2.25 have been taken at different heights above the liquid level corresponding to different film thicknesses and are normalized for each thickness at 1.25 K (*filled circle*). The temperature dependence is mainly determined by the temperature dependence of the superfluid density ϱ_s. Surprisingly, there seems to be a systematic deviation between the experimental data and the theoretical fit with (2.48). The origin of this discrepancy is unknown, but it has been speculated that the excitation process of third sound somehow affects the film thickness.

An interesting variation on such experiments is the investigation of third sound in moving films. In this case, the velocity of third sound is superim-

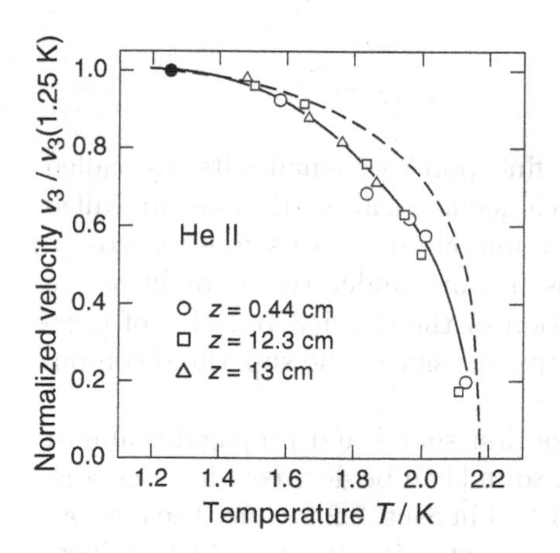

Fig. 2.25. Velocity of third sound normalized to the velocity at 1.25 K versus temperature. The measurements were carried out at different heights z [68]. The *dashed line* represents the variation according to (2.48)

posed upon the flow velocity. An important question studied in connection with these experiments is whether the velocity of third sound is identical to the critical velocity of film flow. If this were the case, third sound should not be observed in the direction in which the film is moving. A schematic representation of the experimental setup and some results are shown in Fig. 2.22 and Fig. 2.26, respectively. The two beakers filled with helium II are connected. Different liquid levels in the beakers result in a film flow, whose velocity depends on the liquid-level height z_1 and thus on the thickness of the film. Using a pulsed laser, third sound is generated and detected optically. Depending on the direction in which the sound wave propagates, one finds that the measured velocity of third sound is enhanced or reduced by the flow velocity of the film. Except for very thick films the flow velocity is identical to the critical velocity (see also Sect. 2.5). From the observed Doppler shift one can conclude that the velocity of third sound is not limited by the critical velocity of film flow.

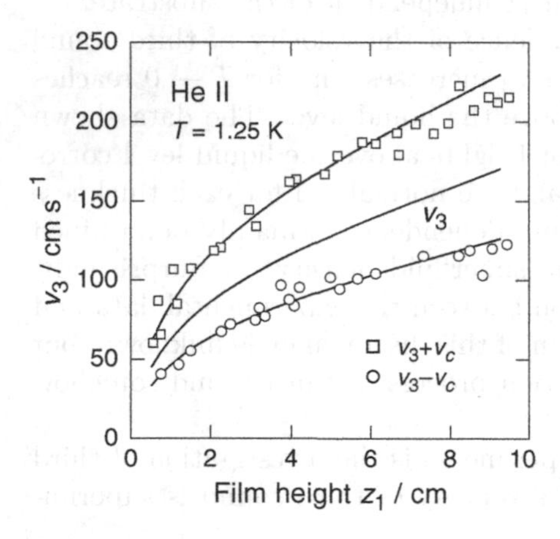

Fig. 2.26. Velocity of third sound in moving helium II films as a function of the height z_1 above the liquid level [69]

Fourth Sound

Compression waves propagating inside fine pores or small slits are called *fourth sound*. As already mentioned, such geometrical restrictions are called superleaks if the normal component is immobile due to its finite viscosity. Since only the superfluid component is moving under these conditions, a compression wave is not just an oscillation of the density, but also of pressure, temperature, entropy and the relative density of the superfluid component ϱ_s/ϱ.

This type of wave can be excited like first sound and propagates almost without damping. The velocity of fourth sound can be derived within nondissipative hydrodynamics, which was described in Sect. 2.2.1. Since the normal-fluid component is immobile ($\boldsymbol{v}_n \approx 0$), the continuity equation (2.4) reduces to $\dot{\varrho} + \varrho_s \operatorname{grad} \boldsymbol{v}_s = 0$, and the entropy conservation (2.6) becomes $\dot{S} = 0$. Using (2.7) to eliminate $\boldsymbol{v}_s$ allows us to construct a wave equation for the sound propagation in fine pores from which the velocity of fourth sound $v_4^2 = \varrho_s(\partial\mu/\partial\varrho)_S$ can be deduced. We find

$$v_4^2 = \frac{\varrho_s}{\varrho}\, v_1^2 \left[1 + \frac{2ST}{\varrho C_p}\left(\frac{\partial\varrho}{\partial T}\right)_p\right] + \frac{\varrho_n}{\varrho}\, v_2^2\,, \tag{2.49}$$

with $\varrho^{-1}(\partial\varrho/\partial T)_p$ being the isobaric expansion coefficient. The second term in the square brackets leads to a correction of only about 2% at T_λ, and is even smaller at lower temperatures. Therefore, to a good approximation the expression

$$v_4 \approx \sqrt{\frac{\varrho_s}{\varrho}\, v_1^2 + \frac{\varrho_n}{\varrho}\, v_2^2} \tag{2.50}$$

can be used for the velocity of fourth sound. As $T \to 0$ all helium atoms belong to the superfluid phase and we thus obtain $v_4 = v_1 \approx 238\,\mathrm{m/s}$. With increasing temperature the velocity of fourth sound decreases and reaches $v_4 = 0$ at $T = T_\lambda$. The contribution of second sound in (2.50) dominates near the lambda point. Under certain experimental conditions only this contribution is observed and is often referred to as *fifth sound* [70, 71]. The velocity of fifth sound is therefore given by $v_5 = \sqrt{(\varrho_n/\varrho)}\, v_2$.

Experimental data showing the temperature dependence of the velocity of fourth sound in helium II are shown in Fig. 2.27. For comparison, the corresponding data of first, second and fifth sound are also included. In this experiment, a cylindrical resonator filled with fine compressed powder was used as a superleak. The average grain size was about 0.5 μm and the porosity was approximately 50%.

Particularly interesting experiments with fourth sound can be performed using experimental arrangements such as those used in persistent-current

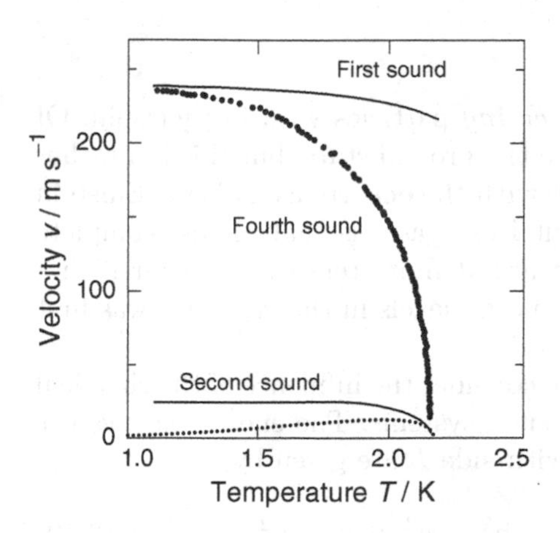

Fig. 2.27. Temperature dependence of the velocity of fourth sound in helium II in comparison with first and second sound [72]. Fifth sound is indicated by the *dotted line* [70]

measurements. A torus filled with fine powder and liquid ^{4}He is set in rotation above the lambda point and is subsequently cooled below T_λ. The rotation of the torus is gently slowed until the torus is at rest. The superfluid component still rotates after this operation. Fourth sound generated under such conditions shows a Doppler shift, given to a good approximation by

$$v_4 \approx v_{4,0} \pm \frac{\varrho_\mathrm{s}}{\varrho} v_\mathrm{D} . \tag{2.51}$$

Here, v_D denotes the flow velocity of the persistent current and $v_{4,0}$ the velocity of fourth sound in stationary helium II. The factor ϱ_s/ϱ takes into account the coupling between the compression wave and second sound.

2.3 Bose–Einstein Condensation

In the previous sections we saw that the superfluid component of liquid helium can flow without any friction. *F. London* suggested in 1938 that this dissipationless flow is related to the frictionless motion of electrons in atomic shells [11]. In atoms, the electrons are in stationary quantum states that are described by the eigenfunctions of the corresponding Hamiltonian. The phases of the wave functions of different atoms in a liquid are not correlated under usual conditions. London assumed that in helium II the wave function is well defined throughout the entire liquid, analogous to the situation in superconductors. We shall see that the superfluid component of helium II can indeed be described by a *macroscopic wave function*. As the possible origin of such a macroscopic wave function in helium II, London discussed the so-called Bose–Einstein condensation. In the following section we will take a brief look at this phenomenon for the case of an ideal Bose gas, although it is clear that the description of liquid ^{4}He as an ideal Bose gas can, at best, be a crude approximation.

2.3.1 Ideal Bose Gas

An ideal Bose gas is a gas of *noninteracting* particles with integer spin. Of course, at $T = 0$ all Bose particles are in the ground state, but this is a rather trivial statement and has nothing to do with the occurrence of Bose–Einstein condensation. The peculiar thing about Bose gases is that almost complete condensation into the ground state occurs at finite temperatures far higher than the corresponding spacing of the energy levels in the gas. This was first realized in 1924 by *Einstein* [73].

To understand this phenomenon, we consider the influence of the chemical potential μ on the level occupation in such systems. The energy eigenvalues of free atoms with mass m in a cube with side L are given by

$$E_n = \frac{\hbar^2}{2m}\left(\frac{\pi}{L}\right)^2 n^2 , \qquad \text{with} \qquad n^2 = n_x^2 + n_y^2 + n_z^2 . \tag{2.52}$$

If we assume a cube of liquid helium with volume $V = 1\,\text{cm}^3$, there are roughly 10^{22} atoms in this cube under normal pressure, and the energy difference between the ground state and the first excited state is about

$$\Delta E/k_\text{B} = (E_{211} - E_{111})/k_\text{B} \approx 2 \times 10^{-14}\,\text{K}. \tag{2.53}$$

This energy difference is extremely small and one would certainly not expect any significant influence on the experimental results at temperatures that can presently be reached. In a classical description using Boltzmann statistics, all levels below 1 K are occupied roughly with equal probability. At low temperatures, a classical description, however, is not adequate, since ^{4}He atoms are Bose particles and therefore obey Bose–Einstein statistics with the distribution function

$$f(E,T) = \frac{1}{\mathrm{e}^{(E-\mu)/k_\text{B}T} - 1}. \tag{2.54}$$

It follows that the chemical potential cannot be larger than the energy of the lowest level because otherwise negative occupation numbers would occur. Furthermore, in the case where the Bose particles are helium atoms the number of particles is conserved, and therefore the chemical potential is nonzero.[3]

First, let us consider the population of the ground state and set the energy scale so that we have $E_{111} = 0$. The occupation of the ground state is then given by

$$f(0,T) = \frac{1}{\mathrm{e}^{-\mu/k_\text{B}T} - 1}. \tag{2.55}$$

Thus, the population of the ground state for $T \to 0$ crucially depends on the chemical potential. If μ goes to zero faster than the temperature, $f(0,T)$ will go to infinity for $T \to 0$. This is also valid in cases in which the chemical potential reaches zero at finite temperatures.

The chemical potential of a gas depends on the ratio of atomic volume $V_\text{A} = V/N$, and quantum volume V_Q. For a gas of noninteracting particles we have

$$\mu = -k_\text{B}T \ln\left(\frac{V_\text{A}}{V_\text{Q}}\right). \tag{2.56}$$

Here, V_Q corresponds to the volume of a cube, whose side length is given by the mean *thermal de Broglie wavelength* $\lambda_\text{B} = h/\sqrt{2\pi m k_\text{B}T}$. It follows that at sufficiently low temperatures the condition $V_\text{A} = V_\text{Q}$ will be reached and thus μ will go to zero. This behavior is schematically shown in Fig. 2.28a. For comparison, we note that the chemical potential of helium gas at room temperature is $\mu/k_\text{B} = -3800\,\text{K}$.

[3] For bosons, for instance photons, whose particle number is not conserved, the chemical potential is exactly zero.

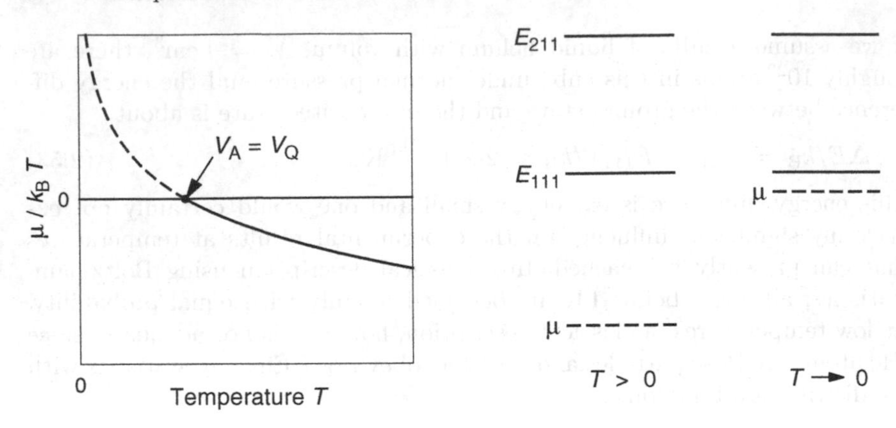

Fig. 2.28. (**a**) Schematic plot of the reduced chemical potential $\mu/k_\mathrm{B}T$ as a function of temperature. The chemical potential of a classical gas is always negative, because $V_\mathrm{Q} < V_\mathrm{A}$. Therefore, the *dashed part* of the curve is not realized. (**b**) Chemical potential μ of ideal ^{4}He atoms relative to E_{111} and E_{211} at finite temperatures, and for $T \to 0$

The ground state will become macroscopically populated for $V_\mathrm{Q} \to V_\mathrm{A}$, meaning that the number N_0 of bosons in the ground state becomes comparable to the total number N of bosons in the system. For the first excited state, we find for $1\,\mathrm{cm}^3$ of helium, $\mu = 0$, and $T = 1\,\mathrm{K}$

$$f(E_{211}, T = 1\,\mathrm{K}) \approx \frac{1}{\exp\left(10^{-14}\right) - 1} \approx 10^{14}\,. \tag{2.57}$$

In spite of this very large number, the population of the first excited state is very small in comparison to the total number of atoms of 10^{22}. The fact that the population of the ground state converges to the total number of particles N for $T \to 0$ and for small values of the chemical potential makes it possible to determine μ:

$$\begin{aligned}\lim_{T\to 0} f(0,T) = N_0(T) &= \lim_{T\to 0}\left(\frac{1}{\mathrm{e}^{-\mu/k_\mathrm{B}T} - 1}\right) \\ &\approx \lim_{T\to 0}\left(\frac{1}{1 - \mu/(k_\mathrm{B}T) + \ldots - 1}\right) \approx -\frac{k_\mathrm{B}T}{\mu}\,.\end{aligned} \tag{2.58}$$

The chemical potential can therefore be approximated by $\mu = -k_\mathrm{B}T/N_0$. This means that it essentially depends on the population of the ground state. For the example of a $1\,\mathrm{cm}^3$ volume of liquid helium we find $\mu/k_\mathrm{B} \approx 10^{-22}\,\mathrm{K}$ at $T = 1\,\mathrm{K}$. Consequently, the absolute value of the chemical potential is, at this point, already much smaller than the energy difference ΔE between the ground state and the first excited state. This situation is depicted in Fig. 2.28b. In this way, the occurrence of macroscopic occupation of the ground state in liquid helium at finite temperatures can be understood.

We shall now calculate the number of bosons N_0 and N_e, in the ground state and the excited states, respectively. We find

$$\sum_i f(E_i, T) = N = N_0(T) + N_e(T)$$
$$= N_0(T) + \int_0^\infty D(E)\, f(E,T)\, \mathrm{d}E\,, \tag{2.59}$$

with $D(E)$ denoting the density of states of the helium atoms. The separate treatment of N_0 and N_e is necessary, because the density of states goes to zero for $E \to 0$ and thus the integral does not include the occupation of the ground state. For $D(E)$ we use the density of states of free particles,

$$D(E) = \frac{V(2m)^{3/2}\sqrt{E}}{4\pi^2\hbar^3}\,, \tag{2.60}$$

which we shall derive in Sect. 3.1. Inserting the Bose–Einstein distribution function in (2.59) we obtain

$$N = N_0 + \frac{V}{4\pi^2}\left(\frac{2m}{\hbar^2}\right)^{3/2} \int_0^\infty \frac{\sqrt{E}}{\mathrm{e}^{(E-\mu)/k_\mathrm{B}T} - 1}\, \mathrm{d}E\,. \tag{2.61}$$

Since we have $|\mu| \ll \Delta E$ under the conditions discussed above, we may set $\exp[(E-\mu)/k_\mathrm{B}T] \approx \exp(E/k_\mathrm{B}T)$. Using this simplification and the substitution $E/k_\mathrm{B}T = x$, we may write

$$N = N_0 + \frac{V}{4\pi^2}\left(\frac{2m}{\hbar^2}\right)^{3/2} (k_\mathrm{B}T)^{3/2} \int_0^\infty \frac{\sqrt{x}}{\mathrm{e}^x - 1}\, \mathrm{d}x\,. \tag{2.62}$$

This definite integral can be evaluated resulting in $\Gamma(5/2) \times \zeta(5/2) \approx 1.783$. Using the expression for the quantum volume V_Q we obtain the following constraints equation for Bose–Einstein condensation

$$N \approx N_0 + 2.6\frac{V}{V_\mathrm{Q}}\,. \tag{2.63}$$

Of course, this equation is only valid as long as the conditions for the approximations are fulfilled, which means that the chemical potential can be expressed by $\mu = -k_\mathrm{B}T/N_0$ and that N_0 is a large number. Equation (2.63) says that the quantum volume has to be at least 2.6 times larger than the atomic volume for Bose–Einstein condensation to occur. Or in other words, the thermal de Broglie wavelength has to be 1.37 times larger than the interatomic distance – and not equal to the dimensions of the container, as F. London originally assumed.

At this point, we will calculate the condensation temperature T_c, which is defined as the temperature above which all particles are excited. We will therefore insert $N_e(T_\mathrm{c}) = N$ and $N_0(T_\mathrm{c}) = 0$ in (2.62) and (2.63). We obtain

$$T_{\mathrm{c}} = \frac{2\pi\hbar^2}{k_{\mathrm{B}}m}\left(\frac{N}{2.6V}\right)^{2/3} . \tag{2.64}$$

Using this result we can write the ratio of the number of excited particles to the total number of particles as

$$\frac{N_{\mathrm{e}}}{N} = \left(\frac{T}{T_{\mathrm{c}}}\right)^{3/2} . \tag{2.65}$$

In the language of the two-fluid model the *condensate* N_0 corresponds to the superfluid component and N_{e} to the normal-fluid component (see Fig. 2.29).

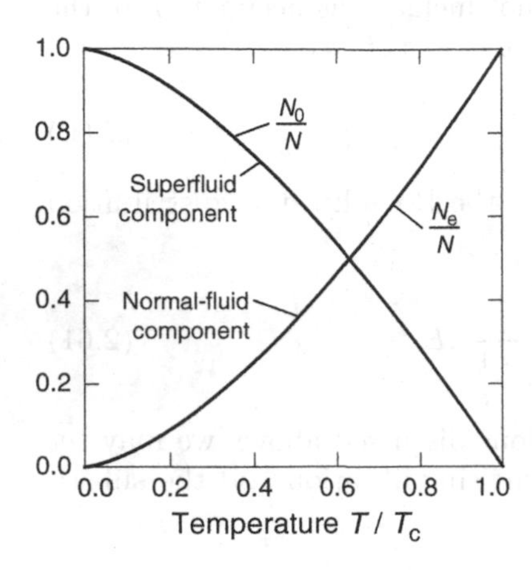

Fig. 2.29. Normalized population of the ground state N_0/N and the excited states N_{e}/N as a function of the reduced temperature T/T_{c}

The condensation of an ordinary gas in real space corresponds to the Bose–Einstein condensation of ^{4}He in momentum space, which means that all atoms have the same wave vector and therefore perform a strictly correlated motion. Because of this, Bose–Einstein condensation can be considered as a disorder-to-order transition. A schematic illustration of the results for an ideal Bose gas is shown in Fig. 2.30. At $T = 0$ (*left*), all particles are in the ground state. For $0 < T < T_{\mathrm{c}}$ (*right*), some particles are excited, but the ground state is still heavily – macroscopically – occupied.

2.3.2 Helium

We now turn to the question of whether (2.64) predicts the condensation temperature T_{c} for helium correctly. Using the number density for ^{4}He gas at saturated vapor pressure at 4.2 K we obtain $T_{\mathrm{c}} \approx 0.5$ K. This temperature obviously lies below the temperature of liquefaction and therefore Bose–Einstein condensation in the gas phase is impossible under equilibrium conditions.[4]

[4] Note that Bose–Einstein condensation of low-density supercooled helium gas has been achieved recently [75].

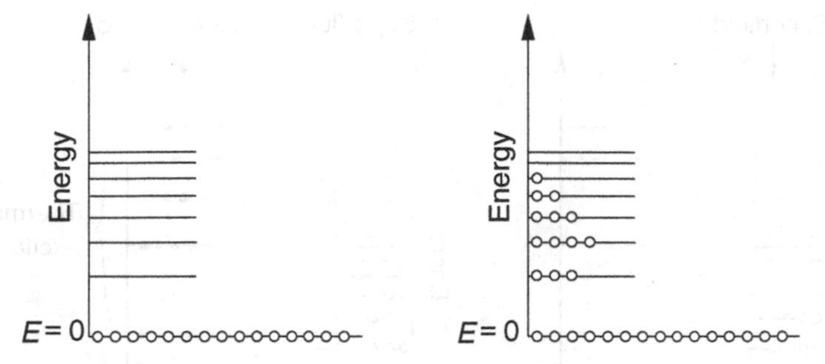

Fig. 2.30. Graphic illustration of the population of the energy levels of an ideal Bose gas. (**a**) $T = 0$, and (**b**) $0 < T < T_c$

Using the parameters for liquid helium one finds a condensation temperature of $T_c = 3.1\,\mathrm{K}$. Considering the simplifications under which (2.64) has been derived, this value is in reasonably good agreement with the measured temperature $T_\lambda = 2.17\,\mathrm{K}$ of the lambda transition.

More distinct differences, however, are visible in the temperature dependence of the specific heat C_V. For an ideal Bose gas a much weaker temperature dependence is expected than experimentally observed for liquid helium. This can be seen in Fig. 2.31 where the specific heat of an ideal Bose gas with the parameters of liquid helium is plotted in comparison with the measured values.

At low temperatures, the rise of C_V should be proportional to $T^{3/2}$ for an ideal Bose gas, whereas a T^3 dependence has been found experimentally for $T < 0.6\,\mathrm{K}$ (see Sect. 2.5.2). The reason for this discrepancy – and others of this kind – lies, of course, in the fact that liquid helium is by no means a noninteracting Bose gas. The interaction between the helium atoms has two

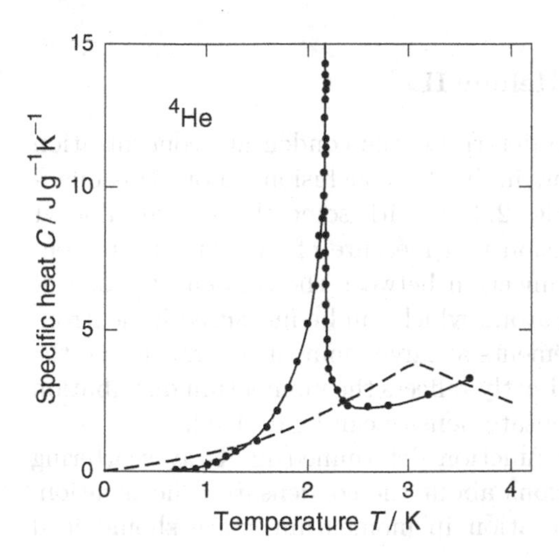

Fig. 2.31. Specific heat of ^{4}He (*points*) after [74] in comparison with the theoretical curve for an ideal Bose gas with the parameters of liquid helium (*dashed line*)

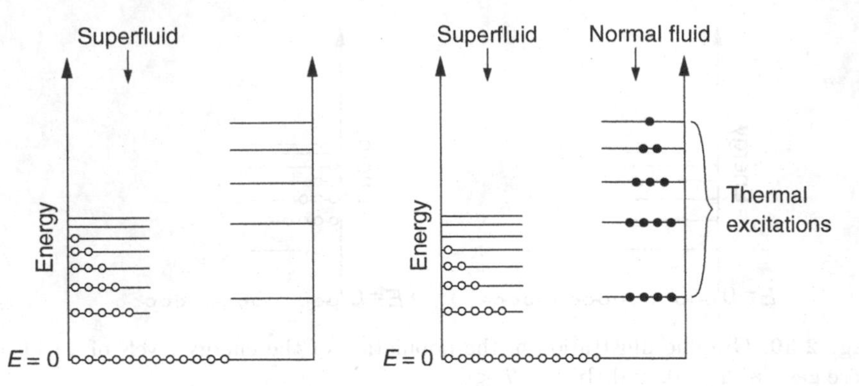

Fig. 2.32. Schematic illustration of the population of the superfluid and normal-fluid component in helium II. (**a**) At $T = 0$, *all* helium atoms belong to the superfluid component. Due to interaction, however, some atoms are scattered into virtual states with $E > 0$. (**b**) At finite temperatures, thermal excitations also occur, which are identical to longitudinal phonons at very low temperatures. As we shall see later, these thermal excitations form the normal-fluid component

important consequences that are schematically shown in Fig. 2.32. First, the ground-state population is somewhat reduced, or in other words, the condensate concentration is lower than one would expect for an ideal Bose gas. One often refers to this as a *depletion* of the ground state. Secondly, the nature of excitations is different for interacting Bose particles. Instead of individually excited atoms, collective excitations occur that have already been investigated theoretically in 1947 by *Bogoliubov* [76]. Despite these differences, the crucial feature of Bose–Einstein condensation in a system of interacting Bose particles is still the fact that a macroscopic number of particles remains in the ground state even at relatively high temperatures, just as in the case of an ideal Bose gas.

2.3.3 Condensate Fraction in Helium II

Although there is no direct way to determine the condensate concentration experimentally it is possible to draw indirectly conclusions about this quantity from different experiments (Fig. 2.33). With some theoretical effort it can be shown that the surface tension is a measure of the condensate concentration. More obvious is the connection between the condensate concentration and the average energy per atom, which can be measured in neutron-scattering experiments. In measurements at large momentum transfer Q, the dynamic structure factor $S(Q, \omega)$ directly reflects the momentum distribution of the atoms from which the condensate density can be derived.

Also, from the pair-correlation function determined in X-ray scattering experiments one can draw conclusions about the condensate concentration. The basic idea is that the condensation in momentum space should lead to broadening of the distribution in real space according to the uncertainty principle. This broadening should be proportional to the number of atoms in the condensate. Therefore, the pair-correlation function below T_λ can be written as

$$g(r) - 1 = (1 - n_0)^2 \left[g^*(r) - 1\right] , \tag{2.66}$$

where n_0 denotes the condensate density and $g^*(r)$ the pair-correlation function of the noncondensate atoms which, in practice, can be taken as being $g(r)$ at a temperature just above T_λ.

As can be seen in Fig. 2.33, the values obtained with different experimental methods agree well. It is remarkable that even for $T \to 0$ the condensate concentration is only about 13%. This means that directly equating ϱ_s with the condensate is not possible. The theoretical values for the condensate fraction lie between 0.09 and 0.12 and are in reasonably good agreement with experimental data.

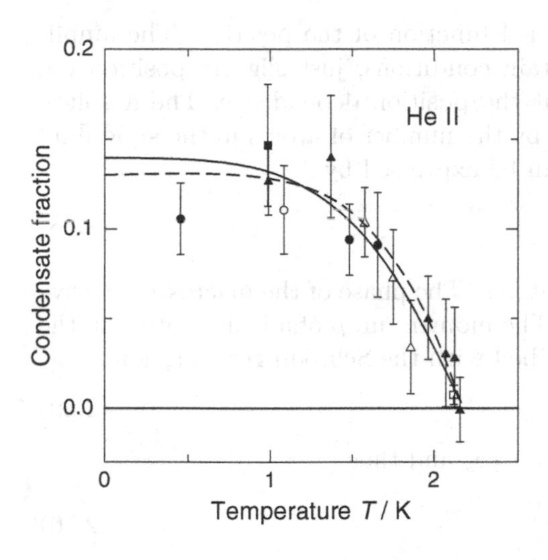

Fig. 2.33. Condensate concentration in helium II as a function of temperature. The experimental data are from X-ray scattering [77], neutron scattering [78–80] and measurements of the surface tension (*dashed curve*) [81]. The *solid line* represents an empirical fit of the data

At the end of this section we note that in recent years Bose–Einstein condensation in dilute gases has been achieved in sophisticated experiments by several groups. The techniques of laser cooling, magnetic trapping and rf evaporative cooling are essential ingredients that have been combined to obtain Bose–Einstein condensation in dilute gases. Meanwhile, not only have such Bose condensates been realized, but many properties, such as sound propagation, have been studied in these systems. Besides the many similarities between the physics of helium II and Bose condensates of dilute gases, these experiments are not considered to belong to traditional low-temperature physics, because of the very different experimental methods involved. We will therefore not discuss them here, but refer to recent reviews about this fascinating subject [82–84].

2.4 Macroscopic Quantum State

F. London repeatedly stressed in different publications that the condensate is a quantum state on a macroscopic scale [11]. Later, this viewpoint was extended to the whole superfluid component since it is assumed that the condensate and the superfluid component are closely related. As we will see, the presence of a macroscopic quantum state has consequences for the properties of helium II. For example, it results in a quantization of circulation and it enables phenomena analogous to the Josephson effect in superconductors.

2.4.1 Wave Function of the Superfluid Component

The macroscopic quantum state present in helium II can be described by the wave function

$$\psi(\boldsymbol{r}) = \psi_0\, \mathrm{e}^{\mathrm{i}\varphi(\boldsymbol{r})}\,, \tag{2.67}$$

where the phase $\varphi(\boldsymbol{r})$ is a real-valued function of the position. The amplitude ψ_0 is constant or, under certain conditions, just slightly position dependent. Henceforth, we shall omit the position dependence. The absolute value of the wave function is given by the number of atoms in the superfluid component per unit volume and can be expressed by

$$\psi^\star\psi = |\psi_0|^2 = \frac{\varrho_\mathrm{s}}{m_4}\,. \tag{2.68}$$

Here, m_4 denotes the mass of ^{4}He atoms. The phase of the macroscopic wave is related to the velocity of atoms. The momentum $\boldsymbol{p}$ of a helium atom in the superfluid component can be described with the Schrödinger equation

$$-\mathrm{i}\hbar\nabla\psi = \boldsymbol{p}\,\psi\,. \tag{2.69}$$

Using (2.67) we find $\boldsymbol{p} = \hbar\nabla\varphi(\boldsymbol{r}) = m_4\boldsymbol{v}_\mathrm{s}$ and thus

$$\boldsymbol{v}_\mathrm{s} = \frac{\hbar}{m_4}\nabla\varphi(\boldsymbol{r})\,. \tag{2.70}$$

The velocity of the superfluid component therefore determines the phase shift of the wave function. The phase is constant for $\boldsymbol{v}_\mathrm{s} = 0$, and changes uniformly for $\boldsymbol{v}_\mathrm{s} = \mathrm{const}$. The phase of the wave function is a well-defined quantity within the entire liquid. We can think of particles being 'rigidly' connected, though it should be emphasized that this rigid coupling takes place in *momentum space* and not in real space. This concept can be verified by investigating helium II under rotation. Corresponding experiments will be discussed in the following section.

2.4.2 Helium II Under Rotation – Quantization of Circulation

In the discussion of the two-fluid model we have assumed that in ϱ_s no turbulence occurs and thus we can write $\operatorname{curl} \boldsymbol{v}_s = 0$. Already in 1941, *Landau* suggested a test of this assumption in experiments with helium II in a rotating vessel [35]. Even before such experiments were conducted *Onsager* speculated whether the assumption of $\operatorname{curl} \boldsymbol{v}_s = 0$ is generally valid and suspected the occurrence of vortices in rotating helium II [85]. In 1955, *Feynman* pointed out that the circulation in helium II should be quantized [86]. The first experimental observation of the quantization of circulation was achieved by *Vinen* in 1961 [87].

First, we consider the normal-fluid component in a rotating vessel. In analogy to the rotation of a rigid body, the velocity of the normal-fluid component is given by $v_n = \omega r$. Here, ω denotes the angular velocity and r the radial distance from the axis of rotation. The shape of the liquid mirror surface, or in other words the liquid meniscus, is described by the parabola

$$z = \frac{\omega^2}{2g} r^2 \,. \tag{2.71}$$

Now we will turn to the question: What happens with the superfluid component ϱ_s in a rotating vessel? We have assumed in Sect. 2.2 that no turbulence occurs in ϱ_s and therefore that the condition $\operatorname{curl} \boldsymbol{v}_s = 0$ is valid. For a single connected region this statement can be transformed by applying Stokes' integral theorem

$$\int_A \operatorname{curl} \boldsymbol{v}_s \cdot \mathrm{d}\boldsymbol{f} = \oint_L \boldsymbol{v}_s \cdot \mathrm{d}\boldsymbol{l} = 0 \,, \tag{2.72}$$

where A is the area enclosed by the contour L. From this it follows directly that $\boldsymbol{v}_s = 0$ and therefore that the superfluid component ϱ_s is at rest. In 1950, experiments with helium II in a rotating vessel were conducted by *Osborne*, in which he investigated the meniscus [88]. In the case where the superfluid component does not participate in the rotation one expects a reduction of the centrifugal force corresponding to the ratio ϱ_n/ϱ. Since gravity acts, of course, on both components, the meniscus of the helium should flatten below the lambda transition. On the contrary, Osborne observed that the meniscus of helium II is identical with that of a classical fluid even well below the lambda point. This means that the superfluid component ϱ_s is not at rest but participates in the rotation, contrary to expectation! As an example, we show in Fig. 2.34 the meniscus curvature $\gamma = \mathrm{d}^2 z/\mathrm{d}r^2$ of helium II under rotation as a function of the angular velocity. Clearly, the data fall onto the classically expected curve (*solid line*).

To understand this result we first discuss a different experiment involving the rotation of helium II in an annular container. In this case, we have a multiply connected region. The circulation κ is given by

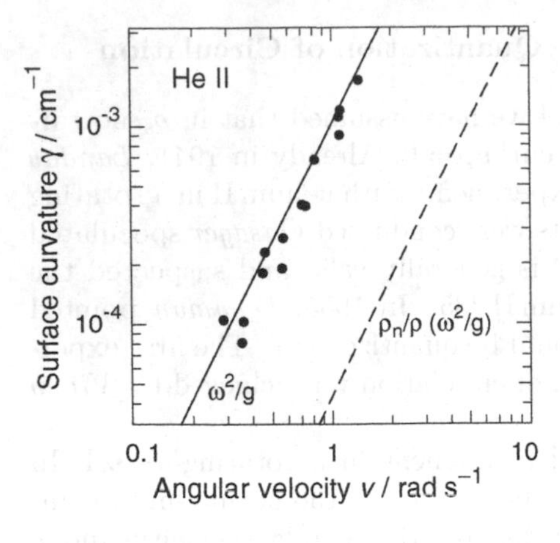

Fig. 2.34. Surface curvature γ of helium II under rotation as a function of angular velocity [89]. The *solid line* represents the expected behavior of a classical liquid $\gamma = \omega^2/g$, and the *dashed line* indicates the prediction for the case that the superfluid component is at rest, i.e., $\gamma = (\varrho_n/\varrho)\omega^2/g$

$$\kappa = \oint_L \boldsymbol{v}_s \cdot d\boldsymbol{l} = \int_A \operatorname{curl} \boldsymbol{v}_s \cdot d\boldsymbol{f} \,. \tag{2.73}$$

The use of $\boldsymbol{v}_s = \hbar \nabla \varphi(\boldsymbol{r})/m_4$ from (2.70) results in

$$\kappa = \frac{\hbar}{m_4} \Delta\varphi_L \,, \tag{2.74}$$

where $\Delta\varphi_L$ denotes the phase difference along the integration path L within the ring. Since the wave function is a unique function, the phase can only differ by integer multiples of 2π, i.e., $\Delta\varphi = 2\pi\, n$, for a complete cycle. Therefore, we have

$$\kappa = \frac{h}{m_4} n \qquad \text{with} \qquad n = 0,\ 1,\ 2,\ 3,\ \ldots \,. \tag{2.75}$$

An experimental proof of this quantization was first obtained by *Vinen* in 1961 [87]. In his experiments, a thin wire (diameter 25 µm, length 5 cm) was placed in the center of a cylindrical vessel filled with helium II. A characteristic transverse vibration of the wire was excited in a constant magnetic field by passing an alternating current through the wire. Without rotation of the surrounding helium, the transverse vibration of the wire can be described by two degenerate oscillations circularly polarized in opposite senses. Under rotation, the degeneracy is lifted by the *Magnus force*. The frequency difference $\Delta\nu$ that is now observed, is given by

$$\Delta\nu = \frac{\varrho_s}{2\pi\,\mathcal{M}} \kappa \,, \tag{2.76}$$

where $\mathcal{M}$ represents the effective mass per length of the wire plus half of the mass of the liquid displaced.

The data shown in Fig. 2.35 are not from the original experiment, but from a more recent investigation similar to that of Vinen. Clearly, quantized values of the circulation are observed. Starting from zero, the rotational velocity was increased slowly, then reduced again and subsequently the direction of rotation was reversed. During this sequence distinctive hysteretic effects were visible. In such experiments, indications for a quantization up to $n = 4$ have been found (not shown in Fig. 2.35).

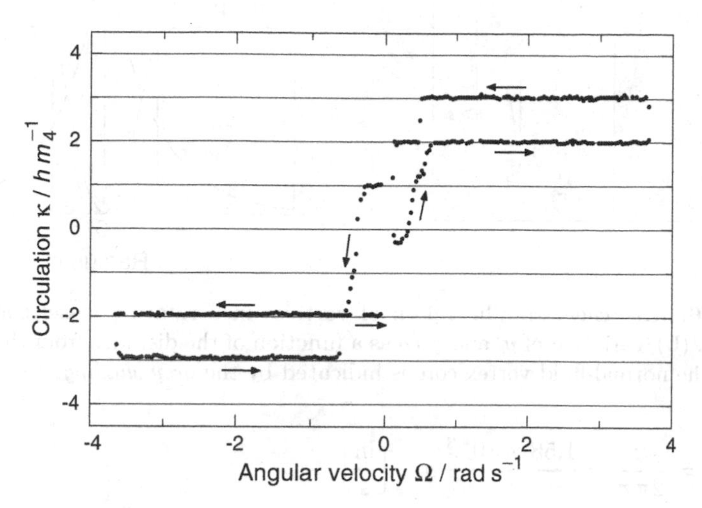

Fig. 2.35. Circulation κ in units of h/m_4 as a function of the angular velocity of the rotating cylinder. The *arrows* indicate the sequence in which the angular velocity was changed [90]

Vortices with Quantized Circulation

We have seen that in a multiply connected region the circulation of the superfluid component can be finite and that its magnitude is quantized. Now we come back to the question why the superfluid component *seems* to participate in the rotation in singly connected regions. The reason is that vortices occur under rotation having a normal-fluid core so that no singly connected region exists within the superfluid component. The occurrence of vortices in helium II therefore provides the explanation for the 'classical' meniscus observed in Osborne's experiments. A schematic illustration of the situation is shown in Fig. 2.36a.

The occurrence of a normal-fluid core in such vortices can be made plausible in the following way. As in a vortex in a classical liquid, the velocity of the superfluid component rises proportional to $1/r$ with decreasing distance r from the center of the vortex. As soon as the critical velocity is exceeded, superfluidity breaks down and a normal-fluid region is formed. The radial dependence of ϱ_s/ϱ and v_s in the vicinity of a vortex core is shown in Fig. 2.36b. Using classical hydrodynamics and (2.75) one finds for the velocity of the superfluid component:

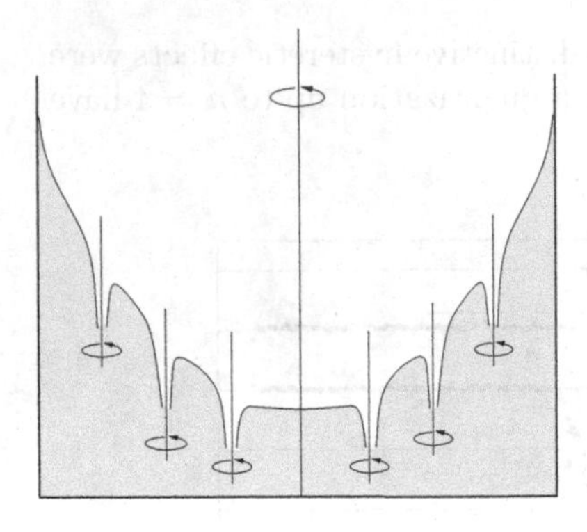

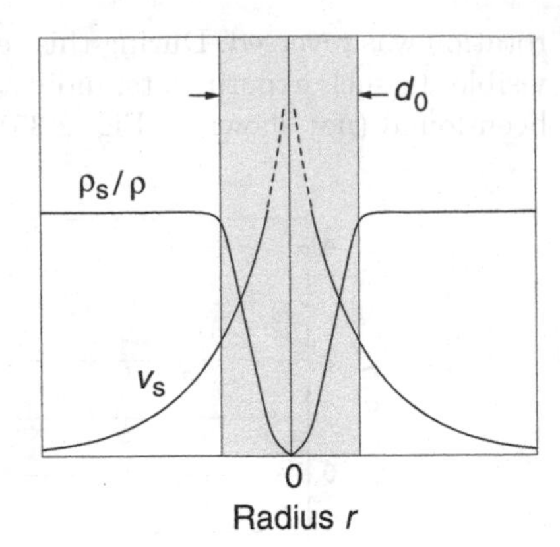

Fig. 2.36. (**a**) Schematic illustration of vortices in a rotating vessel containing helium II. (**b**) Variation of v_s and ϱ_s/ϱ as a function of the distance from the vortex center. The normal-fluid vortex core is indicated by the *grey shading*

$$v_s(r) = \frac{\kappa}{2\pi r} = \frac{1.58 \times 10^{-8}}{r} n \quad \left[\frac{\mathrm{m}}{\mathrm{s}}\right] . \tag{2.77}$$

We can estimate the diameter d_0 of the vortex core if we use as a rough approximation in (2.77) for v_s, the critical velocity for roton formation (see Sect. 2.5.3). In this way, we find the very small value of only a few Å for $T \to 0$. The radius of the vortex core corresponds to the *correlation length* – or *healing length*. This quantity is defined by the length over which the superfluid density falls from its bulk value to zero.

The energy E_v per unit length of a vortex can be calculated by integrating the kinetic energy per unit volume associated with the rotation of ϱ_s, i.e.,

$$E_v = \int_{a_0}^{b} \frac{\varrho_s v_s^2}{2} 2\pi r \, \mathrm{d}r . \tag{2.78}$$

Here, a_0 denotes the radius of the normal-fluid core and b is given either by the radius R of the vessel or by half the distance between the vortices. Using $\kappa = v_s 2\pi r$ we find

$$E_v = \frac{\varrho_s \kappa^2}{4\pi} \ln\left(\frac{b}{a_0}\right) \propto n^2 . \tag{2.79}$$

Because of the quadratic dependence $E_v \propto n^2$, the creation of many vortices with $n = 1$ is energetically more favorable than the creation of a smaller number of vortices with correspondingly higher circulation. The angular momentum L_v per unit length associated with a single vortex is given by

$$L_v = \int_0^R \varrho_s r \, v_s \, 2\pi r \, \mathrm{d}r = \frac{1}{2} \varrho_s \kappa R^2 . \tag{2.80}$$

It can be shown that the critical angular velocity ω_c for the formation of the first vortex is given by the condition $\omega_c = E_v/L_v$ (see, for example, [91]) resulting in

$$\omega_c = \frac{h}{2\pi m_4 R^2} \ln\left(\frac{R}{a_0}\right) . \tag{2.81}$$

For $R = 1\,\text{cm}$, we find the small critical angular velocity $\omega_c \approx 10^{-3}\,\text{s}^{-1}$.

The presence of vortices can be shown via electrons trapped in the vortex cores. As we will see in the following section, free electrons create bubbles in helium. Because of the circulation and the action of Bernoulli forces the electrons are captured by the vortex cores. By applying an electric field parallel to the vortex lines, the electrons are pulled along the vortices to the surface and the charge accumulating at the surface can be measured with an electrometer. Under the assumption that the electrons are captured at constant rate, the voltage at the electrometer is proportional to the number of vortices present at a given time. If a new vortex is formed upon increasing the angular velocity, the electrometer signal jumps to a higher value. The result of such an experiment is shown in Fig. 2.37. The nature of this data is somewhat analogous to the measurement of quantized magnetic flux lines in type II superconductors (see Sect. 10.1.4). In fact, there are a number of other analogies between these two phenomena. Just as in vortices, the cores of the flux tubes in superconductors are normal conducting. In addition, we find in both cases a regular spatial arrangement, the so-called *Abrikosov lattice*.

A very impressive direct visualization of the vortex lattice is possible by accelerating the captured electrons with an electric field and having them hit a phosphorescent screen. In Fig. 2.38 the results of such a measurement

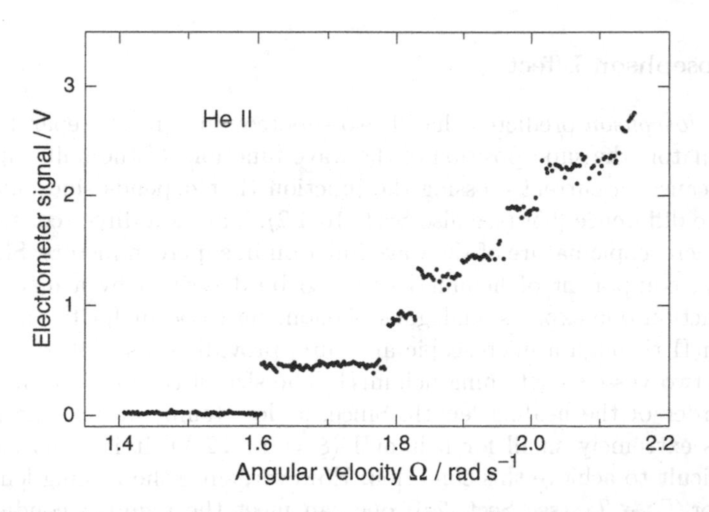

Fig. 2.37. Electrometer signal as a function of angular velocity. The velocity of rotation of the helium II was increased steadily in this experiment [92]

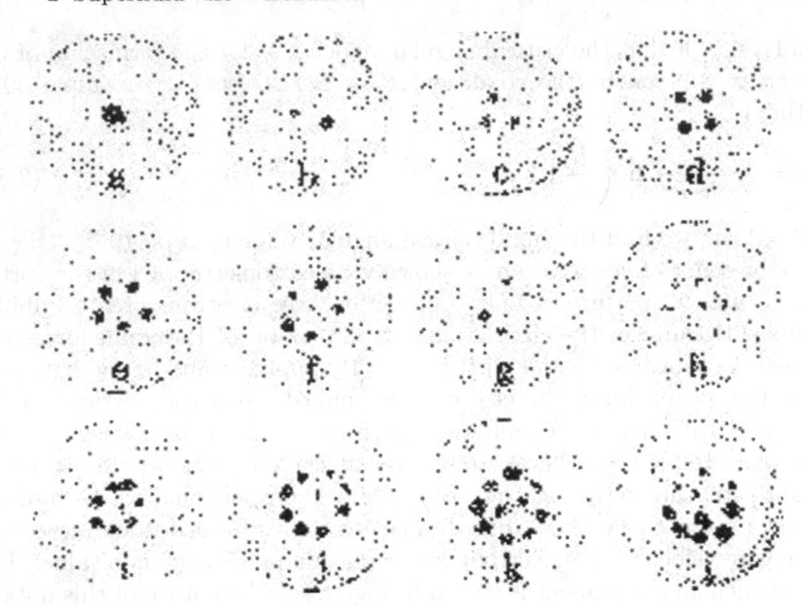

Fig. 2.38. Visualization of quantized vortices in rotating helium II of 0.1 K [93,94]

are shown. In this experiment, the helium II was located in a rotating vessel, 2 mm in diameter and 25 cm long. The angular velocities were in the range between 0.3 and 0.8 rad s^{-1}. The image was transferred out of the cryostat by optical fibers, amplified and photographed. With increasing angular velocity the number of vortices increases as expected. The regular arrangement of the vortices is evident.

2.4.3 Josephson Effect

In 1962, *Josephson* predicted that if two superconductors are separated by a thin insulator, the superposition of the wave functions of the bulk superconductors causes a current crossing the junction that depends sinusoidally on the phase difference [95] (see also Sect. 10.4.2). This is a direct consequence of the macroscopic nature of the wave function in superconductors. Since the superfluid component of helium II can also be described by a macroscopic wave function one expects analogous phenomena to occur for the mass flow of helium II through a microscopic aperture, providing a so-called *weak link* between two vessels containing helium II. The size of the aperture has to be of the order of the healing length. Since, at low temperatures, the healing length is extremely small for helium II ($\xi = 1 \ldots 2\,\text{Å}$), it is experimentally very difficult to achieve this condition. However, since the healing length diverges for $T \to T_\lambda$ (see Sect. 2.6) one can meet the required condition by performing measurements very close to the lambda transition.

In the following, we will consider two vessels containing helium II that are connected by a weak link. A difference in the fill height Δz of the two vessels results in a difference in the chemical potential $\Delta\mu = \mu_2 - \mu_1 = m_4 g \Delta z$. Putting Ψ_1 and Ψ_2 for the macroscopic wave function on either side of the aperture, we can write for the Schrödinger equations of the two vessels

$$\mathrm{i}\hbar\dot{\Psi}_1 = \mu_1\Psi_1 + \mathcal{K}\Psi_2 \quad \text{and} \quad \mathrm{i}\hbar\dot{\Psi}_2 = \mu_2\Psi_2 + \mathcal{K}\Psi_1 \,. \tag{2.82}$$

Here, $\mathcal{K}$ represents the coupling across the weak link. In the absence of coupling the wave functions depend on time as $\Psi(t) = \Psi_0 \exp(-\mathrm{i}\mu t/\hbar)$. In order to solve the coupled equations we use the ansatz

$$\Psi_1 = \sqrt{\varrho_s}\,\mathrm{e}^{\mathrm{i}\varphi_1} \quad \text{and} \quad \Psi_2 = \sqrt{\varrho_s}\,\mathrm{e}^{\mathrm{i}\varphi_2} \,, \tag{2.83}$$

where φ_1 and φ_2 denote the phases of the two wave functions. Inserting these solutions in (2.82) and separating real and imaginary parts we find

$$\frac{\partial \varrho_s}{\partial t} = \frac{2\mathcal{K}}{\hbar}\,\varrho_s \sin(\varphi_2 - \varphi_1) \,, \tag{2.84}$$

$$\frac{\partial}{\partial t}(\varphi_2 - \varphi_1) = -\frac{1}{\hbar}(\mu_2 - \mu_1) = -\frac{1}{\hbar}\, m_4 g \Delta z \,. \tag{2.85}$$

For $\Delta\mu = 0$ we have a constant phase difference $(\varphi_2 - \varphi_1)$ that results in a stationary mass flow j without any pressure applied. This phenomenon is called the *dc Josephson effect*. For $\Delta\mu \neq 0$ we find an oscillating mass flow with frequency $\omega_J = m_4 g \Delta z/\hbar$. This phenomenon is called the *ac Josephson effect*.[5]

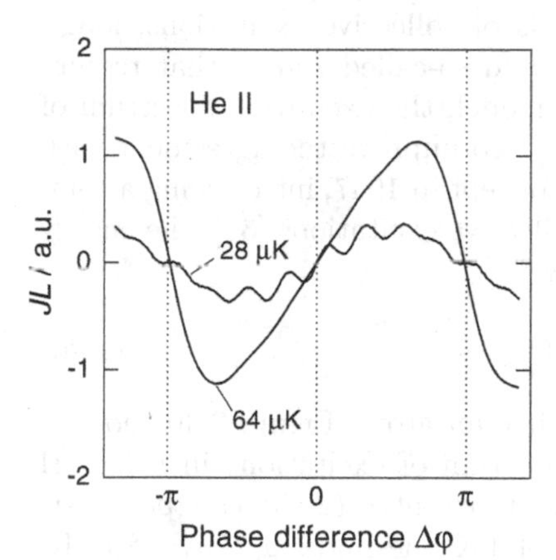

Fig. 2.39. Product JL of mass flow J and hydrodynamic inductance L of helium II passing through a microscopic apertures versus phase difference on the two sides of the weak link [96]. The temperatures noted next to the curves indicate the intervals $(T_\lambda - T)$. The hydrodynamic inductance L is given by the inertia of the fluid in the parallel flow path

In a recent experiment, the ac Josephson effect has been observed [96]. The result is shown in Fig. 2.39 for two different temperatures $(T_\lambda - T)$. Clearly, a 2π-periodic mass flow is observed with almost sinusoidal shape.

[5] The solution of (2.82) is discussed in detail in Sect. 10.4, where the Josephson effect in superconductors is considered.

In this experiment, the weak link consisted of a rectangular array of 24 slit-apertures of dimensions 3 μm by 0.17 μm each, separated by about 10 μm. These apertures were made in a membrane of thickness 0.15 μm. An array of apertures was used in order to increase the total mass current through the weak link. For these dimensions, the Josephson regime is expected to be observable at temperatures no more than 100 μK below the lambda transition. From the observed ac Josephson effect the relation between the phase difference and mass current was calculated.

2.5 Excitation Spectrum of Helium II

A question of fundamental importance is whether the presence of a condensate directly explains the occurrence of superfluidity. The answer is no! The presence of a condensate is a necessary but not a sufficient condition. Of crucial importance is the nature of possible excitations, just as in the case of superconductors (see Chap. 10).

2.5.1 Phonons and Rotons

For an ideal gas the energy of atoms is given by $E = p^2/2m$. In such a system, superfluidity cannot occur, because momentum transfer is allowed at infinitely small velocities. *Landau* therefore proposed in 1941 that only collective excitations exist in superfluid helium and that single-particle excitations are suppressed [35]. He assumed two kinds of collective excitations, longitudinal phonons with a linear dispersion and so-called *rotons* that require a minimum excitation energy Δ_r. In this model, the excitation spectrum of the rotons is given by $E = \Delta_\mathrm{r} + p^2/(2m^*)$. To improve the agreement with experimental data, Landau modified this concept in 1947, introducing a common dispersion curve for both kinds of collective excitations [35], the energy spectrum of the rotons being described by

$$E = \Delta_\mathrm{r} + \frac{(p - p_0)^2}{2m^*} . \tag{2.86}$$

Here, m^* denotes the effective mass of a helium atom. Figure 2.40 shows a graphical representation of the energy spectrum of excitations in helium II as suggested by Landau. The parameters that enter (2.86) can be determined experimentally. At a temperature of 1 K one finds $\Delta_\mathrm{r}/k_\mathrm{B} = 8.67\,\mathrm{K}$, $p_0/\hbar = 1.94\,\text{Å}^{-1}$, and $m^* = 0.15\,m_4$.

In Landau's version of the two-fluid model, phonons and rotons constitute the normal-fluid component. From statistical thermodynamics, the contribution of these excitations to ϱ_n as a function of temperature is given by

$$\varrho_\mathrm{n,ph} = \frac{2\pi^2 k_\mathrm{B}^4}{45\,\hbar^3\,v_1^5}\,T^4 \tag{2.87}$$

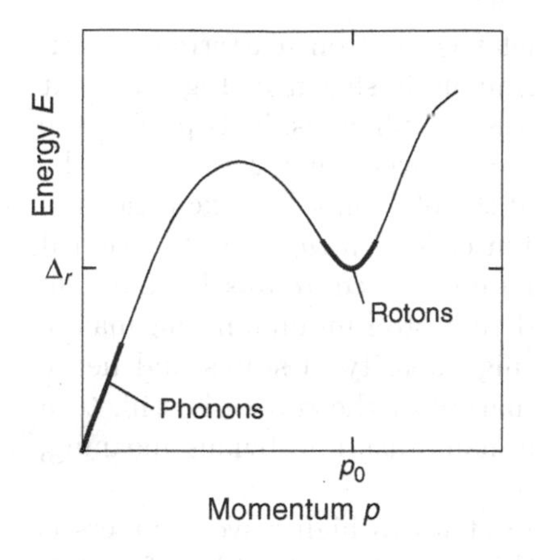

Fig. 2.40. Dispersion curve of helium II, after a suggestion of Landau in 1947

and

$$\varrho_{\mathrm{n,r}} = \frac{2\,p_0^4}{3\,\hbar^3}\sqrt{\frac{m^*}{(2\pi)^3 k_{\mathrm{B}} T}}\,\mathrm{e}^{-\Delta_{\mathrm{r}}/k_{\mathrm{B}}T}\,. \tag{2.88}$$

Figure 2.41 shows the normal-fluid component as a function of temperature. The data have been obtained in second-sound measurements. At low temperatures ($T < 0.6\,\mathrm{K}$) the phonon contribution dominates, and $\varrho_{\mathrm{n}} \propto T^4$ is observed as expected. At higher temperatures, the normal-fluid component rises more steeply due to the roton contribution. Above $T = 1.2\,\mathrm{K}$ the data are identical to the data shown in Sect. 2.2.3. As we have seen, in this range the temperature dependence can be approximated by $\varrho_{\mathrm{n}} \propto (T/T_\lambda)^{5.6}$.

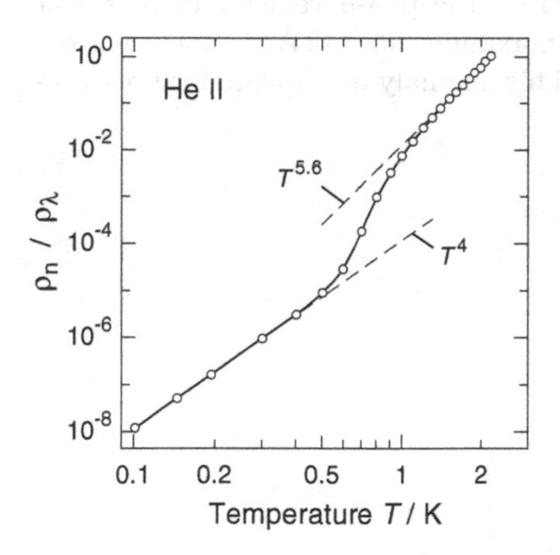

Fig. 2.41. Temperature dependence of the normal-fluid component $\varrho_{\mathrm{n}}/\varrho_\lambda$ of helium II [97]

In 1953 *Feynman* showed that the excitation spectrum postulated by Landau can be derived – at least qualitatively – from quantum-mechanical considerations [36]. In addition, he suggested the investigation of the energy spectrum of excitations in helium II by inelastic neutron-scattering experiments. A spectrum obtained with this technique is shown in Fig. 2.42. At small energies one finds ordinary sound waves, i.e. phonons. This part of the dispersion curve can be approximated by $E = pv$, where $v = 238\,\mathrm{m\,s^{-1}}$ is the sound velocity and p the quasimomentum of the phonons. The excitations in the range of the maximum of $E(p)$ are often called *maxons*. As mentioned above, the excitations near the minimum were named *rotons* by Landau. The regions of the dispersion curve around the roton minimum and maxon maximum, where $\partial E/\partial k$ is small, have a high density of states and hence motivated their naming. The microscopic nature of the rotons is still, to a large extent, unknown today, therefore the name has no obvious meaning, contrary to early suggestions.

The fact that well-defined excitations exist up to high wave numbers is very astonishing, and distinguishes helium II from ordinary liquids. Of course, at low frequencies sound waves do also exist in ordinary liquids, but at high wave numbers the lifetime of phonons becomes very short and collective excitations are overdamped in ordinary liquids. Besides, one finds single-particle excitations in all other liquids, which do not occur in helium II.

If we look at the dispersion curve in more detail, we find that its behavior is remarkable, even at small wave numbers. Helium seems to be the only liquid for which the sound velocity initially increases with increasing frequency. This effect is very small, however, and not visible in Fig. 2.42. Nevertheless, this has considerable consequences for the damping of sound waves, since this anomalous dispersion makes three-phonon scattering processes much easier. For a long time the unusual slope of the dispersion curve was not generally accepted. The proof of its existence was finally established by very precise ultrasonic measurements (Fig. 2.43). First, the phase velocity increases at small wave numbers passes through a maximum at $0.42\,\mathrm{Å^{-1}}$, and then decreases again at higher wave numbers. This anomaly disappears for pressures higher than 20 bar [101].

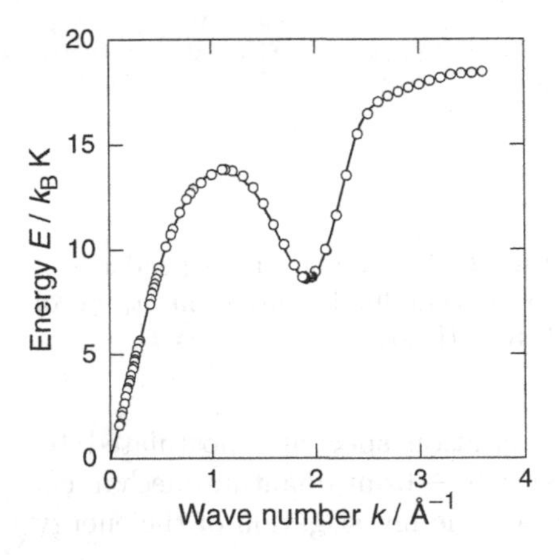

Fig. 2.42. Dispersion curve of helium II as determined experimentally [98]

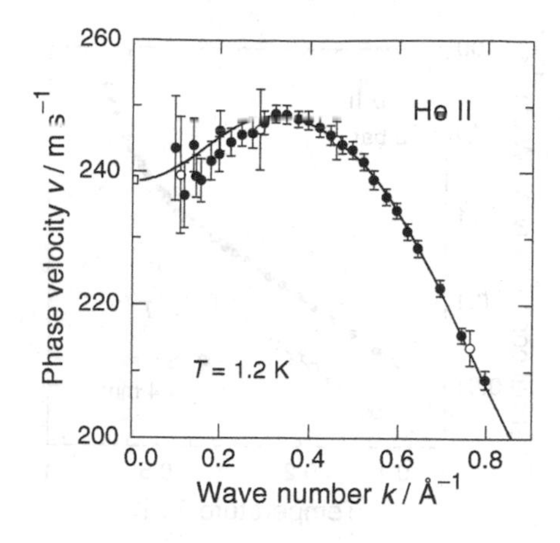

Fig. 2.43. Phase velocity of phonons in helium II at small wave numbers [99, 100]

2.5.2 Specific Heat

The discussion of the specific heat of helium II can be subdivided into three temperature ranges, each being dominated by different excitations.

$T < 0.6$ K In this range, only longitudinal phonons having long wavelengths are excited in liquid helium. These phonons can be described by the Debye model (see Sect. 6.1). According to this theory, the specific heat is expected to follow

$$C_{\mathrm{ph}} = \frac{2\pi^2 k_{\mathrm{B}}^4}{15\varrho\hbar^3 v_1^3}\, T^3 \,, \tag{2.89}$$

i.e., the temperature dependence should be $C_{\mathrm{ph}} \propto T^3$, as in solids. Note that, in contrast to solids, only one phonon branch exists in helium. As shown in Fig. 2.44a this temperature dependence has indeed been observed experimentally below 0.6 K. This conclusion is confirmed by measurements of the thermal conductivity Λ in the so-called *Casimir regime* (see Sect. 6.2.4) where the relation

$$\Lambda = \frac{1}{3}\, C_{\mathrm{ph}}\, v\, d \propto T^3 \tag{2.90}$$

holds. In this regime, the temperature dependence of the conductivity is exclusively determined by the specific heat because the mean free path ℓ of the phonons is given by the diameter d. The results of such measurements on helium II at very low temperatures are shown in Fig. 2.44b. The two data sets belong to measurements with capillaries with different diameter. The T^3 dependence is found in both experiments below $T \approx 0.6$ K in agreement with the specific heat data. The absolute value of the thermal conductivity is determined by the diameter of the capillaries, as expected.

$0.6 < T < 1.2$ K In this temperature range, the specific heat of helium II is dominated by the contribution of rotons. This contribution can be calculated

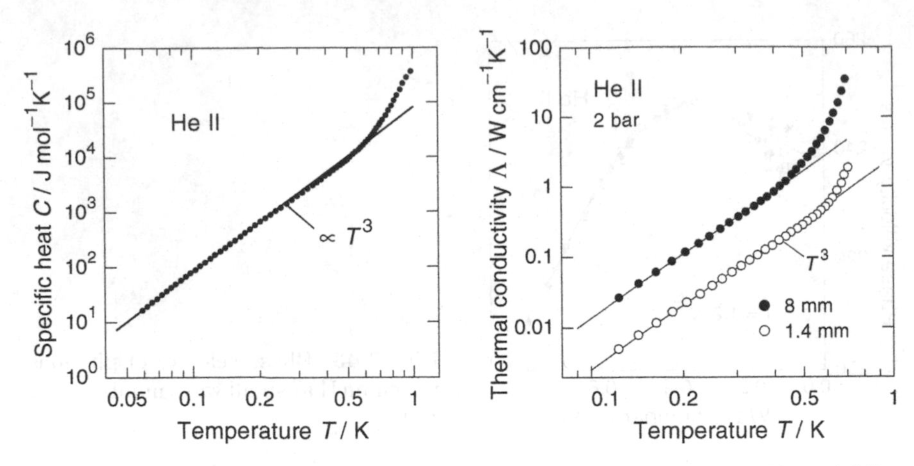

Fig. 2.44. (**a**) Specific heat of helium II as a function of temperature. The *solid line* represents the expected dependence $C_{\mathrm{ph}} \propto T^3$ [102]. (**b**) Thermal conductivity Λ of helium II in thin capillaries with different diameter as a function of temperature measured at a pressure of 2 bar [103]

from the free energy $F_{\mathrm{r}} = -k_{\mathrm{B}}Tn_{\mathrm{r}}$, with n_{r} being the number density of rotons:

$$n_{\mathrm{r}} = \frac{2p_0^2}{3\varrho\hbar^3}\sqrt{\frac{m^*k_{\mathrm{B}}T}{(2\pi)^3}}\,\mathrm{e}^{-\Delta_{\mathrm{r}}/k_{\mathrm{B}}T}\,. \tag{2.91}$$

According to (2.91) the number of rotons increases rapidly with temperature. With the entropy of the roton gas $S_{\mathrm{r}} = -\partial F_{\mathrm{r}}/\partial T$, the specific heat of the rotons $C_{\mathrm{r}} = T\partial S_{\mathrm{r}}/\partial T$ is given by

$$C_{\mathrm{r}} = \frac{2k_{\mathrm{B}}p_0^2}{3\varrho\hbar^3}\sqrt{\frac{m^*k_{\mathrm{B}}T}{(2\pi)^3}}\left\{\frac{3}{4} + \frac{\Delta_{\mathrm{r}}}{k_{\mathrm{B}}T} + \left(\frac{\Delta_{\mathrm{r}}}{k_{\mathrm{B}}T}\right)^2\right\}\mathrm{e}^{-\Delta_{\mathrm{r}}/k_{\mathrm{B}}T}\,. \tag{2.92}$$

This relation describes the experimental results in this temperature range very well.

$1.2\,\mathrm{K} < T < T_\lambda$ At high temperatures, the broadening of the roton states due to lifetime shortening becomes large enough to be comparable with the splitting of the energy levels. Besides that, additional excitations, other than phonons and rotons, appear at these temperatures. The dispersion curve changes and the concepts presented here are not meaningful in this regime.

2.5.3 Concept of a Critical Velocity

For a long time it was generally believed that the occurrence of superconductivity is intimately connected with the presence of an energy gap (see Sect. 10.3). Therefore, the question had to be answered whether a similar situation also exists in helium II. Although rotons cannot be excited with energies below Δ_r, phonons can be excited with arbitrary small energies and, therefore, no energy gap exists in helium II. The reason why helium II is, nevertheless, a superfluid can be explained with the concept of a *critical velocity* introduced by *Landau*.

Consider a macroscopic sphere with mass $\mathcal{M}$ falling with velocity $\boldsymbol{v}$ through a column of liquid helium at temperature $T = 0$. The question is: at which velocity does the sphere cause dissipation by the creation of excitations? In the following, we shall assume that the sphere has created *one* excitation with energy $\mathcal{E}$ and momentum $\boldsymbol{p}$. The energy conservation for this process can be expressed by

$$\frac{1}{2}\mathcal{M}v^2 = \frac{1}{2}\mathcal{M}v'^2 + \mathcal{E}\,. \tag{2.93}$$

Here, $\boldsymbol{v}'$ denotes the velocity of the sphere after creating the excitation. Since momentum conservation is valid for this process, we have a second constraint

$$\mathcal{M}\boldsymbol{v} - \boldsymbol{p} = \mathcal{M}\boldsymbol{v}'\,. \tag{2.94}$$

The crucial point here is that the two conservation laws cannot be fulfilled simultaneously for arbitrary combinations of the parameters $\mathcal{E}$ and $\boldsymbol{p}$. This statement holds even for processes in which the direction of the created excitation is arbitrary. One can convince oneself of this fact by taking the square of (2.94) and dividing by $2\mathcal{M}$. The result is

$$\frac{1}{2}\mathcal{M}v^2 - \boldsymbol{v}\cdot\boldsymbol{p} + \frac{1}{2\mathcal{M}}p^2 = \frac{1}{2}\mathcal{M}v'^2\,. \tag{2.95}$$

Comparing this expression with (2.93) results in

$$\boldsymbol{v}\cdot\boldsymbol{p} - \frac{1}{2\mathcal{M}}p^2 = \mathcal{E}\,. \tag{2.96}$$

This equation can only be fulfilled if the velocity v exceeds a certain minimum value. We can neglect the second term on the left side of this equation if we assume for simplicity that the mass of the sphere is very large. The minimum value of the velocity appears if the momentum of the excitation is parallel to the velocity of the sphere. The critical velocity v_c for creating an excitation with energy $\mathcal{E}$ and momentum p is thus given by

$$v_c = \frac{\mathcal{E}}{p}\,. \tag{2.97}$$

Since in our discussion the nature of the excitations has not been specified, the critical velocity v_c described by (2.97) is valid for all types of excitations. For phonons, rotons and hypothetical single-particle excitations this is schematically illustrated in Fig. 2.45.

Because of the dispersion relation $\mathcal{E} = pv$, we find a critical velocity of $v_c = 238\,\mathrm{m\,s^{-1}}$ for phonons. The critical velocity of rotons is $v_c \approx 60\,\mathrm{m\,s^{-1}}$ and therefore significantly smaller than for phonons. The crucial point for the occurrence of superfluidity is that, in contrast to ordinary liquids in helium II at low energies and small wave numbers, no single-particle excitations with $v_c = 0$ exist.

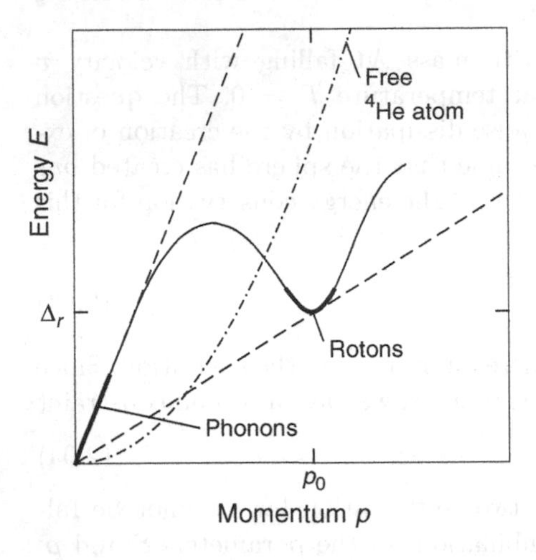

Fig. 2.45. Phonon–roton spectrum in comparison with the dispersion of free ^{4}He atoms (*dashed-dotted line*). The two *dashed lines* are tangents to the dispersion curve and reflect the critical velocities for phonons and rotons, respectively

Landau assumed that the critical velocity that is observable in experiments is determined by rotons. It turned out, however, that under ambient pressure, macroscopic vortices can be created at velocities considerably lower than the critical velocity of rotons. As we shall see in the following section, the creation of vortices and rotons in helium II, and in turn their critical velocities, depends on the detailed experimental conditions and can vary over a wide range. Nevertheless, the occurrence of a critical velocity still means that at low velocities no energy dissipation takes place and the viscosity is zero.

2.5.4 Experimental Determination of the Critical Velocity

The basic concept of a critical velocity as described in the previous section is clear and plausible. However, in general it is very difficult to perform experiments in which the parameters can be chosen in such a way that a simple theory can be applied. Of the large number of experiments that have been carried out in this context, we pick out just two and discuss their results.

Motion of Ions in Liquid Helium

The motion of ions has been used frequently in experiments to investigate the properties of liquid helium. The reason is that the ions can be manipulated and detected easily because of their electric charge. The behavior of negative ions – like electrons – and positive ions – like $^4He_2^+$ – is very different.

$^4He_2^+$-ions attract helium atoms and form a snowball (solid helium) with a radius of $r \approx 7$ Å. In contrast, electrons create a bubble because of the Pauli repulsion. The energy E_b of such a bubble can be described in a simple model as the sum of the zero-point energy of the electron plus the surface energy and volume energy of the displaced helium:

$$E_\mathrm{b} = \frac{h^2}{8mr^2} + 4\pi r^2 \alpha + \frac{4}{3}\pi r^3 p \,. \tag{2.98}$$

Here, α denotes the surface tension, m the mass of the electron, p the pressure, and r the radius of the bubble. The dependence of E_b on the radius is shown in Fig. 2.46. Without external pressure the bubble has a radius of about 19 Å, as indicated by the minimum. With increasing pressure the bubble radius is reduced.

Ions in helium in a static electric field $\mathcal{E}$, move in the stationary state uniformly with a constant drift velocity $\overline{v}_\mathrm{d}$, since the force of the electric field is compensated by viscous drag. The mobility $\mu = \overline{v}_\mathrm{d}/\mathcal{E}$ of the ions can be calculated with Stokes' law for the mobility of a sphere in a laminar flow:

$$\mu = \frac{\overline{v}_\mathrm{d}}{\mathcal{E}} = \frac{q}{6\pi\eta r} \,. \tag{2.99}$$

Here, r denotes the radius of the sphere, q its charge and η the viscosity of the liquid. This equation applies well for snowball-type defects. For electrons, the prefactor $1/6\pi$ must be replaced with $1/4\pi$ to obtain a quantitative agreement. The viscosity seen by ions in helium II is determined by the interaction with phonons, rotons, vortices and the almost unavoidable ^{3}He impurities. In very pure ^{4}He (i.e., without ^{3}He) at pressures above 12 bar, one can reach the condition where the ions are accelerated up to the *Landau velocity* v_L, where pairs of rotons are created. In the temperature range $0.7\,\mathrm{K} < T < 1.8\,\mathrm{K}$ and at moderate electric fields, one finds in this case

$$\mu \propto \frac{1}{n_\mathrm{r}} \propto \mathrm{e}^{\Delta_\mathrm{r}/k_\mathrm{B}T} \,. \tag{2.100}$$

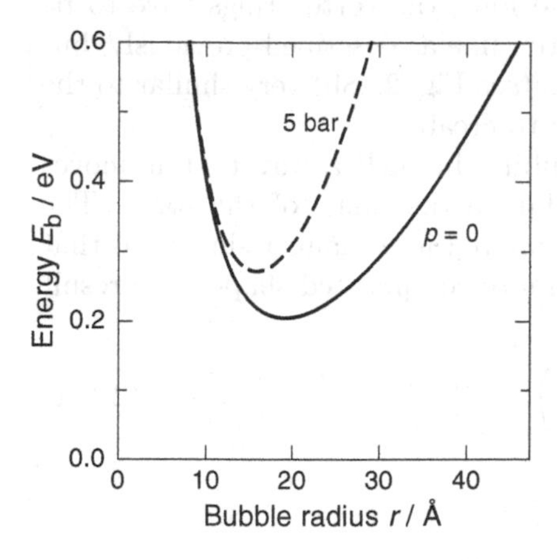

Fig. 2.46. Energy of a bubble in liquid helium created by an electron as a function of the bubble radius at two different pressures

The fact that the average drift velocity and therefore the inverse mean free path is inversely proportional to the density n_r of rotons demonstrates that in this regime the rotons are the dominating scattering centers. Since the energy gap Δ_r of the rotons decreases with increasing pressure one finds a corresponding reduction of the critical velocity. This is shown in Fig. 2.47.

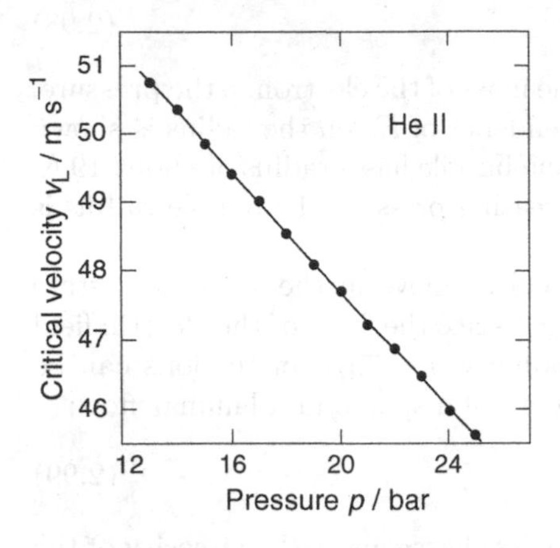

Fig. 2.47. Critical velocity v_L for exciting a roton pair as a function of the applied pressure [104]

At lower temperatures $T < 0.4\,\text{K}$, the mean free path of the ions becomes very large, because there are hardly any rotons. Surprisingly, the observed drift velocities at these temperatures are much lower (typically 10–100 cm s^{-1}) than the ones just discussed. Moreover – and perhaps even more surprising – the velocity of the ions *decreases* with *increasing* energy after reaching a maximum. As an example of this amazing observation we show in Fig. 2.48a data from measurements with both positive and negative ions. Clearly, the velocity of the ions decreases with increasing ion energy independent of the sign of their charge.

This phenomenon can be explained by the creation of macroscopic vortex rings by the ions when reaching a critical velocity. The ions are captured in the vortex rings. In order to move the ions, the vortex rings have to be pulled with them. A vortex ring is a vortex line as described previously, but with both ends connected to form a torus (see Fig. 2.48b) very similar to the smoke rings, smokers are sometimes able to create.

The flow fields of a vortex ring combine in such a way that it moves forward with a velocity v_{vr} perpendicular to the plane of the torus. The kinetic energy of vortex rings can be calculated in a manner similar to that of vortex lines, but accounting for their more complicated shape. The result is

$$E_{\text{vr}} = \int \frac{1}{2}\varrho_s v_s^2 \text{d}V = \frac{1}{2}\varrho_s \kappa^2 r \left[\ln\left(\frac{2r}{a_0}\right) - \frac{7}{4}\right] . \tag{2.101}$$

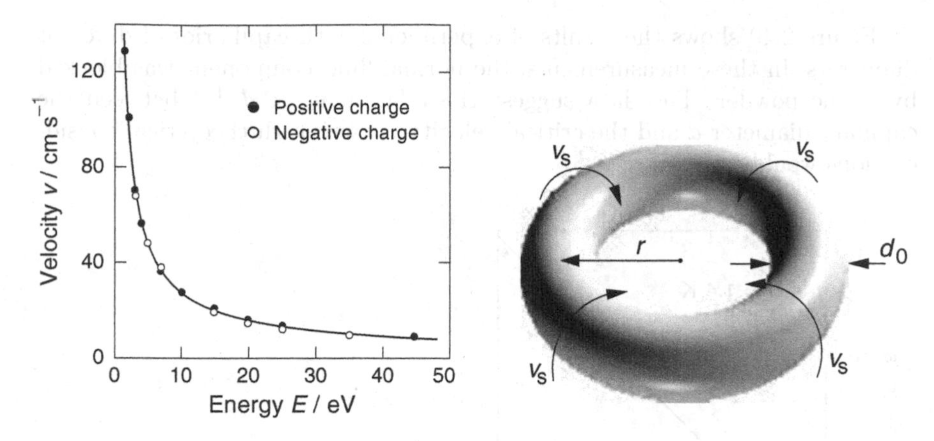

Fig. 2.48. (**a**) Velocity of positive and negative ions in helium II as a function of energy. The *solid line* has been calculated using (2.101) with $\kappa = h/m_4$ and $d_0 = 2a_0 = 2.4\,\text{Å}$. After [105, 106]. (**b**) Schematic illustration of a vortex ring

Here, a_0 refers to the radius of the normal-fluid core of the vortex ring and r to the radius of the torus. The momentum of such a ring is $p_{\text{vr}} = \pi \varrho_{\text{s}} \kappa r^2$ and thus its velocity is given by

$$v_{\text{vr}} = \frac{\partial E}{\partial p_{\text{vr}}} = \frac{\kappa}{4\pi r}\left[\ln\left(\frac{2r}{a_0}\right) - \frac{1}{4}\right] . \tag{2.102}$$

Neglecting the small logarithmic variation of the bracket with r we find for the velocity at a given circulation $v_{\text{vr}} \propto 1/E$ in agreement with the experimental observations shown in Fig. 2.48a.

It is interesting to note that because of the dispersion $E_{\text{vr}} \propto \sqrt{p_{\text{vr}}}$, the critical velocity is minimal for the largest possible ring. For a capillary with diameter d we find the critical velocity

$$v_{\text{c,vr}} = \frac{\hbar}{m_4 d}\left[\ln\left(\frac{d}{a_0}\right) - \frac{1}{4}\right] . \tag{2.103}$$

This result explains qualitatively why in flow experiments the observed flow velocity decreases with increasing capillary diameter (see Sect. 2.1.1).

Flow Experiments

Although it is clear that in a typical flow experiment, the excitation of large vortex rings is of crucial importance, it is not known which velocity in terms of the normal and superfluid components is important for this process: v_{s}, v_{n} or $(v_{\text{n}} - v_{\text{s}})$. Good reasons have been proposed for each one of these possibilities. Despite the fact that theoretical models favor the relative velocity, the experimental observation of a very weak temperature dependence of the critical velocity seems to contradict this option.

Figure 2.49 shows the results of experiments with capillaries of different diameters. In these measurements, the normal-fluid component was blocked by a fine powder. The data suggest the relation $v_c \propto d^{-1/4}$ between the capillary diameter d and the critical velocity v_c, although theoretical considerations would favor $v_c \propto d^{-1}$.

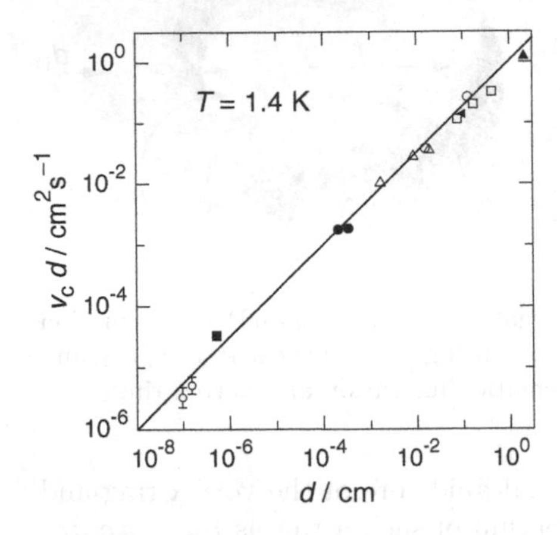

Fig. 2.49. Product $v_c d$ of the critical velocity and the capillary diameter plotted as a function of the diameter d [107]

Finally, we mention that for extended samples, networks of vortices – similar to the dislocation networks in solids – play an important role and have stimulated extensive theoretical work [108].

2.6 Critical Phenomena Near the Lambda Point

We have seen that many of the properties of ^{4}He exhibit a sudden change at the lambda point at $T_\lambda = 2.17\,\mathrm{K}$. As mentioned before, the origin of these changes is the continuous phase transition that occurs in liquid helium. The investigation of phase transitions is of general importance in physics. Since helium is a very clean substance consisting of extremely simple constituents, there has been considerable interest in the investigation of the properties of liquid helium in the vicinity of the phase transition. The hope was that with such a well-defined system, fundamental questions can be investigated that are of relevance in a broader context.

2.6.1 Brief Theoretical Background

The behavior near a critical point is determined by quantities that vanish, such as an order parameter, or by quantities that diverge, such as specific heat or susceptibility. Qualitative descriptions of the critical behavior of some special systems were already given around the turn of the 19th century. Examples are the transition between liquid and gas [109] and the transition between ferromagnetism and paramagnetism [110]. To describe the properties of physical systems in the vicinity of phase transitions, Landau published a generalized theory in 1937 [111]. He introduced the concept of the order parameter and provided, in this way, a mean-field description of the critical behavior. The central aspect of this theory is that the free energy of the system near the critical temperature T_c is described by a series expansion in terms of an order parameter.

The Landau theory predicts power laws with certain *critical exponents* for the temperature dependence of various quantities near the critical point as shown in Table 2.2.

Table 2.2. Predicted power laws according to the Landau model. The reduced temperature is defined by $t = (T - T_c)/T_c$

Quantity	Power Law	Critical Exponent
specific heat	$C_V \propto \|t\|^{\alpha}$	$\alpha = 0$
order parameter	$\Phi \propto \|t\|^{\beta}$	$\beta = 1/2$
susceptibility	$\chi \propto \|t\|^{-\gamma}$	$\gamma = 1$
correlation length	$\xi \propto \|t\|^{-\nu}$	$\nu = 1/2$

Since in the Landau theory no specific assumption is made about the physical systems being described, it should be valid for any kind of phase transition, independent of its specific nature. It was therefore quite surprising that in many cases the measured critical exponents are different from those listed in Table 2.2. Later, it was realized that critical phenomena are different from most other phenomena in physics in that one has to deal with fluctuations of the system over widely different length scales. Fluctuations are not included in Landau's mean-field approach although they become increasingly important near the critical point. One can show by means of the fluctuation-dissipation theorem that for systems with dimensionality $d \leq 3$ any Landau-type theory[6] fails at some point close to T_c. The range of validity of Landau-type theories is determined by the Ginzburg criterion, which states that a Landau-type theory should hold as long as the fluctuations in the order parameter are smaller than the mean value of the order parameter itself.

Despite the failure of Landau's universal approach it has been realized that it is possible to assign different physical systems independent of their specific physical nature to *universality classes*, characterized by a set of critical exponents. The systems are only classified by their dimensionality, by

[6] One example of a Landau-type theory is the Ginzburg–Landau theory for superconductors that we shall briefly discuss in Sect. 10.2.4.

the degrees of freedom of the order parameter and by the length scale of the interaction with respect to the correlation length. Furthermore, certain scaling laws for the critical exponents in different universality classes such as $\alpha + 2\beta + \gamma = 2$ have been suggested. Liquid helium, for example, belongs to the so-called *3D XY class*, because it has dimensionality $d = 3$ and two degrees of freedom for the order parameter (absolute value and phase of the wave function).

A thorough theoretical foundation for the existence of universality classes and scaling laws has been given by *Wilson* in 1971 in the framework of the *renormalization group theory* [112]. In this approach the critical exponents can be calculated. The predicted critical exponents for the lambda transition are: $\alpha = 0.0079 \pm 0.003$, $\beta = 0.3454 \pm 0.0015$ and $\nu = 0.6693 \pm 0.001$ [113].

In the following, we briefly discuss the behavior of liquid helium near the lambda transition and we introduce some experiments in which certain critical exponents of liquid helium have been determined.

2.6.2 Specific Heat

The occurrence of a divergence in the specific heat is characteristic of a second-order phase transition. The behavior of C_V near T_c is almost independent of the specific nature of the interaction between particles and is determined mostly by the presence of fluctuations.

In the following, we shall take a close look at the specific heat of ^{4}He near the phase transition. Figure 2.50 shows the temperature dependence of the specific heat of liquid helium as determined in very careful experiments. It can

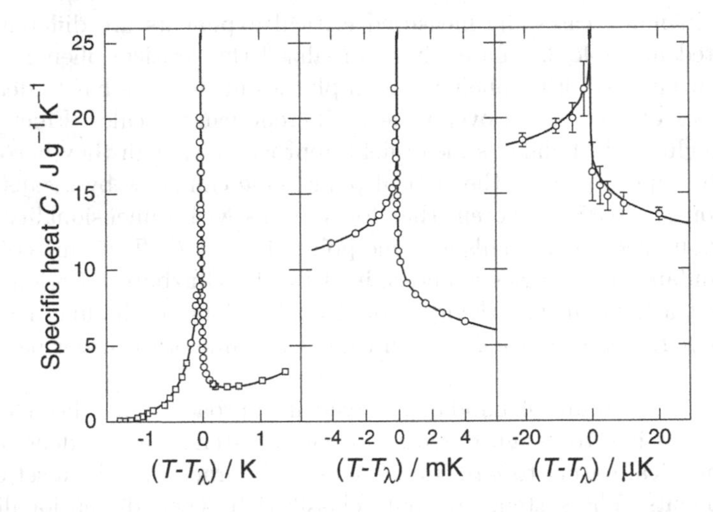

Fig. 2.50. Specific heat of liquid ^{4}He as a function of temperature $(T - T_\lambda)$. The temperature scale is expanded from *left* to *right* each time by almost a factor of 1000. The *solid lines* correspond to an empirical approximation [74]

be seen from this figure that the general shape of the curves are maintained on more expanded temperature scales.

Empirically, the behavior of C_V near the lambda transition can be approximated by $C_V \propto \log t$, with $t = (T/T_\lambda - 1)$. By way of illustration, the specific heat of liquid helium is plotted in Fig. 2.51 on a reduced logarithmic scale $\log t = \log |T/T_\lambda - 1|$. Over several orders of magnitude one finds, both below and above the lambda point, a nearly perfect logarithmic dependence.

At first glance, this logarithmic dependence seems to be in contradiction to the power-law dependence on t that is expected from the theories of phase transition. According to the results of the renormalization group theory, the specific heat of liquid helium should vary as

$$C = B + A\frac{t^{-\alpha}}{\alpha}\left(1 - D\sqrt{t}\right) , \tag{2.104}$$

where A, B and D are numerical constants. Nevertheless, the logarithmic behavior seen in the experiment is understandable. Because the critical exponent $\alpha \approx -0.014$ is small, $t^{-\alpha}$ can be expanded as $t^{-\alpha} = \mathrm{e}^{-\alpha \ln t} \approx 1 - \alpha \ln t$. Note that the magnitude of the experimentally observed critical exponent is somewhat larger that the predicted value.

Studies of the specific heat of liquid helium even closer to the lambda transition are very difficult because effects come into play that seem rather exotic at first glance. The transition temperature T_λ depends on pressure and therefore depends on the depth of the liquid helium in the experimental container. Because of the variation of the hydrostatic pressure in the liquid, the peak of the specific heat is broadened. In addition, the walls of the container can influence the results of the measurement. The first two atomic layers of helium on the walls are solid because of the strong van der Waals forces between the wall and the helium atoms. The superfluid component ϱ_s is reduced near the container wall. The length scale that determines the range over which ϱ_s increases from zero at the wall to the value for an infinitely

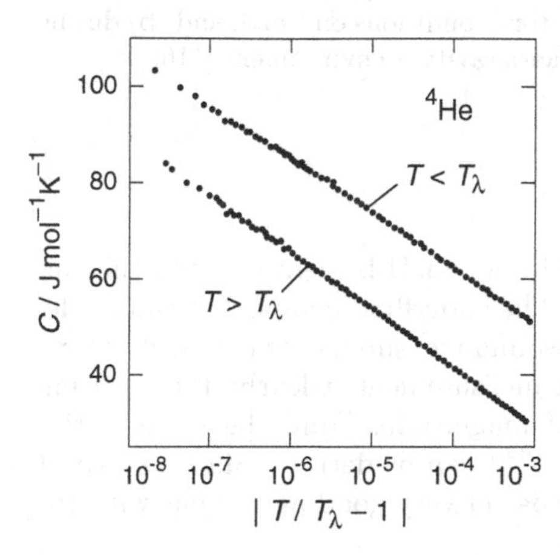

Fig. 2.51. Specific heat of ^{4}He as a function of $\log |T/T_\lambda - 1|$. In this plot the logarithmic variation of the data is clearly visible [114, 115]

large container is the correlation length ξ, which is often referred to as the *healing length*. Size effects become increasingly important near the lambda transition because the healing length diverges for $T \to T_\lambda$.

Figure 2.52a shows the results of a measurement in which the specific heat near the lambda point has been measured with high resolution. The flattening of the experimental curve at the peak, in comparison to the theoretically expected behavior, is clearly visible. This is mainly a result of the influence of gravity. To study the temperature dependence of the specific heat of liquid helium free of the gravity of the Earth in a sufficiently large container, a measurement was performed on a space shuttle mission in 1992. The result of this measurement is shown in Fig. 2.52b. Systematic deviations are clearly reduced, but the data are more noisy than the data obtained on Earth. The larger scatter of the data originated from cosmic rays that caused a fluctuating heat deposition in the extremely sensitive thermometers in this experiment.

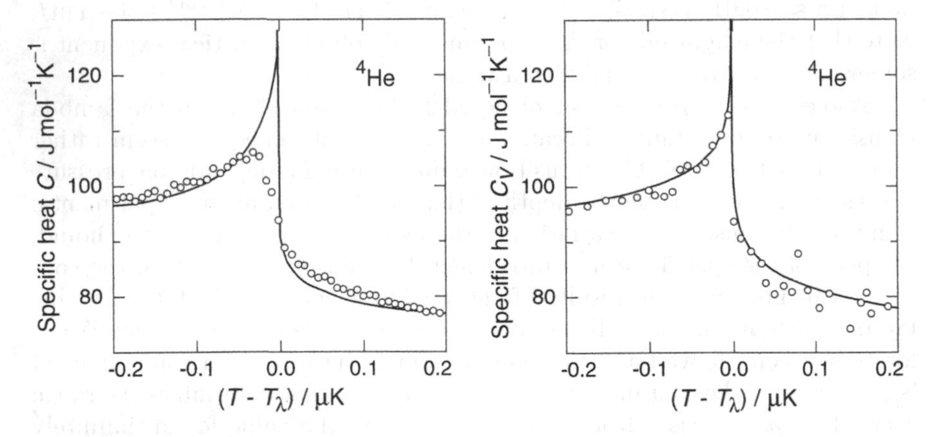

Fig. 2.52. Specific heat of ^{4}He very close to the lambda point as a function of temperature measured (**a**) under normal laboratory conditions on Earth and (**b**) during a space shuttle flight in a gravity-free – microgravity – environment [116]

2.6.3 Order Parameter

The real part of the order parameter Φ in helium II is identical with the amplitude of the wave function $\Psi_0 = \sqrt{\varrho_s}$. The superfluid component can be determined with high precision by second-sound measurements (see Sect. 2.2.8). Figure 2.53 shows the result of such a measurement. Clearly, the data follow a power law over several orders of magnitude. From the slope of this double-logarithmic representation, $\varrho_s = t^{0.67}$ can be derived. Since we expect $\varrho_s = t^{2\beta}$, the critical exponent is $\beta = 0.34$, in very good agreement with the predicted value.

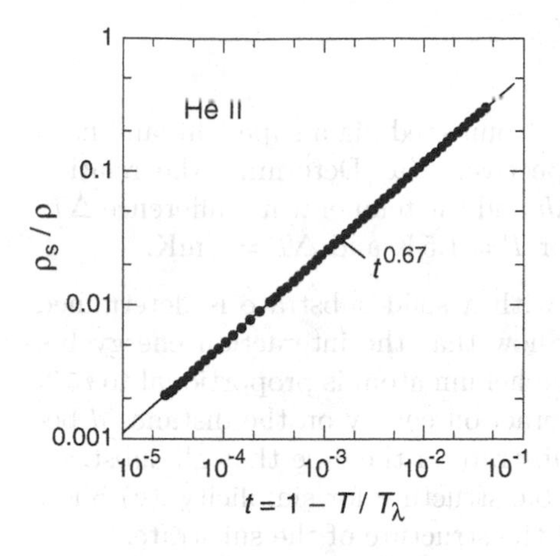

Fig. 2.53. Superfluid density as a function of temperature as determined from second-sound measurements [117]

2.6.4 Correlation Length

Finally, we mention a measurement in which the correlation length has been determined. Near T_λ, the correlation length – or healing length – is expected to vary as $\xi = \xi_0 t^{-\nu}$. This quantity has been determined in an experiment using unsaturated films in porous media such as Vycor glass. In the vicinity of a solid wall, the superfluid component is reduced and recovers to the bulk value over the correlation length. Therefore, the correlation length can be determined by measurements of the superfluid fraction ϱ_s/ϱ as a function of the film height. The results of such measurements for helium films on different materials are shown in Fig. 2.54. From the data, $\nu = 0.63$ and $\xi_0 = 2.8 \pm 0.5$ Å has been deduced. Again, the critical exponent agrees well with the theoretical prediction. The value of ξ_0 is slightly higher than found from other measurements, but is still in fair agreement.

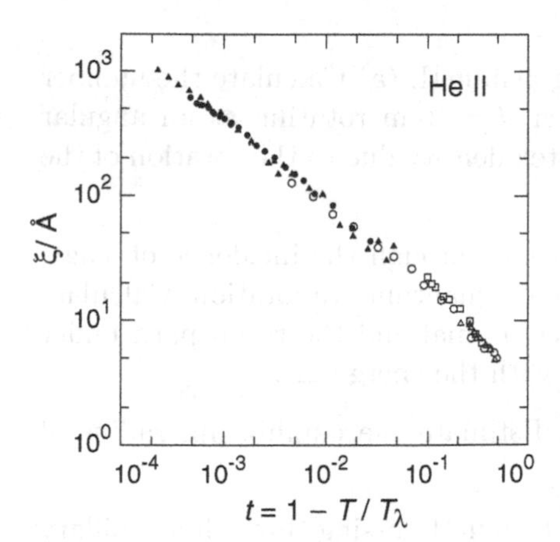

Fig. 2.54. Healing length versus reduced temperature $t = 1 - T/T_\lambda$ [118]

Exercises

2.1 Two vessels containing helium II are connected via a superleak and have the temperature T and $T + \Delta T$, respectively. (**a**) Determine the relation between the resulting level difference Δh and the temperature difference ΔT, and (**b**) calculate the level difference for $T = 1.5\,\mathrm{K}$ and $\Delta T = 1\,\mathrm{mK}$.

2.2 The interaction of helium atoms with a solid substrate is determined by the Van der Waals potential. (**a**) Show that the interaction energy between one atom of the substrate and one helium atom is proportional to r^{-6}. (**b**) What is the dependence of the interaction energy on the distance d between a helium atom and the plane substrate in the case that all substrate atoms are considered? Use a simple cubic structure for simplicity. (**c**) Show that the result of (b) is independent of the structure of the substrate.

2.3 A helium film on a copper wall located 1 cm above the surface of the helium bath ($T = 1\,\mathrm{K}$) has the thickness $d \approx 30\,\mathrm{nm}$. (**a**) Calculate the chemical potential of a helium atom on the film surface. (**b**) Determine the vapor pressure at the film surface. (**c**) Calculate the Hamaker constant for helium on copper.

2.4 First and second sound are weakly coupled. Deduce from thermodynamic identities the relation

$$\left(\frac{\partial p}{\partial S}\right)_{\varrho}\left(\frac{\partial \varrho}{\partial p}\right)_{S}\left(\frac{\partial T}{\partial \varrho}\right)_{S}\left(\frac{\partial S}{T}\right)_{\varrho} = \frac{C_p - C_V}{C_p}\,.$$

2.5 Bose–Einstein condensation leads to a macroscopic occupation of the ground state. (**a**) Which condition must be fulfilled by the chemical potential μ at the phase transition? (**b**) Deduce the relation for the critical temperature from the Bose–Einstein distribution function, and (**c**) for the number of particles in the ground state.

2.6 Quantized vortices occur in rotating helium II. (**a**) Calculate the number of vortices in a vessel with the diameter $d = 2\,\mathrm{cm}$ rotating at an angular velocity $\omega = 1\,\mathrm{s}^{-1}$. (**b**) Estimate the vortex density due to the rotation of the Earth.

2.7 The evaporation of single helium atoms through the incidence of single rotons onto the surface of helium II is called quantum evaporation. Calculate the critical angle Θ_c between the surface normal and the roton path under which this process is allowed for rotons with the energy Δ_r.

2.8 Electrons in helium form bubbles. Estimate the equilibrium radius of such an electron bubble at $T = 0$.

2.9 Calculate the critical velocity v_c for helium II passing through a capillary with an inner diameter $d = 0.1\,\mathrm{mm}$ at $T < 0.3\,\mathrm{K}$.

3 Normal-Fluid ^{3}He

Above 1 K, liquid ^{3}He exhibits many of the characteristics of a dense classical gas. Below this temperature, the quantum nature of the liquid becomes evident. In the temperature range between 1 mK and 100 mK, ^{3}He behaves in many ways like a degenerate Fermi gas. Therefore, we shall look at some basic properties of an ideal, noninteracting Fermi gas at the beginning of this chapter and we shall compare the results of this model to the ^{3}He data. We shall find that the agreement is quantitatively inadequate, although qualitatively it is quite reasonable. A better quantitative description is provided by Landau's theory of a *Fermi liquid* developed between 1956 and 1958 [119, 120]. In contrast to the Fermi-gas model, strongly interacting ^{3}He atoms are the starting point of this description. The interaction between the ^{3}He atoms are treated phenomenologically in a quasiparticle description. In Sect. 3.2 we will introduce the basic idea of the Landau Fermi-liquid theory. One great success of this model was the prediction of the so-called *zero sound* [120]. It was observed experimentally a few years after it was predicted, in full agreement with Landau's theory. We will discuss the phenomenon of zero sound in Sect. 3.3.

3.1 Ideal Fermi Gas – Comparison with Liquid ^{3}He

First, we derive the density of states of an ideal Fermi gas. We use this result not only for a description of normal-fluid ^{3}He, but also in discussion of conduction electrons in metals in Chap. 7. As a starting point we assume that the particles are contained in a box represented by a square-well potential with side length L and volume V. The potential energy is chosen in such a way that it is zero inside the cube. Under these conditions, free particles have only kinetic energy and the stationary Schrödinger equation is simply

$$-\frac{\hbar^2}{2m}\nabla^2\psi(\boldsymbol{r}) = E\psi(\boldsymbol{r})\,. \tag{3.1}$$

This equation can be solved with plane waves of the form

$$\psi(\boldsymbol{r}) = \frac{1}{\sqrt{V}}\,\mathrm{e}^{\mathrm{i}\boldsymbol{k}\cdot\boldsymbol{r}}\,, \tag{3.2}$$

with wave vector $\boldsymbol{k}$. The resulting energy eigenstates are given by

$$E_k = \frac{\hbar^2 k^2}{2m} . \tag{3.3}$$

Using periodic boundary conditions the solutions are restricted to the wave vectors

$$k_x = \frac{2\pi}{L} n_x , \quad k_y = \frac{2\pi}{L} n_y , \quad k_z = \frac{2\pi}{L} n_z , \tag{3.4}$$

where n_x, n_y and n_z are integers. As shown in Fig. 3.1, the allowed wave vectors are uniformly distributed in momentum- or k-space. The density ϱ_k in k-space is thus given by $\varrho_k = (L/2\pi)^3 = V/(2\pi)^3$.

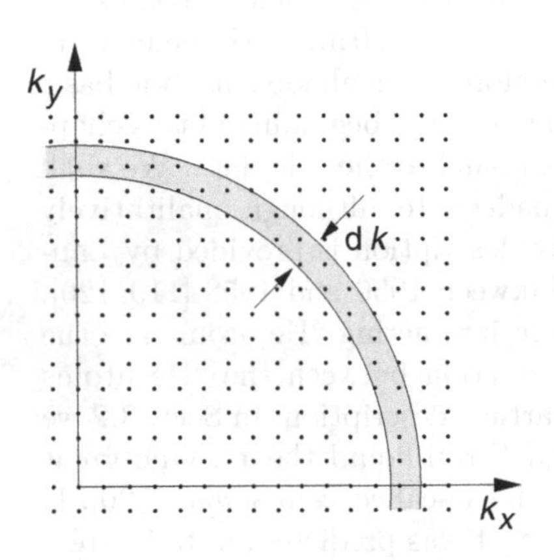

Fig. 3.1. Schematic illustration of the allowed states in k-space. The *points* indicate the allowed wave numbers, which are uniformly distributed

As one can see from Fig. 3.1, the volume element d^3k may be replaced by $4\pi k^2 \mathrm{d}k$, since there is no directional dependence. Therefore, we find for the number of states in the wave-number interval $\mathrm{d}k$ the expression

$$\mathcal{D}(k)\mathrm{d}k = \varrho_k 4\pi k^2 \mathrm{d}k = \frac{V}{2\pi^2} k^2 \mathrm{d}k . \tag{3.5}$$

This result is a consequence of the geometrical boundary conditions and is independent of the dispersion relation of the particles. Since in a Fermi gas each state can be occupied by two particles with opposite spin direction, we find for the density of states in k-space

$$D(k) = \frac{2\mathcal{D}(k)}{V} = \frac{k^2}{\pi^2} . \tag{3.6}$$

Using the dispersion relation (3.3), the density of states for free fermions is given by:

$$D(E) = D(k)\frac{\mathrm{d}k}{\mathrm{d}E} = \frac{(2m)^{3/2}\sqrt{E}}{2\pi^2\hbar^3} . \tag{3.7}$$

In order to calculate the internal energy we also need the distribution function. For fermions, this quantity is given by the *Fermi–Dirac distribution*

$$f(E,T) = \frac{1}{e^{(E-\mu)/k_BT}+1}\,, \tag{3.8}$$

where μ denotes the chemical potential. At $T = 0$ all states up to the *Fermi energy* $E_F \equiv \mu(T = 0)$ are occupied, which is determined by the mass m and the number density n of the fermions, and follows from

$$\begin{aligned} n = \frac{N}{V} &= \int_0^\infty D(k) f(E,T)\,dk = \int_0^\infty D(E) f(E,T=0)\,dE \\ &= \int_0^{E_F} D(E)\,dE = \frac{1}{3\pi^2}\left(\frac{2\,m\,E_F}{\hbar^2}\right)^{3/2} . \end{aligned} \tag{3.9}$$

With the definition $E_F = k_B T_F$, the *Fermi temperature* is given by

$$T_F = \frac{\hbar^2}{2mk_B}\left(3\pi^2 n\right)^{2/3} . \tag{3.10}$$

The Fermi temperature of metals is typically of the order of $10^4 - 10^5$ K because of the small electron mass and the high charge-carrier density $n \approx 10^{23}\,\mathrm{cm}^{-3}$. This means that metals at all experimentally accessible temperatures behave in many ways like they where close to absolute zero. In contrast, a Fermi gas with particles having the mass and the density of liquid ^{3}He has a Fermi temperature of only $T_F \approx 4.9$ K.

3.1.1 Specific Heat

In calculating the specific heat we have to take into account that the thermal energy leads to a blurring of the Fermi surface. This means that the distribution function $f(E,T)$ drops over the range $E_F \pm k_B T$. As a starting point, we consider the internal energy u per unit volume

$$u = \frac{U}{V} = \int_0^\infty D(E)\,f(E,T)\,E\,dE\,. \tag{3.11}$$

This integral is not solvable analytically, but it can be approximated at temperatures $T \ll E_F/k_B$ by the expression

$$u(T) = \frac{3}{5}\,n\,k_B T_F + \frac{\pi^2}{4}\frac{n}{E_F}\,(k_B T)^2\,. \tag{3.12}$$

Using this result, we find for the specific heat of a gas of free fermions

$$C_V = \left(\frac{\partial u}{\partial T}\right)_V = \frac{\pi^2}{2}\frac{n}{E_F}\,{k_B}^2 T = \gamma\,T\,. \tag{3.13}$$

At low temperatures $T \ll T_F$, we expect therefore a linear temperature dependence of the specific heat of liquid ^{3}He. For the specific heat per mole we can write

$$C_V = \frac{\pi^2 R}{2}\left(\frac{T}{T_F}\right) , \tag{3.14}$$

where R represents the universal gas constant. For liquid ^{3}He the constant γ/R is of the order of one.

As mentioned at the beginning of this chapter, ^{3}He behaves like a dense classical gas above 1 K. Below about 0.05 K one indeed finds, to a good approximation, the expected linear temperature dependence of C_V. The corresponding data are shown in Fig. 3.2. Nevertheless, this agreement cannot be regarded as a quantitative success for the free-fermion description, because in this model one would expect the linear regime to occur up to significantly higher temperatures, namely up to $T \approx T_F/10$. There are two reasons for this discrepancy: First, the effective mass of ^{3}He atoms in the liquid is larger than the mass of free ^{3}He atoms, and secondly, there is an additional contribution to the specific heat of liquid ^{3}He at low temperatures due to spin fluctuations. We briefly consider the latter contribution in the following.

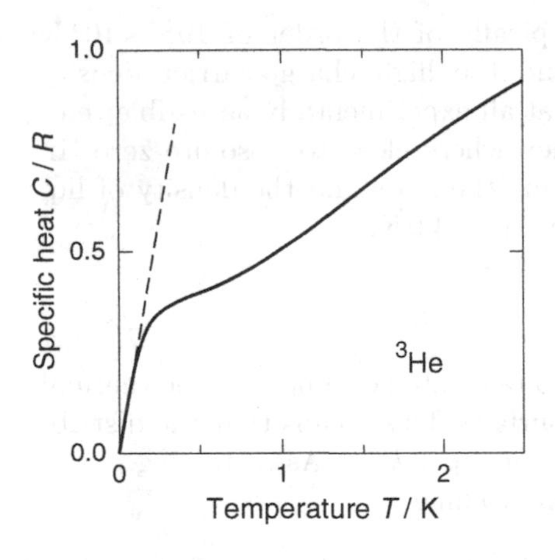

Fig. 3.2. Normalized specific heat C/R of liquid ^{3}He at a molar volume of 36.82 cm^3 mol^{-1} as a function of temperature. The *dashed line* marks the linear temperature dependence of C_V at very low temperatures [121]

There are two conflicting effects in liquid ^{3}He that influence the alignment of the nuclear spins, namely the ferromagnetic exchange interaction and the effect of Fermi statistics. In a system of noninteracting fermions, because of the exclusion principle, antiparallel spin alignment is energetically favorable. Fermi statistics prefers this arrangement since nonoccupied states enhance the total energy of the system. In the case of ^{3}He atoms there is, in addition, a short-range repulsive force between the atoms having the effect of producing an exchange interaction. Because of the strongly repulsive potential at short distances, it is energetically favorable for interacting ^{3}He atoms to occupy a state with an antisymmetric orbital wave function. As a result, the spins tend to orient parallel to each other meaning that the parallel spin orientation of neighboring atoms has a particularly long lifetime. These fluctuating spin domains are called ‘paramagnons’. In a simplified picture, we may assume that fluctuating ferromagnetic domains exist in the liquid and that the concentration of these domains depends on temperature. In this way, they influence the temperature dependence of entropy and specific heat of liquid ^{3}He. In a phenomenological model, the specific heat of normal-fluid ^{3}He below $T < 0.2\,\mathrm{K}$ can be described by the equation

$$C_V = \gamma T + \Gamma T^3 \ln\left(\frac{T}{\Theta_\mathrm{c}}\right). \tag{3.15}$$

The second term on the right side characterizes the contribution of the spin fluctuations. This contribution is a correction to the dominant term that is linear in T and makes up about 20% of the total at 0.1 K. The knowledge of the precise value of the second term in (3.15) is important in deriving an accurate value of γ and thus the effective mass of the ^{3}He atoms. At a molar volume of $36.74\,\mathrm{cm^3\,mol^{-1}}$ one finds the following values for the parameters entering the phenomenological relation (3.15): $\gamma/R = 2.78\,\mathrm{K^{-1}}$, $\Gamma/R = 35.4\,\mathrm{K^{-3}}$, and $\Theta_\mathrm{c} = 0.46\,\mathrm{K}$.

Figure 3.3 shows data for the specific heat of ^{3}He at different pressures, and hence at different densities. In this graph, the quantity $(\gamma - C_V/T)/(RT^2)$ is plotted that would be zero if there were no contribution from nuclear spins. With increasing density the interaction between the ^{3}He atoms becomes increasingly important and, in turn, the contribution of the spin fluctuations to the heat capacity rises.

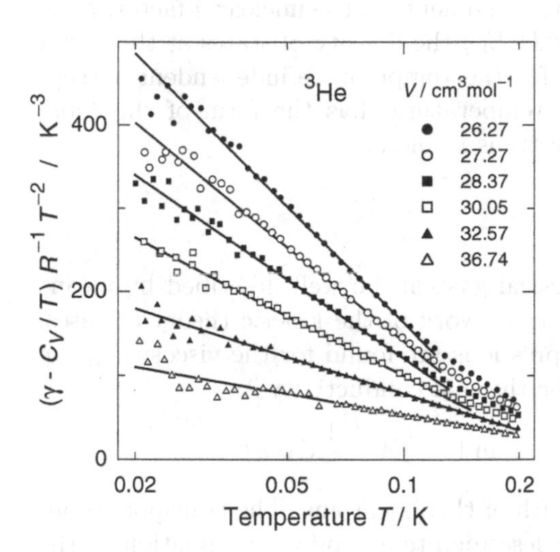

Fig. 3.3. Contribution of nuclear spins to the specific heat of ^{3}He at different molar volumes. Plotted is $(\gamma - C_V/T)/RT^2$ on a logarithmic temperature scale. The *solid lines* correspond to fits using (3.15) [121]

3.1.2 Susceptibility

The temperature dependence of the magnetic susceptibility of liquid ^{3}He is shown in Fig. 3.4. At high temperatures, ^{3}He behaves like a paramagnetic

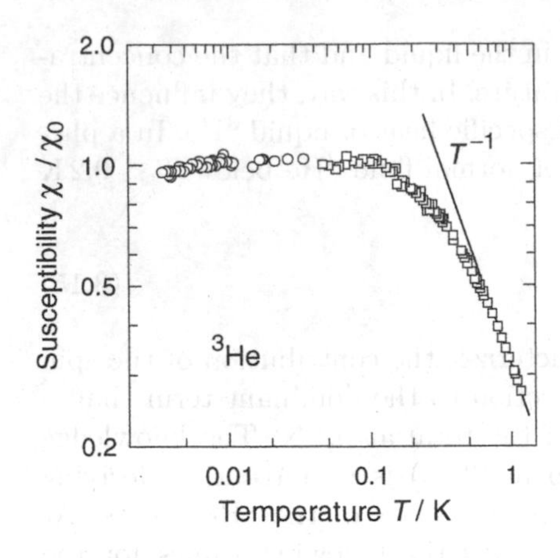

Fig. 3.4. Magnetic susceptibility χ of liquid ^{3}He at 0.5 bar as a function of the temperature normalized to the low-temperature limit χ_0. The *solid line* indicates the proportionality $\chi \propto 1/T$ at high temperatures [122, 123]

liquid, meaning that the magnetic susceptibility due to the nuclear spins varies proportionally to $1/T$, as expected from the *Curie law*. At low temperatures, the susceptibility becomes independent of temperature. This behavior is expected for an ideal Fermi gas. In the free-fermion model the magnetic susceptibility χ at low temperatures is given by

$$\chi = I(I+1)\,\mu_0\,\mu_\mathrm{n}^2\,g_\mathrm{n}^2\,\frac{2}{3}\frac{n}{E_\mathrm{F}} = \beta^2\,D(E_\mathrm{F})\,. \tag{3.16}$$

Here, μ_n denotes the nuclear magnetic moment, g_n the nuclear g-factor, I the nuclear spin of the ^{3}He atoms and $D(E_\mathrm{F})$ the density of states at the Fermi energy. Note that the expression for the temperature-independent susceptibility of liquid ^{3}He at very low temperatures has the form of the *Pauli susceptibility* of the conduction electrons in metals.

3.1.3 Transport Properties

The transport properties of a classical gas can be well described by means of the Boltzmann equation in the framework of the kinetic theory of gases. In this approach, the following expressions are found for the viscosity η, the self-diffusion coefficient D_s, and the thermal conductivity λ:

$$\eta = \frac{1}{3}\,\varrho\,v\,\ell\,, \qquad D_\mathrm{s} = \frac{1}{3}\,v\,\ell\,, \qquad \text{and} \qquad \Lambda = \frac{1}{3}\,C_V\,v\,\ell\,.$$

Here, ℓ represents the mean free path of the gas atoms. The transport properties of an ideal Fermi gas can be described to a good approximation by the same relations, but with the replacement of the thermal velocity v by the Fermi velocity $v_\mathrm{F} = (\hbar/m)(3\pi^2 n)^{1/3}$. The mean free path is limited by the scattering of the fermions, i.e., in our case by the scattering of ^{3}He atoms among each other. The corresponding mean collision time can by expressed by $\tau = v_\mathrm{F}/\ell$. Because of the exclusion principle the phase space for fermion scattering is rather limited. To show this, we consider a system of fermions at

$T = 0$ with a single excited fermion in a level with the energy $E_1 > E_F$ that interacts with another fermion with an energy E_2. The latter fermion has the energy $E_2 < E_F$ since only levels with energies less than E_F are occupied. The exclusion principle requires that these two fermions only scatter into unoccupied levels whose energy E_3 and E_4 must therefore be greater than E_F. Energy conservation requires $E_1 + E_2 = E_3 + E_4$. This means that the other three energies can only vary within a shell of the Fermi sphere of the thickness of order $|E_1 - E_F|$ about the Fermi surface. This leads to a scattering rate proportional to $(E_1 - E_F)^2$. This quantity appears squared rather than cubed, because energy conservation allows no further choice for E_4. In other words: because of the exclusion principle, only the fraction $(E_1 - E_F)^2/E_F^2$ from all the fermions can participate in the scattering event. At finite temperatures we may replace $|E_1 - E_F|$ by the thermal energy $k_B T$, because the accessible energy range is determined by the blurring of the Fermi surface. Therefore, we expect the scattering rate to be proportional to $(k_B T/E_F)^2$. As a result, the collision time is given, to a good approximation, by

$$\tau = \frac{5\hbar^3}{\sigma m^* k_B^2 T^2}, \tag{3.17}$$

where σ represents the scattering cross section. The previous reasoning is, in principle, also valid for the electron–electron scattering in metals. However, for electron–electron scattering, the effect of screening must also be taken into account. This leads to a modification of the prefactor of (3.17), but the temperature dependence remains unaffected.

Viscosity

Inserting the expression $\tau = \ell/v_F$ for the collision time we may write the viscosity of a Fermi gas as

$$\eta = \frac{1}{3}\,\varrho \tau v_F^2\,. \tag{3.18}$$

Since the temperature dependence of the viscosity is governed by the collision time, we expect η to vary proportional to T^{-2} at low temperatures. Figure 3.5 shows experimental data for the viscosity of liquid ^{3}He obtained with an oscillating disc. At temperatures above 1 K, the viscosity is nearly constant and has an absolute magnitude of about 25 μP, which is comparable to that of He I. Water at room temperature, in comparison, has a much higher viscosity of about 10 mP. At low temperatures, the viscosity of liquid ^{3}He increases strongly. As shown in Fig. 3.5b, at very low temperatures one finds the expected $\eta \propto T^{-2}$ dependence. It is remarkable that η increases up to about 0.2 P just above the transition temperature into the superfluid state. This value is comparable to the viscosity of honey.

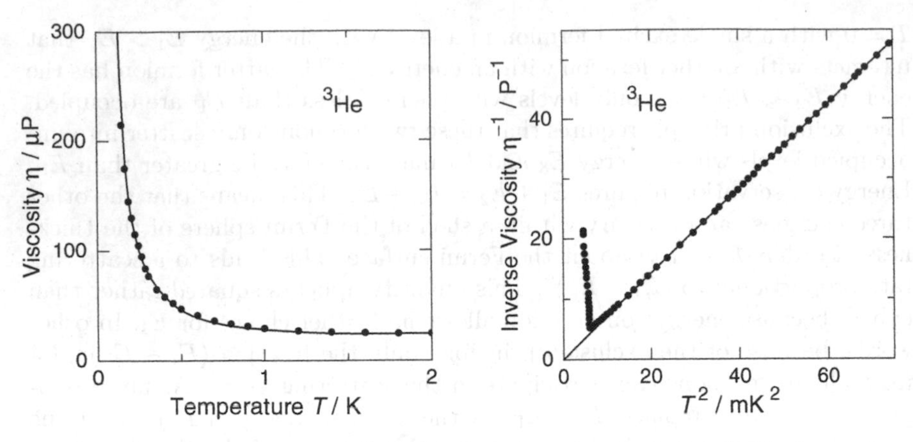

Fig. 3.5. (**a**) Viscosity of liquid ^{3}He as a function of temperature. With decreasing temperature viscosity increases drastically [124, 125]. (**b**) Inverse viscosity of liquid ^{3}He as a function of T^2 at a pressure of 16 bar. Down to the transition into the superfluid state (sharp increase of η^{-1}) one finds $\eta^{-1} \propto T^2$ [126]

Self-diffusion Coefficient

Closely related to the viscosity is the self-diffusion coefficient D_s, which describes the nuclear spin transport. This quantity is usually determined from nuclear spin echo experiments. For a Fermi gas, the self-diffusion coefficient $D_\mathrm{s} = \eta/\varrho$ is given by

$$D_\mathrm{s} = \frac{1}{3}\tau v_\mathrm{F}^2\,. \tag{3.19}$$

Figure 3.6 shows the result of measurements of D_s. As expected, at low temperatures the self-diffusion coefficient decreases with increasing temperature proportional to T^{-2}. At higher temperatures, ^{3}He behaves like a dense

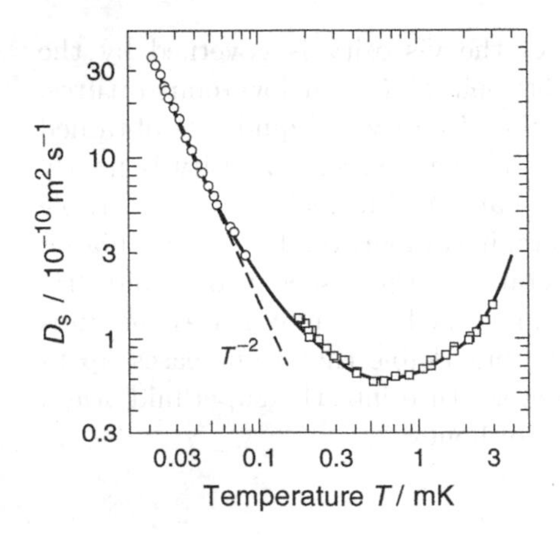

Fig. 3.6. Self-diffusion coefficient D_s of liquid ^{3}He as a function of temperature. The *dashed line* indicates that the self-diffusion coefficient varies proportional to T^{-2} at low temperatures [127–129]

classical gas and the self-diffusion coefficient increases with temperature. The transition from a Fermi gas to a classical gas is marked by a minimum of the self-diffusion coefficient, which is found at about 0.5 K.

Thermal Conductivity

The thermal conductivity Λ of liquid ^{3}He also exhibits a minimum at about 0.2 K. Experimental data for Λ are shown in Fig. 3.7. Normal-fluid ^{3}He is a very poor heat conductor. At 1 K, for example, the thermal conductivity is much lower than that of amorphous materials. We find $\Lambda \approx 10^{-4}\,\mathrm{W\,cm^{-1}K^{-1}}$ for ^{3}He, and in comparison, a typical value for glasses at this temperature is $5 \times 10^{-3}\,\mathrm{W\,cm^{-1}K^{-1}}$ (see Sect. 9.5).

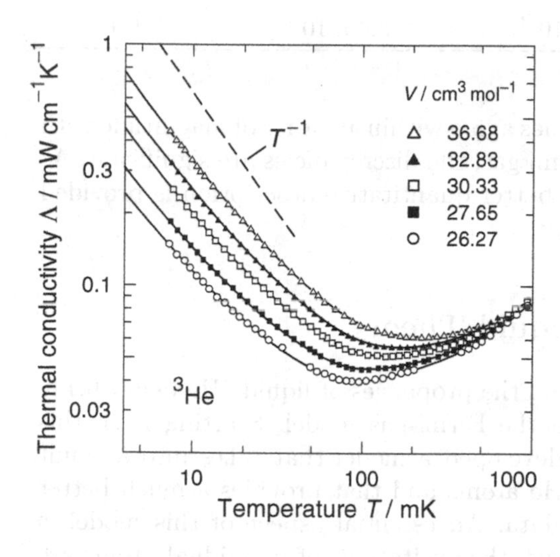

Fig. 3.7. Thermal conductivity Λ of liquid ^{3}He as a function of the temperatures at different molar volumes. At low temperatures and a molar volume of $36.68\,\mathrm{cm^3\,mol^{-1}}$ one finds roughly a $1/T$ dependence, as indicated by the *dashed line* [130]

The thermal conductivity of a Fermi gas is given by

$$\Lambda = \frac{1}{3}\, C_V\, \tau\, v_\mathrm{F}^2 \,. \tag{3.20}$$

Since the specific heat C_V varies proportional to T, and τ proportional to T^{-2}, we expect a $1/T$ dependence for the thermal conductivity. As shown in Fig. 3.7, the thermal conductivity data of liquid ^{3}He in the absence of external pressure ($36.68\,\mathrm{cm^3mol^{-1}}$) indeed do approach such a dependence at very low temperatures. With increasing pressure, the contribution of spin fluctuations causes deviations, as we have also seen in the discussion of the specific heat.

3.1.4 Quantitative Comparison: ^{3}He and Ideal Fermi Gas

We have seen that liquid ^{3}He at low temperatures behaves at least qualitatively as a free Fermi gas. We will now look at the quantitative comparison for a few selected physical properties that we have gathered in Table 3.1.

Table 3.1. Specific heat, sound velocity and magnetic susceptibility of ^{3}He in comparison to an ideal Fermi gas

	^{3}He	Fermi Gas	Ratio
$C_V/\gamma T$	2.78	1.00	2.78
$v = v_F/\sqrt{3}$ $(\mathrm{m\,s^{-1}})$	188	95	1.92
χ/β^2 $(\mathrm{J\,m^3})^{-1}$	3.3×10^{51}	3.6×10^{50}	9.1

Although the experimental values agree within an order of magnitude with the expected ones for an ideal Fermi gas, the discrepancies are significant. As we see in the following section, a better quantitative description is provided by the Landau Fermi-liquid theory.

3.2 The Landau Fermi-Liquid Theory

As we saw in the preceding section, the properties of liquid ^{3}He below 0.1 K can qualitatively be described in the Fermi-gas model. Starting from this simplifying description, *Landau* developed a model that takes into account the strong interaction between ^{3}He atoms and that provides a much better description of the experimental data. An essential aspect of this model is that, due to the strong interactions, the excitations of individual atoms are not the proper means to describe the system. Rather, collective excitations of the atoms must be considered. These elementary excitations can be treated as *quasiparticles* with energies and momenta. Within this model, Landau predicted the so-called *zero sound*, which was later on experimentally discovered by *Keen, Matthews* and *Wilks* in 1963 [131]. We will discuss this property of liquid ^{3}He in Sect. 3.3.

3.2.1 Quasiparticle Concept

Landau assumed that the interaction between ^{3}He atoms changes their energy but not their momentum. This assumption is plausible since the allowed momenta are determined by the geometrical boundary conditions. Therefore, we can still write for the momentum of the quasiparticles at the Fermi surface

$$p_F = \hbar \left(3\pi^2 n\right)^{1/3} . \tag{3.21}$$

Because of the interaction of the ^{3}He atoms the total energy cannot be written as a simple sum $U = \sum_i f_i E_i$, where f_i and E_i represent the occupation number and the energy of the i-th state. The energy E_i of a level cannot unambiguously be related to a certain atom, since its value depends on the occupation of all other states. Therefore, it is not reasonable to consider the energy of single atoms but the energy of so-called *quasiparticles*. Without interaction, the quasiparticles are identical with the atoms. Switching on the interaction between the atoms 'slowly' does not alter the number of levels but shifts their energy, as illustrated in Fig. 3.8. The number of quasiparticles is therefore given by an expression analogous to (3.9)

$$n = 2\varrho_k \int f \, \mathrm{d}^3k = \int D(k) \, f \, \mathrm{d}k \,, \tag{3.22}$$

with f denoting the occupation number, which we will discuss later in more detail. The factor of two accounts for the possible spin orientations. Landau defined the energy of the quasiparticles by

$$\delta u = \int E \, \delta f \, \mathrm{d}^3k \,, \tag{3.23}$$

where δu is the change of the total energy caused by a small change δf in the distribution function. In other words, the energy E is the energy of a single atom that interacts with all other particles in the system. This implies that the quasiparticle states are not eigenstates.

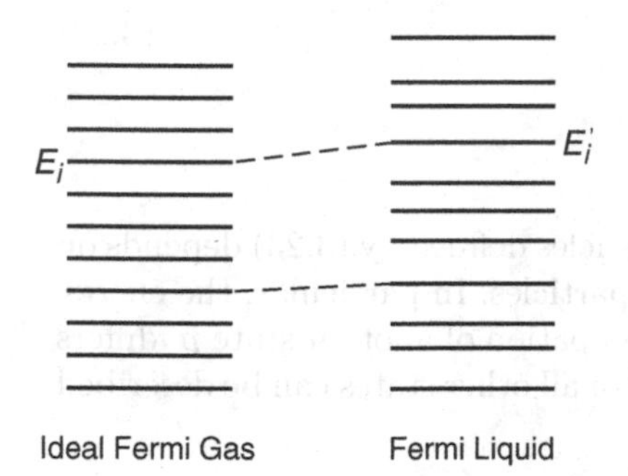

Fig. 3.8. Schematic illustration of the energies E_i of a degenerate ideal Fermi gas (*left*) and for a Fermi liquid of interacting particles (*right*). For simplicity, the levels of the Fermi gas have been drawn equidistantly

At this point, the question is: what does the distribution function f look like? Can we use the Fermi–Dirac distribution as before? This is only the case if the energy states are well defined. Since the quasiparticle states are not true eigenstates, transitions between levels occur, which in turn lead to a level broadening according to the uncertainty principle $\delta E \approx \hbar/\tau$. Here, τ represents the lifetime of the considered state. The energy states are well defined as long as the uncertainty broadening δE is small compared to the thermal broadening $\Delta E \approx k_\mathrm{B}T$. This condition can always be fulfilled at sufficiently low temperatures since $\tau \propto T^{-2}$, and therefore $\delta E \propto T^2$. This means that the distribution function

$$f(E,T) = \frac{1}{e^{(E-\mu)/k_B T} + 1}, \tag{3.24}$$

is also applicable to quasiparticles. Therefore, the complication that E is a function of f is irrelevant at sufficiently low temperatures. In this temperature range, we may approximate the dispersion relation of the quasiparticles at the Fermi surface by

$$E = E_F + \left(\frac{\partial E}{\partial p}\right)_F (p - p_F) . \tag{3.25}$$

Variables with index F relate to quantities at the Fermi surface. For an ideal Fermi gas it follows from (3.3) that

$$\left(\frac{\partial E}{\partial p}\right)_F = \frac{p_F}{m} = v_F . \tag{3.26}$$

This expression is adopted in the Landau theory, but the mass m is replaced by the effective mass m^* of the quasiparticles. Using the modified expression we may rewrite (3.25) in the form

$$E = E_F + \frac{p_F}{m^*}(p - p_F) . \tag{3.27}$$

Therefore, the density of states of quasiparticles at the Fermi surface differs from the expression for an ideal Fermi gas only by the appearance of the effective mass m^*, i.e.,

$$D(E_F) = \frac{m^* k_F}{\pi^2 \hbar^2} = \frac{m^*}{\pi \hbar^2} \sqrt[3]{\frac{3n}{\pi}} . \tag{3.28}$$

3.2.2 Interaction Function

As mentioned above, the energy of quasiparticles defined by (3.23) depends on the configuration of the surrounding quasiparticles. In particular, the energy $E(\boldsymbol{p},T)$ of a quasiparticle changes if the occupation of another state $\boldsymbol{p}'$ differs from that at $T = 0$ by $\delta f(\boldsymbol{p}')$. The influence of all other states can be described by the phenomenological formula[1]

$$E(\boldsymbol{p},T) = E(\boldsymbol{p},0) + 2\varrho_k \int h(\boldsymbol{p},\boldsymbol{p}')\, \delta f'\, d^3p' . \tag{3.29}$$

---

[1] Note that in the theory of Fermi liquids $h(\boldsymbol{p},\boldsymbol{p}')$ and δf are usually denoted as $f(\boldsymbol{p},\boldsymbol{p}')$ and δn. We have not used this notation here to avoid possible confusion with other quantities.

The function $h(\boldsymbol{p},\boldsymbol{p}')$ is an important characteristic quantity in the theory of Fermi liquids. Landau interpreted this function as the quasiparticle scattering amplitude. It can be expressed by

$$h(\boldsymbol{p},\boldsymbol{p}') = \frac{\partial^2 U}{\partial f(\boldsymbol{p})\,\partial f'(\boldsymbol{p}')} \,. \tag{3.30}$$

At low temperatures, only states at the Fermi surface are important. Therefore, we may write, to a good approximation, $p \approx p' \approx p_\mathrm{F}$ for the magnitudes of the momenta. This means that the function $h(\boldsymbol{p},\boldsymbol{p}')$ depends, to a first approximation, only on the angle Θ between $\boldsymbol{p}$ and $\boldsymbol{p}'$. Thus, we use the simplification

$$h(\boldsymbol{p},\boldsymbol{p}') = h(\Theta) \,. \tag{3.31}$$

Handling of the interaction function is somewhat easier if one defines the function

$$F(\Theta) = D(E_\mathrm{F})\, h(\Theta) \,, \tag{3.32}$$

and makes use of a Legendre polynomial expansion

$$F(\Theta) = \sum_i F_i\, P_i(\cos\Theta) = F_0 + F_1\, \cos\Theta + F_2\, \frac{3\cos^2\Theta - 1}{2} + \,\ldots \,. \tag{3.33}$$

The expansion coefficients F_i must be determined experimentally.

So far we have considered an isotropic liquid and have neglected the spin-dependent interaction. If magnetic-field effects have to be taken into account this is no longer admissible since the total energy is then a function of the spin coordinates. To introduce the magnetic exchange interaction into the Landau model, the interaction function $h(\boldsymbol{p},\boldsymbol{p}')$ has to be augmented by a term that contains the spin coordinates $\boldsymbol{s}$ and $\boldsymbol{s}'$. We therefore write

$$\mathcal{F}(\boldsymbol{p},\boldsymbol{s},\boldsymbol{p}',\boldsymbol{s}') = h(\boldsymbol{p},\boldsymbol{p}') + \xi(\boldsymbol{p},\boldsymbol{p}')\; \boldsymbol{s}\cdot\boldsymbol{s}' \,. \tag{3.34}$$

The first term describes the isotropic interaction and the second term accounts for the anisotropic exchange interaction. Analogous to $F(\Theta)$ one can define a function $G(\Theta)$ and its corresponding expansion:

$$G(\Theta) = D(E_\mathrm{F})\, \xi(\Theta) = \sum_i G_i\, P_i(\cos\Theta) = G_0 + G_1 \cos\Theta + \ldots \,. \tag{3.35}$$

In this way, the complex many-body problem is reduced to a determination of the coefficients F_i and G_i. In many cases, only the two leading terms are necessary for a good description of experimental quantities.

3.2.3 Application of Landau's Theory to Normal-Fluid ^{3}He

Finally, we have to link the results of the Landau Fermi-liquid theory with measurable properties of ^{3}He. A somewhat elaborate calculation, which we do not present here, leads to

$$\frac{m^*}{m} = \left(1 + \frac{1}{3} F_1\right) \tag{3.36}$$

for the effective mass [132]. The theoretical treatment of specific heat, sound velocity, susceptibility and transport properties in the Landau model leads to the same temperature dependencies as for a Fermi gas, which we have discussed in Sect. 3.1.3. The following equations contain the numerical corrections computed from the Landau Fermi-liquid theory for the different quantities.

For the specific heat, one finds the expression

$$C = \frac{m^*}{m} C_{\text{FG}} , \tag{3.37}$$

and for the sound velocity the relation

$$v_1^2 = \frac{p_{\text{F}}^2}{3m^2} \frac{1 + F_0}{1 + \frac{1}{3} F_1} . \tag{3.38}$$

In the expression for the magnetic susceptibility, the magnetic exchange interaction enters as well as the effective mass. One obtains

$$\chi = \frac{m^*}{m} \left(\frac{1}{1 + \frac{1}{4} G_0}\right) \chi_{\text{FG}} . \tag{3.39}$$

Here, the quantities C_{FG} and χ_{FG} represent the specific heat and the susceptibility of an ideal Fermi gas.

The specific heat is higher than that of an ideal Fermi gas since the density of states at the Fermi surface is enhanced by the interaction. This is expressed by the ratio $m^*/m > 1$. In principle, this is also true for the magnetic susceptibility, which is also proportional to the density of states $D(E_{\text{F}})$ at the Fermi surface, but it also contains the factor $1/(1 + \frac{1}{4} G_0)$ that results from the spin-dependent exchange interaction. The experimental value is $G_0 = -2.8$. The negative sign indicates that because of the exchange interaction the spins tend to orient parallel to each other, in contrast to the antiparallel orientation expected from Fermi statistics. If the magnitude of the exchange interaction were larger by a factor of two, then $\frac{1}{4} G_0 < -1$, and the ground state of liquid ^{3}He would be ferromagnetic.

The Landau theory is relevant for many different properties of liquid ^{3}He. For example, one finds very good agreement with transport properties such as viscosity, thermal conductivity and self-diffusion coefficient including their dependence on pressure. To obtain the Landau parameters one uses numerous experimental results adjusting the parameters to obtain as good a fit as possible. The first few coefficients of the expansion at different pressures, obtained in this way, are listed in Table 3.2. Using these values, we find for the specific heat and the sound velocity $C_V = 2.78\,\gamma T$ and $v = 180\,\text{m}\,\text{s}^{-1}$, respectively, in excellent agreement with the experimental values listed in Table 3.1.

Table 3.2. Molar volume V_m, experimentally determined values of the Landau parameters F_0, F_1 and G_0 and effective mass m^*/m of liquid ^{3}He at different pressures [133]

p (bar)	V_m (cm^3)	F_0	F_1	G_0	m^*/m
0	36.84	9.30	5.39	−2.78	2.80
3	33.95	15.99	6.49	−2.89	3.16
6	32.03	22.49	7.45	−2.93	3.48
9	30.71	29.00	8.31	−2.97	3.77
12	29.71	35.42	9.09	−2.99	4.03
15	28.89	41.73	9.85	−3.01	4.28
18	28.18	48.46	10.60	−3.03	4.53
21	27.55	55.20	11.34	−3.02	4.78
24	27.01	62.16	12.07	−3.02	5.02
27	26.56	69.43	12.79	−3.02	5.26
30	26.17	77.02	13.50	−3.02	5.50
33	25.75	84.79	14.21	−3.02	5.74

3.3 Zero Sound

Sound propagation in liquid ^{3}He at moderate temperatures is similar to the sound propagation in simple ordinary liquids. At very low temperatures, however, an anomalous sound propagation is observed. This new type of sound occurs when the collision time τ of the quasiparticles is long compared to the period of the sound wave. Because of the relation $\tau \propto T^{-2}$ this condition can be fulfilled for sound waves of any frequency at sufficiently low temperature. At first glance, one would not expect any sound propagation under these circumstances. In the analogous case of dilute gases, for example, where the mean free path is large compared to the wavelength of the sound wave, the damping rises drastically with increasing frequency, and finally sound propagation dies out. This is also true for an ideal Fermi gas. However, at low temperatures liquid ^{3}He is a Fermi liquid of strongly interacting particles, as we have discussed in the previous sections. Within the Landau model, one finds normal hydrodynamic sound propagation for $\omega\tau \ll 1$, but this model also predicts the existence of sound waves for $\omega\tau \gg 1$.

The force that acts on a quasiparticle originates from the interaction with the complete surrounding and not just with one other collision partner. Therefore, density fluctuations that occur in one region are transferred to other regions without direct collisions between quasiparticles. In a theoretical treatment of the problem one first derives the equation of motion for the Fermi liquid from the equations for the number density and the mass current density. The solution of the equation of motion has poles in the plane at complex frequencies, which can be identified as collective oscillations. We shall not go into the details of this rather complicated calculation, but simply state that this theory predicts three different types of waves. Two of them are similar to ordinary sound waves and the third one is identical to a collisionless spin wave. These new types of sound waves, that only exist at very low temperature, were named *zero sound* by Landau.

3.3.1 Longitudinal Sound Propagation

The following expressions for the sound velocity and damping coefficient are not exact but represent first-order approximations using the Landau parameters. For general considerations these expressions are sufficient. In the hydrodynamic regime, $\omega\tau \ll 1$, one finds, normal, or first sound, propagating with the velocity v_1 as described by (3.38). This expression only differs by a numerical constant from the expression for sound waves in an ideal Fermi gas:

$$v_1^2 = \frac{v_F^2}{3} \frac{(1+F_0)}{\left(1+\frac{1}{3}F_1\right)} . \tag{3.40}$$

For $\omega\tau \gg 1$, that is, in the collisionless regime of zero sound, one can express the difference between the velocities of zero sound and normal sound by

$$\frac{v_0 - v_1}{v_1} = \frac{2}{5} \frac{\left(1+\frac{1}{5}F_2\right)}{(1+F_0)} . \tag{3.41}$$

The numerical value is $(v_0 - v_1) \approx 6\,\mathrm{m\,s^{-1}}$. In the intermediate regime one can approximate the sound velocity v and the damping coefficient α with the equations

$$\frac{v}{v_1} = 1 + \frac{v_0 - v_1}{v_1} \frac{\omega^2\tau_s^2}{1+\omega^2\tau_s^2} , \tag{3.42}$$

$$\frac{\alpha v}{\omega} = -2 \frac{v_0 - v_1}{v_1} \frac{\omega\tau_s}{1+\omega^2\tau_s^2} . \tag{3.43}$$

The index s of the relaxation time τ_s indicates that this quantity is not exactly identical with the relaxation time introduced in Sect. 3.1. Essentially, the difference is a numerical factor that does not influence the temperature dependence. The expression for the internal friction, or damping, has the well-known form of a Debye relaxation (see Sect. 9.1.3). In the limiting cases of the hydrodynamic regime and the collisionless regime, the damping coefficient is given by

$$\alpha_1 = A_1\omega^2\tau \propto \omega^2 T^{-2} \qquad \text{for} \quad \omega\tau \ll 1, \quad \text{and} \tag{3.44}$$

$$\alpha_0 = A_0\tau^{-1} \propto T^2 \qquad \text{for} \quad \omega\tau \gg 1, \tag{3.45}$$

respectively. Figure 3.9 shows, for two frequencies in the MHz range, the absorption coefficient α and the velocity v of sound waves propagating in ^{3}He.

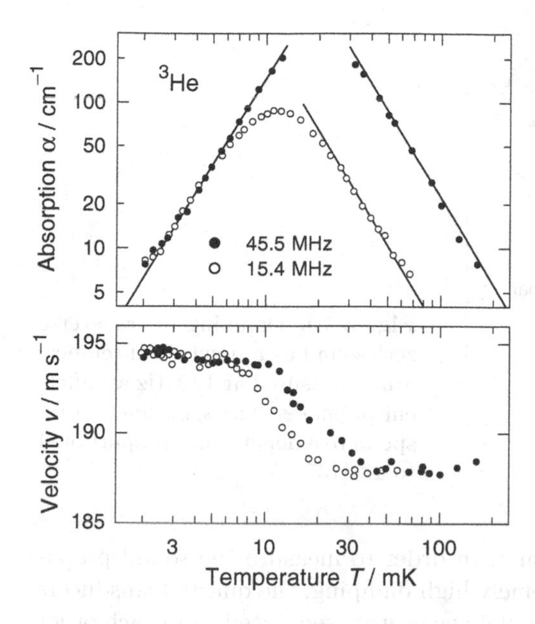

Fig. 3.9. Absorption coefficient α and sound velocity v of liquid ^{3}He as a function of temperature at 15.4 and 45.5 MHz. The data agree very well with the prediction of the Landau theory. The *straight solid lines* represent the proportionalities T^2 and $1/T^2$ at low and high temperatures, respectively [134]

The transition from normal sound to zero sound can be recognized easily. The damping on the low-temperature side of the absorption maximum is independent of frequency, whereas the damping on the high-temperature side is proportional to ω^2, as expected from (3.44) and (3.45). The velocity of zero sound is about 6 m s^{-1} higher than the velocity of normal sound, as predicted by the Landau Fermi-liquid theory.

3.3.2 Transverse Sound Propagation

In classical fluids and gases, the propagation of transverse sound waves is impossible, because no restoring force exists for transverse atomic displacements. In contrast, in liquid ^{3}He such sound modes are observed. In the hydrodynamic regime $\omega\tau \ll 1$ only diffusive shear vibrations are found that decay rapidly as in ordinary liquids. For $\omega\tau \gg 1$, theory predicts the occurrence of transverse sound waves. The rather complicated expression for the dispersion relation has a real solution for $F_1 > 6$. A look at Table 3.2 shows that under normal pressure no transverse zero sound is expected to exist since F_1 is too small. Under pressure, however, the parameter F_1 increases and reaches a value of about 15 at the melting point.

The first experimental confirmation of the existence of transverse zero sound in normal-fluid ^{3}He was obtained by *Roach* and *Ketterson* in 1976 [135]. Figure 3.10 shows that the damping of transversal zero sound rises proportional to T^2, as expected from theory. Note that the absolute magnitude of the damping coefficient is extremely high. Both the prefactor of the sound absorption and the quasiparticle collision time depend on pressure, leading to

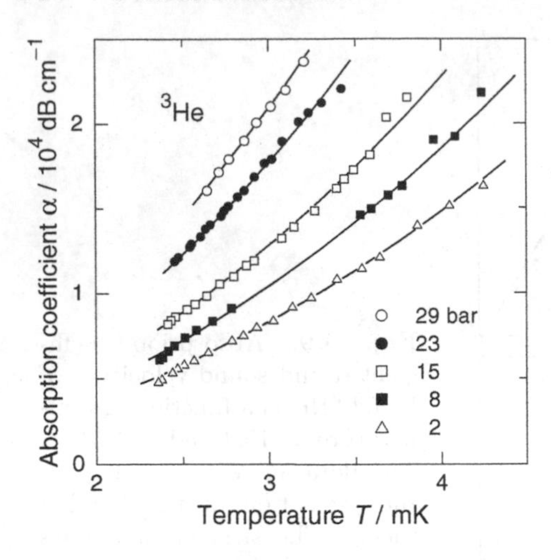

Fig. 3.10. Damping of transverse zero sound as a function of temperature measured at 12 MHz at different pressures. The *solid lines* correspond to a dependence proportional to T^2 [135]

the observed variation with pressure. In order to measure the sound propagation in the presence of the extremely high damping, the quartz transducers for generating and detecting the sound waves were separated from each other by just 25 μm.

3.3.3 Collisionless Spin Waves

For spin transport two different regimes are also observed: For $\omega\tau \ll 1$, in the hydrodynamic regime, one finds normal spin diffusion. In the Landau theory, one has the usual expression for the self-diffusion coefficient (3.19), but with the additional factor $(1 + G_0/4)$. This factor accounts for the exchange interaction resulting in

$$D_{\mathrm{s}} = \frac{1}{3}\,\tau_{\mathrm{D}}\,v_{\mathrm{F}}^2 \left(1 + \frac{1}{4G_0}\right) . \tag{3.46}$$

In the case where the quasiparticle collision time τ_{D} becomes larger than the precession period of the nuclear spins in a magnetic field, collisionless spin waves can propagate. Although this was predicted in 1957 by *Silin* [136], the first experimental observation was made in 1984 by a group from Ohio in NMR studies of normal-fluid ^{3}He [137].

In these experiments, the NMR absorption of ^{3}He was measured in a magnetic field with a linear field gradient. Above 5 mK one obtains a nearly rectangular absorption band, as expected from the field distribution over the sample. At lower temperatures it becomes possible to excite standing spin waves. The absorption maxima related to these spin waves are superposed on the normal rectangular absorption band as shown in Fig. 3.11. The results obtained are in very good agreement with the theoretical expectation for a Fermi liquid.

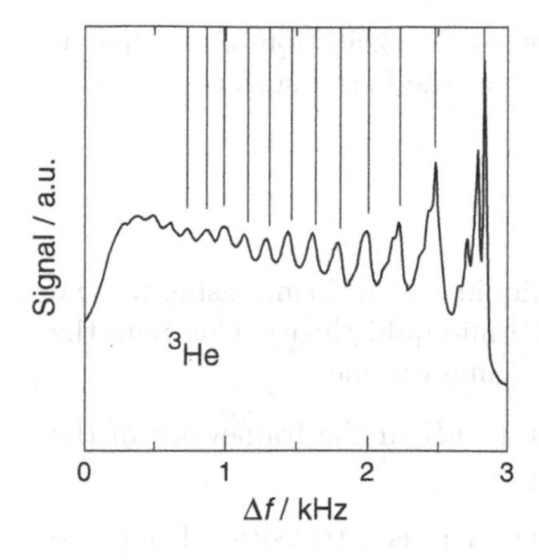

Fig. 3.11. NMR absorption at 2 MHz in a magnetic field gradient of 44 mT m^{-1}. The *vertical lines* indicate the expected positions for absorption maxima of standing spin waves [137]

3.3.4 Final Remarks

The original formulation of the theory of Fermi liquids has been developed further and improved over time. In particular, calculations of the transport properties have been made more precise and spin fluctuations have been included. These spin fluctuations are very strong in liquid ^{3}He because of the competition between antiparallel alignment of the nuclear spins due to the Fermi statistics and the parallel alignment favored by the exchange interaction. This causes long-lived local ferromagnetic fluctuations.

Investigations of the excitations in normal-fluid ^{3}He by neutron scattering experiments have contributed to the further development of the Landau theory. Such experiments are very difficult to perform since they have to be carried out at very low temperatures and, moreover, the capture cross section for neutrons by ^{3}He is extremely large. Results of this kind of measurement are shown in Fig. 3.12. For phonons, one finds an amazingly sharp dispersion curve up to large wave vectors. This behavior distinguishes liquid ^{3}He clearly

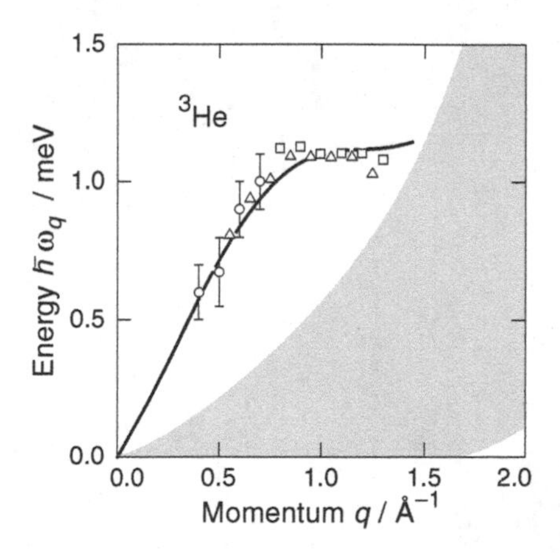

Fig. 3.12. Dispersion curve of ^{3}He. The *data points* were obtained in different neutron scattering experiments. The *line* represents a theoretical calculation of the zero sound dispersion with an improved Landau model. The *grey tinted region* indicates the quasiparticle continuum observed in neutron-scattering experiments [138–140]

from ordinary liquids. In addition, one observes the excitation of quasiparticles and can determine their density of states at the Fermi surface.

Exercises

3.1 Calculate the heat capacity of 1 cm^3 of liquid ^{3}He at 10 mK using (**a**) the free Fermi-gas model and (**b**) the Landau Fermi-liquid theory. Compare the results to that of a piece of copper with the same volume.

3.2 Calculate the viscosity of liquid ^{3}He at 10 mK in the framework of the Landau model.

3.3 At low temperatures, it is expected that there is a transition from zero sound to first sound. Estimate the transition temperature for a sound frequency of 2 MHz.

4 Superfluid ^{3}He

For a long time the existence of a superfluid phase of liquid ^{3}He was considered as being very unlikely. Since ^{3}He atoms obey the Pauli exclusion principle, a Bose–Einstein condensation such as in ^{4}He is impossible. Instead, the absence of a superfluid phase of ^{3}He was taken as conformation of the interpretation that the lambda transition is a Bose–Einstein condensation. This viewpoint remained unchanged until the publication of the revolutionary BCS theory by *Bardeen*, *Cooper* and *Schrieffer* in 1957 that describes the superconductivity of metals on a microscopic basis (see Sect. 10.3). The BCS theory elaborates the fact that the conduction electrons mutually interact via the exchange of virtual phonons. This mechanism leads to the formation of pairs of coupled electrons in the superconducting state. These so-called *Cooper pairs* have a total spin of $S = 0$ and thus have quasibosonic character. Although the occurrence of superfluidity in liquid ^{3}He is also connected with the formation of pairs, there is no one-to-one correspondence between the Cooper pairs in superconductors and the pairs in superfluid ^{3}He. The magnetic exchange interaction favors a parallel alignment of the nuclear spins for the paired quasiparticles in liquid ^{3}He. The superfluid ground state therefore consists of quasiparticle pairs with total spin $S = 1$ and, to satisfy the requirement that the wave function describing a pair of fermions must be antisymmetry to their interchanges, the nonzero orbital angular momentum $L = 1$.

Although the complexity of the pairing was realized long before the experimental discovery of the superfluid phases of ^{3}He, the estimation of the transition temperature proved to be problematic. In 1963 a transition temperature T_c of about 100 mK was expected. In the following years the estimate of T_c was changed several times towards lower values, since no transition was observed in the predicted temperature range. At that time, the search for superfluid ^{3}He was a strong driving force for the development of new and better cooling methods. After many unsuccessful attempts, superfluid ^{3}He was finally discovered in 1971. *Osheroff, Richardson* and *Lee* observed clear indications for two phase transitions in ^{3}He at temperatures around 2 mK in experiments using a Pomeranchuk cell (see Sect. 11.5). From these experiments it was impossible to tell whether the observed phase transitions were related to liquid or to solid ^{3}He, because both phases were present in their Pomeranchuk cell at the same time. However, since they were searching for an antiferromagnetic state of solid ^{3}He, they believed that the phase transitions occurred in solid ^{3}He and published their results in a paper entitled: *'Evidence for a New Phase in Solid ^{3}He'* [141]. But only a few months later they were able to show in beautiful NMR experiments that the phase transitions they had observed were actually taking place in the liquid [142]. Later it became clear that three different superfluid phases of ^{3}He exist.

Superfluid ^{3}He is an extraordinarily complex fluid, that shows a great variety of exotic phenomena (for recent monographs see [133, 143, 144]). In this chapter we will give a brief introduction to some of these properties.

4.1 Basic Experimental Facts

In this section, we briefly discuss some basic experimental observations that have been made on superfluid ^{3}He. It is not possible, however, to cover all the important aspects of the large amount of experimental data available for superfluid ^{3}He. The selection of observations we present in this introductory section has been made with the intention of introducing some basic properties of this fascinating liquid.

4.1.1 Phase Diagram

First, we take a look at the phase diagram of ^{3}He at very low temperatures. As shown in Fig. 4.1, in the absence of magnetic fields, liquid ^{3}He can exist in three different phases, namely in a normal-fluid phase ^{3}He-N and in two superfluid phases ^{3}He-A and ^{3}He-B.

The bcc phase of solid ^{3}He is divided into two regions depending on the behavior of the nuclear spins: a paramagnetic phase and at very low

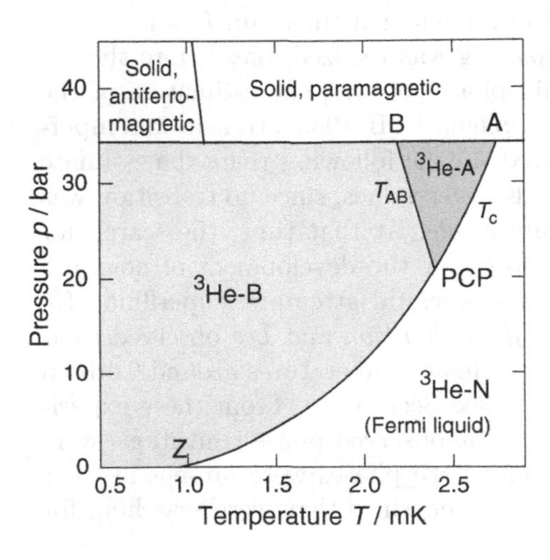

Fig. 4.1. Pressure–temperature phase diagram of ^{3}He below 3 mK in the absence of magnetic fields [145]. The superfluid phases are *grey tinted*

temperatures an antiferromagnetic phase. The transition temperature for the antiferromagnetic transition – the so-called Néel temperature T_N – is about 0.9 mK at melting pressure.

Four points of this phase diagram are specifically indicated by capital letters: A and B are the transition points of the superfluid phases along the melting curve. The polycritical point PCP marks the coexistence point of all three liquid phases. The letter Z labels the transition between ^{3}He-N and ^{3}He-B at zero pressure $p = 0$. In Table 4.1, we have gathered together the values for the pressures and temperatures at which these points occur. The phase transition from normal to superfluid ^{3}He at the line A–PCP–Z is of second order. In contrast, the phase transition from the A to the B phase, indicated by the line B–PCP, is a first-order phase transition.

Table 4.1. Some special points from the temperature–pressure phase diagram of liquid ^{3}He. Note that we have listed the values according to the new Provisional Low Temperature Scale PLTS-2000 (see Table 12.3). However, such an adaptation was *not* performed for the temperatures given in the remaining text of this book

	A	B	PCP	Z
pressure p (bar)	34.3	34.3	21.5	0
temperature T (mK)	2.44	1.90	2.24	0.92

Even small magnetic fields have a great influence on the phase diagram of liquid ^{3}He, as illustrated in Fig. 4.2. An additional superfluid phase, ^{3}He-A_1,

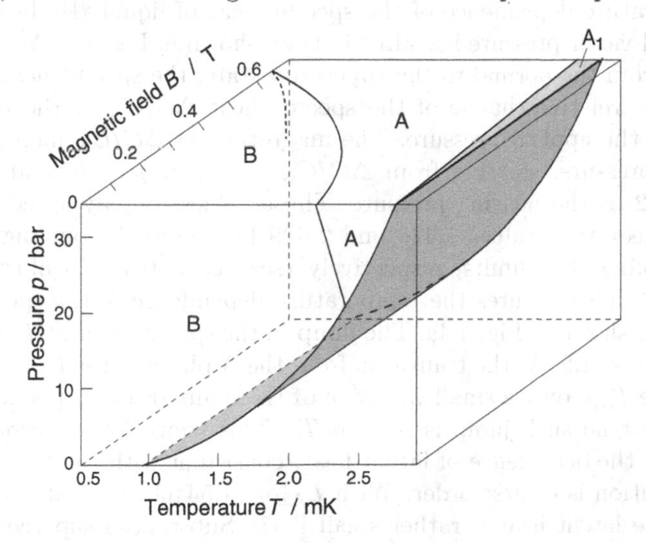

Fig. 4.2. Pressure–temperature–magnetic field phase diagram of liquid ^{3}He in the temperature range 0.5 K to 3 mK. After [133]

occurs. The temperature range in which this phase exists depends on the magnitude of the magnetic field. It increases with the applied field and is 0.5 mK wide at 10 T. The A phase also widens with increasing magnetic field and above 0.65 T it displaces the B phase completely. In addition, the polycritical point PCP disappears in finite magnetic fields and the A phase exists in a small temperature range between ^{3}He-N and ^{3}He-B. At low temperatures and low fields this range is extremely small. As shown in Fig. 4.3, the A phase has a width of only about 20 μK at $B = 38\,\mathrm{mT}$ and $p = 10\,\mathrm{bar}$.

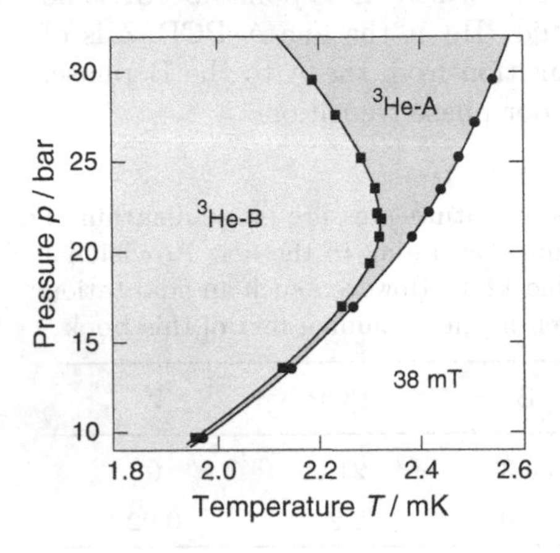

Fig. 4.3. Pressure–temperature phase diagram of liquid ^{3}He below 2.6 mK in a magnetic field of 38 mT [146]

4.1.2 Specific Heat

The temperature dependence of the specific heat of liquid ^{3}He below 2 mK at saturated vapor pressure has already been shown in Fig. 1.3. At the phase transition from the normal to the superfluid state, the specific heat exhibits a jump. The relative change of the specific heat $\Delta C/C_\mathrm{N}$ at the transition depends on the applied pressure. The magnitude of $\Delta C/C_\mathrm{N}$ increases with increasing pressure, starting from $\Delta C/C_\mathrm{N} \approx 1.4$ at $p = 0$, and reaching $\Delta C/C_\mathrm{N} \approx 2$ at the melting pressure. These values roughly agree with the expected theoretical values 1.426 and 2.029 for the weak coupling and the strong coupling BCS limits, respectively (see Sect. 10.3). To illustrate the behavior at high pressures the temperature dependence of the specific heat at 28.7 bar is shown in Fig. 4.4a. The jump in the specific heat at T_c is, in this case, $\Delta C/C_\mathrm{N} \approx 1.9$. At the transition from the A phase to the B phase at the temperature T_AB, only a small variation of the temperature dependence of C is visible, but no such jump is seen at T_c. This transition is accompanied, however, by the occurrence of latent heat, consistent with the fact that this phase transition is of first order. With $L_\mathrm{AB} \approx 1.54\,\mu\mathrm{J\,mol}^{-1}$ at the melting pressure, the latent heat is rather small [147]. Substantial supercooling can occur at the A–B transition, whereas only modest superheating effects have been observed [148]. At pressures below the polycritical point PCP and at

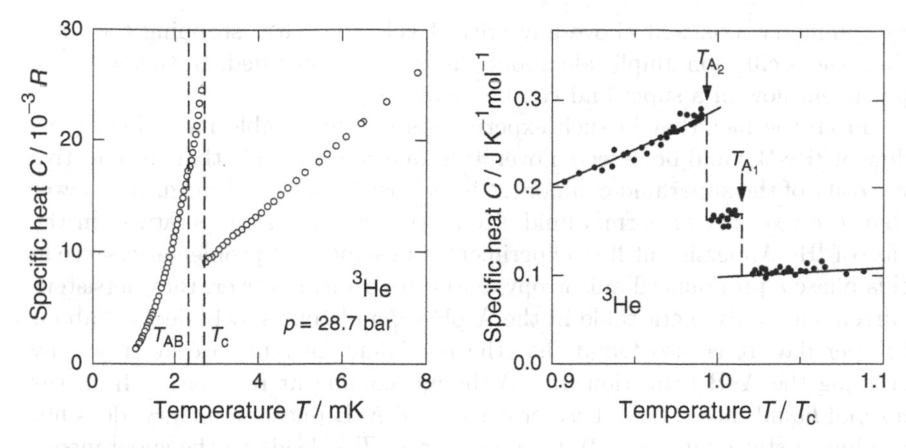

Fig. 4.4. (a) Reduced specific heat C/R of ^{3}He at a pressure of 28.7 bar as a function of temperature. At T_c a clearly visible jump in C exists, while a weak variation of C occurs at T_{AB} [21]. (b) Specific heat of superfluid ^{3}He at the melting pressure as a function of the reduced temperature T/T_c in a magnetic field of 0.88 T [147]. The *solid lines* above T_{A_1} and below T_{A_2} are fits to zero-field data in ^{3}He-N and ^{3}He-A, respectively

zero magnetic field, the transition from normal-fluid to superfluid ^{3}He leads directly to the B phase and the small variation of the specific heat at T_{AB} due to the A–B transition disappears.

As discussed above, the A transition splits into two transitions A_1 and A_2 in an external magnetic field, whereas the B transition moves to lower temperatures. The A_2 phase is identical with the A phase at zero magnetic field. The splitting of the A phase transition is clearly observed in specific heat measurements. Figure 4.4b shows the temperature dependence of the specific heat of liquid ^{3}He in a magnetic field of 0.88 T. A discontinuity is found at each of the transitions, with the A_1 jump $\Delta C_{A_1}/C_N = 0.74$ at the melting pressure being somewhat smaller than the jump at T_{A_2}. The discontinuities add up to the jump observed in zero magnetic field. The width of the A_1 phase is just 56 μK at $B = 0.88$ T.

4.1.3 Superfluidity

Persistent currents and frictionless flow are fundamental properties of a superfluid. The superfluidity of ^{3}He in both the A and the B phase has been investigated in persistent-flow experiments [149–151] using the gyroscope principle [152]. If a superfluid circulates in a ring in the xy-plane, a periodic torque about the x-axis produces oscillations about the y-axis with an amplitude proportional to the angular momentum of the circulating superfluid. One method to generate a persistent supercurrent is to cool the liquid below the superfluid transition temperature and to rotate the cryostat about its symmetry axis well above any critical velocity. After stopping the rotation, the oscillation amplitude about the y-axis is recorded. In this way, the persistent flow of a superfluid can be monitored.

From the fact that in such experiments no measurable reduction of the flow of ^{3}He-B could be detected over 48 h, one can conclude that the effective viscosity of the superfluid component is at least 12 orders of magnitude lower than the viscosity of normal-fluid ^{3}He at the transition temperature. In the case of ^{3}He-A, persistent-flow experiments are somewhat problematic, since in this phase a pronounced anisotropy exists. It seems, however, that persistent currents are only metastable in the A phase, and thus slowly decay – about 1% per day. It is also found that the persistent currents are destroyed by crossing the A–B transition line. Although cooling at rest, either from the normal liquid into the A phase or across the A–B phase boundary, does not produce a state with $\boldsymbol{v}_s \neq 0$, warming across T_{AB} leads to the spontaneous generation of a small persistent current in ^{3}He-A. This flow is independent of the magnitude and the direction of any persistent current that might have been present in the B phase before warming.

The critical velocity at which flow is no longer frictionless is very low in superfluid ^{3}He, as shown in Fig. 4.5. Typically, one finds critical velocities in the range 1 to 100 mm s^{-1}. Although the mechanism responsible for these very low critical velocities has not been unambiguously identified, it seems clear that the formation of vortex rings and the breaking of quasiparticle pairs (see Sect. 4.2) play an important role.

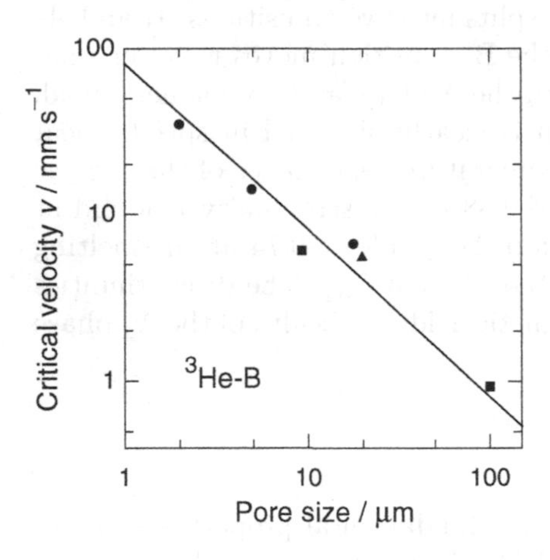

Fig. 4.5. Dependence of the critical velocity of ^{3}He-B in a superleak on the size of the pores. The data are taken at different temperatures and pressures and have been normalized to the value at 29 bar and $T/T_c = 0.5$ [149, 150, 153, 154]

4.1.4 Nuclear Magnetic Resonance (NMR)

An investigation of the nuclear spins of ^{3}He atoms provides a means to obtain detailed information on the dynamical properties of the liquid. In a constant magnetic field $\boldsymbol{B}_0$, a spin-1/2 system has only two possible orientations. In this case, the resonance frequency, or Larmor frequency is given by $\omega_L = \gamma|\boldsymbol{B}_0|$, where γ represents the gyromagnetic ratio.

For isolated ^{3}He atoms and for ^{3}He-N, one finds experimentally the expected frequency ω_L. However, in superfluid ^{3}He several anomalies are observed. We briefly introduce some of the remarkable features. A more detailed discussion of the spin dynamics in superfluid ^{3}He is presented in Sect. 4.5.

Transverse rf Fields

The first NMR experiments on superfluid ^{3}He were carried out in 1972 to clarify whether the newly discovered phases occur in liquid or in solid ^{3}He. Figure 4.6 shows the NMR spectra at different temperatures obtained in these early investigations. The experiments were performed using a Pomeranchuk cell (see Sect. 11.5) that contained both liquid and solid ^{3}He. The large, nearly temperature independent, pair of resonance lines[1] are due to the nuclear spins in solid ^{3}He. At $T = T_c$, the absorption lines in the solid and the liquid phase (*grey tinted*) lie on top of each other. Below T_c, the absorption line of the nuclear spins in the liquid is shifted towards higher frequency. This frequency shift grows with decreasing temperature. The total shift at each temperature is indicated by a *double arrow*. The line shape of the resonance originating in the liquid phase did not change significantly with temperature and is similar to that observed in ^{3}He-N.

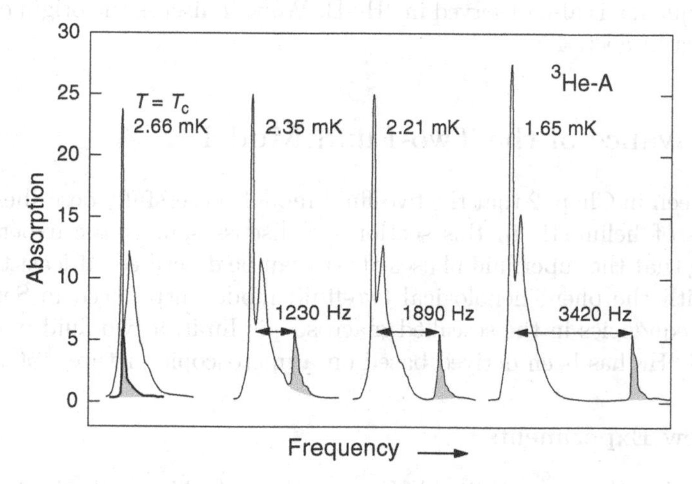

Fig. 4.6. Transverse NMR absorption spectra observed in a mixture of *solid* and *liquid* ^{3}He at the melting pressure for different temperatures [142]. The *grey tinted line* corresponds to the absorption line in the *liquid* phase

[1] The origin of the doublet structure of the resonance line of solid ^{3}He is an experimental artifact and of no interest to our discussion.

The resonant frequency in the superfluid A phase is higher than the one in the solid phase. The temperature-dependent shift is rather large. From this we can conclude that there must be a large additional magnetic field present in the superfluid phase that is absent in the case of normal-fluid ^{3}He. In 1972 *Leggett* showed that the frequency shift originates from the magnetic dipole–dipole interaction between the spins and that the macroscopic coherence in the superfluid state leads to a large enhancement of this additional magnetic field [155]. Note that a corresponding frequency shift of the transverse resonance is absent in the B phase. At the A–B phase transition, the shifted line abruptly disappears. This striking difference has helped to identify the microscopic nature of the two superfluid phases.

Longitudinal rf Fields

A further remarkable effect occurs in the nuclear resonance of ^{3}He. In the superfluid state there exists a resonant absorption at a well-defined frequency with the rf magnetic field applied parallel to the dc magnetic field. In this case, the energy splitting of the spin states is modulated by the rf field. In normal-fluid ^{3}He the application of an rf field results in relaxation processes for the spin system to reach the momentary equilibrium. In ^{3}He-A, however, no relaxation is observed, instead a resonant effect occurs. As we will see in Sect. 4.5, this so-called longitudinal resonance is also a result of the magnetic dipole–dipole interaction. A longitudinal resonance with somewhat higher frequency is also observed in ^{3}He-B. We will discuss the origin of these phenomena in Sect. 4.5.

4.2 Relevance of the Two-Fluid Model

We have seen in Chap. 2 that the two-fluid model successfully describes many properties of helium II. In this section we discuss some basic experiments indicating that the superfluid phases of ^{3}He can be described, at least to some extent, with the phenomenological two-fluid model introduced in Sect. 2.2. For low frequencies in the so-called macroscopic limit, a two-fluid model for superfluid ^{3}He has been derived based on a microscopic picture [156].

4.2.1 Flow Experiments

The anomalous behavior of liquid ^{4}He flow through thin capillaries below T_λ was historically an important indication for the superfluidity of helium II. Figure 4.7 shows the result of corresponding experiments with ^{3}He-B. In this measurement the flow through 1000 parallel channels with a diameter of 0.8 μm and a length of 10 μm was investigated.

According to the Hagen–Poiseuille law (2.1) one would expect the mass current j_s to depend linearly on the pressure gradient along the channels.

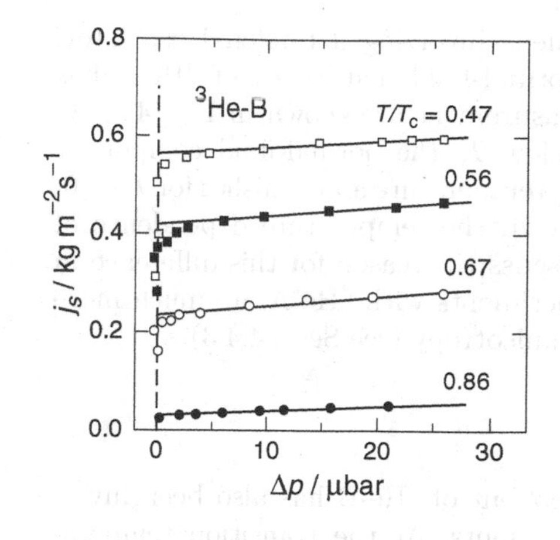

Fig. 4.7. Mass flow density j_s of ^{3}He-B through 1000 thin parallel capillaries with a diameter of 0.8 µm and a length of 10 µm as a function of the pressure gradient Δp at different temperatures [157]

In contrast, the flow of ^{3}He-B is nearly constant, similar to the behavior of helium II. This can be explained by the frictionless flow of the superfluid component, which is limited by a critical velocity. It is remarkable that even at the lowest pressures a significant mass flow occurs. With decreasing temperature the mass flow increases, as expected from the two-fluid model, because the ratio ϱ_s/ϱ becomes larger upon cooling. In addition, the temperature dependence of j_s is influenced by the variation of the critical velocity.

4.2.2 Normal-Fluid Density

The central tenet of the two-fluid model is that the properties of the superfluid can be described in terms of interpenetrating normal-fluid and superfluid components. Numerous studies of superfluid ^{3}He have been performed to determine the densities of these two components. For example, we shall discuss in Sect. 4.8 the determination of ϱ_s/ϱ by fourth-sound measurements.

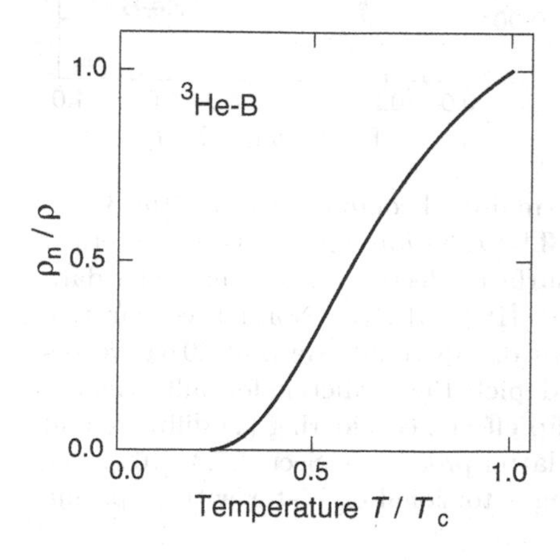

Fig. 4.8. Normalized normal-fluid density ϱ_n/ϱ of ^{3}He-B as a function of the reduced temperature T/T_c [158]

In an *Andronikashvili*-type experiment involving a torsional oscillator, the temperature dependence of the normal-fluid density ϱ_n of ^{3}He-B has been measured. The results of this measurement are shown in Fig. 4.8. As expected from the two-fluid model, below T_c the normal-fluid component decreases monotonically with decreasing temperature and vanishes for $T \to 0$. However, there is an obvious difference in the temperature dependence of ϱ_n/ϱ compared to helium II. We will discuss the reason for this difference in Sect. 4.7. The results of analogous experiments with ^{3}He-A are much more complex, since one finds a pronounced anisotropy (see Sect. 4.4.3).

4.2.3 Viscosity

The viscosity of the normal-fluid component of ^{3}He-B has also been investigated in a number of different experiments. At the transition temperature, the viscosity is very high, but below T_c it drops quite rapidly (see Fig. 3.5). Figure 4.9a shows the temperature dependence of the shear viscosity of the normal-fluid component η_n as determined in different experiments at pressures of about 20 bar. The *solid line* represents a curve calculated for bulk ^{3}He. The data taken with different apparatus agree only at high temperatures $T/T_c > 0.6$. In this range, the theoretical curve also fits very well.

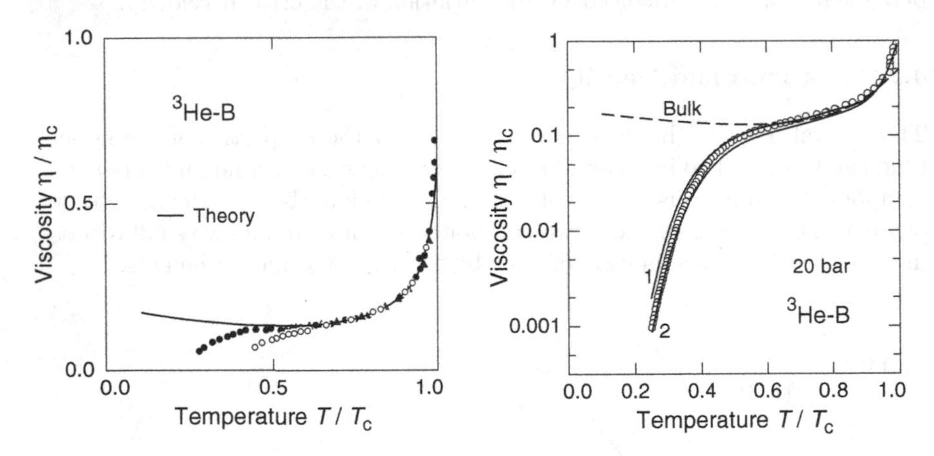

Fig. 4.9. (**a**) Normalized viscosity of the normal-fluid component η_n of ^{3}He-B as a function of the reduced temperature T/T_c. The *solid line* represents a theoretical curve for bulk ^{3}He [159] without taking surface effects into account. The data have been obtained in different experiments [160, 161]. (**b**) Normalized effective shear viscosity of the normal-fluid component of superfluid ^{3}He-B at 20 bar versus reduced temperature T/T_c. The *dashed line* depicts the prediction for bulk ^{3}He, the *solid lines* have been calculated including slip effects, considering (1) diffusive and (2) diffusive *and* Andreev scattering (for the latter process see Sect. 4.7.3) [162]. The data (*open circles*) have been obtained using a torsional oscillator with a spacing of 135 μm [163]

From bulk theory a slight minimum is expected to occur at about $T/T_c \approx 0.4$, which, however, is not seen in the experiments discussed here. The origin of the growing discrepancies between the different sets of data at low temperatures is the increasing influence of boundary effects. Depending on the geometry of the experimental setup, the measured viscosity is an effective viscosity that depends on the details of the interaction of the ^{3}He with the walls. Roughly speaking, this regime is entered when the mean free path of the quasiparticles in the bulk becomes large compared to the mean free path due to scattering at the walls and so-called *slip phenomena* occur.[2]

The effective viscosity explicitly depends on the size and geometry of the measuring cell. To obtain the bulk shear velocity in such experiments at low temperatures one has to correct for slip effects, which has not been done for the data plotted in Fig. 4.9a. As shown in Fig. 4.9b, it is possible to calculate the effective viscosity for a certain experimental setup quite well taking into account slip effects.

Comparable experiments in ^{3}He-A are even more complex to interpret because one finds a pronounced anisotropy in the measured viscosity and also a strong influence by external magnetic fields.

4.2.4 Heat Transport

The measurement of heat transport in ^{3}He provides an additional means to investigate the relevance of the two-fluid model. In particular, one can measure the momentum transfer associated with heat flow in a superfluid. One such experiment involves the measurement of the torque on a so-called *Rayleigh disc.*[3] In normal-fluid media the heat transport is purely diffusive and no torque acts on the disc. In the two-fluid model one expects a counterflow of ϱ_n und ϱ_s that leads to a twist of the Rayleigh disc, because the drag associated with the normal-fluid flow exerts torque on the disc, whereas no drag results from the motion of the superfluid component. Such experiments have been performed in both the A and the B phase and, in fact, a momentum associated with the heat transport has been observed in both phases. Again, the experiments in ^{3}He-A are more complicated to interpret because of strong anisotropy effects.

4.3 Quantum States of Pairs of Coupled Quasiparticles

As we mentioned in the introductory part of this chapter, there is no simple one-to-one correspondence between the pairs of ^{3}He atoms in superfluid ^{3}He and the Cooper pairs in BCS type superconductors (see Sect. 10.3). The main difference lies in the strong magnetic exchange interaction between the

[2] An excellent review on this topic is given in [164].

[3] This is a standard technique to measure momentum flow [165, 166].

^{3}He atoms. As discussed in Sects. 3.2 and 3.3, the Landau Fermi-liquid theory is an adequate description of the normal-fluid phase ^{3}He-N. Although it is not possible to obtain the superfluid ground state by slowly switching on the interactions between the particles – as in the case of the Fermi-liquid theory – it is still possible to describe the superfluid state of ^{3}He as occurring from the formation of pairs of quasiparticles. The strongly repulsive core potential, together with the exchange interaction, favors a parallel alignment of the spins in pairs of quasiparticles. We can roughly picture the attractive interaction between two quasiparticles in the following way: As we have seen in Chap. 3, the nuclear magnetic susceptibility of normal ^{3}He is extraordinarily high at very low temperatures. This means that below about 100 mK, ^{3}He is magnetically soft and a nuclear spin can easily polarize its surrounding ferromagnetically. Thus, parallel alignment of the quasiparticles in pairs is energetically favored.

In contrast to conventional BCS superconductors, in which the Cooper pairs have total spin $S = 0$, quasiparticle pairs are formed with $S = 1$ in superfluid ^{3}He. Since the spin wave function with $S = 1$ is symmetric upon the exchange of the two particles, the exclusion principle only allows for fermions odd quantum numbers for the orbital momentum ($L = 1, 3, \ldots$). In superfluid ^{3}He the pairs have $S = 1$ and $L = 1$. This type of pair formation is known as *spin triplet* or *odd parity pairing*.

4.3.1 Spin-Triplet Pairing

The spin state of the quasiparticle pairs is determined by $|S, S_z\rangle$. Since the z-component of a spin system with $S = 1$ has three possible values, $S_z = 0, \pm 1$, there are three different spin states:

$$|1, +1\rangle = |\uparrow\uparrow\rangle\,, \tag{4.1a}$$

$$|1,\ 0\ \rangle = \frac{1}{\sqrt{2}}\Big[\,|\uparrow\downarrow\rangle + |\downarrow\uparrow\rangle\Big]\,, \tag{4.1b}$$

$$|1, -1\rangle = |\downarrow\downarrow\rangle\,. \tag{4.1c}$$

The most general wave function of the quasiparticle pairs in superfluid ^{3}He can be represented by a linear superposition of all three spin states.

Furthermore, since the angular momentum of the pair is a well-defined quantity, the pair wave function can be expanded in terms of the three states $L_z = 0, \pm 1$ of the $L = 1$ manifold. The general expression for the wave function can thus be written as a linear combination of $3 \times 3 = 9$ terms. The amplitudes of the terms are complex-valued, since each has a magnitude and phase. In other words, the wave function of quasiparticle pairs in superfluid ^{3}He is determined by $2(2S + 1)(2L + 1) = 18$ real-valued parameters. Therefore, the order parameter of superfluid ^{3}He is not described by a complex scalar, as in the case of helium II or in the case of conventional superconductors, but by a 3×3 matrix with complex-valued components.

For mathematical convenience, *Balian* and *Werthamer* introduced a vector representation $\boldsymbol{d}(\widehat{\boldsymbol{k}})$ for this order parameter. Since this notation is widely used in the literature, we also adopt it. The vector $\boldsymbol{d}(\widehat{\boldsymbol{k}})$ represents the pair amplitude for a particular direction $\widehat{\boldsymbol{k}}$ defining a point on the Fermi surface – or in other words, it defines the amplitude of the quasiparticle condensate for a point on the Fermi surface. The components d_x, d_y and d_z are each described by three complex-valued parameters and transform under rotation like a vector. The direction of $\boldsymbol{d}$ is such that its projection onto the direction of the spin of the quasiparticle pairs is zero for any point of the Fermi surface, i.e., $\boldsymbol{d}\cdot\boldsymbol{S}=0$. We would like to point out that this not only defines the plane in which $\boldsymbol{d}$ lies, but fully determines the direction of $\boldsymbol{d}$, because $\boldsymbol{d}$ is a vector with complex components. Furthermore, we note that, in the case where $\boldsymbol{d}$ has only real-valued components, the expectation value of $\boldsymbol{S}$ is zero. The definition of $\boldsymbol{d}$ implies that it represents a unique direction in spin space, as well as the amplitude of the quasiparticle pairs for a particular direction $\widehat{\boldsymbol{k}}$. Using $\boldsymbol{d}$ the general pair wave function can be written as

$$\begin{aligned}|\Psi\rangle &= d_x\Big[\,|\downarrow\downarrow\rangle-|\uparrow\uparrow\rangle\Big]+\mathrm{i}d_y\Big[\,|\downarrow\downarrow\rangle+|\uparrow\uparrow\rangle\Big]+d_z\Big[\,|\uparrow\downarrow\rangle+|\downarrow\uparrow\rangle\Big]\\ &= -\left(d_x-\mathrm{i}d_y\right)|\uparrow\uparrow\rangle+\left(d_x+\mathrm{i}d_y\right)|\downarrow\downarrow\rangle+d_z\Big[\,|\uparrow\downarrow\rangle+|\downarrow\uparrow\rangle\Big]\,. \qquad (4.2)\end{aligned}$$

In principle, the internal degrees of freedom of the spin-triplet pairing allows for many different quasiparticle pair states and hence superfluid phases. Of the different states, the one with the lowest energy for a given set of external parameters will be realized. Using the general wave function (4.2) as a starting point, we will discuss the pair states of the different superfluid phases of ^{3}He that have actually been observed.

^{3}He-A$_1$

The simplest case is that of ^{3}He-A$_1$, which only exists in magnetic fields. The spins are aligned parallel to the applied magnetic field ($S_z=+1$) that means only pairs in the state $|\uparrow\uparrow\rangle$ exist and that the components of $\boldsymbol{d}$ must obey the relations $d_x+\mathrm{i}d_y=0$ and $d_z=0$. In this case, the pair wave function reduces to

$$|\Psi_{\mathrm{A}_1}\rangle = -2\,d_x\,|\uparrow\uparrow\rangle\,. \qquad (4.3)$$

^{3}He-A

The wave function of the A phase is a linear combination of the states $S_z=\pm1$ and thus the d_z component must be zero. This corresponds to a pair state predicted by *Anderson, Brinkman* and *Morel* [167,168] and is therefore often referred to as the ABM state. The wave function of this state is given by

$$|\Psi_{\mathrm{A}}\rangle = -(d_x-\mathrm{i}d_y)\,|\uparrow\uparrow\rangle+(d_x+\mathrm{i}d_y)\,|\downarrow\downarrow\rangle\,. \qquad (4.4)$$

^{3}He-B

The B phase corresponds to a pair state predicted by *Balian* and *Werthamer* in which the wave function is given by the general expression (4.2) that consists of all possible linear combinations [169]. The so-called BW state is quasi-isotropic with vanishing total angular momentum $J = L + S = 0$.

It is noteworthy that both the ABM and the BW model were worked out long before the experimental discovery of superfluid ^{3}He.

4.3.2 Broken Symmetry in Superfluid ^{3}He

The phase transition of liquid ^{3}He into a superfluid state is accompanied by a spontaneous breaking of symmetry, just as it is with other phase transitions of condensed matter into an ordered state. Above the critical temperature T_c, liquid ^{3}He is an isotropic, uniform, nonsuperfluid, nonchiral and nonmagnetic liquid, i.e., it has all the symmetries allowed in condensed matter. This can be expressed by the symmetry group

$$G = SO(3)_L \times SO(3)_S \times U(1)_\varphi \,. \tag{4.5}$$

Here, $SO(3)$ represents a special orthogonal – non-Abelian – rotational group in three-dimensional space. The indices L and S denote orbital and spin degrees of freedom, respectively. The symmetry $U(1)$ is a unitary Abelian rotational group around one axis. The index φ indicates that this group describes the gauge symmetry related to the quantum-mechanical phase.

All these symmetries – except the translational symmetry, which is not included in the above definition of G – are simultaneously broken in liquid ^{3}He below T_c. This results in a complex behavior of the superfluid phases of this fluid. Roughly speaking, liquid ^{3}He below T_c exhibits properties such as those of magnetically ordered systems, liquid crystals and superfluids all at the same time.

This situation can be illustrated by a simple two-dimensional representation of a model system with two distinct degrees of freedom.[4] The two segments of the angled arrow represent the orbital and spin orientation of an individual particle. The different configurations shown in Fig. 4.10 can be identified with certain real physical systems: Fig. 4.10a corresponds to an isotropic paramagnetic liquid, i.e. no ordering of both degrees of freedom. Figure 4.10b is the analogue of a liquid ferromagnet with ideal parallel alignment of the spins, and Fig. 4.10c corresponds to a nematic liquid crystal with perfect order of the orbital degrees of freedom.

Figure 4.10d shows a configuration that schematically illustrates the order of spin and orbital degrees of freedom in ^{3}He-A. In the A phase, both degrees of freedom are strictly correlated resulting in a long-range orientational order.

[4] This instructive visualization was introduced by Liu [170].

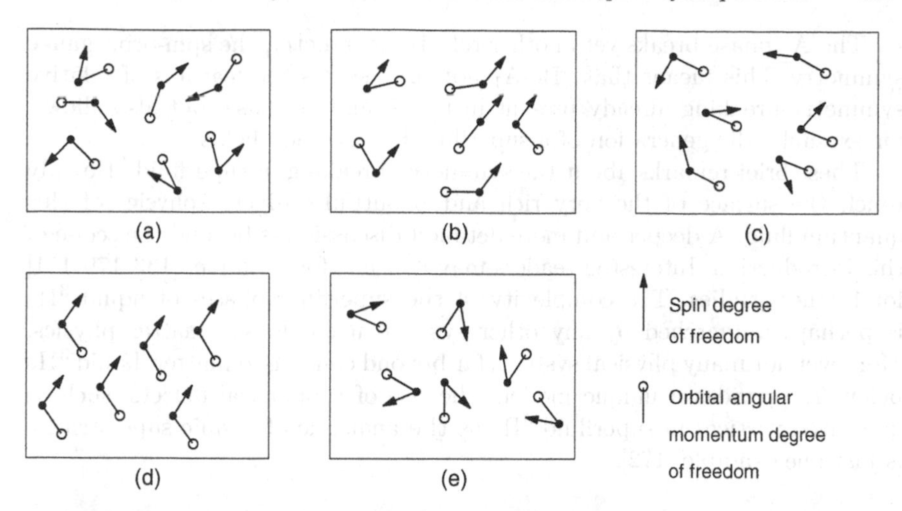

Fig. 4.10. Illustration of disorder and long-range order of a two-dimensional model fluid with an orbital and a spin degree of freedom. After [170]

As we shall see in Sect. 4.4.1, the intrinsically preferred alignment of $\boldsymbol{d}$, representing the spin space, is parallel to the orbital angular momentum vector $\boldsymbol{l}$ of the quasiparticle pairs.

In contrast, as illustrated in Fig. 4.10c, in the B phase none of the degrees of freedom exhibit a long-range order, but the relative orientation of spin and orbital moments of a pair is fixed. Therefore, one refers to this as a broken *relative* spin-orbit symmetry. The fixed angle between the spin and orbital moments is determined by the small magnetic dipole interaction in ^{3}He-B as first pointed out by *Leggett* [155]. This angle is often referred to as the *Leggett angle*. Generally speaking, a broken relative symmetry occurs if a system is invariant under transformations given by a certain linear combination of two symmetry operations, but is not invariant under any deviation from this linear combination. An intriguing point here is that such a system will respond in ways characteristic of a system breaking both symmetries, but mixes up one and the other. For ^{3}He-B, this means that its response to mechanical rotations is, within certain limits, indistinguishable from its response to magnetic fields, representing spin-space rotation. For example, one can mechanically generate spin waves and NMR in the B phase or, vice versa, magnetically induce shear instabilities.

In the A phase, the relative gauge-orbit symmetry is broken, which allows the system, for example, to change the phase of the macroscopic wave function by mechanical rotation. It also has interesting consequences for the mass flow of ^{3}He-A. Superfluid mass current and superfluid velocity $\boldsymbol{v}_\mathrm{s}$ are not necessarily parallel, because the superfluid mass current explicitly depends on the direction of the orbital angular momentum $\boldsymbol{l}$ (see Sect. 4.6.1).

The A_1 phase breaks yet another relative symmetry, the spin-orbit gauge symmetry. This means that ^{3}He-A_1 not only shows the features of relative symmetry breaking already present in the A and B phase, but also allows, for example, the generation of a superflow by magnetic fields.

These brief remarks about the symmetry breaking in superfluid ^{3}He only touch the surface of the very rich and beautifully diverse physics of this quantum fluid. A deeper and more detailed discussion is beyond the scope of this introduction. Interested readers may consult, for example, [133, 170, 171] for further studies. The complexity of the superfluid phases of liquid ^{3}He is perhaps unmatched by any other system in condensed matter physics. Moreover, for many physical systems far beyond condensed matter, liquid ^{3}He below T_c provides a unique model. The use of topological defects, such as quantized vortices in superfluid ^{3}He as the analogue of cosmic superstrings, is just one example [172].

4.3.3 Energy Gap and Superfluidity

In the preceding sections we have seen that the order parameter of the superfluid phases of liquid ^{3}He has an internal structure, but of course it also has an absolute value, which is an actual measure of the order in the system. The phase transition between the normal-fluid and superfluid states takes place only because it is energetically favorable. Analogous to the Cooper-pair formation in superconductors, the formation of quasiparticle pairs in superfluid ^{3}He leads to a reduction of the energy (see Sect. 10.3). The absolute value of the order parameter describes the amount by which the free energy of the system is reduced at a certain temperature compared to the energy at $T = T_c$. This reduction is proportional to the energy gap $\Delta(T)$ that occurs in the excitation spectrum of the quasiparticles in the superfluid state. Therefore, the minimal energy required to break a quasiparticle pair and generate two single quasiparticles is $2\Delta(T)$.

The quasiparticle pairs in superfluid ^{3}He consist of quasiparticles with wave vectors $\pm\boldsymbol{k}$, i.e., the net translational momentum of the pairs is zero. In conventional superconductors, the magnitude of the energy gap does not depend on the direction of $\boldsymbol{k}$ and therefore is isotropic. This is not the case for ^{3}He-A. Because of the anisotropic nature of the order parameter, the energy gap in ^{3}He-A not only depends on the temperature but also on the direction $\widehat{\boldsymbol{k}}$ of the momentum of the quasiparticles relative to the direction $\widehat{\boldsymbol{l}}$ of the orbital angular momentum of the pairs. The energy gap Δ_A of ^{3}He-A is defined by

$$\boldsymbol{d}(\boldsymbol{k}) = \sqrt{3/2}\,\Delta_m(T)\,\sin(\widehat{\boldsymbol{k}},\widehat{\boldsymbol{l}})\,\widehat{\boldsymbol{d}} = \Delta_A(\widehat{\boldsymbol{k}},T)\,\widehat{\boldsymbol{d}}\,, \tag{4.6}$$

where $\Delta_m(T)$ denotes the maximum gap. In the weak coupling limit its value for $T \to 0$ should be $\Delta_m(0) = 2.029\,k_B T_c$. A schematic illustration of the energy gaps of ^{3}He-A and ^{3}He-B is shown in Fig. 4.11. Whereas the energy

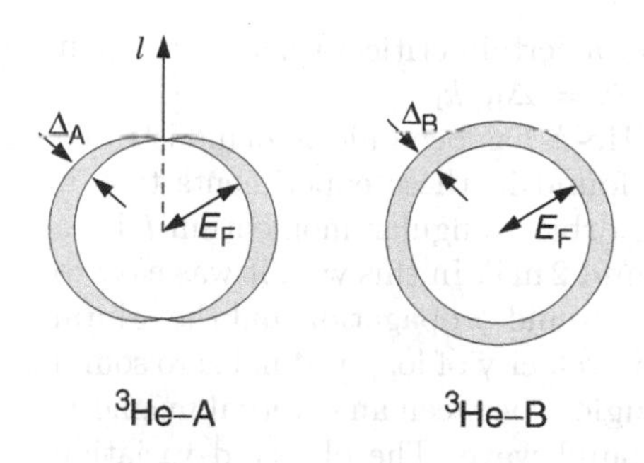

Fig. 4.11. Schematic illustration of the energy gap (*tinted grey*) in ^{3}He-A (*left*) and ^{3}He-B (*right*)

gap of the A phase is anisotropic, the gap Δ_B of the B phase is quasi-isotropic, just as it is in conventional s-wave superconductors. It is referred to as *quasi-isotropic*, because the small magnetic dipole interaction leads to a rotation of the spin coordinates relative to the orbital coordinates by an angle $\Theta \approx 104°$ in ^{3}He-B. This rotation of $\widehat{\boldsymbol{d}}$ about an axis $\widehat{\boldsymbol{n}}$ is described by $\widehat{\boldsymbol{d}} = \overleftrightarrow{\mathbf{R}}(\widehat{\boldsymbol{n}}, \Theta)\widehat{\boldsymbol{k}}$, where $\overleftrightarrow{\mathbf{R}}$ represents the rotation tensor. The definition of the energy gap of the B phase corresponding to (4.6) is given by

$$\boldsymbol{d}(\boldsymbol{k}) = \Delta_B(T)\,\widehat{\boldsymbol{d}} = \Delta_B(T)\,\overleftrightarrow{\mathbf{R}}(\widehat{\boldsymbol{n}}, \Theta)\,\widehat{\boldsymbol{k}}\,. \tag{4.7}$$

In the weak coupling limit the energy gap of the B phase should be given by $\Delta_B \approx 1.76\, k_B\, T_c$ for $T \to 0$.

The most prominent feature of (4.6) is that the energy for breaking pairs becomes zero at two particular points in momentum space, at $\boldsymbol{k} = \pm k_F \widehat{\boldsymbol{l}}$. This has consequences for many properties of ^{3}He-A (see Sect. 4.7). Consider, for example, a massive object moving in the liquid with velocity $\boldsymbol{v}$. It will always be able to break pairs even if it moves with an arbitrarily low velocity. Thus, it experiences a frictional force in the liquid. Nevertheless, the A phase is still superfluid since it can flow without friction! At least metastable persistent mass currents have been observed experimentally in ^{3}He-A.

The possibility of easily breaking pairs does not necessarily prevent superflow. The emitted fermionic quasiparticles of broken pairs, so-called *Bogoliubov quasiparticles*, fill all the negative energy levels $E_{\boldsymbol{k}} - \boldsymbol{k} \cdot \boldsymbol{v} < 0$, and after that, further emission stops due to the Pauli exclusion principle. Here, $E_{\boldsymbol{k}}$ is given by $E_{\boldsymbol{k}} = \sqrt{\eta_{\boldsymbol{k}}^2 + \Delta_A^2(\boldsymbol{k})}$, where $\eta_{\boldsymbol{k}}$ represents the kinetic energy of an unpaired quasiparticle relative to the Fermi energy (see Sects. 10.3.2 and 10.5). For a moving massive object this is not the case, since it occupies only a *finite* part of the volume of the liquid. The quasiparticles created by the moving object cannot fill all the negative energy levels. As a result, the emission process by a massive object is continuous and such an object will experience a frictional force in the liquid.

In contrast, ^{3}He-B – like helium II – is superfluid in both senses: it sustains persistent flow without friction and an external body can also move within this fluid without dissipation below a certain critical velocity. For pair breaking, the critical velocity is given by $v_c = \Delta_B/k_F$.

The anisotropy of the energy gap in ^{3}He-A has been nicely demonstrated in ultrasound experiments [173]. It was found in these experiments that it is possible to orient the direction of the orbital angular momentum $\boldsymbol{l}$ by a rather small magnetic field, typically around 2 mT. In this way, it was easy to vary the angle between the direction of the sound propagation and the orbital angular momentum. Figure 4.12 shows the velocity of longitudinal zero sound and its attenuation as a function of the angle ϕ between an external magnetic field $\boldsymbol{B}$ and the wave vector $\boldsymbol{q}$ of the sound wave. The observed variation agrees with the expected anisotropy.

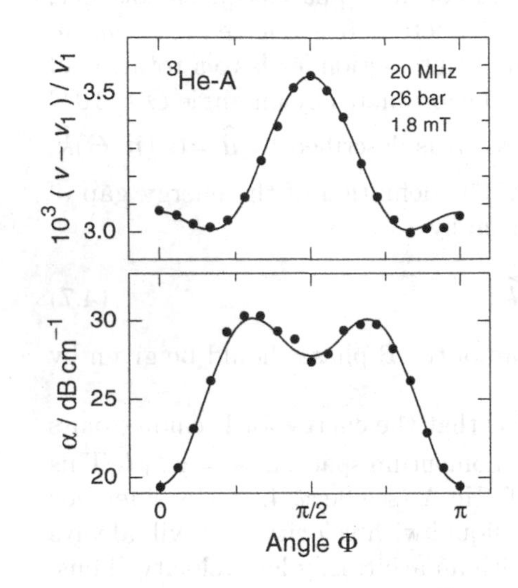

Fig. 4.12. (**a**) Velocity and (**b**) attenuation of longitudinal zero sound in ^{3}He-A at 20.24 MHz as a function of the angle ϕ between the wave vector $\boldsymbol{q}$ of the sound wave and the external magnetic field $\boldsymbol{B}$ [173]. The experiment has been performed at a pressure of 26 bar and a temperature of just 48 μK below T_c

4.4 Order-Parameter Orientation – Textures

As we have already seen, ^{3}He-A shows an anisotropic behavior in many experiments. In the discussion of the orientational dependence of the properties of the A phase, one has to consider not only internal forces that determine the directions of $\boldsymbol{l}$ and $\boldsymbol{d}$, but also external electric and magnetic fields, velocity fields and the influence of the container walls. As a first step, we discuss the relative orientation of $\boldsymbol{l}$ and $\boldsymbol{d}$ without external fields as determined by the magnetic dipole–dipole interaction of the nuclear spins of the quasiparticle pairs. Later, we shall see that $\boldsymbol{d}$ and $\boldsymbol{l}$ are responsive to various external influences such as the boundary conditions that, in general, lead to a nonuniform alignment of the order parameter. The resulting spatial structure of the order parameter is called a *texture*. The concept of textures was originally introduced by *de Gennes* to describe orientational effects in liquid crystals [174].

4.4.1 Intrinsic Alignment

Without external fields the spin $\boldsymbol{S}$ and the orbital angular momentum $\boldsymbol{l}$ of the quasiparticle pairs are oriented in such a way that the magnetic dipole–dipole energy is at its minimum. The resulting configuration for A and B phase will be discussed in the following.

^{3}He-A

As we mentioned previously, the vectors $\boldsymbol{d}$ and $\boldsymbol{l}$ are aligned on a macroscopic scale in ^{3}He-A. It turns out that the dipole–dipole energy is at a minimum when $\boldsymbol{d}$ is parallel to $\boldsymbol{l}$. To illustrate this, let us consider the two possible configurations for rotation of two nuclear dipoles about one another. One involves rotations such that the pair orbital angular momentum is parallel to the nuclear dipole moments and the other involves rotations where the pair orbital angular momentum is perpendicular to the dipole moments. Classically and quantum-mechanically, this latter configuration has a lower energy, and is the favored state. The potential energy of two classical dipoles oriented in the same direction will be minimized when their dipole axes are collinear. If the dipoles are rotating about one another, the most favorable configuration is for the angular momentum vector to be perpendicular to the dipole axes. This means that the magnetic dipole–dipole interaction forces a *parallel* alignment of $\boldsymbol{d}$ and $\boldsymbol{l}$ in the A phase.

To calculate the dipolar interaction taking into account quantum mechanics involves the spatial average of the dipolar interaction over the pair wave function. For the A phase this calculation leads to a dipolar free energy

$$F_{\mathrm{d}} = -\frac{3}{5}\, g_{\mathrm{d}}(T)\left[1-(\widehat{\boldsymbol{d}}\cdot\widehat{\boldsymbol{l}})^2\right] = -\frac{3}{5}\, g_{\mathrm{d}}(T)\,\sin^2\Theta\,. \tag{4.8}$$

The quantity g_{d} is proportional to the number of quasiparticle pairs forming the condensate. It can by approximated by $g_{\mathrm{d}} \approx 10^{-10}(1-T/T_{\mathrm{c}})\,\mathrm{J\,cm^{-3}}$. Therefore, to minimize the free energy, $\widehat{\boldsymbol{l}}$ and $\widehat{\boldsymbol{d}}$ must be parallel for the case of the ABM state, in agreement with the above qualitative argument. Figure 4.13a shows the dependence of the dipolar free energy as a function of the angle Θ between $\boldsymbol{l}$ and $\boldsymbol{d}$ in the A phase.

^{3}He-B

As mentioned before, ^{3}He-B is isotropic in both orbital and spin space, i.e., the spin and orbital moments have no preferred direction on a macroscopic scale (see Fig. 4.10). Therefore, $\widehat{\boldsymbol{d}}$ cannot be constant over the Fermi surface. Instead, the ordering is described by a vector $\widehat{\boldsymbol{n}}$, which originates from the broken relative rotational symmetry of the spin and orbital spaces, caused by the tiny dipole–dipole interaction of the quasiparticle pairs. This interaction results in a rotation of the spin coordinates relative to the orbital coordinates

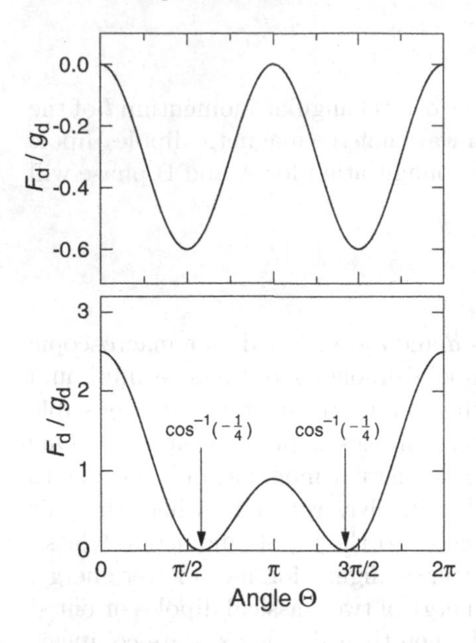

Fig. 4.13. Variation of F_d in (**a**) ^{3}He-A and (**b**) ^{3}He-B

locally at each point of the Fermi surface by an angle Θ about the vector $\widehat{\boldsymbol{n}}$. The dipolar free energy in the B phase is given by

$$F_\mathrm{d} = \frac{8}{5}\, g_\mathrm{d}(T)\left(\cos\Theta + \frac{1}{4}\right)^2 , \tag{4.9}$$

which means that the minimum dipole–dipole energy is for a rotation angle $\Theta = \arccos(-1/4) \approx 104°$. The variation of F_d in ^{3}He-B with Θ is plotted in Fig. 4.13b.

Although this subtle anisotropy caused by the magnetic dipole–dipole interaction leads to textural effects in the B phase, the overall orbital symmetry of the order parameter is still spherical.

4.4.2 Textures in ^{3}He-A

External fields and boundary conditions change the free energy of ^{3}He-A and influence the order-parameter orientation. In the presence of container walls or any type of residual field, the order-parameter alignment will not be uniform, since the perturbing influences tend to orient the order parameter locally in different, often competing, ways. A nonuniform orientation of the order parameter is opposed by the internally preferred alignment, because any bending of the order-parameter field causes an increase in the free energy of the system. The competition between orientational and bending forces leads to a continuous configuration for the order-parameter field, called a *texture*.

In general, the walls of the container play a very important role in determining the configuration of the order parameter. The fact that walls have an influence on the orientation of $\boldsymbol{l}$ seems obvious if one considers the quasiparticle pairs as molecules of two ^{3}He atoms that orbit each other. Clearly, the rotation in a plane parallel to the wall is strongly preferred. That this simplified picture is correct was theoretically shown by *Ambegaokar*, *de Gennes* and *Rainer* in 1974 [175]. Therefore, $\boldsymbol{l}$ will always be oriented perpendicular to the walls of the container.

The importance of the influence of the different external fields varies substantially. The preferred orientation and the energy change ΔE that result from various external fields and boundary conditions are listed in Table 4.2. It is interesting to ask at what magnitude these external effects become comparable with the intrinsic alignment forces. Here, we note the values at which external fields are equally important for the alignment of $\boldsymbol{d}$ and $\boldsymbol{l}$ as the intrinsic dipole–dipole interaction: $\mathcal{E} = 17\,\mathrm{V\,m^{-1}}$, $B = 3.3\,\mathrm{mT}$ and $v_s = 2.4\,\mathrm{mm\,s^{-1}}$. As mentioned before, the walls of the container are particularly important and even very small angles between $\boldsymbol{l}$ and $\boldsymbol{N}$, the normal vector of the container wall, result in an effect comparable with the intrinsic magnetic dipole–dipole interaction.

Table 4.2. Preferred orientation of $\boldsymbol{d}$ and $\boldsymbol{l}$ under the influence of different fields and the energy penalty ΔE when d and l are not optimally aligned in these fields. (After [176])

	Preferred Alignment	$\Delta E/(1 - T/T_c)$ $(\mathrm{J\,m^{-3}})$
magnetic dipole interaction	$\boldsymbol{d} \parallel \boldsymbol{l}$	$-6 \times 10^{-5}\,(\widehat{\boldsymbol{d}} \cdot \widehat{\boldsymbol{l}})^2$
electric field	$\boldsymbol{l} \perp \boldsymbol{\mathcal{E}}$	$2 \times 10^{-7}\,(\widehat{\boldsymbol{l}} \cdot \boldsymbol{\mathcal{E}})^2$
magnetic field	$\boldsymbol{d} \perp \boldsymbol{B}$	$5\,(\widehat{\boldsymbol{d}} \cdot \boldsymbol{B})^2$
mass flow	$\boldsymbol{l} \parallel \boldsymbol{v}_s$	$-10\,(\widehat{\boldsymbol{l}} \cdot \boldsymbol{v}_s)^2$
wall alignment	$\boldsymbol{l} \parallel \boldsymbol{N}$	$-30\,(\widehat{\boldsymbol{l}} \cdot \widehat{\boldsymbol{N}})^2$

Any deviation from a uniform orientation of $\boldsymbol{d}$ and $\boldsymbol{l}$ leads to an increase of the free energy. This energy change is often called *gradient energy*, since it is determined by the gradient of the orientational fields of $\boldsymbol{l}$ and $\boldsymbol{d}$. It is invariant with respect to an inversion $\widehat{\boldsymbol{d}} \to -\widehat{\boldsymbol{d}}$ or $\widehat{\boldsymbol{l}} \to -\widehat{\boldsymbol{l}}$, meaning that states with parallel and antiparallel alignment of $\boldsymbol{l}$ and $\boldsymbol{d}$ are energetically equal. Therefore, it is possible that the ground state differs in different parts of the liquid. For example, in one part a parallel alignment of $\boldsymbol{l}$ and $\boldsymbol{d}$ is realized and in another part an antiparallel alignment. Between these two configurations, an interface must exist in which the transition from one configuration to the other takes place. These transition regions are called *domain walls* analogous to the interface between ferromagnetic domains. Since two different orientation fields are relevant, there are different types of domain walls.

4.4.3 Surface-induced Texture – ^{3}He-A in a Slab

At this point, we discuss one particular experimental example to demonstrate the influence of the container walls on the properties of ^{3}He-A. In an external magnetic field the order parameter tends to orient in a suitable way such that the total free energy is minimal. Depending on the direction of the magnetic field, this preferred orientation will either be the same as or different from the preferred alignment at the walls. Therefore, the properties of the sample depend on the relative orientation of wall and magnetic field. This leads to a directional dependence of the superfluid density ϱ_s/ϱ.

This effect has been demonstrated in an experiment in which the superfluid component of the density was measured using a torsional oscillator as was used in the famous study of helium II by Andronikashvili. A schematic drawing of the setup is shown in Fig. 4.14a. In this experiment, the liquid ^{3}He was located in a slab-like cavity made out of epoxy resin and formed part of the moment of inertia of the oscillator. The thickness of the cavity was 50 µm and its diameter was 8.4 mm. The tube connecting the cavity with the main sample cell served as both the filling tube and the torsional rod. The ^{3}He column in this tube provided a thermal link to the ^{3}He in the cavity. The ^{3}He was cooled by using a sintered copper sponge that was thermally connected via a copper wire link to a Pomeranchuk cell (see Sect. 11.5.1).

As in the original Andronikasvili experiment, the normal-fluid density was determined from the change of the period of the torsional oscillations of the pendulum. Because of its viscosity η_n, the normal-fluid component ϱ_n of the density is immobile in the narrow slab and contributes to the moment of

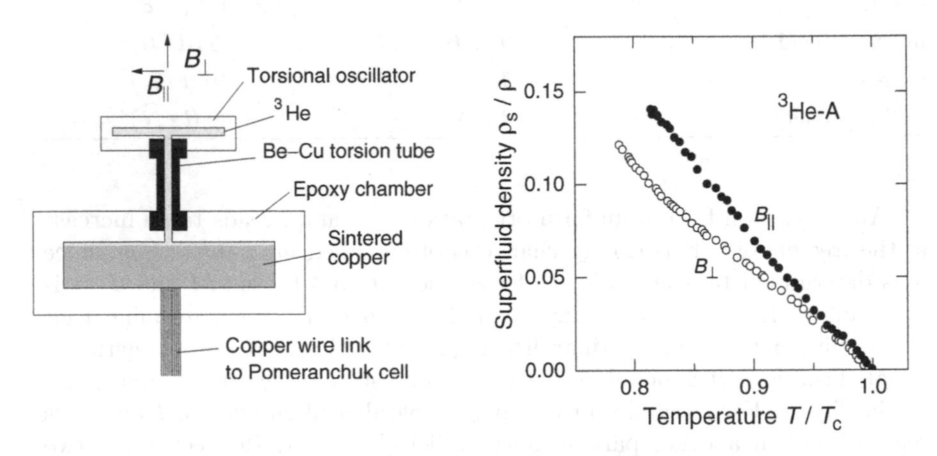

Fig. 4.14. (**a**) Schematic sketch of the experimental setup used to determine ϱ_s/ϱ depending on the orientation of an external static magnetic field. Two relative orientations of the magnetic field with respect to the slab were realized: $\boldsymbol{B}_\parallel$ and $\boldsymbol{B}_\perp$. (**b**) Superfluid density ϱ_s/ϱ of ^{3}He-A, measured at 27 bar in a magnetic field of 29 mT as a function of $(1 - T/T_c)$ for different magnetic-field alignments [177]

inertia of the oscillator, in contrast to the superfluid component. Below T_c, the moment of inertia of the oscillator decreases with temperature, and the ratio ϱ_s/ϱ can be determined. The resolution of this setup in measuring the superfluid fraction was better than 0.1%, despite the fact that the ratio of the moment of inertia of the fluid to that of the bob was only of the order of 10^{-4}.

The results of measurements in the presence of an external magnetic field parallel (•) or perpendicular (∘) to the slab are shown in Fig. 4.14b. Although the magnetic field was small, one finds a strong directional dependence of ϱ_s/ϱ. This can be understood if we consider the preferred order-parameter orientations for the two different situations.

The preferred direction of $\boldsymbol{d}$ is perpendicular to $\boldsymbol{B}$ and that of $\boldsymbol{l}$ parallel to the normal vector $\boldsymbol{N}$ of the walls. For the geometry $\boldsymbol{N} \perp \boldsymbol{B}_{\parallel}$ this means that $\boldsymbol{d}$ and $\boldsymbol{l}$ tend to orient themselves parallel, which is identical with the preferred alignment without external influences. In this case, the order parameter is uniform, except at the circumference of the slab and there is no significant increase of the free energy.

In the vicinity of the walls we still find the alignment of $\boldsymbol{l}$ parallel to $\boldsymbol{N}$. Since $\boldsymbol{d}$ is perpendicular to $\boldsymbol{B}_{\perp}$, this orientation can only occur if $\boldsymbol{d}$ is perpendicular to $\boldsymbol{l}$. Of course, thus is not the preferred alignment. The state of minimum free energy will be one in which $\boldsymbol{l}$ is parallel to $\boldsymbol{N}$ near the walls and turns gradually to an alignment of $\boldsymbol{d} \parallel \boldsymbol{l}$ in the bulk liquid. Therefore, the free energy will be larger than in the situation where the magnetic field direction is parallel to the slab. In general, the superfluid density $\varrho_{s\perp}$ associated with a superflow perpendicular to $\boldsymbol{l}$ is twice as high as $\varrho_{s\parallel}$. Because of this, one expects a reduced superfluid density for the nonuniform texture in the case of $\boldsymbol{B}_{\perp}$. The experimental results shown in Fig. 4.14b clearly confirm this idea.

4.4.4 Textures in ^{3}He-B

In this section, we briefly turn to texture effects in ^{3}He-B. The BW state, in ^{3}He-B, is a quasi-isotropic state consisting of pairs with zero total angular momentum $J = 0$. In ^{3}He-B, the spin and orbital moments have no preferred direction on a macroscopic scale (see Fig. 4.10). As we have seen before, however, the magnetic dipole interaction leads to a rotation of the spin coordinate system, in which the pairs were formed, with respect to the orbital coordinate system about the axis $\widehat{\boldsymbol{n}}$ by an angle of $\arccos(-1/4) \approx 104°$. Alignments of $\widehat{\boldsymbol{n}}$ by external influences such as the container walls or a magnetic field also lead to the formation of textures in the B phase.

These textures can be described completely in terms of $\widehat{\boldsymbol{n}}$, which, in equilibrium, defines the axis for the tiny anisotropy of the B phase, in both the spin and the orbital spaces. The anisotropy is much smaller than that of the A phase by a factor of approximately 10^5; the B phase can therefore still be regarded as an isotropic superfluid. Moreover, all orienting forces on $\widehat{\boldsymbol{n}}$ are at least five orders of magnitude weaker than the corresponding effects on $\hat{\boldsymbol{l}}$ in ^{3}He-A. Nevertheless, they have been clearly seen in various experiments (see, for example, [178]).

4.5 Spin Dynamics – NMR Experiments

Investigations of the spin dynamics of liquid ^{3}He below T_c are very informative regarding the properties of the superfluid phases. Historically, the results of NMR experiments have played a key role in the discovery and identification of the different phases of superfluid ^{3}He.

Nuclear spin resonances are collective, uniform excitations of the spin system. In the A phase, the spin dynamics is determined by the behavior of the order parameter $\boldsymbol{d}$. A thorough description of the spin dynamics of ^{3}He-A was worked out by *Leggett* soon after the discovery of the superfluid phases of ^{3}He [155]. In this section, we mainly focus on some basic aspects of ^{3}He-A, because the spin dynamics of ^{3}He-B is somewhat more complicated.

4.5.1 Leggett Equations

In NMR experiments, one applies a magnetic field oscillating at the precession frequency of the nuclear spins in the static magnetic field $\boldsymbol{B}_0$. The oscillating field causes the magnetization (the expectation value of the spin) to tip away from $\boldsymbol{B}_0$ and to precess about it. The duration and amplitude of the oscillating transverse field determines the so-called tipping angle. In ^{3}He-A, a tipping pulse leads to a relative motion between $\boldsymbol{d}$ and $\boldsymbol{l}$ as depicted in Fig. 4.15 and thus to a deviation from their optimal relative orientation. In this figure, the trajectories of the tips of $\boldsymbol{d}$ and $\boldsymbol{S}$ after various tipping pulses

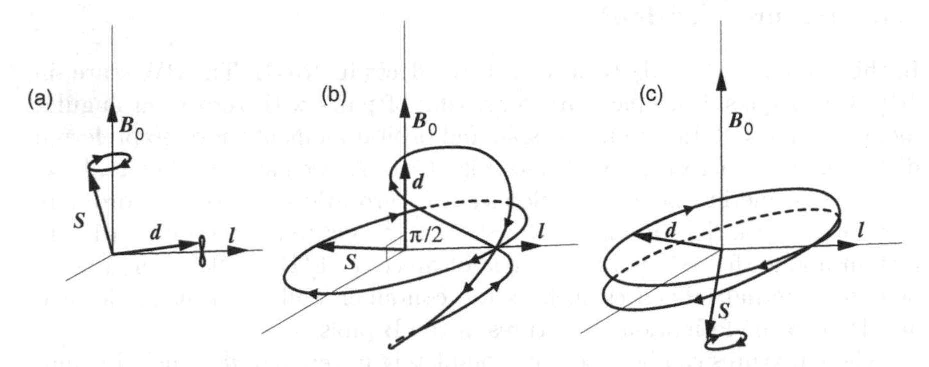

Fig. 4.15. Free precession of $\boldsymbol{S}$ and $\boldsymbol{d}$ after different tipping pulses: (**a**) 10°, (**b**) 90° and (**c**) 170°. In each case, $\boldsymbol{d}$ goes through a 'figure of eight' motion, returning to the axis of $\boldsymbol{l}$ twice in one Larmor precession of $\boldsymbol{S}$. The figure is based upon the motion predicted for ^{3}He-A by the Leggett equations of motion [176]

are shown. Any deviation from the optimal orientation $\boldsymbol{d} \parallel \boldsymbol{l}$ costs energy proportional to $\sin^2(\boldsymbol{d}, \boldsymbol{l})$, which in turn provides a restoring torque.

Taking into account the dipole–dipole interaction (4.8), *Leggett* derived a set of coupled equations giving a complete description of the nondissipative spin dynamics of superfluid ^{3}He [179]. His equations of motion are

$$\frac{\mathrm{d}\boldsymbol{S}}{\mathrm{d}t} = \gamma \boldsymbol{S} \times \boldsymbol{B}_0 + \boldsymbol{R}_\mathrm{d} \tag{4.10}$$

and

$$\frac{\mathrm{d}\boldsymbol{d}}{\mathrm{d}t} = \boldsymbol{d} \times \gamma \boldsymbol{B}_\mathrm{eff} = \boldsymbol{d} \times \gamma \left(\boldsymbol{B} - \frac{\mu_0 \gamma \boldsymbol{S}}{\chi_\mathrm{N}} \right) . \tag{4.11}$$

Here, χ_N denotes the static magnetic susceptibility of ^{3}He-N. The first term on the right side of the first Leggett equation represents the Larmor precession as expected in ordinary NMR. The second term, $\boldsymbol{R}_\mathrm{d}$, corresponds to a restoring torque resulting from the coherent dipole–dipole interaction. The second Leggett equation describes the precession of $\boldsymbol{d}$ in an effective field. The motions of $\boldsymbol{d}$ and $\boldsymbol{S}$ are coupled. All predictions derived from the Leggett equations have been confirmed to high accuracy by experiments.

4.5.2 Transverse Resonance – Frequency Shift

In the introductory part of this chapter, we have seen that a large frequency shift of the NMR resonance line occurs below T_c in the A phase. Here, we discuss the origin of this phenomenon. In the A phase, the restoring torque $\boldsymbol{R}_\mathrm{d}$ is given by

$$\boldsymbol{R}_\mathrm{d}(T) = \frac{6}{5} g_\mathrm{d}(T) \, (\boldsymbol{d} \times \boldsymbol{l}) \, (\boldsymbol{d} \cdot \boldsymbol{l}) \, . \tag{4.12}$$

It is this term that gives rise to the frequency shift in the A phase. Leggett calculated the frequency ω_t of the transverse resonance in the A phase for small tipping angles to be

$$\omega_\mathrm{t}^2 = (\gamma B_0)^2 + \frac{\gamma^2 \mu_0 \langle H_\mathrm{d} \rangle}{\chi_\mathrm{N}} = \omega_\mathrm{L}^2 + \Omega_\mathrm{A}^2(T) \, . \tag{4.13}$$

Here, ω_L represents the Larmor frequency. The second term is determined by the magnetic dipole–dipole interaction, where the quantity $\langle H_\mathrm{d} \rangle$ denotes the spatial average of the dipole–dipole energy. An important point here is that in the A phase, $\langle H_\mathrm{d} \rangle$ is substantially enhanced compared to a classical system. In a classical system, $\langle H_\mathrm{d} \rangle$ is proportional to g_d^2. The dipole–dipole energy of individual dipoles is given by $g_\mathrm{d}/k_\mathrm{B} = \gamma^2 \hbar^2 / (a^3 k_\mathrm{B}) < 10^{-7}\,\mathrm{K}$. Since this energy is very small, its spatial average is negligible compared to the thermal fluctuations at temperatures around 1 mK. In ^{3}He-A, the situation is quite different because all quasiparticle pairs belong to a common ground state with a well-defined orientation of the spins. Thus, one finds a coherent amplification of the interaction energy proportional to the number of quasiparticle pairs N_p. This has an interesting consequence, namely that the strong correlation of the quasiparticle pairs in ^{3}He-A leads to a permanent local field of about 3 mT. It also explains the large shift of the transverse resonance line below T_c. As shown in Fig. 4.16a, the frequency shift Ω_A increases monotonically with decreasing temperature. This temperature dependence is expected because Ω_A is proportional to the superfluid density. The data, obtained by different authors with different techniques, agree very well.

At this point let us briefly return to Fig. 4.15. The motion of $\boldsymbol{d}$ with respect to $\boldsymbol{l}$ depends on the tipping angle. The dipolar torque exerted by the nonequilibrium orientation of $\boldsymbol{d}$ varies throughout the cycle with the maximum torque depending on the size of the pulse tipping angle ϕ. Thus, the frequency of the spin precession changes as the tipping angle is varied. Figure 4.16b shows the relative change of the frequency shift $\Delta\omega(\phi)/\Delta\omega(0)$ observed in pulsed NMR measurements following various tipping pulses.

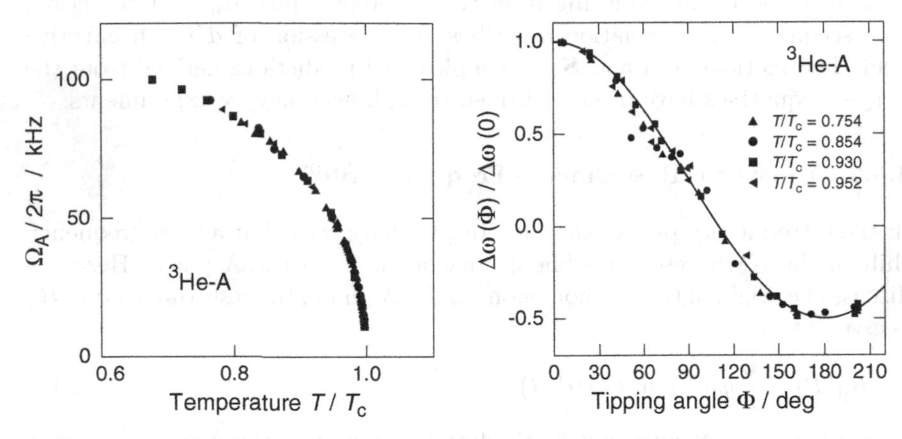

Fig. 4.16. (**a**) Characteristic frequency $\Omega_\mathrm{A}/2\pi$ as a function of the reduced temperature (T/T_c). The data were obtained by different groups with different techniques. After [176]. (**b**) Relative change of the frequency shift $\Delta\omega(\phi)/\Delta\omega(0)$ as a function of the tipping angle ϕ [180]

4.5.3 Longitudinal Resonance

In his theory of the spin dynamics of superfluid ^{3}He, Leggett calculated not only the frequency shift of the transverse resonance in the A phase, but also predicted the existence of a longitudinal resonance in both the A phase and the B phase. As mentioned before, longitudinal resonance experiments are performed by orienting the rf field coil parallel to the applied static magnetic field $\boldsymbol{B}_0$. This leads to a modulation of the static magnetic field with the frequency of the rf field. Therefore, the magnetic field in the z-direction can be written as $B_z = B_0 + B_\mathrm{rf}(\omega)$. This field variation does *not* lead to a tipping of the magnetization, but to a nonequilibrium magnetization along $\boldsymbol{B}_0$. In normal-fluid ^{3}He such a nonequilibrium would cause relaxation processes, but no resonant absorption.

In ^{3}He-A, the field variation in the z-direction leads to an oscillation of $\boldsymbol{d}$ in the plane normal to $\boldsymbol{B}_0$. When $\boldsymbol{d}$ is driven away from its equilibrium orientation, it exerts a torque. Since $\boldsymbol{B}_0$ and $\boldsymbol{B}_{\rm rf}$ both point in the z-direction we have $(\boldsymbol{S} \times \boldsymbol{B})_z = 0$ and thus the first Leggett equation (4.10) results in $\mathrm{d}S_z/\mathrm{d}t = R_{\rm d,z}$. In an ordinary system we would have $\boldsymbol{R}_{\rm d} = 0$ and therefore $\mathrm{d}S_z/\mathrm{d}t = 0$, which means that there is no resonance. In ^{3}He-A, however, $R_{\rm d} \neq 0$ and one finds:

$$\Delta S_z = \frac{\chi}{\mu_0 \gamma} \Delta B_0 \left(1 - \cos \Omega_{\rm A} t\right) . \tag{4.14}$$

One can interpret the longitudinal resonance as transitions between the spin configurations $|\uparrow\uparrow\rangle$ and $|\downarrow\downarrow\rangle$ that are coupled via the magnetic dipole–dipole interaction. The frequency of the longitudinal resonance is identical to $\Omega_{\rm A}$, the extra term in the transverse resonance. This identity has been experimentally confirmed with high accuracy. Note that the data shown in Fig. 4.16 are obtained from both transverse and longitudinal NMR experiments.

Although there is no shifted transverse resonance in ^{3}He-B, the Leggett equations also predict a longitudinal resonance for this phase. If one assumes that at a given pressure the mean energy gap is the same in the A and B phases, $\Omega_{\rm B}$ is related to $\Omega_{\rm A}$ via the equation

$$\Omega_{\rm B}^2(T) = \Omega_{\rm A}^2(T) \, \frac{5}{2} \frac{\chi_{\rm B}}{\chi_{\rm A}} , \tag{4.15}$$

where $\chi_{\rm A}$ and $\chi_{\rm B}$ represent the magnetic susceptibility of ^{3}He-A and ^{3}He-B, respectively. The susceptibility of the A phase is identical with that of normal-fluid ^{3}He, because the spin pairs $|\uparrow\uparrow\rangle$ and $|\downarrow\downarrow\rangle$ can respond directly to any field change in a fashion similar to that of ordinary Fermi liquids. In contrast, the susceptibility of the B phase is reduced, because one third of the spins are involved in zero-spin pairs $|\downarrow\uparrow\rangle + |\uparrow\downarrow\rangle$. In addition, there are Fermi-liquid corrections.

In fact, the longitudinal resonance has been experimentally observed in the A and B phase, fully confirming Leggett's prediction. The temperature dependence of the frequency $\omega_{\rm l}$ of the longitudinal resonance in both phases is shown in Fig. 4.17.

Finally, we remark that despite considerable effort, no longitudinal resonance has been observed in the ^{3}He-A$_1$ [182]. This is consistent with what one would expect because the longitudinal resonance is characteristic of transitions between the two spin populations $|\uparrow\uparrow\rangle$ and $|\downarrow\downarrow\rangle$. In the A$_1$ phase, where only one configuration is present, this process cannot take place.

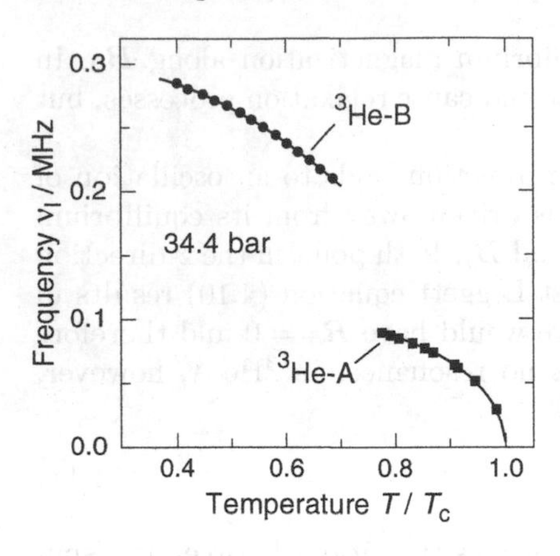

Fig. 4.17. Frequency ω_l of the longitudinal NMR resonance in ^{3}He-A and ^{3}He-B as a function of temperature at the melting pressure [181]

4.6 Macroscopic Quantum Effects

One of the most fascinating aspects of superfluids and superconductors is the existence of quantum effects on a macroscopic scale. Just like helium II, superfluid ^{3}He can be described by a macroscopic wave function. However, as we have already seen, the order parameter is much more complex in liquid ^{3}He below T_c. We can express the macroscopic wave function of superfluid ^{3}He by

$$\Psi_{\alpha\beta}(\boldsymbol{r}) = \mathcal{A}_{\alpha\beta}(\boldsymbol{r})\,\mathrm{e}^{\mathrm{i}\varphi(\boldsymbol{r})}\,, \tag{4.16}$$

where $\mathcal{A}_{\alpha\beta}$ is a 3×3 matrix with complex elements that reflects the angular momentum state of the system. The indices α and β refer to the spin and orbital degrees of freedom, respectively. In total, the order parameter of superfluid ^{3}He has 18 degrees of freedom. This enormous manifold leads to a great variety of macroscopic phenomena, a few selected examples of which will be discussed in the following.

4.6.1 Superflow

Frictionless flow is a fundamental property of all superfluids. In the isotropic B phase, mass currents can be described in the same way as in helium II. As we have already mentioned in Sect. 4.1.3, persistent currents have been observed in ^{3}He-B. The situation in the A phase is more complicated. The density of the superfluid component is not isotropic but depends on the direction of the mass flow relative to the orbital angular momentum $\boldsymbol{l}$. Therefore, ϱ_s is not a scalar but a tensor with 3×3 components. For example, for flow parallel to $\boldsymbol{l}$ the superfluid density is only half as large as for flow perpendicular: $\varrho_{s,\perp} = 2\varrho_{s,\parallel}$. Since the total density must be isotropic we can conclude that the normal-fluid density is also a tensor. The superfluid density can be written as

$$\overleftrightarrow{\varrho}_{\mathrm{s}} = \varrho \overleftrightarrow{1} - \overleftrightarrow{\varrho}_{\mathrm{n}}, \tag{4.17}$$

where $\overleftrightarrow{1}$ represents the unity tensor. This implies that in an anisotropic superfluid, a superflow need not necessarily be parallel to the superfluid velocity. In ^{3}He-A, the situation is even more complicated because the superflow velocity is not only determined by the phase of the macroscopic wave function, but also depends on the orbital angular momentum. A general expression for superflow mass currents in the A phase is given by [183]

$$\boldsymbol{j}_{\mathrm{s}} = \overleftrightarrow{\varrho}_{\mathrm{s}}\, \boldsymbol{v}_{\mathrm{s}} + \frac{\hbar}{2m_3}\, \overleftrightarrow{\mathbf{C}}\ \mathrm{curl}\,\widehat{\boldsymbol{l}}\,. \tag{4.18}$$

Here, $\overleftrightarrow{\mathbf{C}}$ is a tensor whose reference frame – just as for $\overleftrightarrow{\varrho}_{\mathrm{s}}$ – is given by the orbital angular momentum. The first term of (4.18) describes the usual flow of quasiparticle pairs and the second term describes an orbital superflow that is driven by a nonuniform $\widehat{\boldsymbol{l}}$ texture. The tensor $\overleftrightarrow{\mathbf{C}}$ therefore describes the anisotropic bending energy.

The fact that the superfluid velocity is not only determined by the phase in ^{3}He-A has the consequence that a superflow should not be stable under ideal conditions. However, in a real experiment, the direction of $\boldsymbol{l}$ is fixed, at the very least, by the walls of the container. Therefore, it is possible to observe a superflow current, for example, at the interface of ^{3}He-A and the walls. Since this type of superflow is not stable against changes in the texture, one observes a slow decay, due to the tunneling of textures.

4.6.2 Quantization of Circulation

In helium II, the velocity $\boldsymbol{v}_{\mathrm{s}}$ of the superfluid component is given by (2.70). It is proportional to the gradient of the phase of the macroscopic wave function. Being the gradient of a scalar, the superflow is potential flow, i.e. curl-free, or $\mathrm{curl}\,\boldsymbol{v}_{\mathrm{s}} = 0$. However, this is not true in all cases. In vessels with multiply connected regions, or in the presence of vortices, circulation of the superfluid component can exist around any closed contour. To keep the wave function single valued, the circulation of the superfluid component has to be quantized in quanta of h/m_4. In the following two sections, we will discuss the corresponding situation for the A and B phases.

^{3}He-A

Superflow in the A phase need not to be curl-free because the velocity of the superfluid component depends not only on the phase gradient but also on the orbital angular momentum $\boldsymbol{l}$ according to (4.18). This is a consequence of the broken relative gauge-orbit symmetry. In particular, the curl of the superflow velocity is, in cylindrical coordinates (r, ϕ and z), given by

$$\mathrm{curl}\, \boldsymbol{v}_\mathrm{s} = \frac{\hbar}{2m_3 r}\, \widehat{\boldsymbol{l}} \cdot \left(\frac{\partial \widehat{\boldsymbol{l}}}{\partial \phi} \times \frac{\partial \widehat{\boldsymbol{l}}}{\partial r} \right) , \tag{4.19}$$

where m_3 is the mass of a ^{3}He atom. From this, it follows that curl $\boldsymbol{v}_\mathrm{s}$ vanishes only if the texture of $\boldsymbol{l}$ is uniform. Therefore, ^{3}He-A is only irrotational under ideal circumstances. When $\boldsymbol{l}$ changes its direction as a function of position, continuous vortices may be formed, and the fields $\boldsymbol{v}_\mathrm{s}(r)$ and $\boldsymbol{l}(r)$ have no singularities.

It is remarkable that under ideal conditions – without external influences – the circulation in ^{3}He-A is not quantized. This is possible because a change in the phase can be compensated for by a change in the order parameter $\boldsymbol{l}$.

^{3}He-B

In the B phase, the situation is similar to that in helium II. The superflow velocity is, according to

$$\boldsymbol{v}_\mathrm{s} = \frac{\hbar}{2m_3} \nabla\varphi \tag{4.20}$$

directly related to the phase gradient. The flow is therefore expected to be irrotational. As in helium II, quantized circulation of the superfluid component exists in ^{3}He-B in vessels with multiply connected regions. This has been demonstrated in an experiment similar to those carried out by *Vinen* (see Sect. 2.4.2) in which the circulation around a thin vibrating wire was measured. The result is shown in Fig. 4.18. Three different values of the circulation were observed depending on the angular velocity of the vessel. The *grey tinted* regions indicate that at these angular frequencies the circulation was instable. The quantum of circulation in ^{3}He-B is given by

$$\kappa_3 = \frac{h}{2m_3} . \tag{4.21}$$

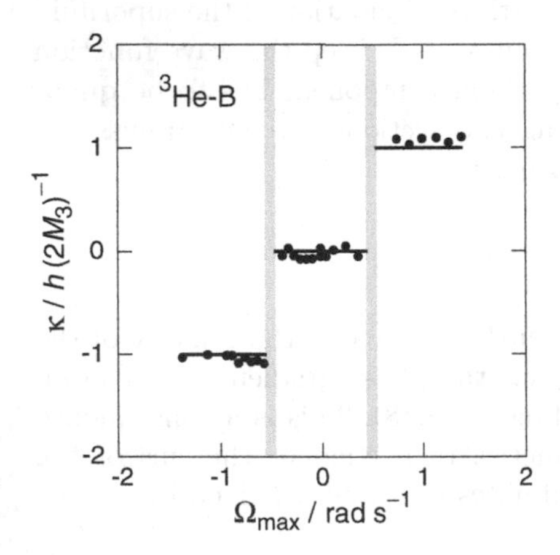

Fig. 4.18. Circulation κ in units of $h/(2m_3)$ as a function of the angular velocity in ^{3}He-B. Within the *grey tinted regions* no stable circulation was observed [184]

It differs from that of helium II, because the relevant mass is twice the mass of a ^{3}He atom, neatly confirming the formation of pairs.

4.6.3 Quantized Vortices

For the same reasons as in superfluid ^{4}He, a regular vortex array is formed in uniformly rotating superfluid ^{3}He. Vortices are also generated easily by liquid flow when the superfluid velocity $\boldsymbol{v}_s$ exceeds a critical value, but in this situation a tangle of vortices is created. To study ordered vortex arrays, one must rotate superfluid ^{3}He smoothly and uniformly. To achieve this, cryostats have been developed that can be rotated axially on air bearings. Most experiments on vortices in rotating superfluid ^{3}He have been performed in the ROTA laboratory in Helsinki [185]. The latest edition of several rotating cryostats – ROTA3 – can by rotated at angular frequencies up to 20 rad s^{-1} at temperatures as low as 0.5 mK. As we will see, most of the investigations on vortices have been performed using nuclear magnetic resonance techniques. Due to the much greater complexity of the order parameter for superfluid ^{3}He compared to helium II, one finds a large variety of different vortex structures in both ^{3}He-A and ^{3}He-B. The usual classification of vortices according to the circulation quantum number is therefore not sufficient for superfluid ^{3}He. Instead, topological considerations are used to classify vortex structures in liquid ^{3}He below T_c [186].

^{3}He-A

As mentioned above, the superflow of ^{3}He-A is not necessarily curl-free. For a texture with uniform $\boldsymbol{l}$ only singular vortices can be formed with single quanta of circulation ($n = 1$). These vortices have two distinct core regions, namely a so-called hard core with radius of the order of the coherence length $\xi_0 \approx 10$ nm,[5] inside of which the order parameter deviates from its bulk value, and a soft core with radius of the order of the dipolar healing length $\xi_d \approx 6$ µm.[6] Because of their hard core they are referred to as singular vortices. When $\boldsymbol{l}$ is 'free' to change its direction, continuous vortices are possible with double-quantum circulation $n = 2$. Their velocity field $\boldsymbol{v}_s$ and the orientational field $\boldsymbol{l}$ have no singularities and no hard core is needed. Both types of vortices obey quantized circulation. Figure 4.19 shows a schematic illustration of the two vortex types described above. Whether a singular or a continuous vortex structure is formed in ^{3}He-A under rotation depends on the external magnetic field, the geometry of the sample and on the angular velocity Ω.

[5] Note that the coherence length of superfluid ^{3}He is roughly two orders of magnitude larger than that of helium II.

[6] If the parallel alignment of $\boldsymbol{d}$ and $\boldsymbol{l}$ is disturbed, e.g., by the container wall, the preferred alignment is restored after a distance given by the dipolar healing length ξ_d.

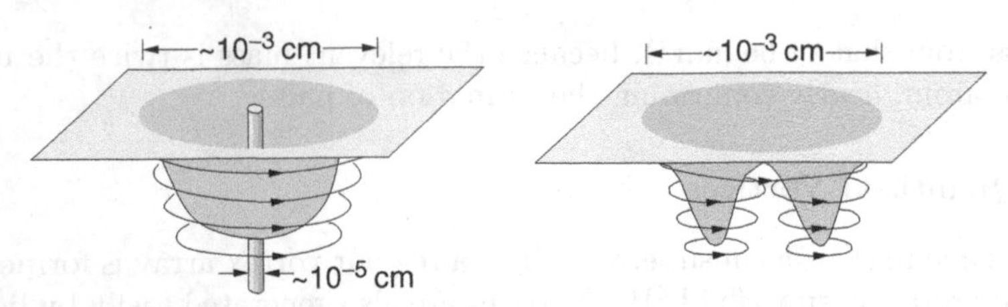

Fig. 4.19. Schematic sketch of a singular vortex (**a**) and a continuous vortex (**b**) in ^{3}He-A. After [187]

Because of the intrinsic coupling between spin and orbital fields of ^{3}He-A, it is, in principle, also possible to have vortices with a half-integer quantum of circulation. This possibility has been pointed out for the case when a vortex in the orbital space is combined with a singularity in the $\boldsymbol{d}$ field [188,189]. In this way, the demand for a single-valued wave function can be fulfilled by a change in phase of π due to both the orbital and the spin part of the order parameter adding up to a phase change of 2π for a full rotation. Because the orbital part of the order parameter changes by π the vortex has a circulation of $\kappa = h/(4m_3) = \kappa_3/2$.

The vector $\boldsymbol{d}$ reverses its direction in this process, producing a disclination in the $\boldsymbol{d}$ field, which acts as a starting point for a planar defect. Since large defects are energetically unfavorable, two half-quantum vortices with opposite disclinations combine to reduce the size of the defect. Such confined pairs of half-quantum vortices are expected to occur, for example, in ^{3}He-A between parallel plates separated by less than the dipolar healing length ξ_d and at temperatures below $0.5\,T_\mathrm{c}$. However, so far these rather exotic vortices have not been observed experimentally.

Nuclear magnetic resonance experiments provide a direct way to study vortices in ^{3}He-A. As we have seen, the coherent magnetic dipole–dipole interaction causes a large frequency shift of the transverse NMR resonance. The size of the shift depends on the order parameter. Large structures, such as the vortex cores in ^{3}He-A, have clear effects on the NMR signal. Under rotation, the shifted peak broadens and an additional small peak appears. Figure 4.20 shows a comparison of the NMR spectrum of ^{3}He-A obtained with and without rotation. The broadening of the main peak, due to rotation, is apparent.

The satellite peak is caused by a bound spin-wave mode localized at the vortex core. The frequency of the spin wave can be calculated from the Leggett equations (4.10) and (4.11), which describe the spin dynamics in superfluid ^{3}He. The frequency is different for different vortex structures that makes it possible to distinguish between them. For the data shown in Fig. 4.20 the satellite peak is caused, for example, by continuous vortices. The height of the satellite peak is proportional to the number of vortices created.

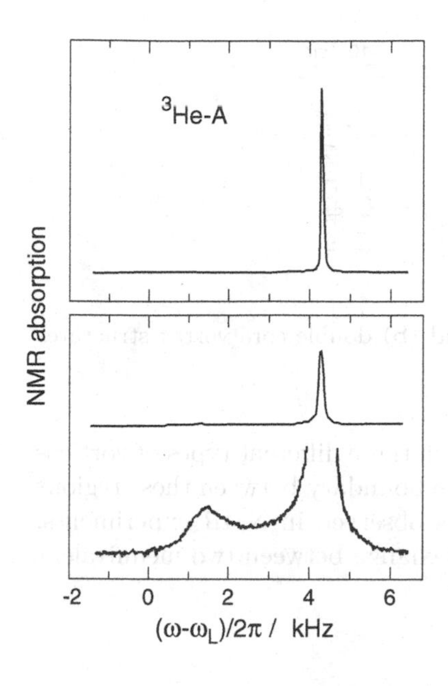

Fig. 4.20. NMR spectra of ^{3}He-A (**a**) at rest and (**b**) under rotation. The lowest curve shows the spectrum vertically magnified by a factor 50 [190]

Experiments indicate that vortices in ^{3}He-A are formed only above a critical velocity of rotation, as in helium II. The critical angular velocity depends on the size of the container, being about 0.1 rad s^{-1} for a cylindrical vessel with diameter 2.5 mm. After the critical speed has been exceeded, continuous vortices enter the liquid within a few seconds. In contrast, the creation of singular vortices in helium II may take several minutes [187].

^{3}He-B

As we have seen above, the superflow velocity in ^{3}He-B is fully determined by the gradient of the phase $\boldsymbol{v}_s = (\hbar/2m_3)\nabla\varphi$ and the circulation of the superflow is quantized. Because $\text{curl}\,\boldsymbol{v}_s = 0$, the vortices in ^{3}He-B exhibit always a core as in helium II. Prior to experiments, it was generally assumed that vortices in ^{3}He-B would be very similar to those occurring in helium II. However, the extended hard core of the vortices in ^{3}He-B permits the formation of different pairing states inside the core, and the core need not necessarily be a normal-fluid. In fact, two different vortex structures with different cores have been identified experimentally. At high pressures and high temperatures one finds an axially symmetrical vortex with ^{3}He-A in the core. At low pressures a nonaxially symmetrical double-core vortex structure composed of two half-quantum vortices is formed. A schematic illustration of these vortex structures is shown in Fig. 4.21. Since the vortices in ^{3}He-B have no soft core, the overall core extension is much smaller than that of the vortices in the A phase.

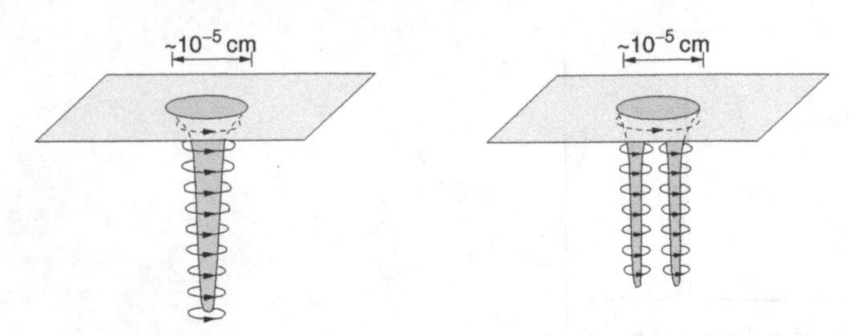

Fig. 4.21. Schematic sketch of (**a**) single- and (**b**) double-core vortex structures occurring in ^{3}He-B. After [187]

The regions of the phase diagram in which these different types of vortices are well defined is shown in Fig. 4.22. At the boundary between these regions a first-order vortex-core phase transition is observed in NMR experiments. This vortex core transition is a topological change between two inequivalent ways to realize a certain vorticity.

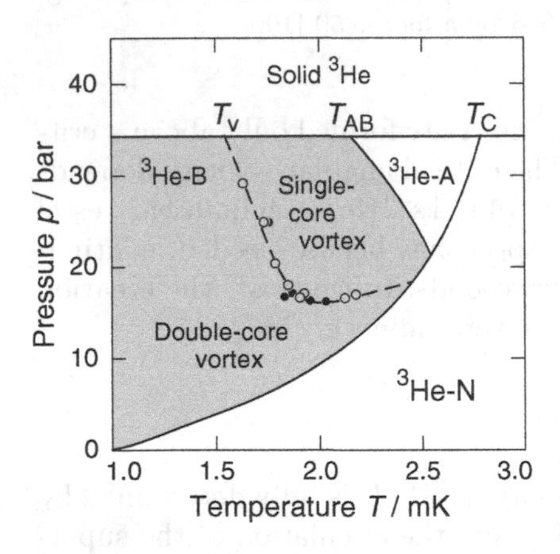

Fig. 4.22. Phase diagram of ^{3}He-B under rotation. The data points represented by *filled* and *unfilled dots* correspond to experiments in magnetic fields of 56.8 mT and 28.4 mT, respectively. After [190]

The phase transition becomes apparent in studies of the spin-wave resonances with the help of NMR experiments. In Sect. 3.3, we already discussed the occurrence of spin waves in the collisionless regime of normal-fluid ^{3}He. Figure 4.23 shows the result of an analogous experiment in ^{3}He-B. Different spin wave resonances are clearly seen. Under rotation the frequencies of these resonances are somewhat shifted and have a larger spacing because additional terms in the free energy enter when ^{3}He-B is rotated.

Figure 4.24 shows the temperature dependence of the resonant frequencies of the three lowest spin-wave modes at rest and under rotation. Whereas no discontinuities are observed when the cryostat is at rest, a sudden drop of the resonant frequency of all three modes occurs under rotation, indicating the vortex-core phase transition. Below the temperature of the discontinuous change, the results of the measurements depend on the direction of rotation.

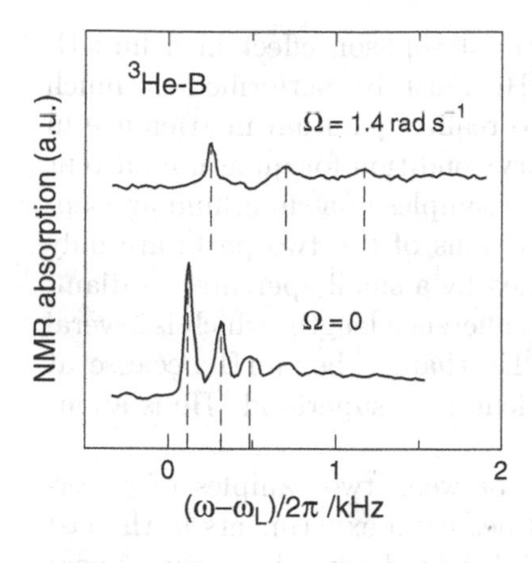

Fig. 4.23. Spin-wave resonances observed in NMR experiments in ^{3}He-B, (**a**) at rest and (**b**) under rotation [191]

The phase-transition temperature is independent of the speed of rotation but shows hysteresis consistent with the interpretation of a first-order phase transition.

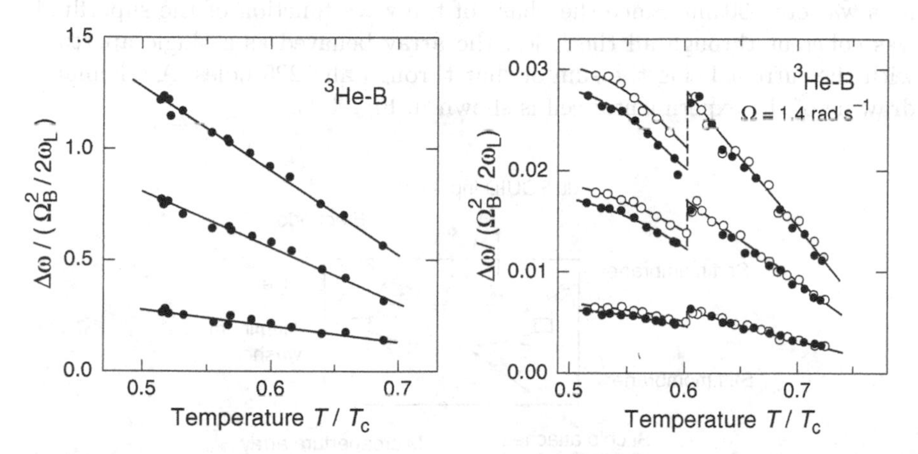

Fig. 4.24. Frequency shift $\Delta\omega = (\omega - \omega_L)$ of different spin wave modes in ^{3}He-B normalized by $\Omega_B^2/2\omega_L$ as a function of the reduced temperature: (**a**) cryostat at rest and (**b**) cryostat rotating. *Filled* and *unfilled* symbols correspond to forward ($\boldsymbol{\Omega} \uparrow\uparrow \boldsymbol{B}$) and reverse ($\boldsymbol{\Omega} \uparrow\downarrow \boldsymbol{B}$) rotations, respectively. The experiments were performed at a pressure of 29.3 bar and in a magnetic field of $B = 28.4\,\text{mT}$. Under these conditions the Larmor frequency is $\omega_L = 922.5\,\text{kHz}$ [191]

4.6.4 Macroscopic Quantum Interference – Josephson Effect

Perhaps the most intriguing property of a macroscopic quantum system is the potential for quantum interference on a macroscopic scale. In Sect. 2.4.3, we have already discussed the occurrence of the Josephson effect in helium II. Although corresponding experiments in ^{3}He must be performed at much lower temperatures it is somewhat easier to realize quantum interference in superfluid ^{3}He than in helium II. A necessary condition for quantum-current oscillations to exist in superfluids is that two samples of a superfluid are separated by such a region that the wave functions of the two parts are only weakly coupled. If such a weak link is realized by a small aperture, its diameter and length must be comparable to the coherence length, which is several orders of magnitude larger in superfluid ^{3}He than in helium II because at low temperatures ($T \ll T_c$) the coherence length in superfluid ^{3}He is about 65 nm.

The existence of quantum interference between two samples of superfluid ^{3}He has been demonstrated in several beautiful experiments in the last few years. The first of two examples we shall briefly discuss, is an experiment in which the Josephson frequency was measured as a function of the pressure difference between two samples of superfluid ^{3}He [192]. The weak link was realized by an array of 4225 holes, each 100 nm in diameter and separated by 3 μm. The holes were etched in a silicon nitride membrane whose thickness was only 50 nm. Since the phase of the wave function of the superfluid was coherent through all the holes, the array behaved as a single aperture with the current being the sum of that through all 4225 holes. A schematic drawing of the experimental cell is shown in Fig. 4.25.

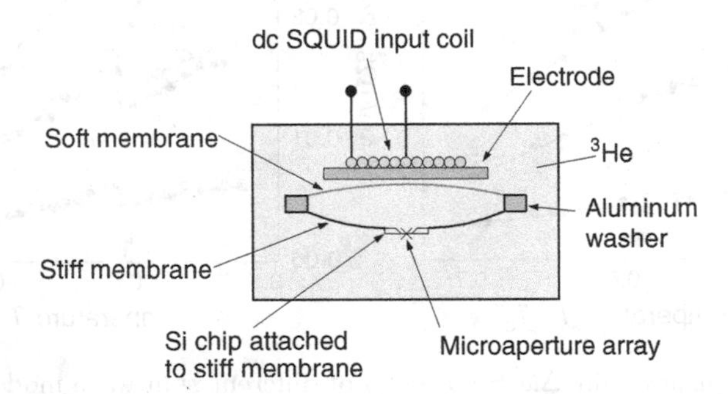

Fig. 4.25. Schematic sketch of the experimental cell used to measure coherent quantum oscillations between two weakly coupled reservoirs of ^{3}He-B. After [192]

The inner cell consisted of a 140 μm thick aluminum washer with one stiff and one very soft diaphragm on the lower and upper side, respectively. A silicon chip containing the silicon nitride membrane with the weak link was glued over a small hole in the stiff diaphragm. The upper diaphragm was metalized on the outside with a thin bilayer of lead and indium. The soft diaphragm was used in the experiment to apply pressure and to detect the oscillating mass current. The pressure on the liquid present between the two diaphragms could be varied by applying a force on the upper diaphragm with an electric field between the diaphragm and the fixed electrode. The frequency of the resulting current oscillations was monitored inductively with a resolution of $10^{-14}\,\mathrm{m\,Hz^{-1/2}}$ by a dc SQUID displacement sensor.[7] As mentioned in Sect. 2.4.3, the Josephson frequency is given by $\omega_\mathrm{J} = m_3\,\Delta p/(\varrho\hbar)$. In Fig. 4.26a, the frequency of the quantum oscillations is shown as a function of the pressure difference. In agreement with the expression for ω_J, the observed frequency is independent of temperature and increases linearly with the pressure difference. The dc Josephson effect – the phase-current relation without pressure drop across the aperture – has also been investigated in the same apparatus. According to (2.84), the mass current density is expected to show the sinusoidal dependence $j = j_\mathrm{c}\sin(\Delta\varphi)$ on the phase difference, where j_c denotes the critical current for the flow through the weak link. The data shown in Fig. 4.26b are in almost perfect agreement with this prediction.

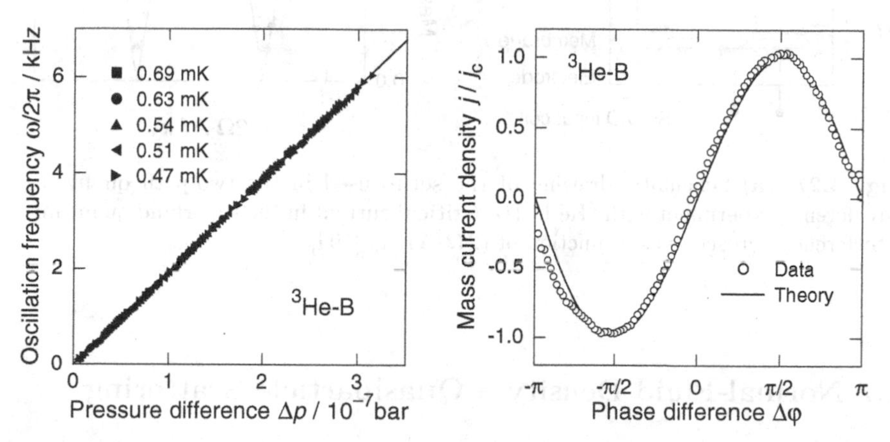

Fig. 4.26. (**a**) Frequency of the quantum oscillations as a function of the pressure difference between two weakly coupled reservoirs of ^{3}He-B at different temperatures [192]. (**b**) Mass current density between these reservoirs versus phase difference $\Delta\varphi$ between the two systems [193]. These data were obtained at $T = 0.85\,T_\mathrm{c}$

Using two arrays of 4225 holes in a closed loop with area A, the analogue of a double-path interference experiment has been performed in ^{3}He-B [194]. A schematic sketch of the experimental setup is shown in Fig. 4.27a. Rotating this quantum interferometer at an angular velocity Ω leads to modulation of the critical current given by

$$j_\mathrm{c} = 2j_\mathrm{c}^0 \left|\cos\left(\pi \frac{2\boldsymbol{\Omega}\cdot\boldsymbol{A}}{\kappa_3}\right)\right| , \tag{4.22}$$

where $\boldsymbol{A}$ represents the area vector of the loop and $\kappa_3 = h/(2m_3)$ the quantum of circulation of superfluid ^{3}He. The variation of the rotation flux $\boldsymbol{\Omega}\cdot\boldsymbol{A}$

[7] A SQUID is a device that allows very sensitive measurements of magnetic flux changes. Its working principle is discussed in Sect. 10.4.4.

was realized by reorienting the normal of the loop plane with respect to the rotation of the Earth. The resulting interference pattern is depicted in Fig. 4.27b. The plot shows the critical current as a function of $(2\boldsymbol{\Omega}\cdot\boldsymbol{A})/\kappa_3$. The *solid line* represents the prediction of (4.22).

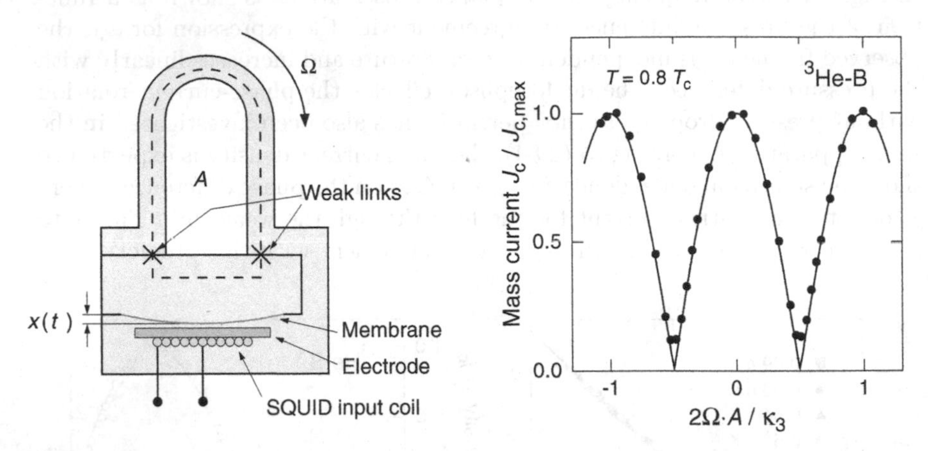

Fig. 4.27. (**a**) Schematic drawing of the setup used in the two-path quantum-interference experiment with ^{3}He-B. (**b**) Critical current in the superfluid quantum-interference gyroscope as a function of $(2\boldsymbol{\Omega}\cdot\boldsymbol{A})/\kappa_3$ [194]

4.7 Normal-Fluid Density – Quasiparticle Scattering

Many properties of superfluid ^{3}He can be described by a system of two interpenetrating fluids, namely the fluid of thermal excitations or normal-fluid component and the pair condensate or superfluid component. In helium II, the normal-fluid density is determined by phonon and roton excitations. In superfluid ^{3}He, thermally excited Bogoliubov quasiparticles make up the normal-fluid density. In this section, we briefly discuss some specific aspects related to the normal-fluid density such as the specific heat below T_c.

4.7.1 Normal-Fluid Density

The normal-fluid density in superfluid ^{3}He is due to thermally excited quasiparticles. The quasiparticle excitation spectrum is given by

$$E_{\boldsymbol{k}} = \sqrt{\eta_{\boldsymbol{k}}^2 + \Delta^2}\,, \tag{4.23}$$

where $\eta_{\boldsymbol{k}}$ represents the kinetic energy of quasiparticles relative to the Fermi level. The normal-fluid density, taking into account Fermi-liquid corrections, has been calculated as

$$\overleftrightarrow{\varrho}_{\rm n} = \frac{m^*}{m}\left(\overleftrightarrow{1} + \frac{1}{3}F_1\frac{\overleftrightarrow{\varrho}_{\rm n,0}}{\varrho}\right)^{-1}\overleftrightarrow{\varrho}_{\rm n,0}\,. \tag{4.24}$$

Here, F_1 denotes the Landau parameter introduced in Sect. 3.2 and $\overleftrightarrow{\varrho}_{\rm n,0}$ the normal-fluid density tensor without Fermi-liquid corrections. The latter is determined by the temperature-dependent Yosida function $Y_0(\widehat{\boldsymbol{k}},T)$, which is a measure of the current response of thermal excitations at the point $\widehat{\boldsymbol{k}}$ on the Fermi surface. For the isotropic BW state this function is independent of $\widehat{\boldsymbol{k}}$ and we can write $Y_0(\widehat{\boldsymbol{k}},T) = Y_0$. Hence the normal-fluid density in the BW state is isotropic and can be expressed by

$$\varrho_{\rm n} = \varrho\frac{(1+F_1/3)\,Y_0}{1+F_1Y_0/3}\,. \tag{4.25}$$

According to the temperature dependence of the Yosida function, $\varrho_{\rm n}$ increases monotonically for $T \to T_{\rm c}$, and vanishes exponentially fast for $T \to 0$.

In the anisotropic ABM state the situation is obviously more complicated because the normal-fluid density depends on the direction of motion relative to the orbital angular momentum $\boldsymbol{l}$. It is an axially symmetric tensor, oriented along $\widehat{\boldsymbol{l}}$. Near $T_{\rm c}$, the parallel and perpendicular components can be written as

$$\varrho_{\rm n,\perp} - \varrho = 2\varrho_{\rm n,\parallel} = -\frac{7}{5}\,\zeta(3)\frac{m}{m^*}\varrho\left(\frac{\Delta_{\rm m}}{\pi k_{\rm B}T_{\rm c}}\right)^2. \tag{4.26}$$

In the low-temperature limit the anisotropy of the ABM state is even stronger. One obtains for the two independent components

$$\varrho_{\rm n,\parallel} = \pi^2\frac{m}{m^*}\varrho\left(\frac{k_{\rm B}T_{\rm c}}{\Delta_{\rm m}}\right)^2, \tag{4.27}$$

and

$$\varrho_{\rm n,\perp} = \frac{7}{15}\pi^4\frac{m}{m^*}\varrho\left(\frac{k_{\rm B}T_{\rm c}}{\Delta_{\rm m}}\right)^4. \tag{4.28}$$

This means that the normal-fluid density obeys a power-law temperature dependence, because of the two gap nodes, rather than an exponential decrease as for the BW state.

4.7.2 Specific Heat

The specific heat of superfluid ^{3}He is determined by thermally excited quasiparticles, just as in ^{3}He-N. However, the quasiparticle spectrum in the superfluid state differs from that in normal-fluid ^{3}He because of the presence of the energy gap. Due to the isotropic energy gap $\Delta_{\rm B}$ of the B phase, the specific heat tends to zero exponentially as $T \to 0$. The low-temperature specific heat of ^{3}He-B is given by

$$C_{\rm B}(T) = \sqrt{2\pi}\, D(E_{\rm F})\, k_{\rm B}\Delta_{\rm B} \left(\frac{\Delta_{\rm B}}{k_{\rm B}T}\right)^{3/2} {\rm e}^{-\Delta_{\rm B}/(k_{\rm B}T)}\,, \tag{4.29}$$

where $D(E_{\rm F})$ represents the density of states at the Fermi surface.

The behavior of ^{3}He-A is quite different, because the energy gap of the A phase is zero along the preferred direction $\hat{l}$. At low temperatures $T \ll \Delta_{\rm m}/k_{\rm B}$, the majority of the quasiparticles are thermally excited in the small region $\Delta k \approx (T/\Delta_{\rm m})\, k_{\rm F}$ on the Fermi surface near $\hat{\boldsymbol{k}} = \pm\hat{l}$. The resulting specific heat can be expressed by

$$C_{\rm A}(T) = \frac{7}{5}\pi^2 \left(\frac{T}{\Delta_{\rm m}}\right)^2 C_{\rm N}(T) \propto T^3\,. \tag{4.30}$$

Here, $C_{\rm N}$ denotes the specific heat of the normal-fluid phase ^{3}He-N, which is given by (3.37). Therefore, the specific heat of the A phase at low temperatures is expected to vary as T^3 rather than exponentially.

4.7.3 Quasiparticle Scattering

For many properties, in particular the transport properties, collisions among quasiparticles are important. We therefore make a few remarks on this subject here. We have to distinguish between the mutual scattering of quasiparticles and scattering processes at boundaries.

Let us consider first the mutual scattering of quasiparticles. In principle, this process is not very different from that occurring in the normal-fluid phase ^{3}He-N, but one has to take into account the differences in the excitation spectrum of the quasiparticles in the two states. Near $T_{\rm c}$, the quasiparticle collision rate – or relaxation rate – in the B phase is given by the same rate as for ^{3}He-N, namely by (3.17). In the low excitation-density limit, $T \ll \Delta_{\rm B}$, the quasiparticle relaxation rate is proportional to the number of excitations available for scattering processes. Therefore, the relaxation rate in the low-temperature limit $T \ll T_{\rm c}$ can be approximated by [156]

$$\tau_{\rm B}^{-1} = \tau_{\rm N}^{-1} \frac{3}{\sqrt{2\pi}}\, W_0 \left(\frac{\Delta_{\rm B}}{k_{\rm B}T}\right)^{3/2} {\rm e}^{-\Delta_{\rm B}/(k_{\rm B}T)}\,, \tag{4.31}$$

where $\tau_{\rm N}^{-1}$ is the quasiparticle collision rate in the normal-fluid state and W_0 is a dimensionless parameter that depends on the quasiparticle scattering amplitude in ^{3}He-N.

In the A phase, the calculation of the relaxation rate is, in general, more complicated because energy and angular integrations are not separable. However, at low temperatures ($T \ll \Delta_{\rm m}/k_{\rm B}$) only quasiparticles with momenta $\boldsymbol{k}$ in the neighborhood of the two gap nodes, are thermally excited, and collide mainly with quasiparticles of the same kind. The energy gap of all these quasiparticles is much smaller than the thermal energy. Therefore, their contribution to the relaxation rate is approximately the same as in the normal-fluid state. By considering the fraction of momenta $\boldsymbol{k}$ accessible in this process one obtains for the collision rate the approximate relation [195]

$$\tau_{\mathrm{A}}^{-1} \propto \tau_{\mathrm{N}}^{-1} \left(\frac{k_{\mathrm{B}}T}{\Delta_{\mathrm{m}}}\right)^2 \approx \tau_{\mathrm{N}}^{-1}(T_{\mathrm{c}}) \left(\frac{T}{T_{\mathrm{c}}}\right)^4, \tag{4.32}$$

where Δ_{m} represents the maximum gap in the A phase.

When considering boundary scattering, for example at the walls, we have to distinguish between specular and diffusive scattering processes. In addition, in superfluid ^{3}He, a special type of scattering process occurs because the order parameter is distorted at the boundary. The thickness of the layer where this distortion occurs is given by the coherence length. This process was first discussed by *Andreev* for the case of a normal-superconducting interface [196] and is known as *Andreev reflection*. In the following, we briefly describe the origin of this process, which leads to important changes in the transport properties of liquid ^{3}He below T_{c}.

Let us describe the situation at a normal metal to superconductor interface in a rather simplified manner: At the interface, an electron incident from the side of the normal metal with an energy smaller than the energy gap of the superconductor cannot enter the bulk superconductor but is converted into a hole that moves backward with respect to the electron. The missing charge $2e$ (an electron has the charge $-e$ and a hole $+e$) propagates as a Cooper pair into the superconductor. The electron–hole conversion is known as Andreev reflection. Therefore, in this scattering process a quasiparticle is turned back by means of an elastic transition from the electron-like excitation branch to the hole-like branch of the excitation spectrum (see also Fig. 10.29), thereby reversing its group velocity. An analogous process of particle–hole conversion at boundaries occurs in superfluid ^{3}He. As in the case described above, a Bogoliubov quasiparticle is reflected as a Bogoliubov quasihole near the suppressed (order parameter) gap at the boundary. This process has been observed in vibrating-wire experiments (see, for example, [197]). In addition, one finds that quasiparticle scattering due to Andreev reflection also occurs in bulk ^{3}He-A that has nonuniform $\hat{\boldsymbol{l}}$ texture. Finally, we remark that Andreev reflection in superfluid ^{3}He leads to an increased slip length and a reduced effective viscosity (see Fig. 4.9b).

4.8 Collective Excitations – Sound Propagation

Because of the complex structure of the superfluid phases of ^{3}He a large variety of collective modes can be excited. In fact, all expected sound modes – zero, first, second, third and fourth sound – and collisionless spin waves have experimentally been observed in liquid ^{3}He below T_{c}. In addition, so-called collective order-parameter modes exist, due to the inner structure of the quasiparticle pairs.

4.8.1 Sound Propagation

Sound propagation in superfluid ^{3}He is rather complex, because it does not only combine the manifold of sound modes of helium II and normal-fluid ^{3}He, but also exhibits a number of phenomena that have no analogue in either fluid. These special modes are a direct consequence of the broken relative symmetries in superfluid ^{3}He.

In general, the two-fluid hydrodynamics discussed in Sect. 2.2.1 for helium II provides a good description for second, third and fourth sound in ^{3}He-B. The anisotropic A and A_1 phases require a more complex theoretical treatment. In this section, we will present a few experimental results that are characteristic of two-fluid sound modes in superfluid ^{3}He. A thorough discussion of the two-fluid hydrodynamics of the superfluid phases of ^{3}He, however, is beyond the scope of this introduction.

Second Sound

The occurrence of second sound is one of the most characteristic features of superfluids, because it is a direct consequence of broken gauge symmetry of the superfluid order parameter. For ^{3}He-B the velocity of second sound is given by (2.43). As an example, we show in Fig. 4.28 the temperature dependence of the velocity of second sound obtained in the B phase at a pressure of 21.3 bar. The *solid line* represents the prediction of the two-fluid model, which agrees with the measured data reasonably well. However, since the entropy density S carried by the normal-fluid component is reduced in ^{3}He by a factor of T/T_F in comparison with ^{4}He, v_2 turns out to be about three orders of magnitude smaller than the Fermi velocity v_F.

As a consequence of the broken relative symmetry of spin rotations and gauge transformations, a propagating 'second sound' mode exists in ^{3}He-A_1 that is a combination of an entropy-density wave and a spin-density wave.

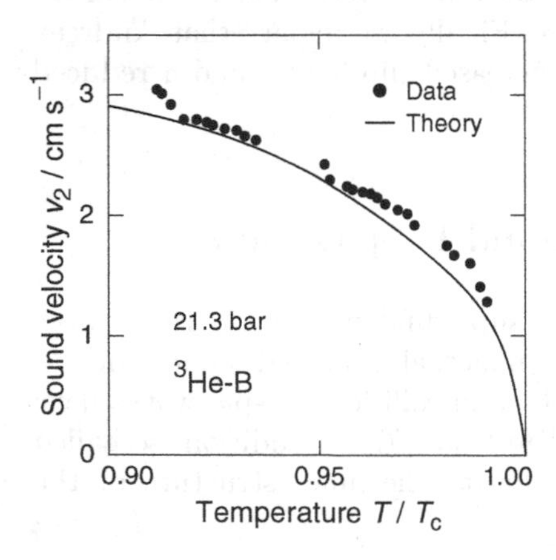

Fig. 4.28. Velocity of second sound in ^{3}He-B as a function of the reduced temperature T/T_c [198]

Since only one spin orientation with respect to the magnetic field exists in ^{3}He-A$_1$, the counterflow of normal and superfluid components involved in second-sound propagation must also carry spins. For the propagation in the direction of the magnetic field, the velocity of second sound in ^{3}He-A$_1$ is given, to a good approximation, by [199]

$$v_2 = \frac{\gamma\hbar}{2m^*}\sqrt{\frac{\varrho}{\chi}\frac{\varrho_{s,\perp}}{\varrho_{n,\perp}}}\,, \tag{4.33}$$

where χ denotes the magnetic susceptibility and m^* the effective mass of the quasiparticles. Figure 4.29 shows the result of a measurement of v_2 in ^{3}He-A$_1$. The experiment was performed at the melting pressure in a magnetic field of 0.85 T, conditions such that the A$_1$ phase exists over a temperature range of 50 μK. The second-sound waves were excited and detected via porous-membrane transducers (5 μm pore diameter) that were driven and monitored capacitively. Since the porous membranes only move the normal-fluid component, it is possible to generate a counterflow of normal-fluid and superfluid density in the liquid.

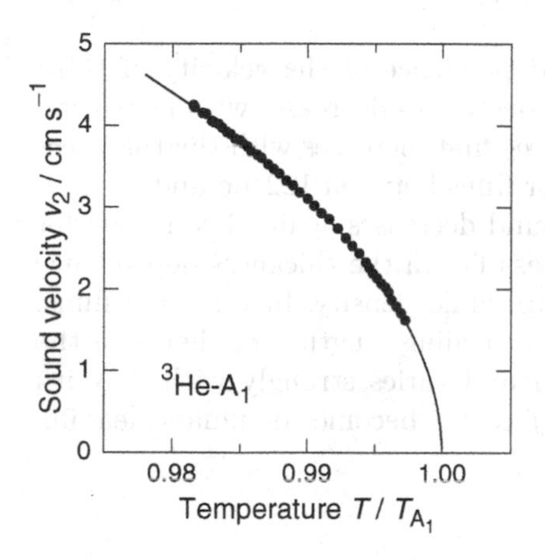

Fig. 4.29. Velocity of second sound in ^{3}He-A$_1$ as a function of the reduced temperature T/T_c [200]

The second-sound velocity in ^{3}He-A$_1$ is much larger than in ^{3}He-B. This is because second sound in ^{3}He-A$_1$ is not only an entropy-density wave but also a spin-density wave. In fact, the velocity of the combined mode is comparable to that of a spin wave in ^{3}He. The temperature dependence of the velocity of second sound is determined by that of the superfluid density. The *solid line* corresponds to the known behavior of $\varrho_s(T) \propto (1-T/T_{A_1})$ and is in very good agreement with the data. Second sound was not detectable near T_{A_1} because either the coherence length exceeded the pore size or the critical velocity dropped below the superfluid velocity within the pores in this temperature range.

Third Sound

As we have seen in Sect. 2.2.8, third sound is a surface wave propagating on a thin superfluid film. Recently, this sound mode was observed in superfluid ^{3}He for the first time [201]. In this experiment, standing waves of third sound in films with thicknesses of about 100 nm were generated by applying an oscillating electric field and detected capacitively. Despite the rather small dielectric constant of the liquid, the resolution for the average film surface displacement was about 30 pm Hz$^{-1/2}$.

Analogous to the expression for helium II, the velocity of third sound in this experiment is given by

$$v_3 = \sqrt{\frac{\langle \varrho_s \rangle}{\varrho} \frac{3\alpha}{d^3}}\,, \tag{4.34}$$

where $\langle \varrho_s \rangle$ represents the superfluid density averaged over the film thickness d. The quantity α denotes the Hamaker constant (see Sect. 2.2.4). The additional temperature-dependent term presented in (2.48) can be neglected in superfluid ^{3}He, because of the low entropy density and the low temperature.

Figure 4.30 shows the temperature dependence of the velocity of third sound for different film thicknesses. As expected, v_3 decreases with increasing temperature. At the lowest temperature, v_3 first increases with the thickness of the film until a maximum is reached for films between 122 nm and 174 nm. For thicker films, the velocity of third sound decreases again. The reason for this nonmonotonic variation with thickness lies in the thickness dependence of the Van der Waals force f and the superfluid density. In very thin films, the thickness becomes comparable to the healing length and therefore the superfluid component is greatly reduced and varies strongly with d. With increasing thickness the proportionality $f \propto d^{-3}$ becomes dominant, leading

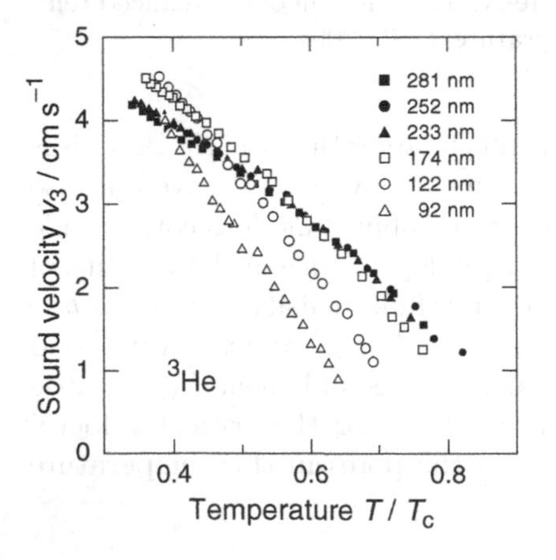

Fig. 4.30. Velocity of third sound in superfluid ^{3}He-B films with different thickness as a function of the reduced temperature T/T_c [201]

to a reduction of v_3. Since the healing length depends on the ratio T/T_c, the cross over from one regime to the other takes place at different temperatures for films with different thicknesses.

Fourth Sound

Density waves propagating through a superleak are called fourth sound. This characteristic mode has been observed in both the A phase and B phase, providing an early proof of the superfluidity of these liquids [202–204]. Whereas the velocity of fourth sound in the isotropic B phase can by described, to a good approximation, by $v_4 = v_1\sqrt{(\varrho_s/\varrho)}$, the propagation of fourth sound in ^{3}He-A depends on the angle Θ between the wave vector $\boldsymbol{q}$ of the sound wave and the orbital angular momentum $\boldsymbol{l}$. Near T_c, the velocity of fourth sound in the A phase can be expressed by

$$v_4 = v_1\sqrt{\frac{\langle\varrho_s\rangle}{\varrho}\frac{5}{3}\left[2-\cos^2\Theta\right)\right]}, \tag{4.35}$$

where $\langle\varrho_s\rangle$ is the average value of the superfluid density and v_1 is the velocity of first sound. The latter is given by (3.38) as in the normal-fluid phase. Fourth-sound experiments have been used to determine the superfluid density and to monitor superflow. Figure 4.31 shows the average superfluid density in liquid ^{3}He below T_c as determined by this type of measurement. In agreement with other methods, one finds a temperature variation $\langle\varrho_s\rangle/\varrho \propto (1-T/T_c)$. No visible change has been observed in measurements of v_4 at the phase transition from the A to the B phase (not shown in Fig. 4.31).

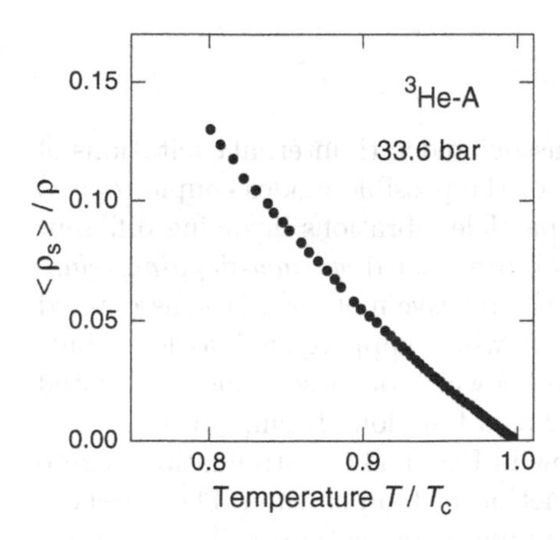

Fig. 4.31. Average superfluid density in liquid ^{3}He below T_c as a function of the reduced temperature T/T_c [202]

4.8.2 Collective Order-Parameter Modes

The quasiparticle pairs in superfluid ^{3}He possess a structure that allows for various types of collective excitations to exist. In a crude analogy, we may view these excitations as being similar in some ways to the stretching mode of a diatomic molecule. In superfluid ^{3}He, the pair excitations are coherently excited and thus vibrate in phase. Since they involve relative motions between $\boldsymbol{d}$ and $\boldsymbol{l}$ they are often referred to as *order-parameter modes*. The entire liquid participates in these collective modes, providing a spectacular example of quantum mechanics on a macroscopic scale. The order-parameter modes are particularly interesting because they reflect the internal degree of freedom of the quasiparticle pairs.

Since some of these modes couple to density fluctuations, they can be excited by sound waves. Measurements of the absorption of longitudinal zero sound provide a possible way to study these modes. One has to take into account, however, that the damping of zero-sound waves is also influenced by pair breaking. As discussed previously, in ^{3}He-A the energy gap has two point nodes and thus pair breaking can take place in the vicinity of these nodes, even far below T_c. This provides an attenuation that masks the pair-vibration resonances to some extent. However, since the phase space for pair-breaking processes occurring at the nodes is rather limited, it is still possible to study pair-vibration resonances in ^{3}He-A in ultrasound experiments at low temperatures.

In ^{3}He-B, pair breaking occurs only above a certain energy threshold, because it has an isotropic energy gap. It is thus possible to investigate the order-parameter modes of ^{3}He-B without the complications of pair breaking, as long as the frequency ω of the sound wave is less than $2\Delta/\hbar$.

^{3}He-A

There are 18 different collective modes associated with internal excitations of the quasiparticle pairs in ^{3}He-A. Three of the possible modes couple to zero sound. They can be visualized as quasiparticle vibrations involving different relative motions between $\boldsymbol{d}$ and $\boldsymbol{l}$. They are named *normal-flapping, clapping* and *superflapping* modes, because the relative motions of the associated vectors reminds one of hands clapping or wings flapping such as for a butterfly. The resonance frequency and the linewidth of these modes calculated numerically [205] are shown in Fig. 4.32 as a function of temperature.

As an experimental example, we show in Fig. 4.33 the attenuation of zero sound at 29.3 bar and 54 MHz as a function of temperature. The observed damping peak is associated with the clapping-mode resonance. The *solid line* represents the result of a theoretical calculation of this mode [206].

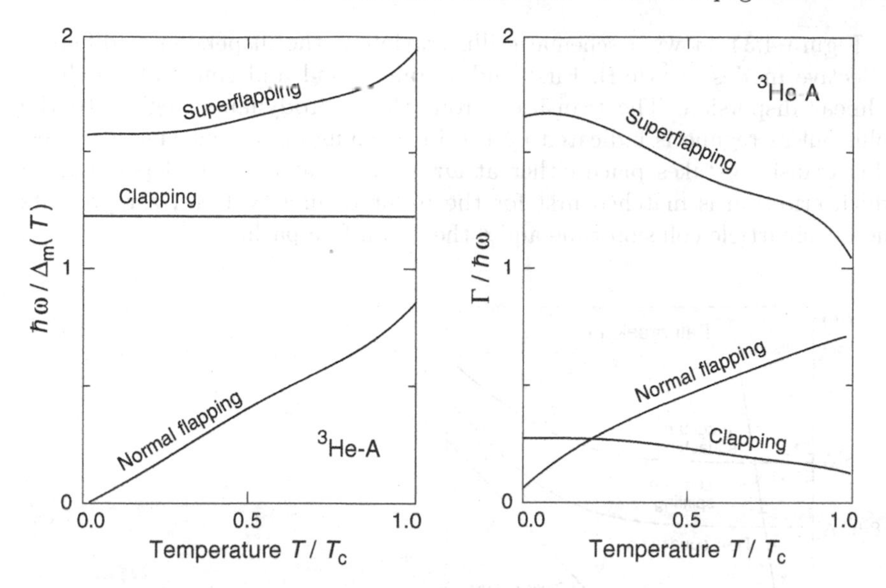

Fig. 4.32. (**a**) Reduced energy $\hbar\omega/[\hbar\Delta_\mathrm{m}(T)]$ and (**b**) reduced linewidths $\Gamma/\hbar\omega$ of normal-flapping, clapping, and superflapping modes as a function of the normalized temperature T/T_c. Here, $\Delta_\mathrm{m}(T)$ denotes the maximum gap [205]

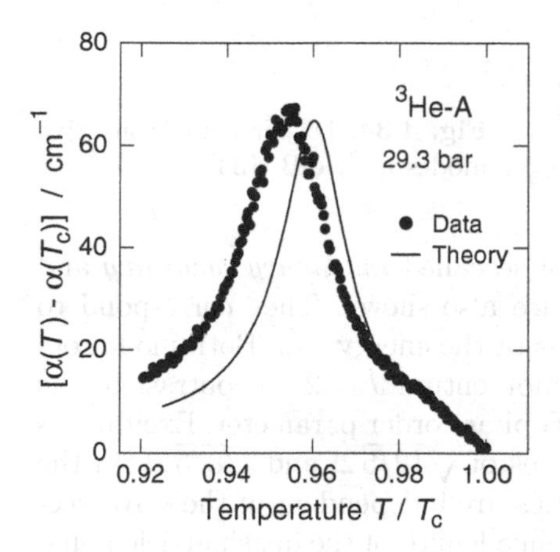

Fig. 4.33. Attenuation of longitudinal zero sound in ^{3}He-A at 54 MHz as a function of the normalized temperature T/T_c at 29.3 bar. The *solid line* represents the theoretical prediction [207]

^{3}He-B

In the B phase, the collective order-parameter modes can be visualized as excited states of the quasiparticle pairs with frequencies proportional to the energy gap $\Delta_\mathrm{B}(T)$. They can be classified by their total angular momenta $\boldsymbol{J} = \boldsymbol{L} + \boldsymbol{S}$ and their azimuthal component J_z. In total, there are 18 possible modes, a few of which couple to zero sound.

Figure 4.34 shows a schematic illustration of the dispersion of different collective modes in ^{3}He-B. First and second sound and spin waves exhibit a linear dispersion. The transition from the hydrodynamic regime to the collisionless regime is indicated by the broadening of the dispersion curves. This transition takes place either at $\omega\tau = 1$, or at $q\ell = 1$, depending on which criterion is matched first for the different modes. Here, τ represents the quasiparticle collision time and ℓ the mean free path.

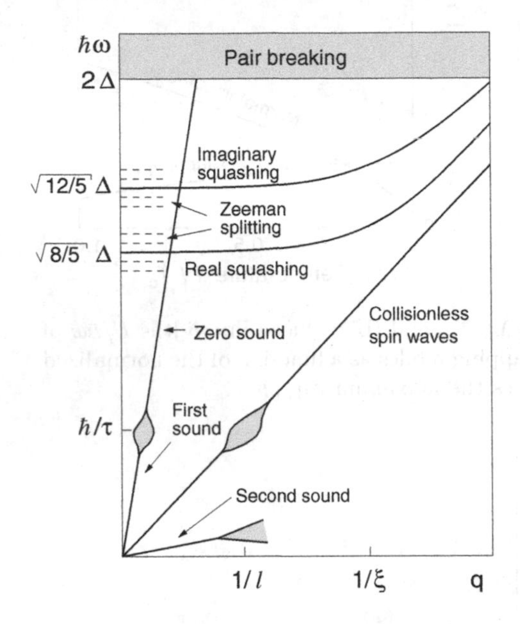

Fig. 4.34. Dispersion of collective modes in ^{3}He-B [133]

Beside these modes, the dispersion of so-called *imaginary squashing* and *real squashing* order-parameter modes are also shown. They correspond to two different types of periodic distortions of the energy gap. Both modes occur in the state with the total angular momentum $J = 2$, in contrast to the $J = 0$ ground state associated with the B phase order parameter. Excitations of these modes have characteristic energies of $\sqrt{12/5}\,\Delta$ and $\sqrt{8/5}\,\Delta$. In the range $q < 1/\xi$ these characteristic energies are independent of the wave vector. The quantity ξ represents the coherence length of the quasiparticle pairs. The *dashed lines* indicate the Zeeman splitting of the imaginary squashing and real squashing mode levels in an external magnetic field at $q = 0$. For phonon energies $\hbar\omega > 2\Delta$ pair breaking takes place.

A peak in the absorption is expected if the sound frequency matches the excitation frequency of a particular mode. Since the energy gap varies with temperature, it is possible to observe different modes even at fixed sound frequency by sweeping the temperature. Figure 4.35 shows the results of a damping measurement of zero sound at a fixed frequency of 44.2 MHz as a function of temperature. Close to T_c, pair breaking dominates the damping.

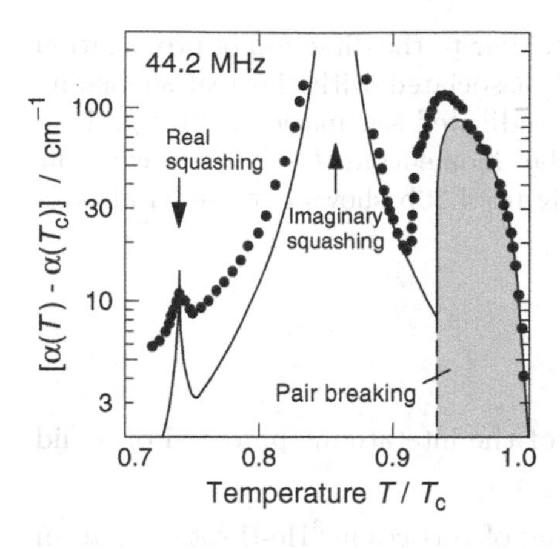

Fig. 4.35. Attenuation of longitudinal zero sound in ^{3}He-B at 44.2 MHz as a function of the reduced temperature T/T_c at 2.44 bar. The *solid line* represents a theoretical curve and the *arrows* indicate the position of the maxima predicted theoretically [208]

The intermediate peak is the imaginary squashing mode. The attenuation is so large for this mode that it could not be measured in the central region in this experiment. The smaller peak on the low-temperature side corresponds to the real squashing mode, which is coupled much more weakly to the sound wave.

Figure 4.36a shows the result of a similar measurement at 60 MHz and 5.3 bar. Pair breaking and the imaginary squashing mode are off scale in this plot. The very narrow absorption line at low temperatures corresponds to the real squashing mode. The width of the peak is determined by the quasiparticle collision rate. Since at low temperatures only very few quasiparticles exist, the peak becomes very sharp.

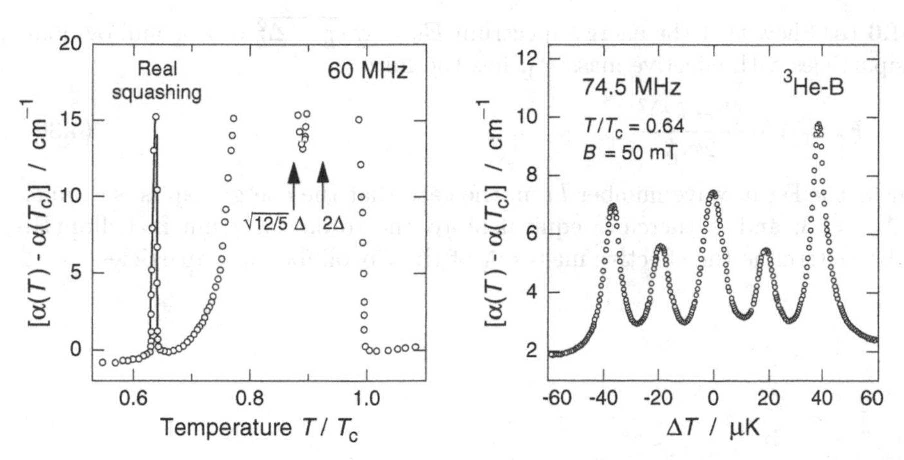

Fig. 4.36. (**a**) Normalized attenuation of longitudinal zero sound in ^{3}He-B at 60 MHz and a pressure of 5.3 bar as a function of the reduced temperature T/T_c [209]. (**b**) Attenuation of zero sound in ^{3}He-B due to the real squashing mode in a magnetic field of 50 mT and at 11 bar. The measuring frequency was 74.4 MHz [210]

In a magnetic field oriented perpendicular to the direction of propagation of the sound waves, the damping peak associated with the real squashing mode shows a spectacular splitting. As indicated schematically in Fig. 4.34, the two squashing modes having angular momentum $J = 2$ are each split into five modes by a magnetic field. Figure 4.36b shows experimental data confirming this prediction.

Exercises

4.1 Show that the hard-core repulsion of the interatomic potential of liquid ^{3}He prevents s-wave pairing.

4.2 Calculate the number and the spacing of vortices in ^{3}He-B rotating at an angular velocity of $2\,\mathrm{rad\,s^{-1}}$. How would these numbers change for continuous vortices in the A phase?

4.3 In a quantum-interference experiment superfluid ^{3}He is rotated in a thin torus with area $A = 10\,\mathrm{cm}^2$ and two weak links. Calculate the Josephson frequency for a pressure difference of 20 μbar. At what angular velocity is the mass current maximal, assuming a rotation axis perpendicular to the plane of the torus?

4.4 Calculate the heat capacity and the normal-fluid density of $1\,\mathrm{cm}^3$ of ^{3}He-B at $T/T_c = 0.1$.

4.5 At what frequency is the transition from first sound to zero sound expected to occur in ^{3}He-A at $T/T_c = 0.1$?

4.6 (**a**) Show that the energy spectrum $E_{\boldsymbol{k}} = \sqrt{\eta_{\boldsymbol{k}}^2 + \Delta_{\boldsymbol{k}}^2}$ of Bogoliubov quasiparticles with effective mass m_{B}^* has the form

$$E_{\boldsymbol{c}} = \Delta + \frac{(k - k_{\mathrm{F}})^2}{2m_{\mathrm{B}}^*} \tag{4.36}$$

near the Fermi wave number k_{F} in the case that the energy gap is isotropic, $\Delta_{\boldsymbol{k}} = \Delta$, and is therefore equivalent to the roton spectrum in helium II. (**b**) Determine the effective mass m_{B}^* of the Bogoliubov quasiparticles.

5 Mixtures of ^{3}He and ^{4}He

Early investigations of the properties of liquid mixtures of ^{3}He and ^{4}He took place in 1947 [211–213] motivated by a suggestion from *Franck* [214] for developing new methods for the separation of the two isotopes. A few years later, in 1950, the first experiments of the propagation of second sound in ^{3}He/^{4}He mixtures showed that the superfluid hydrodynamics of ^{4}He is strongly affected by the presence of ^{3}He [215]. Later, in 1954, studies of the influence of ^{3}He impurities on the lambda transition of ^{4}He [216] were carried out. In particular, the reduction of the magnitude of the specific-heat maximum at the phase transition and the shift of the lambda point to lower temperatures with increasing ^{3}He concentration were investigated. Soon after, in 1956, it was discovered in NMR experiments that the solutions separate into two phases with different mixing ratio below a critical temperature depending on the ^{3}He concentration [217].

Mixtures of ^{3}He and ^{4}He are of technical importance for obtaining very low temperatures, because dilution refrigerators, which are widely used for cooling in the mK range, are based on their properties (see Sect. 11.4). In addition, ^{3}He/^{4}He mixtures serve as model system for various theoretical questions regarding interacting fermion systems. One major advantage is the possibility to vary the ^{3}He concentration and thus the mean interaction energy. One other interesting aspect of ^{3}He/^{4}He mixtures is the occurrence of a tricritical point at which the lambda transition and the phase separation line meet. At this point, a second-order transition line changes into one of first order. Investigations in the vicinity of the tricritical point have shown that the critical exponents are modified compared to those observed at a single-transition line.

In this chapter, we will introduce some basic properties of ^{3}He/^{4}He mixtures starting with the phase diagram and the specific heat and then continue to discuss sound propagation and transport properties. At the end of this chapter we will briefly report on the search for superfluidity of ^{3}He in mixtures of ^{3}He/^{4}He.

5.1 Specific Heat, Phase Diagram and Solubility

In Chap. 2 we saw that the lambda transition of pure ^{4}He occurs at about 2.17 K. As depicted in Fig. 5.1, the lambda point is shifted towards lower temperature upon adding ^{3}He. At the same time, the magnitude of the specific-heat anomaly reduces. The general shape of the temperature dependence, however, appears to be unaltered qualitatively.

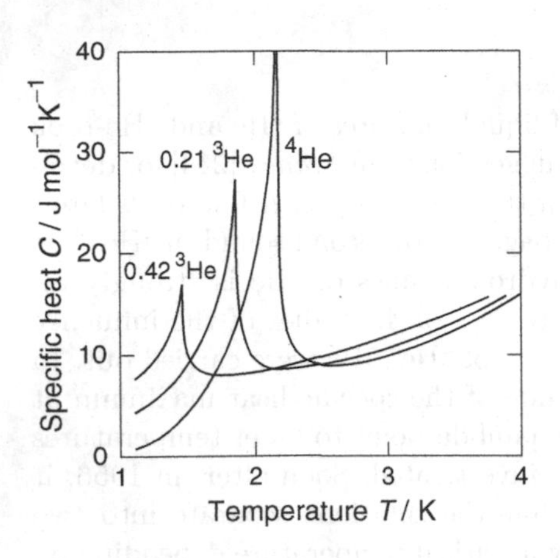

Fig. 5.1. Specific heat of liquid ^{3}He/^{4}He mixtures at saturated vapor pressure for the range of temperature in which the lambda transition takes place. For comparison, the specific heat of pure ^{4}He is also shown [218]

The monotonic decrease of the lambda transition with increasing ^{3}He concentration $c_3 = N_3/(N_3 + N_4)$ continues until the tricritical point is reached at $T = 0.87$ K and $c_3 = 0.67$. Below $T = 0.87$ K, solutions with mixing ratios in a certain range do not exist, or in other words, a *miscibility gap* occurs. For solutions with $c_3 > 0.0648$, one finds a separation into two phases at a temperature that depends on the ^{3}He concentration. A ^{3}He-rich phase, being less dense, floats on a ^{4}He-rich phase. This phase separation is a transition of first order. Figure 5.2 shows the discontinuities of the heat capacities associated

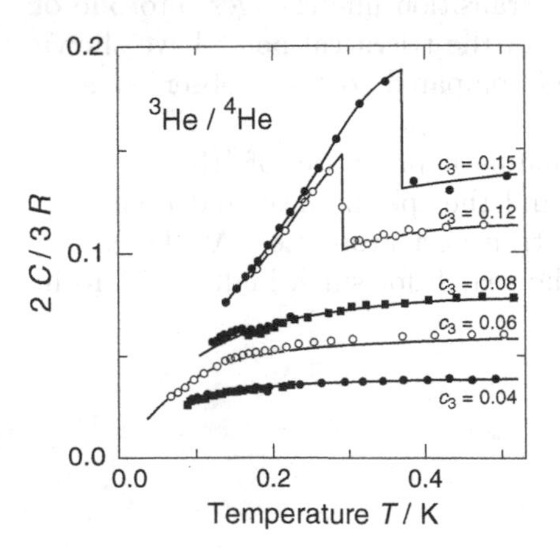

Fig. 5.2. Reduced specific heat $2C/3R$ of ^{3}He/^{4}He mixtures with different mixing ratios as a function of temperature [219]. The *solid lines* are guides for the eye

with the phase separation for some dilute solutions of ^{3}He. Note that for the two solutions with the lowest ^{3}He concentrations no separation is observed. We will come back to this point in Sect. 5.2. The specific-heat data shown in Fig. 5.2 are normalized to the value of classical gases. This means that in the classical limit the normalized heat capacity should directly reflect the concentration of the ^{3}He atoms. For the solutions with low ^{3}He concentration good agreement is found.

5.1.1 Phase Diagram

Figure 5.3 shows the phase diagram of liquid ^{3}He/^{4}He mixtures, as derived from various experiments. Three regions can be distinguished that correspond to mixtures containing normal-fluid ^{4}He, superfluid ^{4}He and the miscibility-gap region. The tricritical point originates from the intersection of the lambda line and the phase-separation line.

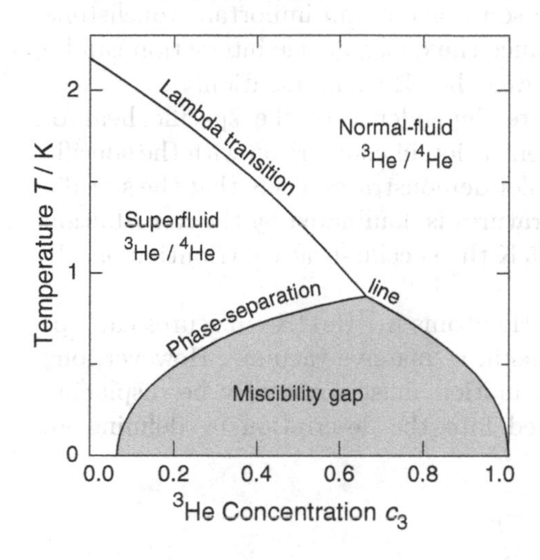

Fig. 5.3. Phase diagram of liquid mixtures of ^{3}He and ^{4}He. The temperature is plotted vertically and the ^{3}He concentration horizontally. The *grey tinted* region marks the extent of the miscibility gap [218, 220–223]

At the phase-separation line, two phases with different ^{3}He content are formed that coexist. A light ^{3}He-rich phase, which consists of pure ^{3}He, for $T \to 0$. The ^{4}He content in this phase can be approximated by the empirical relation

$$c_4 = (1 - c_3) = a\sqrt{T^3}\,\mathrm{e}^{-b/T}\,, \tag{5.1}$$

with $a = 0.85\,\mathrm{K}^{-3/2}$ and $b = 0.56\,\mathrm{K}$. The ^{3}He concentration in the heavier ^{3}He-poor phase is given, to a good approximation, by

$$c_3 = c_{3,0}\,(1 + \tilde{a}\,T^2 + \tilde{b}\,T^3)\,, \tag{5.2}$$

with $\tilde{a} = 8.4\,\mathrm{K}^{-2}$, $\tilde{b} = 9.4\,\mathrm{K}^{-3}$, and $c_{3,0} = 0.0648$, the ^{3}He concentration at $T = 0$. An interesting and essential property for the realization of a dilution refrigerator (see Sect. 11.4), a cooling technique that is based on the properties of ^{3}He/^{4}He, is the fact that even at $T = 0$ a finite solvability of ^{3}He in liquid ^{4}He exists. As we will see in Sect. 5.1.3, the reason for this extraordinary behavior is the interplay of zero-point energy and binding energy. Before we discuss this point further, we shall consider in the following section the specific heat of dilute solutions of ^{3}He in helium II at very low temperatures.

5.1.2 Specific Heat of Dilute Solutions of ^{3}He in Helium II

Mixtures with ^{3}He concentrations up to about $c_3 = 0.15$ are often referred to as dilute solutions of ^{3}He in helium II. The ^{4}He part of these solutions behaves practically like a passive background fluid below about 0.5 K because the normal-fluid component and thus the thermal excitations such as phonons and rotons are negligible. The excitations in mixtures under these conditions stem mainly from ^{3}He atoms, which can be described as an interacting Fermi gas. As mentioned previously, dilute solutions are an important touchstone for theories of interacting fermions, since the quasiparticle interaction can be varied in a well-defined way by changing the ^{3}He concentration.

Figure 5.4 shows the temperature dependence of the specific heat of ^{3}He/^{4}He mixtures with low ^{3}He concentration in comparison with the specific heat of pure ^{3}He and pure ^{4}He. This plot demonstrates nicely that the specific heat of dilute solutions at low temperatures is dominated by the contribution from ^{3}He atoms. Below about $T = 0.5$ K the specific-heat contribution of ^{4}He is negligible.

At very low concentrations, the ^{3}He atoms in ^{3}He/^{4}He mixtures can approximately be described as a free gas in a 'massive vacuum'. However, one has to consider that a ^{3}He atom in motion must constantly be displacing ^{4}He atoms. This can be incorporated into the description by defining an

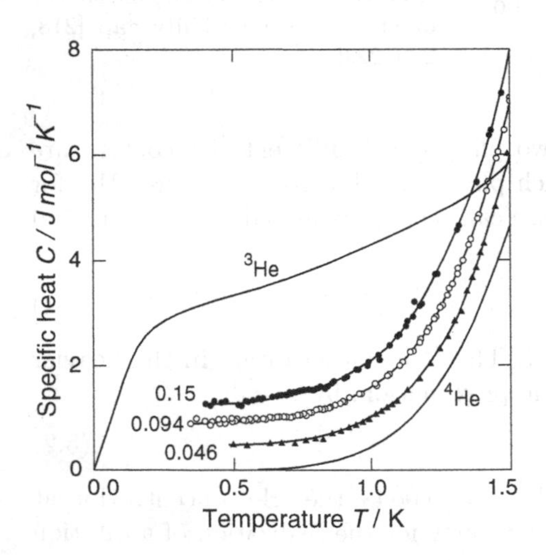

Fig. 5.4. Specific heat of dilute solutions of ^{3}He in ^{4}He as a function of temperature. For comparison, the corresponding data for pure ^{3}He and pure ^{4}He are also drawn [223]

effective mass m_3^* for ^{3}He atoms. The experimental value in the dilute limit, $c_3 \to 0$, is given by $m_3^* = 2.4\,m_3$. This is about 15% lower than the effective mass of the quasiparticles in pure ^{3}He (see Table 3.2). With increasing ^{3}He concentration the interaction between the ^{3}He atoms becomes important and adds to the effective mass leading to $m_3^* \approx 2.5\,m_3$ at $c_3 = 0.08$. Depending on temperature and concentration, ^{3}He in dilute solutions behaves either as a classical gas or as a degenerate Fermi gas. According to (3.10) the Fermi temperature is given by

$$T_{\mathrm{F}} = \frac{\hbar^2}{2m_3^* k_{\mathrm{B}}} \left(3\pi^2 n_3\right)^{2/3} \propto c_3^{2/3}, \tag{5.3}$$

with the number density $n_3 = c_3 N/V$ for the ^{3}He atoms. Figure 5.5 shows the temperature dependence of the specific heat of dilute solutions of ^{3}He in ^{4}He at low temperatures. The transition from the classical behavior to that of a degenerate Fermi gas takes place at roughly the temperature $T \approx T_{\mathrm{F}}/3$. The *solid lines* in Fig 5.5 represent the theoretical prediction for an ideal Fermi gas taking into account the effective mass. The agreement between the fit and the data is very good over the whole temperature range.

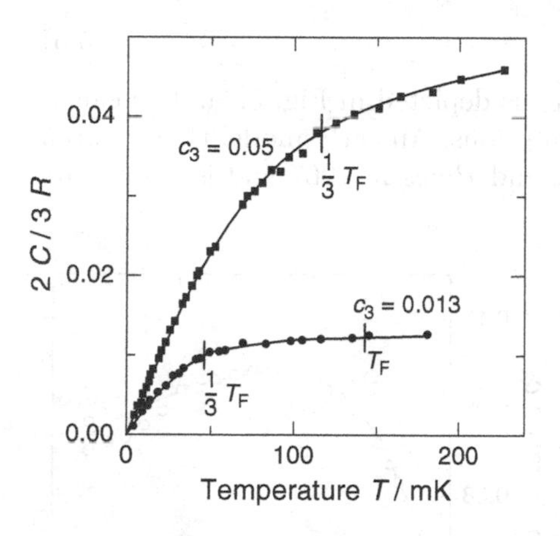

Fig. 5.5. Reduced specific heat $2C/3R$ of two ^{3}He/^{4}He mixtures with different ^{3}He concentration as a function of temperature [224]. The *solid lines* are calculated for an ideal Fermi gas with the corresponding effective masses

5.1.3 Finite Solubility of ^{3}He in Liquid ^{4}He at $T = 0$

The finite solubility of ^{3}He in ^{4}He at absolute zero is ultimately a consequence of the difference in the zero-point energies of ^{3}He and ^{4}He. As we mentioned in Sect. 1.1, the Van der Waals potentials of both isotopes are identical. But due to its smaller mass a ^{3}He atom exhibits a larger zero-point motion than a ^{4}He atom. Therefore, it will be closer to ^{4}He atoms than it would be to ^{3}He atoms, and consequently, its binding to a ^{4}He atom is stronger than a ^{3}He–^{3}He bond.

In equilibrium, the chemical potentials of the concentrated – or ^{3}He-rich – phase and the dilute – or ^{3}He-poor – phase are equal:

$$\mu_{3,\mathrm{d}}(T, c_{3,\mathrm{d}}) = \mu_{3,\mathrm{c}}(T, c_{3,\mathrm{c}})\,. \tag{5.4}$$

Here, the indices d and c refer to the dilute and concentrated phase, respectively. At $T = 0$ the concentrated phase consists of pure ^{3}He. The energy required to move one atom from the concentrated phase into vacuum is given by the latent heat per atom $L_3(T = 0)$ and thus the chemical potential per atom can by expressed by $\mu_{3,\mathrm{c}}(0,1) = \mu_3(0) = -L_3(0)$. The experimentally determined value is $\mu_3(0)/k_\mathrm{B} = -2.473\,\mathrm{K}$ [225].

In the dilute phase, $E_3 = -\mu_{3,\mathrm{d}}(0,0)$ is the binding energy of one ^{3}He atom in liquid ^{4}He at $T = 0$ and $c_{3,\mathrm{d}} \to 0$. The stronger binding of ^{3}He atoms in liquid ^{4}He than in ^{3}He is expressed by $\mu_{3,\mathrm{d}}(0,0) > \mu_3(0)$. With rising ^{3}He concentration the effective binding energy per ^{3}He atom decreases, because of the Pauli exclusion principle. In calculating $\mu_{3,\mathrm{d}}$, we must take into account the Fermi energy $E_\mathrm{F} = k_\mathrm{B}T_\mathrm{F}(c_3)$. In addition, we have to consider the changes in the binding energy with increasing ^{3}He concentration, due to the interaction between the ^{3}He atoms. Using (5.4) we can calculate the equilibrium concentration of ^{3}He in liquid ^{4}He at absolute zero via the equation

$$-L_3(0) = -E_3(0, c_3) + k_\mathrm{B}T_\mathrm{F}(c_3)\,. \tag{5.5}$$

The variation of these quantities with c_3 is depicted in Fig. 5.6a. The calculation of $E_3(0, c_3)$ requires several assumptions. An early model that is often used was developed by *Bardeen, Baym* and *Pines* in 1967 and is commonly

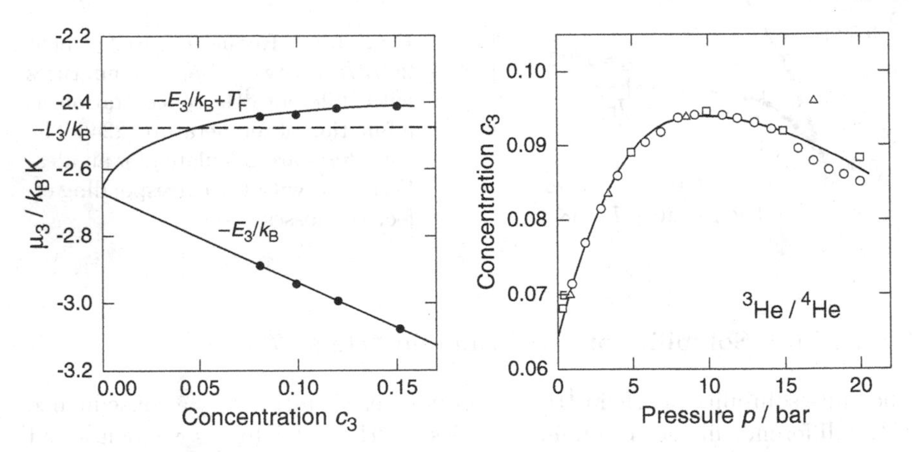

Fig. 5.6. (**a**) Binding energy $-E_3(0, c_3)$ and chemical potential $\mu_3(0, c_3)$ as a function of the ^{3}He concentration [222]. The *dashed line* represents the chemical potential of pure ^{3}He at absolute zero $\mu_3(0) = -L_3(0)$. (**b**) Pressure dependence of the maximal ^{3}He concentration in liquid ^{4}He for $T \to 0$. The data have been obtained with different experimental techniques [226–228] (after [229]). The *solid line* is a guide for the eye

referred to as the BBP theory [230]. The BBP theory leads to a satisfying description at saturated vapor pressure. At low concentrations, the binding energy decreases roughly linearly with c_3. Using the BBP theory together with the effective potential suggested by *Ebner* [231] allows also a qualitatively description to be given of the pressure dependence of the equilibrium concentration of ^{3}He in liquid ^{4}He at absolute zero. One finds a maximal solubility of roughly 9.6% at a pressure of about 8.7 bar. The pressure dependence of the limiting equilibrium concentration of ^{3}He in ^{4}He is shown in Fig. 5.6b.

5.2 Normal-Fluid Component

In the previous sections, we saw that one way to describe the low-temperature properties of dilute solutions of ^{3}He in liquid ^{4}He is to assume that the ^{3}He atoms act as excitations in superfluid ^{4}He. This is the starting point of a model developed by *Landau* and *Pomeranchuk* in 1948 [232]. They treat the ^{3}He atoms as quasiparticles with energy E and momentum $\boldsymbol{p}$, similar to phonons and rotons in helium II. Under the assumption that the interaction between the ^{3}He quasiparticles is negligible compared to their kinetic energy, the excitation spectrum can be approximated by $E(\boldsymbol{p}) = -E_3 + p^2/(2m_3^*)$. In the following, we will discuss two experiments that show the validity of this picture.

5.2.1 Andronikashvili Experiment

In dilute ^{3}He/^{4}He mixtures, ^{3}He atoms behave like the normal-fluid component in superfluid ^{4}He. This can be proven in an experiment similar to that performed by Andronikasvili in helium II. In this experiment, 15 mica discs with a diameter of 4 mm were used. The mica discs were mounted at the end of a torsion pendulum separated by only 190 μm from each other. The oscillation period varied between 3 and 5 s. Figure 5.7 shows the result of such a measurement. In contrast to the behavior of pure helium II, the normal-fluid component ϱ_n does not vanish at low temperatures, but becomes constant below about 1 K. This constant value of ϱ_n is due to the contribution of the ^{3}He atoms. In further experiments, it was shown that the constant contribution of the ^{3}He atoms in dilute solutions is proportional to the ^{3}He concentration, as we would expect for noninteracting quasiparticles (see Sect. 5.3.2). The normal-fluid density can be expressed by

$$\varrho_\mathrm{n} = \varrho_{\mathrm{n},4} + \varrho\,\frac{m_3^*}{m_4}\,c_3\,, \tag{5.6}$$

where $\varrho_{\mathrm{n},4}$ denotes the normal-fluid density of pure helium II.

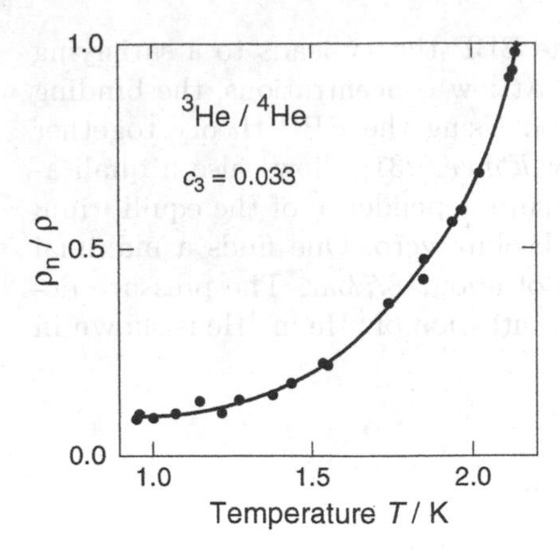

Fig. 5.7. Normal-fluid component ϱ_n/ϱ of a ^{3}He/^{4}He mixture with 3.3% ^{3}He as a function of temperature [233]

5.2.2 Osmotic Pressure

The fact that ^{3}He atoms in dilute solutions behave like the normal-fluid component in helium II, can also be seen in experiments in which two vessels containing ^{3}He/^{4}He mixtures with different ^{3}He concentrations are connected via a superleak.

Starting with equal levels, but with different amounts of ^{3}He, one observes that an osmotic pressure is quickly built up, as illustrated in Fig. 5.8. The reason is that superfluid ^{4}He flows through the superleak to even the ^{3}He concentration in both vessels. The ^{3}He atoms cannot contribute to the reduction of the concentration gradient, because they cannot pass through the superleak due to their viscosity. The osmotic pressure can be viewed as being equivalent to the fountain pressure discussed in Sect. 2.2, where the difference in the normal-fluid density was established by heating. To avoid any influence of the thermomechanical effect in the osmotic pressure experiments both vessels are kept at the same temperature. For $T \to 0$, a concentration difference of one per cent leads to a difference in height of about 20 cm.

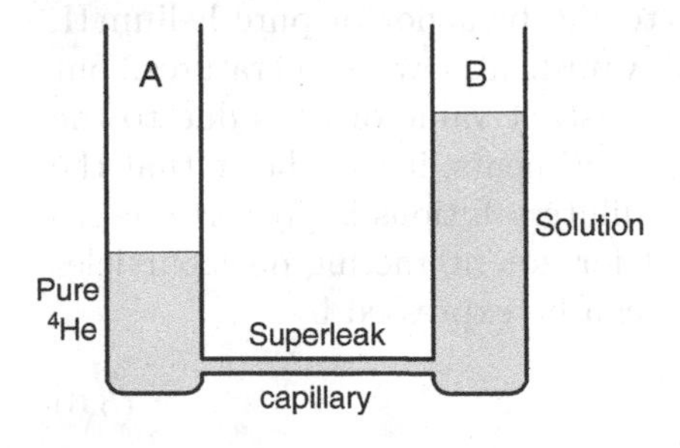

Fig. 5.8. Schematic illustration of an experimental arrangement used to demonstrate the osmotic pressure buildup between two vessels that are connected via a superleak and contain ^{3}He/^{4}He mixtures with different ^{3}He concentrations

The osmotic pressure Π, that ^{3}He atoms produce in helium II, can be described at high temperatures in the classical limit $T \gg T_F$ by the *Van't Hoff law*. Analogous to the ideal gas law, one can write

$$\Pi = n_3 k_B T \propto c_3 T . \tag{5.7}$$

For a degenerate Fermi gas at $T \ll T_F$, the osmotic pressure can be calculated directly from the internal energy given by (3.12). Using the relation $p = -(\partial U/\partial V)_{S,N}$ one obtains

$$\Pi = \frac{2}{5} n_3 k_B T_F \propto c_3^{5/3} = \text{const.} \tag{5.8}$$

Figure 5.9 shows the osmotic pressures of dilute mixtures below 250 mK. For the mixtures investigated, the transition from the classical behavior to that of a degenerate Fermi gas occurs in the temperature range shown in this graph. In addition, the expected strong dependence of the osmotic pressure on the ^{3}He concentration is observed.

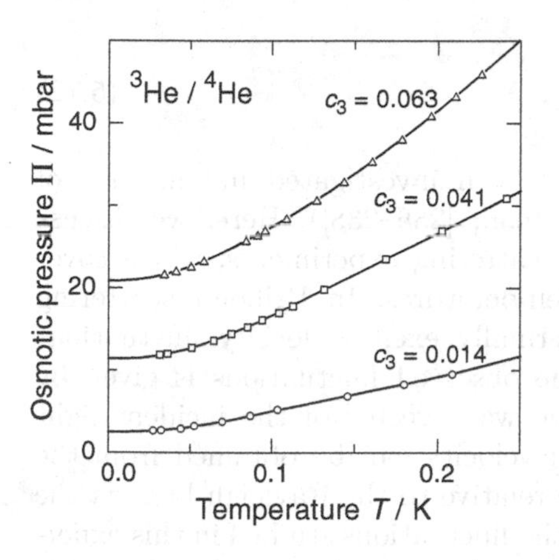

Fig. 5.9. Osmotic pressure of three dilute ^{3}He/^{4}He mixtures with different ^{3}He concentrations as a function of temperature [234]

5.3 Sound Propagation

Without going into the details of the hydrodynamics of mixtures of ^{3}He and ^{4}He we will make some brief remarks on the sound propagation in these mixtures.[1] As we have seen before, the ^{3}He atoms contribute to the normal-fluid component. The properties of first and second sound in mixtures of ^{3}He and ^{4}He depend therefore on the ^{3}He concentration c_3. In the absence of dissipation, sound propagation in ^{3}He/^{4}He mixtures can be described, analogous to (2.26) and (2.27), by three coupled differential equations:

[1] A nice discussion of the hydrodynamics of ^{3}He/^{4}He mixtures can be found in Khalatnikov's book [235].

$$\frac{\partial^2 \varrho}{\partial t^2} = \nabla^2 p , \tag{5.9}$$

$$\frac{\partial^2 S}{\partial t^2} = \frac{\varrho_s S^2}{\varrho_n} \left[\nabla^2 T + \frac{c_3}{S} \nabla^2 (\mu_3 - \mu_4) \right] , \tag{5.10}$$

$$\frac{\partial c_3}{\partial t} = \frac{c_3}{S} \frac{\partial S}{\partial t} . \tag{5.11}$$

Here, the quantities μ_3 and μ_4 denote the chemical potentials of ^{3}He and ^{4}He atoms in the solution. Using this set of equations, first- and second-sound modes in ^{3}He/^{3}He mixtures can be derived in a similar way to the derivation for pure ^{4}He that was presented in Sect. 2.2.8.

5.3.1 First Sound

In the dissipationless regime, the velocity of first sound in ^{3}He/^{4}He mixtures is given by

$$v_1^2 = \left(\frac{\partial p}{\partial \varrho} \right)_{S, c_3} \left[1 + \frac{\varrho_s}{\varrho_n} \left(\frac{\partial \varrho}{\partial c_3} \frac{c_3}{\varrho} \right)^2 \right] . \tag{5.12}$$

The propagation of first sound has been investigated in mixtures of ^{3}He and ^{4}He using different methods (e.g., [236–238]). Here, we discuss data obtained at 600 MHz in Brillouin-scattering experiments, which cover a wide range of concentrations and temperatures. In Brillouin-scattering experiments, laser light couples to thermally excited density fluctuations in the liquid. The wave vector q of the observed fluctuations is given by $q = 2k_0 \sin(\Theta/2)$, where k_0 denotes the wave vector of the incident light and Θ the scattering angle. The sound velocity can be obtained from the frequency shift $\Delta\omega$ of the Brillouin line relative to the Rayleigh line via the relation $v_1 = \Delta\omega/q$. The wavelength of the fluctuations studied in this experiment was about 350 nm. In the temperature range 0.4 K to 1.8 K, where this experiment was carried out, this wavelength is much longer than the mean free path of the thermal excitations. Therefore, one can use the hydrodynamic equations of motion to describe this experiment.

Figure 5.10 shows the temperature dependence of the velocity of first sound for mixtures with different ^{3}He concentration derived in Brillouin-scattering experiments. With increasing temperature the sound velocity in helium II decreases, because of the rising density of phonons and rotons. This general temperature dependence is also found in mixtures of ^{3}He and ^{4}He. However, the presence of the ^{3}He adds to the excitation spectrum and thus lowers the velocity of first sound with increasing ^{3}He concentration.

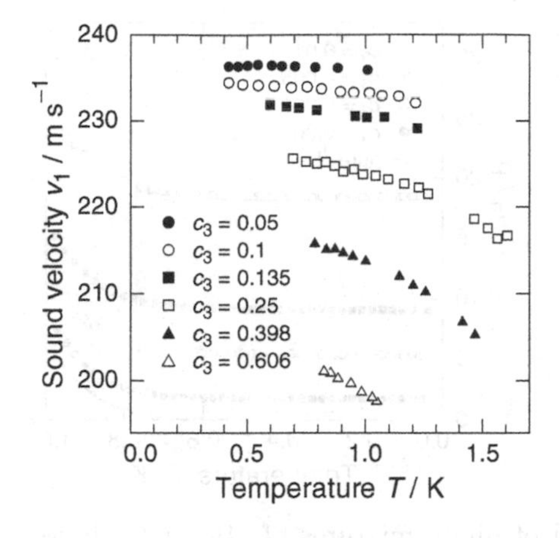

Fig. 5.10. Velocity of first sound in mixtures of ^{3}He and ^{4}He as a function of temperature measured by Brillouin scattering [237]

5.3.2 Second Sound

Second sound in dilute ^{3}He/^{4}He mixtures was predicted by *Pomeranchuk* in 1949 [239]. The first experimental observation of the propagation of second sound in mixtures of ^{3}He and ^{4}He was made by *Lynton* and *Fairbank* in 1950 [215].

Measurements of second sound in ^{3}He/^{4}He mixtures have been used to determine the normal-fluid component. As discussed before, the ^{3}He atoms contribute to the normal-fluid component. At sufficiently low temperatures the only contribution to the normal-fluid component is the ^{3}He content and therefore second-sound modes in dilute solutions are identical to concentration waves. In this case, second sound is equivalent to the propagation of first sound in an ideal ^{3}He gas. Since the variation of the ^{3}He concentration is, in turn, also a density variation, the coupling of first and second sound in mixtures of ^{3}He and ^{4}He is enhanced compared to the situation in helium II.

In the hydrodynamic regime, and in the absence of damping, the velocity of second sound is given by

$$v_2^2 = \frac{\varrho_s}{\varrho_n}\left[\overline{S}\left(\frac{\partial T}{\partial S}\right)_{\varrho,c_3} + c_3^2\frac{\partial(\mu_3-\mu_4)}{\partial c_3}\right]\left[1+\frac{\varrho_s}{\varrho_n}\left(\frac{\partial\varrho}{\partial c_3}\frac{c_3}{\varrho}\right)^2\right]^{-1}, \tag{5.13}$$

with

$$\overline{S} = S_{4,0} - \frac{k_B}{m_4}\left[c_3 + \ln(1-c_3)\right] + \frac{k_B}{m_3}c_3\,. \tag{5.14}$$

Figure 5.11a shows the result of a measurement of the velocity of second sound for two solutions with very low ^{3}He concentration, in comparison with the second-sound velocity in helium II. Clearly, at low temperatures, very small amounts of ^{3}He change the velocity of second sound drastically. At the lowest temperature, in this experiment of 40 mK, the velocity of second sound is only $v_2 \approx 10\,\mathrm{m\,s^{-1}}$ in a 1% solution, compared to about $130\,\mathrm{m\,s^{-1}}$ in helium II.

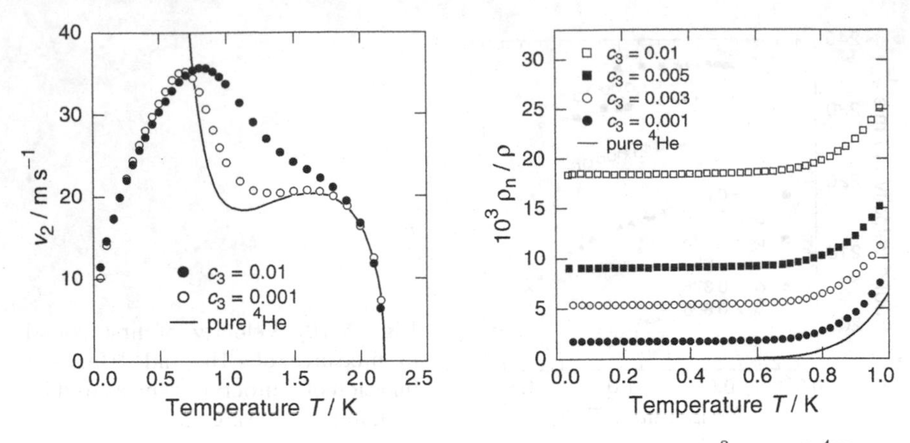

Fig. 5.11. (**a**) Velocity of second sound of dilute mixtures of ^{3}He and ^{4}He as a function of temperature [240]. The *solid line* represents the velocity of second sound in helium II. (**b**) Normal-fluid density of pure helium II (*solid line*) and dilute mixtures of ^{3}He and ^{4}He as a function of temperature [240]

Using (5.13) the normal-fluid component can be derived from the second-sound data. The resulting temperature dependence for mixtures with different ^{3}He concentration is shown in Fig. 5.11b. The normal-fluid component scales with the ^{3}He concentration at low temperatures, as expected. For comparison, the temperature dependence of ϱ_n/ϱ for pure helium II is also plotted as a *solid line*.

5.4 Transport Properties

A thorough treatment of the transport properties of ^{3}He/^{4}He mixtures is rather involved for high temperatures, because phonons, rotons and ^{3}He atoms all have to be considered. At sufficiently low temperatures, the problem reduces to that of interacting ^{3}He quasiparticles and allows us to use Fermi-liquid theory to describe the transport properties.

5.4.1 Heat Transport

Heat transport in ^{3}He/^{4}He mixtures is a rather complex phenomenon. At high temperatures, the roton and phonon contributions dominate the heat transport. In this case, a temperature gradient leads to motion of the normal-fluid component of helium II towards the colder end of the apparatus. The rotons and phonons interact with the ^{3}He atoms and drag them towards the cold end as well. This results in a concentration gradient, which in turn leads to a back diffusion of the ^{3}He atoms, which therefore leads to a strong reduction of the heat current.

At sufficiently low temperatures, a different mechanism of heat transport dominates in dilute solutions. The heat is not carried predominantly by phonons and rotons, but by ^{3}He quasiparticles. As we have seen in Sect. 3.1.3, the quasiparticle interaction limits the heat flow. It is remarkable that the cross-over between heat transport mediated by phonons and quasiparticles takes place at very low temperatures. For example, for a solution with 1% ^{3}He this transition occurs for $T < 25\,\mathrm{mK}$.

According to (3.17), the mean free path of a degenerate Fermi gas is given by $\ell = v_\mathrm{F}\tau \propto (T_\mathrm{F}/T)^2$. The heat capacity varies at low temperatures as $C \propto T/T_\mathrm{F}$ and thus we obtain the following expression for the thermal conductivity due to weakly interacting ^{3}He quasiparticles:

$$\Lambda = \frac{1}{3} C\, v_\mathrm{F}\, \ell \propto \frac{c_3}{T} \,. \tag{5.15}$$

Note that the Fermi velocity is given by $v_\mathrm{F} = (\hbar/m_3^*)(3\pi^2 n_3)^{1/3}$, containing the effective mass of the ^{3}He quasiparticles. According to (3.36) in the Landau model this quantity is related to the mass of a free ^{3}He atom by $m_3^* = (1 + F_1/3)m_3$. Figure 5.12 shows the results of thermal-conductivity measurements of dilute solutions of ^{3}He in liquid ^{4}He. For comparison, the data for pure ^{3}He are shown as well. At low temperatures, one indeed finds the expected $1/T$ dependence. One also can see that with rising concentration the thermal conductivity increases, as expected.

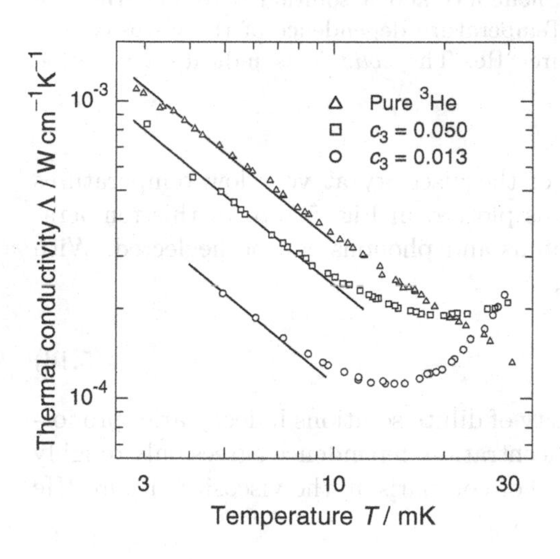

Fig. 5.12. Temperature dependence of the thermal conductivity of ^{3}He/^{4}He mixtures with $c_3 = 0.013$ and $c_3 = 0.05$. In addition, the corresponding data for pure ^{3}He are shown. The *straight solid lines* indicate the $1/T$ dependence [241]

5.4.2 Viscosity

As a further example of the influence of ^{3}He atoms on dynamical processes in dilute solutions, we briefly discuss the viscosity of the normal-fluid component. The viscosity η_n is proportional to the mean free path of the excitations in the liquid (see Sect. 2.2). The presence of the ^{3}He atoms leads to a reduction of the mean free path of the excitations in helium II, because the ^{3}He atoms act as scattering centers for rotons and phonons. Correspondingly, the viscosity η_n of dilute solutions is greatly reduced compared to that of helium II. The result of a measurement using a rotary viscosimeter for a 1% solution is shown in Fig. 5.13a.

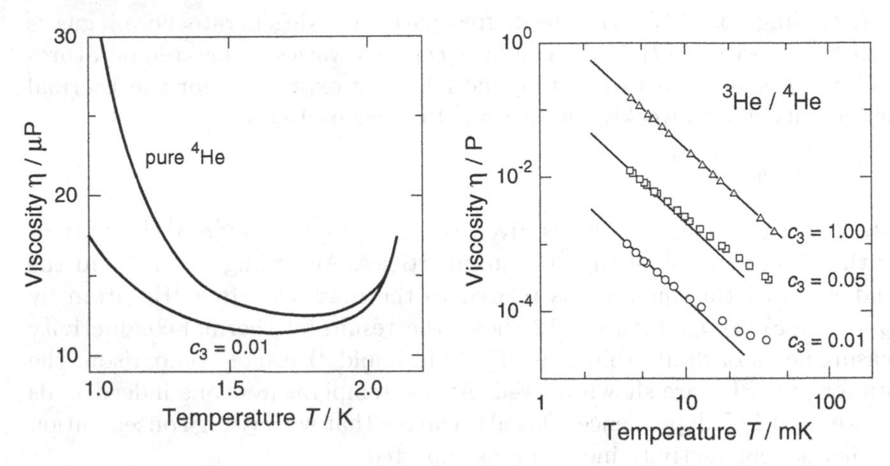

Fig. 5.13. (**a**) Viscosity η of pure helium II and a solution with 1% ^{3}He as a function of temperature [242]. (**b**) Temperature dependence of the viscosity η of dilute ^{3}He/^{4}He mixtures and of pure ^{3}He. The *solid lines* indicate a variation proportional to T^{-2} [243]

The temperature dependence of the viscosity at very low temperatures obtained with a torsion oscillator is plotted in Fig. 5.13b. In this temperature range, the contribution of rotons and phonons can be neglected. With $v_F^2 \propto c_3^{2/3}$ and $\varrho \propto c_3$, one expects

$$\eta = \frac{1}{3}\varrho v_F^2 \tau \propto \frac{c_3^{5/3}}{T^2} . \tag{5.16}$$

As seen from Fig. 5.13b, the viscosity of dilute solutions indeed varies proportional to $1/T^2$. In contrast, the concentration dependence agrees only roughly with the one expected from (5.16). For comparison, the viscosity of pure ^{3}He is also shown in Fig. 5.13b.

5.4.3 Self-Diffusion Coefficient

As we have seen in Sect. 3.1.3, the self-diffusion coefficient D_s is closely related to the viscosity. Figure 5.14 shows the temperature dependence of D_s for two ^{3}He/^{4}He mixtures with different ^{3}He concentration. For comparison, the data for pure ^{3}He are also plotted. At high temperatures ($T > 100\,\mathrm{mK}$),

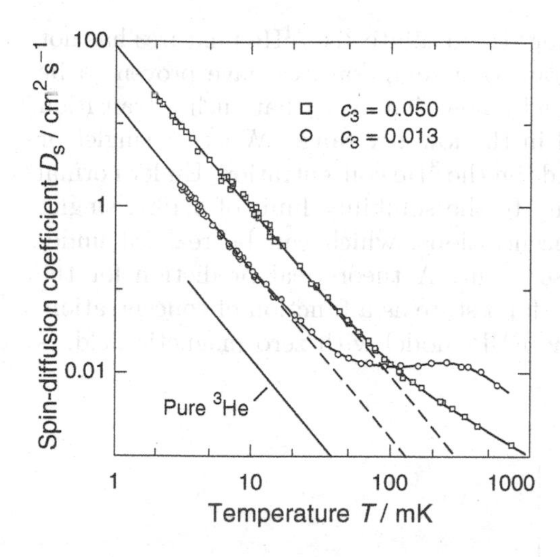

Fig. 5.14. Temperature dependence of the self-diffusion coefficient of ^{3}He/^{4}He mixtures with concentrations $c_3 = 0.013$ and $c_3 = 0.05$ [224]. The data for pure ^{3}He are shown for comparison

the ^{3}He atoms in the mixture behave like a classical gas and the self-diffusion coefficient decreases with increasing ^{3}He concentration. At low temperatures, one expects, according to the Landau model,

$$D_s = \frac{1}{3}\tau_D v_F^2 \left(1 + \frac{1}{4G_0}\right) \propto \frac{c_3^{2/3}}{T^2} . \tag{5.17}$$

This means that in this temperature range the self-diffusion coefficient of ^{3}He atoms in mixtures increases with increasing ^{3}He concentration. The data shown in Fig. 5.14 nicely reflect the transition from the classical behavior to that of a Fermi liquid. The self-diffusion coefficient of pure ^{3}He is also proportional to $1/T^2$ at low temperatures, as we have seen in Sect. 3.1.3. However, the absolute value of D_s for pure ^{3}He is substantially lower than that of dilute mixtures. The origin of this effect is the value of the parameter G_0 that is negative for pure ^{3}He (see Table 3.2). In contrast, for the mixtures shown in Fig. 5.14, one finds roughly $G_0 \approx 0.35$.

5.5 Search for a Superfluid Phase of ^{3}He in Mixtures

There are a number of very interesting questions related to the possibility of superfluidity of ^{3}He in mixtures of ^{3}He and ^{4}He. Such a system gives us a unique opportunity to study a mixture of two superfluids. In addition, the much weaker interaction between the ^{3}He atoms in solution may result in a different realization of Cooper pairs compared to those in pure ^{3}He.

Soon after the publication of the BCS theory, it was speculated that there might exist a possible paired superfluid state in pure ^{3}He, and also for ^{3}He in dilute mixtures of ^{3}He/^{4}He. As we have seen in the previous chapter, in pure ^{3}He a transition into a superfluid p-wave state was discovered in

1972. A corresponding transition for ^{3}He in dilute ^{3}He/^{4}He mixtures has not yet been observed. The precise transition temperatures have proven to be difficult to calculate. It seems clear by now, however, that such a transition in zero magnetic field is expected in the low μK range. Whether singlet or triplet pairing should occur, depends on the ^{3}He concentration. Under normal pressure, for ^{3}He concentrations up to the solubility limit of 6.48%, singlet pairing is favored. At higher concentrations, which can be realized under pressure, triplet paring might also occur. A theoretical prediction for the transition temperature into a superfluid state as a function of concentration, calculated in the framework of the BBP model with zero magnetic field, is shown in Fig. 5.15.

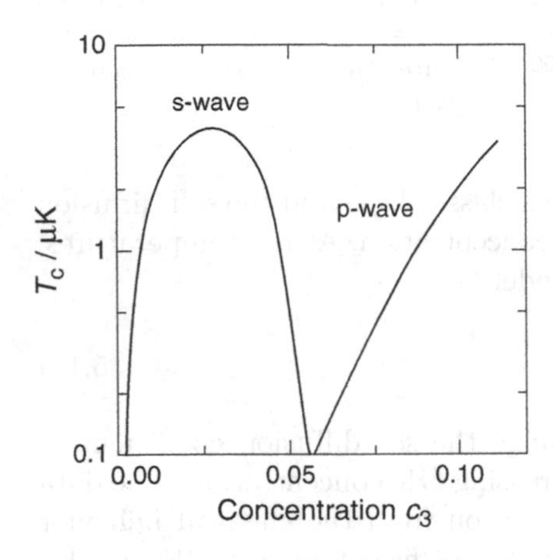

Fig. 5.15. Calculated transition temperatures into a superfluid state of ^{3}He in dilute ^{3}He/^{4}He solutions. After [244]

Note that somewhat higher transition temperatures are predicted for spin-polarized ^{3}He/^{4}He mixtures. However, these temperatures are still lower than those obtained in experiments with liquid helium so far. The lowest temperatures to which mixtures of ^{3}He and ^{4}He have been cooled are just below 100 μK. The difficulty of cooling ^{3}He/^{4}He solution arises mainly from the thermal boundary resistance between the solid container and the mixture. We will discuss the origin of this so-called Kapitza resistance in Sect. 11.4.3. To enhance the thermal flow, large contact areas have been used, which were realized by covering the walls with thick layers of sintered silver. However, the occurrence of various heat leaks prevented the cooling of ^{3}He/^{3}He solutions to temperatures much below 100 μK so far. Nevertheless, the minimum temperature reached by ^{3}He/^{4}He mixtures has been constantly improved in the last 25 years, giving hope that one day superfluidity of ^{3}He in dilute ^{3}He/^{4}He solutions will be discovered. For a review of the experimental efforts to cool ^{3}He/^{4}He mixtures to ultralow temperatures we refer the reader to the recent book by *Dobbs* [143].

Exercises

5.1 Estimate the Fermi temperature for pure ^{3}He and ^{3}He/^{4}He mixtures with concentrations of 1%, 6.5% and 10%.

5.2 A mixture of 10% ^{3}He in ^{4}He is cooled to 10 mK starting from 1.5 K. Calculate the heat capacity at 100 mK and 10 mK.

5.3 At what temperature does the ^{3}He-rich phase of ^{3}He/^{4}He mixtures contain 10 ppm of ^{4}He?

5.4 Calculate the osmotic pressure for a ^{3}He/^{4}He mixture with 5% ^{3}He for $T \rightarrow 0$.

Part II

Solids at Low Temperatures

6 Phonons

Systematic studies of the low-temperature properties of insulating solids were not carried out prior to the end of the 19th century. In the first experiments the specific heat was measured by *Weber*, *Behn* and *Dewar* [245–247]. Surprisingly, the specific heat was not constant but decreased rapidly on cooling. Such behavior was incomprehensible because it violated the well-established equipartition law [248]. In his fundamental paper of 1906, *Einstein* assumed that the vibrations of N atoms in a solid can be represented by $3N$ harmonic oscillators having identical frequency [249]. He then quantized the energy of the oscillators in an approach similar to that used by *Planck* when developing his theory of blackbody radiation. In this way, the general variation of the specific heat with temperature could be described correctly. However, later it turned out that the Einstein model did not lead to an adequate description of the specific heat in the low-temperature limit. It was obvious that the assumption of uncoupled oscillators is an oversimplification since the motion of adjacent atoms can hardly be considered to be independent [250]. Models taking into account the coupling between the atoms were proposed in 1912 by *Born* and *v. Kármán* [251], and by *Debye* [252] as well. Interatomic coupling causes an extension of the vibrational spectrum towards lower frequencies. In their description, Born and v. Kármán started from the discrete atomic structure and the intermolecular forces acting between the atoms, and derived their vibrational spectrum. In contrast, Debye did not consider the vibrations of individual atoms at all but rather the (collective) vibrations of the entire lattice. The energy of the resulting vibrational states turned out to be quantized. In analogy to photons, the quanta of the electromagnetic field, the quanta of the elastic field are called *phonons*.

In this chapter, we use the concept of phonons to describe the specific heat and the heat transport in insulating solids. In addition, we consider other selected low-temperature phenomena that can be understood in terms of the phonon concept.

6.1 Specific Heat – Debye Model

In this section, we briefly introduce the Debye model, which is a simple but elegant approach to describe the specific heat of solids. In this model, solids are treated as homogeneous and isotropic media. Consequently, the phonons can be considered as a gas of free bosons. In contrast to the atoms of a real gas, the number of phonons is not conserved. As a result, their chemical potential vanishes, i.e., $\mu = 0$. Though phonons are not real particles, we may use the density of states $\mathcal{D}(q)$ of free particles (see Sect. 3.1) to calculate the internal energy of solids. There is a simple reason for this correspondence: in both cases the allowed wave vectors $\boldsymbol{q}$ are determined by the geometrical boundary conditions (3.4). Hence, the density of phonons in momentum space is the same as for free particles. The dispersion relation, which depends on the nature of the particles being considered, does not play a role.

We rewrite (3.6) using the abbreviation $\boldsymbol{q}$ for the wave vector of phonons and find $\mathcal{D}(q) = q^2V/2\pi^2$. From this expression, the density of states $\mathcal{D}(\omega)$ in frequency space follows directly from the relation

$$\mathcal{D}(\omega) = \mathcal{D}(q)\frac{\mathrm{d}q}{\mathrm{d}\omega}\,, \tag{6.1}$$

if the dispersion relation $\omega(q)$ is known. In the Debye model, the linear relation $\omega = vq$ is assumed, where v represents the velocity of sound. With this relation the famous ω^2 dependence of the *Debye density of states* follows directly from (6.1):

$$\mathcal{D}(\omega)\,\mathrm{d}\omega = \frac{V\omega^2}{2\pi^2 v^3}\,\mathrm{d}\omega\,. \tag{6.2}$$

In monatomic solids there exist three phonon branches, namely a longitudinal and two transverse branches. In the Debye model, this fact is taken into account by introducing the *Debye velocity* v_D through the relation

$$\frac{3}{v_\mathrm{D}^3} = \frac{1}{v_\ell^3} + \frac{2}{v_\mathrm{t}^3}\,. \tag{6.3}$$

So, we finally obtain for the Debye density of states the expression

$$\mathcal{D}(\omega) = \frac{V\omega^2}{2\pi^2}\left(\frac{1}{v_\ell^3} + \frac{2}{v_\mathrm{t}^3}\right) = \frac{V\omega^2}{2\pi^2}\,\frac{3}{v_\mathrm{D}^3}\,. \tag{6.4}$$

For a rough estimate of v_D we may put $v_\ell \approx 2\,v_\mathrm{t}$ resulting in $v_\mathrm{D} \approx 1.12\,v_\mathrm{t}$, meaning that the Debye velocity and hence the Debye density of states is mainly determined by transverse sound waves.

The density of states is normalized to $3N$, the total number of modes, by introducing the cutoff frequency ω_D, i.e.,

$$3N = \int_0^{\omega_\mathrm{D}} \mathcal{D}(\omega)\,\mathrm{d}\omega\,. \tag{6.5}$$

Through this integral, the *Debye frequency* ω_D is defined leading to the expression

$$\omega_D = v_D \sqrt[3]{\frac{6\pi^2 N}{V}} = \frac{v_D}{a}\sqrt[3]{6\pi^2}\,. \tag{6.6}$$

The last term holds for simple cubic crystals with the lattice constant a, because in this case $N/V = a^3$. In Fig. 6.1 the dispersion relation used in the Debye model is compared with that of a linear chain. Obviously, the maximum values of the wave vector are different in these two models. In the case of a linear chain $q_m = \pi/a$ is determined by the size of the Brillouin zone, whereas for the Debye model $q_D = \sqrt[3]{6\pi^2}/a \approx 3.9/a$ follows from (6.6). Typical numerical values are: $a = 0.2\,\text{nm}$, $v_D = 3000\,\text{m/s}$, and hence $\omega_D \approx 10^{14}\,\text{s}^{-1}$.

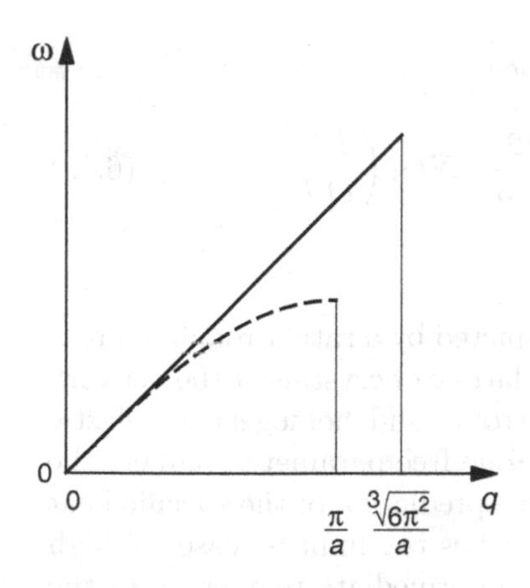

Fig. 6.1. Comparison between the linear dispersion relation of the Debye model (*full line*) and the 'real' dispersion of a linear chain (*dashed line*). Note that the dispersion curves end at different values of the wave vector

The internal energy of the lattice vibrations is given by

$$U(T) = \int_0^{\hbar\omega_D} \hbar\omega\, D(\omega)\, f(\omega,T)\, d\omega\,. \tag{6.7}$$

Inserting the Bose–Einstein distribution function $f(\omega,T)$ and using the abbreviations $x = \hbar\omega/k_B T$ and $x_D = \hbar\omega_D/k_B T$, we find for the specific heat[1] of dielectric crystals the expression

$$C_V = \frac{\partial U}{\partial T} = 9Nk_B\left(\frac{T}{\Theta}\right)^3 \int_0^{x_D} \frac{x^4 e^x}{(e^x-1)^2}\, dx\,, \tag{6.8}$$

where the *Debye temperature* Θ is defined by $k_B\Theta = \hbar\omega_D$.

[1] In general, we do not explicitly distinguish between heat capacity, specific heat, molar specific heat, etc., because in most cases the exact meaning follows from the context in which it is used.

Before we make comparisons between theory and experiment more carefully, let us first discuss three special cases:

- For $T \to \infty$, the variable $x \to 0$, and therefore

$$\lim_{x\to 0} \int_0^{x_\mathrm{D}} \frac{x^4\, \mathrm{e}^x}{(\mathrm{e}^x - 1)^2} \mathrm{d}x \approx \int_0^{x_\mathrm{D}} \frac{x^4 \cdot 1}{x^2} \mathrm{d}x = \frac{x_\mathrm{D}^3}{3} = \frac{1}{3} \left(\frac{\Theta}{T}\right)^3 . \quad (6.9)$$

Thus, in agreement with the law of *Dulong* and *Petit* [248], we obtain

$$C_V = 3Nk_\mathrm{B} . \quad (6.10)$$

- For $T > \Theta$, we find by series expansion

$$C_V = 3Nk_\mathrm{B} \left[1 - \frac{1}{20}\left(\frac{\Theta}{T}\right)^2 + \ldots\right] . \quad (6.11)$$

- For $T \to 0$, the variable $x_\mathrm{D} \to \infty$, and hence

$$C_V = 9Nk_\mathrm{B} \left(\frac{T}{\Theta}\right)^3 \underbrace{\int_0^{\infty} \frac{x^4 \mathrm{e}^x}{\left(\mathrm{e}^x - 1\right)^2} \mathrm{d}x}_{4\pi^4/15} = \frac{12\pi^4}{5} Nk_\mathrm{B} \left(\frac{T}{\Theta}\right)^3 . \quad (6.12)$$

The vibrational states of real solids are depicted by a rather rough approximation. In the Debye model, the anisotropic lattice of crystals or the network of amorphous substances is replaced by isotropic and homogeneous elastic continua. Although this model contains only one free parameter, namely the Debye temperature Θ, it provides a reasonable prediction of the specific heat over a wide temperature range. First, it describes the limiting cases of high and low temperatures correctly, but even at intermediate temperatures the agreement with experiment is surprisingly good.

As an example of the remarkable agreement between the Debye model and experimental data in the low-temperature region, we show in Fig. 6.2 data on the specific heat of solid argon. Indeed, the experimental curve follows the T^3-dependence predicted by the model extremely well. In addition, the magnitude of the Debye temperature Θ, and therefore also the magnitude of C_V, can be calculated without additional assumptions from the elastic data, and good agreement with experiments is found.

It should be pointed out that the density of states (6.4) is an over-simplified model of the real density of states. This fact is demonstrated by Fig. 6.3 where the density of states of diamond is plotted and compared with the Debye model. At low frequencies, the phonon spectrum rises proportional to ω^2 as formulated in the Debye model. But at higher frequencies, the real spectrum exhibits *Van Hove singularities*, where the spectrum changes discontinuously. So, even though the Debye model is successful in describing the broad features of many specific heat data across a wide temperature range,

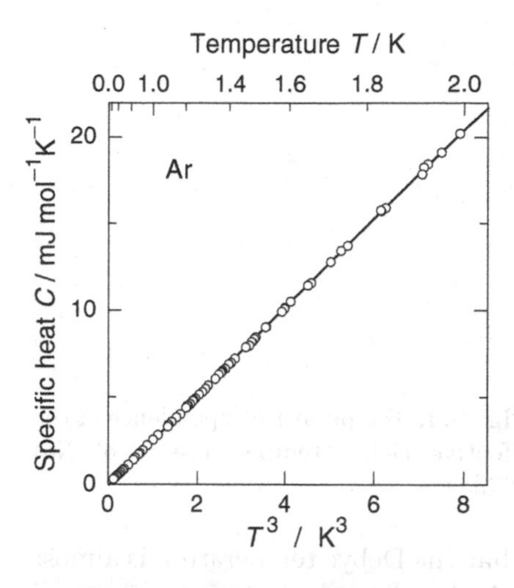

Fig. 6.2. Specific heat of solid argon at temperatures below 2 K plotted versus T^3. The *full line* reflects the prediction of the Debye model with the Debye temperature $\Theta = 92\,\mathrm{K}$ [253]

the comparison between the two spectra makes it clear that the specific heat is rather insensitive to the shape of the density of states. It is determined by the mean behavior of the vibrational states and not by details of the vibrational spectrum.

At what temperatures do we expect noticeable deviations of the experiment from the Debye approximation? At 'very low' temperatures, i.e., for $T < \Theta/100$ only phonons with long wavelength are thermally excited, the dispersion of which is linear. In this case, the density of states of crystalline solids is proportional to ω^2, as assumed in the Debye model. Similarly, at high temperatures ($T > \Theta/4$) details of the phonon spectrum become unimportant and the prediction of the Debye model approaches the classical limit. In the intermediate range significant deviations may exist, but even so, very often the Debye approximation is used. Deviations of the temperature dependence of the specific heat from (6.8) are expressed in terms of Θ varying with temperature. As an example of this procedure, $\Theta(T)$ is shown in Fig. 6.4 for NaI.

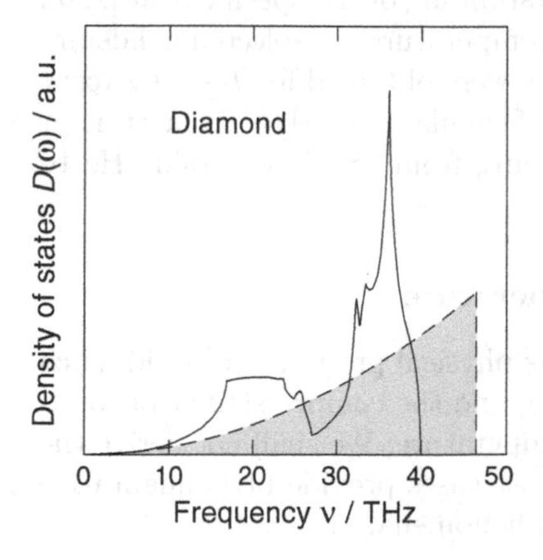

Fig. 6.3. Density of states $\mathcal{D}(\omega)$ of diamond. For comparison, the density of states in the Debye approximation with $\Theta = 2230\,\mathrm{K}$ is also drawn [254]

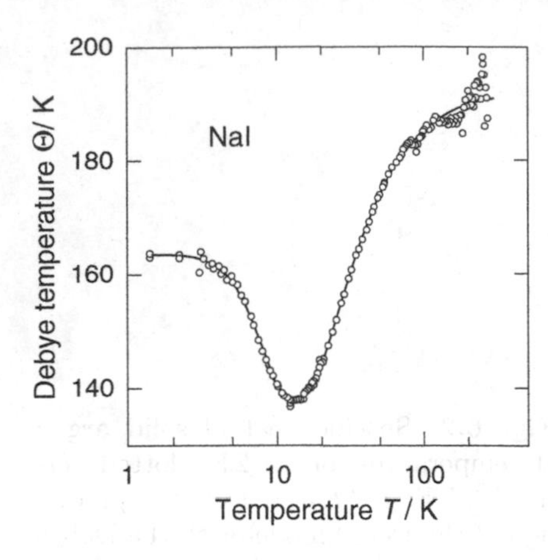

Fig. 6.4. Temperature dependence of the effective Debye temperature Θ of NaI [255]

It is only at low and high temperatures that the Debye temperature is almost constant. In the intermediate range, relatively strong deviations are observed that are due to the specific shapes of the density of states of different materials. In addition, it should be noted that the Debye temperatures deduced in the limit of low and high temperature do not coincide. In order to improve the Debye approximation, the real dispersion, or at least the Van Hove singularities, need to be taken into account. Furthermore, the contribution of the optical phonons must be treated separately.

As already mentioned, in 1912 an alternative calculation of the specific heat of solids was proposed by *Born* and *v. Kármán*. They started from the crystalline atomic structure and the forces acting between the atoms. In fact, with these calculations the authors were laying down the foundations for lattice dynamics. Their approach is mathematically demanding, but in 1912 the agreement between calculation and experimental data was still poorer than the agreement with Debye's theory. Therefore, for a long time this model was 'unpopular'. This attitude changed with the work by *Blackman*, who succeeded in deriving the density of states within the framework of lattice dynamics thereby leading to a better understanding of the specific heat [256].

At the end of this section, the Debye temperatures of selected solids are listed in Tables 6.1 and 6.2. All these values were obtained for $T \rightarrow 0$, except for those substances where the chemical formula is marked by a star. A remarkably wide span is covered by Θ, going from 18.8 K for solid ^{3}He to more than 2000 K for diamond.

6.1.1 Significance of the Debye Temperature

In this section, we will describe how various physical properties of solids such as thermal expansion, melting temperature, and the compressibility of solids can be expressed in terms of the Debye temperature. We shall not derive the relevant mathematical relations that in most cases provide little qualitative understanding of the underlying physical phenomena.

Table 6.1. Debye temperature Θ of various elements in the limit $T \to 0$. The Debye temperature of elements marked by (*) was determined at $T \approx \Theta/2$. After [257,258]

Element	Θ (K)	Element	Θ (K)	Element	Θ (K)	Element	Θ (K)
Ar	92	Cu	347	Mn	409	Sc	346
Ac*	100	Er	118	Mo	423	Se	152
Ag	227	Fe	477	N*	70	Si	645
Al	433	Ga	325	Na	156	Sm	169
Am	121	Gd	182	Nb	276	Sn	199
As	282	Ge	373	Nd	163	Sr	147
Au	162	H (para)	122	Ne	75	Ta	245
B	1480	H (orth)	114	Ni	477	Tb	176
Ba	111	^{3}He	19–33	Np	259	Te	152
Be	1481	Hf	252	O*	90	Th	160
Bi	120	Hg	72	Os	467	Ti	420
C (Dia.)	2250	Ho	190	Pa	185	Tl	78
C (Gra.)	413	I	109	Pb	105	Tm	200
Ca	229	In	112	Pd	271	U	248
Cd	210	Ir	420	Pr	152	V	399
Ce	179	K	91	Pt	237	W	383
Cl*	115	Kr	72	Rb	56	Xe	64
Cm	123	La	145	Rc	416	Y	248
Co	460	Li	344	Rh	512	Yb	118
Cr	606	Lu	183	Ru	555	Zn	329
Cs	40	Mg	403	Sb	220	Zr	290

Table 6.2. Debye temperature Θ of several compounds in the limit $T \to 0$. The Debye temperature of elements marked by (*) was determined at $T \approx \Theta/2$. After [258–260]

Compound	Θ (K)	Compound	Θ (K)	Compound	Θ (K)
$AgBr^*$	140	$Cr_2Cl_3^*$	360	MgO^*	800
$AgCl^*$	180	FeS_2^*	630	MoS_2^*	290
$As_2O_3^*$	140	KBr	173	RbBr	131
$As_2O_5^*$	240	KCl	235	RbCl	165
$AuCu_3$	285	KI	131	RbI	103
BN^*	600	InSb	206	SiO_2 (Quartz)	470
CaF_2	508	LiF	736	TiO_2^* (Rutile)	450
$CrCl_2^*$	80	LiCl	422	ZnS	315

Elastic Properties In the low-temperature limit $T \to 0$, the correlation between thermal and elastic properties of dielectrics becomes very simple. The frequency of thermally excited phonons tends to zero, i.e., thermal phonons and ordinary sound waves become identical. Therefore, the calorimetric Debye temperature $\Theta_{\mathrm{cal}} \equiv \Theta$ and the value Θ_{ela} calculated from elastic data, should be equal. As can be seen from Table 6.3, the two types of Debye temperatures agree remarkably well. This agreement is a substantial success of the Debye model.

Table 6.3. Comparison between Θ_{cal} and Θ_{ela} for $T \to 0$. After [260]

	Sn	Cu	Zn	Al	Ag	Ge	Pb	LiF	CaF_2
Θ_{cal} (K)	200	344	337	428	225	371	107	736	508
Θ_{ela} (K)	201	345	327	431	226	374	105	734	514

Compressibility As noted by *Madelung* and *Einstein*, a relation between the compressibility and the Debye temperature can be derived for simple solids. According to *Blackman*, for ionic crystals with sodium chloride structure, the compressibility κ is related to the Debye temperature Θ_κ via the expression [261]

$$\Theta_\kappa = \frac{\hbar}{k_{\mathrm{B}}}\sqrt{\frac{5\,a}{2m^*\kappa}} \propto \kappa^{-1/2}\,, \tag{6.13}$$

where a represents the lattice constant and m^* the reduced mass of the ions. An equivalent expression relating Θ_κ and κ but with a different numerical coefficient is found for crystals with cesium chloride structure.

Melting Temperature To a very rough approximation, the Debye temperature can be linked with the melting temperature T_{m} through the relation

$$\Theta_{\mathrm{m}} = B\sqrt{\frac{T_{\mathrm{m}}}{\overline{M}\,\overline{V}_{\mathrm{A}}^{2/3}}}\,, \tag{6.14}$$

using *Lindemann's* simple picture of the melting of a solid [262]. Here, $\overline{M}$ stands for the mean atomic mass and $\overline{V}_{\mathrm{A}}$ for the mean atomic volume. The constant B depends on the crystal type and is about $1 \times 10^{-21}\,\mathrm{kg\,m^2\,K^{1/2}}$ for cubic crystals.

Thermal Expansion Specific heat and the coefficient β of thermal expansion are linked via the *Grüneisen relation*

$$\beta = \frac{\gamma\kappa C_V}{V}\,, \tag{6.15}$$

where $\gamma = -\,\partial(\ln\omega)/\partial(\ln V)$ is the *Grüneisen parameter*. It is assumed to be frequency independent and is a measure of the elastic nonlinearities of solids. From the temperature dependence of the thermal expansion, Θ_β can be deduced through C_V, and in many cases rather good agreement with the calorimetric values of Θ is found. But this agreement has to be treated with caution because, in general, different phonon branches i exhibit different values γ_i of the Grüneisen parameter. This can be taken into account by replacing the right-hand side of (6.15) with a sum over the parameters γ_i weighted with the specific heat C_V^i of the individual branches.

Infrared Data The Debye temperature of ionic crystals can be estimated from the wavelength of their 'Reststrahlen', because at high temperatures the main contribution to the specific heat arises from optical phonons. Consequently, their vibrational energy can be approximated by $k_B\Theta$. The Debye temperatures deduced from the reflection maximum or the transmission minimum in the infrared are in good agreement with the calorimetric values of Θ.

Electrical Conductivity As we shall see in Sect. 7.2, the low-temperature variation of the electrical conductivity σ of metals is given by the Bloch–Grüneisen relation

$$\sigma \propto \left(\frac{T}{\Theta_r}\right)^5 \int_0^{\Theta_r/T} \frac{x^5 e^x \, dx}{(e^x - 1)^2} \,. \tag{6.16}$$

The good agreement between experimental results and this prediction is demonstrated in Fig. 7.8. However, a closer inspection shows that Θ_r deduced from such measurements does, in general, not agree very well with the calorimetric Θ. This discrepancy demonstrates that the different phonon branches contribute differently to the two phenomena. The electrical conductivity is predominantly limited by the interaction between conduction electrons and longitudinal phonons, whereas the low-temperature specific heat is dominated by transverse phonons.

Debye–Waller Factor The atomic vibrations affect the reflection of X-rays or other radiations of wavelength comparable with the atomic spacing. The reduction of the intensity I of the Bragg reflections for a monatomic solid depends on temperature in the form $I = I_0 \exp(-2W)$ where the Debye–Waller factor W is related to the Debye temperature through the relation $W \propto (T/\Theta_{DW})^2$. The values of Θ_{DW} deduced from Mössbauer, X-ray or neutron-scattering experiments are in fair agreement with the calorimetric values.

6.1.2 Specific Heat of Finite-Size Systems

Until now we have only discussed the behavior of uniform three-dimensional samples. Since the exact vibrational spectrum depends on the actual shape of the specimen, the specific heat will also be influenced by the sample geometry. As examples, we consider here the specific heat of two-dimensional atomic layers and of fine powders.

Adsorbed Gases

As we have already mentioned, the wave vectors of phonons that are allowed in insulating solids are determined by the boundary conditions that, in turn, depend on the dimensionality d of the system. Within the Debye approximation, the relation

$$D(\omega) \propto \omega^{d-1} \tag{6.17}$$

is found for the phonon density of states. Thus, the proportionality $C_V \propto T^d$ should hold in the low-temperature limit. In particular, for two-dimensional systems, we expect a quadratic temperature dependence of the specific heat, i.e., $C_V \propto T^2$.

As an example of a two-dimensional system, we consider briefly monolayers of helium atoms adsorbed on graphite. The helium atoms are relatively tightly bound to the surface by Van der Waals forces, but motion parallel to the surface is possible. At higher temperatures and small surface densities n, the adsorbed atoms behave like a two-dimensional classical gas exhibiting a specific heat $C_V \approx Nk_B$. With rising vapor pressure the number of adsorbed atoms increases. For $n > 0.078\,\text{Å}^{-2}$, ^{3}He atoms form a two-dimensional solid. On increasing the coverage further the solid becomes more rigid because of the rising interaction between the atoms in the layer. As a consequence, the stiffness of the layer, and hence the Debye temperature increases. The specific heat of ^{3}He layers with different areal densities is shown in Fig. 6.5. As expected, a variation proportional to T^2 is found at low temperatures. At higher temperatures, melting of the two-dimensional crystals occurs, leading to deviations from the quadratic temperature variation of C. Of course, similar effects are also observed with other gases and other substrates.

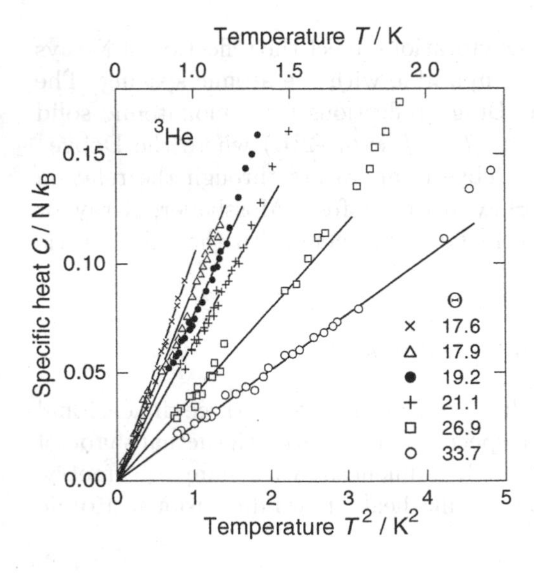

Fig. 6.5. Specific heat of ^{3}He adsorbed on graphite as a function of T^2. Starting from the left side, the density of helium atoms per Å^2 is given by: 0.078, 0.079, 0.080, 0.082, 0.087 and 0.092. The *full lines* correspond to fits with the indicated Debye temperatures [263]

Powders

Let us now consider the specific heat of small particles or powders. So far, we have not taken into account the existence of surfaces for three-dimensional samples or, in other words, we have not yet considered finite-size effects. On the one hand, the restricted geometry of systems is reflected in discrete frequency values of the bulk modes. On the other hand, the deformation of the surface becomes important, leading to an enhancement of the density of states at intermediate frequencies. Depending on temperature, the specific heat can therefore either be reduced due to the discrete spectrum or enhanced due to the existence of surface modes.

In a simple approximation, the particles of a fine powder can be considered as elastically vibrating spheres with radius r and their spectrum can be calculated with the theory of elasticity. The lowest possible frequency of a vibrating sphere will be of the order of $\omega_0 \approx v/r$, where v is an appropriate mean value of the sound velocity. The value of ω_0 allows us to distinguish between two temperature ranges in which one of the two effects will be predominant.

In the low-temperature limit $T < \hbar\omega_0/k_\mathrm{B}$, only the lowest-lying mode has to be considered and the internal energy U of N identical small particles is given by $U = N\hbar\omega_0/[\exp(\hbar\omega_0/k_\mathrm{B}T) - 1]$. For the specific heat, we therefore find the expression:

$$C_V = \frac{\partial U}{\partial T} = Nk_\mathrm{B}\left(\frac{\hbar\omega_0}{k_\mathrm{B}T}\right)^2 \frac{\exp(\hbar\omega_0/k_\mathrm{B}T)}{[\exp(\hbar\omega_0/k_\mathrm{B}T) - 1]^2}$$
$$\approx Nk_\mathrm{B}\left(\frac{\hbar\omega_0}{k_\mathrm{B}T}\right)^2 \exp\left(-\frac{\hbar\omega_0}{k_\mathrm{B}T}\right) . \qquad (6.18)$$

The same result was obtained by Einstein in his theory of the specific heat.

If the temperature is 'high', i.e., if the condition $\hbar\omega_0/k_\mathrm{B} < T < \Theta$ holds, an asymptotic expansion of C_V can be made leading to

$$C_V = C_V^\mathrm{B} + a\frac{T^2}{r^2} + b\frac{T}{r^2} , \qquad (6.19)$$

where C_V^B represents the Debye specific heat of the bulk sample given by (6.12). The last two terms on the right side express the contribution of the surface modes. The positive constants a and b can be calculated [264].

The calculated specific heat of an ensemble of identical spherical particles is shown schematically in Fig. 6.6. Deviations from the Debye behavior are expected to occur in the vicinity of $T_0 = \hbar\omega_0/k_\mathrm{B}$. Below this temperature the specific heat is reduced because of the suppression of phonons of long wavelength. Above T_0 the surface causes additional vibrational states resulting in an increase of C_V. What is the temperature range where finite-size effects should become noticeable? With $v = 3 \times 10^3\,\mathrm{m\,s^{-1}}$ and $r = 5\,\mathrm{nm}$, we obtain $T_0 \approx 5\,\mathrm{K}$. This is a temperature easily accessible in experiments.

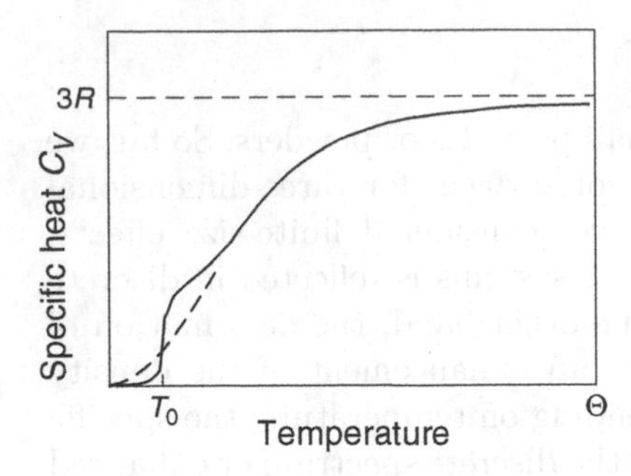

Fig. 6.6. Schematic temperature variation of the specific heat of small spheres. The Debye specific heat and the Dulong–Petit limit are represented by the *dashed lines*. The *full line* displays the calculation taking into account finite-size effects. After [265]

In fact, finite-size effects have been found in measurements of the specific heat of small particles. In all cases, an enhancement of C was observed that is larger than expected from the simple theoretical considerations outlined above. Moreover, to our knowledge no reduction of the vibrational specific heat below the bulk value is reported in the literature although the characteristic temperature T_0 was expected to lie within the accessible temperature range in several experiments. There are several possible explanations for this discrepancy. Most importantly, small particles never exhibit an ideal spherical shape but instead are of somewhat irregular polyhedron shape. In addition, surface atoms, depending on their location are bound with different strengths. In particular, corner atoms will exhibit soft vibrations and thus give rise to additional states in the low-frequency part of the spectrum. As a result, the specific heat can be enhanced even below T_0 [266]. Furthermore, small particles of a real sample always exhibit defects and impurity states that contribute to the specific heat at low temperatures as well. Experimentally, metallic powders provide the advantage that thermal equilibrium is more easily achieved. But in this case, finite-size effects in the electronic system have also to be taken into account. In the particular case of superconducting powders the electronic part of the specific heat can be switched 'on' and 'off' because superconductivity can be suppressed by magnetic fields [267].

6.2 Heat Transport

The heat current density $\boldsymbol{j}$ is related to the temperature gradient by *Fourier's law* $\boldsymbol{j} = -\Lambda \nabla T$. The proportionality constant Λ is known as the *thermal conductivity*. The validity of this equation presupposes that heat transport is a random process because otherwise the thermal flux would not be determined by the gradient but rather by the temperature difference. This also means that temperature (and consequently the phonon occupation number) can be defined locally. Inelastic phonon–phonon collision processes maintain the local equilibrium. In addition, collision processes must occur that change the total momentum of the phonon gas and thus reduce the energy transport. Otherwise, the thermal conductivity of insulating crystals would be infinite.

In our simple description of heat conduction we consider the ensemble of phonons as an ideal gas and describe the energy transport using the equations of the kinetic theory of gases. In this approximation, the thermal conductivity is given by

$$\Lambda = \frac{1}{3} C v \ell \,, \tag{6.20}$$

where C is the specific heat (per unit volume) of the particles carrying the heat, v their mean velocity and ℓ their mean free path.

In the case of a phonon gas, the quantities v and ℓ depend on the wave vector and the right-hand side of (6.20) has to be replaced by an integral. As in the isotropic Debye model, we replace the orientation-dependent integration over the wave vector $\boldsymbol{q}$ by the integration over the frequency $\omega = vq$ and obtain

$$\Lambda = \frac{1}{3} \sum_i \int c_i(\omega)\, v_i(\omega)\, \ell_i(\omega)\, \mathrm{d}\omega \,, \tag{6.21}$$

where $c_i(\omega)\,\mathrm{d}\omega = (\partial C_i/\partial\omega)\,\mathrm{d}\omega$ represents the contribution to the specific heat arising from modes of the phonon branch i and lying in the frequency interval $\mathrm{d}\omega$. The contribution of the different branches is taken into account by the summation.

In most cases, this equation can be simplified in two ways. As in the Debye model, the summation over the phonon branches can be replaced by the introduction of one effective phonon branch with the linear dispersion $v_\mathrm{D} = \omega/q = \text{const}$. The integration is avoided by applying the so-called *dominant phonon approximation*. Instead of the entire frequency spectrum, only that narrow frequency interval is considered that contributes most to the heat transport. Roughly speaking, these are phonons with an energy comparable with the thermal energy $k_\mathrm{B}T$. In most cases, the use of these approximations provides the only possible means to carry out an analysis of experimental data since the exact frequency dependence of the quantities appearing in (6.21) is usually unknown.

The dominant phonon approximation is not only used in the description of the heat conduction but also in treating other physical phenomena. So the specific heat and the ultrasonic attenuation are often analyzed in this way. A more detailed consideration shows that the frequency $\overline{\omega}$ of the dominant phonons can be defined via the first moment of the measured quantity resulting in the relation $\hbar\overline{\omega} = pk_\mathrm{B}T$. The constant p depends on the phenomenon under consideration and is often found to be in the range between two and three.

In the following sections, we first describe an elegant method allowing us to determine thermal conductivity and specific heat in one experiment. Afterwards, we consider in more detail various scattering processes that limit the heat flux in dielectric crystals. Here, we have to distinguish between processes caused by the thermal motion and processes provoked by crystalline imperfections.

6.2.1 Experimental Determination of the Thermal Conductivity

Different sample geometries and experimental methods can be used to measure the thermal conductivity. Here, we do not discuss experimental details but concentrate on a simple and elegant technique that provides a determination of thermal conductivity and specific heat in a single measurement.

The principle of this method is shown in Fig. 6.7a. A sample of length L is glued to the cooling plate of a cryostat at temperature T_0. The temperature rise is measured at $x = x_\mathrm{p}$ while T_0 is kept constant. The black area on top of the sample is irradiated by a short light pulse. The absorbed light produces a heat pulse that travels towards the heat sink at the bottom. The spatial and temporal variation in temperature $T(x,t)$ is described by the equation for heat conduction

$$\frac{\partial^2 T(x,t)}{\partial x^2} = \frac{1}{\kappa}\frac{\partial T(x,t)}{\partial t} . \tag{6.22}$$

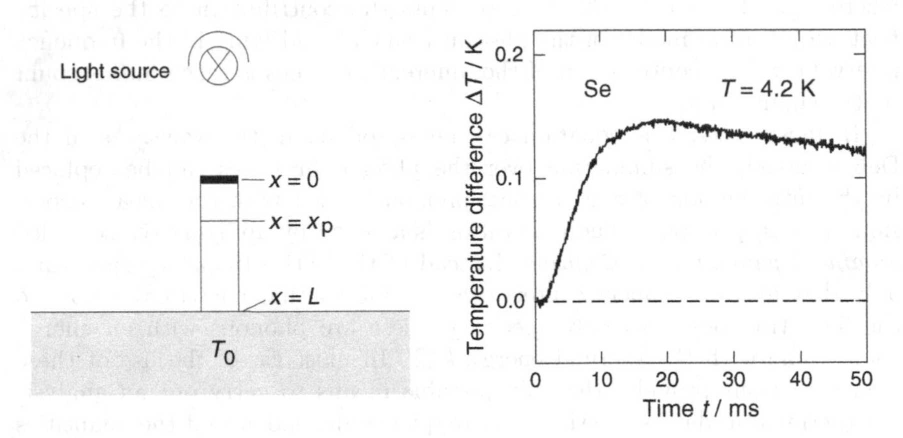

Fig. 6.7. (**a**) Schematic drawing of a setup to determine thermal conductivity and specific heat simultaneously. (**b**) Temporal variation of the temperature increase ΔT at $x = x_\mathrm{p}$ for selenium at $T_0 = 4.2\,\mathrm{K}$ [268]

The two quantities of interest, namely Λ and C are hidden in the thermal diffusivity $\kappa = \Lambda/C$.[2] For simple boundary conditions, the equation can be solved analytically. In particular, if a short heat pulse is applied and the thermometer is fixed remote from the heat sink, the sample can be considered as an infinitely long rod, and the solution of (6.22) can be expressed as a series expansion. Provided that $x_\mathrm{p} \ll L$, the solution for the temperature variation $\Delta T(x,t) = [T(x,t) - T_0]$ can be approximated with sufficient accuracy by

[2] In the definition of the thermal diffusivity the quantity C represents the heat capacity per unit volume.

$$\Delta T(x,t) \approx \frac{Q_{\text{pulse}}}{A\sqrt{\pi \Lambda C t}} \exp\left(-\frac{x^2}{4\kappa t}\right), \tag{6.23}$$

where Q_{pulse} is the heat input due to the light pulse and A the cross-sectional area of the sample.

In Fig. 6.7b, a graph is shown that was obtained in such an experiment with selenium at 4.2 K. As expected, the temperature rises steeply, passes through a maximum and then decreases slowly. Eventually, after a comparatively long time the specimen returns to the bath temperature T_0. Knowing the heat input Q_{pulse}, the specific heat C can directly be deduced from the temperature rise $\Delta T_{\max}$ at the maximum:

$$C = \frac{1}{A\, x_{\text{p}}\, \sqrt{\pi \text{e}/2}} \frac{Q_{\text{pulse}}}{\Delta T_{\max}}. \tag{6.24}$$

With the knowledge of the specific heat the thermal conductivity can be deduced from the thermal diffusivity κ. According to (6.23), κ determines the time $t_{\max}$ at which the temperature reaches its maximum value:

$$\kappa = \frac{x_{\text{p}}^2}{2\, t_{\max}}. \tag{6.25}$$

Of course, with higher accuracy both quantities, κ and Λ, can be determined by taking into account the full time dependence of the response.

6.2.2 Thermal Conductivity of Dielectric Crystals

In dielectric solids, heat is exclusively carried by phonons. Depending on temperature, different mechanisms of phonon scattering are dominant and therefore the temperature variation of Λ also changes. As an example, we show in Fig. 6.8 the conductivity of a very pure sodium fluoride crystal. Starting from high temperatures, the conductivity increases with decreasing temperature, passes through a maximum and finally decreases proportional to T^3. The maximum conductivity observed in this crystal was, for a long period, the highest thermal conductivity measured in a solid. Higher values have since been found in monoisotopic crystals of diamond.

In the following semiquantitative discussion, we make the plausible assumption that scattering processes are independent of each other. In this case, the effective inverse mean free path ℓ^{-1} is given by the sum $\ell^{-1} = \sum_i \ell_i^{-1}$, where ℓ_i stands for the mean free path caused by the process i. Furthermore, we express the inverse phonon mean free path ℓ_i^{-1} by the density $\tilde{n}$ of scattering centers and the total scattering cross section σ_i of the particular process:

$$\ell_i^{-1} = \tilde{n}_i\, \sigma_i\,. \tag{6.26}$$

Depending on the scattering process, $\tilde{n}$ will represent, in this chapter, the defect density or the density of thermally excited phonons.

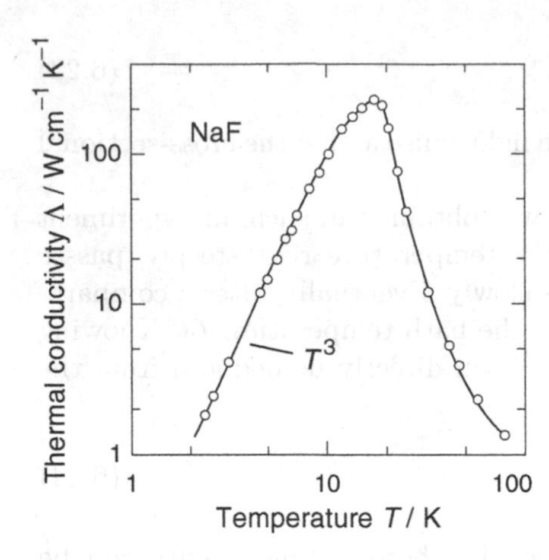

Fig. 6.8. Thermal conductivity of a sodium fluoride crystal as a function of temperature [269]

In the following section we briefly consider mechanisms of phonon scattering that play an important role at low temperatures. However, before we start with this discussion we consider first the interaction between phonons that limits the thermal conductivity at higher temperatures.

6.2.3 Phonon–Phonon Scattering

The most important interaction process between phonons is the *three-phonon process* in which two phonons merge into a single phonon, or a single phonon decays into two phonons. Conservation of energy and quasimomentum requires:

$$\hbar\omega_1 \pm \hbar\omega_2 = \hbar\omega_3 \qquad \text{and} \qquad \hbar \boldsymbol{q}_1 \pm \hbar \boldsymbol{q}_2 = \hbar \boldsymbol{q}_3 + \hbar \boldsymbol{G} \,. \tag{6.27}$$

Depending on the signs, these equations reflect the creation or annihilation of a phonon in the collision process. A characteristic feature of quasimomentum conservation is the occurrence of a reciprocal lattice vector $\boldsymbol{G}$ in this equation. Processes that do not involve a reciprocal lattice vector are called *normal processes*, whereas those that do, are called *umklapp processes*. Frequently, *N-process* and *U-process* are used as abbreviations. In Fig. 6.9, the two types of processes are shown for the case of phonon annihilation.

The scattering event exclusively takes place in the first Brillouin zone if the wave vectors $\boldsymbol{q}_1$ and $\boldsymbol{q}_2$ of the colliding phonons are relatively small. No reciprocal lattice vector is involved, and the sum of quasimomenta stays unchanged. This also holds for the reverse process where a single phonon decays into two phonons. If, however, the resulting vector $\boldsymbol{q}_3$ ends outside the first Brillouin zone, a nonzero reciprocal lattice vector $\boldsymbol{G}$ has to be added. Although the vector $\boldsymbol{q}_3' = \boldsymbol{q}_3 + \boldsymbol{G}$ lies again inside the first Brillouin zone (see Fig. 6.9b), the sum of quasimomenta of the participating phonons is changed.

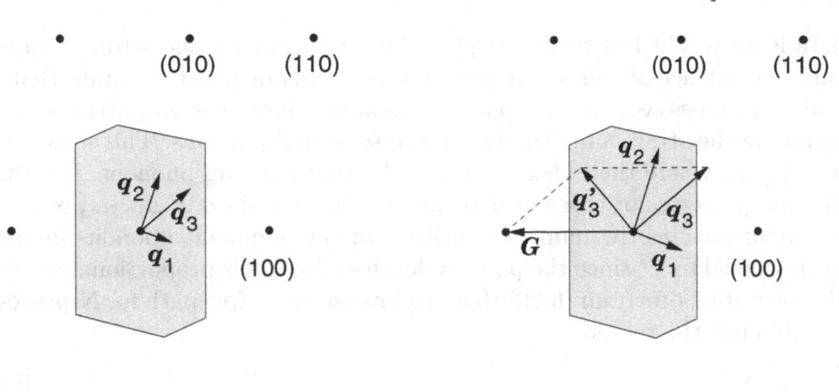

Fig. 6.9. Three-phonon processes. *Full circles* represent the reciprocal lattice. (**a**) Normal process: the wave vectors of all phonons lie within the *grey tinted* first Brillouin zone. (**b**) Umklapp process: the resulting vector $\boldsymbol{q}_3$ ends outside the first Brillouin zone. Addition of the reciprocal lattice vector $\boldsymbol{G}$ leads to the vector $\boldsymbol{q}_3'$ inside the zone

Phonon–phonon collisions are subjected to 'selection rules' because conservation of energy and quasimomentum (6.27) have to be fulfilled simultaneously. As a result, collisions between arbitrary phonons are not possible. The main consequence of this restriction is that only phonons of different polarization and hence with different velocity can take part in collisions [270]. However, for the following quantitative discussion such 'subtleties' are unimportant.

Normal Processes

As already mentioned, the sum of the quasimomenta of the colliding phonons is conserved in N-processes, and consequently the total quasimomentum $\boldsymbol{P}$ of all phonons is also conserved. Therefore

$$\boldsymbol{P} = \sum_{\boldsymbol{q}} n_{\boldsymbol{q}} \hbar \boldsymbol{q} = \text{const.} \tag{6.28}$$

is satisfied, where $n_{\boldsymbol{q}}$ represents the number density of the phonons with the wave vector $\boldsymbol{q}$. Since N-processes influence neither the flux of momentum nor the transport of energy they do not degrade the thermal current. With increasing temperatures, higher-order processes, such as four-phonon processes, become important. Because these processes also conserve the total quasimomentum $\boldsymbol{P}$, they do not give rise to heat resistance either.

Although normal collisions do not diminish the transport of momentum, they do change the frequency of the colliding phonons and thus contribute to the establishment of the local thermal equilibrium. Without going into details we would like to give a plausible argument for the strong temperature dependence of this scattering process. Without proof, we state that the scattering cross section σ for phonon–phonon collisions is proportional to the square of the strain amplitude e_0 of the scattered phonons. This means that $\sigma \propto \prod_i \omega_i$, where the index i denotes the participating phonons. For three-phonon processes of thermal phonons we thus obtain $\sigma \propto \omega_1\omega_2\omega_3 \propto T^3$. At low temperatures the number density $\tilde{n}$ of the dominant phonons increases proportional to T^2 since the phonon density of states is proportional to ω^2. In this way, it follows from (6.26) that the inverse mean free path for N-processes should obey the relation

$$\ell_{\mathrm{N}}^{-1} \propto T^5 . \tag{6.29}$$

Umklapp Processes

It is possible for $\boldsymbol{q}_3$ to lie outside the first Brillouin zone even if three phonons take part, as in the N-process. In this case, the addition of a reciprocal lattice vector brings the wave vector of the generated phonon back into the first zone. However, the total quasimomentum $\boldsymbol{P}$ is changed by these umklapp processes and consequently the heat flow is degraded.

At high temperatures, the overwhelming number of excited phonons are phonons with a frequency close to the Debye frequency ω_{D} and a wave vector comparable with that of the zone boundary. As a consequence, virtually every collision leads to a final state outside the Brillouin zone and is therefore an umklapp process. At $T > \Theta$ the number of thermally excited phonons and hence the density of scattering centers $\tilde{n}$ rises proportional to T. Since the frequency of the dominant phonons is ω_{D} that does not change with temperature, the cross section σ for the phonon–phonon collisions is constant. Thus (6.26) leads to a phonon mean free path ℓ_{U} limited by umklapp scattering that varies as $\ell_{\mathrm{U}} \propto T^{-1}$, and it follows from (6.20) that $\Lambda \propto T^{-1}$ since C_V is approximately constant at high temperatures. It should be noted that there is a gradual transition from N-processes to U-processes, i.e., from $\ell \propto T^{-5}$ to $\ell \propto T^{-1}$.

Umklapp processes are only possible if the colliding phonons carry high enough momenta. Roughly speaking, this condition is fulfilled for phonons with the frequency $\omega > \omega_{\mathrm{D}}/2$. At intermediate temperatures, i.e., below the Debye temperature, the number of phonons with energy $\hbar\omega_{\mathrm{D}}/2$ decreases rapidly on cooling. The probability for finding such phonons follows from the occupation number $f(\omega, T)$. Since $\hbar\omega_{\mathrm{D}} > k_{\mathrm{B}}T$, we may use the approximation $f(\omega,T) = [\exp(\hbar\omega_{\mathrm{D}}/2k_{\mathrm{B}}T) - 1]^{-1} \approx \exp(-\hbar\omega_{\mathrm{D}}/2k_{\mathrm{B}}T) = \exp(-\Theta/2T)$, meaning that the density $\tilde{n}$ of the relevant phonons varies exponentially with temperature. This exponential temperature dependence dominates the weaker dependence on temperature of σ and C_V with the result that

$$\Lambda \propto \ell_{\mathrm{U}} \propto \mathrm{e}^{\Theta/2T} . \tag{6.30}$$

An analysis of the data of Fig. 6.8 shows that the thermal conductivity of NaF ($\Theta_{\mathrm{NaF}} = 492\,\mathrm{K}$) indeed varies roughly exponentially on the high-temperature side of the conductivity peak. A more quantitative analysis of the temperature variation of Λ in the regime where umklapp processes dominate is beyond the scope of this discussion.

6.2.4 Defect Scattering

If the temperature decreases further, the rate of umklapp processes becomes vanishingly small and only N-processes remain. Nevertheless, the thermal conductivity does not continue to increase on cooling because the phonon mean free path is finally limited by defect scattering. In the following section, we consider the influence of different types of defects that are generally present in solids.

Surfaces

The most important 'defect' of a solid is its finite size. The existence of a surface becomes important as soon as the mean free path becomes comparable to the characteristic dimension d of the sample, e.g., the diameter, if the sample has a cylindrical shape. Therefore, we may put $\ell \approx d$ below a certain cross-over temperature and the conductivity is given by $\Lambda \approx \frac{1}{3} C_V v d$. In this low-temperature regime, the so-called *Casimir regime*, the temperature dependence of Λ is determined by C_V leading to the relation $\Lambda \propto T^3$ [271].

The size dependence of Λ is demonstrated by Fig. 6.10a. This graph shows measurements of the thermal conductivity of several LiF crystals with different cross-sectional areas. At high temperatures, Λ is not influenced by the

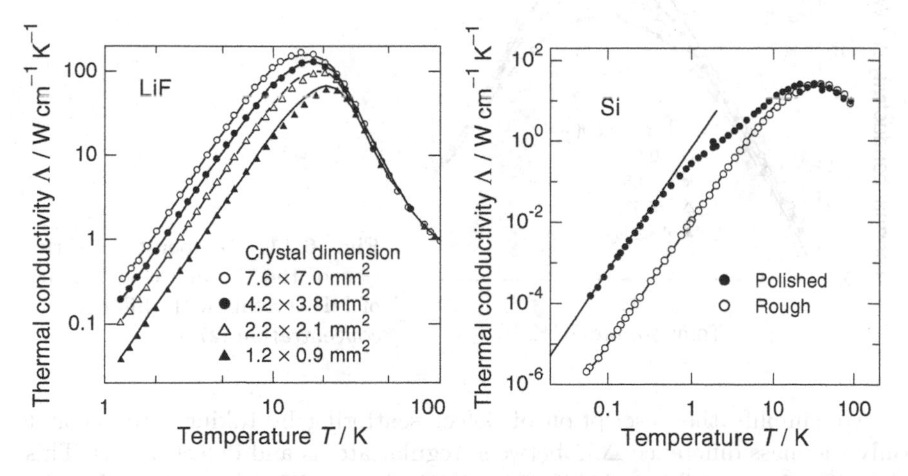

Fig. 6.10. (**a**) Thermal conductivity versus temperature of LiF crystals with different cross-sectional areas [272]. (**b**) Thermal conductivity of silicon crystals with polished (•) and rough (∘) surface, respectively [273]

sample size because ℓ is determined by umklapp processes. In the Casimir regime, however, the expected relation between Λ and d is evident.

The surface roughness has an important influence on the effective mean free path ℓ_{eff}. For well-polished samples, specular phonon reflection occurs, resulting in $\ell_{\text{eff}} > d$. This phenomenon is clearly demonstrated by measurements on silicon crystals with different surface quality. As shown in Fig. 6.10b, below $T = 0.5\,\text{K}$ the effective mean free path in the polished crystal surpasses that of the sample with the rough surface by a factor of 50. The observed mean free path of $\ell_{\text{eff}} \approx 7\,\text{cm}$ corresponds to the length of the sample!

Point Defects

Point defects also give rise to phonon scattering. Here, we have to distinguish between elastic and inelastic processes. In this chapter, we will restrict ourselves to elastic scattering, the influence of resonant scattering by tunneling defects will be discussed in detail in Chap. 9. In Fig. 6.11, the thermal conductivity of LiF samples is shown containing between 0.01 and 49.2% ^{6}Li. The maximum conductivity of the sample with 49.2% ^{6}Li is, by a factor of eight, smaller compared to the isotopically pure sample. At high and low temperatures, the conductivity is virtually the same in all samples. The small differences at low temperatures are caused by the slight variations in the cross-sectional area of the samples.

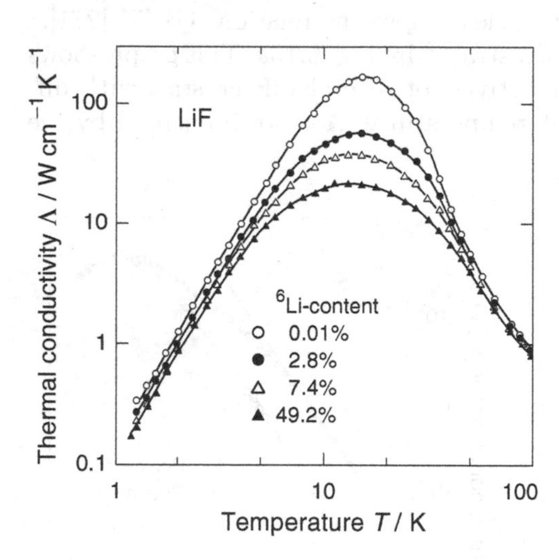

Fig. 6.11. Temperature variation of the thermal conductivity of LiF crystals with different ^{6}Li concentration [274]

We simplify the description of defect scattering by taking into account only the mass difference ΔM between regular atoms and defect atoms. This simplification is well founded if scattering is due to different isotopes. But for arbitrary impurity atoms, changes in the binding to neighboring atoms have to be considered. An elementary analysis utilizing elasticity theory leads to a scattering cross section for isotope scattering analogous to the well-known Rayleigh scattering in optics. For the mean free path, we may write [275]

$$\ell^{-1} = \frac{n_{\mathrm{p}} V_{\mathrm{A}}^2}{4\pi} \left(\frac{\Delta M}{M} \right)^2 q^4 \propto \omega^4 \,, \tag{6.31}$$

where n_{p} denotes the number density of the point defects and V_{A} their atomic volume. From the dominant phonon approximation $\hbar\overline{\omega} = \hbar v \overline{q} \approx 3k_{\mathrm{B}}T$ it follows immediately that scattering by point defects loses its importance at low temperatures. At high temperatures, phonon–phonon collisions are more effective but in the transition region, i.e., close to the maximum of Λ, point defects cause a noticeable reduction of the conductivity. Let us roughly estimate the scattering strength. If 10% of the atoms exhibit a mass difference of 10%, (6.31) predicts a mean free path $\ell \approx 5\,T^{-4}\,\mathrm{cm}$ (temperature expressed in kelvin). This means that at 1 K the mean free path due to isotope scattering is about 5 cm, but at 10 K it is only 5 μm. Therefore, there is, in most cases, a temperature range where this mechanism can give a noticeable contribution to the reduction of the phonon mean free path. If the sample is very small, surface scattering will predominate until umklapp processes take over at higher temperatures.

It should be added that the integral (6.21) diverges at low frequencies if defect scattering is limiting the mean free path. This means that in the case of defect scattering, long-wavelength phonons are important for the heat transport. Since N-processes interchange energy among the different phonon modes they have to be taken into account in a quantitative analysis. We do not consider this problem further but refer to the literature (see, e.g., [274]).

Dislocations

In principle, we may distinguish between two types of phonon–dislocation interactions, namely static and dynamic. The origin of static phonon scattering is the nonlinear elastic response of the strained region surrounding a sessile dislocation. Roughly speaking, this is a structurally deformed cylinder with a radius of a few lattice spacings. It has been shown [275] that its scattering cross section is proportional to the phonon frequency. Since $\omega \propto T$ for thermal phonons, the relation

$$\ell^{-1} = n_{\mathrm{d}}\,\Gamma\omega \propto T \tag{6.32}$$

is obtained, where Γ is a constant and n_{d} is the number of dislocations per unit area. For dielectric crystals, $C_V \propto T^3$ at low temperatures and consequently we expect $\Lambda \propto T^2$. In fact, this variation is observed for covalently bonded crystals like germanium or silicon. As an example, Fig. 6.12 shows the thermal conductivity of germanium crystals with a dislocation density ranging from $10^8\,\mathrm{m}^{-2}$ to $2.2 \times 10^{13}\,\mathrm{m}^{-2}$. Clearly, with increasing n_{d} the conductivity is reduced and the low-temperature slope slightly decreased. The *solid lines* represent fitting curves including dislocations, boundary and isotope scattering, and also phonon–phonon collisions. From this fit, it follows that the value of Γ is in order-of-magnitude agreement with the theory.

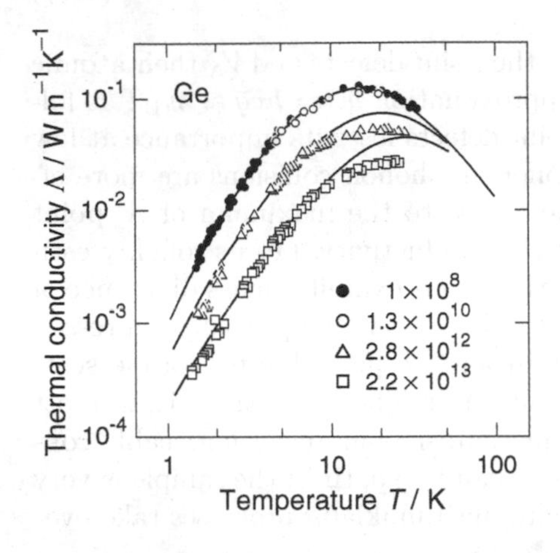

Fig. 6.12. Thermal conductivity of germanium crystals with a dislocation density ranging from $10^8\,\mathrm{m}^{-2}$ to $2.2 \times 10^{13}\,\mathrm{m}^{-2}$. *Solid lines* represent theoretical curves, as described in the text [276]

In metals and ionic crystals, the dynamic interaction is generally more important. At high temperatures, the 'vibrating string model' is used to describe the scattering of phonons by dislocations [277]. Phonons induce travelling waves on the string and are thus absorbed. Subsequently, the energy is reradiated in all directions. At low temperatures, the motion of dislocations is better described in terms of oscillating kinks [278]. Theoretical considerations lead to the following expression for the inverse phonon mean free path:

$$\ell^{-1} = A\, n_\mathrm{d}\, \frac{\omega^3}{(\omega_0^2 - \omega^2)^2 + \beta^2 \omega^4} \,. \tag{6.33}$$

The resonance frequency ω_0 depends on the length of the dislocation and exhibits typical values of the order of $10^{11}\,\mathrm{s}^{-1}$. For the constant β numerical values around 0.1 are found.

In Fig. 6.13, measurements of the thermal conductivity of high-purity single crystals of tantalum are depicted. Since the superconducting transition of Ta occurs at 4.4 K, the contribution of free electrons to thermal conduction can be neglected below 0.8 K, and the heat transport is only brought about by phonons (see Sect. 10.3.4). The two curves show the conductivity of an undeformed and a deformed sample with a dislocation density of $4 \times 10^{14}\,\mathrm{m}^{-2}$. The conductivity Λ of the untreated sample can be described by the relation $\Lambda/T^3 = 38\,\mathrm{W\,m^{-1}\,K^{-4}}$ indicating that the measurement has been carried out in the Casimir regime, i.e., ℓ is determined by the sample diameter. In the second sample, a large number of dislocations was created by plastic deformation, resulting in a reduction of Λ by about a factor of 800 at 0.3 K. The solid line represents a fit based on (6.20). The quantities C_V and v

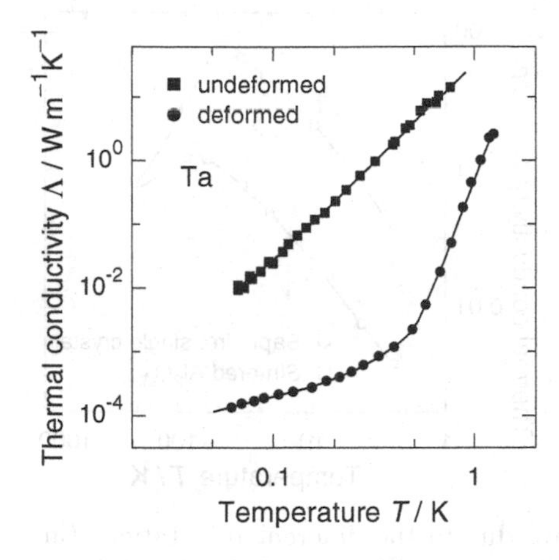

Fig. 6.13. Thermal conductivity of high-purity tantalum single crystals. The conductivity of the undeformed specimen (*full squares*) is much higher than that of the deformed sample (*full circles*) with a dislocation density of $4 \times 10^{14}\,\mathrm{m}^{-2}$ [279]

are known (or can be taken from the fit of the undeformed sample). The mean free path ℓ was calculated from (6.33) using $A = 1.5 \times 10^{-6}\,\mathrm{m\,s}^{-1}$, $\omega_0 = 2 \times 10^{11}\,\mathrm{s}^{-1}$, and $\beta = 0.15$, in agreement with the theory of kink oscillations [279].

Finally, we note that at present no quantum-mechanical description of the dynamics of dislocations exists, although at low temperatures thermally activated processes do not occur and tunneling phenomena should be taken into account.

Grain Boundaries

Although grain boundaries have a strong influence on the thermal conductivity of 'real' solids, no comprehensive theories exist for this phenomenon. There is a simple reason for this. Ordinary grain boundaries are not well-defined structural units but can rather be considered as an accumulation of point defects and dislocations. The scattering from a low-angle grain boundary is an exception for which the thermal resistance has been investigated theoretically. We consider here briefly the refraction of phonons at ordinary grain boundaries. As sketched in Fig. 6.14a, this effect occurs since adjacent crystallites of polycrystalline materials are, in general, orientated differently. At low temperatures, the wavelength of the thermal phonons is large compared to the thickness of the grain boundaries. In this case, the refraction simply leads to a mean free path $\ell \approx L/\langle \sin^2 \alpha \rangle$, where L is the average size of the crystallites and α is the angle of tilt [275].

The thermal conductivity of a single crystal of sapphire and of a sintered powder of Al_2O_3 is shown in Fig. 6.14b. The powder density was about 5% smaller than that of sapphire. The low-temperature variation of the conductivity of both samples is similar, but the magnitude differs considerably below the conductivity maximum. This difference can be understood as resulting from the phonon mean free path in the single crystal being determined by

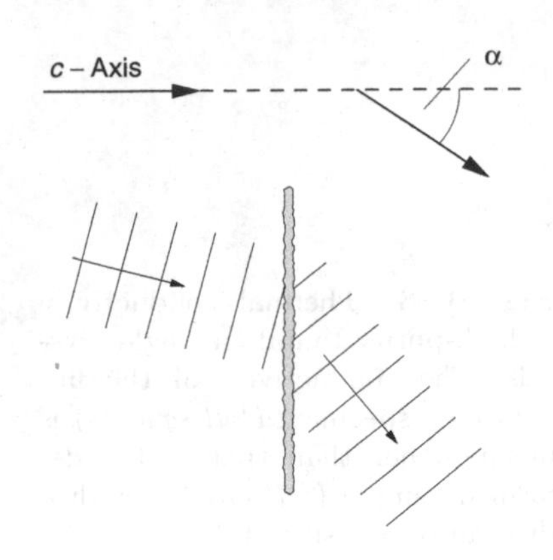

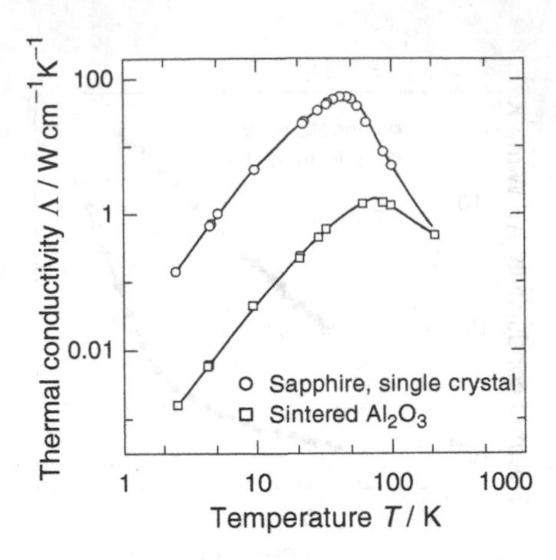

Fig. 6.14. (a) Refraction of a sound wave due to the different orientations (indicated by their c-axis) of adjacent crystallites. The thickness of the tinted drawn grain boundary is neglected since it is, in general, much smaller than the wavelength of low-temperature thermal phonons. **(b)** Comparison between the thermal conductivity of a single crystal of sapphire and a sintered powder of Al_2O_3 with a mean grain size of about 20 μm [280]

the diameter of the sample ($d = 1.5\,\text{mm}$), but in the sintered powder by the size of the crystallites, which ranged from 5 to 30 μm.

6.3 Significance of N-processes in Heat Transport

In the discussion of isotope scattering, it was mentioned that under certain circumstances, normal processes can have an influence on the thermal conductivity. Here, we consider two phenomena where N-processes play a crucial role, namely *Poiseuille flow* and *second sound*.

6.3.1 Poiseuille Flow

If a crystal is cooled to such a low temperature that U-processes and defect scattering can be neglected, there exists a similarity between the motion of gas atoms and the motion of phonons. First, energy and momentum are conserved in atomic collisions as well as in N-processes. Secondly, the number of atoms and the average number of phonons remains unchanged in the respective processes. The last statement sounds surprising at first glance since in a single N-process the number of phonons is always changed. However, in the sum of all phonon-collision processes, creation and annihilation balance one another, and the average number of phonons is a constant. In particular, it is possible to describe the flow of phonons through a 'thin' crystal like the viscous flow of atoms through a capillary, provided that the inequality

$$\ell_{\rm N} \ll d \ll \ell_{\rm R} \tag{6.34}$$

holds. Here, d stands for the diameter of the crystal and the capillary, respectively. $\ell_{\rm R}$ is the mean free path for scattering processes giving rise to thermal resistance. If $\ell_{\rm D}$ stands for the mean free path due to defect scattering, $\ell_{\rm R}$ is given by $\ell_{\rm R}^{-1} = \ell_{\rm U}^{-1} + \ell_{\rm D}^{-1}$.

The mass flow dm/dt of a gas through a capillary is described by the famous *Hagen–Poiseuille law*

$$-\frac{\dot{m}}{\pi r^2}\frac{1}{|\nabla p|} = \frac{\varrho}{8}\frac{r^2}{\eta} = \frac{3}{8}\frac{1}{\overline{v}_{\rm th}}\frac{r^2}{\ell_{\rm g}}, \tag{6.35}$$

where ∇p denotes the pressure gradient, η the viscosity, and r the radius of the capillary. Inserting the expression $\eta = \frac{1}{3}\varrho\,\overline{v}_{\rm th}\,\ell_{\rm g}$ for the viscosity leads to the right side of the equation. Here, $\overline{v}_{\rm th}$ is the average thermal velocity of the gas atoms and $\ell_{\rm g}$ their mean free path. The mass transport is limited by viscosity or, in other words, by the transfer of momentum to the capillary wall. The momentum transfer becomes increasingly effective with rising $\ell_{\rm g}$.

Let us now consider the Poiseuille flow in a gas of phonons. If the condition (6.34) is satisfied, we may use the analogy and replace the mass flow dm/dt in (6.35) by the heat flow $\dot{Q}$ and the pressure gradient by the temperature gradient, and obtain for the thermal conductivity the expression

$$-\frac{\dot{Q}}{\pi r^2}\frac{1}{|\nabla T|} = \Lambda = \frac{1}{3}C_V\, v\, \ell_{\rm eff}\,. \tag{6.36}$$

A comparison with the Hagen–Poiseuille equation shows that, as a first approximation, we may write $l_{\rm eff} \approx r^2/\ell_{\rm N}$ for the effective mean free path. As in ordinary heat conduction in the Casimir regime, the heat resistance is the result of surface scattering. However, because of the frequently occurring N-processes, phonons perform a random walk and cover on average a distance $r^2/\ell_{\rm N}$ before they reach the crystal surface. This means that the mean free path becomes much larger than the sample diameter for scattering events that do not conserve quasimomentum. A more precise calculation leads to

$$\Lambda = \frac{1}{3}C_V v\,\frac{5\,d^2}{32\,\ell_{\rm N}}, \tag{6.37}$$

i.e., the effective mean free path is enlarged by a factor $5d/32\ell_{\rm N}$ in comparison with ordinary heat transport. Surprisingly, under this condition the thermal conductivity varies *inversely* proportional to $\ell_{\rm N}$. Since $\ell_{\rm N} \propto T^{-5}$ (see (6.29)), the thermal conductivity in the Poiseuille regime should exhibit the strong temperature variation $\Lambda \propto T^8$.

Because of the different temperature dependence of the mean free path of N- and U-processes ($\ell_{\rm N} \propto T^{-5}$ and $\ell_{\rm U} \propto {\rm e}^{\Theta/2T}$) it is, in principle, always possible to satisfy condition (6.34) in a limited temperature interval if a suitable sample diameter is chosen. However, for real crystals the restriction (6.34) is difficult to fulfill, and until recently Poiseuille flow has only been observed in solid helium.

Here, the question arises why Poiseuille flow is not found in 'ordinary' crystals like silicon, which can be produced with nearly perfect quality. The transport of heat is accompanied by the flow of quasimomentum from the core of the sample to its surface. If p is the average number of scattering events taking place before the surface is reached, the inequality $\ell_R > p\,\ell_N$ has to be satisfied in order to have a negligible influence of resistive processes. As indicated above, the momentum transfer is caused by a random walk of the heat-carrying phonons, i.e., we may write $r \approx \sqrt{p}\,\ell_N$. Consequently, $\ell_R > r^2/\ell_N$ has to be satisfied. From (6.34) it follows that $\ell_N \ll r$, meaning that ℓ_N should be as small as possible, and ℓ_R as large as possible.

Because the two inequalities are difficult to meet simultaneously, Poiseuille flow has only been found in solid ^{4}He [281, 282] and ^{3}He [283], and recently in the quasi-one-dimensional single crystals $Ta_{1-x}Nb_xSe_4I$ with $x = 0$, 0.008 and 0.01 [284]. Measurements of the thermal conductivity of ^{4}He and ^{3}He crystals are shown in Fig. 6.15. In ^{4}He crystals it was found that between 0.5 K and 0.8 K the conductivity varies as $\Lambda \propto T^6$, i.e., the temperature variation is much stronger than the T^3-dependence known from the Casimir regime. Measurements on solid ^{3}He also indicate the occurrence of Poiseuille flow (curves A to C) just below the conductivity maximum, although the effect is less pronounced. In the measurement of curve D, Poiseuille flow was suppressed by the presence of 100 ppm ^{4}He impurities, i.e., by isotope scattering.

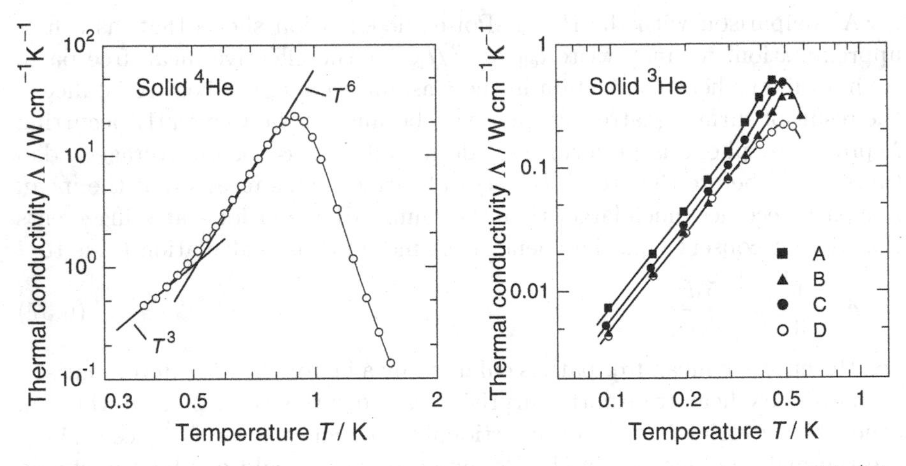

Fig. 6.15. Thermal conductivity of solid helium. (**a**) Measurement on solid ^{4}He [285]. (**b**) Measurement on solid ^{3}He at different pressures (A to C). Curve D was obtained on ^{3}He doped with 100 ppm ^{4}He [283]

From an analysis of the ^{4}He data, not only ℓ_{eff} but also ℓ_N can be determined [285]. The result is shown in Fig. 6.16. It demonstrates unambiguously that within a certain temperature range, ℓ_{eff} exceeds the diameter of the sample considerably. Below 0.5 K, $\ell_{\rm eff}$ approaches the Casimir limit. Although the temperature variation of Λ is much stronger than T^3, it is weaker than expected. From $\Lambda \propto T^6$, it follows that $\ell_{\rm N}$ is proportional to T^3 rather than T^5. Because experimental data are only available in a narrow temperature range (0.5 to 0.8 K), it is difficult to deduce the correct power law. In addition, the crystals investigated in these studies were possibly not completely free of dislocations. Scattering by dislocations reduces the temperature dependence of the conductivity and hence masks the temperature variation of $\ell_{\rm N}$. In fact, in different experiments the temperature dependence of $\ell_{\rm N}$ deduced from the data, was found to vary within the range $\ell_{\rm N} \propto T^{3-5}$ [286].

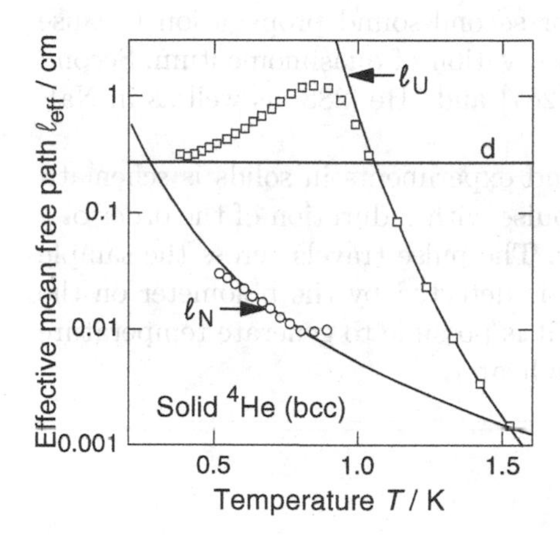

Fig. 6.16. Temperature dependence of the mean free paths. $\ell_{\rm eff}$ (*squares*) and $\ell_{\rm N}$ (*circles*) deduced from the data on the thermal conductivity of solid ^{4}He. *Full lines* represent the temperature variations of $\ell_{\rm N} \propto T^{-3}$ and $\ell_{\rm U} \propto {\rm e}^{\Theta/2T}$, respectively [285]

6.3.2 Second Sound

In Sect. 2.1.2, we discussed the occurrence of second sound in superfluid helium. As we shall see, it can also be observed in solids. Sound waves in gases are density waves propagating with the velocity $v_{\rm s}$, which can be estimated via the simple relation

$$v_{\rm s} \approx \frac{1}{\sqrt{3}} \overline{v}_{\rm th} , \tag{6.38}$$

where $\overline{v}_{\rm th}$ is the average thermal velocity of the gas atoms. Similarly, second sound may be considered as a density wave in a phonon gas. Because of this analogy the velocity v_2 of second sound can also be expressed by (6.38) after replacing the velocity of the gas atoms by the velocity of phonons:

$$v_2 \approx \frac{1}{\sqrt{3}} v_{\rm s} . \tag{6.39}$$

It should be pointed out that this picture is only applicable if resistive scattering processes can be neglected.

Experiments investigating the propagation of second sound are generally carried out with heat pulses such that the collision time $\tau_N = \ell_N/v_s$ for N-processes is much shorter than the duration t_p of the heat pulse. Therefore, the inequality

$$\ell_N \ll v_s t_p \ll \ell_R \tag{6.40}$$

has to be satisfied. From a comparison of (6.34) and (6.40) it is obvious that the latter condition is less restrictive. If $\ell_R \gg \ell_N$, in principle, only the duration of the heat pulse has to be adjusted to allow the observation of second sound. However, it turns out that the temperature range over which these conditions can be met, is rather narrow. Only a few elements and compounds are potential candidates for second-sound propagation because isotope scattering also destroys the conservation of quasimomentum. Second sound has been observed in solid ^{3}He [287] and ^{4}He [288] as well as in NaF [289] and Bi [290].

A typical setup used in second-sound experiments in solids is schematically shown in Fig. 6.17. A short heat pulse with a duration of the order of a microsecond is generated with a heater. The pulse travels across the sample with the characteristic velocity v_2 and is detected by the bolometer on the other side of the sample. Alternatively, it is possible to generate temperature waves using a periodic excitation of the heater.

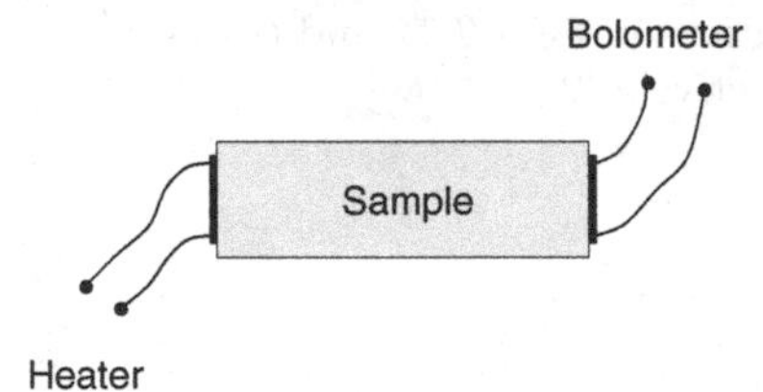

Fig. 6.17. Schematic drawing of an experimental setup to investigate second sound in solids

As an example of second-sound propagation in solids, we show in Fig. 6.18 a measurement on the same NaF crystal that exhibited the especially high maximum thermal conductivity of 240 W cm^{-1} K^{-1} shown in Fig. 6.8. The 'temperature window' for the observation of second sound was between 13 K and 17 K in this experiment. At 9 K the propagation of phonons is ballistic, i.e., N-processes do not occur frequently enough and phonons travel from the heater to the bolometer virtually without interaction. At low temperatures, the detector signal exhibits two maxima that can be ascribed to longitudinal and transverse phonons (see the following Sect. 6.4). With increasing temperature the first maximum gradually disappears, and above 11 K the second maximum shifts towards longer delay times. Within a narrow temperature range the heat pulse travels with the characteristic velocity v_2. At higher temperatures ($T > 17$ K) U-processes become important, the propagation of phonons becomes diffusive, and the second-sound signal disappears.

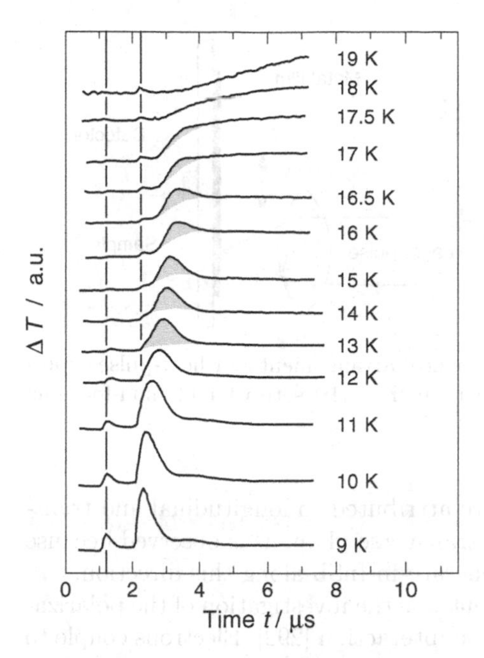

Fig. 6.18. Bolometer signal registered in a heat-pulse experiment on a NaF crystal. The *tinted regions* observed at temperatures between 13 K and 17.5 K are attributed to second sound [291]

6.4 Ballistic Propagation of Phonons

If crystals are cooled to sufficiently low temperatures, N-processes die out and finally phonons propagate ballistically. In Fig. 6.19, two possible experimental arrangements for the investigation of *ballistic phonons* are depicted schematically. In one case (see Fig. 6.19a), a small metal film is evaporated onto a disc-shaped sample and serves as a heater. Because the film has a low heat capacity, short heat pulses can be generated and detected on the other side of the sample by a fast superconducting bolometer or tunnel junction (see Sect. 10.3.6). In Fig. 6.19b, a setup is shown that is used in so-called *phonon-focusing* experiments. One side of the sample is covered by a thin metal film that is heated by a focused laser pulse. Again, a bolometer is used for the heat-pulse detection on the other side. While the detector is fixed, the laser beam scans over the surface, thus providing a movable heat source. As we shall see, in such an experiment two-dimensional images can be generated that reflect the elastic anisotropy of crystals.

6.4.1 Time-Resolved Measurements of Phonon Propagation

The time $t_i = d/v_i$ needed by phonons to travel through the sample of thickness d from the heater to the detector (see Fig. 6.19a) depends on their polarization, where the index i labels the phonon branch. The detector signal produced by heat pulses propagating along the [111] direction of InSb crystals is shown in Fig. 6.20. In pure samples (upper trace) two heat pulses

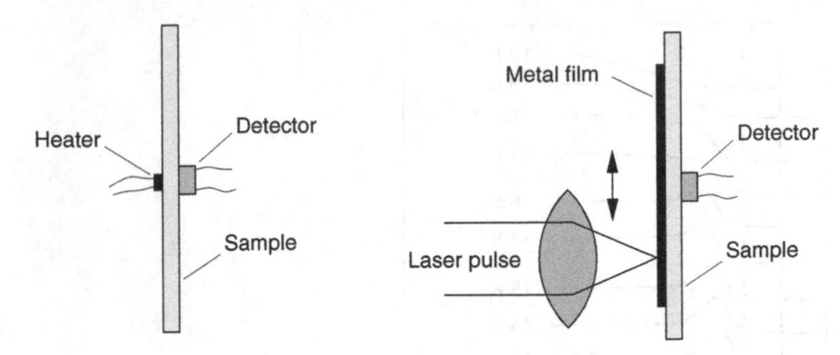

Fig. 6.19. Schematic drawings of experimental arrangements for heat-pulse experiments. (**a**) Setup for time-resolved measurements, (**b**) setup for phonon-focusing experiments

can clearly be distinguished, which are attributed to longitudinal and transverse phonons. Only a single pulse of transverse phonons is observed because the two transverse branches are degenerate in InSb along this direction.

The main purpose of this experiment was the investigation of the polarization dependence of the electron–phonon interaction [292]. Electrons couple to lattice vibrations via density changes. Since pure transverse phonons produce no change in density, it is expected that in metals or semiconductors with a spherical Fermi surface, only longitudinal phonons couple to electrons. To test this prediction, heat-pulse propagation was first studied in a nominally pure InSb sample (impurity content $n \approx 2 \times 10^{14}\,\mathrm{cm}^{-3}$). A strong signal due to longitudinal (L) phonons was found. In the n-doped sample (donor concentration $n \approx 5 \times 10^{17}\,\mathrm{cm}^{-3}$), however, the longitudinal heat pulse was strongly attenuated because of the strong electron–phonon interaction. In contrast, transverse (T) phonons are hardly influenced by the presence of electrons.

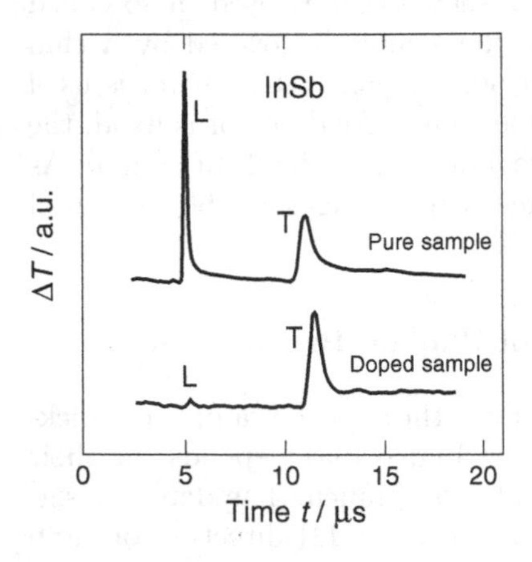

Fig. 6.20. Propagation of heat pulses in pure and n-doped InSb at 1.68 K. Heat pulses were generated in a gold film of 450 Å thickness and detected by a superconducting Al-Sn bolometer [292]

As a further example, we show in Fig. 6.21 data from the propagation of ballistic phonons in GaAs crystals. The samples, discs 2.6 mm thick, were doped with oxygen and chromium as indicated. Heat pulses were generated by Joule heating of a constantan film and detected with a superconducting bolometer. In the [100] and [111] directions, the fast and the slow transverse phonon branch (denoted by FT and ST, respectively) are degenerate. The relative amplitudes of the signals depend on the direction of propagation, on the density of states of the respective phonon branches and on 'focusing effects' due to the elastic anisotropy of GaAs. Scattering by free electrons was unimportant in these experiments.

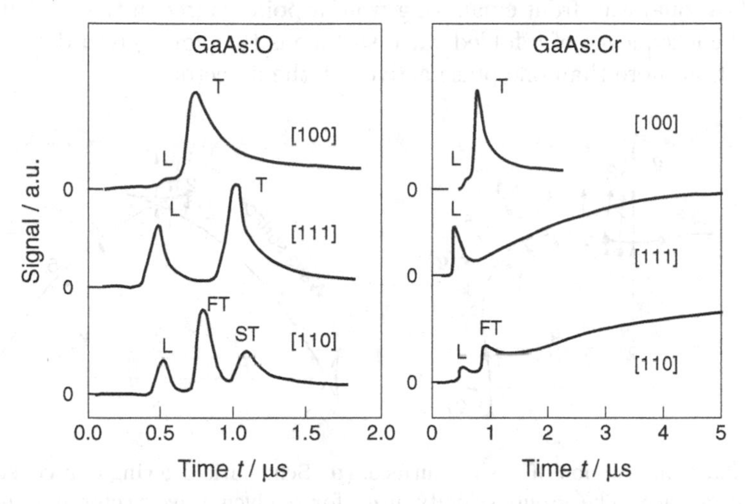

Fig. 6.21. Propagation of ballistic phonons in GaAs at 1.5 K. The specimens were doped with oxygen and chromium [293]

Heat-pulse experiments are well suited for the investigation of point defects. In the oxygen-doped samples, the phonons propagate ballistically to the bolometer. But in the chromium-doped samples ($n \approx 1 \times 10^{17}\,\text{cm}^{-3}$) the propagation of transverse (T) phonons in the [111] direction, and the propagation of slow transverse (ST) phonons in the [110] direction is particularly impeded. These phonons, along with the weaker scattered longitudinal phonons, end up in a diffusive pulse observed after a few microseconds. A closer analysis shows that the dominant scattering mechanism is not Rayleigh scattering but resonant interaction between phonons and chromium impurities. Since the cross section for this process depends on the polarization and propagation direction, information on the site symmetry of the defects can be obtained.

6.4.2 Phonon Focusing

Crystals are elastically anisotropic. As a consequence, the vibrational energy of a (plane) sound wave does not, in general, flow in the direction of its wave vector but in the direction of its group velocity. The locus of all possible group velocities is known as the group-velocity surface, or wave surface. This concept is illustrated in Fig. 6.22. The left figure shows the schematic drawing of a constant-frequency curve in q-space. The group velocity $\boldsymbol{v} = \mathrm{d}\omega(\boldsymbol{q})/\mathrm{d}\boldsymbol{q}$ for a given wave vector is normal to this curve. In the figure on the right, the wave surface is constructed by connecting the tails of all group-velocity vectors. This leads to the folded curve shown in Fig. 6.22b. It defines the shape of a vibrational wavefront emanating from a point source in the crystal. One unusual consequence of a folded wave surface is that for a given direction of propagation more than one pulse arrives at the detector.

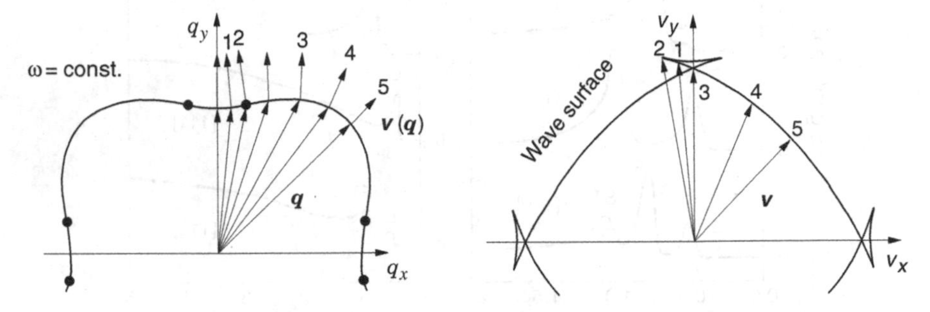

Fig. 6.22. Construction of a wave surface. (**a**) Schematic drawing of a constant-ω curve in q-space. The group velocity $\boldsymbol{v}(\boldsymbol{q})$ for a given wave vector $\boldsymbol{q}$ is normal to this curve. (**b**) Construction of the wave surface by connecting the tails of all group-velocity vectors [294]

An experiment carried out with a setup of the type drawn schematically in Fig. 6.19b produces the images shown in Fig. 6.23. The four crystals have cubic symmetry and were cut perpendicularly to their [111] direction, resulting in the observed three-fold symmetry. To understand the details of the intensity distribution the exact shape of the wave surface of the crystals has to be known. We do not discuss further this interesting topic in low-temperature physics because here we only wanted to give a brief introduction to this field and refer the reader to [295] for more details.

An elegant alternative method of demonstrating the occurrence of phonon focusing is shown schematically in Fig. 6.24. This method makes use of the properties of superfluid helium films. Because of the fountain effect (see Sect. 2.2) the thickness of the saturated superfluid helium film on a surface above a liquid bath depends upon its temperature. Focusing of phonons generated at a point source on one surface of a crystal leads to a spatial vari-

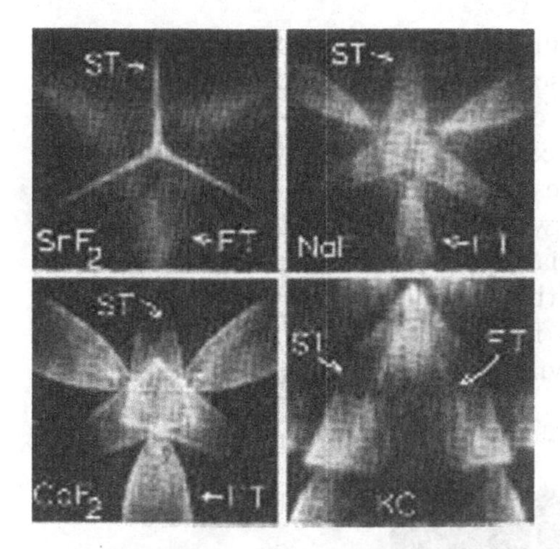

Fig. 6.23. Phonon image of four cubic crystals. Bright lines are caustics in ballistic heat flux due to phonon focusing [295]. The abbreviations ST and FT stand for 'slow transverse' and 'fast transverse', respectively

ation of the density of phonons incident on the second surface. This variation in phonon density produces local heating with a variation in the film temperature and film thickness. Thus, a variation of the thickness of the helium film on top of the crystal occurs that is a direct measure of the local energy density of the lattice vibrations.

In contrast to the pulsed experiments discussed above, the crystal is steadily heated. In equilibrium, the local gain in helium due to the inflow of the superfluid component to the heated regions is compensated for by a higher evaporation rate. In other words, in steady state the film thickness in a heated region is determined by the balance between the heat flow, limited by the critical velocity, and the enhanced evaporation. This method of observing phonon focusing has the advantage that no scanning of the crystal surface is necessary.

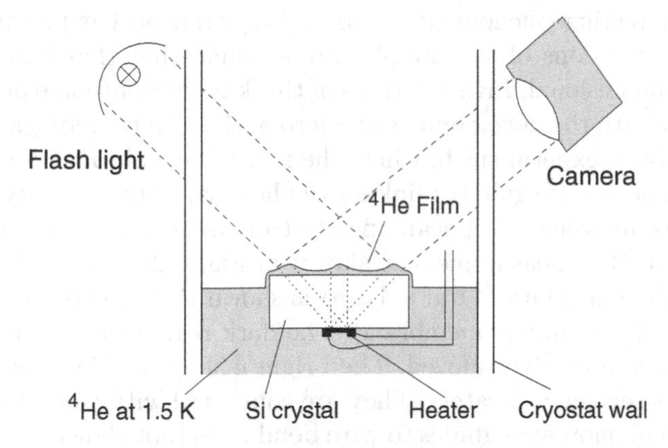

Fig. 6.24. Schematic drawing of the setup used to demonstrate phonon focusing by the thickness variations of a helium film

In Fig. 6.25, the result of a measurement of this type is shown with the helium bath at $T = 1.5\,\mathrm{K}$. A silicon disc was used in this experiment, 30 mm in diameter and 7 mm thick. The surface of the helium bath was 5 mm below the upper surface of the crystal. To enhance the critical velocity and thus increase the attainable thickness variation, the surface of the silicon crystal had been contaminated with air before filling with liquid helium. In the experiment shown in Fig. 6.25, the thickness variation was estimated to be of the order of 100 nm. Although the quality of the reproduction of the photograph is not very good, the effect of phonon focusing is clearly visible. Again, the phonon density exhibits a threefold symmetry since the cubic crystal was cut perpendicular to the [111] direction.

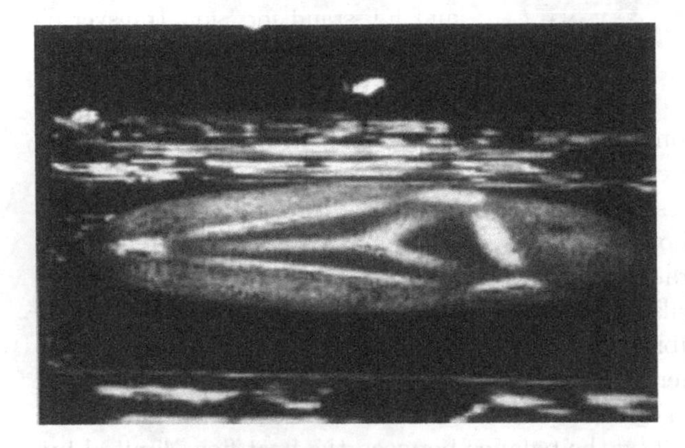

Fig. 6.25. Phonon focusing in silicon made visible by the thickness variation of a superfluid helium film [296]

6.5 Thermal Conductivity of One-Dimensional Samples

New and interesting phenomena occur in heat-transport experiments when the lateral dimensions of the samples are so small that they can be considered as one-dimensional. Investigations of this kind have become possible only very recently with the development of microfabrication technologies. Here, we discuss briefly an experiment in which the flow of heat through four tiny connections ('phonon waveguides') linking the heater ('phonon cavity') with the heat sink was investigated. A scanning electron micrograph of the mesoscopic device used in this measurement is shown in Fig. 6.26. On the left side the suspended phonon cavity ($4\,\mu\mathrm{m} \times 4\,\mu\mathrm{m}$) is shown that was patterned from a 60 nm thick silicon nitride membrane. The dark regions are areas where the membrane is completely removed. The bright c-shaped objects on the cavity are gold films serving as heaters. They are connected with thin niobium leads on top of the phonon waveguides to wire bond pads (not shown in this figure). The right photograph shows a close-up of one of the catenoidal waveguides. The width w of the neck is smaller than 200 nm.

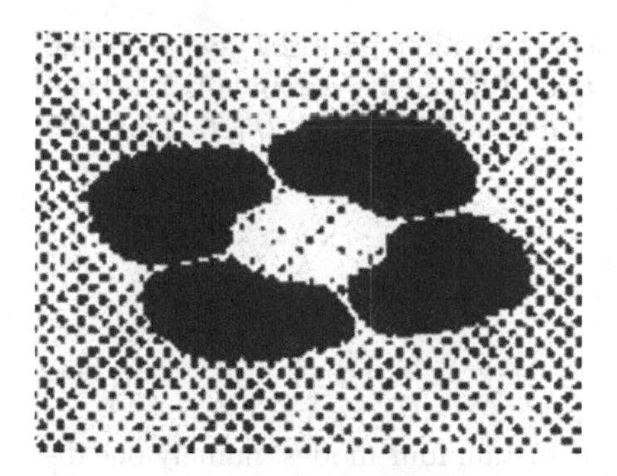

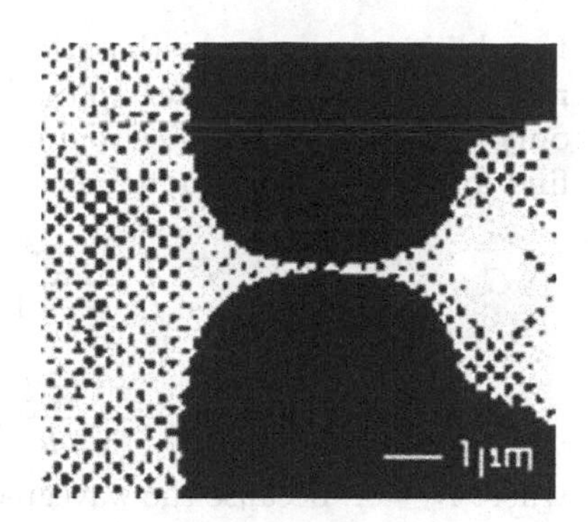

Fig. 6.26. View of experimental arrangement. (**a**) The suspended device consists of a 4 μm × 4 μm 'phonon cavity' in the center patterned from the silicon nitride membrane. The *dark regions* are areas where the membrane has been completely removed. The *bright 'c'-shaped* objects on the cavity are gold films serving as heaters. They are connected with thin niobium leads on top of the phonon waveguides to wire bond pads (not shown in this figure). (**b**) Close-up of one of the catenoidal waveguides with a neck smaller than 200 nm [297]

Before we look at the experimental results let us consider the theoretical predictions [298]. The energy current J flowing through a one-dimensional sample of length L is given by

$$J = \frac{1}{L} \sum_{\boldsymbol{q}} \hbar\omega_{\boldsymbol{q}} v_{\boldsymbol{q}} \, , \tag{6.41}$$

where $v_{\boldsymbol{q}}$ is the phonon velocity. The summation indicates that all thermally excited phonons contribute to the heat flow. After replacing the sum by an integral, the energy current between a heat source on the right side and a heat sink on the left side can be expressed by

$$J = \sum_i \frac{1}{L} \int_0^\infty \mathcal{D}_i^1(q) \, \hbar\omega_i \, v_i \, [f_\mathrm{h}(\omega,T) - f_\mathrm{c}(\omega,T)] \, \mathrm{d}q \, . \tag{6.42}$$

Here, i denotes the index of the phonon mode taking part in the heat transport, $\mathcal{D}_i^1(q)$ the one-dimensional density of states in momentum space, and v_i the group velocity of the phonons. The thermal occupation of the phonon states is expressed by the Bose–Einstein factor $f(\omega,T)$, where the indices 'h' and 'c' stand for the hot and cold thermal reservoir, respectively. The lower limit $q = 0$ for the integral follows from the fact that the sign of the wave number of the 'cold' phonons coming from the heat sink has already been taken into account by the sign of $f_\mathrm{c}(\omega,T)$. Furthermore, in our consideration we have put the transmission coefficient characterizing the coupling of the waveguide modes to the reservoirs equal to unity.

We insert $\mathcal{D}_i^1(q) = L/2\pi$, and change the variables from q to ω. This leads to a factor $\partial q/\partial\omega$, which just cancels the phonon group velocity $\partial\omega/\partial q$.

In addition, we assume that the temperature differences ΔT between the reservoirs is small so that $[f_{\rm h}(\omega,T) - f_{\rm c}(\omega,T)]$ can be expanded. We keep only the term linear in ΔT and use the abbreviation $x = \hbar\omega/k_{\rm B}T$. Thus, we find for the thermal conductance

$$G = \frac{J}{\Delta T} = \frac{k_{\rm B}^2 T}{h} \sum_i \int_0^\infty \frac{x^2 {\rm e}^x}{({\rm e}^x - 1)^2} {\rm d}x$$
$$= N_i G_0 = N_i \frac{\pi^2}{3} \frac{k_{\rm B}^2 T}{h} , \tag{6.43}$$

where $N_i = 4$, because the waveguide can sustain four modes, namely one dilatational, one torsional, and two flexural modes. It is remarkable that under ideal conditions, each mode contributes the same amount to the conductance, namely $G_0 = (9.456 \times 10^{-13}\,{\rm W\,K^{-2}})\,T$. Because of the cancellation of the one-dimensional density of states and the group velocity, this result does not depend on particle statistics. It is universal for fermions, bosons and anyons.[3]

The data shown in Fig. 6.27 confirm this concept. In this graph, the normalized thermal conductance of the four waveguides is plotted. At high temperatures, the conductivity is proportional to T^3, as expected in the Casimir regime. From this set of data an effective mean free path $\ell_{\rm eff} \approx 0.9\,\mu$m can be deduced. One-dimensional behavior is expected when the wave number $q_{\rm th} \approx k_{\rm B}T/(\hbar v)$ of the thermal phonons becomes smaller than the spacing between the lowest-lying modes that is roughly given by $\Delta q \approx \pi/w$.

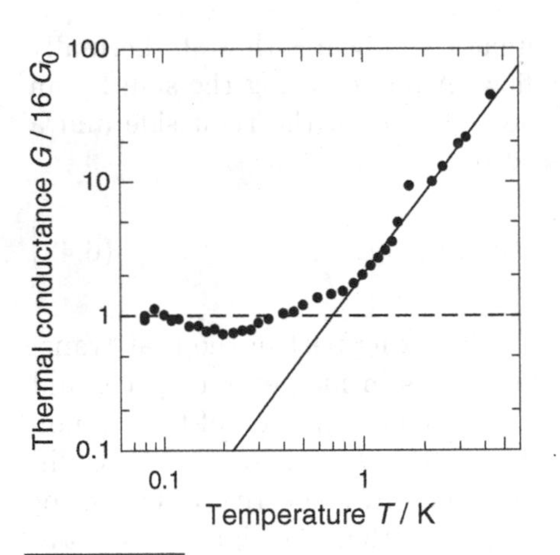

Fig. 6.27. Reduced thermal conductance G of four one-dimensional 'waveguides'. Below 0.8 K the waveguides behave like one-dimensional samples. Above that cross-over temperature the conductivity increases proportional to T^3, as expected in the Casimir regime for a three-dimensional sample [297]

[3] It is worth mentioning that the electrical conduction of a thin, one-dimensional wire follows directly from (6.43) if we use the Wiedemann–Franz law $\Lambda_{\rm el}/\sigma = (\pi^2/3)(k_{\rm B}/e)^2 T$ (see (7.26)). Taking into account that in the electrical case two 'channels' exist because of the spin degeneracy, we immediately obtain $G_{\rm el} = 2e^2/h$ for the quantized unit of the electric conductance.

Assuming an average sound velocity $v \approx 6000\,\mathrm{m\,s^{-1}}$ and a width of the neck $w \approx 200\,\mathrm{nm}$, the cross-over temperature T_{co} is estimated to be about $T_{\mathrm{co}} \approx hv/(2wk_{\mathrm{B}}) \approx 0.8\,\mathrm{K}$. In fact, below 1 K the conductance starts to saturate and approaches $16\,G_0$ as expected since the phonon cavity is cooled via four waveguides, each carrying just four populated modes.

Exercises

6.1 At high temperatures ($T \geq \Theta$) the specific heat (per atom) is comparable for all solids (Dulong–Petit law). However, at low temperatures the specific heat of different materials can differ by orders of magnitude. Which solid shows the largest and which one the smallest specific heat at low temperatures?

6.2 The low-temperature specific heat of crystalline ethanol is well described by the Debye model. At 0.7 K the values $0.26\,\mathrm{mJ\,mol^{-1}\,K^{-1}}$ and $0.31\,\mathrm{mJ\ mol^{-1}\,K^{-1}}$ were measured for ordinary ethanol (C_2H_6O) and deuterated ethanol (C_2D_6O), respectively. Calculate the Debye temperatures. Is the observed isotope effect in agreement with the mass dependence of the sound velocity?

6.3 In an experiment, the ballistic propagation of phonons through a 5 mm thick plate of LiF is studied at 0.5 K and 4 K. The temperature of the heat pulse is 20% higher than the equilibrium temperature. Has the natural abundance of 7.5% ^{6}Li an influence on the attenuation of the heat pulse? Use the parameters $c_{11} = 112\,\mathrm{GPa}$, $c_{12} = 46\,\mathrm{GPa}$, $c_{44} = 63.5\,\mathrm{GPa}$, $\varrho = 2.65\,\mathrm{g\ cm^{-3}}$, and the lattice constant $a = 4.03\,\mathrm{Å}$.

6.4 Show that the total number of thermally excited phonons is proportional to T^3 at low temperatures ($T \ll \Theta$), and proportional to T at high temperatures ($T \gg \Theta$). Calculate the number of phonons existing in silicon at 1 K.

6.5 The lower ends of 1 cm long cylindrical samples are thermally anchored at a heat sink with $T_0 = 0.5\,\mathrm{K}$. Their upper ends are heated with $10\,\mu\mathrm{W}$ resulting in a temperature difference of 1 mK between the two ends. (**a**) What are the cross-sectional areas of the samples if they are made of silicon, vitreous silica, and copper? Use the following values for the thermal conductivity at 0.5 K: $\lambda_{\mathrm{Si}} = 1 \times 10^{-2}\,\mathrm{W\ cm^{-1}K^{-1}}$, $\lambda_{\mathrm{a-SiO_2}} = 5 \times 10^{-5}\,\mathrm{W\ cm^{-1}K^{-1}}$, and $\lambda_{\mathrm{Cu}} = 4\,\mathrm{W\ cm^{-1}K^{-1}}$. (**b**) Now the heater power is increased to 10 mW. Under this condition the samples exhibit different temperature gradients. Which sample shows the highest and which the lowest temperature gradient? Note that $\lambda_{\mathrm{a-SiO_2}} \propto T^2$, and $\lambda_{\mathrm{Cu}} \propto T$ (see Sect. 9.5.2 and Sect. 7.3).

7 Conduction Electrons

Around 1900, only a few years after *Thomson* discovered the electron, *Drude* put forth his theory of electrical and thermal conductivity based on the idea that conduction electrons in metals behave like atoms of a gas. He assumed that the compensating positive charge was attached to much heavier particles, which he considered to be immobile. In this way, he was able to develop a theory that successfully described electrical and thermal conductivity, as well as the origin of several optical properties [299]. The valence electrons were assumed to be highly mobile and to move with the speed given by the equipartition law. The interaction of electrons with the lattice was summarily taken into account by the introduction of a characteristic collision time. Later, Drude's theory was refined by *Lorentz* (1905) who assumed that the electrons, like the atoms of an ideal gas, move with a velocity given by the Maxwell–Boltzmann distribution. Although the theory was able to describe many properties of metals, the assumption of a classical velocity distribution was incompatible with other observations, e.g., with measurements of the specific heat to which electrons hardly contribute. An understanding of all these observations is only possible within a theory based on the laws of quantum mechanics. The cornerstone to such a theory was laid by *Sommerfeld* (1927) who also started from the free-electron gas but took into account the Pauli exclusion principle [300], i.e., he used Fermi–Dirac statistics rather than Maxwell–Boltzmann statistics.

In this chapter, we first consider some fundamental properties of electrons in metals such as specific heat, electrical and thermal conductivity, and then go on to discuss some specific phenomena observed in so-called *Kondo alloys* and *heavy-fermion systems*. We restrict ourselves to a discussion of the behavior of normal metals, i.e., we will not consider electrons in semiconductors or superconductors.

7.1 Specific Heat

For a long time the specific heat of metals appeared to be rather mysterious. Since, in Drude's theory, the electrons were assumed to be in thermal equilibrium with the lattice, the specific heat was expected to be given by $C_V = 3R/2$ in accordance with the equipartition theorem. As a consequence, the specific heat of all metals should be higher than that of insulators. However, this expectation was in conflict with the Dulong–Petit law that tells us that at higher temperatures all solids exhibit the specific heat $C_V = 3R$ regardless of whether they are metals or insulators. As mentioned above, the explanation was finally given by Sommerfeld in his famous theory.

7.1.1 Conduction Electrons in Simple Metals

Electrons in simple metals are well described as a gas of free fermions although the Coulomb interaction of the conduction electrons with the ion cores is definitely of importance. At first glance, this approach seems to be too coarse a simplification and thus unable to reflect the actual physical situation. However, the modulation of the potential experienced by the conduction electrons in simple metals is relatively weak. To a first approximation, the potential variation can be neglected and the actual potential can be considered to be approximately constant.

As shown in Sect. 3.1, a linear increase of the specific heat with temperature is expected for a free Fermi gas as long as $T \ll T_F$. In metals, this condition holds for all accessible temperatures since T_F lies well above 10^4 K. Taking into account the lattice contribution as well, the total low-temperature specific heat of metals may be expressed by

$$c_V = \gamma T + \beta T^3 \,. \tag{7.1}$$

The Sommerfeld coefficient

$$\gamma = \frac{\pi^2 n k_B^2}{2E_F} \propto n^{1/3}\, m_{th}^* \tag{7.2}$$

characterizes the electronic contribution and $\beta = 12\,\pi^4 n k_B/(5\,\Theta^3)$ the lattice part (see (3.13) and (6.12), respectively). Here, m_{th}^* represents the effective mass defined below.

The specific heat of copper is depicted in Fig. 7.1. As shown in Fig. 7.1a the observed specific heat (*open circles*) consists of the linear contribution from the electrons (*dashed line*) and the T^3-contribution from the lattice (*dashed-dotted line*). At about 4 K the two parts are of equal size. Below this temperature, the electronic contribution dominates, while above it phonons are more important. To separate the two contributions it is convenient to plot C/T versus T^2 (see Fig. 7.1b). The value of γ can directly be deduced from the intersection of the straight line with the ordinate.

Of course, the question then arises of whether a description of the specific heat in terms of a free Fermi gas leads to good agreement with experimental data. From the data in Table 7.1 it follows that the ratio $\gamma_{exp}/\gamma_{theo}$ between the experimental and the theoretical value of γ is close to unity for simple metals, i.e., the agreement is rather good.

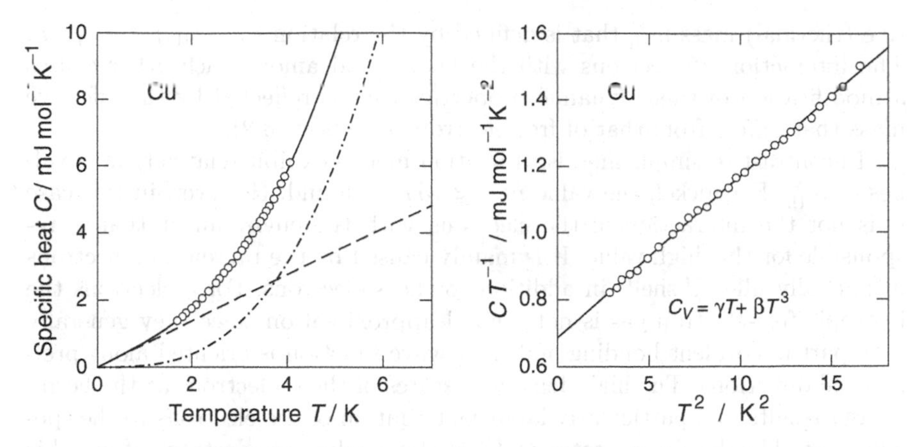

Fig. 7.1. Low-temperature specific heat of copper [301]. (**a**) C plotted linearly versus temperature. The contribution of the electrons and of the phonons are represented by *dashed* and *dashed-dotted lines*, respectively. The *full line* reflects the sum of the two contributions. (**b**) C/T versus T^2. This plot permits the separation of the electronic and lattice contributions by graphical means

Since the calculated value of γ is proportional to the electronic mass m, the fit to the experimental data is often improved by the introduction of an effec-

Table 7.1. Electron density n, Fermi temperature T_F, experimental and theoretical Sommerfeld coefficient γ_exp and γ_theo, and the ratio $\gamma_\mathrm{exp}/\gamma_\mathrm{theo}$. Apart from the densities of the alkali metals Li (77 K), Na (5 K), K (5 K), Rb (5 K), and Cs (5 K), the electron densities in the table were obtained at room temperature. After [302]

Element	n ($10^{22}\,\mathrm{cm}^{-3}$)	T_F (10^4 K)	γ_exp [mJ $(\mathrm{K}^2\mathrm{mol})^{-1}$]	γ_theo [mJ $(\mathrm{K}^2\mathrm{mol})^{-1}$]	$\gamma_\mathrm{exp}/\gamma_\mathrm{theo}$
Li	4.70	5.51	1.75	0.75	2.3
Na	2.65	3.77	1.46	1.08	1.3
K	1.40	2.46	1.92	1.67	1.2
Rb	1.15	2.15	2.42	1.92	1.3
Cs	0.91	1.84	3.22	2.21	1.5
Cu	8.47	8.16	0.67	0.50	1.3
Ag	5.86	6.38	0.65	0.63	1.1
Au	5.90	6.42	0.65	0.63	1.1
Mg	8.61	8.23	1.34	1.00	1.7
Ba	3.15	4.23	2.72	1.96	1.4
Al	18.10	13.60	1.25	0.91	1.4
In	11.50	10.00	1.79	1.21	1.5
Sn	14.80	11.80	1.83	1.37	1.3
Pb	13.20	11.00	2.92	1.50	1.9

tive (thermal) mass $m^*_{\rm th}$ that is defined by the relation $\gamma_{\rm exp}/\gamma_{\rm theo} = m^*_{\rm th}/m$. The interaction of electrons with the lattice and among each other causes a modification of their dynamic properties that is reflected by the effective mass that differs from that of free electrons (see Sect. 3.2).

In contrast to simple metals, transition metals exhibit relatively high values of $m^*_{\rm th}$. For nickel, the value $m^*_{\rm th} \approx 15\,m$ is found. However, in this case it is not the interaction of the electrons with the environment that is responsible for this high value. It is mainly caused by the presence of electrons in partially filled d-shells in addition to the s-electrons. For d-electrons the isotropic free-electron gas is not a good approximation since they generally take part in covalent bonding and their wave function is oriented along preferential directions. The high density of states of the d-electrons at the Fermi energy results in a particularly large contribution of the electrons to the specific heat. The density of states $D(E)$ of the conduction electrons of nickel is shown in Fig. 7.2 illustrating this particular situation. At the Fermi energy the density of states is dominated by d-electrons, whereas in the free-electron gas approximation only s-electrons are considered.

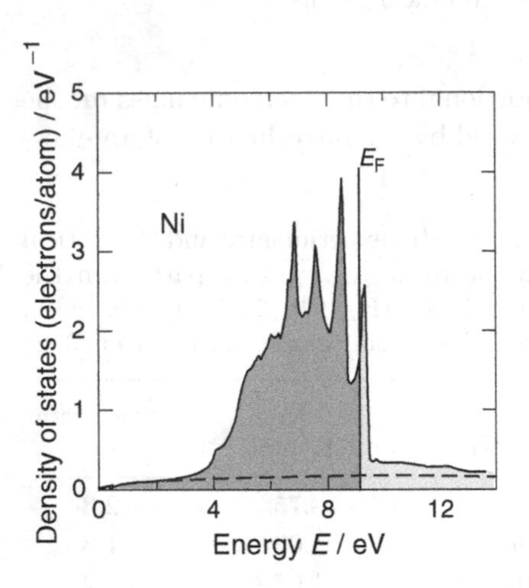

Fig. 7.2. Density of states $D(E)$ of the conduction electrons of nickel. At the Fermi energy a large contribution of the d-band is superimposed on the s-band (*dashed line*). The *dark area* reflects the occupied states [303]

7.1.2 Heavy-Fermion Systems

The specific heat shows astonishingly high values in metals with so-called *heavy electrons*, i.e., electrons with a very large effective mass. These metals – usually called *heavy-fermion systems* – are intermetallic compounds such as $CeCu_2Si_2$. They contain elements with $4f$- or $5f$-electrons that carry a magnetic moment. It is worth mentioning that $CeCu_2Si_2$ becomes superconducting below 0.6 K even though magnetic moments usually destroy superconductivity. Here, we do not discuss these unusual superconducting properties of heavy-fermion systems but postpone this discussion until Sect. 10.5.

The specific heat divided by temperature C/T of a $CeCu_2Si_2$ crystal in its normal conducting state is depicted in Fig. 7.3a. Below 20 K, γ is not constant but rises with decreasing temperature. An extrapolation of the high-temperature data to $T = 0$ leads to $\gamma \approx 30\,\mathrm{mJ\,(mol\,K^2)^{-1}}$. Extrapolating the low-temperature data shown in the insert results in the astonishingly high value $\gamma \approx 1050\,\mathrm{mJ\,(mol\,K^2)^{-1}}$. In Fig. 7.3b, the specific heat of $CeAl_3$ is shown. The very high value $\gamma \approx 1800\,\mathrm{mJ\,(mol\,K^2)^{-1}}$ is found at 0.35 K. For $CeAlCu_4$, a Sommerfeld coefficient as high as $\gamma \approx 2200\,\mathrm{mJ\,(mol\,K^2)^{-1}}$ is reported [304]. From the relation $\gamma \propto n^{1/3} m^*_{\mathrm{th}}$ it follows that the exceptionally high values of γ are caused by the extremely large effective electron masses. The reason for this anomaly and further interesting properties of this class of substances will be discussed in Sect. 7.5 at the end of this chapter after our discussion of the Kondo effect.

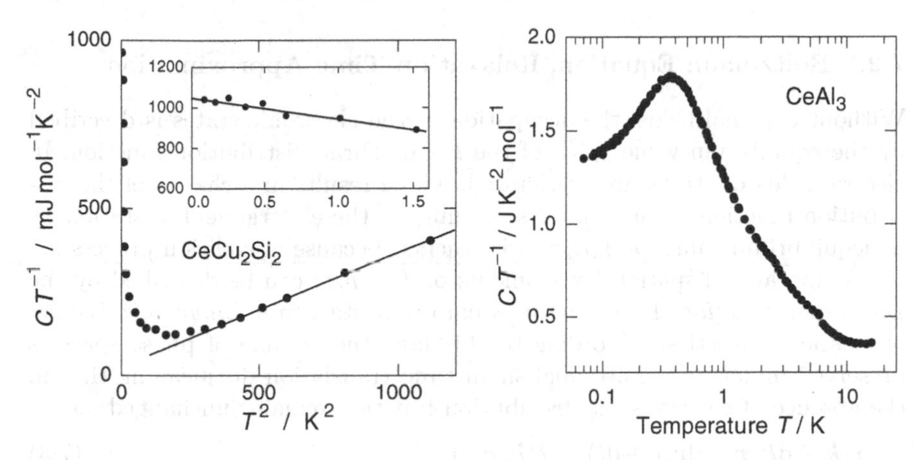

Fig. 7.3. Specific heat of heavy-fermion systems. (**a**) Specific heat divided by temperature C/T of $CeCu_2Si_2$ versus T^2. The low-temperature behavior is shown in the insert in the upper part of the figure with spread scale [305]. (**b**) C/T of $CeAl_3$ versus T [306]

7.2 Electrical Conductivity

In this section, we discuss some interesting low-temperature aspects of the electrical conductivity of metals. Superconductivity will be excluded from our discussion because we will treat this phenomenon separately in Chap. 10. To avoid unnecessary conceptual complications we mainly pay attention to simple metals. Concerning transition metals we only make a very brief remark since they often stand out due to peculiarities in their physical properties. As mentioned before, in these metals not only s-electrons but also d-electrons are present in the conduction band. Although the density of states of d-electrons at the Fermi energy exceeds that of s-electrons they hardly contribute to conduction because of their low mobility. This poor mobility is mainly due to the high effective mass of the electrons in the narrow subbands. In transition metals the mobility of the s-band electrons is generally lower than in simple metals. This reduction is caused by the scattering of s-electrons into d-band states. As a result, the conductivity of transition metals is, in general, smaller than that of simple metals although both s-band and d-band electrons are present.

The application of a dc electric field causes a shift of the Fermi sphere as a whole. But after a very short time a stationary state is reached, since the electrons located at the front of the sphere are scattered to its back by various processes. Before we consider some of these scattering mechanisms in more detail, and compare theoretical predictions with experimental results, we briefly introduce the relaxation-time approximation that is usually applied in theoretical treatments of the electrical conductivity.

7.2.1 Boltzmann Equation, Relaxation-Time Approximation

Without external fields, the occupation of the electronic states is described by the equilibrium value $f_0(\boldsymbol{k})$ of the Fermi–Dirac distribution function. In electric fields electrons are accelerated, which results in a change of the distribution function. Shortly after switching on the electric field, a stationary nonequilibrium value for $f(\boldsymbol{k},\boldsymbol{r},t)$ is reached because of collision processes.

Temporal and spatial developments of $f(\boldsymbol{k},\boldsymbol{r},t)$ can be described by the *Boltzmann equation*. To set up this equation we start from *Liouville's theorem* of classical statistics. According to this law, the volume of phase space is preserved under an infinitesimal small time translation dt, meaning that in the absence of collisions the distribution function remains unchanged, i.e.,

$$f(\boldsymbol{k}+\mathrm{d}\boldsymbol{k},\boldsymbol{r}+\mathrm{d}\boldsymbol{r},t+\mathrm{d}t) = f(\boldsymbol{k},\boldsymbol{r},t)\,. \tag{7.3}$$

Expansion to linear order in dt leads to

$$f(\boldsymbol{k}+\mathrm{d}\boldsymbol{k},\boldsymbol{r}+\mathrm{d}\boldsymbol{r},t+\mathrm{d}t) - f(\boldsymbol{k},\boldsymbol{r},t) = \\ = \frac{\partial f}{\partial \boldsymbol{k}}\cdot \mathrm{d}\boldsymbol{k} + \frac{\partial f}{\partial \boldsymbol{r}}\cdot \mathrm{d}\boldsymbol{r} + \frac{\partial f}{\partial t}\,\mathrm{d}t = 0\,. \tag{7.4}$$

Collisions may summarily be taken into account by introducing a 'correction term' $(\partial f/\partial t)_{\mathrm{coll}}$. This leads to the celebrated Boltzmann equation

$$\dot{\boldsymbol{k}}\cdot\frac{\partial f}{\partial \boldsymbol{k}} + \dot{\boldsymbol{r}}\cdot\frac{\partial f}{\partial \boldsymbol{r}} + \frac{\partial f}{\partial t} = \left.\frac{\partial f}{\partial t}\right|_{\mathrm{coll}}, \tag{7.5}$$

which is a starting point often used for the treatment of transport problems. The first two terms are called the *field term* and the *diffusion term*, respectively. On the right-hand side we find the *scattering term* or *collision term*. The diffusion term vanishes, since a constant electrical field does not change the homogeneous spatial distribution of electrons. Therefore, we omit this term henceforth.

It is possible to find quantum-mechanical expressions for specific scattering processes. However, inserting these expressions in (7.5) often leads to a nonlinear integrodifferential equation that is difficult to treat [307]. Therefore, we do not pursue this idea further but exploit the simple *relaxation time approximation*

$$\left.\frac{\partial f(\boldsymbol{k})}{\partial t}\right|_{\text{coll}} = -\frac{f(\boldsymbol{k}) - f_0(\boldsymbol{k})}{\tau(\boldsymbol{k})} \,. \tag{7.6}$$

This is based on the assumption that a nonequilibrium distribution $f(\boldsymbol{k}, \boldsymbol{r}, t)$ gradually returns to its equilibrium value within a characteristic time, the relaxation time $\tau(\boldsymbol{k})$, by the scattering of electrons with the wave vector $\boldsymbol{k}$ into states $\boldsymbol{k}'$, and vice versa.

The dc electrical conductivity follows from the current density $\boldsymbol{j}$ caused by the constant electrical field $\boldsymbol{\mathcal{E}}$. We disregard transients and consider only the stationary state where $\partial f / \partial t$ vanishes. In this case, the distribution function does not vary either with space or time, i.e., it is only a function of the wave vector $\boldsymbol{k}$.

In an electric field $\boldsymbol{\mathcal{E}}$, the electrons are subjected to the force $\hbar \dot{\boldsymbol{k}} = -e\boldsymbol{\mathcal{E}}$. Inserting $\dot{\boldsymbol{k}}$ and the collision term (7.6) in (7.5), we find

$$-\frac{e}{\hbar}\boldsymbol{\mathcal{E}} \cdot \frac{\partial f(\boldsymbol{k})}{\partial \boldsymbol{k}} = -\frac{f(\boldsymbol{k}) - f_0(\boldsymbol{k})}{\tau(\boldsymbol{k})} \,. \tag{7.7}$$

Rearranging the terms in this equation we obtain the simple-looking expression

$$f(\boldsymbol{k}) = f_0(\boldsymbol{k}) + \frac{e\tau(\boldsymbol{k})}{\hbar}\boldsymbol{\mathcal{E}} \cdot \frac{\partial f(\boldsymbol{k})}{\partial \boldsymbol{k}} \,. \tag{7.8}$$

For small deviations from equilibrium, the gradient $\partial f(\boldsymbol{k})/\partial \boldsymbol{k}$ of the actual distribution function can be replaced by the gradient of $f_0(\boldsymbol{k})$. Through this simplification we obtain the *linearized Boltzmann equation*

$$f(\boldsymbol{k}) \approx f_0(\boldsymbol{k}) + \frac{e\tau(\boldsymbol{k})}{\hbar}\boldsymbol{\mathcal{E}} \cdot \frac{\partial f_0(\boldsymbol{k})}{\partial \boldsymbol{k}} \,. \tag{7.9}$$

For the sake of simplicity, henceforth we consider an isotropic or cubic solid in which the current is only flowing in the field direction, i.e., $j_y = j_z = 0$ for a field in the x-direction. Since each electron gives the contribution $-ev_x(k)$ to the electrical current, we obtain j_x by integrating over all occupied states. With (3.6) for the density in k-space we find

$$j_x = -e \int D(k)\, v_x(k)\, f(k)\, \mathrm{d}k = -\frac{e}{\pi^2} \int k^2\, v_x(k)\, f(k)\, \mathrm{d}k \,. \tag{7.10}$$

Since the current is only driven by the gradient of the Fermi distribution, the first term in (7.9) may be omitted. Inserting $\partial f_0(k)/\partial k_x = [\partial f_0(k)/\partial E]\,\hbar v_x$, the current density can be expressed by

$$j_x = -\frac{e^2 \mathcal{E}_x}{\pi^2} \int k^2 v_x^2(k)\, \tau(k)\, \frac{\partial f_0(k)}{\partial E}\, \mathrm{d}k\,. \tag{7.11}$$

The Fermi–Dirac distribution is virtually constant except at the Fermi energy where it exhibits a step-like decrease. Therefore, we may replace its derivative by a delta function and write $[\partial f_0(\boldsymbol{k})/\partial E] \approx -\delta(E - E_\mathrm{F})$. Furthermore, from the dispersion relation $E = \hbar^2 k^2/2m$ of free electrons it follows that $\mathrm{d}k = \mathrm{d}E/\hbar v$. Inserting this relation and the delta function in (7.11) we find for the electrical conductivity σ the expression

$$\begin{aligned}\sigma = \frac{j_x}{\mathcal{E}} &\approx \frac{e^2}{\pi^2 \hbar} \int \frac{v_x^2(k)}{v(k)}\, k^2\, \tau(k)\, \delta(E - E_\mathrm{F})\, \mathrm{d}E \\ &\approx \frac{e^2}{\pi^2 \hbar} \frac{v(k_\mathrm{F})}{3}\, k_\mathrm{F}^2\, \tau(k_\mathrm{F})\,.\end{aligned} \tag{7.12}$$

The second line of this equation is obtained by carrying out the integration after replacing v_x^2 by $v^2/3$ because of the isotropic properties of the free-electron gas. Note that only electrons at the Fermi surface contribute to the charge transport, since it is only there that the gradient of the distribution function differs distinctly from zero.

Inserting the density of states at the Fermi surface $D(E_\mathrm{F}) = 3n/(2E_\mathrm{F})$ in (7.12) we finally find for the electrical conductivity the equation

$$\sigma = \frac{1}{3} e^2\, D(E_\mathrm{F})\, v_\mathrm{F}^2\, \tau(E_\mathrm{F})\,. \tag{7.13}$$

With the Fermi velocity $v_\mathrm{F}^2 = 2E_\mathrm{F}/m$, the expression can be simplified further, and we obtain finally

$$\sigma = \frac{n\, e^2}{m}\, \tau(E_\mathrm{F})\,. \tag{7.14}$$

Therefore, a full description of the electrical conductivity of metals can be traced back to the treatment of the relevant collision processes. It should be mentioned that the same expression was found by *Drude* at the turn of the nineteenth century under the erroneous assumption that *all* electrons contribute to the electrical current.

Before we discuss the collision processes determining the resistance of metals, let us make a short remark regarding the electron–electron scattering. At first glance it seems that this process should be of great importance because of the high electron density of metals and the long range of the Coulomb interaction. However, there are two reasons why this process can be neglected under most circumstances. On the one hand, electrons are screened by the lattice ions, resulting in a strong reduction of the range of the Coulomb potential. On the other hand, only a small proportion of the electrons can participate in collision processes because of the exclusion principle. As already mentioned in the discussion of the ideal Fermi gas, and described by (3.17), the rate of electron–electron collisions is reduced by a factor $(T/T_F)^2$ due to this latter effect. Estimates of the electron–electron scattering rate result in values much lower than the rates measured experimentally. Therefore, electron–electron scattering processes can be neglected and we do not discuss them further.

7.2.2 Residual Resistivity of Metals – Matthiessen's Rule

In most cases, the relaxation time τ of the conduction electron is limited by scattering through defects or through lattice vibrations. If the two mechanisms are independent of each other, the effective scattering rate τ^{-1} is, to a good approximation[1], just given by the sum of the individual rates, τ_{de}^{-1} and τ_{ph}^{-1}, caused by defects and phonons, respectively:

$$\tau^{-1} = \tau_{\text{de}}^{-1} + \tau_{\text{ph}}^{-1} . \tag{7.15}$$

Consequently, the electrical resistivity $\varrho = 1/\sigma$ can be split into two independent components and we obtain

$$\varrho = \frac{m}{ne^2\tau} = \frac{m}{ne^2\tau_{\text{de}}} + \frac{m}{ne^2\tau_{\text{ph}}(T)} = \varrho_{\text{de}} + \varrho_{\text{ph}}(T) . \tag{7.16}$$

This empirical relationship is called *Matthiessen's rule* (1862). As indicated, the electron–phonon scattering leads to a temperature-dependent contribution to the resistance, while defect scattering is temperature independent.

According to (6.26) scattering rate τ^{-1} and inverse mean free path ℓ^{-1} are connected via the relation

$$\tau^{-1} = v\ell^{-1} = v n \Sigma . \tag{7.17}$$

Here, v stands for the electron velocity, n for the density of the scattering centers, and Σ for the total scattering cross section. In this chapter, we use the symbol Σ for the cross section in order to avoid confusion with the abbreviation for the conductivity σ.

For vanishing temperatures, the phonon scattering is 'frozen out' and the electrical resistivity is determined by defect scattering only. Consequently, a temperature-independent *residual resistivity* is observed that depends on the density of scattering centers and therefore on purity and quality of the sample. As an example, we show in Fig. 7.4 the conductivity of relatively pure copper and the alloy Manganin ($Cu_{0.86}Mn_{0.12}Ni_{0.02}$). For the entire temperature range the resistivity of pure copper is considerably lower than that of Manganin. At low temperatures, the resistivity of both materials is only due to defect scattering and is constant, as expected. At temperatures above 20 K a pronounced temperature dependence is observed for copper. The relatively weak temperature dependence of Manganin indicates that in this alloy, defect scattering also dominates at high temperatures.

[1] This assumption is critically discussed in [302].

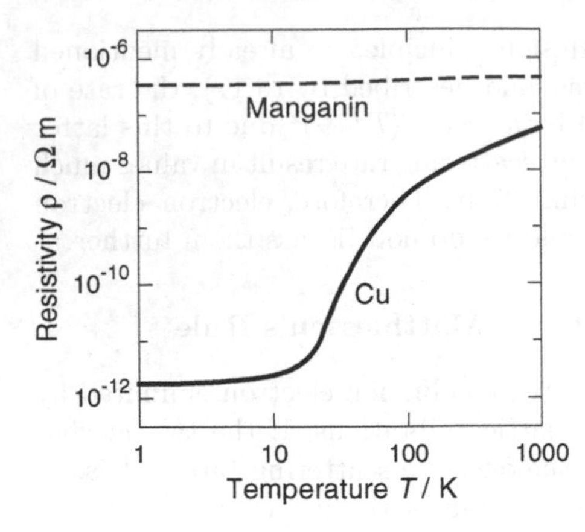

Fig. 7.4. Temperature dependence of the electrical resistivity of copper (*full line*) and the alloy Manganin (*dashed line*)

In the following sections, we will discuss scattering by impurities, phonons and also by spin waves that contribute to the resistivity of ferromagnets.

7.2.3 Impurity Scattering

Impurities influence the mean free path of charge carriers because they destroy the lattice periodicity. The resulting electron–impurity scattering is very effective if local variations in the charge density accompany the defects. This happens if the number of valence electrons of the regular atoms is different from the number in the impurity atoms. Although defect scattering is not a typical low-temperature phenomenon, it is also important in this temperature range. Therefore, we will also consider examples of this phenomenon.

For the sake of completeness we want to mention that large impurity concentrations can change the concentration of the charge carriers noticeably and thus the shape and position of the Fermi surface. As a consequence, the scattering probabilities and hence the collision rates are changed. However, in the following discussion we will disregard the repercussions of this effect.

Influence of Local Charge Fluctuations

If the charge of the ion core of the impurities differs from that of the host lattice, the resulting Coulomb field will give rise to scattering, as mentioned above. This scattering may be treated similarly to the well-known Rutherford scattering that results in a cross section $\Sigma \propto (\Delta Z)^2$, where ΔZ represents the charge difference. Consequently, the residual resistivity ϱ_D should increase proportional to $(\Delta Z)^2$ provided the impurity content is small enough.

As an example, the residual resistivity of various copper samples is shown in Fig. 7.5. These samples were alloyed with about one atomic per cent of an element with a higher number of valence electrons. In fact, the residual resistivity ϱ_de, normalized to the impurity content, increases proportional to $(\Delta Z)^2$, as predicted.

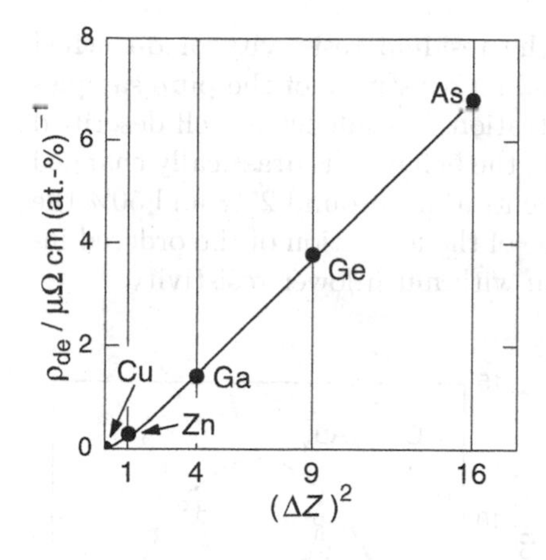

Fig. 7.5. Residual resistivity $\varrho_{\rm de}$ of copper alloyed with elements having a different number of valence electrons. The resistivity is normalized to an impurity content of one atomic per cent [308]

Concentration Dependence

Alloying two metals usually leads to an increase of the resistivity because of the loss of periodicity. Provided that the two components are completely miscible, as is the case in disordered alloys, the concentration dependence of the resistivity can be described with *Nordheim's rule* [309].

This rule can be made plausible by simple considerations: We start with the reasonable assumption that for the binary mixture A_xB_{1-x} an average potential with amplitude $U_0 = xU_{\rm A} + (1-x)U_{\rm B}$ exists, where U_A and U_B represent the potentials of the atoms A and B, respectively. Deviation from the average potential is given by $(U_0 - U_A) = (1-x)(U_{\rm B} - U_{\rm A})$ at the sites of atoms A, and $(U_0 - U_{\rm B}) = x(U_{\rm A} - U_{\rm B})$ at the atoms B. These deviations give rise to the above-mentioned electron scattering. As a crude approximation, the probabilities $w_{\rm A}$ and $w_{\rm B}$ for the scattering of the conduction electrons are given by

$$w_{\rm A} = (1-x)^2 \left| \int \psi^*(\boldsymbol{k})\,(U_{\rm B} - U_{\rm A})\psi(\boldsymbol{k}')\,{\rm d}^3k' \right|^2 \tag{7.18}$$

and

$$w_{\rm B} = x^2 \left| \int \psi^*(\boldsymbol{k})\,(U_{\rm A} - U_{\rm B})\psi(\boldsymbol{k}')\,{\rm d}^3k' \right|^2 = \frac{x^2}{(1-x)^2}\,w_{\rm A}\,. \tag{7.19}$$

From (7.16) and (7.17) it follows that the resistivity can be expressed by $\varrho_{\rm de} \propto xw_{\rm A} + (1-x)w_{\rm B}$ since scattering probability and scattering cross section are proportional to each other. Inserting (7.18) and (7.19) leads directly to Nordheim's rule

$$\varrho_{\rm de} \propto x(1-x)\,. \tag{7.20}$$

This concentration dependence has been confirmed by measurements of the resistivity of many disordered alloys. In particular, a maximum is observed at $x = 0.5$, since at this composition the disorder has its maximum.

As an example, we show in Fig. 7.6a the residual resistivity of quenched copper-gold alloys. In this graph, the residual resistivity of the pure samples has been subtracted. The overall concentration dependence is well described by Nordheim's rule. As shown in Fig. 7.6b, the behavior is drastically changed after annealing the samples. At gold concentrations around 25% and 50% the resistivity is considerably reduced because of the formation of the ordered intermetallic compounds CuAu and Cu_3Au with much lower resistivity.

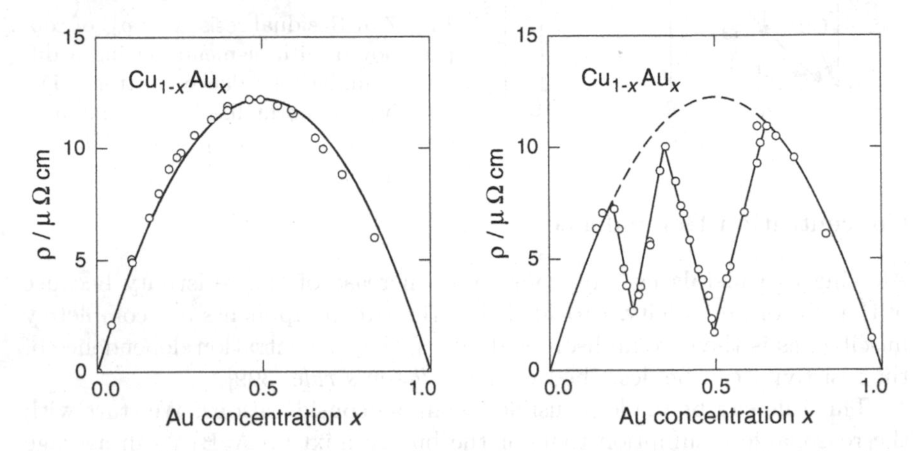

Fig. 7.6. (**a**) Residual resistivity of quenched copper-gold alloys. The *full line* represents the relation $x(1-x)$ predicted by Nordheim's rule. (**b**) Residual resistivity of copper-gold alloys after annealing. The *dashed line* shows the prediction of Nordheim's rule [310]

7.2.4 Electron–Phonon Scattering

As mentioned in the introduction to this section, in an electric field the Fermi sphere is shifted and scattering of electrons from the front to the empty states in the back takes place thus stabilizing the position of the sphere. Besides impurities, phonons give rise to such scattering processes (see Fig. 7.7). Since phonon energies are small compared to the Fermi energy only electrons close to the Fermi surface can participate in scattering events. This means that the magnitude $|\boldsymbol{k}|$ of the electronic wave vector is hardly changed. Therefore, we may put $|\boldsymbol{k}| \approx |\boldsymbol{k}'|$, where $\boldsymbol{k}$ and $\boldsymbol{k}'$ represent the wave vectors before and after the collision.

In Fig. 7.7 electron–phonon scattering processes are illustrated. The Fermi sphere is drawn together with the Brillouin zone of a simple cubic lattice. The shift δk of the spheres is caused by the applied electric field. The blurring of the Fermi surface due to thermal excitations is indicated by tinted areas. At high temperatures, the wave number q of the dominant thermal phonons

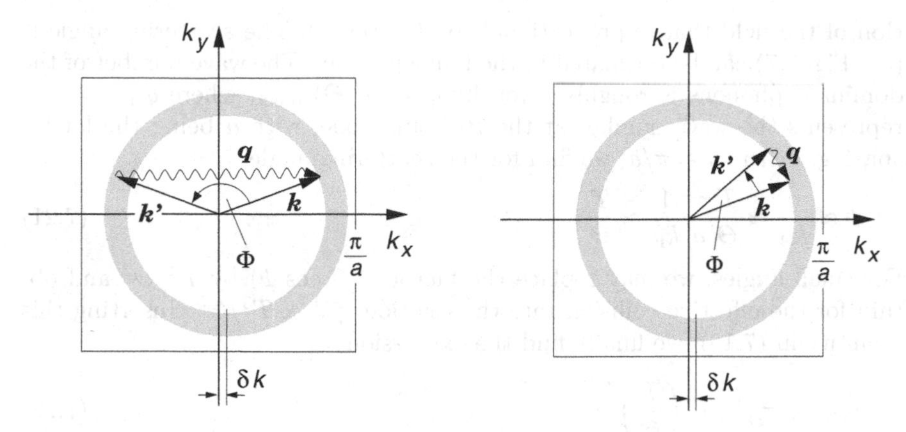

Fig. 7.7. Illustration of electron–phonon scattering processes in a free-electron gas. The electronic wave vectors are represented by *straight arrows*, those of the phonons by *wavy arrows*. The blurring of the Fermi surface is indicated by *tinted areas*. The collision of an electron with (**a**) a short-wavelength phonon (high temperature), and with (**b**) a long-wavelength phonon (low temperature) are shown

is relatively large meaning that the scattering angle Φ is also large. Thus, one scattering event is sufficient to scatter the electrons from the front to the back of the Fermi sphere. These scattering events are the main source of the electric resistance at high temperatures. With decreasing temperature, the wave number of the dominant phonons becomes smaller, and scattering events only give rise to small angle changes. Therefore, many scattering events are necessary to transport electrons along the Fermi surface from the front to the back of the sphere meaning that phonon scattering becomes less effective.

A thorough quantitative theoretical treatment of the electron–phonon scattering is rather involved. Therefore, we will discuss the temperature dependence of the electrical conductivity only qualitatively: The mean free path of free electrons and hence their scattering rate is given by the density of the colliding phonons and the cross section Σ. Phonons with energy $\hbar\omega \approx k_\mathrm{B}T$ are dominant and are the main contributors to the scattering rate. At low temperatures, i.e., below the Debye temperature Θ, the phonon density n is proportional to T^2/Θ^2. As in the case of phonon–phonon scattering (see Sect. 6.2), the cross section Σ is proportional to the square of the amplitude e_0 of the strain fields caused by the phonon participating in the collision process. Since, for thermal phonons $e_0^2 \propto \omega/\Theta \propto T/\Theta$, we find for the relaxation rate the proportionality $\tau^{-1} \propto n\Sigma \propto T^3/\Theta^3$. However, this is not the quantity that should be inserted in the Boltzmann equation. The relaxation time introduced by (7.6) is the time needed by nonequilibrium electrons to travel from the front to the back of the Fermi sphere. As pointed out above, at low temperatures a single collision is not sufficient to achieve this transition. The crucial quantity is the momentum transfer in the direction of the field that is proportional to $(1 - \cos \Phi)$. The scattering angle Φ (see Fig. 7.7) can be estimated in the following way: The wave number of the dominant phonons is roughly given by $q \approx (T/\Theta)\, q_{\text{max}}$, where $q_{\text{max}} = \pi/a$ represents the wave number at the Brillouin zone, with a being the lattice spacing. With $k_F \approx \pi/a$, we find for the scattering angle

$$\Phi \approx \frac{q}{k_F} \approx \frac{T}{\Theta}\frac{\pi}{a}\frac{1}{k_F} \approx \frac{T}{\Theta}. \tag{7.21}$$

For small angles, we may replace the factor $(1 - \cos \Phi)$ by T^2/Θ^2 and obtain for the effective collision rate the relation $\tau_{\text{eff}}^{-1} \propto T^5/\Theta^5$. Inserting this quantity in (7.16), we finally find the expression

$$\varrho_{\text{ph}} \propto \tau_{\text{eff}}^{-1} \propto \left(\frac{T}{\Theta}\right)^5 \tag{7.22}$$

for the low-temperature electrical resistivity caused by lattice vibrations. In a more careful treatment, an integration over the whole Fermi sphere has to be carried out, but the qualitative arguments are unchanged. The exact expression is known as the *Bloch–Grüneisen relation*. In Fig. 7.8, the temperature dependence of the resistivity of several metals is plotted after normalization to the value ϱ_Θ at the Debye temperature. Clearly, the relation (7.22) is well obeyed by these metals.

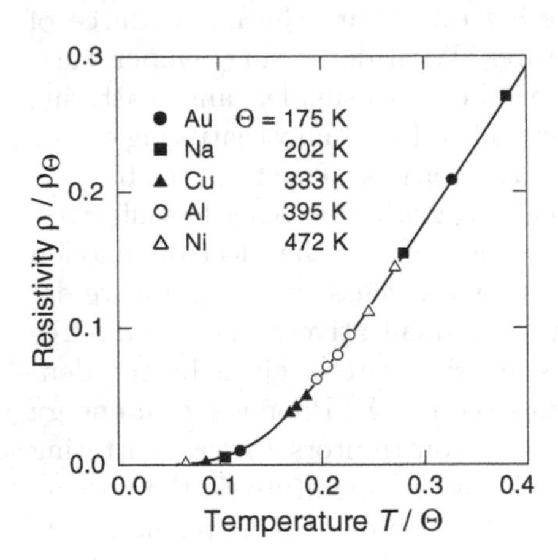

Fig. 7.8. Normalized resistivity ϱ/ϱ_Θ plotted as a function of the reduced temperature T/Θ. The *full line* depicts the prediction of the exact calculation [311]

7.2.5 Electron–Magnon Scattering

The electrical conductivity of *ferromagnetic materials* is a very interesting example of low-temperature physics. Besides defect scattering, the scattering of electrons by spin waves is the limiting mechanism for the electrical conductivity of ferromagnets at low temperatures. Spin waves are the collective excitations of the aligned spins in ferromagnets that will be discussed in more detail in Sect. 8.2. The quasiparticles assigned to these excitations are called *magnons*. The dispersion curve of spin waves in ferromagnets is schematically depicted in Fig. 7.9a. The energy gap Δ_{ma} at $q = 0$ is caused by the anisotropy of the exchange interaction and depends therefore on the structural characteristics of the ferromagnet under consideration.

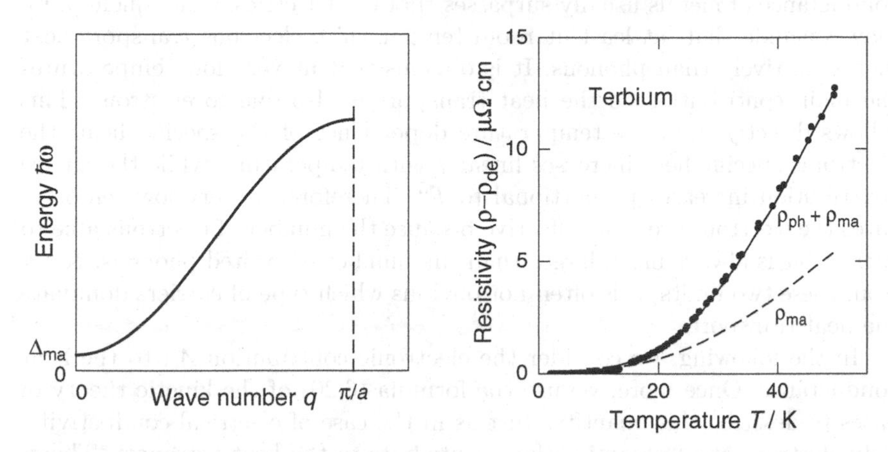

Fig. 7.9. (**a**) Illustration of the dispersion relation of spin waves. (**b**) Electrical resistivity of terbium as a function of temperature. The temperature-independent residual resistivity ϱ_{de} has been subtracted. The *full line* depicts the sum of the resistance caused by phonon and magnon scattering, the *dashed line* shows the calculated magnon contribution [312]

The scattering rate and hence the magnon contribution ϱ_{ma} to the resistivity is proportional to the number of magnons present in the sample. This number depends on temperature and on the shape of the dispersion curve. However, the exact shape plays a minor role if the gap is comparable to or even larger than $k_{\mathrm{B}}T$, because in this case the magnon density n_{ma} increases exponentially with temperature. Thus, we expect for the scattering rate the relation

$$\tau^{-1} \propto n_{\mathrm{ma}} \propto \mathrm{e}^{-\Delta_{\mathrm{ma}}/k_{\mathrm{B}}T}\,, \tag{7.23}$$

and for the electrical resistivity

$$\varrho_{\mathrm{ma}} \propto \mathrm{e}^{-\Delta_{\mathrm{ma}}/k_{\mathrm{B}}T}\,. \tag{7.24}$$

Both magnons and phonons contribute to the resistance of ferromagnets, i.e., $\varrho = \varrho_{\mathrm{ph}} + \varrho_{\mathrm{ma}}$. While the magnon part increases exponentially with temperature, the phonon part is proportional to T^5 according to (7.22). In Fig. 7.9b, the temperature variation of the resistivity of the ferromagnet terbium is shown. The *dashed line* shows the magnon part, the *full line* the sum of the magnon and the phonon contribution, which agrees well with the experimental data. In the case of terbium, the excitation gap is about $\Delta_{\mathrm{ma}}/k_{\mathrm{B}} \approx 20\,\mathrm{K}$.

7.3 Thermal Conductivity of Metals

We will first establish whether, in metals, the heat is mainly transported by electrons or phonons. There are two limiting temperature ranges where the answer is relatively simple. Our daily experience tells us that the thermal conductance of metals usually surpasses that of dielectrics. Consequently, we may conclude that, at least at room temperature, electrons transport heat more effectively than phonons. It is obvious that at very low temperatures the main contribution to the heat transport is also due to electrons. This follows directly from the temperature dependence of the specific heat: the electronic specific heat increases linearly with temperature, while the lattice contribution increases proportional to T^3. Therefore, at very low temperatures the electrons are more effective because the number of electrons able to carry heat is always much larger than the number of excited phonons. Away from these two limits, it is often not obvious which type of carriers dominate the heat transport.

In the following, we consider the electronic contribution Λ_{el} to the heat conductance. Once more, we use the formula (6.20) of the kinetic theory of gases to describe this quantity. Just as in the case of electrical conductivity, only electrons at the Fermi surface contribute to the heat transport. Therefore, we insert the electronic specific heat c_V^{el} and the Fermi velocity v_{F} in (6.20), and obtain

$$\Lambda_{\mathrm{el}} = \frac{1}{3}\, c_V^{\mathrm{el}}\, v\, \ell = \frac{1}{3}\, \frac{\pi^2 n k_{\mathrm{B}}^2 T}{m v_{\mathrm{F}}^2}\, v_{\mathrm{F}}\, \ell\,. \tag{7.25}$$

We assume that electric and thermal conduction have the same microscopic origin and eliminate the electronic mean free path ℓ by dividing the expression for the thermal conductivity (7.25) by that for the electrical conductivity (7.14):

$$\frac{\Lambda_{\mathrm{el}}}{\sigma} = \frac{\pi^2}{3}\left(\frac{k_{\mathrm{B}}}{e}\right)^2 T = \mathcal{L} T\,. \tag{7.26}$$

This relation is the famous *Wiedemann–Franz law* (1853). The ratio between the two transport quantities is exclusively determined by the universal *Lorenz number* $\mathcal{L} = 2.45 \times 10^{-8}\,\mathrm{V^2\,K^{-2}}$, and by the temperature. This remarkable law is valid if the electronic mean free path determining these two transport quantities is limited by the same process.

Before we continue with the discussion of the Wiedemann–Franz law, we briefly consider the different sensitivity of the electric and thermal transport to the scattering processes. The application of a constant electric field across a metallic sample changes the wave vector of the electrons. Because of their negative charge, electrons moving opposite to the field are accelerated, while those at the back of the Fermi sphere are slowed down. Altogether, the origin of the Fermi sphere is shifted by the amount $\boldsymbol{\delta k} = -e\tau\boldsymbol{\mathcal{E}}/\hbar$, and intimately connected, the electronic energy is changed by $\delta E \approx -e\tau \boldsymbol{v}(\boldsymbol{k})\cdot\boldsymbol{\mathcal{E}}$. This means that electric fields change the argument of the Fermi–Dirac distribution function by the amount $\boldsymbol{\delta k}$, but do not alter the functional form. Accordingly, we may write

$$f(\boldsymbol{k}) = f_0\left(\boldsymbol{k} + \frac{e\tau\boldsymbol{\mathcal{E}}}{\hbar}\right) = f_0\Big(E(\boldsymbol{k}) + e\tau\boldsymbol{v}(\boldsymbol{k})\cdot\boldsymbol{\mathcal{E}}\Big)\,. \tag{7.27}$$

This effect is illustrated in Fig. 7.10. It should be pointed out that the shift is the result of a dynamic process in which electrons at the front of the Fermi surface are scattered until they reach the back of the sphere.

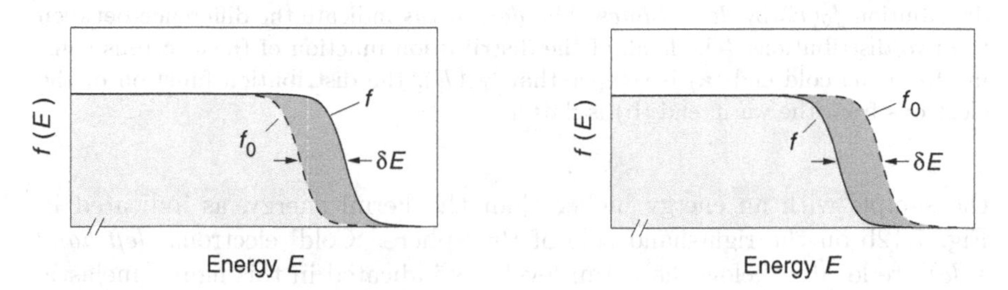

Fig. 7.10. Fermi–Dirac distribution $f(E)$ in a dc electric field. The equilibrium distribution $f_0(E)$ is represented by *dashed lines*, the distribution $f(E)$ in the electric field by *full lines*. The *dark areas* indicate the difference between the two distributions. The energy of the electrons on the front (**a**) of the Fermi sphere is increased by δE, on the back (**b**) the energy is reduced by the same amount

The situation is different for the transport of heat. The distribution $f(\boldsymbol{k},\boldsymbol{r})$ is different at the hot and cold end of the sample, but there is no net flow of electrons. Electrons coming from the hot end and moving with the heat current are faster than those travelling in the opposite direction. While an electric field causes a change of the wave vector, as discussed above, a temperature gradient leads to a temperature difference δT given by an analogous expression: $\delta T = -\tau\boldsymbol{v}(\boldsymbol{k})\cdot\nabla T$. The corresponding transformation of the Fermi distribution function thus reads:

$$f(T) = f_0\Big(T - \tau\boldsymbol{v}(\boldsymbol{k})\cdot\nabla T\Big)\,. \tag{7.28}$$

This result is illustrated in Fig. 7.11. The distribution function of the electrons coming from the cold end of the sample has a steeper flank than that from the warmer end.

Clearly, the relaxation processes have to meet different requirements in the electrical and thermal cases. Electric fields cause a shift of the Fermi sphere and relaxation processes compensate for the momentum gain. This process is shown in Fig. 7.12a, where electrons from the front of the Fermi sphere are scattered to the back. At high temperatures, a single collision is sufficient because of the high momentum carried by the colliding phonon. In a heat current there are 'hot' electrons coming from the warm end of

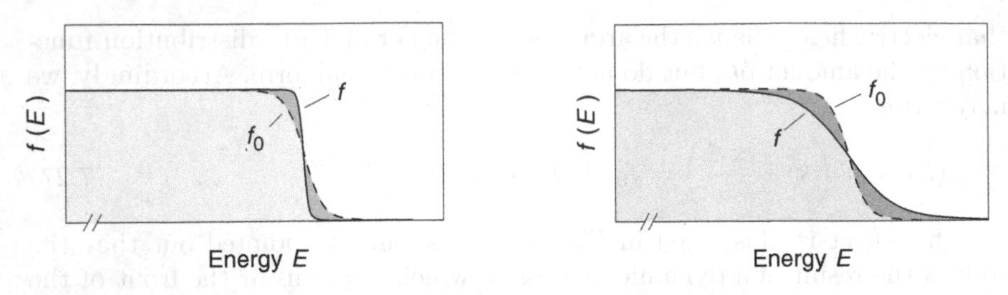

Fig. 7.11. Fermi–Dirac distribution $f(E)$ in the presence of a temperature gradient. The distribution function $f(E)$ is represented by *full lines*, the equilibrium distribution $f_0(E)$ by *dashed lines*. The *dark areas* indicate the difference between the two distributions. The flank of the distribution function of the electrons coming from the cold end (**a**) is steeper than $f_0(E)$, the distribution function of the electrons from the warm end (**b**) is flatter

the sample with an energy higher than the Fermi energy, as indicated in Fig. 7.12b on the right-hand side of the sphere. 'Cold' electrons (*left-hand side*) are located below the Fermi level. As indicated in this figure, inelastic electron scattering gives rise to relaxation at the Fermi surface. However, in this case the momentum change is unimportant because it is the energy relaxation that counts.

We finish our discussion by once more taking a quick look at the Wiedemann–Franz law. At high temperatures, scattering processes with large

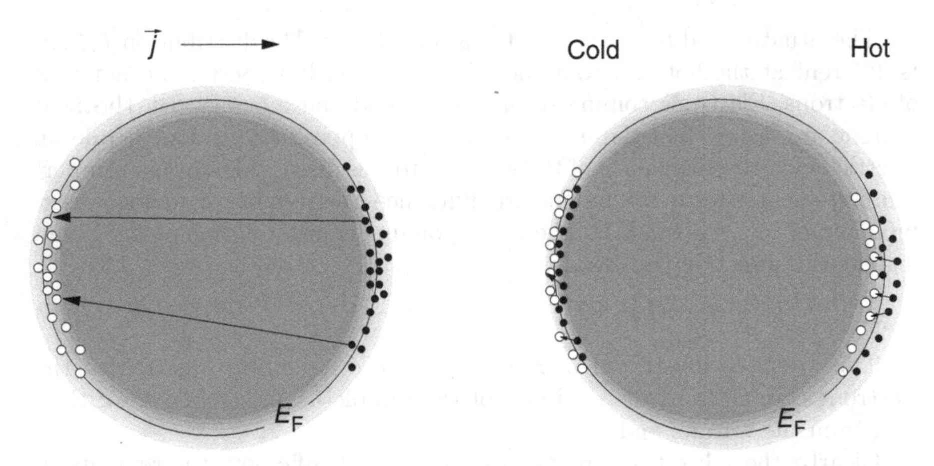

Fig. 7.12. Fermi sphere and scattering processes. The *tinted areas* represent the Fermi spheres, the *circles* the Fermi levels in equilibrium. Charge carriers exceeding the equilibrium occupation are represented by *dots* (electrons), and by *open circles* (holes or empty states). (**a**) Scattering processes stabilizing the position of the Fermi sphere in an electric field. (**b**) Scattering processes leading to a stationary state in the presence of a thermal gradient

momentum transfer limit both the electrical and the thermal conductivity. The mean free path can be eliminated as we showed in (7.26), since the basic assumptions of the Wiedemann–Franz law are fulfilled. At moderately low temperatures, processes with small momentum changes dominate. These processes cause the energy to change to the relaxed state but the momentum is hardly altered. This leads to a greater degradation of the thermal current than the electrical current. Consequently, the ratio $\Lambda_{\rm el}/\sigma T$ decreases and differs from the Lorenz number $\mathcal{L}$. Finally, at very low temperatures impurity scattering dominates both electrical and thermal transport, and the ratio $\Lambda_{\rm el}/\sigma T$ generally recovers.

In Fig. 7.13, experimental low-temperature data on the thermal conductivity of a copper foil are compared with the prediction (7.26). In this experiment, the sample was annealed repeatedly, resulting in a decreasing residual resistivity, and hence in an increasing thermal conductivity. As can be seen, the agreement between theoretical expectation and experiment is nearly perfect in this case. However, it should be mentioned that for other materials deviations from the Wiedemann–Franz law are also reported at low temperatures.

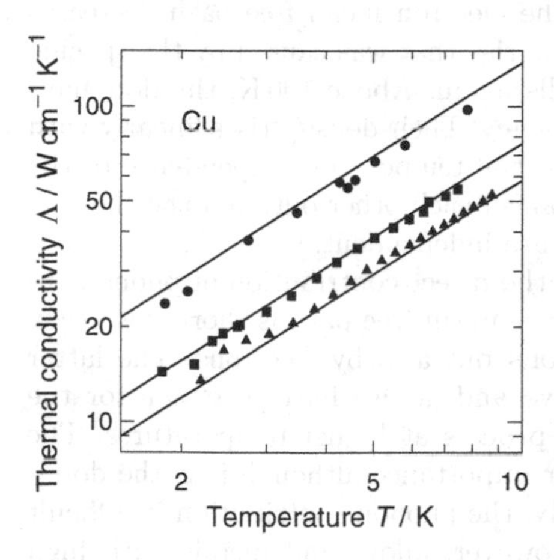

Fig. 7.13. Thermal conductivity of a copper foil 60 μm thick and 20 mm wide. By heat treatment, the residual electrical resistivity and the thermal conductivity were varied. The calculated thermal conductivities are shown by *full lines* [313]

As we already mentioned, at room temperature the thermal conductivity of most metals exceeds that of dielectric solids. An example of this is the thermal conductivity of copper plotted in Fig. 7.14. The overall temperature variation is similar to that of dielectric crystals. However, the temperature dependence at high and low temperatures is different. Note that the maximum conductivity of metals is considerably lower than that of pure dielectric crystals.

Qualitatively, the temperature variation of the thermal conductivity of pure metals can be explained by considering just the electronic contribution. At low temperatures, electrons are predominantly scattered by impurities,

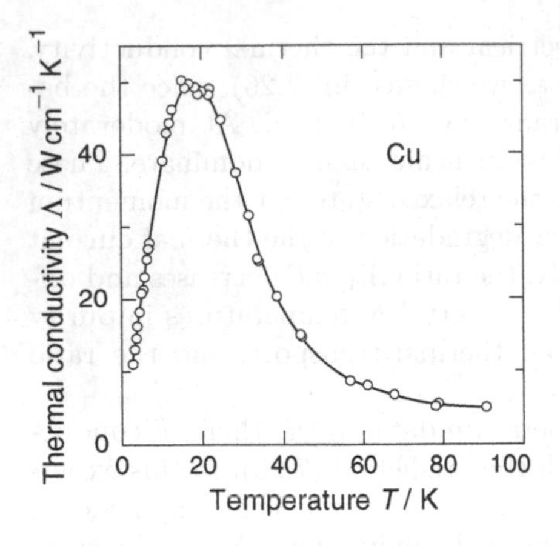

Fig. 7.14. Thermal conductivity of copper. As in dielectric crystals, a pronounced maximum is observed [314]

resulting in a constant electron mean free path. In this case, the linear temperature dependence of the specific heat gives rise to a linear increase of the thermal conductivity. With increasing temperature, the electron–phonon interaction becomes more and more important because of the growing number of high-frequency thermal phonons. The electron mean free path decreases rapidly, more than compensating for the rise that was caused by the specific heat, and the thermal conductivity falls again. Above 100 K, the dominant phonons are those with the Debye frequency. Their density rises linearly with temperature leading to $\ell \propto 1/T$. Thus, the temperature dependence of the specific heat and the mean free path cancel each other out, and the thermal conductivity becomes almost temperature independent.

Finally, we make a few remarks on the direct contribution of phonons to the thermal conductivity of metals. Their mean free path is short since they are not only scattered by other phonons but also by electrons. The latter scattering mechanism is rather effective and, as we have just seen for the electric conductivity, is the dominant process at higher temperatures. The phonon–phonon interaction is of minor importance although it is the dominant interaction in dielectrics. Typically, the phonon contribution is difficult to separate from that of electrons. However, alloys and metals with high impurity content are exceptions since in these solids electrons are strongly scattered by impurities and defects. Thus, their contribution to the heat transport is heavily reduced and the lattice contribution becomes dominant again.

7.4 Kondo Effect

As early as 1930, *Meissner* and *Voigt* observed that electrical resistivity does not become constant at low temperatures in all metals [311]. In particular, simple metals such as gold or copper with magnetic impurities exhibit a resistivity minimum, i.e., the resistivity rises again at low temperature. As an example, we show in Fig. 7.15 the resistivity of copper containing 440 ppm Fe. While the residual resistivity of pure copper stays constant at low temperatures, a pronounced minimum is found at about 27 K in the iron-doped sample. Above that minimum, the temperature variation of the resistivity of the two samples is nearly identical.

In 1964, it was shown by *Kondo* that this phenomenon reflects the spin-dependent scattering of the conduction electrons by the magnetic moments of impurity atoms [316]. Typical systems that show this effect are simple metals containing a small amount of a transition metal. With decreasing temperature the exchange interaction between the conduction electrons and the localized d-electrons of the impurity atoms becomes more and more significant, resulting in a rising electrical resistivity. Together with the T^5-dependent resistivity due to the electron–phonon interaction this specific scattering mechanism leads to the above-mentioned minimum. In this section, we want to sketch the theory describing this effect and show experimental results supporting these ideas.

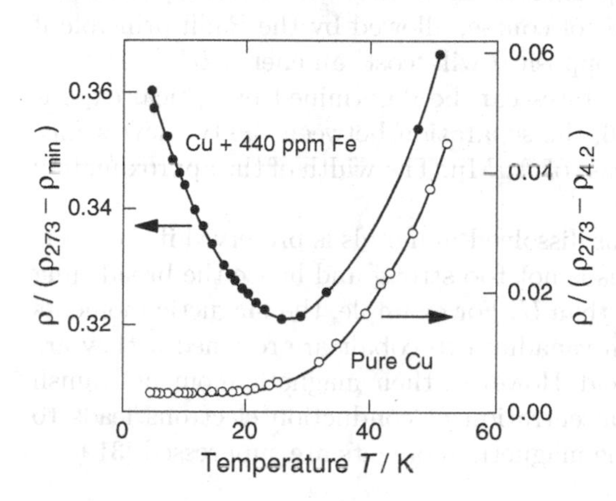

Fig. 7.15. Reduced electrical resistivity of pure copper and copper doped with 440 ppm Fe. $\varrho_{4.2}$, ϱ_{273}, and $\varrho_{\min}$ represent the resistivity at helium temperature, room temperature, and at the minimum of the resistance curve, respectively. The *full lines* connect the data points [315]

7.4.1 Localized Magnetic Moments

When small amounts of transition metals or rare-earth elements are dissolved in ordinary metals their localized magnetic moments are sometimes retained. Depending on the position of the ion levels with respect to the Fermi energy of the host, electrons from the ions might join the conduction electrons or, vice versa, electrons from the conduction band might drop into lower-lying ionic levels. In this way, the magnetic moment of the ion may be altered or even quenched. Furthermore, there is a mixing of the d- and f-levels of the ions with the degenerate continuum of the conduction-band levels. As a result, the d-electrons of the impurities are less localized and the charge distribution of the nearby conduction band levels is changed.

The starting point for a theoretical treatment of this phenomenon is usually the Hamiltonian introduced by *Anderson* in 1961 [317]. However, since we are not going to address the full theoretical aspects of this phenomenon here, we proceed now to discuss only some of the results. In a metallic environment, the d-levels of impurity atoms with the energy E_d become polarized and split by the d–d interaction. As mentioned above, the interaction between localized moments and conduction electrons leads to a hybridization that results in broadening of the localized levels. With the help of the golden rule their width W may be expressed by

$$\frac{W}{\hbar} = \frac{\pi}{\hbar}\mathcal{V}^2 D_s(E_{d\sigma}) , \tag{7.29}$$

where $\mathcal{V}$ represents the matrix element for transitions between s- and d-states and thus reflects the coupling strength. $D_s(E_{d\sigma})$ stands for the density of states of the conduction electrons that have an energy $E_{d\sigma}$, which is the energy of the d-resonances with the spin orientation σ. Because of the rather small spatial extent of the d-orbitals compared to the s-orbitals, there is an appreciable on-site Coulomb repulsion at the d-site, i.e., the occupation of a d-orbital with a second electron (of course, allowed by the Pauli principle if the spins of the d-electrons are opposite) will 'cost' an energy U.

The energy of the localized states can be determined by optical experiments. As indicated in Fig. 7.16, the separation between the two levels $E_{\mathrm{d}+}$ and $E_{\mathrm{d}-}$ is about 4.8 eV in the case of **Ag**Mn. The width of the approximately Lorentzian lines is about 0.5 eV.

The magnetic moment of ions dissolved in metals is preserved if the interaction with conduction electrons is not too strong and hence the broadening of the resonances much smaller than U. For example, the magnetic moments of the iron group elements from vanadium to cobalt are retained if they are dissolved in copper, silver, or gold. However, their magnetic moments vanish in aluminum since the high concentration of conduction electrons leads to such a large broadening that the magnetic moments are suppressed [319].

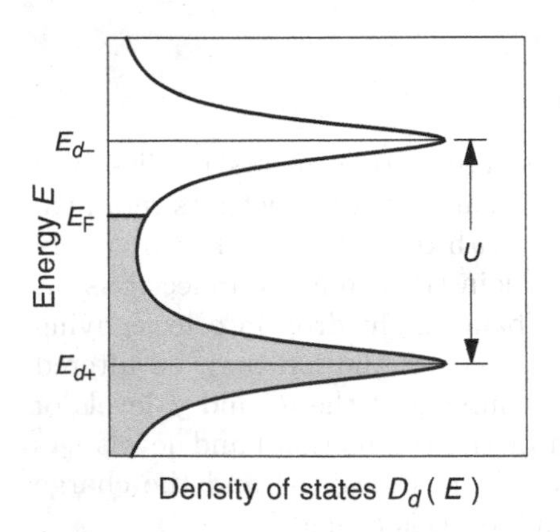

Fig. 7.16. Density of states $D_{\mathrm{d}}(E)$ of the d-resonances in **Ag**Mn. The level splitting $U = 4.8$ eV and the linewidth $W = 0.5$ eV were measured optically [318]

As mentioned above, the resonant states also act on the conduction-band electrons and cause a change of their charge distribution. For instance, suppose that all d-states are occupied with 'up-spins'. In this case, states with that orientation would no longer be accessible to s-electrons. Interactions would only be possible with 'down-spin' electrons. Thus, the system behaves as if there is a spin-dependent interaction between the localized spin $\boldsymbol{S}$ of the impurity ion at the position $\boldsymbol{R}$ and the spin $\boldsymbol{s}$ of the conduction electron at $\boldsymbol{r}$. This fact may be expressed by a Hamiltonian containing only the exchange term of the form

$$\mathcal{H}_{sd} = -J\,\boldsymbol{S}\cdot\boldsymbol{s}\;\delta(\boldsymbol{r}-\boldsymbol{R})\,. \tag{7.30}$$

Without justification, we state that in this so-called *s-d model* the coupling factor J is approximately given by $J \approx -\mathcal{V}^2/U$. The minus sign indicates that the antiparallel spin orientation is favored independent of the sign of $\mathcal{V}$.

Using the Hamiltonian $\mathcal{H}_{sd}$, the generalized susceptibility of the system can be calculated. In the free-electron model an analytic expression is obtained with a singularity of the second derivative at $q = 2k_\mathrm{F}$, where q is the wave number of the perturbation. A Fourier transformation gives us the spatial distribution of the magnetization in the neighborhood of localized moments. As shown in Fig. 7.17, the singularity gives rise to a spatial variation of the magnetization of the form $\cos(2k_\mathrm{F}r)/r^3$, i.e., to a magneti-

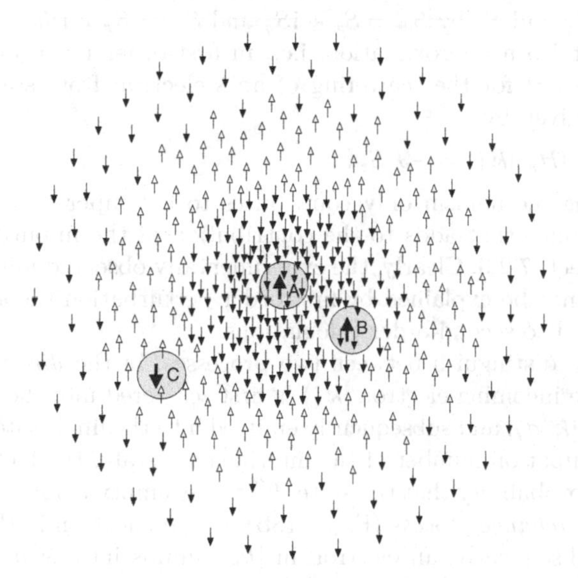

Fig. 7.17. Friedel oscillation of the spin polarization. The localized magnetic moment A causes an oscillation of the spin polarization of the neighboring conduction electrons, resulting in an indirect exchange interaction between neighboring impurities. Depending on the distance, a ferro- (B) or antiferromagnetic (C) alignment of the magnetic moments of the impurities is favored. For clarity, only the spin polarization due to the magnetic moment A is shown

zation oscillating with a period given by the wave number $2k_\mathrm{F}$ and decaying proportional to $1/r^3$. This effect is analogous to the so-called *Friedel oscillations* of the charge density that are seen near a charged impurity [320]. The induced magnetization causes an indirect exchange interaction between impurities that has antiferromagnetic or ferromagnetic character depending on the distance. The so-called *RKKY interaction* is named after *Ruderman, Kittel, Kasuya* and *Yosida* [321] and is of great importance in our discussion of the properties of magnetic compounds at the end of this chapter.

7.4.2 Electron Scattering by Localized Moments

We now consider the Kondo problem, i.e., we investigate the scattering of conduction electrons by localized magnetic moments. In principle, both elastic and inelastic scattering contribute to the electrical resistivity, but only the elastic scattering is relevant to the Kondo effect. To calculate the scattering amplitude we use the so-called N representation in which the Hamiltonian $\mathcal{H}_{sd}$ (7.30) has the form

$$\mathcal{H}_{sd} = -J \sum_{\boldsymbol{k}\boldsymbol{k}'} S_z(c^+_{\boldsymbol{k}'\uparrow} c_{\boldsymbol{k}\uparrow} - c^+_{\boldsymbol{k}'\downarrow} c_{\boldsymbol{k}\downarrow}) + S_+ c^+_{\boldsymbol{k}'\downarrow} c_{\boldsymbol{k}\uparrow} + S_- c^+_{\boldsymbol{k}'\uparrow} c_{\boldsymbol{k}\downarrow} \,. \tag{7.31}$$

Here, $c^+_{\boldsymbol{k}}$ and $c_{\boldsymbol{k}}$ depict the creation and annihilation operators that act on the free electrons with wave vector $\boldsymbol{k}$. Furthermore, we have replaced the spin components S_x and S_y by $S_+ = S_x + \mathrm{i}S_y$ and $S_- = S_x - \mathrm{i}S_y$.

In the first Born approximation, i.e., in first-order perturbation theory, the amplitude $t^{(1)}$ for the scattering of an s-electron from state $|\boldsymbol{k}\uparrow\rangle$ into state $|\boldsymbol{k}'\uparrow\rangle$ is given by

$$t^{(1)} = \langle \boldsymbol{k}'\uparrow|\mathcal{H}_{sd}|\boldsymbol{k}\uparrow\rangle = -J\, S_z \,. \tag{7.32}$$

This scattering mechanism only contributes to a temperature-independent residual resistance that adds to the contribution of the impurity scattering discussed in Sect. 7.2.3. Clearly, the experimentally observed minimum of the resistivity cannot be explained by first-order perturbation theory and so we have to proceed to second-order calculations.

We need to distinguish between two processes. In the *direct process* (see Fig. 7.18a) the incoming electron $|\boldsymbol{k}\uparrow\rangle$ is first scattered into the empty intermediate state $|\boldsymbol{k}''\sigma\rangle$, and subsequently emitted into the final state $|\boldsymbol{k}'\uparrow\rangle$. Since $f_{\boldsymbol{k}''}$ is the occupation number of the intermediate state, the factor $(1 - f_{\boldsymbol{k}''})$ describes the probability that the state $|\boldsymbol{k}''\sigma\rangle$ was empty before the scattering event. In the *exchange process* (Fig. 7.18b) first an electron in $|\boldsymbol{k}''\sigma\rangle$ scatters into $|\boldsymbol{k}'\sigma\rangle$. Subsequently, an electron in $|\boldsymbol{k}\sigma\rangle$ jumps into $|\boldsymbol{k}''\sigma\rangle$ that is now empty. These two processes lead to the scattering amplitude

$$\begin{aligned} t^{(2)} = \sum_{\boldsymbol{k}'',\sigma} \frac{1}{E(\boldsymbol{k}) - E(\boldsymbol{k}'')} \Big[&(1 - f_{\boldsymbol{k}''}) \langle \boldsymbol{k}'\uparrow|\mathcal{H}_{sd}|\boldsymbol{k}''\sigma\rangle \langle \boldsymbol{k}''\sigma|\mathcal{H}_{sd}|\boldsymbol{k}\uparrow\rangle \\ &+ f_{\boldsymbol{k}''} \langle \boldsymbol{k}''\sigma|\mathcal{H}_{sd}|\boldsymbol{k}\uparrow\rangle \langle \boldsymbol{k}'\uparrow|\mathcal{H}_{sd}|\boldsymbol{k}''\sigma\rangle \Big] \,. \end{aligned} \tag{7.33}$$

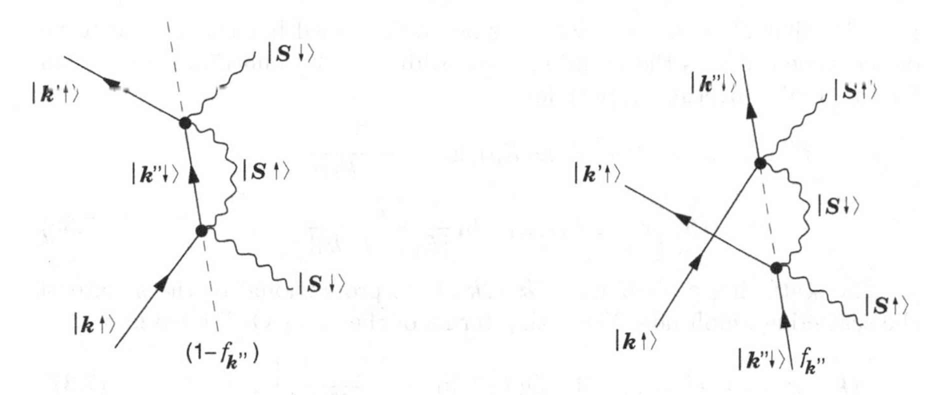

Fig. 7.18. Schematic diagrams of second-order scattering processes. (**a**) Direct process and (**b**) exchange process. Note that in both cases, events are drawn with a spin-flip in the intermediate state

Without spin-flip processes, i.e., taking into account only those terms of the Hamiltonian (7.30) that depend on S_z, the product of the matrix elements does not depend on their order, and the terms containing $f_{\boldsymbol{k}''}$ cancel each other out. The remaining contribution to the scattering cross section is only a small and temperature-independent correction to the first-order term (7.32). If spin-flip processes of the type shown in Fig. 7.18 are included, terms such as S_x and S_y, or S_+ and S_- have to be considered, which do not commute. As a consequence, terms containing $f_{\boldsymbol{k}''}$ do not cancel each other and give rise to a relatively strong and temperature-dependent scattering amplitude. Thus, spin-flip processes determine the electrical resistivity at low temperatures.

7.4.3 Kondo Resistance

The Kondo resistance can be calculated by evaluating the scattering amplitude (7.33). Neglecting the insignificant contribution of the direct processes, the second-order scattering amplitude is given by

$$t^{(2)} = J^2\, S_z \sum_{\boldsymbol{k}''} \frac{2 f_{\boldsymbol{k}''} - 1}{E(\boldsymbol{k}) - E(\boldsymbol{k}'')}\,. \tag{7.34}$$

We can replace the summation by an integration and assume that within the energy range $\delta E < |E_\mathrm{F} \pm \mathcal{D}|$ the density of states can be approximated by $D(E) \approx D(E_\mathrm{F})$. The integration of the second term, namely of the antisymmetric function $1/[E(\boldsymbol{k}) - E(\boldsymbol{k}'')]$, leads again to a small and unimportant temperature-independent contribution. $f_{\boldsymbol{k}''}$ is approximately constant, except in the vicinity of E_F, where its value changes abruptly. For simplification, we replace $f_{\boldsymbol{k}''}$ by a step function, carry out the integration, and obtain

$$t^{(2)} \approx 2\, J^2\, S_z\, D(E_\mathrm{F}) \ln \frac{\mathcal{D}}{|E_\mathrm{F} - E(\boldsymbol{k})|}\,, \tag{7.35}$$

provided that $\mathcal{D} \gg |E_F - E(\boldsymbol{k})|$. $\mathcal{D}$ is not well defined but is expected to be of the same order as the Fermi energy. Adding (7.32) and (7.35) we obtain for the total scattering amplitude

$$t^{(1)} + t^{(2)} = -JS_z + 2\,J^2\,S_z\,D(E_F)\,\ln\frac{\mathcal{D}}{|E_F - E(\boldsymbol{k})|}$$
$$= -JS_z\left[1 - 2\,J\,D(E_F)\,\ln\frac{\mathcal{D}}{|E_F - E(\boldsymbol{k})|}\right]. \qquad (7.36)$$

The scattering probability $w(\boldsymbol{k}\uparrow, \boldsymbol{k}'\uparrow)$ is proportional to the square of the scattering amplitude. Neglecting terms of the order $O(J^4)$, we find

$$w(\boldsymbol{k}\uparrow, \boldsymbol{k}'\uparrow) \propto J^2\,S_z^2\left[1 - 4\,J\,D(E_F)\,\ln\frac{\mathcal{D}}{|E_F - E(\boldsymbol{k})|}\right]. \qquad (7.37)$$

To obtain a final expression for the scattering probability an integration over all wave vectors or over all energies should be carried out. Since only electrons with the energy $E(\boldsymbol{k}) \approx E_F \pm k_B T$ can be scattered we may simply write for the resistivity $\varrho(T)$ owing to magnetic impurities in an otherwise nonmagnetic metal:

$$\varrho(T) \propto \varrho_0\left[1 - 4J\,D(E_F)\,\ln\frac{\mathcal{D}}{k_B T}\right]. \qquad (7.38)$$

Here, we have used the abbreviation ϱ_0 for the temperature-independent residual resistivity obtained in the first-order calculation. Since J is negative the second term leads to an increase of the resistivity with decreasing temperature. Adding the lattice contribution $\varrho_{ph} = aT^5$, we obtain for the total resistivity the expression

$$\varrho = aT^5 + c\varrho_0 + c\varrho_1\,\ln\frac{\mathcal{D}}{k_B T}, \qquad (7.39)$$

where c stands for the concentration of the magnetic impurities, and ϱ_1 is a positive constant. As observed in experiments, a characteristic minimum is predicted at

$$T_{min} = \left(\frac{c\varrho_1}{5a}\right)^{1/5}. \qquad (7.40)$$

As can be seen in Fig. 7.19a, the position of the minimum depends rather weakly on the impurity concentration. Although the iron concentration varies by a factor four, the resistivity minimum is only shifted by about 35%. The predicted logarithmic temperature variation of the resistivity below the minimum is clearly visible in Fig. 7.19b for the three gold samples with different impurities.

Although the theory presented here was an important step towards an understanding of the resistivity of metals with magnetic impurities, the description is unsatisfactory in several respects. In particular, the logarithmic divergence of the resistivity for $T \to 0$ not only contradicts experimental observations, it is also 'nonphysical'. At first glance, it seems that it could

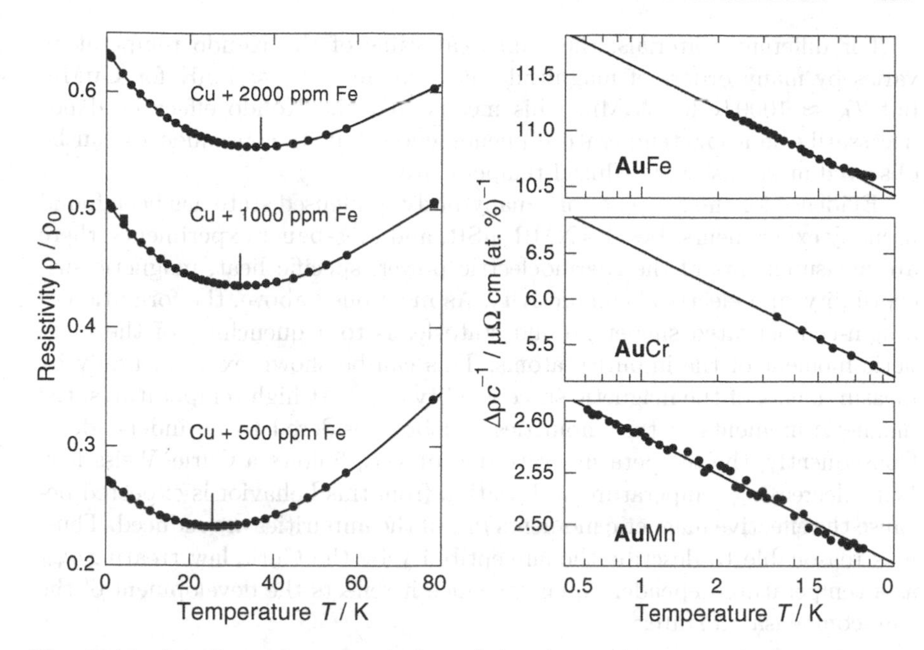

Fig. 7.19. (**a**) Temperature variation of the electrical resistivity of copper containing different concentrations of iron. The *full lines* represent the theoretical results [322]. (**b**) Electrical resistivity of gold samples with different magnetic impurities. The logarithmic variation of the resistivity below the resistance minimum is clearly visible [323]

be improved by taking into account higher-order corrections in the calculation of the scattering amplitude or accounting for the mutual polarization of neighboring impurity atoms. However, it turns out that the behavior of dilute magnetic alloys changes qualitatively in going below the so-called *Kondo temperature* T_{K}.

Until now, we have discussed the weak coupling regime ($T > T_{\mathrm{K}}$) in which the impurities carry well-defined moments. Below T_{K} we enter the strong coupling regime where many-body effects play an important role in causing a breakdown of perturbation theory. The magnetic moments of the impurities become screened by the spin polarization of the surrounding conduction electrons. Thus, a spatially extended correlation between impurity spins and the spins of the conduction electron develops resulting in a spin-compensated singlet ground state. The reason for the transition from a magnetic to a nonmagnetic state is the existence of an energy gain per magnetic impurity given by

$$k_{\mathrm{B}} T_{\mathrm{K}} = \mathcal{D}\, \mathrm{e}^{-1/J D(E_{\mathrm{F}})} \approx k_{\mathrm{B}} T_{\mathrm{F}}\, \mathrm{e}^{-1/J D(E_{\mathrm{F}})} \,. \tag{7.41}$$

Conversely, this internal binding energy must be overcome in order to strip off the spin-compensation cloud. For this reason, a maximum in the specific heat of Kondo systems is observed at temperatures around T_{K}.

For different materials, the numerical value of the Kondo temperature varies by many orders of magnitude. For example, $T_K \approx 1\,\text{mK}$ for **Cu**Mn, but $T_K \approx 1000\,\text{K}$ for **Al**Mn. This means that the Kondo effect need not necessarily be a low-temperature phenomenon, but its consequences can be observed most easily at reduced temperatures.

Evidence for the existence of a many-body condensed state has been found in many experiments. Besides NMR, μSR, and Mössbauer experiments, there are measurements of the thermoelectric power, specific heat, magnetic susceptibility, and electrical conductivity. As mentioned above, the formation of a spin-compensated singlet ground state leads to a quenching of the magnetic moment of the impurity atoms. This can be shown experimentally by measurements of the magnetic susceptibility χ_{imp}. At high temperatures, the magnetic moments of the impurities can be considered to be independent. Consequently, the temperature variation of χ_{imp} follows a Curie–Weiss law. With decreasing temperature, a deviation from this behavior is expected because the effective magnetic moments μ_{eff} of the impurities are reduced. Thus, it is reasonable to describe the susceptibility by the Curie law treating μ_{eff} as a temperature-dependent quantity since it reflects the development of the spin-compensated state:[2]

$$\chi_{\text{imp}} = \frac{\mu_{\text{eff}}^2(T)}{3k_B T} \,. \tag{7.42}$$

Figure 7.20 displays this development in the dilute alloy **Au**V with a Kondo temperature of about 300 K. The data convincingly demonstrates the disappearance of the magnetic moments with decreasing temperature.

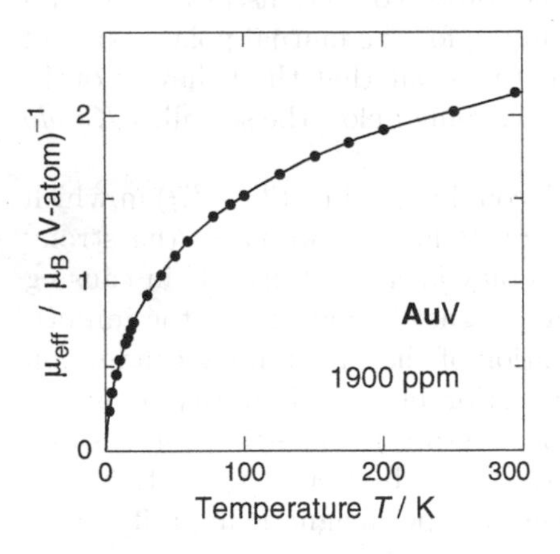

Fig. 7.20. Temperature dependence of the effective magnetic moment of vanadium in gold, derived from measurements of the magnetic susceptibility [324]

[2] In general, the atomic magnetic moment is given by $\mu_{\text{eff}}^2 = g^2 \mu_B^2 J(J+1)$, where g is the Landé factor and J, as customary in atomic physic, the angular momentum quantum number.

A clear, though indirect indication of the formation of a new state is found in resistivity measurements down to lower temperatures. As shown in Fig. 7.21, with decreasing temperature the resistivity caused by iron impurities increases logarithmically in a certain temperature range but finally flattens out. At still lower temperatures, the resistivity stays constant down to 35 mK [325]. It is a remarkable feature of this graph that data from different host materials fall on top of each other. This has been achieved by subtracting out the known resistivity due to the AuCu alloy and normalizing the data with respect to the concentration c and an appropriate temperature T_x that is approximately the Kondo temperature T_K.

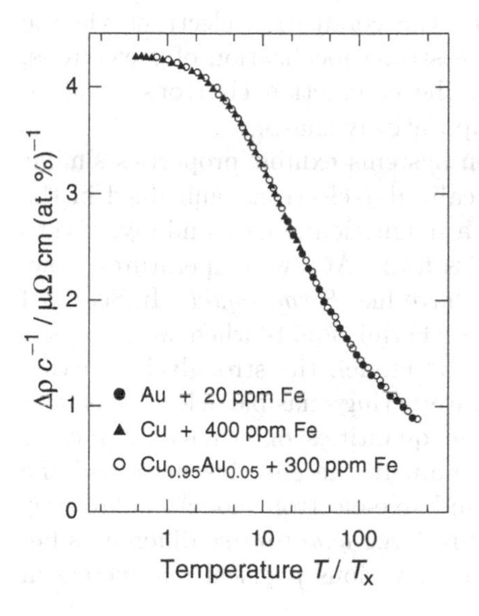

Fig. 7.21. Normalized electrical resistivity $\Delta\varrho/c$ of different Kondo systems plotted against the reduced temperature T/T_x. The values $T_x = 26$ mK, 690 mK, and 1 K were used for **Au**Fe, **CuAu**Fe, and **Cu**Fe, respectively [323, 326]

The Kondo effect, and in particular the many-body aspect of this phenomenon, has attracted many theoreticians and the fundamental problems associated with the Kondo effect are now solved. In particular, as $T \to 0$ the many-body effect can be described by the *local Fermi-liquid theory* involving strongly renormalized quasiparticles. Here, we do not go deeply through the extensive literature that exists, but refer to review articles [326–328].

7.5 Heavy-Fermion Systems

During the last 25 years, a new class of solids has attracted the attention of many low-temperature physicists. These intermetallic compounds contain rare-earth or actinide elements with partially filled $4f$- or $5f$-electron shells. They are called *heavy-fermion systems* or heavy-electron systems because of the high thermal effective mass m^* of their conduction electrons. This heavy mass manifests itself, for example, not only in the high specific heat but also in a large Pauli susceptibility, and in peculiarities of the de Haas–van Alphen oscillations. Prominent examples are the cerium compounds $CeCu_6$, $CeAl_3$, $CeCu_2Si_2$, or $CeRu_2Si_2$, and the uranium compounds UBe_{13} and UPt_3. It is a fascinating aspect of heavy-fermion systems that a wealth of ground states can occur that result in a variety of different low-temperature properties. Within the bounds of this book only a superficial presentation of the complex properties of this class of substances is possible. For a more detailed discussion we refer to relevant review articles (see, e.g., [329–332]).

It is important to note that f-states are localized and retain, to a large extent, their atomic character even in solids. Like the d-electrons of the transition metals, f-electrons hybridize with the conduction electrons via the exchange interaction. But because of the strong localization of f-electrons, the overlap of the wave functions with the conduction electrons is rather weak, and $4f$- or $5f$-resonances are comparatively sharp.

At high temperatures, heavy-fermion systems exhibit properties similar to conventional metals with strongly localized f-electrons embedded in the sea of conduction electrons. They carry a magnetic moment and give rise to a susceptibility that is of the Curie–Weiss form. At low temperatures, however, heavy-fermion systems generally behave like *Fermi liquids*. In Sect. 3.3 we already considered Landau's theory of Fermi liquids when we discussed the properties of normal fluid ^{3}He. In this model, the strongly interacting helium atoms were replaced by weakly interacting quasiparticles. Thus, this model allows us to describe macroscopic quantities of heavily interacting Fermi particles with a few parameters, namely the effective mass and the Landau parameters. The Fermi liquid and free-electron model *qualitatively* agree to a great extent. However, there are large *quantitative* differences between the predicted and observed values of various physical parameters in heavy-fermion systems.

As in dilute Kondo alloys, at low temperatures screening of local magnetic moments becomes energetically more favorable. The transition from one regime to another occurs around a *characteristic temperature* T^* with a typical value between 5 K and 50 K. The fact that T^* is comparable to the Kondo temperature T_K indicates that the thermodynamic properties, such as the high effective thermal mass, are mainly determined by single-ion effects. On the other hand, there are properties such as electrical conductivity, superconductivity and long-range magnetic order that are strongly influenced by coherence effects and cooperative phenomena.

In Kondo systems, i.e., in dilute alloys, the impurities with their local moments are randomly distributed. In heavy-fermion systems, however, the moments are periodically arranged and systems of this type are therefore often called *Kondo lattices*. As a consequence of the periodic structure, a real band of heavy quasiparticles is formed (see Fig. 7.22). Owing to the large concentration of localized magnetic moments, intersite RKKY interaction (see Sect. 7.4.1) also plays an important role. The energy associated with this type of interaction is proportional to $J^2 D(E_F)\cos(2k_F r)/r^3$ for sufficiently large distances r between the moments. Its strength can be expressed by the characteristic temperature T_{RKKY} that satisfies the relation $k_B T_{RKKY} \propto J^2 D(E_F)$.

In Kondo lattices, the competition between the demagnetizing on-site Kondo interaction and the intersite RKKY exchange interaction is believed to be the key factor that determines whether the heavy-fermion ground state occurs. It should be mentioned that there are also other effects that contribute to the complexity of this problem that we do not consider here for the sake of simplicity. For small values of J, the relation $T_{RKKY} > T_K$ holds because of the exponential dependence of T_K on J (see (7.41)). In this case, magnetic ordering is more favorable. For large values of J the energy gained by the formation of local Kondo singlets will surpass the gain due to magnetic ordering. Consequently, a nonmagnetic ground state without long-range magnetic order will develop, although short-range order and short-time correlations between the moments may also exist. However, it should be mentioned that at the lowest temperatures most heavy-fermion systems become either magnetic or superconducting.

Close to the critical value J_c, where the two characteristic temperatures T_K and T_{RKKY} are equal, *non-Fermi liquid* properties are observed that we will consider at the end of this chapter. But first, we will discuss a few examples of the large number of surprising anomalies found in heavy-fermion systems.

7.5.1 Specific Heat

We have already described examples of the specific heat of heavy-fermion systems in Sect. 7.1.2, where data on the compound $CeCu_2Si_2$ and $CeAl_3$ were presented. Starting from room temperature, the specific heat divided by temperature C/T first drops with decreasing temperature. The slope is not only determined by the properties of phonons but also by the contribution of crystal-field excitations, i.e., by thermally activated transitions between crystal-field levels. Around the characteristic temperature T^*, the quantity C/T increases steeply with decreasing temperature. In $CeCu_2Si_2$, the curve finally flattens out and a constant but extremely high value is found at very low temperatures. In other cases, such as $CeAl_3$, C/T goes through a maximum and falls again.

Inserting experimental values in the expression (7.2) for the Sommerfeld coefficient γ, a huge value for the density of states $D(E_F)$ is typically found at the Fermi energy. The pronounced rise of γ with decreasing temperature indicates that $D(E_F)$ varies strongly with temperature. This behavior is illustrated in Fig. 7.22. It shows schematically the density of states of heavy-fermion systems for the limiting cases $T \gg T^*$ and $T \ll T^*$. At high temperatures, the conduction band is virtually constant in the neighborhood

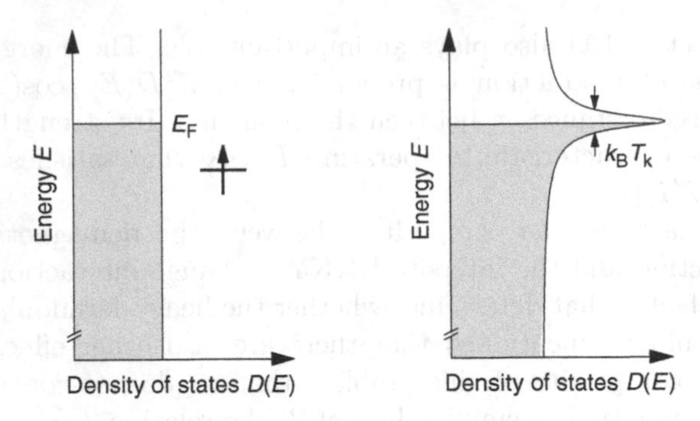

Fig. 7.22. Density of states at the Fermi energy. (**a**) The density of states of the conduction electrons is constant at $T \gg T^*$. The energy level of the $4f$- or $5f$-electrons is indicated by a line with a spin symbol. (**b**) At $T \ll T^*$ a narrow band of the heavy quasiparticles is formed by the f-electrons

of the Fermi energy. The f-electrons are localized and are indicated in the schematic drawing as a line with the symbol of a spin. The density of states at the Fermi energy is exclusively determined by the conduction electrons with an effective mass that is slightly renormalized due to the coupling with localized moments. At low temperatures, the quasiparticle density of states is characterized by a pronounced peak at the Fermi energy that is caused by the hybridized states. The bandwidth is determined by the binding energy and is therefore roughly given by $k_B T_K$. Well below T_K, the thermodynamic properties can be described by the Fermi-liquid model.

The narrow band at E_F gives rise to the extraordinarily high effective mass m^* of the quasiparticles that is deduced from the specific heat. Direct evidence for the existence of such high values of m^* was found in studies of the *de Haas–van Alphen effect*. For example, for the compound $Ce_2Ru_2Si_2$ the mass $m^* = 120\,m$ was deduced [333]. But it should be pointed out that m^* determined in this manner, and by calorimetric means, do not always agree. The specific heat cannot always exclusively be traced back to the excitation of quasiparticles since magnetic correlation effects, for instance, can become important at low temperatures. We shall briefly discuss this phenomenon below.

In Fig. 7.23, we show a survey of measurements of the specific heat of several heavy-fermion systems. Again, C/T is plotted as a function of T^2. This figure demonstrates that the properties of heavy-fermion systems vary considerably. The reason is that different types of heavy-fermion systems exist. There are systems that become superconducting and/or magnetically ordered at low temperatures. In these cases, C/T will go through a maximum. And there are systems that become neither superconducting nor magnetically ordered even at very low temperatures such as $CeCu_6$. In such Fermi-liquid paramagnets C/T will eventually stay approximately constant.

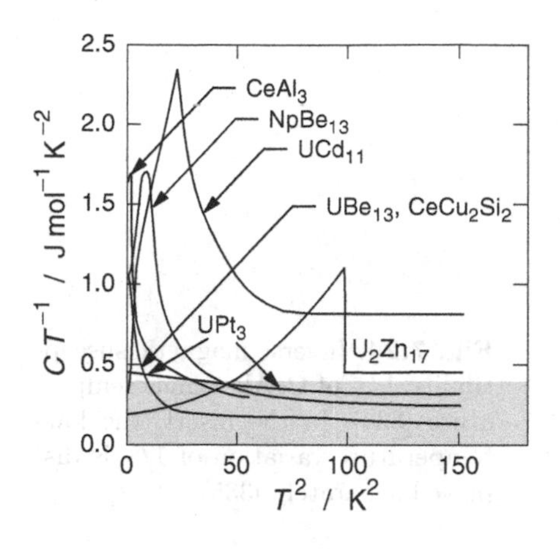

Fig. 7.23. Specific heat C/T of different heavy-fermion compounds as a function of T^2. Substances with a Sommerfeld coefficient $\gamma = C/T$ above the *tinted area* are considered to be heavy-fermion systems. For comparison: this limit corresponds to $\gamma \approx 270\,\gamma_{\mathrm{Na}}$ [329]

7.5.2 Susceptibility

Besides the enormously high specific heat, heavy-fermion systems also exhibit a strongly enhanced susceptibility. In ordinary nonmagnetic metals there is an almost temperature-independent susceptibility, the so-called Pauli susceptibility

$$\chi = \mu_0 \mu_{\mathrm{B}}^2 D(E_{\mathrm{F}}) \propto n^{1/3} m^* . \tag{7.43}$$

However, at high temperatures, heavy-fermion systems exhibit a Curie–Weiss like behavior that can be attributed to the localized moment of f-electrons. At temperatures below T^*, the temperature variation of χ flattens and finally the susceptibility remains approximately constant. The enhanced, temperature-independent susceptibility can be interpreted as the Pauli susceptibility of the heavy quasiparticles.

As an example, the susceptibility of $CeAl_3$ is shown in Fig. 7.24. At high temperatures, the expected Curie–Weiss behavior of systems with magnetic moments is found. From these data it follows that all Ce-ions carry a magnetic moment of $2.56\,\mu_{\mathrm{B}}$, corresponding roughly to that of free trivalent Ce-ions.[3] As shown in the insert of the figure, the susceptibility becomes approximately constant below 1.5 K and $\chi \approx 3.6 \times 10^{-2}\,\mathrm{cm^3\,mol^{-1}}$ is found. This value is about two orders of magnitude greater than that of free electrons. Just like the specific heat, the susceptibility (7.43) is proportional to the density of states and we conclude that the high effective mass of the quasiparticles is the origin of this unusual behavior.

This conclusion is supported by the clear correlation between γ and χ displayed in Fig. 7.25. In the range $T \ll T^*$, the strongly enhanced value of the magnetic susceptibility and the specific heat of heavy-fermion systems

[3] In a quantitative comparison it has to be taken into account that the crystal field modifies the magnetic moments.

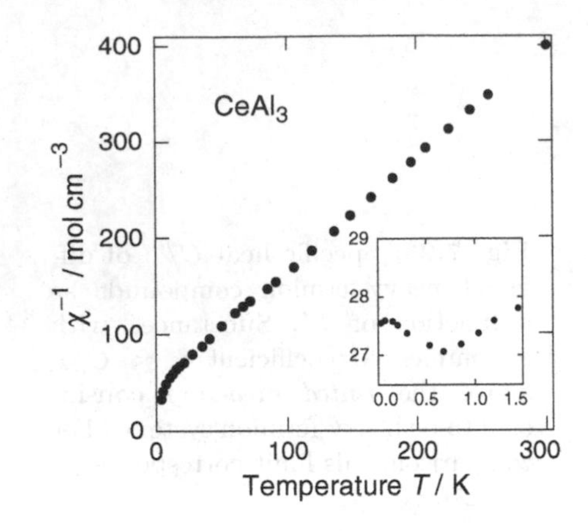

Fig. 7.24. Inverse magnetic susceptibility $1/\chi$ of $CeAl_3$ versus temperature [334]. In the insert, the low-temperature variation of $1/\chi$ is displayed separately [335]

are, as in simple metals, proportional to each other. This fact can be expressed by the *Wilson ratio*

$$R = \frac{\chi}{\gamma} \frac{\pi^2 k_B^2}{\mu_0 \mu_{\text{eff}}^2} . \tag{7.44}$$

For free electrons this ratio is one, but is between two and five in heavy-fermion systems (see Fig. 7.25). The deviation from the free-electron value can easily be understood within the framework of Fermi-liquid theory (see Sect. 3.2) because of the negative value of the parameter G_0. In addition, Fig. 7.25 shows that there is a continuous transition from systems with comparatively small effective mass, i.e., only weakly enhanced mass to the heavy-fermion systems.

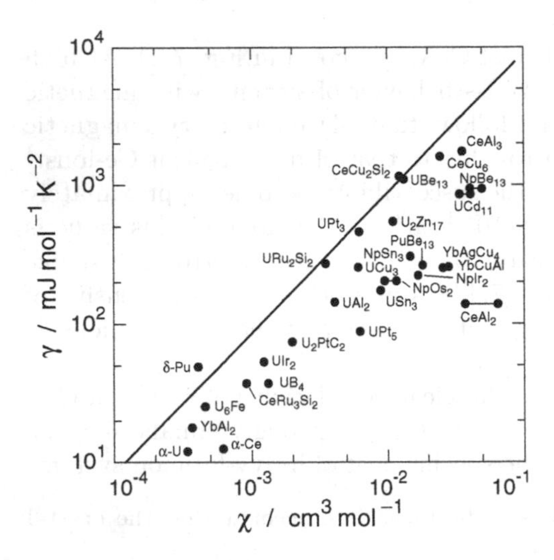

Fig. 7.25. Sommerfeld coefficient γ plotted as a function of the low-temperature value of the susceptibility $\chi \to 0$ of different compounds. The *straight line* represents the Wilson ratio $R = 1$ [336]

7.5.3 Electrical Resistivity

While in normal metals the electrical conductivity is limited by the scattering of conduction electrons by phonons and impurities, heavy-fermion systems exhibit a more complex behavior. As a typical example of the temperature variation of the electrical resistivity ϱ, data from $CeAl_3$ is shown in Fig. 7.26. Around a few hundred kelvin, ϱ is relatively large and depends only weakly on temperature. Besides phonons and defects, crystal-field excitations also contribute to the scattering of the conduction electrons.

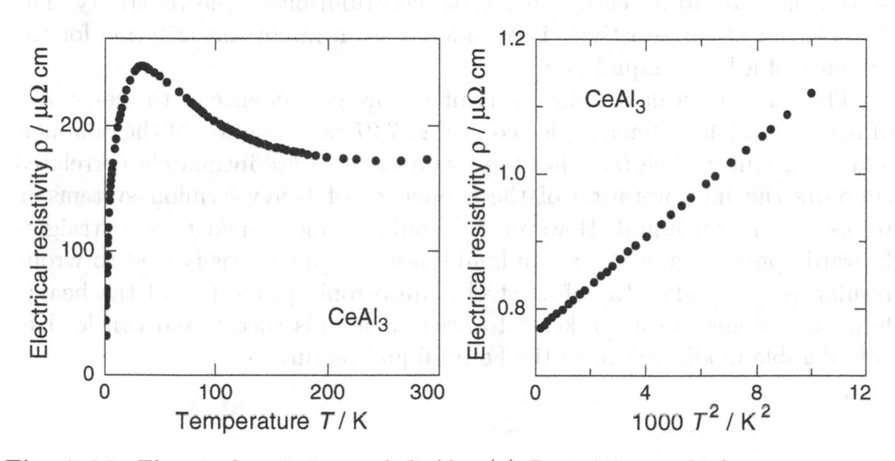

Fig. 7.26. Electrical resistivity of $CeAl_3$. (**a**) Resistivity at high temperatures versus T [337]. (**b**) Resistivity below 100 mK as a function of T^2 [335]

At about 30 K, the resistivity passes through a maximum and drops upon cooling again. The increase of the electrical resistance with decreasing temperature is due to electron scattering by the cerium ions. For a theoretical description, the formalism that was used to explain the logarithmic temperature variation of the resistivity of dilute alloys can be applied. As in Kondo systems, the curve flattens out at lower temperatures as the screening of the magnetic moments by the conduction electrons becomes more effective.

At still lower temperatures, the periodic arrangement of the magnetic moments becomes important and coherence effects come into play. In this low-temperature regime, the resistivity is given by

$$\varrho(T) = \varrho_0 + AT^2 , \tag{7.45}$$

where A is a constant and ϱ_0 is the temperature-independent residual resistivity. In high-quality samples this residual resistivity ϱ_0 is found to be comparable to that of pure transition metals without magnetic moments. This low value indicates that the rather complex wave functions of the heavy-fermion systems are eigenstates of the periodic Kondo lattice [330, 331].

The temperature-dependent part of the resistivity can, to a good approximation, be described by the Fermi-liquid model. The resistivity in this temperature range is due to the scattering of the heavy quasiparticles amongst each other. Analogous to electron–electron scattering (see Sect. 7.2.1) and in agreement with experimental results, a temperature variation $\varrho(T) \propto AT^2$ is expected. In ordinary metals, electron–electron scattering causes a resistivity $\varrho \propto (T/T_\mathrm{F})^2$ that can generally be neglected and can only be observed in extremely pure samples at very low temperatures. In heavy-fermion systems the relevant temperature is not T_F but T^*. Thus, the electron–electron scattering leads to an easily observable contribution to the resistivity. The T^2-behavior of the resistivity is considered as an important criterion for the presence of a Fermi liquid [338].

The constant A describing the temperature dependence of the resistivity of heavy-fermion systems is plotted in Fig. 7.27 as a function of the Sommerfeld coefficient γ. The fact that the two quantities are intimately correlated supports the interpretation of the properties of heavy-fermion systems in terms of a Fermi liquid. However, it should be mentioned that a straightforward application of the Fermi-liquid description can easily lead to wrong results. In particular, the often strong anisotropic properties of the heavy-fermion systems are not taken into account in this theory and can lead to considerable modifications of the Fermi-liquid picture.

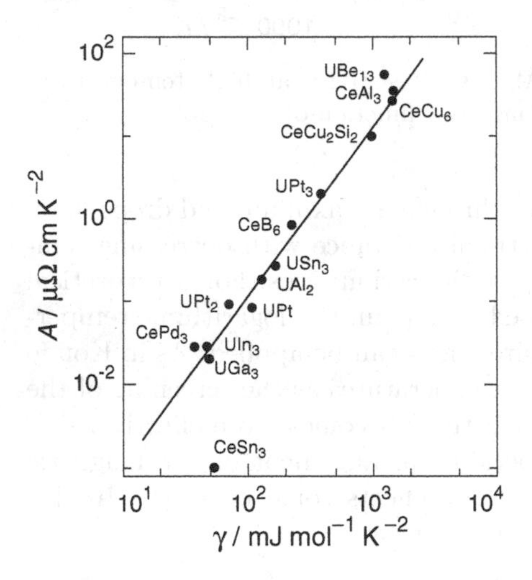

Fig. 7.27. Coefficient A of (7.45) plotted as a function of the Sommerfeld coefficient γ of different heavy-fermion systems. The *straight line* reflects the relation $A \propto \gamma^2$ [339]

In Sect. 7.3 we briefly discussed the Wiedemann–Franz law, relating the electrical and thermal conductivity of ordinary metals. As we have shown, in ordinary metals, the two quantities fulfill the relation $\Lambda_\mathrm{el}/\sigma = \mathcal{L}T$, where $\mathcal{L}$ is the Lorenz number. Such a behavior was also found for $CeAl_3$, supporting the assumption that both charge and heat are transported by quasiparticles and that the same scattering mechanism limits the transport [337].

7.5.4 Non-Fermi Liquids

In our discussion of the properties of heavy-fermion systems, the impression was given that at low temperatures either perfect magnetic order exists or else complete screening occurs resulting in a nonmagnetic ground state. However, often a tiny static magnetic moment of the order of $0.01\mu_B$ per atom or even less remains. This is caused by the so-called heavy-fermion band magnetism. Without further justification, we add that it can be ascribed to the exchange splitting of the heavy quasiparticle band.

One of the few systems where long-range magnetic order has not been observed down to very low temperatures is $CeCu_6$.[4] However, this compound can be driven towards magnetic order by alloying. For example, adding gold to the compound leads to the formation of an antiferromagnetic phase. In Fig. 7.28, the Néel temperature of $CeCu_{6-x}Au_x$ is plotted as a function of the gold concentration. Above the critical threshold of $x = 0.1$, the Néel temperature increases linearly with x. This rise is mainly caused by the lattice expansion due to the presence of gold atoms. This expansion causes a decrease of the hybridization and hence a reduction of J. As a consequence, the magnetic moments are stabilized and now interact via the RKKY interaction. The sharp maximum at $x = 1$, and the decrease beyond that composition, are not related to the problems discussed here and will not be considered further. In fact, it has been shown by elastic neutron scattering that the effective magnetic moment increases with the gold concentration.

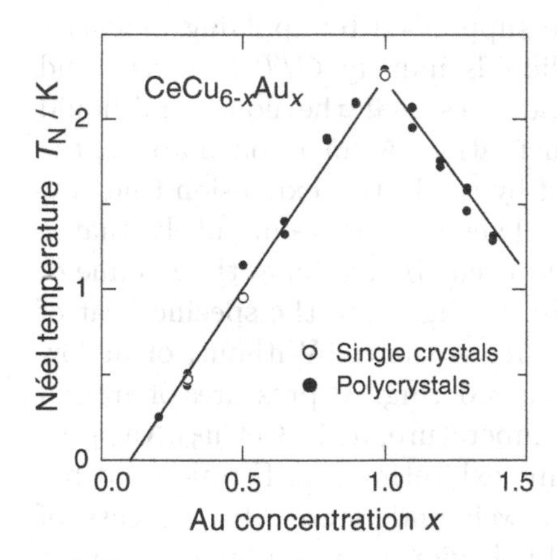

Fig. 7.28. Néel temperature T_N of $CeCu_{6-x}Au_x$ versus gold concentration x determined by measurements of the specific heat (*triangles*) and magnetic susceptibility (*circles*) [341]

In Fig. 7.29, the specific heat and the magnetic susceptibility of pure $CeCu_6$ and two alloys are shown as a function of temperature. The properties of the pure sample can approximately be described by the Fermi-liquid model.

[4] There are indications that around 3 mK magnetic ordering occurs in this system but this temperature is considerably lower than the transition temperature found in other heavy-fermion systems [340].

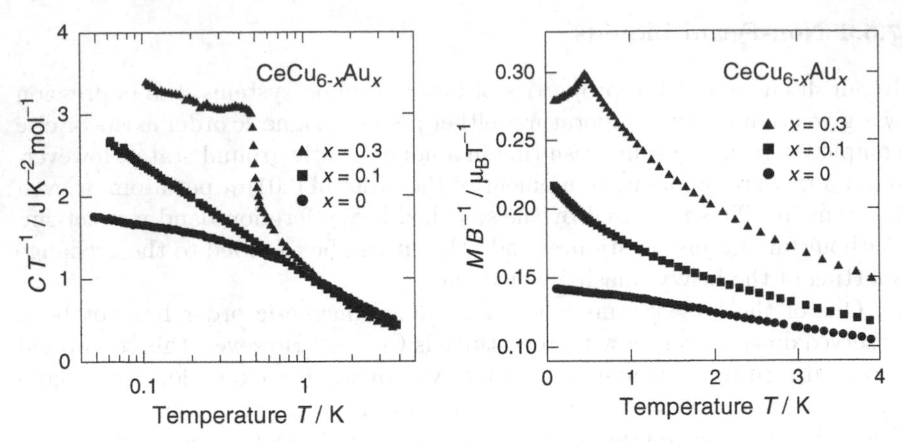

Fig. 7.29. Specific heat and susceptibility of $CeCu_{6-x}Au_x$. (**a**) Specific heat divided by temperature C/T versus temperature. (**b**) Susceptibility $\chi/\mu_0 = M/B$ measured at $B = 0.1\,T$, and plotted as a function of temperature [342]

The sample with $x = 0.3$ shows features at 0.48 K that indicate the onset of magnetic order. But most intriguing is that one finds $C/T \propto -\ln(T/T_0)$ over a wide temperature range for $x = 0.1$. At that composition, the susceptibility follows the relation $\chi \propto (1 - \alpha\sqrt{T})$, where α is a constant. Furthermore, the electrical resistivity increases linearly with temperature, in contrast to the quadratic increase observed in pure $CeCu_6$ [343].

The non-Fermi liquid behavior can be suppressed by applying magnetic fields, i.e., the characteristics of Fermi liquids namely $C/T \approx$ const. and $\varrho = \varrho_0 + AT^2$ can be recovered. This indicates that the non-Fermi-liquid behavior is related to some sort of spin fluctuation. As mentioned above, the magnetic order in $CeCu_{6-x}Au_x$ is induced by the lattice expansion that was caused by incorporating the gold atoms. Therefore, it seems likely that a transition back to the ordered state should occur by reducing the volume of the alloy with the help of applied pressure. In Fig. 7.30, the specific heat of $CeCu_{5.7}Au_{0.3}$ measured at different pressures is shown. Without, or at low pressure, C/T exhibits the onset of magnetic ordering. At pressures of around 8 kbar, C/T varies logarithmically with temperature, as in $CeCu_{5.9}Au_{0.1}$.

Although there are many other systems exhibiting non-Fermi-liquid behavior, the microscopic origin is not yet well understood. In the case of $CeCu_{6-x}Au_x$, it is likely that the unusual behavior is caused by an incipient quantum phase transition between the magnetic and nonmagnetic ground states taking place at $T = 0$ ('the quantum critical point'). However, other scenarios of non-Fermi-liquid behavior may be important in other materials. A comprehensive discussion of this interesting phenomenon would, however, go beyond the scope of this introductory treatment (see, for example, [345]).

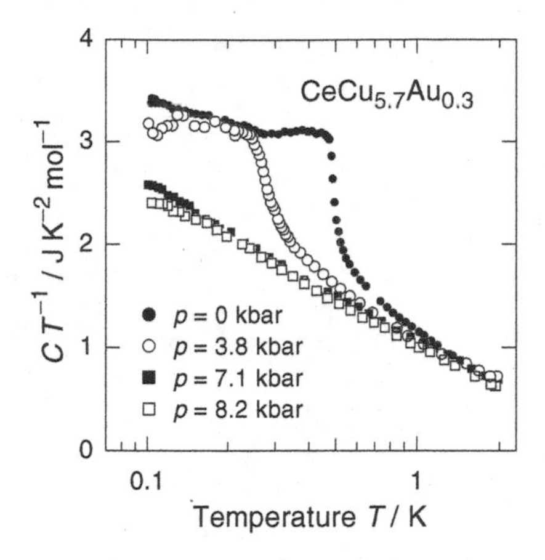

Fig. 7.30. Specific heat C/T of $CeCu_{5.7}Au_{0.3}$ versus $\ln T$, measured at different pressures [344]

Exercises

7.1 In metals, electrons and lattice contribute to the specific heat. At what temperature are the two contributions in potassium equal? What would be the ratio at room temperature?

7.2 The residual resistivity of the alloy 'Manganin' ($Cu_{0.86}Mn_{0.12}Cu_{0.02}$) is about $4.5 \times 10^{-7}\,\Omega\,\mathrm{m}$. Show that the mean free path of the conduction electrons is only of the order of a few lattice spacings.

7.3 In a low-temperature experiment, a 0.5 mm thick silicon plate covered by a 100 nm thick copper film is exposed to a heat gradient. The room-temperature resistivity $\varrho = 1.7 \times 10^{-8}\,\Omega\,\mathrm{m}$ of bulk copper is due to electron–phonon scattering. Electrons are also scattered at the film surface and the interface, leading to an additional resistance. (**a**) At what temperature is the contribution of the two scattering mechanisms equal? (**b**) The sample is cooled down to 0.1 K. What is the expected electrical and thermal conductivity of the copper film? (**c**) Is the main contribution to the heat transport due to the silicon plate or due to the copper film?

8 Magnetic Moments – Spins

To describe consistently the fine structure of the optical spectra of hydrogen atoms and to explain the anomalous Zeeman effect, *Uhlenbeck* and *Goudsmit* postulated in 1925 that electrons carry an intrinsic angular momentum – called 'spin' – and thus possess an intrinsic magnetic moment [346]. Their model was based on the classical idea of a charged sphere rotating on an axis through its center. Although this point of view led to inherent contradictions, it provided a formal answer to many open questions of atomic physics. Some years later, in 1928, *Dirac* showed that the existence of an intrinsic angular momentum is an inevitable consequence of the relativistic formulation of quantum mechanics, and that the analogy with a classically rotating particle is not sensible [347].

The spin is not only an indispensable ingredient in atomic physics it is also responsible for many phenomena observed in solid-state physics. As we have seen in the previous chapter, magnetic moments can have a strong influence on electron scattering, provoking interesting low-temperature phenomena. Moreover, spins determine to a great extent the magnetic properties of solids. As well-known examples, we mention here Pauli paramagnetism of metals, ferromagnetism and antiferromagnetism, and Van-Vleck paramagnetism. Although spins play an important role in the low-temperature behavior of many solids, we pick out here only a few typical examples. In particular, we discuss relatively simple problems related with isolated spins, touch on the properties of collective excitations in ferro- and antiferromagnets and give a short introduction to the wealth of phenomena observed in spin glasses. Finally, we discuss the magnetic order of nuclear spins that can be investigated at positive and negative spin temperatures.

8.1 Paramagnetic Systems – Isolated Spins

In the following section, we consider the low-temperature specific heat and susceptibility of paramagnetic substances. In these systems, applied magnetic fields cause a preferential alignment of magnetic moments. In our discussion, we neglect the interaction of the magnetic moments with the environment and the contribution of neighboring magnetic moments to the local field.

8.1.1 Magnetic Moments

Both spin and orbital motion of the electrons contribute to the magnetic moment $\boldsymbol{\mu}$ of free atoms or ions that is given by

$$\boldsymbol{\mu} = -g\mu_\mathrm{B}\boldsymbol{J}\,, \quad \text{with} \quad g = 1 + \frac{J(J+1)+S(S+1)-L(L+1)}{2J(J+1)}\,, \tag{8.1}$$

where g is the Landé factor, and S, L, and J refer to spin, orbital angular momentum, and total angular momentum, respectively.

In solids, the moment of magnetic ions is not only determined by the electronic configuration of the ions but also by their chemical bonding and the crystal structure. In particular, the nonspherical crystal field caused by neighboring atoms has a remarkably strong influence on the formation of the magnetic moment. Fortunately, the situation is especially simple if *rare-earth ions* are incorporated in insulators. As shown in Fig. 8.1a, in this case, the observed magnetic moments agree well with those calculated for free ions using *Hund's rules*. The crystal field has only a minor effect because the $4f$-electrons of the rare-earth elements are localized within the ion core and are thus screened by the outer $5s$- and $5p$-electrons. In fact, the radii of trivalent rare-earth ions lie between 0.9 Å and 1.1 Å, while $4f$-electrons stay within a shell of only 0.3 Å in radius.

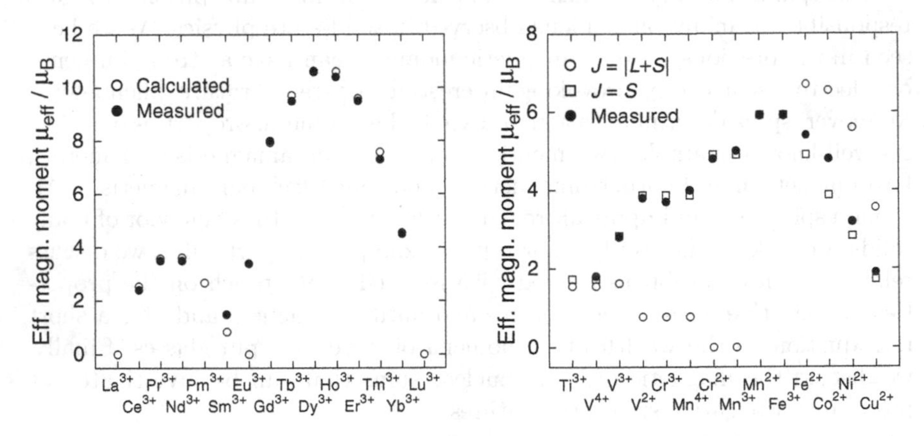

Fig. 8.1. Experimental (*full circles*) and calculated (*open circles*) effective magnetic moments of (**a**) rare-earth ions and (**b**) ions of the iron group in dielectric crystals. *Open squares* are calculated under the assumption $J = S$ [348]

In contrast, the magnetic moments of ions of the *iron group* differ in solids substantially from those of free ions. In this case, the contribution of the orbital motion is usually 'quenched' by the crystal field, and only the spins remain effective. This different behavior originates from the fact that the $3d$-electrons are unscreened by outer electrons. Therefore, the crystal field has a strong influence on the electronic motion and lifts even the LS-coupling. The magnitude of the orbital angular momentum is conserved even in strong noncentrosymmetric fields, but L_z is no longer a constant of motion. The time average of L_z vanishes and hence the contribution of the orbital angular motion to the magnetic moment. Thus, only the contribution of the spins has to be taken into account, in agreement with the experimental data shown in Fig. 8.1 b for salts of the iron group.

The behavior of magnetic ions is considerably more complex if a metal serves as a host. In this case, the conduction electrons exert a strong influence on the magnetic properties, which we have already discussed in some detail in Sect. 7.4.

8.1.2 Susceptibility

In the absence of a magnetic field, the magnetic sublevels of atoms or ions are degenerate and the mean value of the magnetic moments vanishes. In magnetic fields, the levels are split into $(2J+1)$ sublevels being separated by $g\mu_B B$. The thermal occupation of the individual levels and hence the associated magnetization can be calculated via the relation

$$M = n\frac{\sum_{-J}^{J} m_J g\mu_B \exp[-g\mu_B m_J B/(k_B T)]}{\sum_{-J}^{J} \exp[-g\mu_B m_J B/(k_B T)]}\,, \tag{8.2}$$

where the summation includes the contribution of the $(2J+1)$ sublevels with the quantum numbers m_J. In this way, the expression

$$M = ng\mu_B J\mathcal{B}(h) \tag{8.3}$$

is found, with $\mathcal{B}(h)$ being the *Brillouin function* defined by

$$\mathcal{B}(h) = \frac{2J+1}{2J}\coth\left[\frac{(2J+1)h}{2J}\right] - \frac{1}{2J}\coth\left(\frac{h}{2J}\right)\,. \tag{8.4}$$

The argument $h = g\mu_B JB/(k_B T)$ reflects the ratio between magnetic and thermal energy. At very low temperatures, all moments are aligned in the field direction and only the ground state is occupied, resulting in $\mathcal{B}(h) = 1$. However, in many experiments the Zeeman splitting is small compared to $k_B T$, and hence $h \ll 1$. In this case, the Brillouin function may be approximated by the first term of the Taylor expansion, leading to the well-known *Curie law* of paramagnetic substances:

$$\chi = \frac{M}{H} = \frac{\mu_0 n g^2 J(J+1)\mu_B^2}{3k_B T} = \frac{C}{T}\,. \tag{8.5}$$

In Fig. 8.2, the Brillouin function $\mathcal{B}(h)$ for $J = 1/2$ is compared with the Curie approximation, depicted in this figure by a *dashed line*.

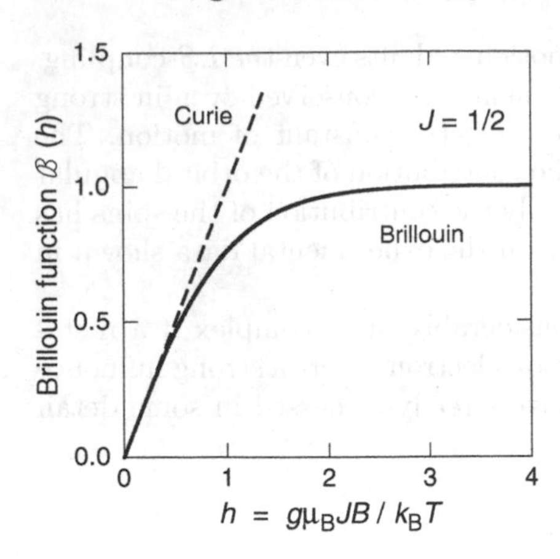

Fig. 8.2. Brillouin function for $J = 1/2$ as a function of $h = g\mu_B JB/(k_B T)$. The Curie approximation is shown by a *dashed line*

Paramagnetic Salts

As an example of spin magnetism we consider the behavior of paramagnetic salts at low temperatures. In Fig. 8.3, the temperature and magnetic field dependence of the normalized magnetization per ion $M/n\mu_B$ of gadolinium sulfate octahydrate ($S = J = 7/2$), ferric ammonium alum ($S = J = 5/2$), and potassium chromium alum ($S = 3/2, J = 3$) are displayed. The measurements were performed in magnetic fields up to 5 T at helium temperature and below. At $T = 1.3$ K and $B = 5$ T, a 99.3% saturation of the magnetization was achieved. The *full lines* in Fig. 8.3 represent the theoretical prediction (8.3). These data demonstrate the perfect agreement between theory and experiment. In particular, they show that in potassium chromium alum the contribution of the orbital angular momentum is quenched.

In Fig. 8.4, the temperature variation of the magnetic susceptibility $\chi = \partial \mu_0 M/\partial B$ of cerium magnesium nitrate (CMN) is shown. In this paramagnetic salt the Curie approximation $\chi = C/T$ is well satisfied down to very

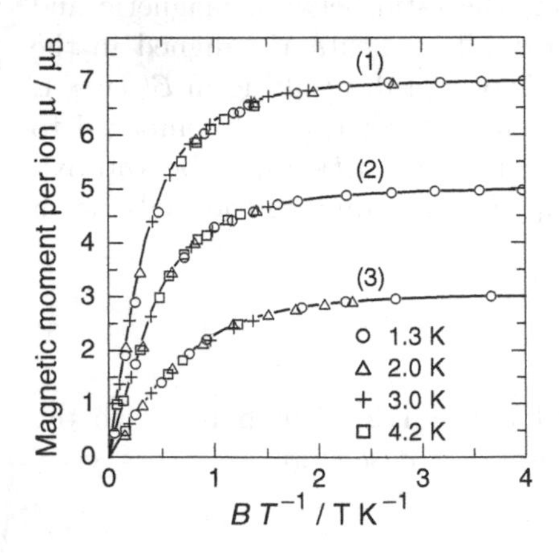

Fig. 8.3. Normalized mean magnetic moment per ion $M/n\mu_B$ versus B/T of the paramagnetic salts (1) gadolinium sulfate, (2) ferric ammonium alum, and (3) potassium chromium alum. The *full lines* represent $(gJ\mathcal{B})$ for the spin quantum numbers given in the text [349]

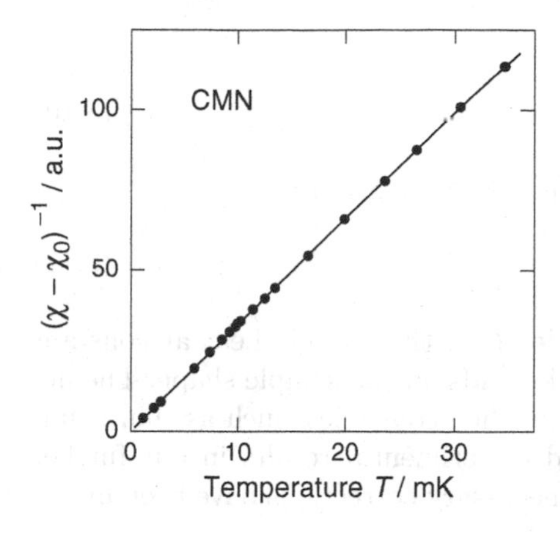

Fig. 8.4. Inverse magnetic susceptibility $(\chi - \chi_0)^{-1}$ of CMN as a function of temperature. A constant χ_0 has been subtracted that depends on the details of the experimental setup [350]

low temperatures. It should be mentioned that CMN is a well-known material in low-temperature physics. It is used for the production of low temperatures and for low-temperature thermometry (see Chaps. 11 and 12).

8.1.3 Specific Heat

Magnetic moments contribute to the free energy and consequently also to the specific heat. Therefore, we include the term $\boldsymbol{m}\mathrm{d}\boldsymbol{B}$ in the expression for the incremental heat change and write

$$\mathrm{d}Q = T\mathrm{d}S = \mathrm{d}U + p\mathrm{d}V + \boldsymbol{m}\mathrm{d}\boldsymbol{B}\,, \tag{8.6}$$

where $\boldsymbol{m} = N\overline{\boldsymbol{\mu}}$ is the magnetic moment of the sample given by the number N of the magnetic dipoles and their mean moment $\overline{\boldsymbol{\mu}}$. In our further discussion we assume for simplicity that $\boldsymbol{\mu}$ and $\boldsymbol{B}$ are parallel to each other and consider only the magnitude of these quantities. We pay attention to the fact that the magnetization can be provoked either by an external field as assumed in (8.6), or by a spontaneous magnetization. In the latter case, the magnetic part of the energy is generally included in the internal energy U'. With the definition $U' = U + mB$, (8.6) reads

$$\mathrm{d}Q = \mathrm{d}U' + p\mathrm{d}V - B\mathrm{d}m\,. \tag{8.7}$$

In the problems we are going to consider, the mechanical work $p\,\mathrm{d}V$ is much smaller than the magnetic energy, and will therefore be neglected. Thus, (8.7) can be simplified and may be replaced by

$$\mathrm{d}Q = \mathrm{d}U' - B\mathrm{d}m\,. \tag{8.8}$$

In analogy to C_V and C_p, we may define C_B and C_m, the specific heat at constant field and at constant magnetic moment, by the following relations

$$C_B = \left(\frac{\mathrm{d}Q}{\mathrm{d}T}\right)_B = T\left(\frac{\partial S}{\partial T}\right)_B \tag{8.9}$$

and

$$C_m = \left(\frac{\mathrm{d}Q}{\mathrm{d}T}\right)_m = T\left(\frac{\partial S}{\partial T}\right)_m , \tag{8.10}$$

where the difference between the two quantities is given by

$$C_B - C_m = -T\left(\frac{\partial B}{\partial T}\right)_m \left(\frac{\partial m}{\partial T}\right)_B . \tag{8.11}$$

In the following, we will always consider C_m, the specific heat at constant magnetic moment. Note that $\partial m/\partial B$ depends on the sample shape. The demagnetization field only vanishes for specific geometries, such as thin cylinders. When comparing theoretical and experimental results in our further discussions we always assume that the necessary corrections have been made.

Two-level Systems

There are many physical systems that are readily described in terms of two energy levels because they either exhibit, in fact, only two levels, or they can be described to a good approximation by two-level systems. A prominent and particularly simple example are free electrons in a homogeneous magnetic field.

Of course, electrons in solids are not really free. Very often the coupling to electric and magnetic fields, and the interaction between them, leads to multi-level systems with partially degenerate levels. Well-known examples are paramagnetic salts, the magnetic properties of which we have discussed above. Despite the fact that their level scheme is, in general, relatively complicated, the overall behavior of the specific heat differs only slightly from that of genuine two-level systems. Therefore, we consider here the basic aspects of the specific heat and the entropy of two-level systems.

Nondegenerate Two-level Systems

We start with the discussion of specific heat and entropy of nondegenerate two-level systems. The general expression for the partition function of a single system with the energy levels E_s reads: $Z = \sum_\mathrm{s} \exp(-E_\mathrm{s}/k_\mathrm{B}T)$. If the energy of the lower level coincides with the energy zero and the upper level is separated by the energy E, we obtain for the partition function of a two-level system the simple expression

$$Z = 1 + \mathrm{e}^{-E/k_\mathrm{B}T} . \tag{8.12}$$

Using the definition of the occupation probability

$$P(E_\mathrm{s}) = \frac{1}{Z}\exp(-E_\mathrm{s}/k_\mathrm{B}T) , \tag{8.13}$$

the occupation difference $\Delta P = P(0) - P(E)$ is given by

$$\Delta P = \frac{1}{Z}\left(1 - \mathrm{e}^{-E/k_\mathrm{B}T}\right) = \frac{1 - \mathrm{e}^{-E/k_\mathrm{B}T}}{1 + \mathrm{e}^{-E/k_\mathrm{B}T}} = \tanh\left(\frac{E}{2k_\mathrm{B}T}\right). \tag{8.14}$$

For the internal energy, i.e., the mean energy of N noninteracting system, we may write

$$U = \langle E \rangle = N \sum_s P(E_s)\, E_s = \frac{NE}{\mathrm{e}^{E/k_\mathrm{B}T} + 1}, \tag{8.15}$$

and for the specific heat $C_V = \partial U/\partial T$:

$$C_V = N k_\mathrm{B} \left(\frac{E}{k_\mathrm{B}T}\right)^2 \frac{\mathrm{e}^{E/k_\mathrm{B}T}}{(\mathrm{e}^{E/k_\mathrm{B}T} + 1)^2}. \tag{8.16}$$

The temperature variation of the specific heat is depicted in Fig. 8.5. Maxima of this type are often referred to as *Schottky anomalies*. The energy splitting of the two-level systems can directly by deduced from the temperature T_max at which C_V has a maximum via the relation $E \approx 0.42\, k_\mathrm{B} T_\mathrm{max}$.

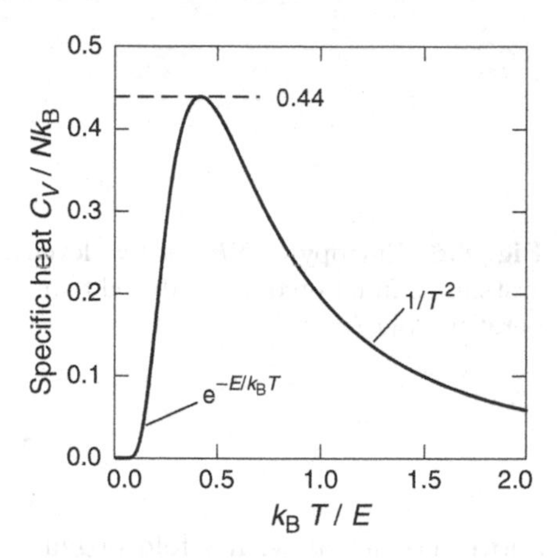

Fig. 8.5. Specific heat C/Nk_B of two-level systems plotted as a function of the reduced temperature $k_\mathrm{B}T/E$

In the limits of high and low temperatures, the temperature dependence of the specific heat can be approximated by the following expressions:

- For $T \to \infty$, the ratio $E/k_\mathrm{B}T \to 0$, leading to the expression

$$C_V \approx \frac{Nk_\mathrm{B}}{4}\left(\frac{E}{k_\mathrm{B}T}\right)^2 \propto \frac{1}{T^2}. \tag{8.17}$$

  This is the temperature dependence found in experiments when the lowest attainable temperature lies well above the maximum temperature.
- For $T \to 0$, the ratio $E/k_\mathrm{B}T \to \infty$, and C_V is approximately given by

$$C_V \approx N k_\mathrm{B} \left(\frac{E}{k_\mathrm{B}T}\right)^2 \mathrm{e}^{-E/k_\mathrm{B}T}. \tag{8.18}$$

  The predicted exponential increase is often covered by contributions to the specific heat arising from other sources.

The temperature variation of the entropy of two-level systems is plotted in Fig. 8.6. It follows directly from the specific heat via the relation

$$S = \int_0^T \frac{C_V}{T'}\,\mathrm{d}T' = Nk_\mathrm{B}\left[\ln\left(1+\mathrm{e}^{-E/k_\mathrm{B}T}\right) + \frac{E}{k_\mathrm{B}T}\,\frac{\mathrm{e}^{-E/k_\mathrm{B}T}}{1+\mathrm{e}^{-E/k_\mathrm{B}T}}\right]. \tag{8.19}$$

Of course, the entropy vanishes for $T \to 0$, and saturates at high temperatures because the number of energy levels is restricted. For $T \gg E/k_\mathrm{B}$, the entropy approaches the value $S_\infty = Nk_\mathrm{B}\ln 2$, or more generally, the value $S_\infty = Nk_\mathrm{B}\ln z$ for systems with z energy levels.

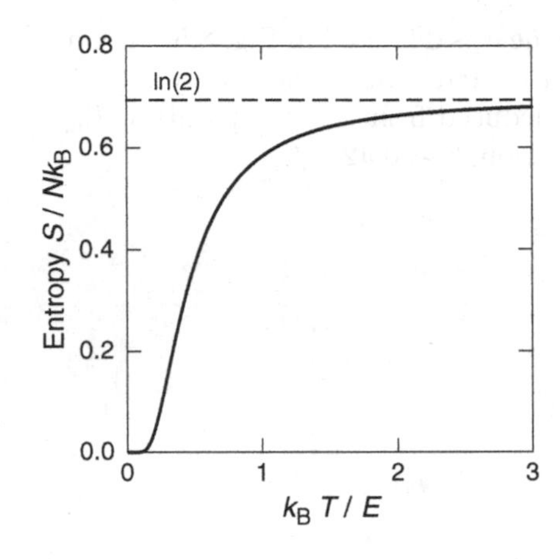

Fig. 8.6. Entropy S/Nk_B of two-level systems as function of the reduced temperature $k_\mathrm{B}T/E$

Degenerate Two-level Systems

Now we generalize the results obtained above. For levels with s-fold degeneracy, the general expression for the internal energy reads

$$U = N\sum_s P(E_s)\,E_s = N\frac{\sum_s E_s g_s \mathrm{e}^{-E_s/k_\mathrm{B}T}}{\sum_s g_s \mathrm{e}^{-E_s/k_\mathrm{B}T}}\,, \tag{8.20}$$

where g_s reflects the degree of degeneracy of the level s. For two-level systems with a g_0-fold degeneracy of the ground state, and a g_1-fold degeneracy of the excited state, these sums simplify to

$$U = N\frac{g_1\,E\,\mathrm{e}^{-E/k_\mathrm{B}T}}{g_0 + g_1\,\mathrm{e}^{-E/k_\mathrm{B}T}} = \frac{NE}{1+\left(\frac{g_0}{g_1}\right)\mathrm{e}^{E/k_\mathrm{B}T}}\,, \tag{8.21}$$

resulting in the specific heat

$$C_V = Nk_\mathrm{B}\left(\frac{g_0}{g_1}\right)\left(\frac{E}{k_\mathrm{B}T}\right)^2 \frac{\mathrm{e}^{E/k_\mathrm{B}T}}{\left[1+\left(\frac{g_0}{g_1}\right)\mathrm{e}^{E/k_\mathrm{B}T}\right]^2}\,. \tag{8.22}$$

The influence of the degeneracy on the specific heat is depicted in Fig. 8.7. As can be seen, both the height of the maximum and its position depend on the ratio g_1/g_0.

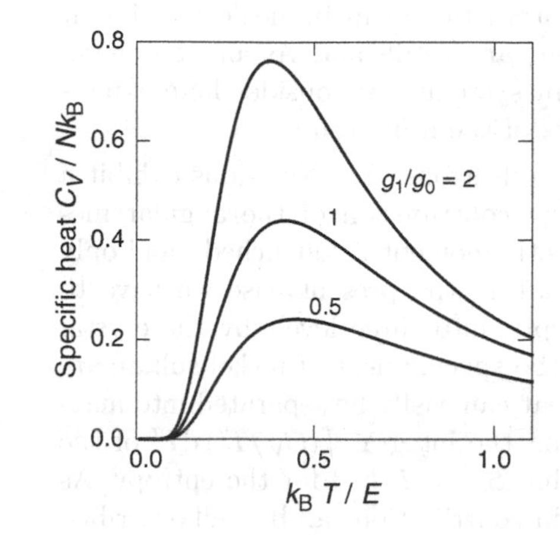

Fig. 8.7. Specific heat C_V/Nk_B versus k_BT/E of two-level systems with different degrees of degeneracy

Multilevel Systems

The corresponding expressions are much more complicated if the number of irregularly spaced energy levels is larger than two. Of course, in this case the specific heat can be deduced from the partition function Z via the thermodynamic relation

$$C_V = Nk_BT\,\frac{\partial^2(T\ln Z)}{\partial T^2}\,, \tag{8.23}$$

but the resulting equation for the specific heat is rather complicated even for three-level systems, and is not particularly illuminating.

Electron Spins – Paramagnetic Salts

As already mentioned, salts containing elements of the iron group or rare-earth metals are typical representatives of paramagnetic substances. As we will see in Chaps. 11 and 12, they are also of great technical importance in low-temperature physics, where they are used for thermometric purposes and for the production of low temperatures. In the latter case, it is an important point that the low-temperature specific heat of spin systems is very large in comparison to contributions from other sources, for example, of phonons or electrons.

In deriving the formula for the specific heat, we have neglected the interaction between the spins, which has to be taken into account as soon as the interaction energy becomes comparable with the energy splitting. Experimentally, its influence can be studied by diluting the spin system, e.g., by a partial replacement of magnetic ions by nonmagnetic ions, and thus reducing the strength of interaction. In particular, the specific heat is proportional to the spin concentration if interactions can be neglected, but it becomes nonlinear as soon as interactions are significant. As an example of the low-temperature specific heat of spin systems, we consider here experimental results on two paramagnetic salts of the iron group.

α–NiSO$_4 \cdot$ 6H$_2$O Because of their $3d^8$-electrons, free Ni^{2+}-ions exhibit a $^3F_4$-ground state. As in most solids, the contribution of the angular momentum of the $3d$-electrons to the magnetic moment is 'quenched', and only spin states are relevant. This means that in the present case we have to consider only the spin states that are split into three levels by the crystal field. In Fig. 8.8, experimental data on the specific heat of nickel sulfate are shown. As indicated, the total specific heat can easily be separated into magnetic and lattice contribution. Evaluating the integral $\int(C_V/T')\,dT'$ of the magnetic part leads to the expected value $S_\infty = R\ln 3$ for the entropy. As shown in Fig. 8.8 by the *full line*, the spin contribution can be well described by two-level systems with a degeneracy factor $g_1 = 2$ and a level splitting of $E/k_B = 6.85\,K$. A more careful analysis and comparison with other experimental data, however, leads to the conclusion that the two upper levels are nearly degenerate and are separated from the ground state by $E_1/k_B = 6.4\,K$, and $E_2/k_B = 7.3\,K$, respectively. This result tells us that the analysis of specific heat is more suitable for qualitative than for quantitative statements on the level scheme.

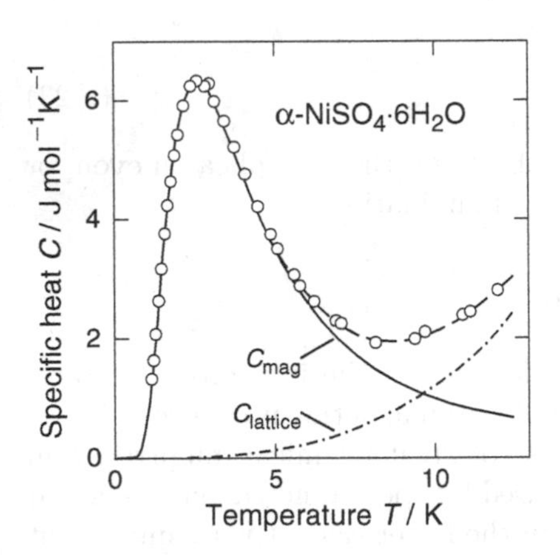

Fig. 8.8. Specific heat of the paramagnetic salt α-NiSO$_4 \cdot 6$H$_2$O. The *full line* represents the fit discussed in the text. The *dashed-dotted line* reflects the lattice, the *dashed line* the sum of both contributions [351]

Fe(NH$_3$CH$_3$)(SO$_4$)$_2$ $\cdot$ 12H$_2$O The $^6S_{5/2}$-ground state of the Fe^{3+}-ions of ferri-methyl-ammonium-sulfate has six levels resulting from the five electrons of the $3d^5$-configuration. From paramagnetic resonance it was known that three levels are present, each one being twofold degenerate. The splitting E/k_B among the levels was determined to be 1.05 K and 0.58 K, respectively [352]. However, the paramagnetic resonance experiments did not distinguish between the two sequences of the energy levels shown in Fig. 8.9.

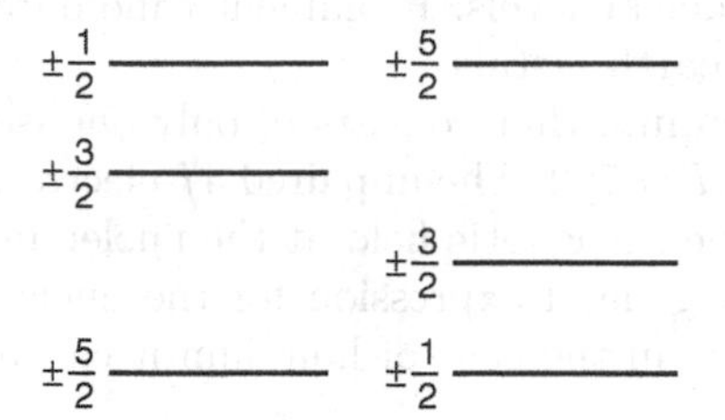

Fig. 8.9. Level schemes of the Fe^{3+}-ions of $Fe(NH_3CH_3)(SO_4)_2 \cdot 12H_2O$ being compatible with the paramagnetic resonance experiments

An unambiguous answer could be given by measurements of the specific heat. In Fig. 8.10, the experimental data are shown together with fitting curves based on the two different level schemes. Clearly, the right scheme of Fig. 8.9 describes the data much better than the left one. The small deviation probably originates from experimental uncertainties and the fact that the magnetic dipole–dipole interaction was neglected in the analysis.

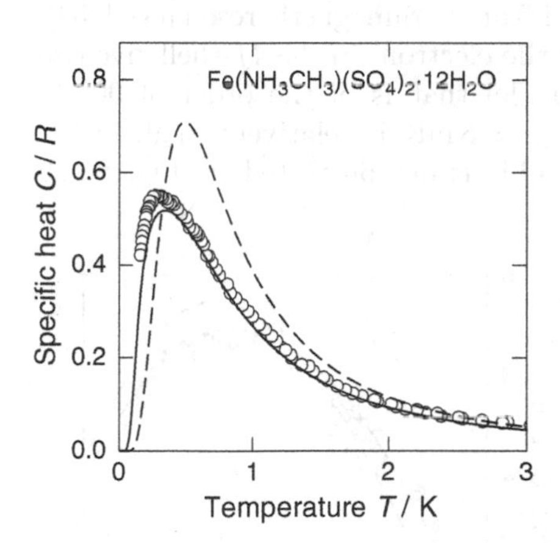

Fig. 8.10. Specific heat C/R of ferri-methyl-ammonium-sulfate as a function of temperature. The *dashed line* represents the fitting curve based on the left, the *full line* on the right the level scheme of Fig. 8.9 [353]

Specific Heat due to Atomic Nuclei

At first glance, it is surprising that nuclei also contribute to the specific heat at low temperatures. But as we shall see, under certain circumstances their contribution can even be dominant. The nuclear specific heat can arise either because of the interaction of the nuclear magnetic moment with magnetic fields caused by the motion of electrons, or by the interaction of the nuclear electric quadrupole moment with electric-field gradients.

In general, nuclear splittings are rather small, meaning that the corresponding Schottky anomalies are found at very low temperatures, typically below 10 mK. The situation is different in ferromagnetic or antiferromagnetic substances because of the presence of strong internal magnetic fields that can lead to an appreciable splitting of the nuclear levels. Prominent candidates for such an effect are ferromagnetic rare-earth metals.

A 'classical' example is the metal holmium that consists of only one isotope, namely ^{195}Ho with the nuclear spin $I = 7/2$. The unpaired $4f$-electrons present in this material give rise to a strong magnetic field at the nuclei and to an electric field gradient as well. The general expression for the nuclear energy levels E_i is rather complicated but in the case of holmium it can be approximated by

$$E_i/k_\mathrm{B} = i\,A + P\left[i^2 + \frac{1}{3}I(I+1)\right] , \tag{8.24}$$

where A and P represent the magnetic hyperfine constant and the quadrupole constant, respectively. The magnetic quantum number i may adopt the values $-I$, $-I+1$, ..., $+I$. The good overall agreement between theory and experiment is demonstrated by Fig. 8.11, where the specific heat of holmium is plotted. The hyperfine constant $A/k_\mathrm{B} \approx 0.32\,\mathrm{K}$, obtained from the fit of the data, agrees well with the values deduced from paramagnetic resonance [354]. From the magnitude of A it follows that the electrons in the $4f$-shell give rise to an effective magnetic field at the nuclei that is of the order of 900 T. The quadrupole coupling constant $P/k_\mathrm{B} \approx 8\,\mathrm{mK}$ is relatively small but it is required for a good fit of the data. This is demonstrated by Fig. 8.11b

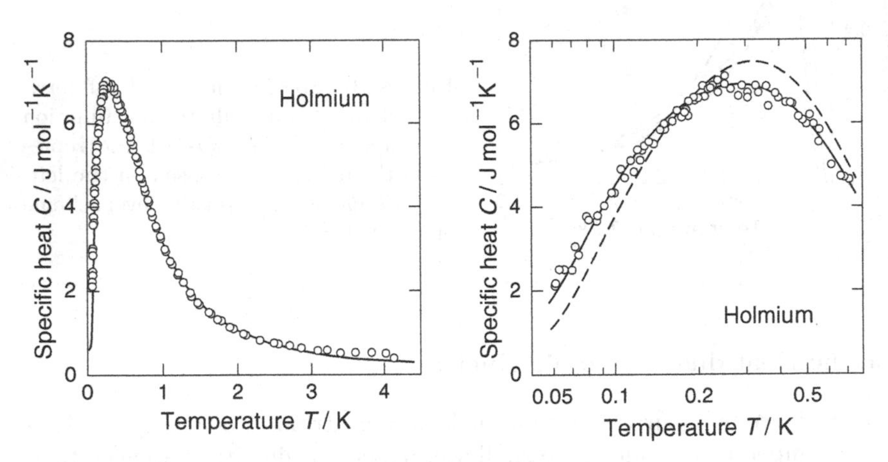

Fig. 8.11. Specific heat of holmium versus temperature. *Full lines* are fitting curves based on (8.24). (**a**) Total temperature variation [355,356] and (**b**) variation close to the Schottky maximum. Neglecting the relatively weak quadrupole coupling leads to the *dashed line* [356]

where fitting curves are drawn taking into account the electrical quadrupole coupling and ignoring it.

It is remarkable that atomic nuclei exhibit a very large specific heat, although their magnetic moments are rather small. At very low temperatures, their contribution is dominant and exceeds all other contributions. Like the electron spins of paramagnetic salts, the nuclear spins of some specific metals play an important role in the generation of very low temperatures. We will discuss this technical aspect in Chaps. 11 and 12.

8.2 Spin Waves – Magnons

As pointed out in Sect. 7.2.5, the coupling between spins leads to collective excitations in ferromagnets and antiferromagnets. In this section, we consider the dispersion relation of the spin waves and discuss their contribution to magnetization and specific heat.

8.2.1 Ferromagnets

The starting point of our considerations is the *Heisenberg model* of ferromagnets. It is based on the Hamiltonian

$$H = -\sum_{i,j} \mathcal{J}_{ij} \boldsymbol{S}_i \cdot \boldsymbol{S}_j \,, \tag{8.25}$$

that describes the isotropic exchange interaction between neighboring spins [357]. Here, $\mathcal{J}_{ij}$ represents the exchange coefficient, and $\boldsymbol{S}_i$ and $\boldsymbol{S}_j$ are the spin operators. In the ground state, all spins of a ferromagnet are aligned, as schematically shown in Fig. 8.12a for a one-dimensional array. One might guess that the flip of a single spin (see Fig. 8.12b) is the excitation with the lowest energy.[1] However, the energy $\delta E = z\mathcal{J}S^2$ is necessary for a flip, where z is the number of neighbors. This means that spin flips hardly occur at temperatures well below the Curie temperature since the coupling energy is comparable with the thermal energy at the Curie temperature T_c, i.e., $\mathcal{J} \approx k_\mathrm{B} T_\mathrm{c}$.

Already in 1930, *Bloch* realized that collective excitations exist in ferromagnetic materials that can be described as the precessing motion of coupled spins [358]. The energy necessary for the excitation of such spin waves or *magnons* is much smaller than the energy required for the flip of a single spin. In Fig. 8.12c, the precession of the spins in such a spin wave is visualized.

[1] This simple argument only holds for spin-1/2 systems where only two possible orientations exist.

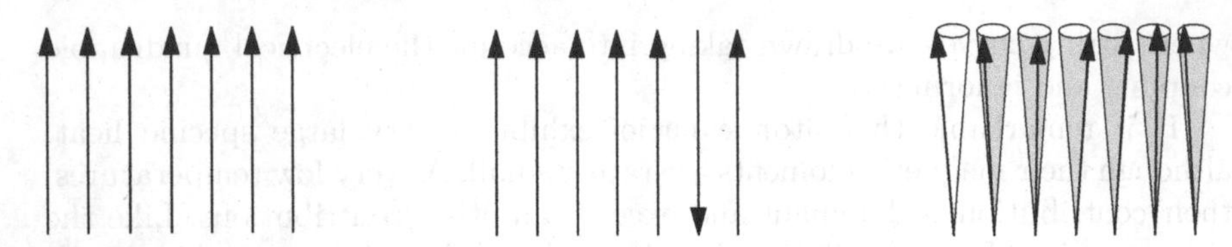

Fig. 8.12. Schematic representation of the spin orientation of an array of spins. (**a**) Ground state, all spins are aligned, (**b**) one-electron excitation, flip of a single spin by 180°, (**c**) precession of the spins in a spin wave

Dispersion of Spin Waves

In the following simple derivation of the dispersion relation of spin waves we use a semiclassical approach, i.e., we treat the spin vectors $\boldsymbol{S}$ like classical vectors $\widetilde{\boldsymbol{S}}$. In a further simplification, we restrict ourselves to a linear chain of spins with spacing a. But it should be pointed out that the conclusions drawn here are qualitatively also correct for three-dimensional solids.

Because of the exchange interaction between neighboring spins, small deviations from the perfect alignment in the z-direction give rise to a torque proportional to the deviation. Equating the rate of change of the angular momentum to the torque acting on the spin $\widetilde{\boldsymbol{S}}_m$ leads to

$$\frac{\mathrm{d}\widetilde{\boldsymbol{S}}_m}{\mathrm{d}t} = \frac{\mathcal{J}}{\hbar}\widetilde{\boldsymbol{S}}_m \times (\widetilde{\boldsymbol{S}}_{m-1} + \widetilde{\boldsymbol{S}}_{m+1}) \,. \tag{8.26}$$

This differential equation is nonlinear since products of the spin vectors occur. For small deflections in the xy-plane, i.e., for $|\widetilde{S}_x|, |\widetilde{S}_y| \ll |\widetilde{S}_z|$, the equation may be linearized by neglecting products of the terms $\widetilde{S}_x$ and $\widetilde{S}_y$, and neglecting $\mathrm{d}\widetilde{S}_z/\mathrm{d}t$. In this case, the equation of motion can be solved with the simple ansatz of a travelling wave with the amplitude A and the wave vector q of the form

$$\widetilde{S}_{m,x} = A\cos(mqa - \omega t) \,, \tag{8.27a}$$

$$\widetilde{S}_{m,y} = A\sin(mqa - \omega t) \,, \tag{8.27b}$$

$$\widetilde{S}_{m,z} = S \,, \tag{8.27c}$$

resulting in the dispersion relation

$$\hbar\omega = 2\mathcal{J}S\Big[1 - \cos(qa)\Big] = 4\mathcal{J}S\sin^2\left(\frac{qa}{2}\right) . \tag{8.28}$$

For small wave numbers, we may replace this equation by

$$\hbar\omega \approx \mathcal{J}Sa^2q^2 \,, \tag{8.29}$$

meaning that in the long-wavelength limit, the frequency of spin waves is proportional to q^2.

The corresponding calculation for cubic crystals leads to a dispersion relation very similar to that of the linear chain, namely to

$$\hbar\omega = \mathcal{J} S \sum_m \left[1 - \cos(\boldsymbol{q} \cdot \boldsymbol{r}_m)\right] . \tag{8.30}$$

The summation runs over all neighbors with the position vectors $\boldsymbol{r}_m$. In the limit $qa \ll 1$, this equation can be simplified and the dispersion relation (8.29) of the linear chain is recovered, where a now stands for the lattice constant of the cubic crystal.

In Fig. 8.13, the dispersion curve of spin waves is shown as measured by inelastic neutron scattering in cobalt alloyed with 8 % iron. As expected, a parabolic variation is observed independent of the crystallographic direction. In addition, a gap is found at $q = 0$ caused by the anisotropy of the exchange interaction already mentioned in Sect. 7.2.5.

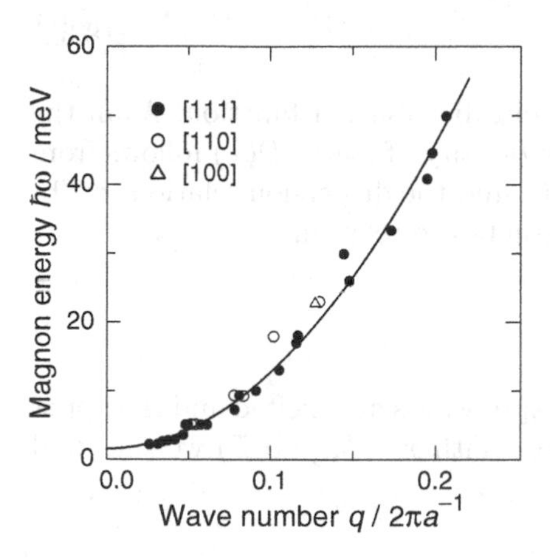

Fig. 8.13. Dispersion of magnons in cobalt alloyed with 8 % iron, measured by inelastic neutron scattering [359]. The curve is parabolic with a gap at $q = 0$ due to the anisotropy of the exchange interaction [359]

Finally, note that there is a deep analogy to lattice vibrations. A quantum-mechanical calculation leads to the result that the energy of the modes of wave number q is quantized and can only take the values $E = (n_q + 1/2)\hbar\omega$. The n_q quanta associated with the spin wave modes are called *magnons*. The excitation of a magnon is equivalent to the flip of one spin, i.e., to a change of the angular momentum of the spin ensemble by $\hbar$. Magnons play the same role in magnetic phenomena as phonons do in lattice dynamics. Both magnons and phonons obey Bose–Einstein statistics and are subjected to the same boundary conditions reflecting the finite size of the crystal (see Sects. 3.1 and 6.1).[2]

[2] It should be mentioned that in contrast to lattice waves, spin waves do not rigorously obey the superposition principle because of the nonlinearity of the equation of motion.

Magnetization

Magnons destroy the perfect alignment of spins in ferromagnets, resulting in a lessening of the total angular momentum in the z-direction. Thus, the spontaneous magnetization of the sample is reduced with increasing temperature owing to the increasing number of thermally excited magnons. As mentioned above, each magnon reduces the total angular momentum by $\hbar$, and thus the magnetization by the amount $g\mu_\mathrm{B}$. Therefore, we may write for the temperature dependence of the spontaneous magnetization

$$M_\mathrm{s}(T) = M_\mathrm{s}(0) - g\mu_\mathrm{B}\frac{N_\mathrm{mag}}{V}\,. \tag{8.31}$$

Here, N_mag stands for the number of excited magnons and $M_\mathrm{s}(0) = ng\mu_\mathrm{B}S$ for the magnetization at $T = 0$ that follows from (8.3) and (8.4) under the assumption $J = S$. The number of excited magnons is given by the integral

$$N_\mathrm{mag} = \int \mathcal{D}(\omega)\, f(\omega, T)\, \mathrm{d}\omega\,, \tag{8.32}$$

where $f(\omega, T)$ represents the Bose–Einstein distribution function. As in the case of phonons or electrons, the magnon density of states $\mathcal{D}(\omega)$ follows from the density $\mathcal{D}(q)$ in momentum space (3.6) and the dispersion relation (8.29). In this way, we obtain for cubic ferromagnets the relation

$$\mathcal{D}(\omega) = \frac{V}{4\pi^2}\left(\frac{\hbar}{\mathcal{J}Sa^2}\right)^{3/2}\sqrt{\omega}\,. \tag{8.33}$$

At low temperatures, only low-frequency spin waves are excited and the upper limit of the integral can be set to infinity. With $x = \hbar\omega/(k_\mathrm{B}T)$ we thus find for the number of excited magnons

$$N_\mathrm{mag} = \frac{V}{4\pi^2}\left(\frac{k_\mathrm{B}T}{\mathcal{J}Sa^2}\right)^{3/2}\int_0^\infty \frac{\sqrt{x}}{\mathrm{e}^x - 1}\mathrm{d}x\,. \tag{8.34}$$

The numerical value of the integral is $\Gamma(3/2)\,\zeta(3/2) \approx 4\pi^2 \times 0.0587$. In cubic crystals $n = r/a^3$, where $r = 1, 2, 4$ for the lattice sc, bcc, and fcc, respectively. Thus, we finally obtain for the temperature variation of the spontaneous magnetization:

$$\frac{M_\mathrm{s}(0) - M_\mathrm{s}(T)}{M_\mathrm{s}(0)} = \frac{0.0587}{rS}\left(\frac{k_\mathrm{B}T}{\mathcal{J}S}\right)^{3/2}\,. \tag{8.35}$$

This is the famous *Bloch* $T^{3/2}$ *law* for the magnetization of ferromagnets. In Fig. 8.14, the experimental data for nickel are compared with the theoretical prediction. At low temperatures it agrees favorably with experiment. The deviation at higher temperatures stems mainly from the fact that the approximation (8.29) for magnons with small wave numbers is no longer valid.

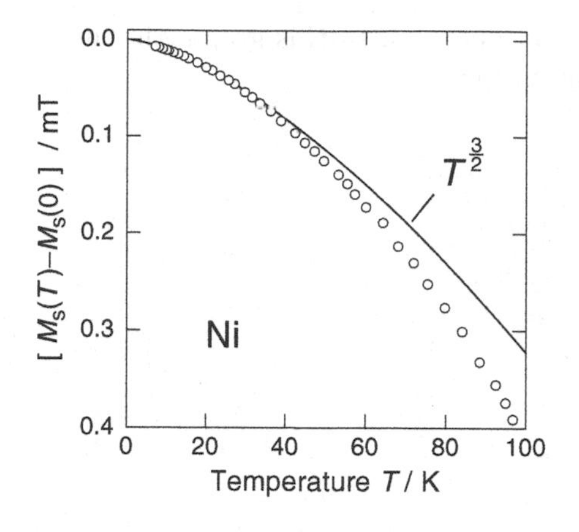

Fig. 8.14. Spontaneous magnetization of nickel plotted as a function of temperature. At low temperatures, the magnetization follows the expected $T^{3/2}$- variation shown by the *full line* [360]

Specific Heat

The internal energy due to magnons can be calculated in a manner similar to that of phonons or free electrons. Using the density of states (8.33) we find for the internal energy

$$U_{\mathrm{mag}} = \frac{V}{4\pi^2} \frac{(k_{\mathrm{B}}T)^{5/2}}{(\mathcal{J}Sa^2)^{3/2}} \int\limits_0^\infty \frac{x^{3/2}}{\mathrm{e}^x - 1} \mathrm{d}x \,, \tag{8.36}$$

where the integral has the value $\Gamma(5/2) \cdot \zeta(5/2) \approx 1.783$. Thus, the magnon contribution to the specific heat (per unit volume) is given by

$$C_{\mathrm{mag}} = 0.113 \frac{nk_{\mathrm{B}}}{r} \left(\frac{k_{\mathrm{B}}T}{\mathcal{J}S} \right)^{3/2} \tag{8.37}$$

for cubic lattices. To get an idea of the significance of the spin-wave contribution we compare C_{mag} with the lattice specific heat C_{Debye}. With the typical Debye temperature $\Theta \approx 300\,\mathrm{K}$ and with $\mathcal{J}S \approx 50\,\mathrm{meV}$, the phonon and magnon contributions are roughly the same at about $1\,\mathrm{K}$, where $C \approx 10^{-5}R$.

Electrons, phonons and magnons contribute to the specific heat of conducting ferromagnets:

$$C_V = \gamma T + \beta T^3 + \delta T^{3/2} \,. \tag{8.38}$$

Because of the different exponents (and prefactors) the respective contributions dominate in different temperature ranges. At high and low temperatures, the contributions of phonons and electrons are larger than the contribution of magnons, but at intermediate temperatures the magnon component is important. The situation is simpler in insulating ferromagnets because of the missing electron contribution. In Fig. 8.15 the specific heat of yttrium-iron-garnet ($Y_3Fe_5O_{12}$) is shown where the predicted $T^{3/2}$-law was observed. In this figure, $C/T^{3/2}$ is plotted as a function of $T^{3/2}$ in order to separate directly the contribution of magnons and phonons. In this graph, the intersection at $T = 0$ reflects the coefficient δ of (8.38).

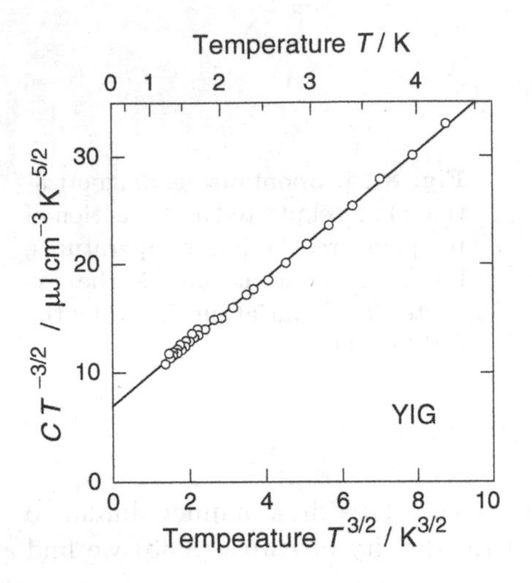

Fig. 8.15. Low-temperature specific heat $C/T^{3/2}$ of yttrium-iron-garnet as a function of $T^{3/2}$ [361]

8.2.2 Antiferromagnets

Spin waves also exist in antiferromagnets, but their dispersion relation differs qualitatively from that of ferromagnets. Therefore, their contribution to the specific heat is also different. The dispersion relation can be derived in the same way as outlined above for one-dimensional ferromagnetics. Again, we consider a linear chain but now assign the spins to the two sublattices A and B. Spins with an even index $2m$ are oriented along the z-direction, i.e., they belong to the sublattice A where the relation $S_z = S$ holds. Spins of the sublattice B have an odd index $2m + 1$ and are oriented antiparallel ($S_z = -S$). Taking into account next-neighbor interaction only, the interaction between the spins can be characterized by the exchange coefficient $\mathcal{J}'$ that is negative in the case of antiferromagnets. In this way, we obtain the equations of motion for the two types of spins that resemble (8.26):

$$\frac{\mathrm{d}\boldsymbol{S}_{2m}}{\mathrm{d}t} = \frac{\mathcal{J}'}{\hbar}\boldsymbol{S}_{2m} \times (\boldsymbol{S}_{2m-1} + \boldsymbol{S}_{2m+1}), \tag{8.39a}$$

$$\frac{\mathrm{d}\boldsymbol{S}_{2m\pm1}}{\mathrm{d}t} = \frac{\mathcal{J}'}{\hbar}\boldsymbol{S}_{2m\pm1} \times (\boldsymbol{S}_{2m} + \boldsymbol{S}_{2m\pm2}). \tag{8.39b}$$

For small displacements the differential equations can be linearized, as in the case of ferromagnets, and an ansatz is made analogous to (8.27a). Solving the equations, the dispersion relation of a one-dimensional antiferromagnet is found to be

$$\hbar\omega = 2S|\mathcal{J}'||\sin qa|. \tag{8.40}$$

Qualitatively, the same result is obtained for three-dimensional systems. Remarkably, magnons in antiferromagnets and phonons exhibit qualitatively the same dispersion relation. In particular, $\omega \propto q$ for small wave numbers q. The theoretical result is confirmed by inelastic neutron-scattering experiments. In Fig. 8.16, experimental data are presented for magnons in $RbMnF_3$ propagating in [100], [110] and [111] directions. Since in this cubic crystal the dispersion curves differ hardly at all for the three directions, we have plotted the three sets of data together in one figure.

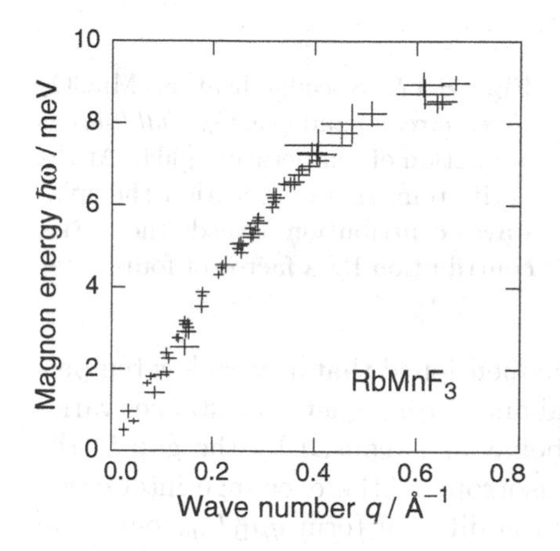

Fig. 8.16. Magnon dispersion curve in the antiferromagnetic crystal $RbMnF_3$ measured by inelastic neutron scattering [362]

The contribution of antiferromagnetic magnons to the low-temperature specific heat can be calculated as described in Sect. 6.1 for the lattice vibrations. The expression

$$C_{\mathrm{mag}} = nk_{\mathrm{B}}A\left(\frac{k_{\mathrm{B}}T}{2S|\mathcal{J}'|}\right)^3 \tag{8.41}$$

is found, where the coefficient A is characteristic for the particular crystal structure. Since the temperature dependence of the magnon contribution is identical with that of phonons, it is difficult to show the existence of C_{mag} by simple specific-heat measurements. However, interesting information is obtained by a comparison of the specific heat of two substances with similar lattice, like $CaCO_3$ and $MnCO_3$. While no magnetic moments are present in $CaCO_3$, $MnCO_3$ is antiferromagnetic at low temperatures. As shown in Fig. 8.17, at low temperatures the specific heat of $MnCO_3$ is considerably higher than that of $CaCO_3$ because of the existence of magnons in the antiferromagnetic substance. Slightly above 30 K, a maximum is observed caused by the transition from the antiferromagnetic to the paramagnetic phase. The position of the maximum agrees well with the Néel temperature $T_{\mathrm{N}} \approx 32.4\,\mathrm{K}$ deduced from measurements of the susceptibility [363]. Both materials exhibit approximately the same specific heat at high temperatures. Subtracting the phonon part deduced from data on $CaCO_3$, the magnon contribution can easily be estimated.

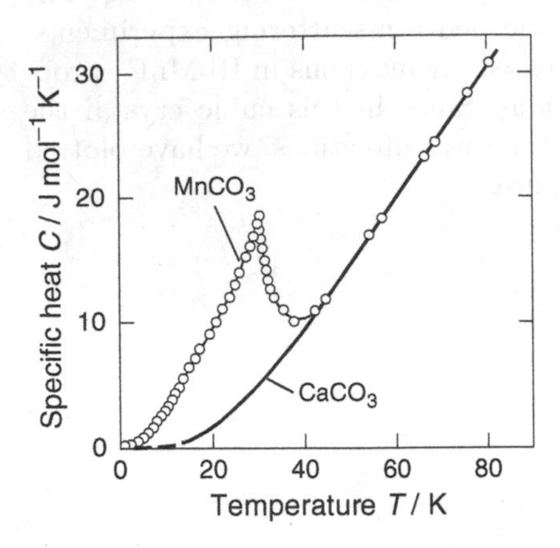

Fig. 8.17. Specific heat of $MnCO_3$ (*open circles*) and $CaCO_3$ (*full line*) as a function of temperature [364]. At the antiferromagnetic transition the spin-waves contribution exceeds the lattice contribution by a factor of four

At the end of this section, it should be mentioned that at very low temperatures the specific heat of most ferro- and antiferromagnetic substances varies exponentially with temperature. This behavior is caused by the gap in the dispersion curve at $q = 0$ due to the anisotropy of the exchange interaction mentioned in Sect. 7.2.5. As a result, an additional term $g\mu_B B_{int}$ occurs in the dispersion relations (8.28) and (8.40), where B_{int} is the internal magnetic field.

8.3 Spin Glasses

The expression *spin glasses* was originally used to describe some magnetic alloys in which nonperiodic 'freezing' of the orientation of magnetic moments was observed.[3] This freezing was accompanied by a slow response and a linear low-temperature heat capacity, such as in conventional glasses. Prominent examples of this class of materials are the alloys **Au**Fe or **Cu**Mn with a concentration of magnetic ions of a few per cent. The magnetic ions are randomly distributed and coupled via the RKKY exchange interaction (see Sect. 7.4). The magnetic moments experience ferromagnetic and antiferromagnetic ordering instructions due to the oscillatory nature of this interaction. Since not all instructions can be satisfied simultaneously, the spins are said to be 'frustrated'. Thus, the two essential ingredients leading to spin-glass ordering are *randomness* and *frustration*. As mentioned above, on cooling, the dynamics of the spins is slowed down and finally the spins are frozen in random

[3] In this section on spin glasses, we use mainly the term 'spin' although 'magnetic moment' would be more accurate in most cases.

directions. This randomness can be demonstrated by neutron-scattering experiments. The broad spectrum of timescales involved in the freezing process can be studied by magnetic-susceptibility measurements, which we will consider in more detail below. Finally, no anomaly in the specific heat is observed at the spin-glass transition.

The basis for most numerical and analytical modelling of spin glasses is the Heisenberg Hamiltonian (8.25). In many theoretical treatments the *Ising model* [365] is used in which this Hamiltonian is simplified by the assumption that only the spin orientations 'up' and 'down' are possible. Depending on the range of interaction, two different models have been developed, namely the EA model based on short-range interaction, and the SK model assuming long-range interaction.[4] Two main theories have been worked out to describe the properties of spin glasses: the *droplet theory* [368] based on renormalisation group arguments for the Ising EA model, and the *replica symmetry breaking theory* [369] providing a mean-field solution of the SK model. Here, we do not discuss these theories but rather review a few interesting experiments and refer the reader to recent monographs [370, 371] for more details.

8.3.1 Structural Properties

We visualize the problem of spin-glass formation by considering the insulating spin glass $Eu_xSr_{1-x}S$ where the nature of the exchange interaction is simpler than in conventional metallic spin glasses. In EuS the exchange constant between neighboring Eu^{2+}-ions is positive but negative between next-nearest neighbors. Therefore, depending on their separation there is a tendency to parallel or antiparallel alignment of the interacting spins as in the case of the RKKY interaction. The magnitude of the exchange constant $\mathcal{J}_1$ for nearest neighbors is roughly twice the constant $\mathcal{J}_2$ for next-nearest neighbors, i.e., $\mathcal{J}_1/\mathcal{J}_2 \approx -2$. Since the interaction with nearest neighbors is dominant in pure EuS, this substance is ferromagnetic below 16.5 K. Replacing Eu^{2+}-ions by nonmagnetic Sr^{2+}-ions causes a shift of the Curie temperature to lower temperatures. The phase diagram of the compound $Eu_xSr_{1-x}S$ is shown in Fig. 8.18. For sufficiently high dilution of the Eu^{2+}-ions by Sr^{2+}-ions, a transition to the spin-glass phase is observed below 2 K.

The reason for the development of the spin-glass state in $Eu_xSr_{1-x}S$ can be understood by considering a two-dimensional Ising system. In order to simplify the calculation of the exchange energy we take into account only the interaction between nearest and next-nearest neighbors. Nevertheless, the arguments given here also hold for three-dimensional systems and for spin systems with more than two orientational degrees of freedom.

Let us first consider the exchange energy of the two spins in Fig. 8.19a being oriented opposite to the ferromagnetic order of the remaining sample.

[4] The EA model is named after *Edwards* and *Anderson* [366], the SK model after *Sherrington* and *Kirkpatrick* [367].

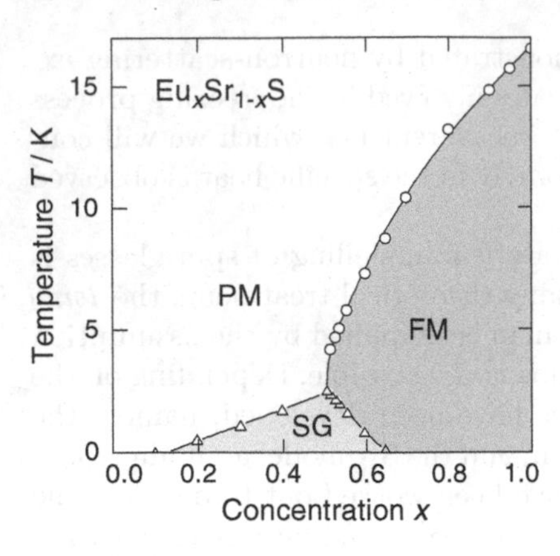

Fig. 8.18. Magnetic phase diagram of $Eu_xSr_{1-x}S$. The abbreviations PM, FM, and SG stand for paramagnetic, ferromagnetic, and spin-glass phase, respectively [372]

With the relative strength of the exchange interactions $\mathcal{J}_1/\mathcal{J}_2 = -2$ it follows directly that the configuration as drawn has the lowest energy for the given arrangement of Sr^{2+}-ions. If, however, the nonmagnetic ions are located as shown in Figs. 8.19b and c, the exchange energy of the spin pair S_1 and S_2 is independent of the orientation, i.e., the two configurations are energetically degenerate. An alignment of the framed spins parallel to the majority of the spins is prevented by the presence of aligned spins S_2 and S_1, respectively. This inability is a consequence of the frustration of the spins involved in the interaction.

With an increasing number of nonmagnetic ions, more and more regions are formed in $Eu_xSr_{1-x}S$ samples, within which the ferromagnetic order of pure EuS is destroyed and a nonperiodic spin orientation is frozen in at low temperatures. In this spin-glass phase, the exchange energy and hence the alignment of spins depends on the orientation of the spins in the environment in a complicated way.

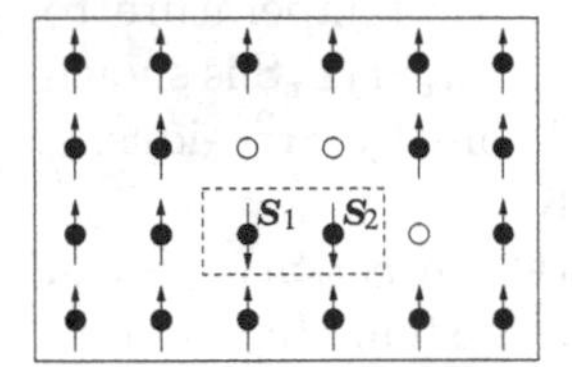

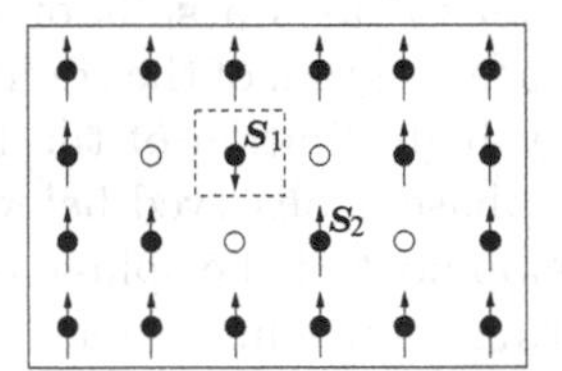

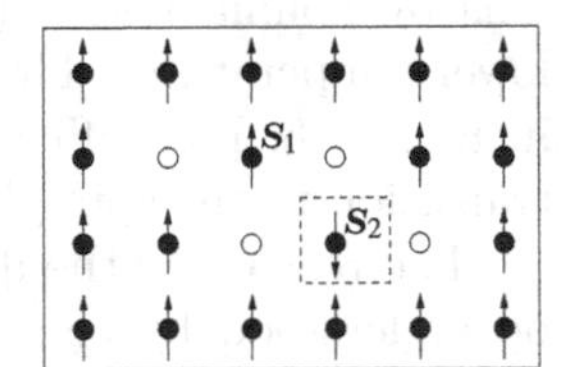

Fig. 8.19. Two-dimensional visualization of the effect of frustration. *Full circles* represent magnetic, *open circles* nonmagnetic ions. The framed spins are oriented opposite to the ferromagnetic order. (**a**) Spins S_1 and S_2 occupy the energetically lowest state. (**b**) Spin S_2 prevents the ferromagnetic alignment of spin S_1. The opposite situation is depicted in (**c**). Both configurations in (**b**) and (**c**) are energetically equivalent

8.3.2 Dynamic Behavior

Close to a characteristic temperature, called the *spin-glass temperature* T_g, the response of spin glasses to external perturbations slows down considerably, resulting in pronounced preparation-dependent effects. This interesting phenomenon can be investigated by measuring the susceptibility. The result of such an experiment on **CuMn** is shown in Fig. 8.20a. The measurement was carried out in a magnetic field of 0.59 mT. As expected, at high temperatures the dc susceptibility χ follows the Curie–Weiss law. However, below the spin-glass temperature the susceptibility depends on the measuring process. In one run, the field was applied above T_g and maintained during the measurement. In this experiment the *upper* curve was found, which is labelled FC for 'field cooling'. In the other run, the sample was first cooled down to the lowest temperature in zero field ($B < 5\,\mu$T). Afterwards, the field was applied and the susceptibility measurement was performed. In this experiment the *lower* curve was obtained, labelled ZFC for 'zero field cooling'. The observed *cusp* in the susceptibility is a characteristic feature of spin glasses.

The response of spin glasses to finite frequencies is illustrated in Fig. 8.20b. The measurements were carried out on the insulating $Fe_{0.5}Mn_{0.5}TiO_3$ that behaves like a 3d Ising spin glass with spins aligned along the hexagonal c-axis. First, let us have a look at the *full line*. It reflects the field-cooled susceptibility χ_{FC} measured with a dc magnetometer. In this experiment, the sample was cooled in temperature decrements of 0.1 K and equilibrated for 1000 s after each temperature step. The temperature variation of the real part χ' of the ac susceptibility is shown for various frequencies. The interesting observa-

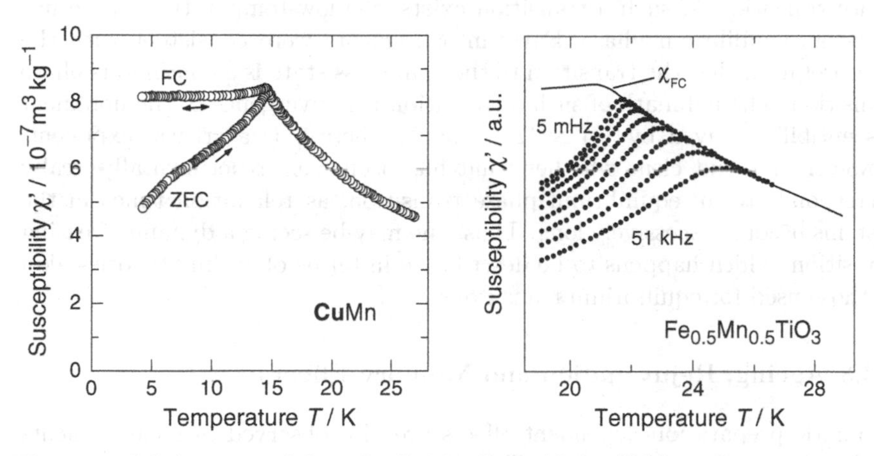

Fig. 8.20. (**a**) dc susceptibility of a **CuMn** alloy with 2 at % Mn versus temperature. At low temperatures two data sets were found: The *upper* curve (FC) was obtained by field cooling, the *lower* curve (ZFC) by cooling in zero field [373]. (**b**) Real part χ' of the ac susceptibility of $Fe_{0.5}Mn_{0.5}TiO_3$ versus temperature. The different curves correspond to frequencies differing by a factor of ten. In addition, the field-cooled dc susceptibility χ_{FC} is depicted by a *full line* [374]

tion is that the cusp shifts to lower temperatures with decreasing frequency. It should be mentioned that in the vicinity of the cusp a maximum in the imaginary part χ'' of the susceptibility is found, which shifts with frequency as χ' does. The imaginary part of the susceptibility reflects the losses due to the motion of spins in the oscillating magnetic field. As we shall discuss in detail in Sect. 9.1.3, this process is most effective if the relation $\omega\tau \approx 1$ is fulfilled, where τ represents the corresponding relaxation time.

We may qualitatively understand the observed phenomena in the following way: With decreasing temperature, more and more spins or clusters of spins are freezing in certain configurations and are no longer able to reorientate themselves on further cooling. Roughly speaking, we may divide the spins into two groups. The spins in the first group are able to follow the external perturbation, i.e., $\omega\tau < 1$. The spins in the second group (with $\omega\tau > 1$) are not able to change their orientation on the timescale given by the period of the applied field. Therefore, the observed loss maximum is mainly due to those spin for which $\omega\tau \approx 1$. With decreasing frequency, spin configurations are probed with longer and longer relaxation times. Since a large change in the frequency of the measurement only leads to a rather small shift of the cusp, this result demonstrates that there is a rapid slowing down of the spin dynamics on approaching the spin-glass transition temperature.

A comparison with the field-cooled susceptibility makes it clear that χ_{FC} corresponds to the equilibrium susceptibility at temperatures $T > 21.5\,\mathrm{K}$. The 'true' freezing temperature, or the 'true' spin-glass temperature, can only be measured in the limit of vanishing frequency. However, it is not clear to date whether spin-glass ordering is associated with a true thermodynamic phase transition. If such a transition exists, the low-temperature phase has to be an equilibrium phase. Early measurements were consistent with the interpretation that the transition to the spin-glass state is a continuous phase transition. The hallmark of such a transition is a divergence of the nonlinear susceptibility[5] given by $\chi_{\mathrm{nl}} \propto |T - T_{\mathrm{g}}|^{-\gamma}$, where γ is a critical exponent. However, it is not clear whether what has been seen experimentally, really corresponds to an equilibrium phase transition, as relaxation times in the systems become excessively long. Thus, one may be seeing a dynamic freezing transition, which happens to be describable in terms of scaling theories akin to those used for equilibrium transitions.

8.3.3 Ageing, Rejuvenation and Memory Effects

Dramatic preparation-dependent effects are also observed in measurements of the remanent magnetization. These observations demonstrate that in the spin-glass phase there are many metastable states whose relative free energies vary in different ways with external perturbations and that have energy barriers impeding motion from one state to another. To describe the complex

[5] The nonlinear susceptibility χ_{nl} is defined by the relation $\mu_0 M = \chi B + \chi_{\mathrm{nl}} B^3$.

ageing phenomena, processes have to be considered that occur on different timescales. In ageing experiments, the age t_a of the system plays an important role. It consists of two parts and is defined by $t_a = t_w + t$. The wait time t_w is the time during which the spin glass is kept at constant temperature before the magnetic field is applied, and the observation time t is the time elapsed after field application.

Crossing the spin-glass temperatures T_g in a magnetic field leads to a remanent magnetization of the sample. After removing the field this magnetization decays with time thus allowing the investigation of effects caused by spin reorientation. An alternative procedure is, first to cool the sample below the freezing temperature in zero field, and then to apply the magnetic field after the wait time t_w. The temporal development of the magnetization $M(t_w, t)$ recorded in such a measurement for different wait times t_w is shown in Fig. 8.21a for the system **CuMn** at $T = 0.85\,T_g$. A significant feature is the inflection point in $M(t_w, t)$ since the derivative of the magnetization with respect to time reflects the corresponding relaxation rate. In Fig. 8.21b, the relaxation rate $S(t) \propto \partial M/\partial(\ln t)$ is plotted. The data clearly demonstrate that the dominant influence of ageing occurs in the time window where the observation time t is of the order of t_a.

Finally, we show the result of an experiment in which the *rejuvenation* and the *memory effect* are demonstrated. In the measurement shown in Fig. 8.22a, the insulating spin glass $CdCr_{1.7}In_{0.3}S_4$ was quenched from a temperature above $T_g = 16.7$ K to 12 K. Here, the imaginary part χ'' of the ac susceptibility at 1 Hz is plotted since this quantity is more sensitive to the effects discussed here than the real part χ'. After cooling the sample down to 12 K, it was left to relax for 350 min. During that time, reorientation of spins in metastable configurations took place resulting in a reduction of the free energy. As a

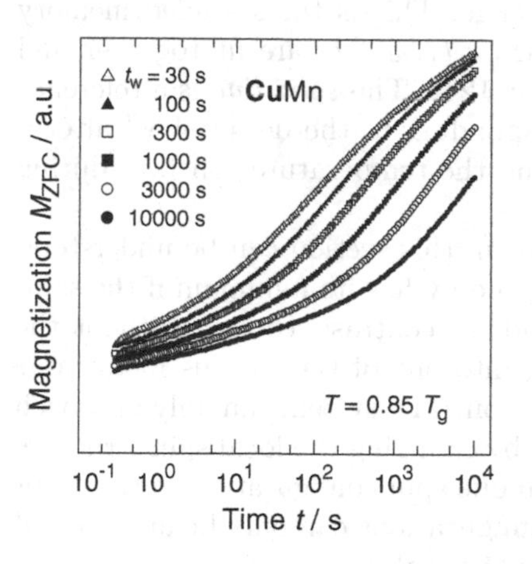

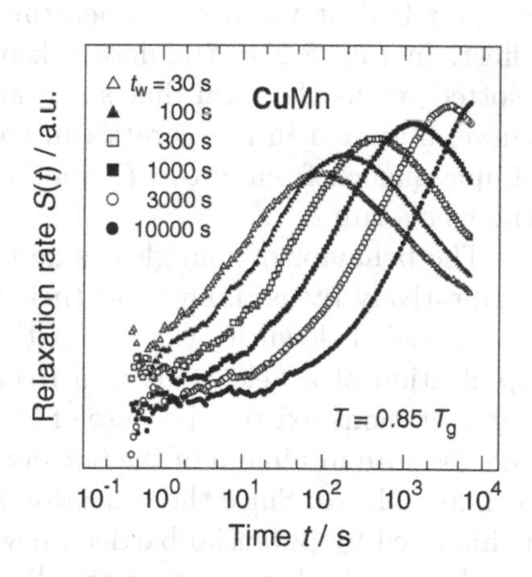

Fig. 8.21. Ageing effects on the magnetization of **CuMn** at $T = 0.85T_g$. (**a**) Zero-field magnetization $M_{ZFC}(t_w, t)$ versus $(\log t)$ for various wait times t_w, (**b**) corresponding relaxation rates $S(t)$ [374, 375]

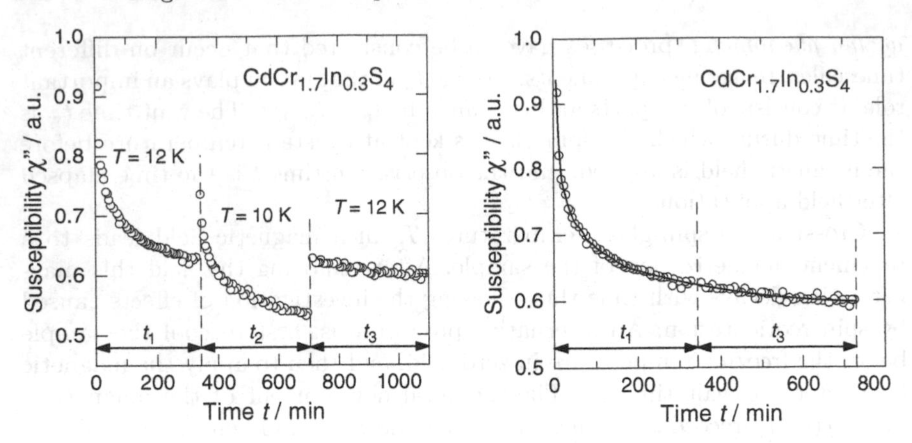

Fig. 8.22. Time variation of χ'' of the insulating spin glass $CdCr_{1.7}In_{0.3}S_4$ measured at 1 Hz. (**a**) After $t_1 = 350$ min the temperature was suddenly lowered by 2 K during $t_2 = 350$ min, and raised to the initial value after a further 350 min. (**b**) The data taken during t_1 and t_3 are fitted together and plotted versus the total time spent at $T = 12$ K [376]

consequence, the number of spins able to reorient on the given timescale was reduced leading to a decay of χ''. After 350 min the temperature was dropped to 10 K. Surprisingly, χ'' restarts to decay with time from a higher value, despite the reduced thermal energy. This is the so-called rejuvenation effect. Obviously, at the new temperature a different subset of spins is contributing to the magnetic loss at 1 Hz. The magnitude of χ'' decreases with time because of the global long-time rearrangement of the spin orientation. After another 350 min the temperature was brought back to 12 K, and χ'' resumes from the value it had at the first temperature change. This is the so-called memory effect. In Fig. 8.22b, the data taken during t_1 and t_3 are fit together and plotted versus the total time spent at $T = 12$ K. The *solid line* is a reference curve, obtained in a separate run corresponding to the decay of χ'' after a simple quench from above T_g and keeping the temperature constant during the whole run.

The behavior of spin glasses described in this section can be understood qualitatively by assuming that their free energy has its minimum if the sample is cooled down in a dc magnetic field. In contrast, cooling without the application of a field leads to a freezing into one of the various metastable configurations existing because of frustration. The system can only approach the absolute minimum of the free energy by changing the local spin structure at many places. Since the transition from one spin configuration to the other is hindered by potential barriers, new configurations can only be approached slowly, even if they are energetically more favorable.

The attempt to understand the cooperative physics of such alloys has exposed many previously unknown concepts and led to new analytical, experimental and computer-simulational techniques. These have had major ramifications throughout the whole field of study of problems involving assemblies of strongly interacting individual entities in which competitive forces yield a complex cooperative behavior.

8.4 Nuclear Magnetic Ordering

Spontaneous nuclear magnetic ordering effects due to nuclear dipole–dipole interactions are expected to take place in the nK or μK range since nuclear magnetic moments are about three orders of magnitude smaller than that of electrons. Therefore, the observation of such phenomena requires, in general, ultralow temperatures.

Ordering of nuclear spins was first observed in 1969 by *Chapellier, Goldmann, Chau* and *Abragam* in insulating materials [377]. By measuring the magnetic susceptibility of CaF_2 they were able to prove the existence of an antiferromagnetically ordered phase of fluorine nuclei. In this experiment, ^{19}F-nuclear spins were polarized in a rather ingenious way at about 0.7 K and in a magnetic field of 2.7 T using paramagnetic impurity ions U^{3+} as a 'mediator'. At the beginning of the experiment the electron spins were almost completely polarized, and the nuclear spins nearly completely unpolarized. The sample was then irradiated by microwaves of frequency $(\omega_e - \omega_n)$, i.e., with the difference between the Larmor frequency ω_e of the electrons, and the Larmor frequency ω_n of the nuclei. In this way, *simultaneous transitions* of the two types of spins were induced, namely flip-flop transitions, where electronic and nuclear spins flip in opposite directions.[6] Since the electron spins were initially polarized, the microwave-induced flipping occurred only in one direction. Because of their short relaxation time, electrons relax back to the polarized state immediately, whereas the nuclei maintain their new orientation. In this way, the nuclei were driven up and a polarization of 90% was reached in the experiment with CaF_2. The corresponding nuclear spin temperature was about 4 mK. By demagnetization, i.e., by decreasing the applied magnetic field, the nuclei were subsequently cooled further to temperatures below 1 μK, where nuclear ordering occurred.[7]

Until now, spontaneous nuclear magnetic ordering in *thermal equilibrium* has been observed only in a few materials. There are two particular types of systems with exceptionally high ordering temperatures, namely solid ^{3}He and so-called *Van-Vleck paramagnets*. In ^{3}He, the strong direct exchange interaction (see Sect. 3.1.1) gives rise to an antiferromagnetic phase with a

[6] For flip-flip transitions with both spins flipping in the same direction, microwaves with the frequency $(\omega_e + \omega_n)$ would be required.

[7] Nuclear demagnetization will be discussed in more detail in Chap. 11, where the generation of ultralow temperatures by demagnetization of nuclear moments will be considered.

Néel temperature of about 1 mK [378]. In Van-Vleck paramagnets such as $PrCu_6$, $PrNi_5$, or $Pr_{1-x}Y_xNi_5$, large hyperfine fields can be induced at the nuclei by moderate external fields. Without an applied field the ground state of the $4f$-shell of the Pr^{3+}-ions is a nonmagnetic singlet ground state. But magnetic fields change the $4f$-atomic wave function and induce an electronic magnetic moment on the ground state. This induced moment, in turn, causes a hyperfine field that can exceed the external field by up to a factor 100. Consequently, the transition to ferromagnetic nuclear order is observed in $PrCu_6$ at $T_c = 2.5$ mK [379].

With decreasing temperature, the exchange of thermal energy between different systems becomes slower and slower. For example, in the μK range the nuclear spin-lattice relaxation time can be of the order of weeks. Since the heat exchange between phonons and nuclear spins is generally much slower than the spin-spin relaxation among the spins, the nuclear spins form a well-defined subsystem that is in internal thermal equilibrium. The decoupling of phonons and spins makes it possible to define the so-called *nuclear spin temperature* T_n, which may differ by orders of magnitude from that of the lattice.

Since nuclear spins couple much more strongly to conduction electrons than to phonons, the relaxation of nuclei is faster in metals. Nevertheless, under certain circumstances it is possible to cool nuclear spins well below the *electron temperature* T_e. The reason for this surprising effect is that the strong coupling not only accelerates the energy exchange between nuclei and electrons but also leads to an enhancement of the coupling between the nuclear spins via the *Ruderman–Kittel interaction*, i.e., the indirect exchange interaction.[8]

Frequently, a distinction is made between systems with weak and systems with strong nucleus–electron coupling. In the case of weak coupling the interaction strength is comparable with the nuclear magnetic dipole–dipole interaction. As examples of weak coupling systems we mention Cu, Au, Ag, or Rh. If the indirect exchange interaction via the conduction electrons predominates, as in Tl, In, Sn, and Sc, the coupling is called 'strong'.

8.4.1 Strong Nucleus–Electron Coupling

Strongly coupled systems are particularly well suited for the observation of nuclear ordering. In this case, electrons and nuclei are in equilibrium because of the relatively short relaxation time, i.e., $T_e = T_n$. The nucleus-electron relaxation time is given by the *Korringa relation* [380]

$$\tau = \frac{\kappa}{T_e}, \tag{8.42}$$

[8] We have already discussed the indirect exchange interaction between localized magnetic moments via conduction electrons in Chap. 7, when the Kondo effect was discussed. However, in that case we considered the localized magnetic moments of electrons.

where κ is the *Korringa constant*. The value of κ not only determines the nucleus-electron relaxation time, it is also a measure for the strength of the indirect exchange interaction *between* the nuclei.

Apart from solid ^{3}He and Van-Vleck paramagnets, spontaneous nuclear ordering in thermal equilibrium has, to our knowledge, only been observed in the intermetallic compound $AuIn_2$. At first glance, pure indium metal seems to be a very good candidate for investigations of spontaneous nuclear ordering. Because of its large magnetic moment $\mu = 5.5\mu_n$, large nuclear spin $I = 9/2$, and small Korringa constant $\kappa = 0.09\,\mathrm{K\,s}$, ordering is expected to take place at relatively high temperatures. However, the strong electric quadrupole interaction of the nuclei in the tetragonal indium crystal and the relatively high critical field of 28 mT of this superconductor suppresses ordering. To avoid these problems the cubic compound $AuIn_2$ with a critical field of only 1.45 mT was used in the experiment considered here [381]. In addition, the Korringa constant $\kappa = 0.11\,\mathrm{K\,s}$ of $AuIn_2$ does not differ much from that of pure indium.

In Fig. 8.23, measurements of the specific heat of $AuIn_2$ at different external magnetic fields are displayed. The magnetic field seen by the nuclei consists of two parts. In addition to the external field B_{ext}, there is an internal field B_{int} produced by the nuclear moments. The resultant effective field is $B_{\mathrm{eff}} = \sqrt{B_{\mathrm{ext}}^2 + B_{\mathrm{int}}^2}$. The internal field B_{int} can be neglected at high

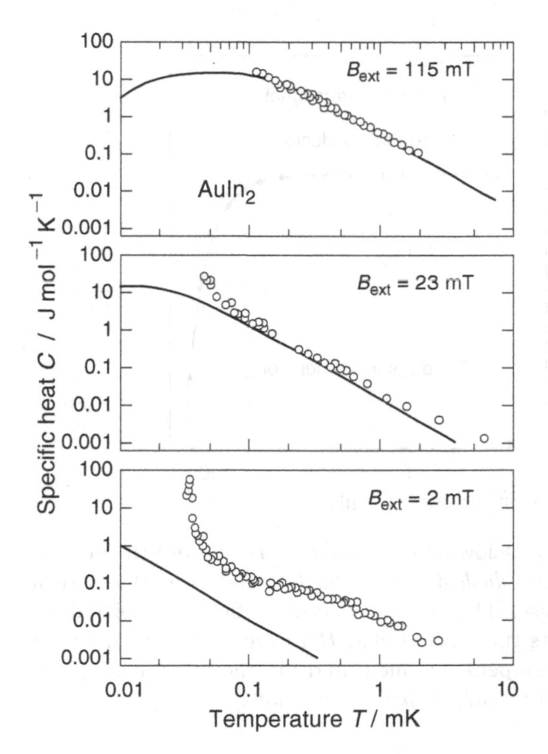

Fig. 8.23. Specific heat of $AuIn_2$ versus temperature measured at different magnetic fields. The *full lines* reflect the contributions expected for noninteracting indium nuclei [381]

external fields. In this case, the specific heat varies with temperature as for noninteracting spin systems, and can therefore be described by a Schottky anomaly. With decreasing field, deviations from this behavior are observed, and at $B_{\mathrm{ext}} = 2\,\mathrm{mT}$ a sharp maximum of the specific heat is found at 35 μK, indicating the occurrence of a phase transition to ferromagnetic order. It is worth noting that the specific heat is very large in this temperature range.

It is remarkable that superconductivity and ferromagnetic nuclear order are simultaneously present in $\mathrm{AuIn_2}$. This effect is clearly visible in the phase diagram drawn in Fig. 8.24, which shows the interdependence of these two quantities. At 'higher temperatures', i.e., between 0.2 mK and 207 mK, the critical magnetic field follows the well-known relation $B_{\mathrm{c}}^*(T) = B_{\mathrm{c}}^*(0)[1-(T^2/T_{\mathrm{c}}^2)]$ of type I superconductors (see (10.2)) where the asterisks indicate that the magnetization of the sample has not been taken into account. Below 0.2 mK, the critical field decreases slightly on cooling by an amount proportional to the nuclear magnetization, i.e., according to $B_{\mathrm{c}}(T) = B_{\mathrm{c}}^* - \mu_0 M(B_{\mathrm{c}}, T)$. Finally, at $T_{\mathrm{c,n}} \approx 35\,\mu\mathrm{K}$, a transition from the nuclear paramagnetic phase to the ordered phase occurs, which is accompanied by a distinct reduction of the critical field. Below $T_{\mathrm{c,n}}$, the sample remains superconducting in weak magnetic fields, indicating that superconductivity and nuclear ferromagnetism coexist. In this temperature range, the measured value of the critical field depends on the precooling conditions implying that domains are present in the nuclear ferromag-

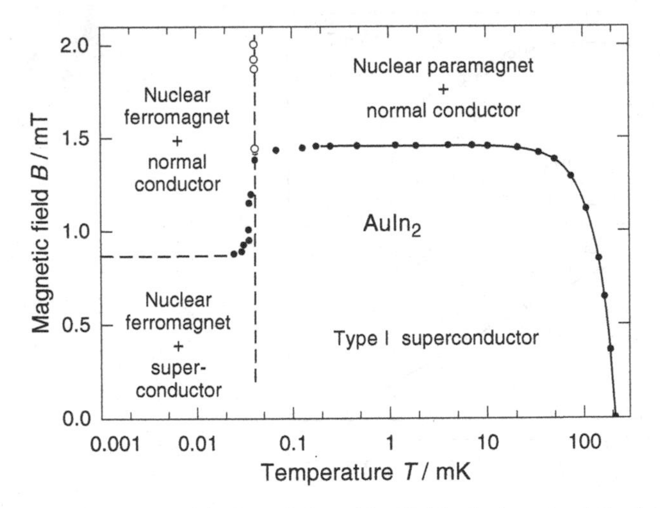

Fig. 8.24. Phase diagram of $\mathrm{AuIn_2}$. Below 35 μK (*dashed vertical line*) the nuclear magnetic moments are aligned. The *dashed horizontal line* indicates the critical magnetic field in the ordered phase. The measured critical field is depicted by *full circles*, the *solid line* represents the critical field $B_{\mathrm{c}}^*(T)$ according to the BCS theory. In addition, the ordering temperature measured via the nuclear magnetic susceptibility is plotted for fields $B \leq 2\,\mathrm{mT}$ (*open circles*) [382]

netic phase. The calculated saturation magnetization of the nuclear spins $\mu_0 M_s(0) = \mu_0 n \mu_n g_n I = 1.01\,\mathrm{mT}$ is smaller than the critical field above the phase transition. Hence, superconductivity is not completely destroyed in the ferromagnetic phase but survives, although with a reduced critical field.

8.4.2 Weak Nucleus–Electron Coupling

In the case of weak Ruderman–Kittel interaction, and hence of weak coupling between the nuclei, spontaneous magnetic nuclear ordering is expected to occur in the range of nanokelvins. The weak coupling between nuclei and conduction electrons also leads to a prolongation of the Korringa relaxation time τ. Therefore, it becomes feasible to reduce the spin temperature T_n well below the electron temperature T_e for a limited time by fast demagnetization. In this way, studies of cooperative phenomena within the ensemble of nuclear spins become possible. The first material to be successfully cooled in this manner was copper. By measurments of the magnetic susceptibility it was demonstrated in 1982 that an antiferromagnetic phase with a Néel temperature of 58 nK exists in this material [383].

These experiments were carried out in a two-stage nuclear demagnetization cryostat (see Sect. 11.7), the second stage being also the copper sample under investigation. In a field of 7 T the sample was precooled to a temperature between 50 and 100 μK by the first stage. Then the field at the sample was reduced to zero in 20 min. Since the Korringa constant of copper is $\kappa = 1.2\,\mathrm{K\,s}$, the relaxation time is about three hours at $T_e \approx 100\,\mu\mathrm{K}$. Obviously, the nuclei were not in thermal equilibrium with the conduction electrons under these conditions. Because of the strong thermal anchoring of the sample to the first nuclear stage, the electron temperature remained virtually constant during the whole experiment.

With a SQUID magnetometer the static magnetic susceptibility of the sample was measured during the warming phase of the experiment. The time after demagnetization can be converted to entropy and also to temperature.[9] In this way, the measured susceptibility can be plotted versus entropy, as shown in Fig. 8.25. The susceptibility first rises, passes through a maximum and decreases again. Since this variation is the typical behavior of antiferromagnets in the vicinity of their phase transition, these data were considered to be a strong indication for antiferromagnetic nuclear ordering in copper.

This interpretation was confirmed by a series of additional experiments. As shown in Fig. 8.26, in measurements without an external magnetic field, an entropy jump is found at $T_{c,n} = 58\,\mathrm{nK}$. At this temperature, the latent

[9] Briefly, the relation between time and entropy can be found by a measurement of the susceptibility in high magnetic fields where equations for the paramagnetic state apply. Then the temperature can be determined via the second law of thermodynamics by giving the spins a heat pulse of known magnitude and finding the ensuing entropy increase.

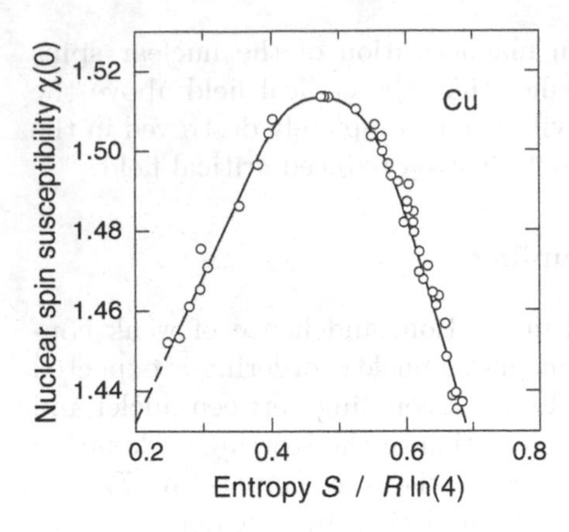

Fig. 8.25. Static nuclear spin susceptibility $\chi(0)$ of copper as a function of entropy [384]

heat $L = T_{\mathrm{c,n}}\Delta S_{\mathrm{c,n}} = 0.09\,\mu\mathrm{J\,mole}^{-1}$ was observed, where $\Delta S_{\mathrm{c,n}}$ is the measured entropy jump. This observation clearly indicates that a first-order phase transition occurs in the spin system. In further investigations, three different antiferromagnetic phases were found. In Fig. 8.27, the phase diagram is shown that was constructed from measurements of the susceptibility and NMR experiments on single crystals with a magnetic field applied in the [001] direction. The shaded areas mark the regions in which first-order transitions take place. In the upper-right corner of this figure the spin arrangements that were originally proposed, are depicted schematically. In later experiments, it was shown that the suggested structures are not fully consistent with neutron-diffraction data. For example, the phases AF1 and AF2 consist of four sublattices rather than two. In particular, in the phase AF2 the observed Bragg reflections are consistent with a so-called 'up-up-down spin configuration' [386]. This kind of order had not been observed before in any fcc antiferromagnet.

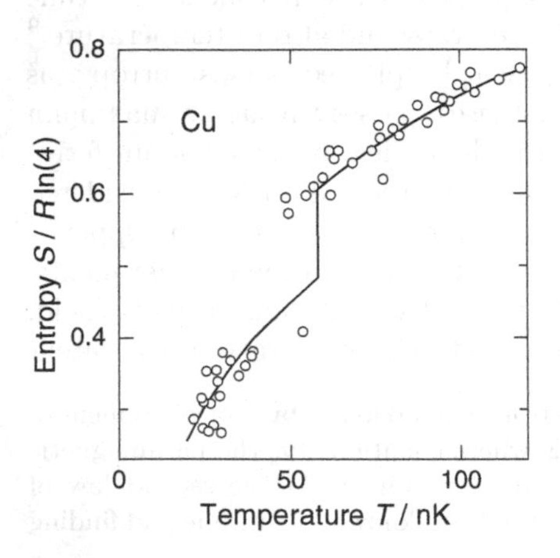

Fig. 8.26. Reduced entropy $S/(R\ln 4)$ of copper in zero field as a function of temperature. Close to the entropy jump at 58 nK, namely in the range $S = (0.38 - 0.56)R\ln 4$ it was not possible to determine the temperature with a high enough accuracy [384]

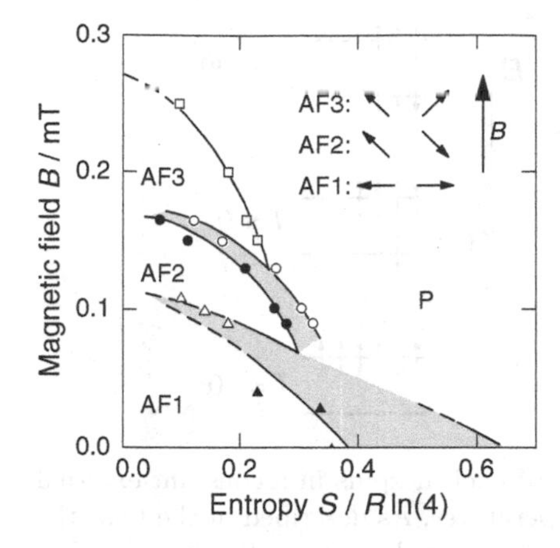

Fig. 8.27. External magnetic field versus entropy for nuclear spins in copper. The *shaded areas* mark the regions in which first-order phase transitions take place. The spin arrangements originally proposed for AF1, AF2, and AF3 are schematically illustrated in the *top-right* corner [384, 385]

In later experiments, nuclear magnetic ordering has also been observed in silver and rhodium by different techniques. Because of the smaller magnetic moment of these nuclei, ordering occurs at much lower temperatures. In measurements of the susceptibility the transition temperatures were determined to be $T_{c,n} = 560\,\text{pK}$ and $T_{c,n} = 280\,\text{nK}$ for Ag and Rh, respectively.

As we shall see, reversal of the external field gives rise to interesting new effects provided the nuclear relaxation is sufficiently slow. In this case, negative occupation temperatures can temporarily be generated. In the following section we will discuss some measurements of this type.

8.5 Negative Spin Temperatures

The concept of negative occupation temperatures was introduced in 1938 by *Casimir* and *Du Pré* [387]. The main idea associated with negative spin temperatures is illustrated in Fig. 8.28. The schematic diagram reflects the occupation of the levels of noninteracting spin-1/2 systems in an external magnetic field $\boldsymbol{B}$. The same ideas readily apply to interacting spins.

At the absolute $T = +0$, all nuclear spins are in the ground state, i.e., they are aligned parallel to the external field. With increasing temperature a growing number of spins is excited, and finally at $T = +\infty$ both levels are equally populated. If the energy of the spin system is further increased by some means, an inverted occupation can be obtained, which can be described by the Boltzmann factor, but now with a temperature $T < 0$. Finally, only the upper levels will be populated, i.e., the temperature $T = -0$ is reached. The transition from positive to negative temperatures takes place smoothly at $T \pm \infty$.

In principle, the experimental production of negative temperatures is rather simple because a sufficiently fast field reversal leads to an inversion of the two energy levels. From an experimental point of view it is important to

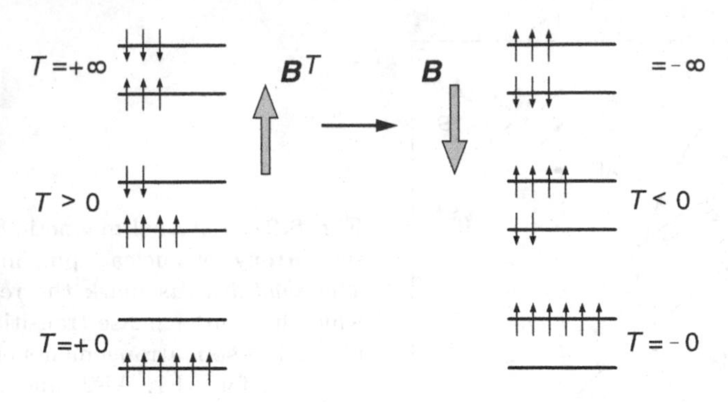

Fig. 8.28. Occupation of the energy levels of nuclear spins in a constant external magnetic field at positive and negative temperatures. As described in the text, the transition from positive to negative temperatures can be achieved by a reversal of the applied magnetic field [388]

carry out the field reversal rapidly in comparison with the spin-spin relaxation time τ_2 by which thermal equilibrium of the spin system is achieved. If the field change is slow, the spins will follow the field reversal adiabatically, and negative temperatures will not be obtained. During the quick field flip, the Boltzmann distribution of the spins breaks down and for a moment, the spin system cannot be assigned a temperature. Once the field is reversed and the system is in internal equilibrium, it can remain in a state of negative temperature for a long time. The spins relax towards positive temperatures by exchanging energy with phonons or conduction electrons, and pass, in this way, the temperature $T \pm \infty$.

The first experimental realization of a negative spin temperature was achieved in 1951 by *Purcell* and *Pound* in investigations of the system LiH [389]. Evidence for the occurrence of a negative occupation temperature was provided in this experiment by NMR measurements where stimulated emission was observed.

8.5.1 Thermodynamics at Negative Temperatures

A necessary condition for the occurrence of negative temperatures is that the number of allowed energy states of the system is limited. Otherwise, the Boltzmann factor $\exp(-E_m/k_B T)$ would not converge for $T < 0$ and the internal energy would tend to infinity. Therefore, the lattice of solids or the conduction electrons cannot be brought to negative temperatures. From the viewpoint of thermodynamics, a further essential requirement is that the entropy S does not monotonically increase with the internal energy U. Since $T = 1/(\partial S/\partial U)_B$, it follows that for positive and negative temperatures the sign of $(\partial S/\partial U)_B$ must be different. As a consequence, the entropy decreases with increasing internal energy for a system at $T < 0$. Furthermore, the spin ensemble must be sufficiently weakly coupled to the environment, but the spins must be in thermodynamic equilibrium among themselves. Therefore, the relaxation time τ_2 for establishing thermal equilibrium among the nuclear spins must be much shorter than the relaxation time τ_1, the time in which the spin system comes into equilibrium with its surroundings. For silver at $T_e = 200\,\mu K$, the condition $\tau_1 \gg \tau_2$ is fulfilled, since $\tau_1 = 5 \times 10^4$ s, and $\tau_2 = 10$ ms.

Systems at negative temperatures exhibit unusual properties. Since energy has to be added to go from positive to negative temperatures, systems with $T < 0$ are always hotter than systems with $T > 0$. This fact leads to interesting consequences. Suppose there is a spin system with a negative temperature and another one with a positive temperature. After making thermal contact, heat will flow from the system with negative temperature to the system with positive temperature! Adiabatic demagnetization of a spin system at $T < 0$ leads to heating and not to cooling, as it does at $T > 0$. Accordingly, a spin system at $T < 0$ has to be heated to increase its polarization.

The order of temperatures on the absolute Kelvin scale, from the coldest to hottest is the following: $+0\,K \rightarrow 300\,K \rightarrow \pm\infty\,K \rightarrow -300\,K \rightarrow -0\,K$. This means that a system at $T = +0\,K$ cannot be cooled further because it cannot give up more of its energy, and vice versa, a system at $T = -0\,K$ cannot be heated further because it cannot absorb more energy. The variation of the entropy $S/(R\ln 2)$, the specific heat C_B/R and the internal energy $U/(N|\mu|B)$ of an ensemble of two-level systems are shown in Fig. 8.29 as a function of $1/T$ for positive and negative temperatures. The quantities are normalized in such a way that common axes without units can be used. From this graph, the symmetry of the thermodynamic quantities with respect to positive and negative temperatures becomes evident.

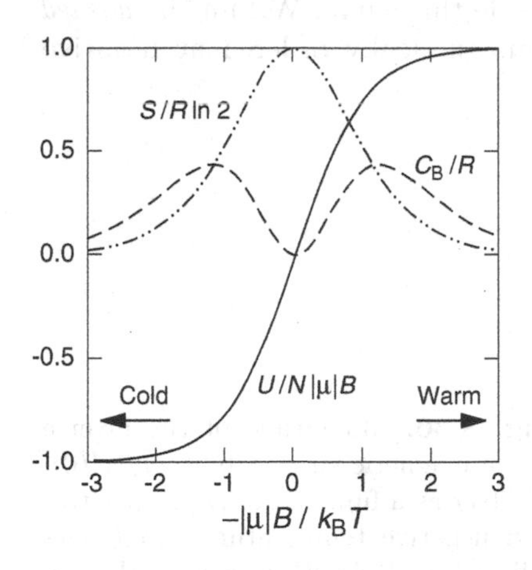

Fig. 8.29. Reduced entropy, specific heat, and internal energy of an ensemble of two-level system versus $-|\mu|B/(k_B T)$ [388]

8.5.2 Nuclear Ordering

At ordinary temperatures, the Gibbs free energy $G = H - |T|S$ possesses a minimum since, in equilibrium, the entropy S exhibits a maximum. In our notation, $H = U - BM$ represents the magnetic enthalpy. At $T < 0$, Gibbs free energy $G = H + |T|S$ reaches its maximum in equilibrium. This has a profound effect on the spontaneous magnetic order of nuclear spins. For example, in silver the Ruderman–Kittel interaction favors an antiparallel orientation of next-nearest neighbors resulting in an antiferromagnetic order as T approaches $+0$. At negative temperatures, thermodynamics predict the occurrence of a ferromagnetic order since the energy must now be maximized.

In Fig. 8.30, the magnitude of the inverse static susceptibility $|\chi'(0)|^{-1}$ of silver is shown for positive and negative temperatures. This set of data has been deduced from measurements of the imaginary part χ'' in the frequency range from 30 to 180 Hz by applying the Kramers–Kronig relation $\chi'(0) = (2/\pi) \int (\chi''/\omega) d\omega$. Clearly, the susceptibility at negative temperatures is much higher than at positive temperatures. This is due to the fact that at $T < 0$ the spin system tries to maximize its energy at constant entropy. Since the exchange interaction is antiferromagnetic in silver, the state with maximum energy has ferromagnetic order with a higher susceptibility. Although these data clearly reflect the different behavior of nuclear spins at positive and negative temperatures, they also demonstrate that in these experiments spontaneous nuclear order of silver was not reached. No deviation from the Curie–Weiss behavior was observed. In later experiments, the existence of an antiferromagnetic phase was demonstrated for $T > 0$ with a Néel temperature $T_N = 560 \pm 60$ pK and a ferromagnetic phase with $T_c = -1.9 \pm 0.4$ nK. The magnetic field–entropy phase diagram of silver is shown in Fig. 8.31. At $T > 0$ the spins are antiferromagnetically ordered inside the *solid curve* and paramagnetic outside this curve. Within the *dashed curve* the spins are expected to be ferromagnetically ordered at negative temperatures.

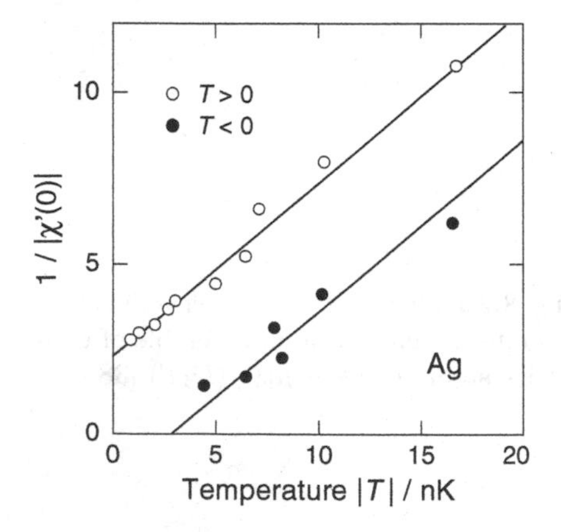

Fig. 8.30. Magnitude of the inverse static magnetic susceptibility $1/|\chi'(0)|$ of silver as a function of $|T|$ at positive and negative temperatures. *Full lines* reflect the Curie–Weiss law for the ferromagnetic ($T < 0$), and antiferromagnetic ($T > 0$) phase [390]

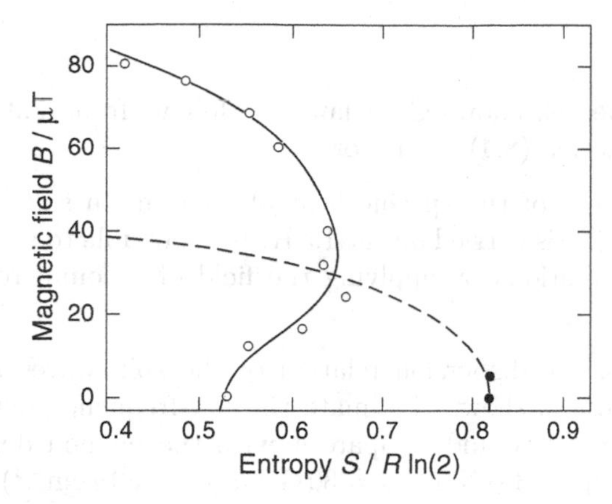

Fig. 8.31. Phase diagram of the nuclear spins of silver at positive and negative temperatures. The *full line* connects the data for $T > 0$. Inside this line the spin system is antiferromagnetically ordered. The *dashed curve* represents the phase boundary between the ferromagnetic (inside) and paramagnetic state at $T < 0$. It is determined by the two *full circles* and the intercept with the $S = 0$ axis. The shape is based on the mean-field theory assuming a linear relationship between S and T [388]

8.5.3 Stimulated Emission

The effect of stimulated emission in NMR experiments provides a confirmation of the occurrence of negative temperatures. As an example, we show in Fig. 8.32 the result obtained in a measurement of the imaginary part of the magnetic susceptibility of silver as a function of frequency. At $T > 0$, an NMR absorption line is observed, but at $T < 0$ emission takes place, i.e., the imaginary part of the NMR signal has opposite sign. In the case of negative temperatures the resonance is shifted towards higher frequencies because of the higher susceptibility of silver in the ferromagnetic phase.

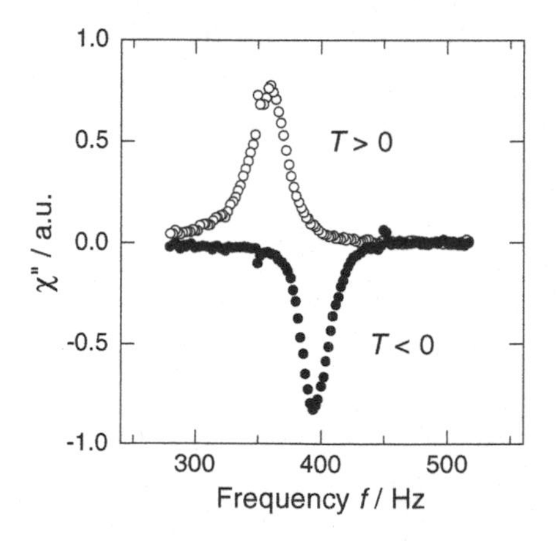

Fig. 8.32. Imaginary part of the magnetic susceptibility of silver at positive and negative temperatures. The absorption line at $T > 0$ is depicted by *open circles*, the emission spectrum at $T < 0$ by *full circles* [391]

Exercises

8.1 Show that the well-known Curie law (8.5) follows from (8.3) by expanding the Brillouin function (8.4) to first order.

8.2 A measurement of the specific heat of gadolinium sulfate octahydrate (total spin $S = 7/2$) is carried out at 1.5 K. Is it possible to separate the spin and lattice contributions by applying the field of a commercially available magnet?

8.3 In Fig. 8.16, the dispersion relation of the spin waves in the antiferromagnet $RbMnF_3$ is shown. Estimate the low-frequency density of states from the group velocity and compare it with the phonon density of states ($c_{11} = 105\,\mathrm{GPa}$, $c_{12} = 33\,\mathrm{GPa}$, $c_{44} = 30\,\mathrm{GPa}$, $\varrho = 4.30\,\mathrm{g\,cm^{-3}}$).

8.4 Nickel is a body-centered cubic ferromagnet with the lattice constant $a = 3.52\,\text{Å}$ and the Debye temperature $\theta = 450\,\mathrm{K}$. The magnon energy is given by $\hbar\omega = Dq^2$ with $D = 6.4 \times 10^{-40}\,\mathrm{J\,m^2}$. Calculate the exchange coefficient and the contribution of the spin waves to the specific heat at 5 K under the assumption that the spin quantum number $S = 1/2$. At what temperature are the contributions of the spin waves and the lattice equal?

9 Tunneling Systems

Already in 1927, the tunneling of atoms was considered by *Hund* shortly after the development of quantum mechanics to explain the vibrational spectrum of NH_3 molecules [392].[1] The nitrogen atom of NH_3 molecules can occupy two energetically equivalent positions located on either side of the plane formed by the hydrogen atoms. These two positions are separated by an energy barrier of 0.3 eV. Transitions between the potential minima are possible via tunneling, resulting in a removal of the degeneracy of the ground state. In 1934, the corresponding energy splitting of 24 GHz could be determined experimentally by measurements of the microwave absorption [393]. Tunneling of atoms in solids was first considered in 1930 by *Pauling* [394]. But it took thirty years until an experimental proof for this type of tunneling could by given. It was not until 1962 that *Känzig* was able to demonstrate the tunneling motion of oxygen ions in alkali halide crystals by dielectric measurements [395].

In general, a certain degree of structural disorder is a prerequisite for the occurrence of atomic tunneling. Therefore, tunneling states are not usually found in perfect crystals but in crystals containing defects and also in amorphous solids. Defect atoms in crystals can occupy several equivalent sites because of the symmetry of the lattice potential, and can tunnel from one site to another at low temperatures. In Sect. 9.1, we consider the theoretical background. In order to avoid unnecessary complications we restrict ourselves to tunneling systems exhibiting only two levels. This introduction is followed by the discussion of the properties of tunneling states in crystals in Sects. 9.2 to 9.4. Furthermore, we report on the low-temperature behavior of amorphous solids in Sect. 9.5, and finally, we consider some examples of echo phenomena based on the coherent motion of tunneling systems at very low temperatures.

9.1 Two-Level Tunneling Systems

In this section, we assume that the considered tunneling atoms can occupy only two distinct sites of equilibrium. This simplification considerably reduces

[1] From a historical point of view, it is interesting that the tunneling effect was first discussed for atoms, even before tunneling of electrons and α-particles was considered.

mathematical effort. Of course, this assumption will not generally be correct since multiwell potentials are present in crystals due to the symmetry of crystalline structures. However, the two-level approximation is sufficient for understanding the fundamental properties of this type of excitation. We will broaden the theoretical description when necessary. In addition, it is generally accepted that the low-temperature properties of amorphous solids are, in fact, determined by two-level systems, so that the concept can be applied directly to this class of solids.

9.1.1 Double-Well Potentials

Formally, a two-level tunneling system can be described by a particle of mass m moving in a double-well potential. For the sake of simplicity we assume that this potential is composed of two identical harmonic wells, as shown in Fig. 9.1. The two wells are shifted with respect to each other by a small amount, namely the asymmetry energy Δ. In crystals with a low concentration of defects, Δ is irrelevant, but at higher concentrations and in glasses, the depths may differ noticeably because of structural disorder. It should be mentioned that the tunneling motion is not necessarily a pure translational motion. Therefore the abscissa in Fig. 9.1 is thought to be a configurational coordinate.

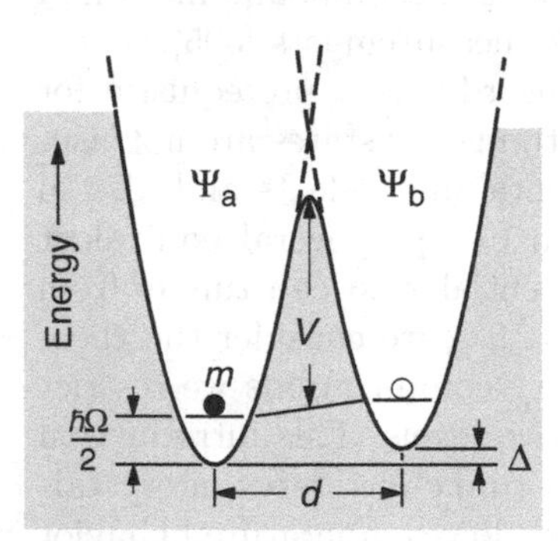

Fig. 9.1. Schematic representation of a particle in a double-well potential with the asymmetry energy Δ, the well distance d, the ground state energy $\hbar\Omega/2$ of the particle in an isolated well, and the potential barrier V

In the discussion of the low-temperature properties of solids only the ground state of the defect systems is of interest. To calculate the energy splitting we solve the Schrödinger equation $H\psi = E\psi$ with the ansatz $\psi = a\psi_\mathrm{l} + b\psi_\mathrm{r}$, where ψ_l and ψ_r are the normalized wave functions of the particle in the isolated 'left' and 'right' wells. The coefficients a and b are assumed to be real quantities. The eigenvalue E is thus given by

$$E = \frac{\int \psi^* H \psi \,\mathrm{d}^3x}{\int \psi^* \psi \,\mathrm{d}^3x} = \frac{a^2 H_\mathrm{ll} + b^2 H_\mathrm{rr} + 2ab H_\mathrm{lr}}{a^2 + b^2 + 2abS} \,. \tag{9.1}$$

The abbreviations have the following meaning: H_ll and H_rr are the eigenvalues of the particle in the isolated wells, i.e., $H_\mathrm{ll} = \int \psi_\mathrm{l}^* H \psi_\mathrm{l} \,\mathrm{d}^3x$, and

$H_{\rm rr} = \int \psi_{\rm r}^* H \psi_{\rm r}\, {\rm d}^3x$. The overlap of the wave functions is expressed by $S = \int \psi_{\rm l}^* \psi_{\rm r}\, {\rm d}^3x$. Furthermore, $H_{\rm lr} = \int \psi_{\rm l}^* H \psi_{\rm r}\, {\rm d}^3x$ is the exchange energy that is of particular interest. If the wells are far apart, the overlap of the wave functions vanishes, i.e., $S = 0$ and $H_{\rm lr} = 0$.

The true eigenvalue is always smaller than the eigenvalue E calculated with the ansatz given above. Therefore, we minimize E by requiring $\partial E/\partial a = 0$, and $\partial E/\partial b = 0$, and obtain the characteristic equations

$$a\,(H_{\rm ll} - E) + b(H_{\rm lr} - E\,S) = 0\,, \tag{9.2}$$

$$a\,(H_{\rm lr} - E\,S) + b(H_{\rm rr} - E) = 0\,, \tag{9.3}$$

leading to the solution

$$(H_{\rm ll} - E)(H_{\rm rr} - E) - (H_{\rm lr} - E\,S)^2 = 0\,. \tag{9.4}$$

Choosing the middle between the two potential minima as zero energy, the eigenvalues of the particle in the isolated wells may be expressed by $H_{\rm ll,rr} = (\hbar\Omega \pm \Delta)/2$. In addition, we assume that the overlap of the wave functions is weak, and neglect the quantity ES. Thus, we obtain the eigenvalues

$$E_\pm = \frac{1}{2}\left(\hbar\Omega \pm \sqrt{\Delta^2 + 4\,H_{\rm lr}^2}\right)\,, \tag{9.5}$$

and hence the energy splitting of the ground state

$$E = E_+ - E_- = \sqrt{\Delta^2 + 4H_{\rm lr}^2} = \sqrt{\Delta^2 + \Delta_0^2}\,. \tag{9.6}$$

Using the WKB method,[2] $H_{\rm lr}$ can be calculated, and $-2H_{\rm lr} = \Delta_0 \approx \hbar\Omega\,{\rm e}^{-\lambda}$ is found. The quantity Δ_0 is called the *tunnel splitting*, and λ is the *tunneling parameter* determined by the potential shape and the mass m of the tunneling particle. For the simple potential we have chosen here, λ is given approximately by

$$\lambda \approx \frac{d}{2\hbar}\sqrt{2mV}\,. \tag{9.7}$$

As we will see in the following sections, the mass dependence leads to an isotope effect that is readily observed in crystalline systems.

9.1.2 Coupling to Electric and Elastic Fields

Without perturbation, the Hamiltonian in the basis $(\psi_{\rm l}, \psi_{\rm r})$ is given by

$$H_0 = \frac{1}{2}\begin{pmatrix} \Delta & -\Delta_0 \\ -\Delta_0 & -\Delta \end{pmatrix}\,. \tag{9.8}$$

Because of the tunneling, $\psi_{\rm l}$ and $\psi_{\rm r}$ are not true eigenstates. In the orthogonal basis the Hamiltonian reads

[2] This method is named for Wentzel, Kramers, and Brillouin [396].

$$\mathcal{H}_0 = \frac{1}{2}\begin{pmatrix} E & 0 \\ 0 & -E \end{pmatrix}. \tag{9.9}$$

Strain fields and electric fields alter the parameters Δ and Δ_0. In the following, we assume that these changes are sufficiently small to be treated by first-order perturbation theory. The total Hamiltonian $H = H_0 + H_S$ is the sum of H_0 and the perturbation Hamiltonian

$$H_S = \frac{1}{2}\begin{pmatrix} \delta\Delta & -\delta\Delta_0 \\ -\delta\Delta_0 & -\delta\Delta \end{pmatrix}. \tag{9.10}$$

In general, it is assumed that the change $\delta\Delta$ in the asymmetry energy is much bigger than the change $\delta\Delta_0$ in the tunnel splitting. This sounds plausible since the shape of the potential, i.e., the separation of the wells and the barrier height, and consequently also the value of Δ_0, remains essentially unchanged by external fields. But external fields influence the environment of the tunneling systems and thus change their asymmetry energy Δ. For sufficiently small perturbations, the asymmetry energy $\delta\Delta$ will vary linearly with the strength of the elastic or electric field, and we may write

$$\delta\Delta = 2\,\overleftrightarrow{\gamma}\cdot\overleftrightarrow{\mathrm{e}} \qquad \text{or} \qquad \delta\Delta = 2\boldsymbol{p}\cdot\boldsymbol{F}. \tag{9.11}$$

Here, $\overleftrightarrow{\gamma}$ denotes the deformation potential and $\overleftrightarrow{\mathrm{e}}$ the applied strain field. In the following discussion, we neglect their tensorial character and simply write γ and $\tilde{e}$ instead. Of course, we will take the anisotropy of these quantities into account when necessary. Furthermore, $\boldsymbol{p}$ stands for the (permanent) electric dipole moment, and $\boldsymbol{F}$ for the strength of the applied electric field. The transformation of the perturbation Hamiltonian into the basis of $\mathcal{H}_0$ leads to the expression

$$\mathcal{H}_S = \frac{1}{E}\begin{pmatrix} \Delta & -\Delta_0 \\ -\Delta_0 & -\Delta \end{pmatrix}(\gamma\,\tilde{e} + \boldsymbol{p}\cdot\boldsymbol{F}). \tag{9.12}$$

The diagonal elements describe the change in the energy splitting caused by external fields, thus reflecting the energy modulation in the case of periodic perturbations. The off-diagonal elements give rise to transitions between the two levels.

The dynamics of tunneling systems in external fields can be discussed with the help of the Bloch equations. This set of equations was introduced in 1946 by *Bloch* to describe the time evolution of spin systems in magnetic fields [397]. We will use the Bloch equations in the discussion of NMR-based thermometers (see Sect. 12.2.5) but use less compact, more illustrative descriptions in the treatment of other phenomena.

9.1.3 Relaxation

Relaxation phenomena play an important role in many areas of physics and lead to interesting low-temperature properties of solids containing defects.

We consider here the dynamic properties of tunneling defects in alternating elastic and electric fields. According to (9.12) the change in the level splitting caused by external fields is given by

$$\delta E = 2\gamma\,\tilde{e}\,\frac{\Delta}{E} \qquad \text{or} \qquad \delta E = 2\boldsymbol{p}\cdot\boldsymbol{F}\,\frac{\Delta}{E}\,. \tag{9.13}$$

Since the occupation of the energy levels depends on the energy splitting, external fields drive the defect states out of thermal equilibrium. They then try to re-establish equilibrium by exchanging energy with the heat bath. Because dielectric and elastic relaxation have the same origin, we consider here only the dielectric relaxation, and adapt the corresponding constants to the description of the elastic relaxation afterwards.

In the treatment of the dielectric relaxation we do not take into account the vectorial character of the polarization $\boldsymbol{P}$ and the electric field $\boldsymbol{F}$, and the tensorial character of the susceptibility $\overleftrightarrow{\chi}$. Although this is a radical simplification it already allows an appropriate description of the low-temperature properties of glasses and a qualitatively correct description of the properties of crystals. However, in the latter case, anisotropy has to be considered for a quantitative comparison.

Without external fields the polarization vanishes in all materials considered here. In an electric field, the tunneling systems give rise to the polarization δP. Of course, there is also a background polarization due to the host solid. However, it does not depend on frequency or temperature in the parameter range of interest. In the linear approximation, the polarization and electric field are connected via the susceptibility χ, i.e.,

$$\delta P = \varepsilon_0 \chi F\,, \tag{9.14}$$

where ε_0 represents the vacuum permittivity. If the applied electric field varies so slowly with time that all systems can follow adiabatically, thermal equilibrium will always be maintained. In the following discussion, we use symbols with a 'hat' for quantities in the quasistatic limit, i.e., we write

$$\delta\widehat{P} = \varepsilon_0 \widehat{\chi} F\,, \tag{9.15}$$

where $\widehat{\chi}$ represents the *static susceptibility* of the sample.

At higher frequencies, the thermal equilibrium of the tunneling systems is disturbed and relaxation effects occur. For the description of this phenomenon we use the *relaxation time approximation* already introduced in Sect. 7.2.1. For the polarization due to tunneling systems the corresponding equation has the form

$$-\frac{\partial(\delta P)}{\partial t} = \frac{\delta P - \delta\widehat{P}}{\tau}\,, \tag{9.16}$$

where the relaxation time τ is of particular interest. It should be pointed out that the perturbed ensemble of tunneling systems does not relax towards the static equilibrium $\delta P = 0$ of the unperturbed state, but towards the 'instantaneous' equilibrium $\delta\widehat{P}$ that would be reached after a sufficiently long waiting time in the 'instantaneous' field. In other words, the system is relaxing towards the state that would be approached in the limiting case $\tau \to 0$.

The finite value of τ leads to a phase shift between driving force and displacement, resulting in a *complex susceptibility* $\chi = \chi' + \mathrm{i}\chi''$. For a periodically varying electric field of the form $F = F_0 \mathrm{e}^{-\mathrm{i}\omega t}$, if we insert (9.14) and (9.15) into (9.16) for the susceptibility χ, we find the relation

$$\chi = \frac{\widehat{\chi}}{1 - \mathrm{i}\omega\tau} \,. \tag{9.17}$$

The real and imaginary parts of the susceptibility of linear systems are linked by the *Kramers–Kronig relations* [3]

$$\chi'(\omega) = \frac{1}{\pi} \fint_{-\infty}^{\infty} \frac{\chi''(\widetilde{\omega})}{\widetilde{\omega} - \omega} \, \mathrm{d}\widetilde{\omega} \,, \tag{9.18}$$

and

$$\chi''(\omega) = -\frac{1}{\pi} \fint_{-\infty}^{\infty} \frac{\chi'(\widetilde{\omega})}{\widetilde{\omega} - \omega} \, \mathrm{d}\widetilde{\omega} \,, \tag{9.19}$$

where $\fint$ denotes the principal value of the integral. It should be noted that, in principle, the calculation of the real part of the susceptibility requires knowledge of the imaginary part at *all* frequencies, and vice versa.

In the following discussion, we consider the response of an ensemble of identical two-level systems. In equilibrium, the difference in occupation of the two levels is given by

$$\Delta N(E,T) = N \tanh\left(\frac{E}{2k_\mathrm{B}T}\right) . \tag{9.20}$$

The application of an electric field leads to a partial alignment of the electric dipoles and to a change in the energy splitting, and consequently to a change $\delta(\Delta N)$ in the population of the two levels. To simplify the treatment further, we assume that the electric dipole moments of the tunneling systems are aligned either parallel or antiparallel to the applied field. Otherwise, the component of the dipole moment in the field direction has to be considered. For alignment in the field direction we may write $\delta P = \delta(\Delta N) p_\mathrm{eff}$, where the effective dipole moment $p_\mathrm{eff} = p\varDelta/E$ is defined by (9.12). Using the definition (9.15), we may express the static susceptibility by

$$\widehat{\chi} = \frac{\partial(\delta P)}{\partial(\Delta N)} \, \frac{\partial(\Delta N)}{\partial E} \, \frac{\partial E}{\partial F} = p\frac{\varDelta}{E}\,(-2N)\frac{\partial f}{\partial E}\, 2p\frac{\varDelta}{E} \,. \tag{9.21}$$

[3] These relations were derived and published independently by Kronig (1926) and Kramers (1927) [398].

To shorten the expression we have introduced the derivative of the Fermi–Dirac distribution $f = (e^{E/k_BT} + 1)^{-1}$ as an abbreviation.[4]

The dielectric function ε and dielectric susceptibility χ are connected via the relation $\varepsilon = \varepsilon' + i\varepsilon'' = 1 + \chi$. Accordingly, the contribution of the defect systems to the variation $\delta\varepsilon$ of the dielectric function can be expressed with the help of (9.17) and (9.21) by

$$\delta\varepsilon = \frac{-4N}{\varepsilon_0}\left(\frac{p\Delta}{E}\right)^2 \frac{\partial f}{\partial E}\frac{1}{1 - i\omega\tau}\,. \tag{9.22}$$

Splitting $\delta\varepsilon$ into real and imaginary parts yields the variation $\delta\varepsilon'$ of the dielectric constant, namely,

$$\delta\varepsilon' = \frac{-4Np^2}{\varepsilon_0}\left(\frac{\Delta}{E}\right)^2 \frac{\partial f}{\partial E}\frac{1}{1 + (\omega\tau)^2}\,, \tag{9.23}$$

and the loss angle

$$\tan\delta = \frac{\varepsilon''}{\varepsilon'} = \frac{-4Np^2}{\varepsilon_0\varepsilon'}\left(\frac{\Delta}{E}\right)^2 \frac{\partial f}{\partial E}\frac{\omega\tau}{1 + (\omega\tau)^2}\,. \tag{9.24}$$

Note that the sign of $\partial f/\partial E$ is negative, meaning that $\delta\varepsilon'$ and $\tan\delta$ are positive quantities. The frequency-dependent terms describe the so-called *Debye relaxator*, well known in the physics of dielectrics. The typical frequency variation of the dielectric function is illustrated in Fig. 9.2.

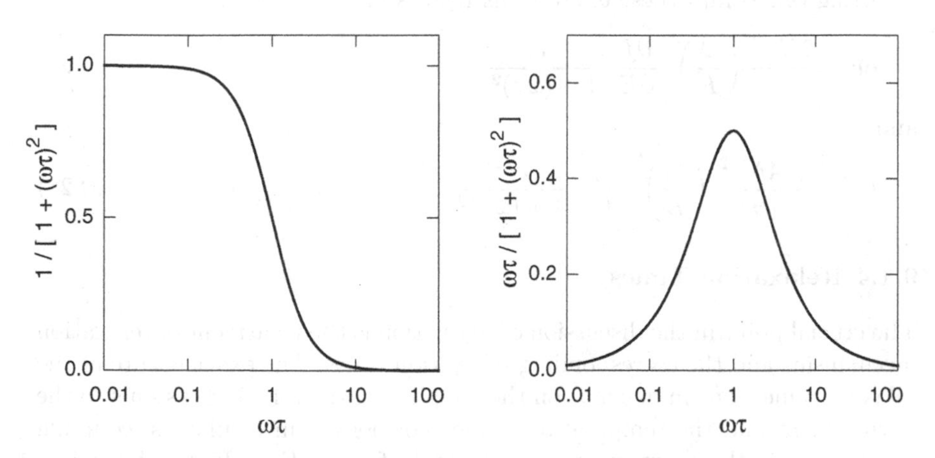

Fig. 9.2. Response of a Debye relaxator versus logarithm of $\omega\tau$. (**a**) Real part with an inflexion point at $\omega\tau = 1$. (**b**) Imaginary part with a maximum at $\omega\tau = 1$

[4] The relation $\partial(\Delta N)/\partial E = -2N\,\partial f/\partial E$ holds for an ensemble of two-level systems. More complicated expressions have to be used if more levels are involved. The general treatment of the relaxation phenomenon, however, does not depend on the specific details of the level scheme.

Now we turn to the elastic relaxation. Formally, the polarization has to be replaced by the elastic stress, and the electric field by the strain field. The 'acoustic susceptibility' S is usually called *elastic compliance*. The elastic compliance of the sample is given by $\widetilde{S} = S_0 + S$, where S_0 represents the host material and S the contribution of the tunneling systems. The same derivation as in the electric case leads to the expression

$$S = \frac{-4N}{\varrho^2 v^4} \left(\frac{\gamma\Delta}{E}\right)^2 \frac{\partial f}{\partial E} \frac{1}{1 - \mathrm{i}\omega\tau}\,, \tag{9.25}$$

where ϱ is the mass density, and v the velocity of sound. In acoustic experiments, the change δv in the sound velocity and the absorption coefficient α is usually measured. They are linked with the compliance S via the relation $\varrho\widetilde{S} = v^{-2} = k^2/\omega^2$, where the wave number $k = k' + \mathrm{i}k'' = k' + \mathrm{i}\alpha/2$ is a complex quantity. The factor 1/2 arises from the fact that k'' describes the decrease in the amplitude, while α reflects the decay of the intensity. Equation (9.25) holds for longitudinal and transverse sound waves in isotropic solids. In the case of crystals, the relation between sound velocity and compliance is more complicated because of the tensorial character of the elastic properties. The separation of the complex compliance into real and imaginary parts leads to

$$\delta v = -\frac{\varrho v^3}{2}\, S' \qquad \text{and} \qquad \alpha \equiv \ell^{-1} = \varrho v\omega\, S''\,. \tag{9.26}$$

Inserting (9.25) into these expressions results in

$$\delta v = \frac{2N\gamma^2}{\varrho v} \left(\frac{\Delta}{E}\right)^2 \frac{\partial f}{\partial E} \frac{1}{1 + (\omega\tau)^2}\,, \tag{9.27}$$

and

$$\ell^{-1} = \frac{-4N\gamma^2}{\varrho v^3} \left(\frac{\Delta}{E}\right)^2 \frac{\partial f}{\partial E} \frac{\omega^2\tau}{1 + (\omega\tau)^2}\,. \tag{9.28}$$

9.1.4 Relaxation Times

The crucial point in the discussion of relaxation is the treatment of relaxation mechanisms and the corresponding relaxation times. The exact nature of the relaxation mechanism depends on the coupling of the relaxing systems to the environment and the temperature range. For defects in insulators, coupling to phonons is the most important channel of relaxation. In the following, we consider briefly the dominant relaxation processes in such materials at different temperatures.

Relaxation occurs at high temperatures via thermally activated jumps of the moving particle over the potential barrier V (see Fig. 9.3). Since many phonons are needed to raise the thermal energy necessary for the jump, this 'classical' process is most effective at high temperature. The relaxation time for this process may be expressed by the *Arrhenius law*

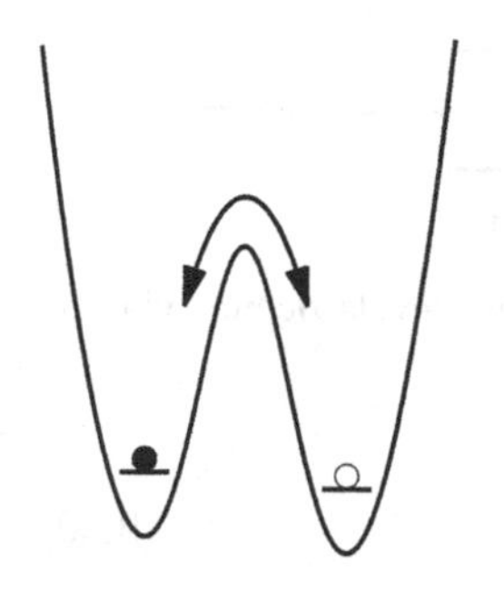

Fig. 9.3. Illustration of thermally activated relaxation. The 'support' of many phonons is needed to enable the particle to jump over the barrier

$$\tau = \tau_0 \, e^{V/k_B T} , \tag{9.29}$$

where τ_0 denotes the period of oscillation of the particle in one of the potential wells with typical values between 10^{-13} s and 10^{-12} s.

At intermediate temperatures, roughly speaking in the range of a few kelvin, the thermal energy is not generally high enough for activated jumps over the barrier, and tunneling processes come to the fore. As in optical Raman scattering, transitions between the levels of the split ground state take place via intermediate states that may be real or virtual. Such many-phonon processes lead to a relaxation rate proportional to T^α, where $\alpha \geq 5$. The value of α depends on the number of phonons participating in the process, on their density of states, and on the coupling mechanism. The simplest process of this type is illustrated in Fig. 9.4. Only two phonons with frequencies ω_1 and ω_2 participate in this so-called *Raman process*, for which the relation $\hbar\omega_1 = \hbar\omega_2 \pm E$ follows from energy conservation.

Fig. 9.4. Two-phonon relaxation process (*Raman process*). The transition between levels occurs via an intermediate state (*dashed line*) that can be either real or virtual

On cooling, the number of excited phonons declines further. If the thermal energy becomes comparable with, or even smaller than the energy splitting of the tunneling systems, i.e., if $k_B T \lesssim E$, the transition between the levels is brought about by the absorption or emission of a single thermal phonon. This is the so-called *one-phonon process* or *direct process* (see Fig. 9.5). With *Fermi's Golden Rule* the probability W_{12} for the transition between level 1 and level 2 may be expressed by

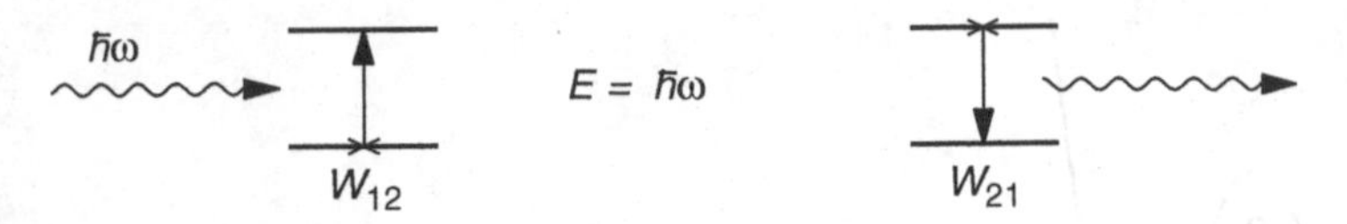

Fig. 9.5. One-phonon process. The transition between the levels occurs via the absorption or emission of a single thermal phonon

$$W_{12} = \frac{2\pi}{\hbar}|\langle\psi_2|\mathcal{H}_\mathrm{s}|\psi_1\rangle|^2\, D(E)\, f(E)\, \delta(\hbar\omega = E)\,. \tag{9.30}$$

In this expression, $D(E)$ denotes the phonon density of states, and $f(E)$ the Bose–Einstein distribution function reflecting the occupation of the phonon states. Since the one-phonon process is relevant only at low temperatures we may use the Debye density of states (6.4). Transitions in the opposite direction are accompanied by the emission of a phonon, and $f(E)$ has to be replaced by $[1 + f(E)]$. With $\tau^{-1} = W_{12} + W_{21}$, the relaxation rate

$$\tau^{-1} = \left(\frac{\gamma_\ell^2}{v_\ell^5} + 2\frac{\gamma_\mathrm{t}^2}{v_\mathrm{t}^5}\right)\left(\frac{\Delta_0}{E}\right)^2 \frac{E^3}{2\pi\varrho\hbar^4}\coth\left(\frac{E}{2k_\mathrm{B}T}\right) \tag{9.31}$$

is found for the one-phonon process. In this expression, the indices ℓ and t refer to the longitudinal and the two transverse phonon branches interacting with the tunneling systems. For $k_\mathrm{B}T \geq E$ the hyperbolic cotangent may be approximated by the first term of the Taylor expansion, resulting in $\tau^{-1} \propto T$.

9.1.5 Resonant Interaction

At low temperatures, the resonant interaction between the applied field and the defect states is of particular importance. As depicted in Fig. 9.6, absorption and induced emission have to be considered. In some sense, this process is the 'reverse' of the direct process just described. As before, the Golden Rule may be used to calculate the transition probability and hence the repercussion of this process on the properties of the sample. As mentioned above, the density of the final states has an important influence. As *spectral function* $g(\omega)$, the Lorentzian line function

$$g(\omega) = \frac{\tau_2}{\pi}\frac{1}{1 + (\omega - \omega_0)^2\tau_2^2} \tag{9.32}$$

Fig. 9.6. Resonant interaction of photons or phonons with two-level systems. Absorption (*left*) and induced emission (*right*) of a photon or phonon with energy $\hbar\omega$

is often used for the upper level with maximum at ω_0, width τ_2^{-1}, and maximum height $g(\omega_0) = \tau_2/\pi$. The physical significance of the quantity τ_2 will be briefly discussed in Sect. 9.6 when we consider coherent phenomena in amorphous solids.

Taking into account absorption and induced emission, the expressions

$$\ell^{-1} = \frac{N\gamma^2}{\varrho v^3}\left(\frac{\Delta_0}{E}\right)^2 \frac{\pi\omega}{\hbar}\, g(\omega)\tanh\left(\frac{E}{2k_\mathrm{B}T}\right) \tag{9.33}$$

and

$$\tan\,\delta = \frac{Np^2}{\varepsilon_0\varepsilon'}\left(\frac{\Delta_0}{E}\right)^2 \frac{\pi\omega}{\hbar}\, g(\omega)\,\tanh\left(\frac{E}{2k_\mathrm{B}T}\right) \tag{9.34}$$

are found for the acoustic attenuation or the dielectric loss angle, respectively. A glance at (9.20) reveals that both quantities are proportional to the difference in the occupation of the two levels, which vanishes for $k_\mathrm{B}T \gg E$. Furthermore, the loss increases linearly with frequency and is most pronounced for symmetric tunneling systems with $E = \Delta_0$, because of the factor $(\Delta_0/E)^2$.

The influence of the coupling on the acoustic absorption (9.33) demands some explanatory remarks. We have to distinguish between homogeneous and inhomogeneous broadening of the spectral function $g(\omega)$. If the line is homogeneously broadened, $g(\omega)$ is determined by the lifetime τ of the excited state. In this case $\tau_2 \approx \tau$, and the maximum height is given by $g(\omega_0) \propto \tau$. According to (9.31), $\tau \propto \gamma^{-2}$, implying that in (9.33) the dependence on the deformation potential γ just cancels. Surprisingly, the maximum attenuation is independent of the coupling strength. The situation is different if the absorption line is inhomogeneously broadened. Since an inhomogeneous line is a superposition of individual lines, its width does not depend on the lifetime of the excited state. In fact, in this case the observed absorption maximum is proportional to γ^2.

Measurements of ultrasonic absorption and dielectric loss at microwave frequencies have been carried out on a variety of defect systems. Since these processes have been studied most extensively in glasses, we postpone the detailed discussion of experimental results and resume this subject in Sect. 9.5 when we consider the low-temperature properties of amorphous solids. In addition, it should be mentioned that the resonant absorption becomes saturated at higher intensities. We have ignored this effect up to now but shall consider it when discussing the properties of tunneling systems in glasses.

The behavior of the real part S' of the susceptibility close to the resonance is more complicated. For brevity, we omit the discussion of S' in the vicinity of the resonance frequency ω_0 and consider only the low-frequency limit, i.e., frequencies $\omega \ll \omega_0$, because most experiments were performed in this frequency range. In this limit, (9.33) together with the Kramers–Kronig relation (9.18) leads to the expression

$$\delta v = \frac{N\gamma^2}{\varrho v}\left(\frac{\Delta_0}{E}\right)^2 \frac{1}{E}\,\tanh\left(\frac{E}{2k_\mathrm{B}T}\right) \qquad (\omega \ll \omega_0) \tag{9.35}$$

for the sound velocity. The corresponding expression for the dielectric constant follows from (9.34) and (9.18):

$$\delta\varepsilon' = \frac{Np^2}{\varepsilon_0}\left(\frac{\Delta_0}{E}\right)^2 \frac{1}{E}\tanh\left(\frac{E}{2k_{\mathrm{B}}T}\right) \qquad (\omega \ll \omega_0)\,. \tag{9.36}$$

9.2 Isolated Tunneling Systems in Crystals

As we have already seen, many properties of real crystals are governed by the presence of point defects. Substitutional atoms or molecules replacing regular atoms on their lattice sites belong to this category of defects. Over the last few decades the influence of point defects on the behavior of crystals at low temperatures has been studied extensively.

In general, impurity atoms or impurity molecules differ in size and shape from the regular atoms of the lattice. Therefore, they can, in most cases, occupy several positions of equilibrium in the host crystals. The number of these energetically equivalent sites depends on the nature of the impurities and the symmetry of the host crystal. At sufficiently high temperatures, the potential barrier between the sites can be surmounted by the defect atom via thermally activated processes. At low temperatures, configurational changes are often enabled by quantum-mechanical tunneling. This process removes the degeneracy of the energy levels, resulting in a splitting of the ground state.

Since impurity atoms do not generally fit exactly into the host lattice, they cause elastic strain fields and/or electric dipole moments at the defect site, and these can give rise to an interaction between the defects. This interaction can be neglected as long as the defect concentration is very small. In this case, individual tunneling systems contribute to the low-temperature properties independently and we speak of *isolated tunneling systems*.

Because of the cubic lattice and the simple ionic binding, the defect structure is especially simple in alkali halides. Therefore, alkali halides have been studied extensively in the past and were often considered as model substances. For this reason, we start this section with the consideration of tunneling systems in this class of materials.

9.2.1 Level Schemes

Three kinds of substitutional defects are compatible with the cubic symmetry of alkali halides. They differ in the number and location of potential minima. There are systems with six energetically degenerate potential minima along the $\langle 100 \rangle$ direction. Well-known examples are OH^- ions or OD^- ions incorporated in KCl. Twelve equivalent sites are found for defects having their potential minima in the $\langle 110 \rangle$ direction. As examples, we mention NaBr:F and RbCl:Ag. Finally, defects with ⟨111⟩ symmetry can occupy eight different sites. Well-known examples are KCl:Li and KCl:CN.

In the following, we will mainly consider KCl:Li because it is a system that has been studied in great detail experimentally and theoretically. Therefore, we restrict our theoretical introduction to ⟨111⟩ systems, although the two other defect types can be treated along the same lines. Differences will be mentioned where necessary during the presentation of experimental data.

⟨111⟩ Systems

In the system KCl:Li, small lithium ions replace relatively large potassium ions. Because of the difference in size, the energetically most favorable position for Li^+ ions is not at the center of the lattice site, but at a position shifted in the ⟨111⟩ direction. This results in a so-called *off-center position*, illustrated in Fig. 9.7, where the radii of the circles reflect the actual ratio of the ionic radii of the three elements: $r_{Cl^-} \approx 1.81\,Å$, $r_{K^+} \approx 1.33\,Å$, and $r_{Li^+} \approx 0.60\,Å$. Of course, the eight equivalent lithium sites do not lie in the plane of the potassium and chlorine ions but slightly above and below it.

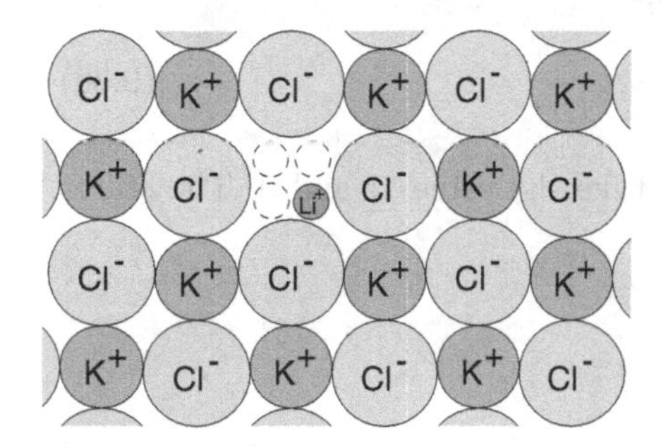

Fig. 9.7. Schematic representation of a (001) plane of KCl after the replacement of a K^+ ion by a Li^+ ion. Eight sites are accessible to the lithium ion, namely four above and four below the plane of the drawing

The displacement of the lithium ion towards the chlorine ions leads to the relatively large dipole moment $p \approx 2.6$ debye (8.7×10^{-30} A s m) without Lorentz correction for the local field. Fortunately, the elastic strain caused by the Li^+ defect is relatively small and will be neglected in our discussion. This fact considerably facilitates the theoretical treatment of lithium tunneling systems and is the main reason why KCl:Li is often considered as a model system. Calculations of the potential landscape have shown that the off-center position in ⟨111⟩ direction is energetically most favorable. The potential minima are displaced approximately 0.2 lattice constants from the central symmetric position and are separated by a potential barrier of at least $V/k_B \approx 100\,K$.

The potential wells are located at the corners of a cube with the edge length d. The positions of the corners are given by the vectors

$$\boldsymbol{r} = \frac{d}{2}(\alpha, \beta, \gamma), \tag{9.37}$$

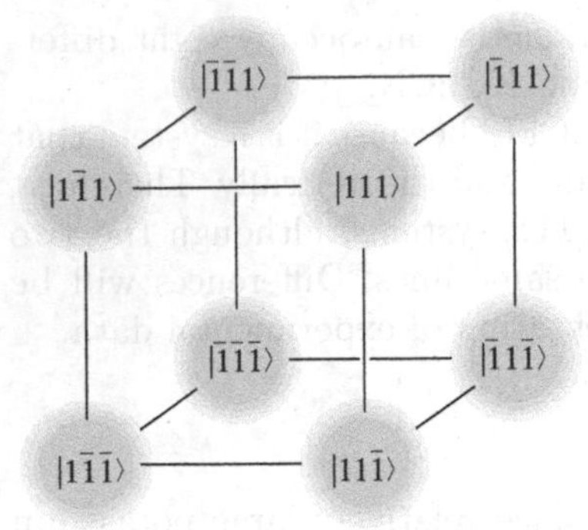

Fig. 9.8. Illustration of the eight localized states $|\alpha\beta\gamma\rangle$. The centers of the wave functions coincide with the corners of a cube

with $\alpha, \beta, \gamma = \pm 1$. The corresponding localized states $|\alpha\beta\gamma\rangle$ are depicted in Fig. 9.8.

The potential landscape allows tunneling between the localized states along the edges of the cube, the face diagonals, and the space diagonals. The energy eigenstates $|\boldsymbol{m}\rangle$ are constructed as a superposition of the localized states with an appropriate phase factor. They are irreducible representations of the point group $\mathrm{O_h}$ and have the form

$$|\boldsymbol{m}\rangle = \frac{1}{\sqrt{8}} \sum_{\alpha\beta\gamma} \exp\left(\frac{\mathrm{i}\pi}{d}\, \boldsymbol{m} \cdot \boldsymbol{r}\right) |\alpha\beta\gamma\rangle\,, \tag{9.38}$$

where the entries of the vector $\boldsymbol{m}$ may take the values 1 and 0. The states are arranged in the four levels:

$$\begin{aligned}
\mathrm{A_{2u}} &: \quad (1,1,1)\,, \\
\mathrm{T_{2g}} &: \quad (1,1,0); (1,0,1); (0,1,1)\,, \\
\mathrm{T_{1u}} &: \quad (1,0,0); (0,1,0); (0,0,1)\,, \\
\mathrm{A_{1g}} &: \quad (0,0,0)\,.
\end{aligned} \tag{9.39}$$

As expected, the ground state $\mathrm{A_{1g}}$, and the state $\mathrm{A_{2u}}$ with the highest energy, are not degenerate and have spherical symmetry. The two levels in the middle are threefold degenerate.

As in most other systems, the potential minima of KCl:Li are approximately spherical. In this case, tunneling along the shortest path, i.e., tunneling along the edges of the cube, is most favorable. The two other tunneling motions hardly contribute to the tunnel splitting [399]. In this simple case, the energy eigenvalues are arranged equidistantly, as shown in Fig. 9.9. They are separated by the tunnel splitting Δ_0, which can be calculated with the help of the WKB approximation as mentioned in Sect. 9.1.1.

Because of the exponential mass dependence of the tunneling splitting, a pronounced isotope effect is expected. Indeed, the ratio ${}^6\Delta_0/{}^7\Delta_0 \approx 1.5$ is found experimentally for the two stable isotopes ^{6}Li and ^{7}Li. Inserting the known parameters of the potential of the lithium defect into (9.7), the

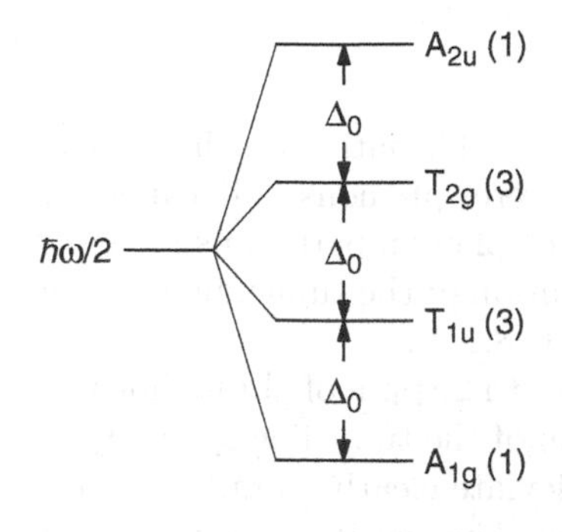

Fig. 9.9. Level scheme of ⟨111⟩ tunneling systems taking only edge tunneling into account. The number of degenerate levels is given in brackets

agreement with expectation is only rough. The main reason for this lack of agreement is the assumption that the potential landscape seen by the tunneling ion is rigid. In reality, the lattice in the immediate neighborhood of the tunneling ion relaxes, resulting in a drag on neighboring ions. In other words, a *dressing effect* occurs. Consequently, the effective mass of the tunneling particles differs from the mass of the 'naked' lithium ions, leading to a modification of the ratio of the tunnel splitting [400].

⟨100⟩ and ⟨110⟩ Tunneling Systems

The level schemes of the two other types of tunneling system are also relatively simple. The separation of the levels is a multiple of Δ_0, provided that the potential wells are spherical, and tunneling between localized states only occurs along one path. The shortest path is equivalent to a rotational motion through 90° for ⟨100⟩ systems, and 60° for ⟨110⟩ systems. The level schemes of the two types of system are displayed in Fig. 9.10.

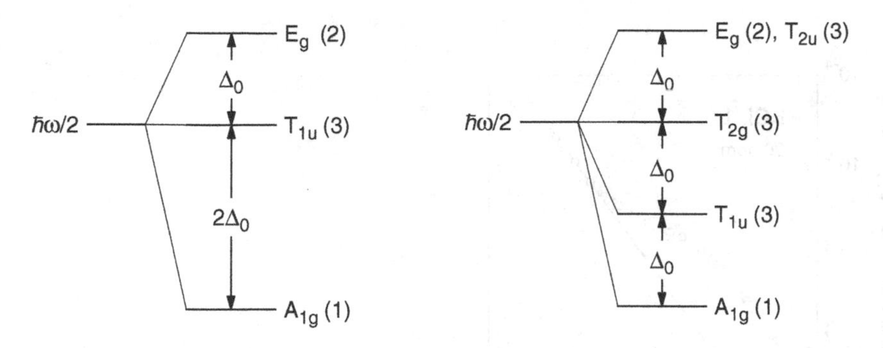

Fig. 9.10. Level schemes of ⟨100⟩ tunneling systems (**a**) and ⟨110⟩ tunneling systems (**b**). The degree of degeneracy of the levels is given in brackets

9.2.2 Specific Heat

As pointed out repeatedly, the specific heat gives information on the degrees of freedom. In the systems we are discussing here, phonons and tunneling systems contribute to the specific heat. While the phonon part vanishes as T^3 with decreasing temperature, the main contribution of the tunneling systems is found at temperatures $T \approx \Delta_0/2k_B$ (see Sect. 8.1.3).

The specific heat C of a KCl crystal doped with 20 ppm of ^{7}Li is shown in Fig. 9.11. Obviously, below 1 K the contribution of the tunneling systems becomes important, and the experimental data deviate clearly from the T^3 behavior of pure crystals indicated by the full line. The specific heat caused by the tunneling systems can be deduced from their internal energy U. For N isolated $\langle 111 \rangle$ tunneling systems the internal energy is given by

$$U = \frac{3}{2} N \Delta_0 \left[1 - \tanh\left(\frac{\Delta_0}{2k_B T} \right) \right] , \tag{9.40}$$

and accordingly the specific heat per unit mass by

$$C_{TS} = \frac{3nk_B}{\varrho} \left(\frac{\Delta_0}{2k_B T} \right)^2 \operatorname{sech}^2 \left(\frac{\Delta_0}{2k_B T} \right) . \tag{9.41}$$

Here, $n = N/V$ is the number density of the tunneling systems and ϱ the mass density of the sample. Because of the similarity with the specific heat of two-level systems (see Sect. 8.1.3), it is often referred to in the literature as the *Schottky anomaly*. Interestingly, the expression (9.41) differs from that for simple two-level systems only by the numerical factor three. Accordingly, the contribution of the tunneling systems to the specific heat has its maximum at $T \approx \Delta_0/2k_B$. Since the temperature of the maximum is determined by the tunnel splitting, its position depends on the isotope.

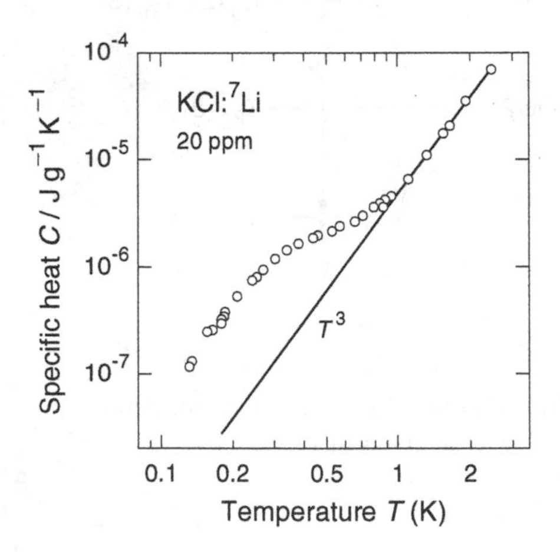

Fig. 9.11. Specific heat of a lithium-doped KCl crystal versus temperature. The *full line* indicates the T^3 contribution of the host lattice [401]

The contribution C_{Li} of lithium tunneling systems in KCl is shown in Fig. 9.12. To demonstrate that the same temperature variation is observed for both isotopes, ^{7}Li and ^{6}Li, and that the isotope effect only causes a shift of the curves along the (logarithmic) temperature axis, the two sets of data are plotted with different temperature scales. The lower scale is valid for the ^{7}Li-doped sample, the upper one for the sample containing ^{6}Li. The temperature scales are shifted by such an amount that the maxima of the two curves coincide. The data sets exhibit a great resemblance and agree well in the maximum region. The full line in Fig. 9.12 follows from (9.41) after inserting the known values $^6\Delta_0/k_{\mathrm{B}} = 1.65\,\mathrm{K}$ and $^7\Delta_0/k_{\mathrm{B}} = 1.1\,\mathrm{K}$. The data obtained above 1 K and below 200 mK deviate from the theoretical prediction. These deviations are probably due to experimental problems. At $T > 1\,\mathrm{K}$, the contribution of the defects is much smaller than that of the host lattice, so that the difference between the doped and undoped samples cannot be determined with great accuracy. The deviations at the lowest temperatures are probably caused by the finite sensitivity of the measurement. From a historical point of view, the experimental observation of an isotope-dependent specific heat constituted unambiguous proof for the tunneling of lithium ions in KCl.

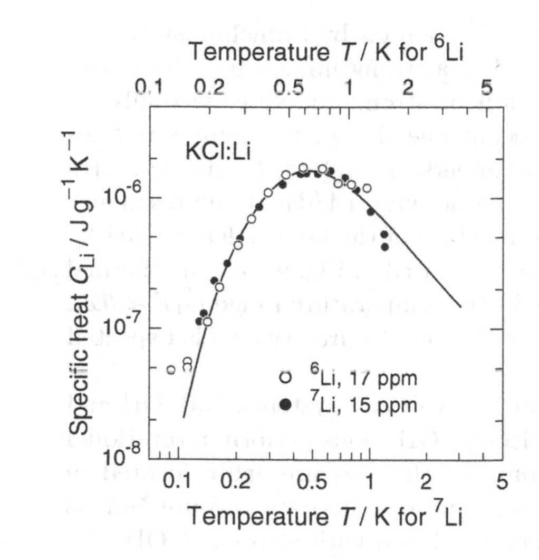

Fig. 9.12. Specific heat of lithium-doped KCl after subtracting the lattice contribution. The *upper* temperature scale is valid for ^{6}Li, and the *lower* scale for ^{7}Li. The *full line* represents the prediction of (9.41) [401]

As an example of a molecular point defect we show the specific heat of a KCl crystal doped with 27 ppm KCN. The data plotted in Fig. 9.13 were obtained after subtracting the lattice contribution. As mentioned above, CN^- ions also form $\langle 111 \rangle$ tunneling systems. The full line is the predicted Schottky curve (9.41) after inserting $\Delta_0/k_{\mathrm{B}} = 1.55\,\mathrm{K}$ for the tunnel splitting. Theory and experiment agree in a nearly perfect manner and confirm once

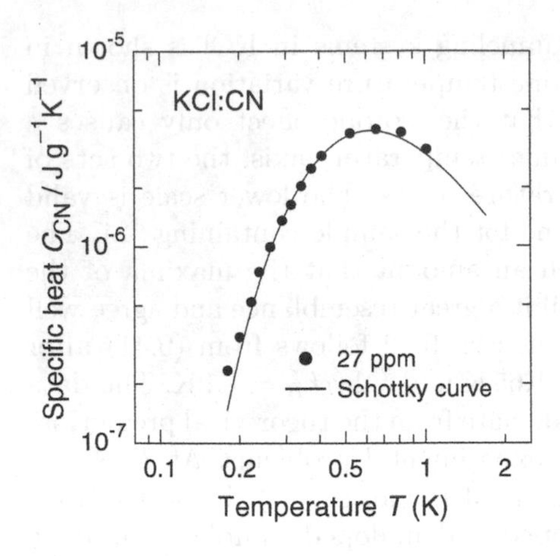

Fig. 9.13. Specific heat of KCl:CN after subtraction of the lattice contribution. The *full line* indicates the fit with (9.41) [402]

again that the specific heat at small defect concentrations can be described by the model of isolated tunneling systems.

9.2.3 Thermal Conductivity

The resonant absorption and re-emission of phonons by tunneling systems – usually called 'resonance fluorescence' – has a strong influence on heat conduction. In the case of strong interaction or strong coupling, virtually no resonantly interacting phonon generated at one side of the sample reaches the other side in heat conduction experiments. Therefore, tunneling states 'burn a hole' into the differential heat conductivity $\mathrm{d}\Lambda/\mathrm{d}\omega$ at the resonance frequency $\omega_\mathrm{r} = E/\hbar$ (see Fig. 9.14a). The width of the hole is determined by the strength of the scattering process. As indicated in Fig. 9.14b, the thermal conductivity is reduced by this process in the temperature range $k_\mathrm{B}T \approx \hbar\omega_\mathrm{r}$. Consequently, a deviation from the T^3 variation of pure crystals is expected in the Casimir limit.

We consider here the thermal conductivity of two systems, NaF:OH and KCl:Li. In the case of NaF:OH, the molecular $\mathrm{OH^-}$ ions perform a rotational tunneling motion. The potential minima for the ions are again located in the $\langle 111 \rangle$ direction. In Fig. 9.15a, the thermal conductivity of pure NaF is compared with the conductivity of a crystal doped with 50 ppm NaOH. Besides a reduction in the region of the maximum due to Rayleigh scattering (see Sect. 6.2.4), a drastic decrease in Λ is found at low temperatures. At 0.2 K the thermal conductivity of the doped crystal is reduced by a factor of 500, although the concentration of tunneling systems is rather small! Experimental results on KCl:Li are shown in Fig. 9.15b. Because of the different value of the tunnel splitting in the case of the two lithium isotopes, the maximum deviation from the conductivity of a pure sample is found at different temperatures.

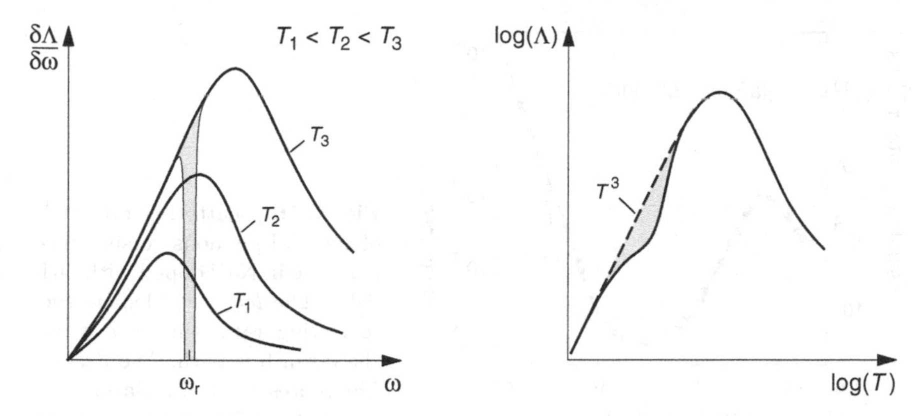

Fig. 9.14. (**a**) Differential thermal conductivity $\mathrm{d}\Lambda/\mathrm{d}\omega$ versus phonon frequency for three temperatures ($T_1 < T_2 < T_3$). Tunneling systems 'burn a hole' into $\mathrm{d}\Lambda/\mathrm{d}\omega$ at the resonance frequency $\omega_\mathrm{r} = E/\hbar$. (**b**) Influence of tunneling systems on thermal conductivity. Resonant scattering causes a reduction of Λ and therefore a deviation from the T^3 law in the Casimir regime

From the data in Fig. 9.15a the scattering rate τ^{-1} and hence the mean free path ℓ of thermal phonons can be deduced using the dominant phonon approximation. The resulting temperature variation is graphed in Fig. 9.16. Coming from high temperatures, the scattering rate first decreases, since the contribution of Umklapp processes vanishes as indicated by the dashed line.

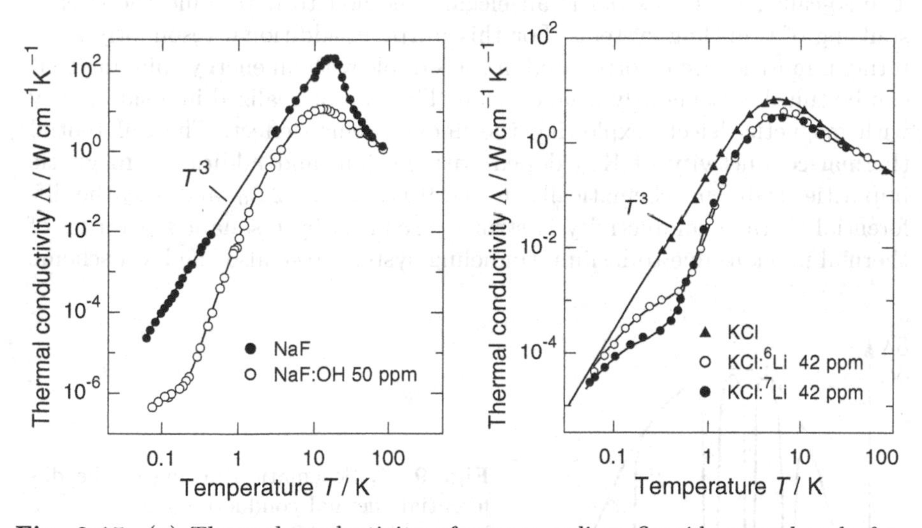

Fig. 9.15. (**a**) Thermal conductivity of a pure sodium fluoride crystal and of a crystal doped with 50 ppm OH^- ions [403]. (**b**) Comparison of the thermal conductivity of pure KCl crystal with the conductivity of crystals doped with ^{6}Li and ^{7}Li, respectively [404]

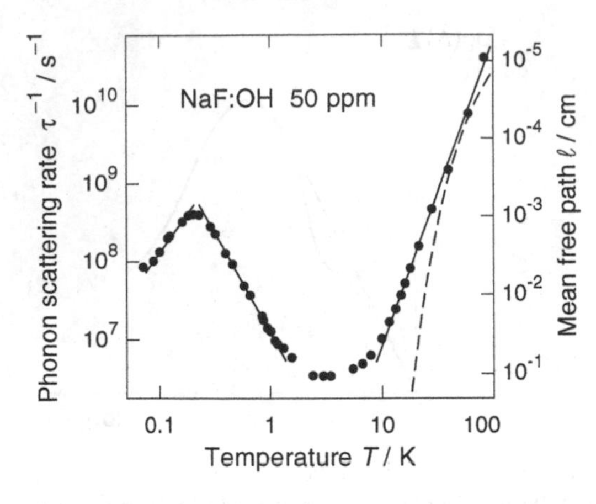

Fig. 9.16. Scattering rate τ^{-1} of thermal phonons versus temperature in NaF doped with OH [403]. The *left scale* depicts the scattering rate, the *right scale* the mean free path. The *dashed line* represents the variation expected from Umklapp processes

Around 3 K the scattering rate τ^{-1} becomes minimal, i.e., the mean free path ℓ passes through a maximum and is roughly given by the diameter of the sample of about 1 mm. Below 1 K, the scattering rate rises again with decreasing temperature because of the resonant interaction of phonons with the tunneling states. The pronounced maximum at 0.2 K follows from the energy splitting $\Delta_0/k_\mathrm{B} \approx 0.5\,\mathrm{K}$ of the OH^- tunneling systems.

9.2.4 Level Crossing

The so-called *level crossing* is an elegant method to determine the energy splitting of tunneling systems. For this purpose, additional resonantly scattering impurities are incorporated in the sample with an energy splitting that can be tuned continuously from outside. This can be realized in a skilful way with magnetic defects exploiting the linear Zeeman effect. The differential thermal conductivity of KCl doped with lithium and additional magnetic impurities is drawn schematically in Fig. 9.17. At Δ_0, $2\Delta_0$, and $3\Delta_0$ the differential thermal conductivity is strongly reduced by resonant scattering of thermal phonons due to lithium tunneling systems (see also the level scheme

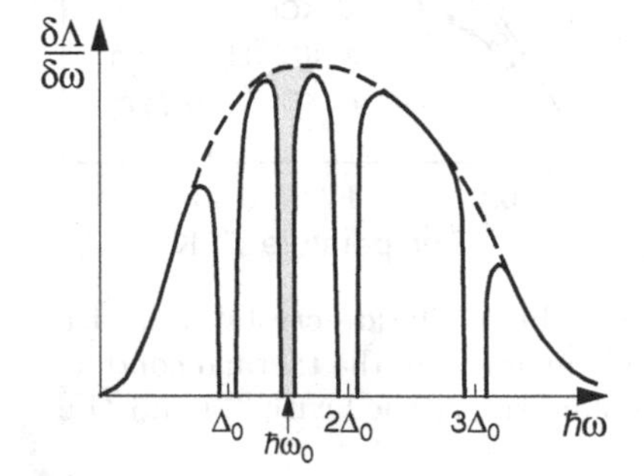

Fig. 9.17. Schematic drawing of the differential thermal conductivity $\mathrm{d}\Lambda/\mathrm{d}\omega$ of a lithium-doped KCl crystal versus phonon energy $\hbar\omega$. The *shaded* hole is caused by resonant scattering of the additionally incorporated magnetic impurities

shown in Fig. 9.9). As depicted in the figure, the magnetic impurities give rise to an additional hole in $\mathrm{d}\Lambda/\mathrm{d}\omega$ at $\hbar\omega_0$, where the magnetic impurity scatters resonantly because of Zeeman splitting.

In order to determine the level scheme of the tunneling systems, the thermal conductivity is measured at constant temperature as a function of the applied magnetic field. The conductivity exhibits a maximum if the hole due to the magnetic impurities coincides with one of the holes caused by the tunneling systems. In this case, the impurity does not give rise to an additional reduction in the conductivity. The result of such a measurement on KCl:Li is depicted in Fig. 9.18. In this experiment, color centers served as 'magnetic impurities' that were generated by irradiation of the crystal by a ^{60}Co source. In this diagram, the relative change in the thermal conductivity $\delta\Lambda(B)/\Lambda$ is plotted as a function of the applied magnetic field. Clearly, maxima of $\delta\Lambda(B)/\Lambda$ were found at 0.7, 1.45, and 2.2 T. From this result the tunnel splitting $\Delta_0 = 1.1\,\mathrm{K}$ can be deduced for ^{7}Li, in agreement with the value determined by other methods and simultaneously confirming the equidistant level scheme shown in Fig. 9.9.

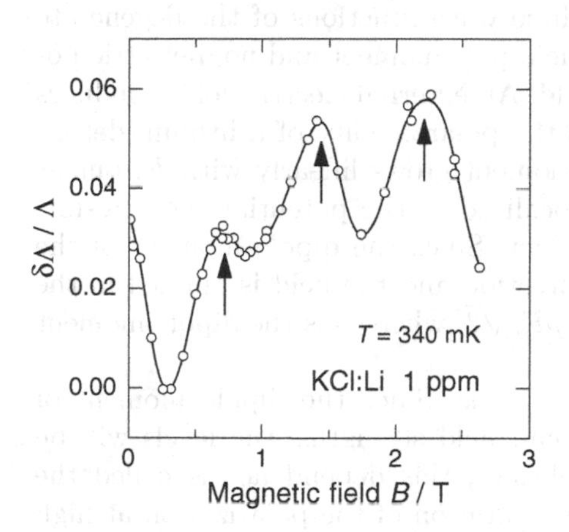

Fig. 9.18. Relative change in the thermal conductivity of KCl:Li with color centers versus external magnetic field. *Arrows* mark the magnetic fields at which Zeeman splitting of the 'magnetic impurities' coincides with the level splittings of the lithium tunneling systems [405]

9.2.5 Dielectric Susceptibility

To understand the experiments on the dielectric susceptibility presented here, it is advisable to look first at the selection rules and the dependence of the energy levels on electric fields. To keep the discussion simple we consider here only data on KCl:Li. In this material, electric dipole transitions between eigenstates are allowed, if the vectors $\boldsymbol{m}$ and $\boldsymbol{m}'$ introduced in (9.38) fulfill the condition $|\boldsymbol{m} - \boldsymbol{m}'| = 1$. In an overwhelming number of experiments the electric field was applied in the $\langle 100 \rangle$ direction. In this particular case, only the x-components are relevant in the selection rules, i.e., $|m_x - m_x'| = 1$. These transitions are indicated by arrows in the level scheme of KCl:Li drawn in Fig. 9.19.

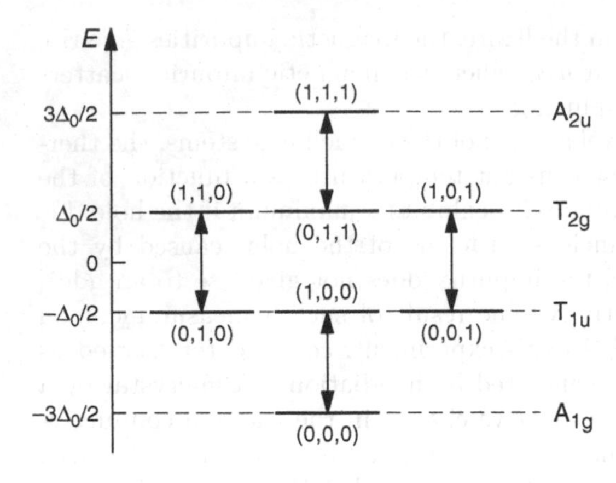

Fig. 9.19. Electric dipole transitions in KCl:Li. *Arrows* mark the allowed transitions for an ac electric field applied in the ⟨100⟩ direction

Without an external electric field, the wave functions of the four energy levels have no distinguished direction, i.e., they have cubic symmetry. This is also valid for the wave functions of the levels in the middle because they are constructed by superposing the three wave functions of the degenerate eigenstates. This means that no electric dipole moment and no dielectric polarization exists without an applied field. An external electric field F induces a dipole moment $p = \alpha F$, where α is the polarizability of a lithium defect. For small field strengths, the dipole moment grows linearly with F, but at high fields the lithium ions become localized in the potential wells, resulting in a field-independent dipole moment. Since the dipole moment of the lithium ions is oriented in the ⟨111⟩ direction and the field is applied in the ⟨100⟩ direction, it follows that $\boldsymbol{p} \cdot \boldsymbol{F} = pF/\sqrt{3}$, where $\boldsymbol{p}$ is the dipole moment of the localized states.

The electric field shifts the energy levels. Since the dipole moment of the defects is proportional to F at small field strengths, the levels will be shifted quadratically. As in atomic physics, this dependence is called the quadratic Stark effect. Because of the saturation of the polarization at high fields, a transition from quadratic to linear dependence is observed. To a good approximation, the field variation of the energy levels is given by

$$\begin{aligned} E_{(1,1,1)} &= -E_{(0,0,0)} = \Delta_0 + \sqrt{\frac{1}{4}\Delta_0^2 + \frac{1}{3}p^2F^2}\,, \\ E_{(0,1,1)} &= -E_{(1,0,0)} = \Delta_0 - \sqrt{\frac{1}{4}\Delta_0^2 + \frac{1}{3}p^2F^2}\,, \\ E_{(1,0,1)} &= E_{(1,1,0)} = -E_{(0,1,0)} = -E_{(0,0,1)} = \sqrt{\frac{1}{4}{\Delta_0}^2 + \frac{1}{3}p^2F^2}\,. \end{aligned} \tag{9.42}$$

The dielectric susceptibility χ can be deduced from the partition function Z taking into account the field dependence of the energy levels. Starting from the free energy $\mathcal{F}$ of tunneling systems, and using $\mathcal{F} = -k_{\mathrm{B}}T \ln Z$, the dielectric susceptibility is found to be

$$\chi = -\frac{1}{V}\left.\frac{\partial^2 \mathcal{F}}{\partial F^2}\right|_{F=0} = \frac{k_\mathrm{B}T}{V}\left.\frac{\partial^2 \ln Z}{\partial F^2}\right|_{F=0} . \tag{9.43}$$

The calculation of the partition function of an ensemble of tunneling systems with the level scheme (9.42) is straightforward. Evaluating its second-order derivative leads to the simple expression

$$\chi_\mathrm{iso} = \frac{2}{3}\frac{np^2}{\varepsilon_0 \Delta_0}\tanh\left(\frac{\Delta_0}{2k_\mathrm{B}T}\right) \tag{9.44}$$

for isolated $\langle 111\rangle$ tunneling systems. As expected, the susceptibility is proportional to the number density of tunneling systems. At high temperatures $T > \Delta_0/k_\mathrm{B}$, this expression merges with the classic *Langevin–Debye formula.* In this regime, the susceptibility is proportional to T^{-1}, and independent of the tunnel splitting. Interestingly, (9.44) corresponds exactly to the susceptibility of an ensemble of simple two-level systems with the tunnel splitting Δ_0.

In experiments, it is usually the dielectric constant ε that is measured rather than the susceptibility. Converting the data, one has to take note of the fact that the field at the defect is not identical with the applied field. When calculating the local field at a certain defect, the polarizability of the host crystal and field contribution of the neighboring defect ions has to be taken into account. The latter can be neglected for small concentrations. In this case, an expression analogous to the *Clausius–Mossotti relation* is found:

$$\chi = \frac{3\,(\varepsilon - \varepsilon_\mathrm{h})}{\varepsilon_\mathrm{h} + 2} . \tag{9.45}$$

Here, ε_h denotes the dielectric constant of the host crystal, and ε the measured relative permittivity of the sample.

The dielectric susceptibility of two KCl crystals containing very small amounts of ^{6}Li and ^{7}Li is shown in Fig. 9.20. The full lines are theoretical curves calculated for the static susceptibility of the two samples. Although the measurement was carried out at $f = 10\,\mathrm{kHz}$, the expression for the static susceptibility is applicable because (9.44) is a very good approximation as long as $f \ll \Delta_0/h$. The theoretical curves have been calculated using values for the dipole moment and for the tunnel splitting from measurements of the *paraelectric resonance*, namely $p = 2.63\,\mathrm{D}$, $^7\Delta_0/k_\mathrm{B} = 1.1\,\mathrm{K}$, and $^6\Delta_0/k_\mathrm{B} = 1.65\,\mathrm{K}$ for the two isotopes [406, 407]. Since the number density n follows directly from the known lithium concentration, there is no free fitting parameter. The excellent agreement of the theory with the experimental data demonstrates that the behavior of lithium tunneling systems in KCl at sufficiently small concentrations can be described quantitatively by the model of isolated tunneling systems. The influence of the isotope mass is clearly visible in this measurement. The total effect scales with $1/\Delta_0$, and the curve of the lighter isotope is shifted to higher temperatures.

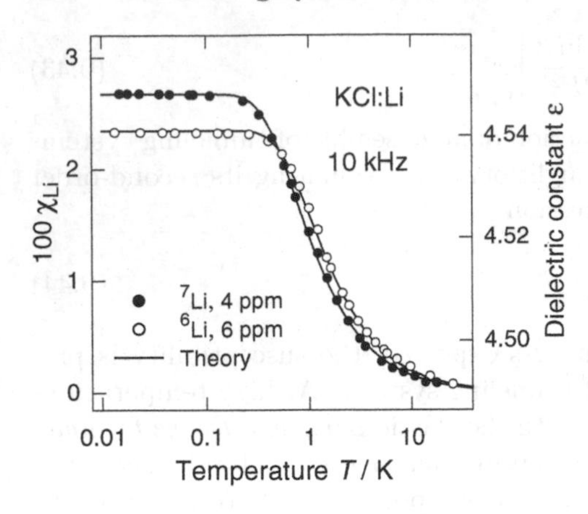

Fig. 9.20. Real part of the susceptibility χ_{Li} of two KCl:Li samples doped with different isotopes of lithium versus temperature. The *right* scale depicts the measured relative permittivity. *Full lines* represent the theoretical prediction according to (9.44) [408]

9.2.6 Sound Velocity

Substitutional defects do not usually fit ideally into the host lattice. They cause local strain fields that mediate the coupling to phonons. The effect of tunneling systems on the acoustic susceptibility can be derived from the strain dependence of the free energy, as in the case of the dielectric susceptibility, where the electric field dependence had to be considered. However, the resulting relations are considerably more complicated in the elastic case since the deformation potential $\overleftrightarrow{\gamma}$ defined in (9.11) and the strain field $\overleftrightarrow{\mathbf{e}}$ are not scalars but tensors. The effect of the tunneling systems depends not only on the direction of propagation but also on the polarization of the sound wave. On the other hand, the sensitivity to different parameters can give selective information on the microscopic structure of the tunneling systems.

Figure 9.21 shows the variation of the sound velocity with the inverse temperature, measured in a KCl crystal doped with 100 ppm lithium. The experiment was carried out at 30 MHz with transverse waves travelling along the [110] direction. Sound waves polarized in the $[1\bar{1}0]$ direction are unaffected, while the velocity of the waves with [001] polarization exhibits a strong temperature dependence. From this result, it can be concluded that lithium ions in KCl occupy a $\langle 111 \rangle$ off-center position.

A comparison of the dielectric and acoustic susceptibility supplies information on differences between the electric and elastic coupling. For shear waves propagating in the [100] direction the relative variation of the sound velocity due to $\langle 111 \rangle$ tunneling systems is expected to vary as

$$\frac{\delta v}{v} = -\frac{2n\gamma^2}{\varrho v^2 \Delta_0} \tanh\left(\frac{\Delta_0}{2k_{\mathrm{B}}T}\right) - \frac{n\gamma^2}{\varrho v^2 k_{\mathrm{B}}T} \operatorname{sech}^2\left(\frac{\Delta_0}{2k_{\mathrm{B}}T}\right) , \tag{9.46}$$

in the static limit. In contrast to the relation derived for the dielectric susceptibility, this expression contains two terms. The reason is that the reaction of the eight energy levels is different for strain and electric fields. The energy of four levels varies quadratically with strain, whereas the other four levels

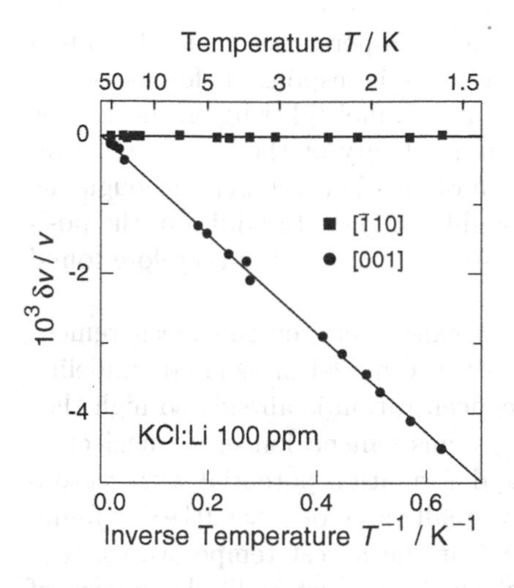

Fig. 9.21. Variation of the velocity of sound at 30 MHz in a KCl crystal doped with 100 ppm lithium. The transverse waves were propagating along the [110] direction and were polarized as indicated in the drawing [409]

experience a linear shift. The first term is caused by the levels with quadratic coupling and corresponds exactly (apart from the negative sign) to the expression (9.44) for the dielectric susceptibility. The second term is due to the levels with linear coupling.

Figure 9.22 shows the temperature variation of the sound velocity of KCL samples. It was measured at 2 kHz with torsion oscillators made of three differently doped KCl crystals. The sound velocity of the nominally pure KCl crystal decreases monotonically with rising temperature. The exact variation is determined by dislocations and lattice vibrations. Both lithium-doped KCl crystals exhibit a pronounced reduction in the velocity at low temperatures. Above 10 K, the contribution of the tunneling systems vanishes, the properties of the host crystal dominate, and the curves merge into the curve

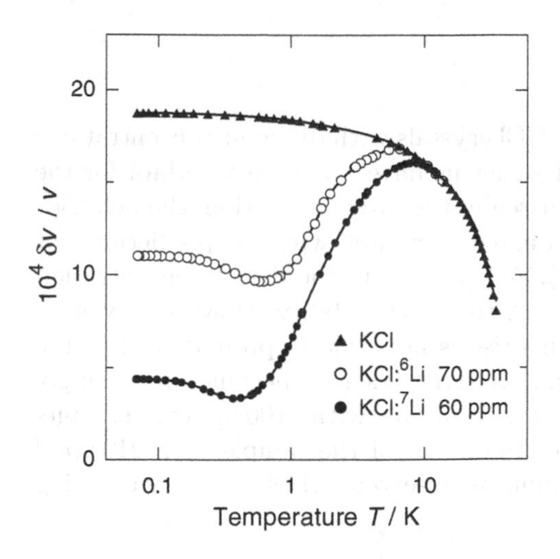

Fig. 9.22. Relative variation of the sound velocity $\delta v/v$ versus temperature in pure and lithium-doped KCl at 2 kHz. *Full lines* connect the data points [410]

for pure KCl. The significant reduction at low temperatures is mainly caused by the first term of (9.46), while the second term is responsible for the occurrence of the minimum. The dependence of the tunnel splitting on the isotope mass is obvious: the reduction in the sound velocity at the lowest temperatures is smaller in the ^{6}Li sample because of the factor $1/\Delta_0$, although the lithium concentration in this sample is slightly higher. In addition, the position of the minima is influenced by the isotope mass and is therefore found at different temperatures.

Although qualitatively good agreement exists between this measurement and (9.46), a quantitative comparison with the model of isolated tunneling systems is not feasible since the defect concentration is already so high that the interaction between the tunneling systems can no longer be neglected. Nevertheless, it is possible to deduce the deformation potential with reasonable accuracy from (9.46). The relatively small value of $\gamma \approx 0.04\,\text{eV}$ is found from the reduction in the speed of sound at the lowest temperatures. It is due to this small value of γ, together with the large electric dipole moment of the lithium tunneling systems in this material, that the electric dipole–dipole interaction dominates and the elastic interaction can be neglected. As a consequence, a simplified theoretical treatment of interacting tunneling systems is possible, and we discuss this in the following section.

9.3 Interacting Tunneling Systems in Crystals

With rising defect concentration the interaction between the tunneling systems becomes increasingly important, and the description based on isolated tunneling systems breaks down. Before presenting a brief theoretical description of interacting tunneling systems we discuss experimental results in order to show how the dielectric properties change with defect concentration.

9.3.1 Dielectric Properties

The dielectric susceptibility of four KCl crystals with different concentrations of ^{6}Li is depicted in Fig. 9.23a. First, let us have a look at the data for the two weakly doped crystals. As discussed in the previous section, the behavior of the sample doped with 6 ppm Li agrees very well with the predictions for isolated defects (see also Fig. 9.20). An increase in the defect concentration to 70 ppm leads to a distinct rise in the susceptibility. However, even at this relatively low concentration this rise is not exactly proportional to the defect concentration. The deviation from strict linearity becomes increasingly pronounced at higher concentrations. The sample with 1100 ppm ^{6}Li exhibits a susceptibility that is even smaller than that of the sample with 210 ppm! Simultaneously, a relaxation maximum is observed, which grows with rising defect concentration.

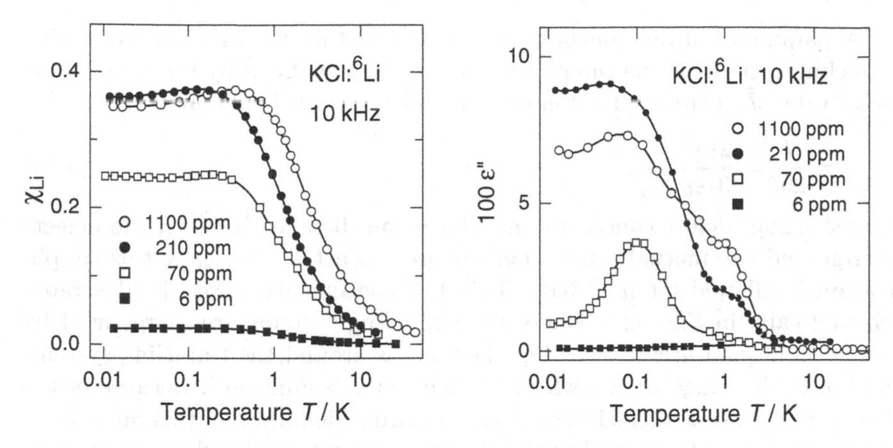

Fig. 9.23. Dielectric susceptibility (**a**) and dielectric absorption (**b**) of KCl crystals with different lithium doping [408]. *Full lines* connect the data points

Figure 9.23b shows the dielectric absorption ε'' of the same crystals. Since no absorption at all is expected for isolated defects, the dielectric loss at high concentrations has to be a direct consequence of interaction between the tunneling systems. Clearly, a high relaxation absorption is observed at low frequencies.

9.3.2 Theoretical Description

As mentioned in Sect. 9.2, many properties of isolated lithium tunneling systems can be reasonably well described in the framework of the so-called *two-state approximation*, where the multilevel structure of real tunneling systems is replaced by two levels. In this way, an essential simplification is achieved, and it also makes feasible a description of interacting tunneling systems in alkali halides.

As already pointed out, in systems like KCl:Li, the elastic interaction is small in comparison with the electric one. We therefore assume in the following discussion that the interaction energy J_{ij} between the defects i and j is exclusively determined by the electric dipole–dipole interaction

$$J_{ij} = -\frac{1}{4\pi\varepsilon_0\varepsilon_{\mathrm{KCl}}}\left[\frac{\boldsymbol{p}_i \cdot \boldsymbol{p}_j}{r_{ij}^3} - \frac{3(\boldsymbol{r}_{ij}\cdot\boldsymbol{p}_i)(\boldsymbol{r}_{ij}\cdot\boldsymbol{p}_j)}{r_{ij}^5}\right] . \tag{9.47}$$

To simplify the theoretical treatment further, the angular dependence of this interaction is replaced by the assumption that only parallel and antiparallel ordering of the dipoles is possible. This means that we replace (9.47) by the much simpler expression

$$J_{ij} = \pm\frac{p^2}{4\pi\varepsilon_0\varepsilon_{\mathrm{KCl}} r_{ij}^3} . \tag{9.48}$$

A parameter of fundamental significance in the treatment of interacting tunneling systems is the *interaction parameter* μ, defined by the ratio of the mean value $\overline{J}$ of the interaction energy and the tunnel splitting Δ_0:

$$\mu = \frac{\overline{J}}{\Delta_0} = \frac{2np^2}{3\varepsilon\varepsilon_0}\frac{1}{\Delta_0}\,. \tag{9.49}$$

At very small defect concentrations the mean distance between the defects is large and the mutual interaction can be neglected. Typically this simplification is allowed for $\mu < 0.01$. This low-concentration case is illustrated schematically in Fig. 9.24, where the tunneling systems are represented by points on a quadratic lattice. The dark areas around the tunneling systems symbolize the range within which interaction is 'significant'. Because of the slow $1/r^3$ variation of the interaction, a rigorous definition of this range is, of course, not possible. Regarding the thermal energy, the significance of the interaction increases steadily with decreasing temperature, but this subtlety is of no importance for the basic statement made by Fig. 9.24, i.e., that defects with nonoverlapping shaded areas may be considered to be isolated tunneling systems. In the low-concentration limit the majority of the tunneling systems fall into this category.

Nevertheless, for randomly distributed defects there is a certain probability, even at small concentrations, that two defects will be so close that their interaction cannot be neglected. Such a defect pair is also present in Fig. 9.24. The properties of pairs of tunneling defects are not well understood in general, but in the case of strong coupling, i.e., for $J_{ij} \gg \Delta_0$, the two defect ions move in their potential in a correlated manner, simplifying the description considerably. The number of such pairs increases quadratically with growing defect concentration, and their influence on the macroscopic properties of the host crystal rises rapidly. Therefore, for crystals with an interaction parameter $\mu \approx 0.1$, pairs of tunneling systems also have to be included in the considerations. The obvious approach in this case seems to be a cluster expansion. In fact, several theories have been worked out based on this idea. Since the expansion was usually truncated after the second term, these descriptions (see, e.g., [411, 412]) are often called *pair models*.

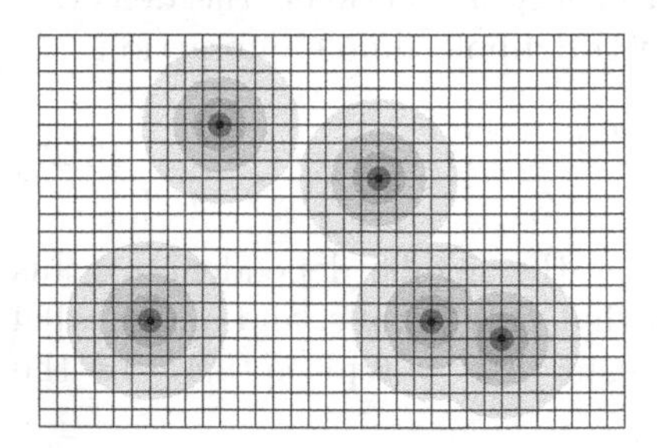

Fig. 9.24. Schematic illustration of defects (*full circles*) and their 'effective' radius of interaction (*dark areas*) at small defect concentration

At high defect concentrations, i.e., for $\mu > 0.5$, cluster expansions are no longer adequate. The main reason for their failure is the fact that inclusion of clusters of increasing size leads to an improvement only within a rather limited concentration range. A schematic illustration is given in Fig. 9.25. Obviously, a many-body problem exists that requires new theoretical concepts. Such a theory has been successfully developed by *Würger* [413], describing the dynamics of tunneling states in model systems. In this novel theoretical treatment, the whole experimentally accessible concentration range is covered, and quantitative predictions have been made for a variety of experiments. This theory is based on the projection method developed by *Mori* and *Zwanzig* [414, 415] and the mode-coupling approximation. The projection method is used to separate known and unknown contributions to the dynamics of complex systems. The unknown contributions are then treated in an appropriate approximation. The mode-coupling theory was originally introduced by *Kawasaki* [416] and developed further by *Götze* [417] in his description of the glass transition. The basic feature of this theory is the separation of processes developing on different time scales. Expressing this idea in a simple manner, we may say that slowly relaxing modes are only relevant on long time scales.

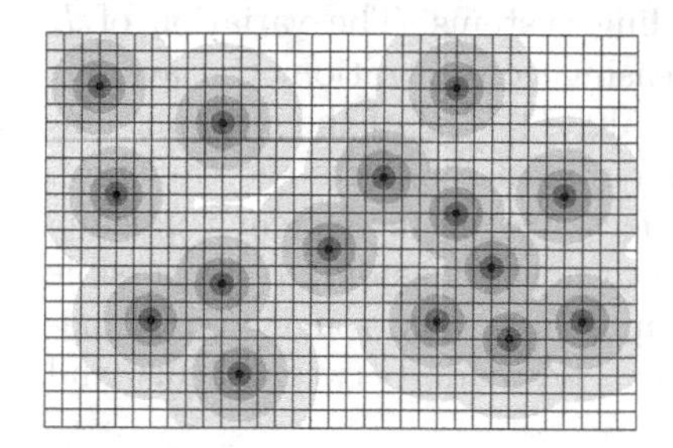

Fig. 9.25. Schematic illustration of defects (*full circles*) and their interaction radii (*dark areas*) at high defect concentration

The most important result of Würger's theory is that, at high concentrations, a transition occurs from the coherent tunneling of isolated defects to the incoherent tunneling of interacting defects. Incoherent tunneling means that, at high concentrations, the tunneling systems are so strongly disturbed by the motion of neighboring systems that, during the tunneling process, the phase information of the wave function is lost. As a direct consequence, the oscillator strength of the tunneling systems is reduced and a new relaxation mechanism appears.

9.3.3 Dielectric Susceptibility

We do not want to reproduce the theory here in all its detail but discuss its outcome in a simple form. The basic result is that relaxation processes occur due to the interaction that do not take place in systems with isolated defects. The strength of these processes is reflected by the relaxation amplitude R. As relaxation becomes more significant, the resonant interaction loses its weight. This fact is reflected by the equations describing the relaxation part χ_{rel} and the resonant part χ_{res} of the dielectric susceptibility, namely by

$$\chi_{\text{rel}} = \frac{p^2}{3\varepsilon_0 V} \frac{1}{k_B T} \sum_i R_i \frac{1}{1 - \mathrm{i}\omega\tau_i} \tag{9.50}$$

and

$$\chi_{\text{res}} = \frac{2p^2}{3\varepsilon_0 V} \sum_i (1 - R_i) \frac{1}{E_i} \tanh\left(\frac{E_i}{2k_B T}\right) . \tag{9.51}$$

Here, R_i, E_i, and τ_i denote the relaxation amplitude, the energy, and the relaxation time of the tunneling system i. The average values $\overline{R}$ and $\overline{E}$ are given by the two expressions

$$\overline{R} = \mu \left[\frac{\pi}{2} - \arctan(\mu)\right] , \tag{9.52}$$

and

$$\overline{E} = \Delta_0 \sqrt{1 + \mu^2} , \tag{9.53}$$

which only depend on the parameter μ. It is remarkable that the interaction changes the energy splitting of the tunneling systems. The variation of $\overline{R}$ with μ is shown in Fig. 9.26. With increasing concentration, the average relaxation amplitude rises while the mean value $(1 - \overline{R})$ of the resonant amplitude decreases to the same extent. The curve also reflects the transition from the coherent tunneling of isolated defects to the incoherent tunneling of interacting defects. Unfortunately, up until now it has only been possible to calculate the spectrum of relaxation times in a very rough approximation. Therefore, only a crude comparison can be made between theory and experiment.

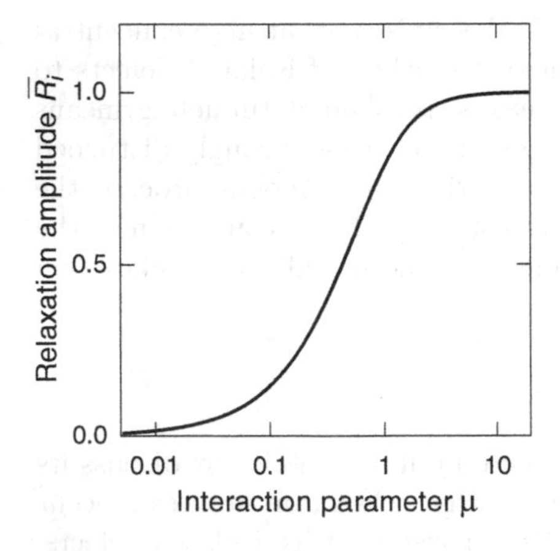

Fig. 9.26. Mean relaxation amplitude $\overline{R}$ as a function of the interaction parameter μ

However, in the limit of very low temperatures, i.e., for $T \to 0$, relaxation becomes very slow. Therefore, χ_{rel} vanishes and only the resonant interaction contributes to the dielectric susceptibility. Furthermore, since $\tanh(E/2k_{\mathrm{B}}T) \to 1$, we may write

$$\chi'(T \to 0) \approx \chi'_{\mathrm{res}}(T = 0) = \frac{2}{3} \frac{np^2}{\varepsilon_0 \Delta_0} \frac{\left(\sqrt{1+\mu^2} - \mu\right)^2}{\sqrt{1+\mu^2}} . \tag{9.54}$$

A comparison of this expression with the dielectric susceptibility χ'_{iso} of isolated tunneling systems (9.44) leads to the ratio

$$\frac{\chi'_{\mathrm{res}}}{\chi'_{\mathrm{iso}}}\,(T \to 0) = \frac{\left(\sqrt{1+\mu^2} - \mu\right)^2}{\sqrt{1+\mu^2}} , \tag{9.55}$$

which only depends on the dimensionless interaction parameter μ, thus allowing a rigorous comparison between theory and experiment.

Figure 9.27 depicts the theoretical prediction (full line) for the ratio $\chi'_{\mathrm{res}}/\chi'_{\mathrm{iso}}$ over a wide range of values of the parameter μ, together with experimental data on a variety of defect systems. The curve demonstrates once again that the amplitude of the resonant effect decreases with growing defect concentration. In particular, the transition from coherent to incoherent tunneling takes place at $\mu \approx 0.5$. Since (9.55) does not contain any free parameters, the outstandingly good agreement with experiment for a great number of systems can be considered as a confirmation of the model.

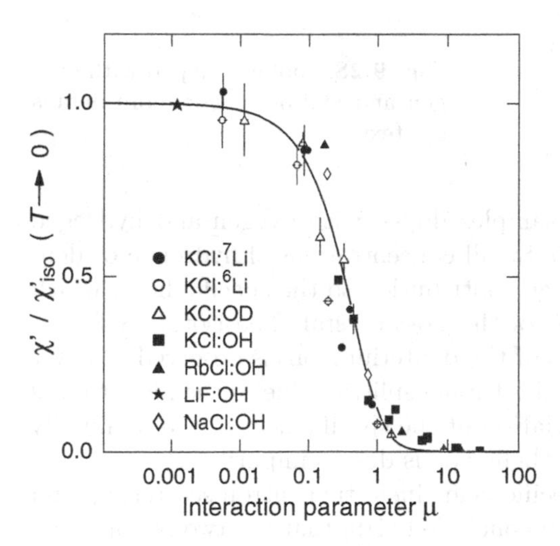

Fig. 9.27. Low-temperature limit of $\chi'/\chi'_{\mathrm{iso}}$ versus interaction parameter μ for a variety of alkali halide crystals with different tunneling systems. The *full line* represents the prediction from (9.55) [418]

9.4 Asymmetric Tunneling Systems in Crystals

9.4.1 Nb:O,H and Nb:O,D

Tunneling systems in niobium can easily be generated by doping with oxygen (or nitrogen) and loading the metal with hydrogen afterwards. Incorporated oxygen or nitrogen atoms are virtually immobile and serve as trap centers for hydrogen ions. By appropriate doping, one hydrogen ion is captured per trap center.[5]

The atomic configuration of such defects in niobium is depicted schematically in Fig. 9.28. For clarity, only the surface of the bcc cell is drawn. Hydrogen or deuterium ions are located at tetrahedral sites indicated by small circles. There are always *pairs* of equivalent sites, implying that the defect states in Nb:O,H or Nb:O,D are 'genuine' two-level systems. Oxygen ions occupy interstitial lattice sites with octahedral symmetry. Their strain field lifts the degeneracy of the tetrahedral sites apart from those in the (100) plane. From a series of investigations using different techniques, it is known that, under the experimental conditions discussed here, only those sites represented in our drawing by a pair of full circles are actually occupied by a hydrogen or deuterium ion.

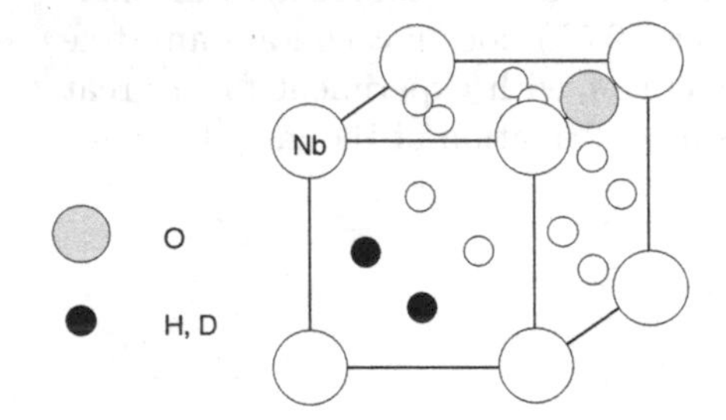

Fig. 9.28. Niobium doped with oxygen and hydrogen. For more details see text

The specific heat of niobium samples doped with oxygen and hydrogen or deuterium is shown in Fig. 9.29. Small concentrations of hydrogen or deuterium obviously give rise to a large contribution to the specific heat, which can easily be separated from that of the host crystal. The isotope effect is well pronounced. The contribution of the deuterium ions is observed at lower temperatures because of their smaller tunnel splitting due to the higher mass of the tunneling particle. The deviation of the specific heat of the nominally pure sample from the expected T^3 behavior is due to impurities.

From measurements of the specific heat, inelastic neutron scattering, and ultrasonic experiments, it has been concluded [419] that the two isotopes exhibit the tunnel splitting $\Delta_0^{\mathrm{H}}/k_{\mathrm{B}} = 1.4\,\mathrm{K}$, and $\Delta_0^{\mathrm{D}}/k_{\mathrm{B}} = 0.18\,\mathrm{K}$, respectively.

[5] It should be mentioned that protons or deuterons attract free electrons and modify the structure of the host lattice locally, resulting in a dressing effect that we do not consider here.

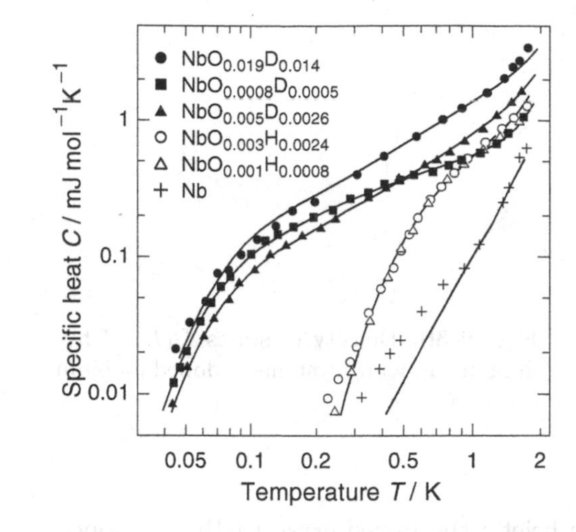

Fig. 9.29. Specific heat of Nb, Nb:O,H, and Nb:O,D [419]

At first, it seems that hydrogen-doped niobium is an especially simple system because of the true two-level nature of the tunneling states. However, the oxygen impurities cause strain fields that decay only slowly with distance and thus give rise to asymmetry energies Δ in neighboring two-level systems. This effect leads to a wide distribution of the asymmetry energies that can be approximately described by the Lorentzian distribution function

$$P(\Delta) = \frac{N_0}{\pi} \frac{\widetilde{\Delta}}{\widetilde{\Delta}^2 + \Delta^2}, \tag{9.56}$$

where N_0 is the number of trap centers. The width of the function can be deduced from ultrasonic experiments, and the value $\widetilde{\Delta}/k_B \approx 3\,\mathrm{K}$ has been found.

The temperature variation of the specific heat is not determined by $P(\Delta)$ but by the density of states of the tunneling systems. With the help of the relation $E^2 = \Delta^2 + \Delta_0^2$, the distribution $P(\Delta)$ can be converted into the density of states $D(E)$, and

$$D(E)\,\mathrm{d}E = P(\Delta)\frac{\partial \Delta}{\partial E}\,\mathrm{d}E = \frac{N_0}{\pi}\,\frac{E}{\sqrt{E^2 - \Delta_0^2}}\,\frac{\widetilde{\Delta}}{\widetilde{\Delta}^2 + E^2 - \Delta_0^2}\,\mathrm{d}E \tag{9.57}$$

is found. As shown schematically in Fig. 9.30, the density of states $D(E)$ diverges at the lower cutoff energy Δ_0, but stays integrable.

The vibrational energy of the hydrogen ions has been determined by inelastic neutron scattering. Using the value $\hbar\Omega/k_B \approx 1230\,\mathrm{K}$, the distance $d = 1.17\,\text{Å}$ between the two ion sites, and the (bare) mass of protons, and assuming parabolic potentials, the tunnel splitting Δ_0 can be estimated, and satisfactory agreement with the measured value is found.

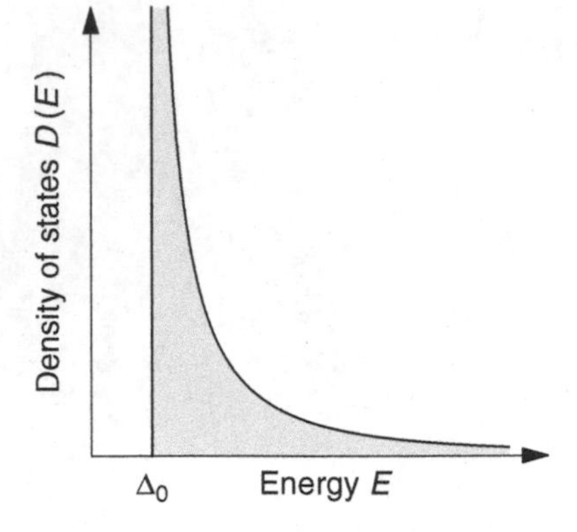

Fig. 9.30. Density of states $D(E)$ of hydrogen tunneling systems in doped niobium crystals

9.4.2 CN^- Ions in KBr:KCl

As a second example, we mention briefly the mixed crystal KBr:KCl doped with KCN. CN^- ions form rotational tunneling states in both host components. The distinctive feature of this system is that, because of the different size of chlorine and bromine ions, the mixing of KBr and KCl gives rise to strong fluctuating internal strain fields. As a consequence, tunneling systems are found with a broad distribution of the two parameters Δ_0 and Δ.

The peculiarity of this system is reflected by the specific heat. As shown in Fig. 9.31, the specific heat of the pure mixed crystal follows the Debye model with a minor deviation at the lowest temperatures. The mixed crystal with nearly the same composition but containing additional CN^- ions exhibits a specific heat that is markedly enhanced at low temperatures and approaches a linear variation with temperature. As we shall see in the following section, this is a characteristic property of glasses. For this reason, mixed crystals with tunneling systems are often considered as model systems for the low-temperature properties of amorphous solids.

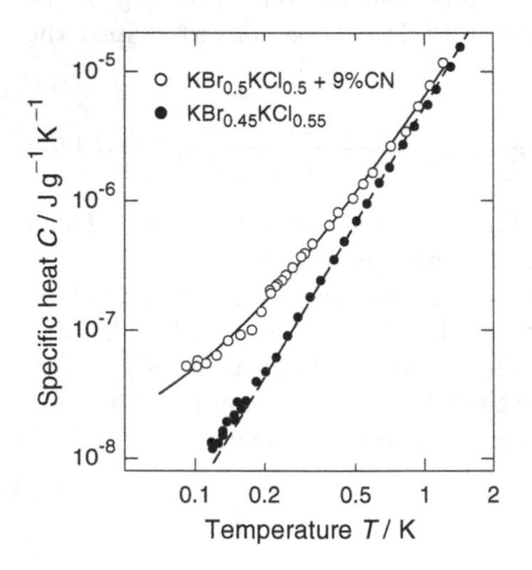

Fig. 9.31. Specific heat of pure KBr_xKCl_{1-x} and the same doped with KCN. The *dashed line* depicts the Debye behavior of pure dielectric crystals. The *full line* serves as an optical guide [420]

9.5 Amorphous Dielectrics

The low-temperature properties of amorphous solids are mainly governed by the presence of tunneling systems. They determine the thermal, elastic and dielectric properties to a large extent. The two-dimensional analogue of the structure of crystalline quartz is depicted schematically in Fig. 9.32 and compared with two-dimensional fused quartz. The crystal is built of regular six-membered rings with well-defined bonding angles resulting in a 'honeycomb' structure. The basic structural unit of the amorphous modification is identical, but the bonding angles at the oxygen atoms vary slightly. As a consequence, the long-range order is lost and not all rings exhibit the same size. The structure of real, three-dimensional SiO_2 is similar. The basic units of the crystalline and glassy phase are SiO_4 tetrahedra. They are weakly distorted in the amorphous modification, resulting in a loss of long-range order in the glass due to the slight variations of the bonding angles.

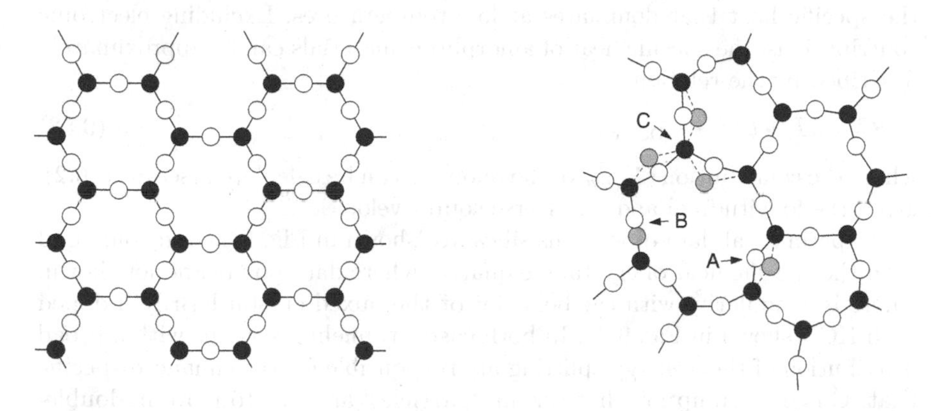

Fig. 9.32. Two-dimensional illustration of the structure of the crystalline (**a**) and glassy (**b**) modification of SiO_2. *Full circles* represent silicon atoms and *open circles* oxygen atoms. *Arrows* indicate parts of the network that could be candidates for tunneling systems

As already stated, the precondition for the occurrence of tunneling systems is the presence of some kind of irregularity. In perfect crystals, all atoms occupy well-defined sites so that there is no possibility for the formation of tunneling systems. In contrast, in amorphous solids, certain atoms or small clusters of atoms occupy energetically nearly equivalent sites in the irregular network. At higher temperatures, transitions between the potential minima take place via thermally activated processes. At low temperatures, such transitions are possible via quantum tunneling. In Fig. 9.32b, three configurations, A, B and C, identified by arrows, are candidates for double-well potentials. Although this schematic drawing is a rather naive caricature of the reality, it illustrates the basic idea.

9.5.1 Specific Heat

In 1971, *Zeller* and *Pohl* [421] demonstrated in their pioneering work that the low-temperature behavior of glasses differs drastically from that of perfect crystals, and that their thermal properties are almost universal. This was a surprising result since, at low temperatures, the thermal properties of pure dielectrics are generally determined by phonons of long wavelength. Phonons in turn are characterized by the elastic behavior of the medium averaged over a distance comparable with their wavelength. Consequently, it was expected that, at low temperatures, the atomic disorder of amorphous solids would be unimportant and glasses would behave as predicted by the Debye model. In reality, it was found that dielectric glasses, amorphous polymers, and superconducting amorphous metals exhibit an additional contribution to the specific heat that dominates at low temperatures. Excluding electronic contributions, the specific heat of amorphous materials can be approximately described by the relation

$$C = aT + bT^3 + C_{\text{Debye}} \,, \tag{9.58}$$

where the contribution C_{Debye} of the phonons can be calculated (see Sect. 6.2), given the longitudinal and transverse sound velocities.[6]

Experimental data of vitreous silica are shown in Fig. 9.33 and compared with the specific heat of crystalline quartz, where data points are not shown. There is a similarity with the behavior of the mixed crystal KBr:KCl doped with KCN shown in Fig. 9.31. In both cases, tunneling systems with a broad distribution of their energy splitting are responsible for the enhanced specific heat. On the assumption that certain 'particles' are able to move in double-well potentials, a phenomenological model, the so-called *tunneling model*, was developed in 1972 by *Phillips* [422], and *Anderson*, *Halperin*, and *Varma* [423].

Despite the great similarity with the defect states in alkali halides discussed in Sect. 9.2, there are also fundamental differences. Of course, the tunneling 'particles' vary from material to material, but even within one system they are not identical. The tunneling particles are thought to be composed of several atoms performing different motions [425,426]. Since these motions will involve translations and rotations through small angles, the separation between the two minima of the double-well potential in Fig. 9.1 symbolizes

[6] The specific heat can also be approximated by replacing the term bT^3 by one proportional to T^5. Since (9.58) only holds in a rather limited temperature range, both expressions give a fair approximation. In addition, the temperature variation of the specific heat also deviates from strict linearity at the lowest temperatures. We will discuss this observation later.

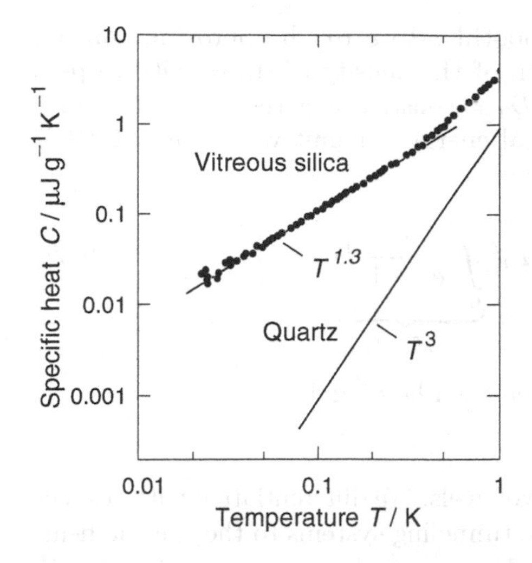

Fig. 9.33. Specific heat of vitreous silica and crystalline quartz versus temperature. The specific heat of the glass exceeds that of the corresponding crystal over the whole temperature range. From [424]

a distance in configurational space. The restriction to only two equilibrium sites is, of course, a simplification, but the occurrence of many nearly equivalent wells is unlikely in the irregular structure of amorphous solids. The assumption of a broad distribution of the asymmetry energy Δ and also of the tunneling parameter λ is a further basic difference with defect centers in crystals. In the tunneling model, it is assumed that Δ and λ are independent of each other and uniformly distributed according to

$$P(\Delta, \lambda)\,\mathrm{d}\Delta\,\mathrm{d}\lambda = P_0\,\mathrm{d}\Delta\,\mathrm{d}\lambda\,, \tag{9.59}$$

where P_0 is a constant. For comparison with experiment, it is more convenient to use the distribution function $P(E, \Delta_0)$ that follows from $P(\Delta, \lambda)$ by a Jacobian transformation. With $E^2 = \Delta^2 + \Delta_0^2$, and $\Delta_0 = \hbar\Omega \mathrm{e}^{-\lambda}$, the transformation leads to

$$\begin{aligned} P(E, \Delta_0)\,\mathrm{d}\Delta_0\,\mathrm{d}E &= P(\Delta, \lambda) \left|\frac{\partial \lambda}{\partial \Delta_0}\right| \left|\frac{\partial \Delta}{\partial E}\right| \mathrm{d}\Delta_0\,\mathrm{d}E \\ &= P_0 \frac{E}{\Delta_0\sqrt{E^2 - \Delta_0^2}}\,\mathrm{d}\Delta_0\,\mathrm{d}E\,. \end{aligned} \tag{9.60}$$

We integrate over all values of Δ_0 to evaluate the density of states $D(E)$, and find

$$D(E) = \int_{\Delta_0^{\min}}^{E} P(\Delta_0, E)\,\mathrm{d}\Delta_0 = P_0 \ln \frac{2E}{\Delta_0^{\min}}\,. \tag{9.61}$$

To avoid a nonintegrable divergence of the distribution function for vanishing Δ_0, a minimum value $\Delta_0^{\min}$ for the tunnel splitting has been introduced. This is a consequence of the fact that an upper limit $\lambda_{\max}$ exists for the tunneling parameter. Of course, in real systems the function $P(\lambda)$ will not end abruptly but will drop away smoothly. To a rough approximation, we may neglect the logarithmic variation of the density of states with respect to E and replace (9.61) by $D(E) \approx D_0 = \text{const}$. Under the assumption of a constant density of states, the internal energy per unit volume $u = U/V$ is easily calculated as

$$u = \int E\, D(E)\, f(E)\, \mathrm{d}E = D_0 (k_\mathrm{B} T)^2 \underbrace{\int_0^\infty \frac{x}{\mathrm{e}^x + 1}\, \mathrm{d}x}_{\pi^2/12} . \tag{9.62}$$

The abbreviation $x = E/k_\mathrm{B}T$ has been used here, and

$$f(E) = [\exp(E/k_\mathrm{B}T) + 1]^{-1}$$

is the occupation probability of the two levels. We differentiate u with respect to T to obtain the contribution of the tunneling systems to the specific heat:

$$C_V = \left(\frac{\partial u}{\partial T}\right)_V = \frac{\pi^2}{6} D_0 {k_\mathrm{B}}^2\, T . \tag{9.63}$$

This result is in fair agreement with the experimentally observed variation of C below 1 K. However, a more careful analysis of the experimental data shows that the data for vitreous silica are better described by $C \propto T^{1.3}$ [427]. Similar deviations from strict linearity have also been found in other glasses. Two types of explanation have been proposed. On the one hand, it has been argued that the density of states is not strictly constant. On the other hand, the deviation has been attributed to time effects. We consider the latter effect more thoroughly here because it is a natural consequence of the distribution of the tunneling parameter λ.

According to (9.31), the low-temperature relaxation time τ depends on the ratio (Δ_0/E). Therefore, the relaxation time of an ensemble of tunneling systems will vary between a minimum value τ_min and infinity, even if the energy splitting is kept constant. Symmetric systems with $(\Delta_0/E) = 1$ exhibit the shortest relaxation time, while highly asymmetric systems with $(\Delta_0/E) \to 0$ relax very slowly. This fact is conveniently expressed by rewriting (9.31) in the form

$$\tau(E,T) = \left(\frac{E}{\Delta_0}\right)^2 \tau_\mathrm{min}(E,T) . \tag{9.64}$$

In many experiments, such as measurements of specific heat or internal friction, the main contribution is due to tunneling systems with an energy splitting $E \approx k_\mathrm{B}T$. The minimum relaxation time $\tau_\mathrm{m} = \tau_\mathrm{min}(E = k_\mathrm{B}T)$ of these systems is an important quantity in the discussion of the dynamics of amorphous solids. Since $\tau_\mathrm{min}^{-1} \propto E^3 \coth(E/2k_\mathrm{B}T)$ according to (9.31), we may write

$$\tau_\mathrm{m}^{-1} = \tau_\mathrm{min}(E = k_\mathrm{B}T) = AT^3 . \tag{9.65}$$

From ultrasonic experiments it is known that typically $A \approx 10^8\,\mathrm{s^{-1}K^{-2}}$, meaning that at 1 K, the fastest tunneling systems relax within a few nanoseconds.

We take the distribution of relaxation times into account by carrying out another Jacobian transformation and replace $P(E, \Delta_0)$ by

$$P(E,\tau) = \frac{P_0}{2\tau\sqrt{1-\tau_{\min}/\tau}}\,. \tag{9.66}$$

The density of states D_{eff} effectively seen in a conventional measurement of the specific heat is then given by integrating over all relaxation times from the minimum time $\tau_{\min}$ up to the time scale t_0 of the experiment:

$$D_{\text{eff}}(E,t_0) = \int_{\tau_{\min}}^{t_0} P(E,\tau)\,\mathrm{d}\tau = \frac{P_0}{2}\ln\frac{4t_0}{\tau_{\min}}\,. \tag{9.67}$$

The occurrence of extremely long relaxation times makes a measurement of the specific heat in the thermodynamic sense impossible. Using the fact that the main contribution to the specific heat is due to tunneling systems with $E \approx k_\mathrm{B}T$, we replace $\tau_{\min}$ by τ_m defined by (9.65). With this simplification, the specific heat C_V is easily calculated and we obtain

$$C_V = \frac{\pi^2}{12} P_0 {k_\mathrm{B}}^2\, T \ln(4At_0T^3)\,. \tag{9.68}$$

The logarithmic factor results in a weak additional temperature dependence. Inserting typical values, like $t_0 = 10\,\mathrm{s}$ and $A = 10^8\,\mathrm{s^{-1}K^{-3}}$ for vitreous silica, satisfactory agreement with the observed temperature variation is found.

There is another interesting type of experiment that allows the investigation of extremely long relaxation times. In the so-called *heat-release experiment*, the amorphous sample is kept for a long time at the charging temperature T_1. It is then brought into contact with a heat sink and rapidly cooled down to T_0. Within a short time the phonons of the sample reach the new thermal equilibrium determined by the heat sink, but the tunneling systems are unable to follow because of the wide spread of relaxation times. The heat release $\dot{Q}$ per volume can be calculated, taking into account the distribution function (9.66). Here, we only quote the result without going into the details of the calculation [428]:

$$\dot{Q} = \frac{\pi^2 k_\mathrm{B}^2}{24} P_0 (T_1^2 - T_0^2)\frac{1}{t}\,. \tag{9.69}$$

Figure 9.34 displays data obtained in a measurement on vitreous silica. In this experiment, the sample was rapidly cooled from nitrogen temperatures down to $T_0 = 20\,\mathrm{mK}$. In fact, the observed heat release follows the predicted t^{-1} law up to very long times. Since (9.69) is a consequence of the distribution function (9.60), this result supports the assumption of the tunneling model that the tunneling parameter λ is uniformly distributed.

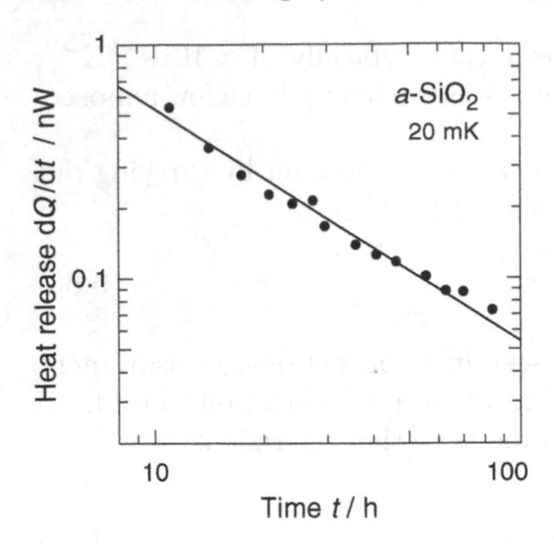

Fig. 9.34. Heat release of vitreous silica. The sample was rapidly cooled from nitrogen temperature to the measuring temperature of 20 mK. The expected t^{-1} dependence is indicated by a *straight line* [429]

Now we come back to the specific heat of amorphous solids given by (9.58). This equation contains not only a linear term caused by the tunneling systems, but also a T^3 term that is due neither to tunneling systems nor to phonons. The occurrence of this additional term follows in a natural way from the so-called *soft-potential model* [430–432], which affords reliable predictions of the properties of glasses, even above 1 K. In the tunneling model only double-well potentials, consisting of two identical harmonic wells, are taken into account. In the soft potential model, a more general expression for the potential landscape is used, leading to the so-called *quasiharmonic potentials* in addition to the double-well potentials discussed so far. Particles moving in these single wells give rise to more or less harmonic vibrational *soft modes*. Below 1 K, both models, the tunneling model and the soft-potential model, lead to virtually the same predictions because tunneling particles in doublewells govern the behavior. At higher temperatures, the quasiharmonic vibrations come into play and determine the properties of glasses, because the great majority of low-energy excitations belong to this class. The existence of these low-frequency vibrations has been demonstrated by inelastic neutron scattering [432]. The additional T^3 term (or T^5 term) in (9.58) is caused by these quasiharmonic soft modes and is well accounted for by the soft-potential model.

Finally, we want to touch briefly upon the magnitude of the specific heat, i.e., the total number of tunneling systems existing in glasses. A unique statement is not possible since, above a few kelvin, the specific heat is mainly determined by quasiharmonic soft modes and phonons. Although it is difficult to separate the contribution of the tunneling centers unambiguously, a comparison with experimental data tells us that, in the range $0 < E/k_{\mathrm{B}} \approx 1\,\mathrm{K}$, roughly 10^{17} to 10^{18} tunneling systems/cm^3 exist in amorphous solids, i.e., the concentration is roughly 10 ppm to 100 ppm. Within rather narrow limits this number is a characteristic quantity of amorphous solids. The few exceptions, e.g., amorphous silicon, can be traced back to the overconstrained random network of these materials. At higher temperatures, tunneling becomes increasingly incoherent. It eventually vanishes completely and particles in double-well potentials behave like classical two-state systems.

It is remarkable that, not only the temperature dependence, but also the magnitude of the specific heat is comparable for most amorphous solids. As we shall see, this statement is also correct for the thermal conductivity and elastic behavior. This so-called 'universality' of the low-temperature properties of amorphous solids is not understood theoretically. However, it indicates that the tunneling particles are not single atoms because in this case the specific structural and chemical properties of the solid would be of importance, leading to pronounced differences in the low-temperature properties known from crystals. Consequently, tunneling particles in glasses very likely consist of a greater number of atoms, as has been demonstrated in simulations [425,426].

9.5.2 Thermal Conductivity

The thermal conductivity of amorphous solids differs fundamentally from that of crystals. This becomes obvious from a comparison of the conductivity of vitreous silica and crystalline quartz depicted in Fig. 9.35. The two curves are characteristic for the two different classes of solids. In pure dielectric crystals, the thermal conductivity goes through a pronounced maximum for reasons we have already discussed in Sect. 6.2. In amorphous solids, however, the thermal conductivity increases steadily with temperature. At all temperatures its magnitude is distinctly below the value of the corresponding crystals. We want to mention once again that amorphous dielectrics not only exhibit the same low-temperature variation of their thermal conductivity but also the magnitude of the conductivity of different materials is similar. This is one reason why the thermal properties of glasses are often referred to as

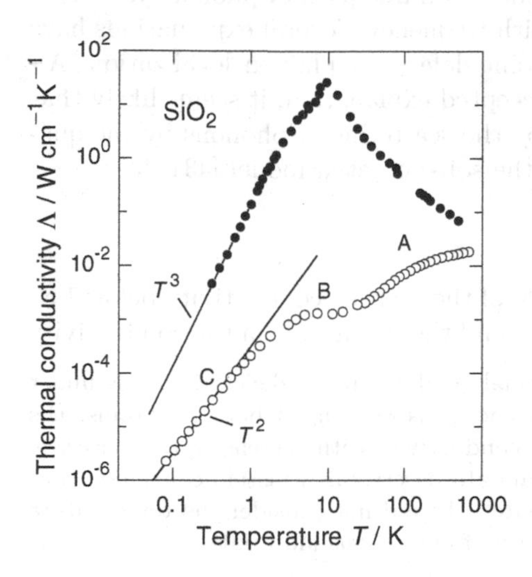

Fig. 9.35. Comparison of the thermal conductivity of a quartz crystal (•) and of vitreous silica (○). Three temperature regimes can be distinguished in the case of vitreous silica denoted by A, B and C [421]

'universal'. As indicated in Fig. 9.35, it is possible to distinguish between three temperature regimes that we discuss briefly.

High-Temperature Regime A

Although it is beyond the scope of this book we make a very brief remark on the high-temperature regime. At temperatures above 20 K, the frequency of the dominant phonons is of the order of 5×10^{13} Hz. Their wavelength is about 5 Å and remarkably, the mean free path is of the same order as well. This means that short-wavelength phonons, if they exist at all in amorphous solids, are overdamped, i.e., the original definition of phonons as elementary excitations is questionable in this case. Under such circumstances the description of the thermal conductivity in terms of phonon propagation becomes doubtful and it might be more useful to develop the idea that the vibrational energy diffuses from atom to atom. Besides the old treatment by *Einstein* in 1911 [250] there are only a few recent theoretical studies of the thermal conductivity of amorphous materials at higher temperatures.[7] Recently, molecular dynamic simulations have been carried out and good agreement was found with the experiment [433].

Intermediate-Temperature Regime B (Plateau Region)

In the temperature range around 10 K, the so-called 'plateau' is found in virtually all amorphous solids. This phenomenon is not related to the maximum of the thermal conductivity of crystals, although Fig. 9.35 could give the impression that they have a common origin. While the maximum observed for crystals shifts with the dimension of the sample, the geometry has no influence on the position of the plateau. Because of the rapid increase in the specific heat in this temperature range, an almost temperature-independent thermal conductivity indicates that the mean free path of phonons decreases steeply either with temperature or with frequency. Several explanations have been proposed, e.g., scattering by point defects or phonon localization. Although there is not yet a generally accepted explanation, it seems likely that the thermal conductivity is limited by the scattering of phonons by the quasiharmonic potentials introduced in the soft-potential model [431, 432].

Low-Temperature Regime C

Two features characterize the behavior of the thermal conductivity below 1 K. Its temperature variation is quadratic, and the magnitude of the conductivity

[7] In 1911, Einstein calculated the thermal conductivity of dielectric solids under the assumption that the vibrational energy is exchanged between atoms. His treatment predicted a decrease in the conductivity with decreasing temperature. From a comparison with experimental data on crystals Einstein concluded that his theory was not correct. Unfortunately, he did not consider the temperature dependence of the thermal conductivity of amorphous materials.

of different amorphous substances is similar. As in crystals, the heat is transported at low temperatures by phonons of long wavelength. However, even at the lowest temperatures the Casimir regime is not reached in conventional conductivity measurements. Phonons are not scattered at the sample surface, but interact resonantly with tunneling systems with the energy $E = \hbar\omega$. This resonant interaction will be considered in Sect. 9.5.4. We anticipate the result (9.75) of this discussion and write for the inverse mean free path of phonons

$$\ell^{-1} \propto \omega \tanh\left(\frac{\hbar\omega}{2k_\mathrm{B}T}\right) . \tag{9.70}$$

Resonant interaction takes place at all phonon frequencies because of the widely distributed energy splitting of the tunneling systems in amorphous solids. In order to deduce the temperature dependence of the conductivity we apply the dominant phonon approximation (see Sect. 6.2). Since the frequency $\overline{\omega}$ of the dominant phonons is roughly given by $\hbar\overline{\omega} \approx k_\mathrm{B}T$, we find $\ell^{-1} \propto \overline{\omega} \propto T$ for the inverse mean free path. In the Debye model, the phonon specific heat at low temperatures is proportional to T^3. From $\Lambda \propto C_\mathrm{Debye}\ell$, we thus finally obtain

$$\Lambda \propto T^2 , \tag{9.71}$$

in good agreement with observation. Given (9.75) it is straightforward to calculate the magnitude of Λ. Inserting the parameters deduced from acoustic measurements, very good agreement between theory and experiment has been obtained.

9.5.3 Relaxation Absorption

Acoustic and dielectric losses vanish in pure dielectric crystals below helium temperature. In contrast, relatively high losses with a characteristic temperature variation are observed in amorphous solids. At low frequencies, i.e., up to the MHz range, relaxation is the dominant loss mechanism. In Sect. 9.1.3, we have written down the relevant expressions for identical tunneling systems. In amorphous solids the distribution of the relaxation times comes into play, and we have to take the distribution function (9.60) or (9.66) into account.

In the following, we discuss acoustic and dielectric properties more or less simultaneously since both quantities are caused by the same dynamics. We consider here the *internal friction* Q^{-1} instead of the acoustic absorption coefficient α, because it is the elastic analogue of the electrical loss angle $\tan\delta$. The absorption coefficient and internal friction are connected via the relation

$$Q^{-1} \equiv \frac{\alpha v}{\omega} \equiv \frac{v}{\ell\omega} . \tag{9.72}$$

Because we restrict our discussion to low temperatures, we neglect higher-order processes and take into account only one-phonon processes given by

(9.31). In typical low-temperature acoustic or dielectric measurements the frequency is kept constant and the loss is measured as a function of temperature. In glasses, with their broad distribution of energy splittings E, the main contribution is due to systems with energy splitting $E \approx k_{\mathrm{B}}T$, because of the factor $\partial f/\partial E$ in (9.22). In addition, we have to take into account the distribution of relaxation times given by (9.31).

We replace the number N of relaxing systems in (9.28) by the distribution function (9.66) and carry out the integration over τ and E numerically. Fortunately, for two limiting cases the expected loss can also be calculated analytically, i.e., for $\omega\tau_{\mathrm{m}} \gg 1$ and $\omega\tau_{\mathrm{m}} \ll 1$, where the minimum relaxation time τ_{m} is defined by (9.65).

In the low-temperature limit the condition $\omega\tau_{\mathrm{m}} \gg 1$ holds, implying that only the 'fastest' systems with $E \approx \Delta_0$ contribute noticeably to the absorption. The temperature dependence of the absorption follows directly from the expression for the 'Debye relaxator' given by (9.28):

$$Q^{-1} \propto \tan\delta \propto \frac{\omega\tau}{1+\omega^2\tau^2} \propto \frac{1}{\omega\tau_{\mathrm{m}}} \propto \frac{AT^3}{\omega}\,. \tag{9.73}$$

In the high-temperature limit, there are always systems fulfilling the conditions $\omega\tau = 1$ and $E \approx k_{\mathrm{B}}T$. For a qualitative assessment, it is therefore sufficient to insert $\omega\tau = 1$ into (9.28), leading to the prediction of a frequency- and temperature-independent loss. Analytically, the expressions

$$Q^{-1} = \frac{\pi}{2}C_{\mathrm{a}} \qquad \text{and} \qquad \tan\delta = \frac{\pi}{2}C_{\mathrm{e}} \tag{9.74}$$

are found for internal friction and electrical loss, respectively. The constants C_{a} and C_{e} are given by $C_{\mathrm{a}} = P_0\gamma^2/\varrho v^2$ and $C_{\mathrm{e}} = P_0 p^2/3\varepsilon_0\varepsilon'$ in the acoustic and dielectric cases respectively. Since the loss is constant, this temperature range is generally referred to as the 'plateau' region. The theoretically predicted loss is shown schematically in Fig. 9.36.

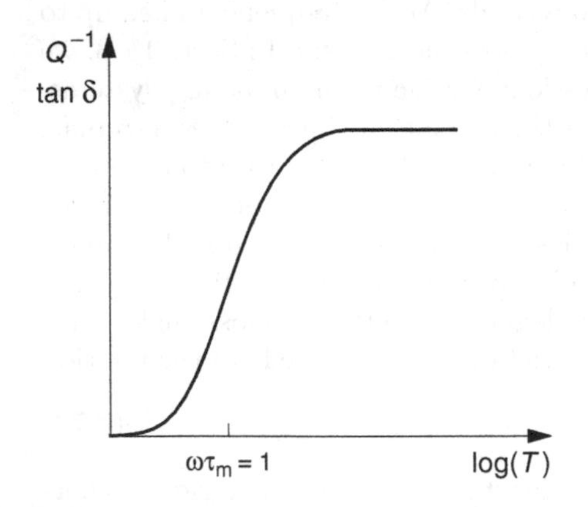

Fig. 9.36. Internal friction Q^{-1} and dielectric loss $\tan\delta$ of glasses. A steep rise at low temperatures is followed by a plateau

The temperature variation of the internal friction of vitreous silica and of the dielectric loss of the borosilicate glass BK7 is shown in Figs. 9.37a and b, respectively. The steep rise of Q^{-1} at low temperatures and the transition to a temperature-independent plateau is clearly visible. At higher temperatures ($T > 5\,\mathrm{K}$), thermally activated processes come into play, which we do not consider further here. Good agreement with the predictions of the tunneling model is also found in the dielectric case (Fig. 9.37b). Again at low temperatures a large increase is observed, ending in an almost temperature-independent plateau at higher temperatures. The great similarity in the shape of the dielectric and acoustic loss curves of these two different glasses demonstrates once again that many low-temperature properties of glasses are almost universal.

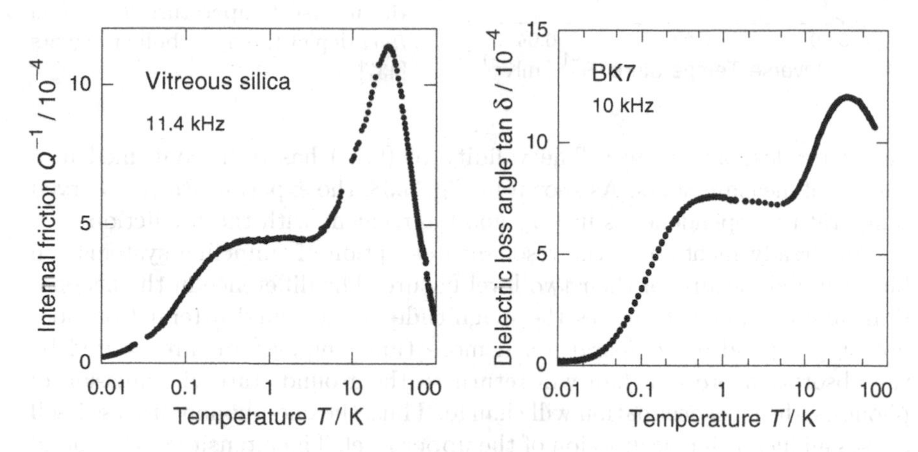

Fig. 9.37. (**a**) Internal friction of vitreous silica (a-SiO_2) [434], and (**b**) dielectric loss angle $\tan\delta$ of borosilicate glass (BK7) versus temperature [435]

9.5.4 Resonant Absorption

Besides the relaxation absorption, the resonant absorption has also been observed in amorphous solids. The corresponding mathematical expression can be obtained by replacing the quantity $Ng(E)$ in (9.33) by the distribution function $P(E, \Delta_0)$ and integrating over all values of Δ_0. One thus finds the following relation for the inverse mean free path ℓ_0^{-1} due to the resonant interaction [436]:

$$\ell_0^{-1} = C_\mathrm{a}\frac{\pi\omega}{v}\tanh\left(\frac{\hbar\omega}{2k_\mathrm{B}T}\right). \tag{9.75}$$

To obtain the corresponding expression for the dielectric loss, we replace C_a by C_e and take into account the fact that Q^{-1} is the quantity correspond-

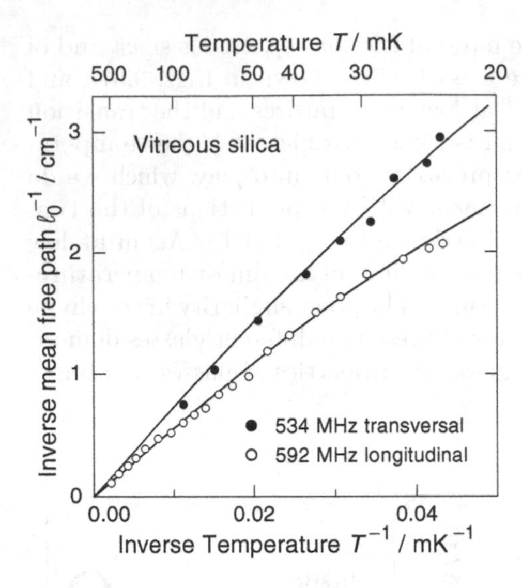

Fig. 9.38. Inverse mean free path of longitudinal and transverse sound waves in vitreous silica at 550 MHz plotted as a function of the inverse temperature T^{-1}. *Full lines* depict the hyperbolic tangents [437]

ing to the loss angle $\tan \delta$. The validity of (9.75) has been confirmed in a variety of measurements. As shown in Fig. 9.38, the experimentally observed temperature dependence is in very good agreement with the prediction.

As already mentioned, the resonant absorption of tunneling systems can be saturated because of their two-level nature. The difference in the occupation number that determines the magnitude of the sound attenuation does not only depend on temperature. If more tunneling systems are excited by the absorption process than will return to the ground state via emission of phonons, the level occupation will change. Thus, the one-phonon process itself causes an increasing occupation of the upper level. The intensity of the sound wave at which this effect becomes noticeable is called the *critical intensity* I_c. Without deriving the corresponding equation we only want to mention that the intensity dependence of the absorption can be described by [436]

$$\ell^{-1} = \ell_0^{-1} \frac{1}{\sqrt{1 + I/I_c}} . \tag{9.76}$$

Here, ℓ_0 denotes the mean free path in the limit of small intensities, and is consequently identical with the quantity given in (9.75). Experimentally, it is difficult to observe the saturation behavior because of the small critical intensity I_c. At 1 K, it is typically of the order of $10^{-7}\,\mathrm{W\,cm^{-2}}$ and decreases further upon cooling. Experiments of this type gave the first evidence for the two-level character of the low-energy excitations of glasses.

Experimental data obtained in measurements on the borosilicate glass BK7 are shown in Fig. 9.39. The decrease in attenuation with increasing intensity is clearly visible. The full line reflects the intensity dependence predicted by (9.76). The temperature dependence of the absorption and its saturation behavior are displayed in Fig. 9.40. On the left-hand side, a dielectric

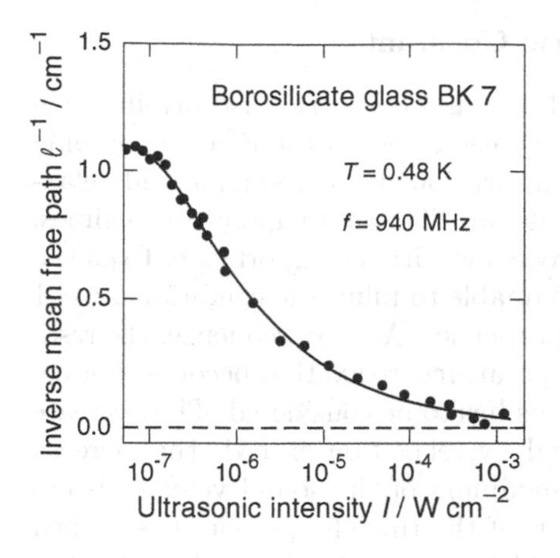

Fig. 9.39. Resonant absorption in the borosilicate glass BK7 at $T = 0.48\,\mathrm{K}$ versus acoustic intensity I. The *full line* reflects the prediction of (9.76) [438]

measurement is shown, and on the right-hand side, an acoustic measurement. The curves demonstrate once again that the observed phenomena have a common origin. The differences in the shape of the curves are due to the fact that the measurements were carried out at frequencies differing by a factor of 10. The intensity dependence at low temperatures is caused by the above-mentioned saturation effect. At higher temperatures, the relaxation absorption dominates, and this cannot be saturated. For comparison, the dielectric and acoustic absorption of quartz crystals are also shown. They are negligibly small in this temperature range!

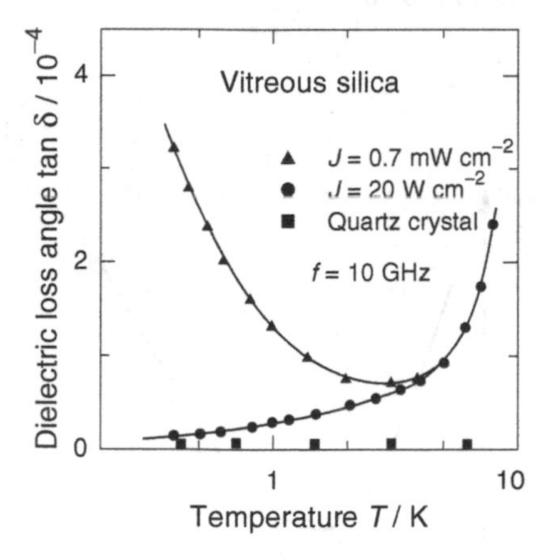

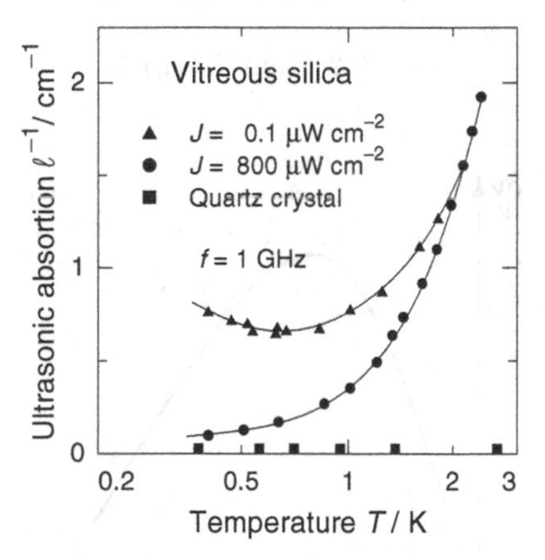

Fig. 9.40. Temperature dependence of the dielectric (**a**) and acoustic absorption (**b**) in vitreous silica at two different intensities [424, 439]

9.5.5 Sound Velocity and Dielectric Constant

The variation of the real part of the elastic and dielectric susceptibility can be deduced either directly from Bloch equations, or via the Kramers–Kronig relation, from the expression for the absorption. Both resonant and relaxation processes contribute, but dominate in different temperature regimes. At low temperatures, the relaxation process is without importance. Even the 'fastest' relaxing tunneling systems are unable to fulfill the condition $\omega\tau = 1$ and hardly contribute to the dynamic properties. As a consequence, the resonant process dominates. At higher temperatures, relaxation becomes noticeable and the sum of the two contributions has to be considered. The decrease in the velocity with temperature caused by relaxation exceeds the increase due to the resonant effect. Hence, a maximum of the sound velocity is expected around $\omega\tau_\mathrm{m} = 1$. The prediction of the tunneling model is sketched in Fig. 9.41a. Analogous arguments hold for the variation of the dielectric constant. As depicted in Fig. 9.41b, $\delta\varepsilon'$ should decrease logarithmically at low temperatures, pass through a minimum, and rise again logarithmically. The slope ratio should now be minus two to one.

Inserting the expression (9.75) for the resonant absorption into the Kramers–Kronig relation and carrying out the integration over all frequencies leads to the following prediction: At low temperature, when $\omega\tau_\mathrm{m} \gg 1$, the velocity is expected to increase according to

$$\frac{\delta v}{v} = C_\mathrm{a} \ln \frac{T}{T_0}\,, \tag{9.77}$$

where T_0 is an arbitrary reference temperature. At higher temperatures, relaxation causes a decrease of the velocity of the form

$$\frac{\delta v}{v} = -(3/2) C_\mathrm{a} \ln(T/T_0)\,.$$

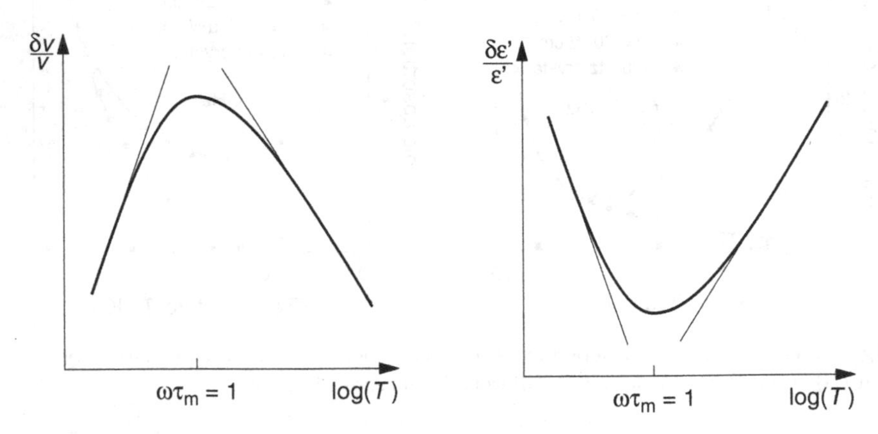

Fig. 9.41. Schematic course of (**a**) the sound velocity, and (**b**) the dielectric constant in glasses versus the logarithm of the temperature

Adding the resonant part, the ratio of the slopes of the logarithmically plotted variation of the sound velocity at low and high temperatures should be two to minus one.

Measurements on vitreous silica and on the borosilicate glass BK7 are shown in Fig. 9.42, and demonstrate the overall agreement with the prediction. However, a more careful analysis shows immediately that the behavior of glasses is more complex than expected. In contrast to the theoretical predictions based on the tunneling model, the slope ratio is rather one to minus one in the acoustic measurements. The reason for this deviation is not yet understood. Probably, the interaction between the tunneling systems plays an important role, as in crystals with a relatively high concentration of defect states. Deviations from the logarithmic decrease of the sound velocity at higher temperatures can be traced back to relaxation processes involving many phonons. As shown in Fig. 9.42b, qualitative agreement with theory is also found in dielectric measurements, but the slope ratio once again deviates significantly from the theoretical value, being minus one to one.

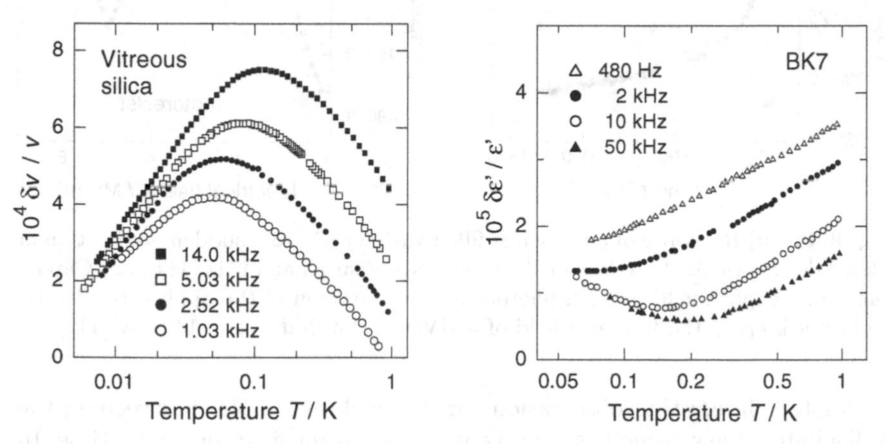

Fig. 9.42. (**a**) Relative change in the sound velocity of vitreous silica as a function of temperature [440]. (**b**) Relative change in the dielectric constant of the borosilicate glass (BK7) versus temperature [441]

There is an additional observation that we would like to mention without going into a detailed discussion. Below 100 mK, the real and imaginary parts of the elastic and dielectric susceptibility exhibit pronounced nonlinearities. The reason is that, with decreasing temperature, tunneling systems with smaller and smaller energy splitting become important. For these systems, the assumption originally made in Sect. 9.1.2, namely, that changes caused by the external fields are small compared to the energy splitting in zero field, is no longer fulfilled. Fields of moderate strength already cause changes $\delta\Delta \gtrsim E$. We do not discuss this interesting aspect further but refer the reader to the literature [442, 443].

Nonequilibrium Phenomena – Memory Effect

The temporary application of a strong dc electric field across a capacitor with a glassy dielectric results in a sudden jump of the capacitance (which is proportional to the dielectric constant of the glass in this measurement) and a subsequent slow relaxation towards its starting value [444–446]. As an example, we show in Fig. 9.43a the time evolution of the capacitance of a capacitor filled with a-SiO_x after the sudden application of a large dc electric field at 50 mK. The slow relaxation observed in this experiment is logarithmic in the time elapsed since the field was applied.

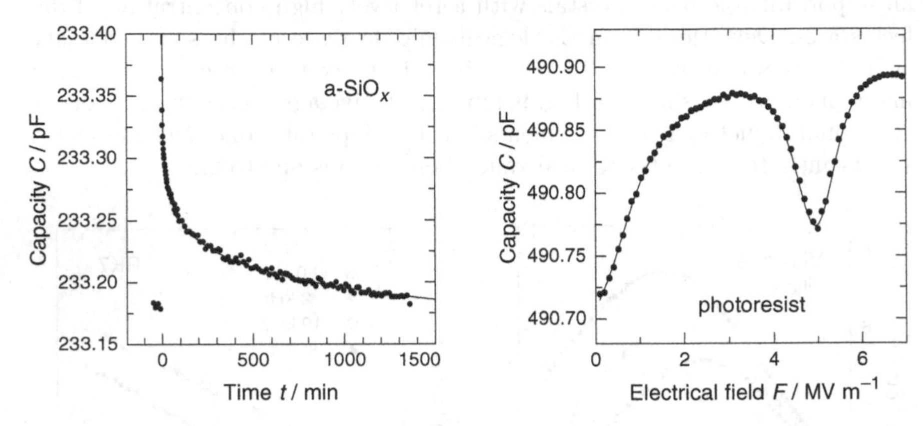

Fig. 9.43. (**a**) Response of a capacitor filled with a-SiO_x to a sudden application of a large dc electric field (10 MV m^{-1}) at 50 mK measured at 1 kHz [444]. (**b**) Capacitance of a capacitor filled with photoresist as a function of the applied dc electric field after keeping the film at a field of 5 MV m^{-1} for 2 at a $T = 140$ mK [444]

Another interesting observation can be made when slowly sweeping the dc field after the sample has been kept in a constant field for a long time. In this case, the capacitance shows a minimum at the field at which the sample was kept prior to the sweep experiment. Figure 9.42b shows the response to such a sweep of a capacitor filled with photoresist (an amorphous polymer) that initially had been cooled down without an electric field and then kept at a field of $F = 5\,\mathrm{MV\,m^{-1}}$ for about 2 h. Clearly, a minimum of the capacitance was found at zero field, at which the sample was kept for a long time before the sweep. In addition, a second minimum is seen at the field of $F = 5\,\mathrm{MV\,m^{-1}}$, at which the sample was kept for 2 h. Obviously, the dielectric constant of the photoresist 'remembers' its history. This is referred to as a *memory effect*. Further investigations have shown that the new minimum grows logarithmically with the length of time for which the new dc field was applied, and at the same time the minimum at zero field disappears slowly.

The theoretical explanation of the nonequilibrium phenomena presented in this section is rather involved [447] and will only be briefly summarized here. The theory is based on the assumption that in the glass network statistically a small fraction of tunneling systems are close neighbors forming pairs with coupling energies larger than the energy splitting of both, if considered uncoupled. This causes a small change of the flat distribution function and is known as the formation of a *dipole gap* in a system of interacting dipoles. In equilibrium, these pairs do not contribute to the dielectric constant. However, a strong dc electric field leads to a shift in the distribution of the asymmetry energies proportional to pF and thus to a nonequilibrium situation in which the pairs contribute to the dielectric response. This explains the sudden increase of the dielectric constant as soon as a dc electric field is applied. This change is followed by a slow relaxation to equilibrium. The memory effect occurs because under the influence of a dc field a new dipole gap is formed and the original dipole gap vanishes.

9.6 Metallic Glasses

In metallic glasses, tunneling systems not only interact with phonons but also with free electrons. Nevertheless, the overall behavior does not differ fundamentally from that of dielectric glasses (see Fig. 9.44a). The relaxation rate due to phonons is proportional to E^3, but linear in the electronic case. This difference is caused by the different energy dependence of the density of states $D(E)$ of phonons and electrons. While for phonons $D(E) \propto E^2$, the density of states of interacting electrons is virtually constant, since only electrons within the narrow energy interval $(E_\mathrm{F} \pm k_\mathrm{B}T)$ at the Fermi level are able to interact with tunneling systems. According to (9.73), the absorption

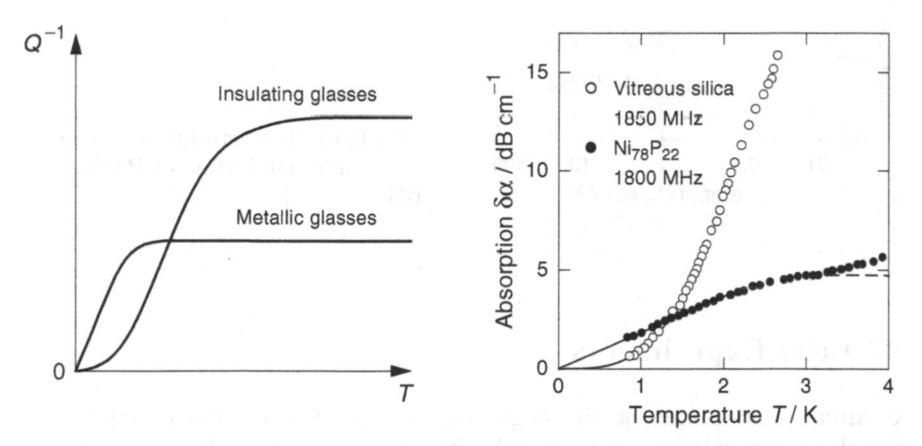

Fig. 9.44. (**a**) Schematic variation of the acoustic loss in metallic and dielectric glasses. Arbitrary values are chosen for the plateau heights and the minimum relaxation time. (**b**) Acoustic relaxation absorption of vitreous silica in comparison with the absorption of the metallic glass $Ni_{78}P_{22}$ [448]

at low temperatures is proportional to the maximum relaxation rate τ_m^{-1}. Therefore, the absorption in metallic glasses is expected to rise linearly with temperature. For comparison, the acoustic absorption of vitreous silica and $Ni_{78}P_{22}$ are shown in Fig. 9.44b. Clearly, the absorption rises linearly with temperature in $Ni_{78}P_{22}$, but in proportion to T^3 in vitreous silica.

Because of the high relaxation rate mediated by free electrons, the condition $\omega\tau_m = 1$ is already fulfilled at much lower temperatures than in dielectric glasses. Although the frequency of measurement of 2 GHz is relatively high, a plateau seems to be reached in the metallic glass at about 2.5 K, whereas in vitreous silica the absorption is still rising in this temperature range. As in most metallic glasses, the attenuation of $Ni_{78}P_{22}$ in the 'plateau range' is considerably smaller than in vitreous silica.

A further example is shown in Fig. 9.45, where the internal friction of PdSiCu at 1 kHz is plotted. The loss depends only weakly on temperature and can therefore be considered to be the 'plateau range'. Because of the short relaxation times the plateau extends to the lowest temperatures attained in this experiment. The maximum seen at 12 K can be attributed to thermally activated processes. Finally, we should mention that, at temperatures below 100 mK, significant nonlinear effects are also observed in metallic glasses.

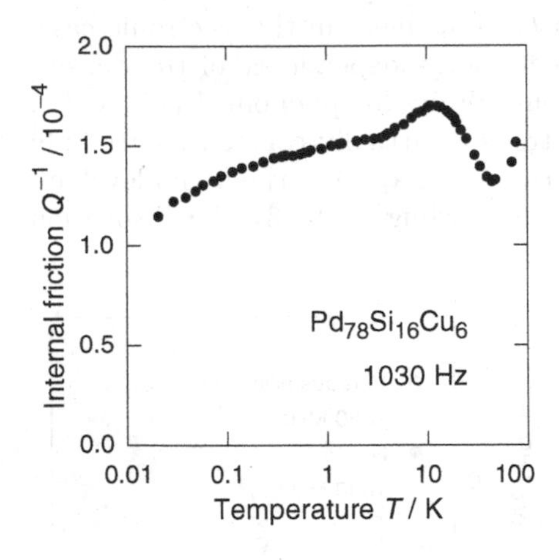

Fig. 9.45. Temperature dependence of the internal friction of PdSiCu at 1030 Hz [449]

9.7 Echo Experiments

As already mentioned at the beginning of this chapter, the dynamics of tunneling systems can be described with the help of the Bloch equations [397, 436]. Obviously, a formal analogy exists between particles with spin 1/2 in a constant magnetic field and two-level systems. We do not work out this aspect but discuss the behavior of tunneling systems at very low temperatures in a semiquantitative manner. For further discussion we refer the reader to the literature [436, 450].

With decreasing temperature, the interaction between the tunneling systems and the heat bath becomes weaker and weaker. Finally, at very low temperatures, the tunneling systems can be considered to be quasi-isolated and coherent effects come into play that are analogous to those known from nuclear magnetic resonance.

The application of an alternating acoustic or electric field causes a forced oscillation of the tunneling systems and thus the generation of an oscillating elastic or electric polarization. After switching the field off, the systems continue to oscillate with their eigenfrequency E/h. Because of the distribution of the energy splitting, tunneling systems exist whose phase $Et/\hbar$ develops faster than the mean value, and of course, there are also systems whose phase lags behind. Therefore, the phase coherence of the tunneling systems is lost as time goes on, and the macroscopic polarization vanishes ('free induction decay'). As in nuclear magnetic resonance, the polarization can be turned back for a short time. In such an experiment, a microwave pulse is applied with a duration chosen in such a way that the resonating systems are brought into the intermediate state. After switching off the applied field they quickly lose their common phase. After the time t_{12}, a second pulse is applied with such a duration that the phase of the wave function of the tunneling systems is changed by 180°. As a consequence the development of the phase in time changes its sign, i.e., time is reversed. After the time t_{12} has elapsed a second time, all systems will be in phase again for a short while. A detectable macroscopic polarization thus develops. This additional pulse is called a *spontaneous echo* or *two-pulse echo*.

As an illustration, the sequence of applied pulses and also the induced macroscopic polarization of the sample is drawn schematically in Fig. 9.46. As indicated, the echo appears after the time $2t_{12}$. Figure 9.47 exploits the analogy with magnetic resonance. The temporal change in polarization is

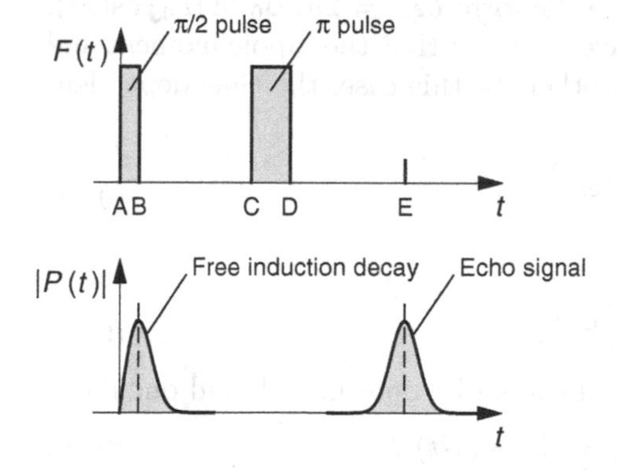

Fig. 9.46. Pulse sequence for the generation of a spontaneous echo (*top*), and the corresponding dielectric polarization (*bottom*). *Capital letters* refer to particular times for which the polarization vector is depicted in Fig. 9.47

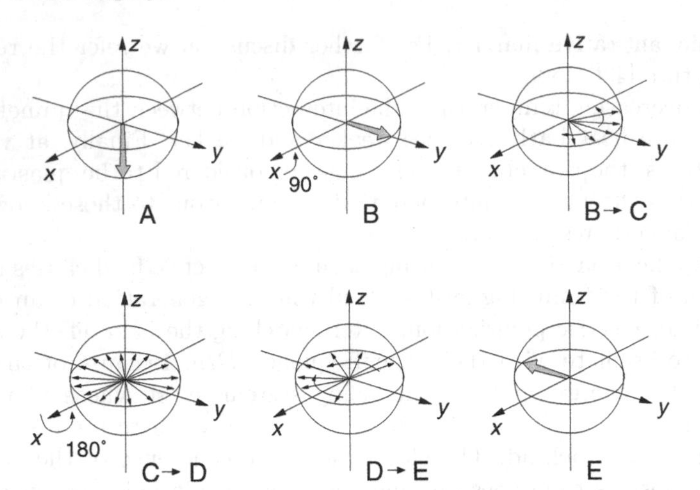

Fig. 9.47. Schematic representation of various stages in the development of a spontaneous echo. The z-component represents the occupation of the energy levels. The x- and y-components reflect the phase of the systems with respect to the applied field. It is assumed that, at the beginning, all tunneling systems are in the ground state. The capital letters refer to particular points in time as indicated in Fig. 9.46

reflected by the motion of the 'pseudopolarization vector' in a coordinate system rotating with the frequency of the applied electric field.

We now discuss the occurrence of the dielectric echo in a more formal way. For this purpose, we consider two-level systems with the energy levels E_1 and E_2, and the corresponding eigenfunctions ϕ_1 and ϕ_2. For the energy difference between the two levels, we write $E = E_2 - E_1 = \hbar\omega_2 - \hbar\omega_1 = \hbar\omega$. The alternating field $F = F_0[\exp(\mathrm{i}\omega t) + \exp(-\mathrm{i}\omega t)] = 2F_0 \cos(\omega t)$ acts on the two-level systems and causes an oscillation of the resonating systems. The off-diagonal elements of the perturbation Hamiltonian (9.12) give rise to transitions between the two levels and have the form $\delta\Delta_0 = 2p\,(\Delta_0/E)F_0 \cos(\omega t)$. To avoid awkward expressions we have assumed that the dipole moment and the electric field are parallel to each other. In this case, the time-dependent Schrödinger equation has the form

$$\mathrm{i}\hbar\frac{\partial\psi}{\partial t} = \left[H_0 + p\frac{\Delta_0}{E}F_0\left(\mathrm{e}^{\mathrm{i}\omega t} + \mathrm{e}^{-\mathrm{i}\omega t}\right)\right]\psi\,. \tag{9.78}$$

With the ansatz

$$\Psi(t) = a_1(t)\,\phi_1\mathrm{e}^{-\mathrm{i}\omega_1 t} + a_2(t)\,\phi_2\mathrm{e}^{-\mathrm{i}\omega_2 t}\,, \tag{9.79}$$

the time-dependent coefficients a_1 and a_2 can be determined, and one finds

$$a_1 = \sin(\Omega t) \qquad \text{and} \qquad a_2 = -\mathrm{i}\,\cos(\Omega t)\,. \tag{9.80}$$

Here, Ω is the so-called *Rabi frequency* given by

$$\Omega = \frac{\mu_{12} F_0}{\hbar} = \frac{p F_0}{\hbar} \frac{\Delta_0}{E} , \tag{9.81}$$

where the quantity μ_{12} represents the expectation value of the off-diagonal element of the perturbation operator, i.e., $\mu_{12} = \mu_{21} = \langle 2|\mathcal{H}_S|1\rangle$.

The coefficients a_1 and a_2 describe the transition between the two energy levels giving rise to the oscillating dipole moments $\langle p\rangle$:

$$\langle p\rangle = a_1^* a_2\, \mu_{12}\, \mathrm{e}^{-\mathrm{i}\omega t} + a_1 a_2^*\, \mu_{21}\, \mathrm{e}^{\mathrm{i}\omega t} = -\mu_{12} \sin(\Omega t)\, \sin(\omega t) . \tag{9.82}$$

A macroscopic polarization is generated parallel to the applied field, since all tunneling systems oscillate in phase.

We now consider the temporal development of the polarization leading to the generation of the echo. As shown in Fig. 9.46, we assume that a short pulse with *pulse area* $\Omega t = \pi/2$ is applied at time $t = t_0$. This pulse is usually called a '$\pi/2$ pulse'. At the end of this pulse, the oscillating dipole moment is given by

$$\langle p\rangle_0 = -\mu_{12} \sin(\pi/2) \sin(\omega t_0) = -\mu_{12} \sin(\omega t_0) . \tag{9.83}$$

After the time t_{12} has elapsed between the two pulses, the dipole moment exhibits the value $\langle p\rangle_{12} = -\mu_{12} \sin(\pi/2) \sin[\omega(t_0 + t_{12})]$. The application of the 'π pulse' generates the polarization

$$\begin{aligned}\langle p\rangle_{12} &= -\mu_{12} \sin(\pi/2 + \pi) \sin[\omega(t_0 + t_{12})] = \mu_{12} \sin[\omega(t_0 + t_{12})] \\ &= -\mu_{12} \sin[\omega(-t_0 - t_{12})] .\end{aligned} \tag{9.84}$$

The polarization has suffered a phase jump by 180°, and the further temporal development follows the relation

$$\langle p\rangle = -\mu_{12} \sin[\omega(-t_0 - t_{12} + t)] . \tag{9.85}$$

Astonishingly, after another time interval t_{12} has elapsed, a macroscopic polarization develops again. Independently of the separation of the two pulses, the polarization $-\langle p\rangle_0$ appears after the time $2t_{12}$. This consideration is also correct if the applied frequency does not coincide exactly with the resonance frequency of the tunneling systems, i.e., if $\omega \approx \omega_2 - \omega_1$ holds.

Such coherent effects are not only observed in glasses but also in crystals with higher concentrations of point defects. As an example, for such a measurement, we show in Fig. 9.48 the signal observed in a dielectric two-pulse echo experiment carried out on a KBr crystal doped with KCN. The applied electric pulse had a frequency of 800 MHz, the width of the pulses was 250 ns for the first and 500 ns for the second, and the pulse separation was 5 μs. Because of the saturation of the amplifier, the signals caused by the applied pulses are cut off at a certain amplitude and broadened. In fact, an echo developed 10 μs after the application of the first pulse. It should be mentioned that the echo observed in this specific experiment was not due to single CN^- ions. At the relatively high concentration of 5%, the tunneling systems form a large number of strongly coupled pairs. The two lower levels of these pairs have a small energy splitting and give rise to the observed echo.

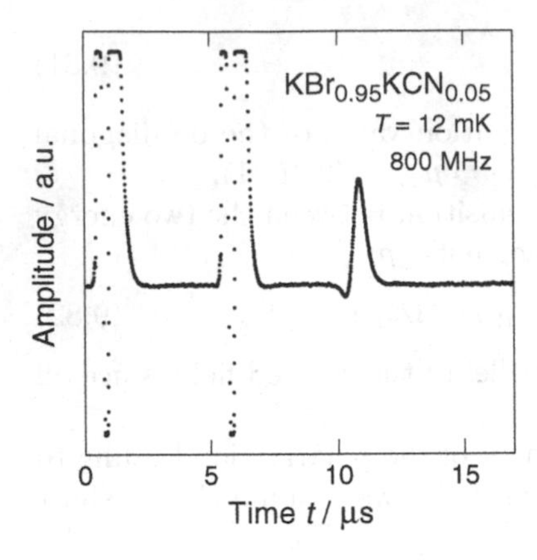

Fig. 9.48. Dielectric two-pulse echo in a KBr crystal doped with KCN [451]

The equations developed above are only correct as long as no processes occur that destroy the phase of the tunneling systems during the time elapsing between the first pulse and the echo. At finite temperatures, the echo amplitude decays with time. The characteristic time is called the *phase memory time* τ_2, or more often the *transverse relaxation time* τ_2. The latter name is commonly used in magnetic resonance experiments. We should point out here that the relaxation time τ_2 also determines the width of the Lorentz function (9.32) introduced in Sect. 9.1.5.

Of course, uncorrelated thermal transitions between the upper and lower levels of the tunneling systems destroy the phase coherence. If this process were dominant, the corresponding transverse relaxation time τ_2 would be comparable to the relaxation time caused by one-phonon processes. Actually, it was found that in the materials discussed here the phase coherence is lost much faster than expected from the coupling to the phonon bath. Therefore, another process must exist that changes the phase of the tunneling systems. The most likely mechanism is the elastic interaction *between* the tunneling systems. To make this mechanism plausible, we consider a tunneling system (A) with the energy splitting E, in resonance with the external field, and a second tunneling system (B) in its neighborhood with the energy splitting $E \lesssim k_{\mathrm{B}}T$, which undergoes thermal transitions between the ground state and the excited state, and vice versa. Each transition will change the asymmetry energy Δ of the system (A) by a small amount through the change in the local strain field caused by the system (B). Since the temporal development of the phase is intimately connected with the energy splitting, the phase of the system (A) will develop differently from that of a resonating tunneling system that is undisturbed. After the time $2t_{12}$, the phase will differ by $2t_{12}\,\delta E/\hbar$, and the contribution of the system (A) to the echo amplitude will be reduced. This means that thermal fluctuations of neighboring tunneling systems reduce the contribution of the resonating systems to the polarization echo. This process is called *spectral diffusion* and is supposed to limit the phase coherence of the tunneling systems in glasses.

Recently, it was found that in most materials the amplitude of the spontaneous echo depends strongly on magnetic fields, although the samples investigated in these experiments were nominally nonmagnetic. Moreover, experiments with different delay times t_{12} showed that certain features in the plot of the echo amplitude as a function of the magnetic field B scale with the product $B\,t_{12}$. Surprisingly, the quantity $h/(B\,t_{12})$, which has the dimensions of a magnetic moment, shows values of the order of a nuclear magnetic moment. At first glance, an influence of nuclear spins on a microsecond timescale seems very unlikely because of the extremely long nuclear spin-lattice relaxation times in insulating materials. However, it is conceivable that the coupling of the tunneling motion to the nuclear degrees of freedom is brought about by the nuclear quadrupole moment experiencing the electric field gradients in the two localized states of the double-well potential. This coupling would give rise to a quadrupole splitting of the energy levels of the tunneling systems. The applied magnetic field would then lead to an additional Zeeman splitting of the nuclear levels. On the basis of this picture, a detailed model has been worked out to explain the magnetic field effects in polarization echo experiments [452].

This idea is strongly supported by the observation that no magnetic field effects occur in vitreous silica. This glass consists of silicon and oxygen atoms, the nuclei of which do not carry a magnetic spin. Recently, the magnetic-field dependence of the echo amplitude has been investigated in glassy samples of ordinary glycerol $C_3H_8O_3$ and deuterated glycerol $C_3D_8O_3$. Carbon and oxygen nuclei carry no nuclear spin. Deuterons with the nuclear spin $I = 1$ possess a magnetic moment and a quadrupole moment, while protons with $I = 1/2$ possess a magnetic moment but no quadrupole moment. The echo amplitude is plotted as a function of the applied magnetic field for both samples in Fig. 9.49. No influence of the magnetic field was detected in the case of natural glycerol,[8] but a dramatic field dependence was found for the deuterated sample. Without going into details we would like to mention that, for certain experimental settings, the decay of the echo amplitude with t_{12} exhibits an oscillatory behavior ('quantum beating').

The chemical binding of deuterium in glycerol has axial symmetry. Due to the quadrupole moment of the deuteron, the electric-field gradient associated with chemical binding causes a splitting of the nuclear spin states into two levels with $m_I = 0$, and $m_I \pm 1$. Magnetic fields lift the degeneracy of the levels with $m_I \pm 1$ because of the Zeeman splitting. At high magnetic fields, the Zeeman splitting exceeds the quadrupole splitting and the nuclear levels become approximately equidistant.

If a deuteron is part of a tunneling system, its nuclear levels cause a fine splitting of the tunneling levels. In general, the electric-field gradients in the

[8] The origin of the small change around $B = 0$ is not yet understood.

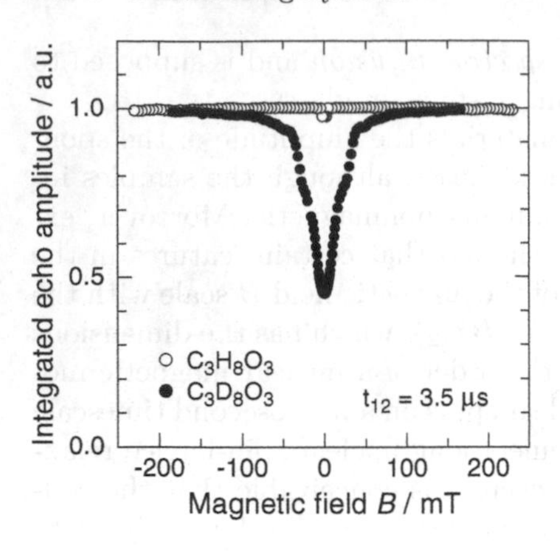

Fig. 9.49. Amplitude of the spontaneous echo generated in fully deuterated glycerol and ordinary glycerol as a function of the magnetic field. Data were taken at 13.4 mK [453]

two wells of a tunneling potential do not point in the same direction. Thus, the appropriate basis for describing the nuclear spin states in a tunneling system is a mixture of the pure spin states. Even in a zero magnetic field, the intermediate state of such a system is a superposition of several states, resulting in a reduction of the echo amplitude in comparison to that of a genuine two-level system. In large magnetic fields, the quantization axis of the nuclear spins is determined by the direction of the magnetic field. Therefore, no mixing of the different nuclear spin states occurs, and the echo amplitude is identical with that of a simple two-level system. Consequently, the echo amplitude is expected to increase with magnetic field and to saturate at high magnetic fields. This is exactly what is seen in Fig. 9.49 for the sample made of deuterated glycerol.

The exact variation of the echo amplitude with the magnetic field depends on the temporal development of the intermediate state and consequently on the delay time t_{12}. Therefore, the shape of the curve reflecting the echo amplitude as a function of the magnetic field can vary considerably and detailed information on the tunneling motion can be extracted from such experiments.

Finally, we mention that the dielectric function of most glasses also reveals the significant influence of the magnetic field. It is tempting to try to explain this phenomenon on the same basis. This would mean that the low-frequency dielectric behavior would also be significantly influenced by the nuclear properties. However, the details of the response are not yet understood.

Exercises

9.1 Calculate the tunnel splitting of a particle of 100 amu in a 'harmonic' double-well potential (see Fig. 9.1) with the parameters $d = 0.1\,\text{Å}$ and $V/k_\mathrm{B} = 200\,\mathrm{K}$. The ground-state energy is assumed to be $\hbar\Omega/k_\mathrm{B} = 20\,\mathrm{K}$.

9.2 Tunneling systems in glasses respond to electric and elastic fields. Typical coupling parameters are $p \approx 1\,\mathrm{debye}$ ($3.2 \times 10^{-30}\,\mathrm{A\,s\,m}$) for the electric dipole moment and $\gamma \approx 1\,\mathrm{eV}$ for the deformation potential. In which temperature range is the assumption $\delta E < k_\mathrm{B}T$ fulfilled for an electric field $F_0 = 10^4\,\mathrm{V\,m^{-1}}$ and a strain field $\tilde{e}_0 = 10^{-7}$?

9.3 Figure 9.35 shows the thermal conductivity of vitreous silica. Show that in the plateau range the mean free path of the thermal phonons is comparable with their wavelength. The Debye temperature of this part of the specific heat is $\theta = 460\,\mathrm{K}$.

9.4 The low-temperature internal friction of amorphous dielectrics rises steeply at very low temperatures and finally becomes temperature independent ('plateau'). The characteristic transition temperature T^* is defined by $\omega\tau_\mathrm{m} = 1$. Show that the relation $T^* \propto \omega^{1/3}$ holds.

10 Superconductivity

Superconductivity was discovered in 1911 by *Kamerlingh Onnes* while studying the resistivity of pure metals [454]. During the investigation of a mercury sample, he noticed that below 4.2 K the resistance suddenly dropped.[1] Later, the same observation was made with other substances, though the drop occurred at different temperatures. In further experiments, many interesting discoveries were made and phenomenological theories by *Gorter* and *Casimir* [455], by *F.* and *H. London* [456], and by *Ginzburg* and *Landau* [457] contributed to a deeper understanding of this phenomenon. However, it took half a century until the mystery of the microscopic origin of superconductivity was resolved. In 1957, *Bardeen, Cooper*, and *Schrieffer* succeeded in developing the BCS theory, named after these three scientists [458]. A new chapter in the history of superconductivity was opened in 1986 by *Bednorz* and *Müller*, when they observed a transition temperature above 30 K in the compound La_2SrCuO_4 [459]. Later, transition temperatures up to 135 K were found in these so-called *high-T_c superconductors*.

In this chapter, we first consider properties that are fundamental to superconductivity. Subsequently, we discuss various theoretical concepts emphasizing BCS theory. Later, we report on experiments that illuminate the properties of the macroscopic wave function that can describe the superconducting state. At the end of this chapter, we report on superconductors with unusual properties. In this context, we discuss the influence of magnetism and consider *unconventional superconductors*.

10.1 Experimental Observations

The historic measurement of superconductivity in mercury is shown in Fig. 10.1. As in many other metals, the resistance of mercury decreased steadily upon cooling, but dropped suddenly at 4.2 K, and became undetectably small. So the important question arose whether the resistance below the critical temperature T_c is indeed zero or only very small.

Extremely small values of the resistance can be measured via the decay of a current generated in a closed loop. As shown schematically in Fig. 10.2,

[1] This experiment was carried out by Kamerlingh Onnes and his coworker *van Holst*.

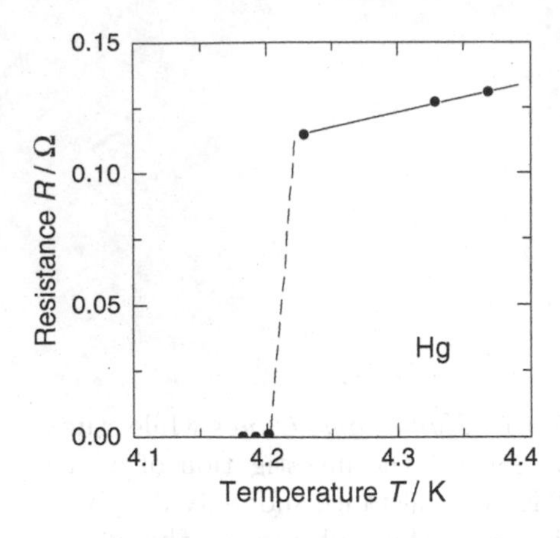

Fig. 10.1. Discovery of superconductivity. Historical data on the temperature dependence of the electrical resistance of a thread of mercury [454]

above the transition temperature a ring made of a superconducting material can be exposed to a magnetic field, e.g., caused by a bar magnet. As described in more detail in Sect. 10.2, a shielding current will then be generated in the superconducting ring below T_c that expels the field from the superconductor. An understanding of this phenomenon is, however, not needed here. Removing the magnet, a current in the ring is induced proportional to magnetic field. Its field can be used to investigate the decay of the current given by $I(t) = I_0 e^{-Rt/L}$, where L is the inductance, and R the resistance of the ring. From measurements of this type, it was concluded that the resistance drops by at least 16 orders of magnitude while passing through the superconducting transition [460]. This means that the current in the superconducting ring could easily be detected a million years after its generation. This result demonstrates that in the superconducting state the resistance is not only very small, but in fact zero.

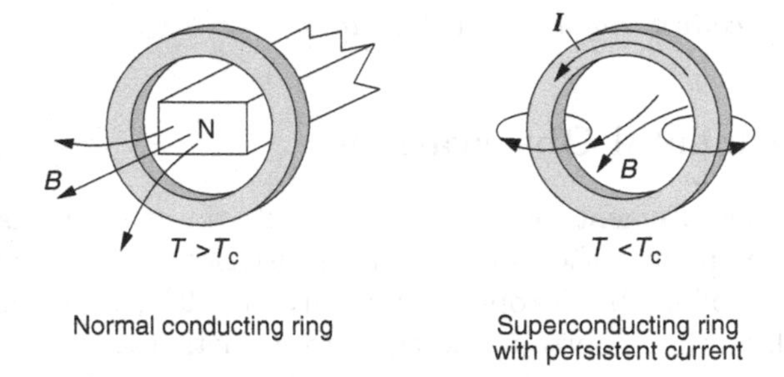

Fig. 10.2. Generation of a persistent current in a superconducting ring. The bar magnet is removed after cooling the specimen below the transition temperature

10.1.1 Transition Temperature

As already indicated, the transition temperature depends on the nature of the superconductor and varies over a wide range. In Table 10.1, the transition temperature is listed of those elements that become superconducting under normal pressure. In addition, the transition temperatures of some superconducting compounds and high-T_c superconductors is given. At first glance, it seems that there is hardly any relation between the transition temperature and other properties. Nevertheless, some general statements are possible and give the first hints on the nature of superconductivity:

- The structural order of the superconducting materials is not important, i.e., pure crystals, alloys, and amorphous metals become superconducting. Pressure may enable superconductivity. Even nonmetallic solids like silicon may become superconducting under pressure.
- Many metallic elements, but not all, become superconducting below 10 K. In intermetallic compounds, transition temperatures even above 20 K are

Table 10.1. Transition temperature of elements that become superconducting under atmospheric pressure. In addition, the transition temperature of selected compounds and high-T_c superconductors are given (from different authors)

Element	T_c (K)	Element	T_c (K)	Element	T_c (K)
Al	1.19	Nb	9.2	Tc	7.8
Be	0.026	Np	0.075	Th	1.37
Cd	0.55	Os	0.65	Ti	0.39
Ga	1.09	Pa	1.3	Tl	2.39
Hf	0.13	Pb	7.2	U	0.2
Hg	4.15	Re	1.7	V	5.3
In	3.40	Rh	0.0003	W	0.012
Ir	0.14	Ru	0.5	Zn	0.9
La	4.8	Sn	3.75	Zr	0.55
Mo	0.92	Ta	4.39		
Compound	**T_c (K)**	**Compound**	**T_c (K)**	**Compound**	**T_c (K)**
Nb_3Sn	18.1	MgB_2	39	UPt_3	0.5
Nb_3Ge	23.2	$PbMo_6S_8$	15	UPd_2Al_3	2
Cs_3C_{60}	19	YPd_2B_2C	23	$(TMTSF)_2ClO_4$	1.2
Cs_3C_{60}	40	$HoNi_2B_2C$	7.5	$(ET)_2Cu[Ni(CN)_2]Br$	11.5

High-T_c superconductor	T_c (K)	High-T_c superconductor	T_c (K)
$La_{1.83}Sr_{0.17}CuO_4$	38	$Tl_2Ba_2Ca_2Cu_3O_{10+x}$	125
$YBa_2Cu_3O_{6+x}$	93	$HgBa2Ca_2Cu_3O_{8+x}$	135
$Bi_2Sr_2Ca_2Cu_3O_{10+x}$	107	$Hg_{0.8}Tl_{0.2}Ba_2Ca_2Cu_3O_{8.33}$	134

observed. Exceptionally high values were found for MgB_2 ($T_c = 39\,K$) and Cs_3C_{60} ($T_c = 40\,K$ under pressure). Most interesting are the extremely high transition temperatures up to 135 K in the so-called high-T_c superconductors.

- 'Good' metals with a high conductivity at room temperature like Ag, Au, Cu, Na, or K stay normal conducting down to the lowest temperatures. Similarly, ferromagnets like Fe or Ni do not exhibit superconductivity.
- In general, impurities do not change the behavior of 'conventional' superconductors qualitatively. However, magnetic impurities turn out to be an exception, they may even suppress superconductivity.
- The transition from the normal to the superconducting state is accompanied by a phase transition without structural changes.
- Compounds may show superconductivity although their constituents stay normal down to the lowest temperatures.

Until now it is impossible to predict the transition temperature, even if the structure and characteristic properties of the materials are well known. However, there are some empirical rules, two of which we mention here briefly. It has been known for a long time that the atomic volume plays an important role. In Fig. 10.3, the atomic volume of the elements is plotted as a function of their atomic number. Normal conducting and superconducting elements are labelled by *open* and *full circles*, respectively. Obviously, large atomic volumes are unfavorable for the development of superconductivity. This conclusion

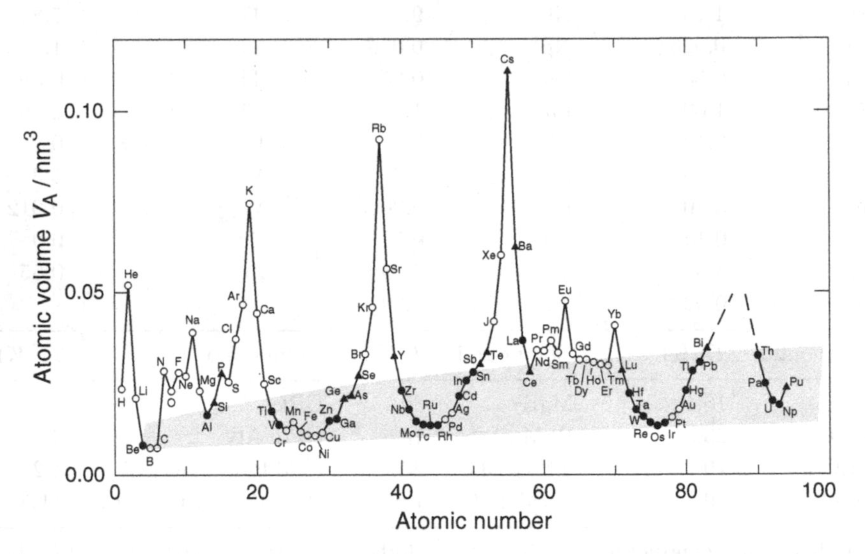

Fig. 10.3. Atomic volume of the elements versus atomic number. *Full circles* represent superconductors, and *open circles* normal conductors. Elements depicted by *triangles* are superconducting under pressure. The *shaded* area shows the region within which superconducting elements are overwhelmingly found

is supported by experiments where pressure is applied. Normal-conducting elements like Cs or Ba become superconducting under high pressure. In addition, elements like Si, Ge or Bi with a metallic high pressure phase exhibit superconductivity under pressure and are represented by *triangles*.

Another empirical relation, the so-called *Matthias rule*, tells us that the mean number N_V of valence electrons has a strong influence on the transition temperature [461], all delocalized electrons being considered as valence electrons. In addition, we have to distinguish between simple metals and transitions metals. In simple metals, the transition temperature increases steadily with the number of valence electrons. In transition metals, well-pronounced maxima are found. An example of the strong influence of N_V is shown in Fig. 10.4 where the number of valence electron is continuously varied by alloying.

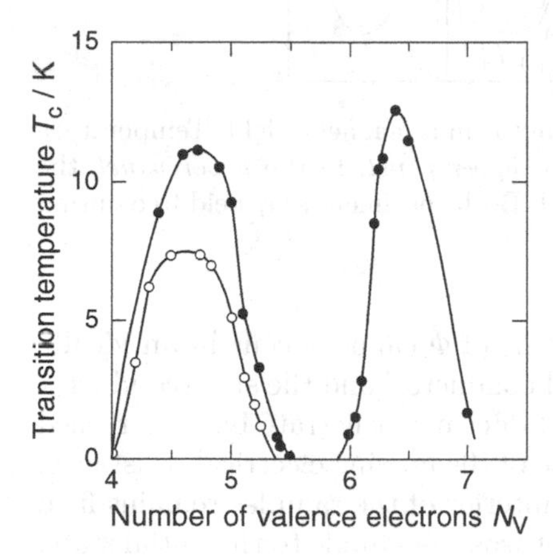

Fig. 10.4. Transition temperature T_c of two series of alloys of transition metals (• Zr-Nb-Mo-Re, ○ Ti-V-Cr) versus the mean number N_V of valence electrons [462]

10.1.2 Meissner–Ochsenfeld Effect

In 1933, *Meissner* and *Ochsenfeld* discovered that superconductors are not only characterized by the absence of resistance but also by their *ideal diamagnetism* [463]. The different behavior of a superconductor and a hypothetical 'ideal conductor' that loses its resistance at T_c may be demonstrated by the thought experiments depicted in Figs. 10.5 and 10.6. The upper panels of these figures show the variation of temperature and magnetic field assumed in the thought experiments. In the lower panels, the (elliptical) samples are sketched together with the spatial variation of the magnetic field. In Fig. 10.5, it is assumed that both samples are cooled in zero field to a temperature below T_c before a magnetic field is applied. According to Faraday's law $\oint \boldsymbol{\mathcal{E}} \cdot \mathrm{d}\boldsymbol{s} = -\partial\Phi/\partial t$, a temporally varying magnetic flux Φ enclosed by an arbitrarily chosen loop causes an electric field $\boldsymbol{\mathcal{E}}$. However, in an 'ideal conductor' and in superconductors the resistance is zero and the electric field $\boldsymbol{\mathcal{E}}$

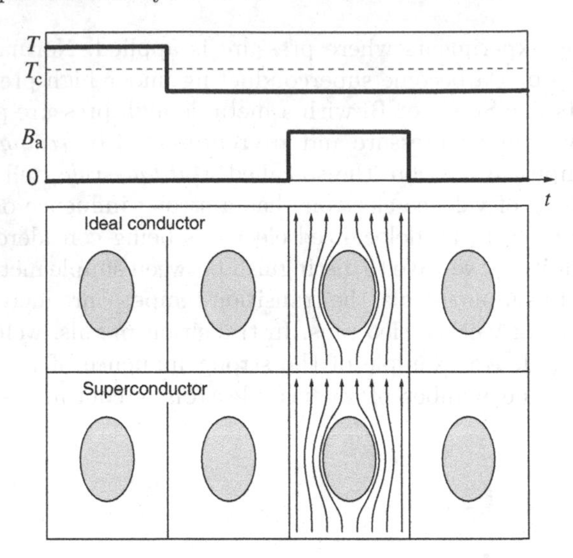

Fig. 10.5. 'Ideal conductor' and superconductor in a magnetic field. Temperature and magnetic field changes are shown in the *upper panel*. In the *lower panel,* the magnetic field next to the samples is sketched. Both specimens stay field free during the whole experiment

vanishes. Consequently, temporal variations of Φ cannot occur in an ideally conducting material. Therefore, the 'ideal conductor' and the superconductor react in the same way: The magnetic field does not penetrate but is screened by currents at the sample surface. Because of the missing electrical resistance, shielding currents do not decay, and the interior of the samples remains field free for all times. Switching off the field returns the sample to the initial state. Obviously, the experiment described does not distinguish between an 'ideal conductor' and a superconductor.

In the second 'experiment', the magnetic field is applied before crossing the transition temperature T_c. As shown in Fig. 10.6, the field penetrates both specimens since shielding currents decay in materials with finite electrical resistance. Passing through T_c does not change the situation in the 'ideal conductor'. Switching off the external field induces currents that prevent changes of the magnetic flux inside the sample, as explained above. As a consequence the 'ideal conductor' now carries a magnetic moment. The behavior of superconductors is completely different: When passing through the transition temperature the magnetic field is expelled from the sample. This means that not only field changes are suppressed, as predicted by Faraday's law, but the field vanishes inside the sample. Energy is needed to expel the magnetic field that is provided by the exothermic superconducting transition. Switching off the external field returns the superconducting sample to the initial state. Depending on the experimental path, two different final states are

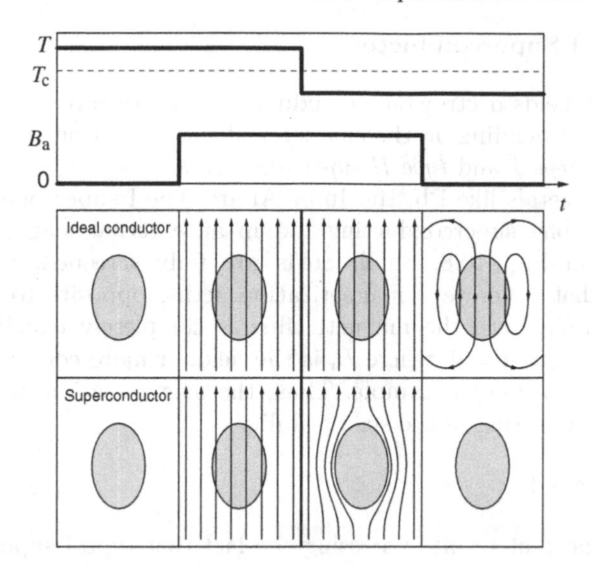

Fig. 10.6. 'Ideal conductor' and superconductor in a magnetic field. The temporal variation of temperature and applied magnetic field is shown in the *upper panel*. The magnetic field is applied above T_c and penetrates the samples. As shown in the *lower panel,* the magnetic flux becomes trapped in the 'ideal conductor' at T_c while it is expelled from the superconductor. After switching off the external field the final states of the two samples differ

expected for the 'ideal conductor' while the final state of the superconductor is independent of the experimental path. The expulsion of the magnetic flux from the superconductor is known as the *Meissner–Ochsenfeld effect.*

The situation is different for superconducting samples that are not singly connected. As shown in Fig. 10.7 for a ring, the flux cannot be expelled from the interior of the annular sample since the magnetic field lines would have to cross the superconductor. Thus, the flux through the ring becomes frozen, while the superconducting material of the ring is field free below T_c.

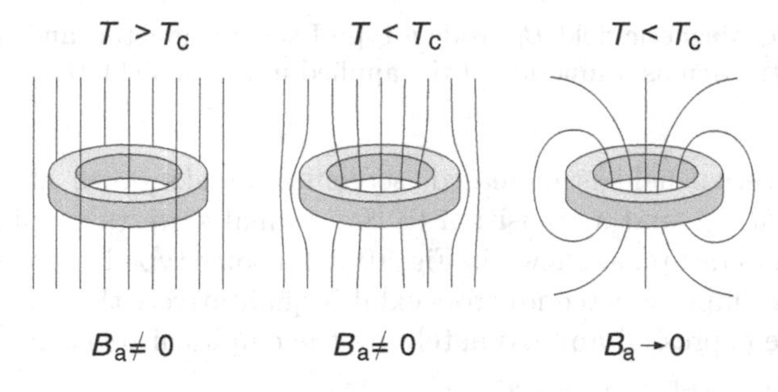

Fig. 10.7. Annular superconductor in a magnetic field. The magnetic field is expelled from the superconducting material but the magnetic flux through the ring becomes trapped

10.1.3 Type I Superconductor

High magnetic fields destroy superconductivity and restore the normal conducting state. Depending on the character of this transition, we may distinguish between *type I* and *type II superconductors.*

Most pure metals like Pb, Hg, In or Al are type I superconductors. The behavior of a long superconducting rod in an external magnetic field B_a is visualized in Fig. 10.8. Small fields are fully screened, i.e., currents are induced that generate a magnetization acting opposite to the applied field $B_a = \mu_0 H$. Since the magnetic flux is completely expelled we may write $B_i = B_a + \mu_0 M = 0$, where B_i is the field, or more correctly the magnetic induction, inside the sample. Thus, the magnetization M is given by $M = -B_a/\mu_0$, resulting in the susceptibility

$$\chi = \frac{\mu_0 M}{B_a} = -1\,. \tag{10.1}$$

This is an important result expressing the fact that type I superconductors are *ideal diamagnets.*

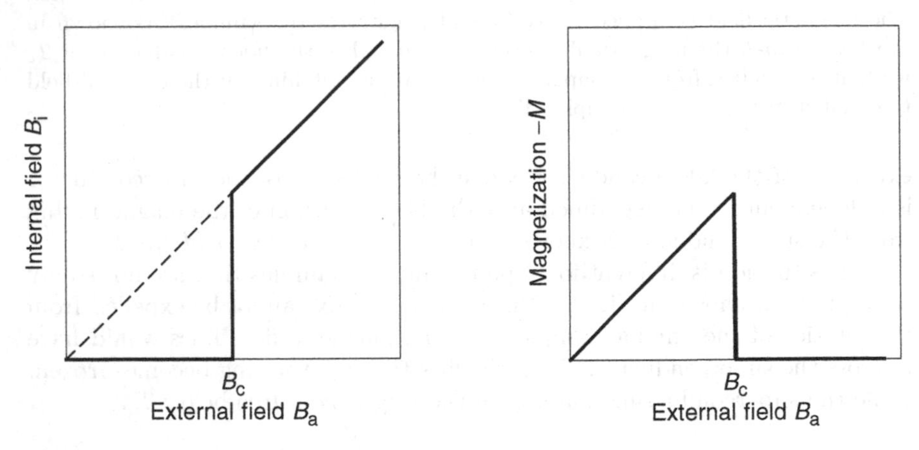

Fig. 10.8. (**a**) Magnetic field B_i inside a type I superconductor, and (**b**) its (negative) magnetization as a function of the applied magnetic field B_a

If the external field is increased, screening breaks down at the *critical magnetic field* B_c, and a transition to the normal state takes place. B_c depends on temperature, as shown in Fig. 10.9, for some type I superconductors. Note that all simple superconductors exhibit qualitatively the same variation, which can be expressed approximately by the empirical relation

$$B_c(T) = B_c(0)\left[1 - \left(\frac{T}{T_c}\right)^2\right]\,. \tag{10.2}$$

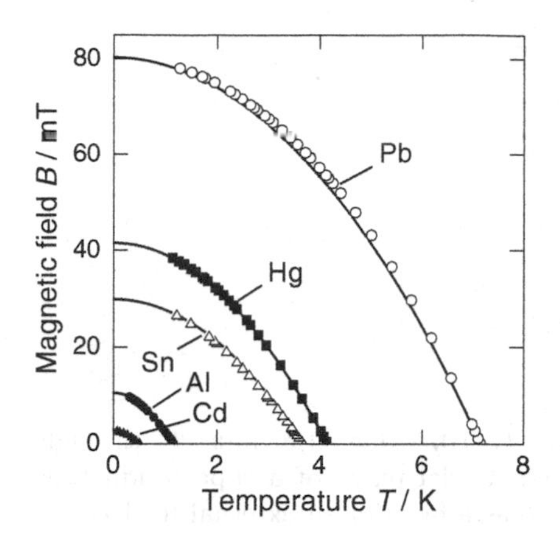

Fig. 10.9. Critical magnetic field B_c versus temperature of type I superconductors (from different authors)

Intermediate State

The shape of the specimen has an important influence on the magnetic behavior of superconductors. For a long rod oriented parallel to the applied field, the magnetization of the rod has no appreciable effect on the magnetic field. However, for samples with different shape, the magnetization alters the magnetic field seen by the specimen. For the 'effective' magnetic field B_{eff} we write

$$B_{eff} = B_a - D\mu_0 M\,, \tag{10.3}$$

where the demagnetization factor D depends on the sample shape. It is zero for long cylinders parallel to the field, $\frac{1}{3}$ for a sphere, $\frac{1}{2}$ for a cylinder perpendicular to the field, and unity for a plate perpendicular to the field.

Using the fact that type I superconductors are ideal diamagnets, i.e., that $M = -B_{eff}/\mu_0$, we find for the effective field $B_{eff} = B_a/(1 - D)$. In particular, for spherical samples the effective field is given by $B_{eff} = \frac{3}{2}B_a$. This means that already at $B_a = \frac{2}{3}B_c$, the field on the equator reaches the critical value B_c, and magnetic flux will penetrate the sphere. In fact, the sphere breaks up into alternating regions of superconducting and normal material, the magnetic flux penetrating the normal conducting regions. In this state, normal and superconducting domains are oriented parallel to the applied field, with $B = B_c$ in the normal domains. This peculiar state of superconducting specimens is known as the *intermediate state*.

When the applied field reaches the critical value, there is very little normal material. Increasing B_a further, the normal domains grow in such a way that the critical field B_c is always present at the equator. In Fig. 10.10, the magnetic flux through the equatorial plane is drawn as a function of B_a. Starting at $B_a = \frac{2}{3}B_c$, the magnetic flux increases linearly and reaches the value of the normal conducting material at B_c.

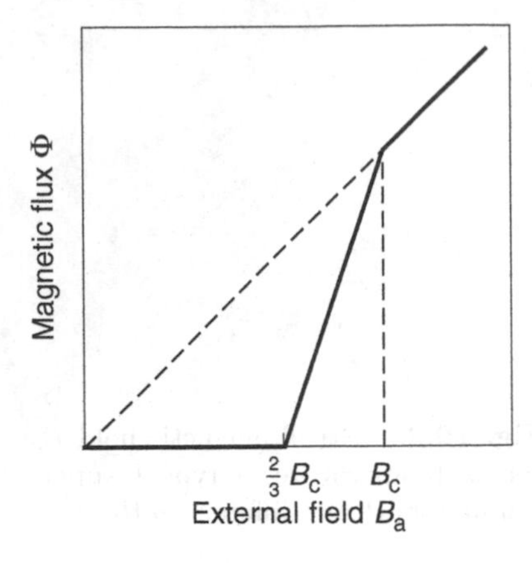

Fig. 10.10. Magnetic flux Φ through the equatorial plane of a superconducting sphere as a function of external field B_a

The intermediate state can be directly visualized with the help of decoration methods. An experimental result is shown in Fig. 10.11, where fine niobium particles were spread onto a superconducting plate of indium. Since the superconducting niobium grains are ideally diamagnetic they are pushed away from the normal conducting areas that are penetrated by the magnetic field. Thus the grains accumulate in the superconducting areas. The different reflectivity of decorated and undecorated areas is used to map superconducting and normal conducting areas of the sample in the intermediate state.

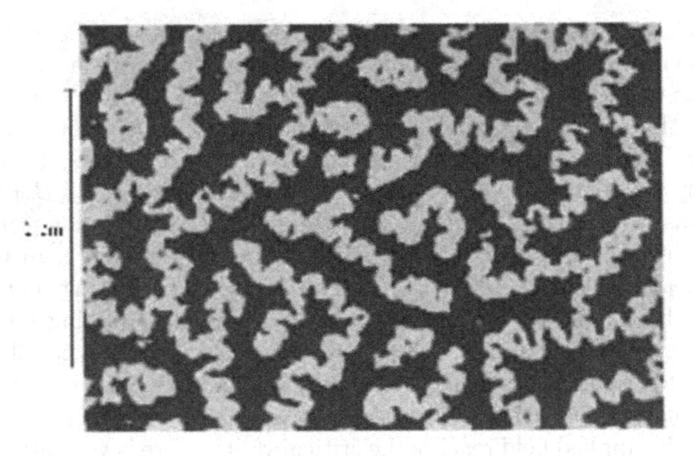

Fig. 10.11. Intermediate state of an indium plate. The whole plate was originally decorated with small niobium particles that were pushed away by the magnetic field in the normal conducting areas. Thus, the *bright areas* mark the normal conducting parts of the plate, while the superconducting areas are *dark* [464]

Critical Current

For technical applications, such as for example in a magnetic coil, the flow of high currents through superconducting wires is of particular interest. In calculating the critical current we have to take into account that in no region of the superconductor may the current density exceed the critical value. As worked out in more detail in Sect. 10.3.7, it is insignificant whether the current is due to shielding or caused by an external source.

For an estimate of the critical current, we consider a wire with radius R. The field at the surface of the wire is given by

$$B = \mu_0 \frac{I}{2\pi R}, \tag{10.4}$$

and consequently, the critical current reads

$$I_c = \frac{2\pi R}{\mu_0} B_c . \tag{10.5}$$

Using this relation, we find for a tin wire with $R = 1$ mm the relatively low critical current $I_c = 150$ A since the critical magnetic field of tin is only 30 mT. As $I_c \propto B_c$, critical current I_c and critical field B_c exhibit the same temperature dependence. It should be noted that the critical current does not rise with the cross-sectional area but linearly with radius of the wire.

What happens with the wire at the transition from the superconducting to the normal conducting state? At first glance, it seems plausible that the transition could take place by the superconductivity retreating to an inner core of the wire. But we immediately get into difficulties because in this case the current would move into the superconducting core, and consequently produce a field at the boundary of the core that is greater than that originally present at the surface. Thus, the retreat would continue until the whole wire was in the normal state. This is obviously impossible since then the current could be uniformly distributed over the cross section of the wire, and the field would be less than B_c over the greater part of the cross section, and this part could not be normal conducting.

As was also in the case of the intermediate state of a sphere, the transition must take place in a more complicated manner. As the current is increased and passes the critical value, normal regions are formed at the surface of the wire and flow towards the center. The wire becomes filled with a dynamic intermediate state whose moving domains are long in the direction of the field and encircle the wire. With increasing current, this state shrinks and the wire consists of a core of intermediate state with a critical radius R_c and a normal conducting sheath outside. The magnetic field will then be equal to B_c inside the normal domains of the core, but greater in the normal sheath.

The flux pattern inside the core is approximately rotationally symmetric. Different models have been developed to determine the pattern along the axis. In the *London model,* the superconducting phase consists of lamellae that are oriented perpendicular to the axis of the wire. The lamellae are static and not connected. The wire will exhibit a finite resistance because current is also flowing in the normal conducting regions. In the, *Gorter model,* it is assumed that superconducting tubes are nucleated at the critical radius R_c, and flow inwards towards the axis. The motion of the superconducting regions causes temporally varying magnetic fields that give rise to electric fields in the normal regions, thus leading to a finite resistance. The true behavior is likely in between these two models.

Without a detailed discussion, we would like to point out that using Ampère's and Faraday's law we can conclude the following: The current density $j(r)$ inside the core will decrease with the radius as $j(r) \propto 1/r$. An electric field parallel to the wire axis will exist in the normal regions. In particular, in the normal sheath, the field will take the constant value $\mathcal{E}_0$ fixed by the applied voltage. The critical radius will be given by $R_c = I_c/2\pi R\sigma\mathcal{E}_0$, where σ is the conductivity of the normal conductor. By setting $R_c = R$, we find that immediately above the critical current the field at the surface of the wire jumps from zero to $\mathcal{E}_c = I_c/2\pi R^2\sigma$. This is just half of the field that would be found if the wire was normal conducting and the current uniformly distributed. In other words, the resistance should rise discontinuously to half of its full value as soon as the critical current is reached, and then continue to rise with a further increase of the current. This prediction is in fair agreement with experimental observations. However, it should be noted that conclusive experiments are hampered by the fact that the transition to the resistive state is accompanied by heat production, and consequently the effects of temperature are difficult to exclude.

10.1.4 Type II Superconductors

Alloys, transition metals, metallic glasses and the novel high-T_c superconductors are members of the family of type II superconductors. As in type I superconductors, the magnetic field is expelled until the *lower critical field* B_{c1} is reached. At higher fields, normal conducting islands are formed allowing the penetration of the magnetic field in the form of thin filaments, usually called *flux lines* or *vortices*. In Sect. 10.4.1, it will be shown that the magnetic flux through these filaments is quantized, each filament carrying a so-called *flux quantum* $\Phi_0 = h/2e = 2.07 \times 10^{-15}\,\mathrm{T\,m^2}$. In contrast to the intermediate state, this so-called *mixed state* or *Shubnikov phase* is not a consequence of the demagnetization field and therefore occurs in all samples independent of their shape. The occurrence of two types of superconductors is due to the fact that the energy needed for the formation of a normal-superconducting interface is positive for type I superconductors, but negative for type II superconductors. We will return to this point in Sect. 10.2.4, where the surface energy is discussed in more detail.

The magnetic field in a type II superconductor as a function of the applied field is shown in Fig. 10.12. As mentioned above, below B_{c1}, both types of superconductors behave in the same way. Above B_{c1}, magnetic flux penetrates in the form of vortices, and the field is only partially screened. After passing through the *upper critical field* B_{c2}, the material becomes normal conducting. Therefore, the magnetization rises with the field at low fields as in type I superconductors, decreases above B_{c1}, and vanishes completely at B_{c2}. In this figure, the *thermodynamic critical field* $B_{c,th}$ is also drawn. It is determined by the equation

$$B_{c,th}^2 = -2\mu_0 \int_0^{B_{c2}} M \, dB_a \tag{10.6}$$

as indicated in the figure by the *shaded areas*. As we will see, the thermodynamic critical field is directly related to the energy gain occurring at the superconducting transition.

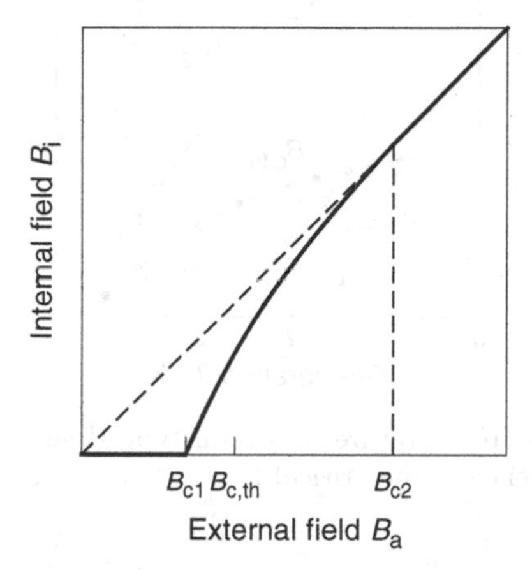

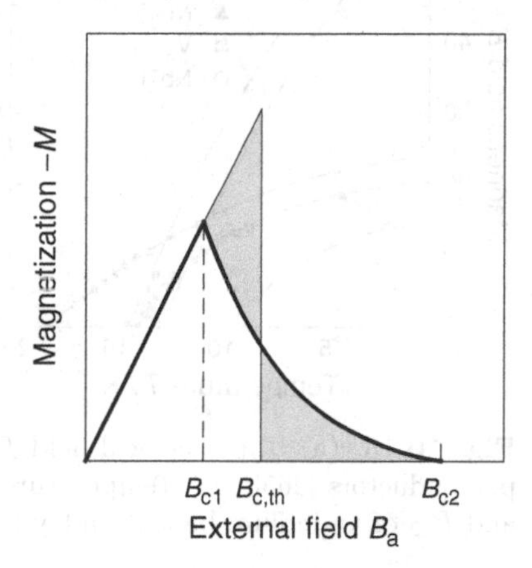

Fig. 10.12. (**a**) Internal field in a type II superconductor, and (**b**) the corresponding magnetization as a function of the applied field B_a. The Meissner phase is found below B_{c1}, the mixed state between B_{c1} and B_{c2}. The thermodynamic critical field $B_{c,th}$ is defined via the equality of the two *shaded areas*

The upper critical field B_{c2} of type II superconductors may be several hundred times larger than the critical field B_c of type I superconductors. For example, a critical field of $B_{c2} \approx 60\,T$ is found in the Chevrel compound[2] $PbGd_{0.3}Mo_6S_8$. Another interesting class of type II superconductors are A15 compounds.[3] A member of this class is Nb_3Sn that has the

[2] Chevrel compounds have the composition MMo_6X_8, where M stands for a rare-earth metal, and X for sulfur or selenium.

[3] A15 compounds are represented by the formula A_3B, and crystallize in the β-W structure. The niobium atoms in Nb_3Sn are arranged in chains and are closer to each other than in metallic niobium.

relatively large critical parameters $B_{c2}(0) = 25\,\mathrm{T}$ and $T_c(B = 0) = 18.7\,\mathrm{K}$. This is technically important because this compound can easily be processed and used as the material for wires in superconducting magnets. The novel high-T_c superconductors with their extremely high values of T_c and B_{c2} will be considered separately in Sect. 10.5.4.

Both the critical field strengths B_{c1} and B_{c2} depend on temperature. They have their largest value at $T = 0$, and vanish at T_c. In Fig. 10.13a, the temperature variation of B_{c2} of several type II superconductors is depicted. In Fig. 10.13b, the critical fields B_{c1} and B_{c2}, and the thermodynamic critical field $B_{c,th}$ of an indium-bismuth alloy are shown. Obviously, all types of critical fields exhibit a similar temperature dependence.

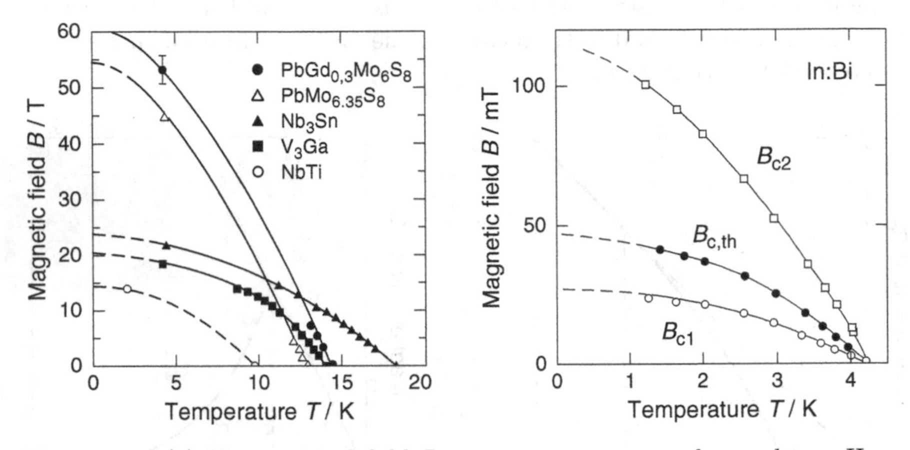

Fig. 10.13. (**a**) Upper critical field B_{c2} versus temperature of several type II superconductors [465]. (**b**) Temperature dependence of the critical fields B_{c1}, $B_{c,th}$, and B_{c2} of an indium-bismuth alloy [466]

In perfect crystals, the flux lines are regularly arranged in the so-called *Abrikosov lattice*, as is schematically shown in Fig. 10.14a. The first evidence for the existence of flux lines and their regular arrangement was obtained by decorating them on the surface of a sample with small particles of colloidal iron and making them visible with an electron microscope [467]. More recently, scanning tunneling microscopes have been used for the investigation of vortex lattices. A typical observation from this type of experiment is shown in Fig. 10.14b. Clearly, the flux lines in $NbSe_2$ are arranged in a hexagonal lattice. The contrast is due to the different work functions of the normal and superconducting phases. Based upon the different tunneling-current characteristics of the two phases and the high spatial resolution of this technique, it is possible to study the electronic density of states even inside the flux lines and in the superconducting neighborhood. In realistic materials, the regular

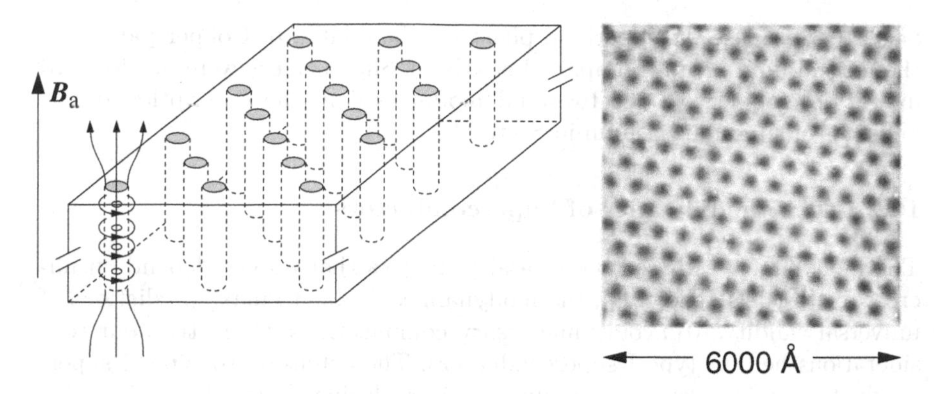

Fig. 10.14. (**a**) Arrangement of flux lines in the mixed state. The normal conducting vortices are depicted by *shaded areas*. Magnetic field and shielding current are schematically drawn for one flux line. (**b**) Direct observation of the Abrikosov structure in $NbSe_2$ with a scanning tunneling microscope. The experiment was carried out at 1.8 K and a magnetic field of 1 T [468]

arrangement of the flux lines is prevented by *flux line pinning* effects. We will briefly discuss this subject further in Sect. 10.2.4.

10.2 Phenomenological Description

Shortly after the discovery of the Meissner–Ochsenfeld effect, *F. London* and *H. London* developed a phenomenological theory of superconductivity that also included magnetic field effects [456]. Based on his experimental results, *Pippard* concluded in 1953 that the London theory is not complete, and an important ingredient had to be added, namely the nonlocal character of superconductivity [469]. A completely different approach to superconductivity was worked out by *Ginzburg* and *Landau* in 1950 [457]. They were able to give an explanation for the existence of the two types of superconductors. Based on this theory, *Abrikosov* predicted in 1957 the existence of the mixed state, in which superconductors contain a finite density of flux lines [470]. Finally, in 1959 *Gorkov* was able to prove that the Ginzburg–Landau theory can be traced back to BCS theory [471]. In addition, he showed that the theory developed by Ginzburg and Landau is not only applicable close to T_c, as originally assumed, but at all temperatures. The general concept developed by this group is often named GLAG theory.

Before we start with the phenomenological description of superconductors, we make a brief remark on the two-fluid model [455] developed by *Gorter* and *Casimir* in 1934. In this model, the electrons were divided into a *normal fluid*, carrying entropy and subjected to scattering, and a *superfluid* condensate, carrying no entropy and not subjected to scattering. This model has some connections to the microscopic theory to be discussed in

Sect. 10.3 because the superfluid phase can be related to Cooper pairs, and the normal-fluid phase to unpaired electrons. Similar ideas were put forward in the development of the two-fluid model used in the description of the behavior of superfluid helium in Sect. 2.2.

10.2.1 Thermodynamics of Superconductors

Thermodynamics is based on general principles that do not depend on microscopic models. Therefore, thermodynamics is able to make predictions of universal validity. To avoid unnecessary complications, we restrict our considerations here to type I superconductors. The extension to type II superconductors is, in most cases, easily done in replacing B_{c} by $B_{\mathrm{c,th}}$.

In the following section, we chose the external field $\boldsymbol{B}$ to be an independent variable as well as T and p. The thermodynamic potential of interest is the *Gibbs free energy* $G(T, p, \boldsymbol{B})$ that is given by

$$G = U - TS + pV - \boldsymbol{m} \cdot \boldsymbol{B} \, . \tag{10.7}$$

Because $\boldsymbol{m}$ and $\boldsymbol{B}$ are either parallel or antiparallel to each other, we can neglect their vector properties and only take into account the sign of the magnetic moment $\boldsymbol{m}$. For the variation $\mathrm{d}G$ of the Gibbs free energy we may write

$$\mathrm{d}G = -S\,\mathrm{d}T + V\mathrm{d}p - m\,\mathrm{d}B \, , \tag{10.8}$$

since the variation of the internal energy is given by $\mathrm{d}U = T\,\mathrm{d}S - p\,\mathrm{d}V + B\,\mathrm{d}m$.

First, we will compare the Gibbs free energy of the normal and superconducting phases because their difference is intimately connected with the nature of the superconducting phase transition. To avoid complications caused by demagnetization effects, we consider specimens with a demagnetization factor $D = 0$. Furthermore, we ignore effects caused by changes of the pressure and use the fact that superconductors behave like ideal diamagnets, i.e., we write $M = m/V = -B/\mu_0$. Thus, we find for the variation $\mathrm{d}G_{\mathrm{s}}$ of Gibbs free energy of a superconductor in a magnetic field the simple relation

$$\mathrm{d}G_{\mathrm{s}} = -S\,\mathrm{d}T + \frac{V}{\mu_0}\,B\,\mathrm{d}B \, . \tag{10.9}$$

Integration leads to

$$G_{\mathrm{s}}(B,T) = G_{\mathrm{s}}(0,T) + \int_0^B \frac{V}{\mu_0}\,B'\,\mathrm{d}B' = G_{\mathrm{s}}(0,T) + \frac{VB^2}{2\mu_0} \, . \tag{10.10}$$

The last term in this equation is positive, indicating that energy is necessary for the expulsion of the magnetic flux.

Changes in magnetization due to superconductivity are large in comparison with the Pauli paramagnetism and the Landau diamagnetism of normal conductors. Therefore, we may neglect the magnetic field variation of the Gibbs free energy of normal conductors, i.e., we write for the normal conductor $G_\mathrm{n}(B,T) = G_\mathrm{n}(0,T)$. According to (10.10), the Gibbs free energy of superconductors increases with the applied field and finally reaches the value G_n of the normal state. At this point, the energy gained by the superconducting transition is used up by the expulsion energy. The superconducting phase becomes unstable, and the normal state is re-established. Using the equality of the Gibbs free energy at the critical field expressed by $G_\mathrm{s}(B_\mathrm{c},T) = G_\mathrm{n}(B_\mathrm{c},T) = G_\mathrm{n}(0,T)$, we find the relation

$$G_\mathrm{n}(0,T) - G_\mathrm{s}(0,T) = \frac{VB_\mathrm{c}^2}{2\mu_0}\,. \tag{10.11}$$

Since G_n and G_s are equal at T_c, this equation implies that B_c approaches zero at T_c, as shown in Fig. 10.9. The energy difference $G_\mathrm{con} = VB_\mathrm{c}^2/2\mu_0$ is an important quantity in the theory of superconductivity, and for reasons we will discuss later, it is also called *condensation energy*. It should be noted that G_con is rather small. For a critical field of 50 mT the value $G_\mathrm{con} \approx 10^3\,\mathrm{J\,m^{-3}}$ is found.

From the difference between the free energy of the normal and superconducting state, the entropy difference can also be deduced via the relation $S = -(\partial G/\partial T)_{B,p}$. Thus we obtain

$$\Delta S = S_\mathrm{n} - S_\mathrm{s} = -V\,\frac{B_\mathrm{c}}{\mu_0}\,\frac{\mathrm{d}B_\mathrm{c}}{\mathrm{d}T}\,. \tag{10.12}$$

The temperature variation of both the free energy and the entropy is schematically depicted in Fig. 10.15.

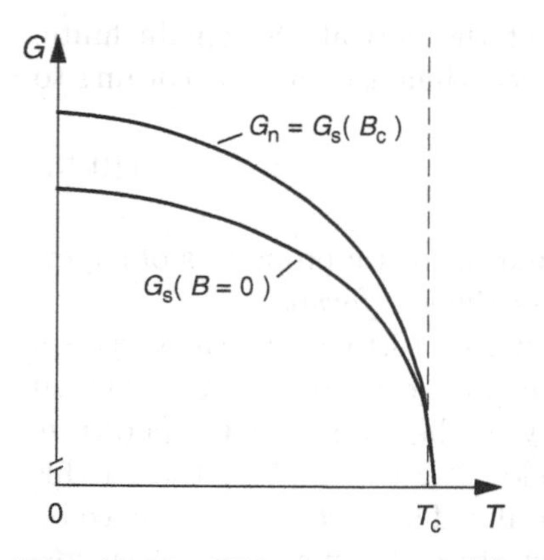

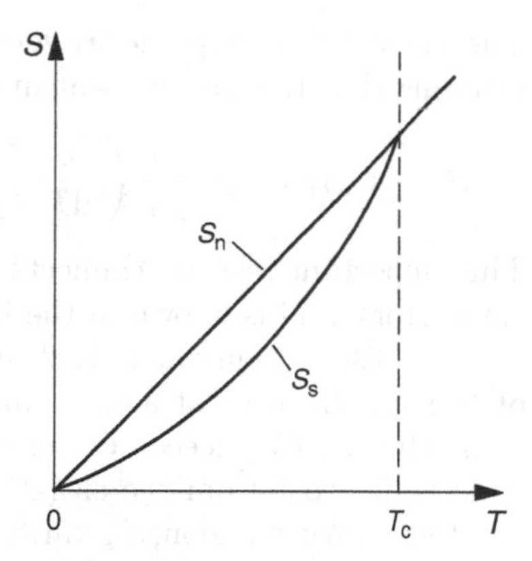

Fig. 10.15. Gibbs free energy (a) and entropy (b) versus temperature for a metal in its normal and superconducting state

We briefly discuss the behavior of these two quantities in different temperatures ranges:

- $T \to T_c$: The critical field B_c vanishes at the transition temperature, and so does the entropy difference ΔS. As a consequence, the latent heat $\Delta Q = T_c(S_n - S_s) = 0$, meaning that the transition to the superconducting state is not a phase transition of first, but of second order.
- $T \to 0$: The derivative dB_c/dT vanishes at absolute zero and consequently the entropy difference also vanishes, as expected from the third law of thermodynamics.
- $0 < T < T_c$: In the intermediate temperature range the derivative $dB_c/dT < 0$, and hence $\Delta S < 0$. From the sign of the entropy difference it follows that in the superconducting state the order is higher than in the normal state. Since $\Delta S \neq 0$, the superconducting transition in a magnetic field is accompanied by a latent heat, which means that the phase transition is of first order. The presence of a latent heat has an interesting consequence: Passing adiabatically through the phase boundary from higher magnetic field leads to a cooling of the specimen. In fact, this effect has occasionally been used in the past for cooling purposes.

The specific heat at constant pressure is defined by

$$C_p = T\left(\frac{\partial S}{\partial T}\right)_{p,B} = -T\left(\frac{\partial^2 G}{\partial T^2}\right)_{p,B} . \tag{10.13}$$

We differentiate (10.12), drop the index p, and find the difference of the specific heats in the superconducting and normal state to be

$$C_s - C_n = \frac{VT}{\mu_0}\left[\left(\frac{dB_c}{dT}\right)^2 + B_c\left(\frac{d^2 B_c}{dT^2}\right)\right] . \tag{10.14}$$

The critical field approaches zero at T_c, but the derivatives remain finite, meaning that the specific heat must vary discontinuously at T_c according to

$$(C_s - C_n)_{T=T_c} = \frac{VT}{\mu_0}\left(\frac{dB_c}{dT}\right)^2_{T=T_c} . \tag{10.15}$$

This important relation connects thermal and magnetic properties of superconductors, and is known in the literature as *Rutgers formula*.

This discontinuity at T_c is clearly visible in Fig. 10.16, where the variation of the specific heat of aluminum with temperature is shown. As expected from (10.15), C_s exceeds C_n in the vicinity of the transition temperature. This is different from the situation well below T_c, where $C_s < C_n$. In this low-temperature region, C_s varies exponentially. This strong dependence on temperature is caused by the increasing density of 'normal fluid' electrons, i.e., by the increasing number of thermally generated unpaired electrons (see Sect. 10.3).

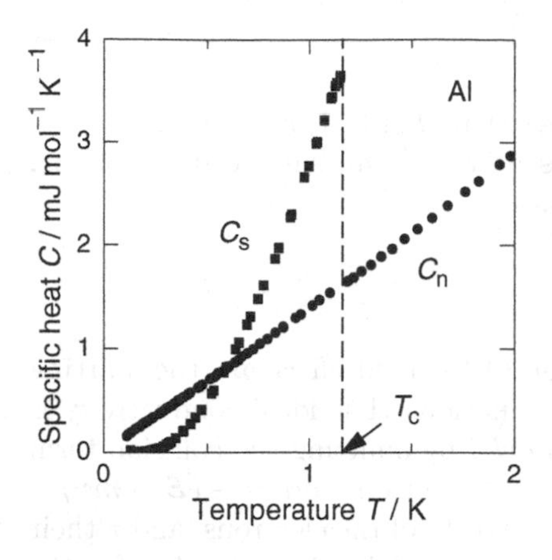

Fig. 10.16. Specific heat C_s and C_n of aluminum versus temperature. For the measurement in the normal state, a magnetic field $B > B_c$ was applied. The phonon contribution is negligible in both cases [472]

In Table 10.2, data obtained by calorimetric measurements are compared with data deduced from the temperature dependence of the critical field. The experimental situation is described well by (10.15) in most cases. However, there is a striking deviation in the case of lead. Similar deviations are also found in some other superconductors. The deviations are a consequence of the strong coupling between electrons and lattice in these metals. We will return to this observation in the discussion of the microscopic theory of superconductivity in Sect. 10.3.

Table 10.2. Abrupt change $\Delta C = C_s - C_n$ of the specific heat of some superconductors at T_c. Calorimetric data ΔC_{cal} are compared with values ΔC_{mag} deduced from the temperature dependence of the critical field [473,474]

Element	T_c (K)	ΔC_{cal} (mJ mole^{-1} K^{-1})	ΔC_{mag} (mJ mole-1 K^{-1})
Tl	2.39	6.2	6.15
In	3.4	9.75	9.62
Sn	3.72	10.6	10.6
Ta	4.39	41.5	41.6
Pb	7.2	52.6	41.8

Finally, we want to investigate the relation between the low-temperature value of $(C_s - C_n)$ and the specific heat of normal conductors. For $T \to 0$, the specific heat C_s rapidly approaches zero and can therefore be neglected. Inserting (10.2) in (10.15) we thus find

$$C_n \approx \frac{2V}{\mu_0} \frac{B_c^2(0)}{T_c^2} T\,. \tag{10.16}$$

The specific heat of normal conducting metals was discussed in Sect. 7.1 and is given by

$$C_n = \frac{2}{3}\pi^2 k_B{}^2 V D(E_F) T . \tag{10.17}$$

A comparison of the two expressions shows that $(B_c(0)/T_c)^2 \propto D(E_F)$, meaning that the electronic density of states at the Fermi surface has a strong influence on superconducting properties.

10.2.2 London Equations

The ideal conductivity and the Meissner–Ochsenfeld effect are the starting points for the derivation of the London equations. The ideal conductivity is easily taken into account in the *Drude model* by omitting the collision term in the equation of motion of the conduction electrons: $m\dot{\boldsymbol{v}} = -e\boldsymbol{\mathcal{E}} - m\boldsymbol{v}/\tau$. Here, $\boldsymbol{\mathcal{E}}$ denotes the electrical field, $\boldsymbol{v}$ the velocity of the electrons, and τ their collision time. By setting $\tau = \infty$, we obtain the following equation for the time derivative of the density $\boldsymbol{j}_s$ of the supercurrent

$$\frac{d\boldsymbol{j}_s}{dt} = \frac{n_s e_s^2}{m_s}\boldsymbol{\mathcal{E}} . \tag{10.18}$$

This is the *first London equation*. In the two-fluid model mentioned at the beginning of this section, the supercurrent is carried by the superfluid condensate. For the time being, we leave open the nature of the carriers that are responsible for $\boldsymbol{j}_s$. We use the subscript 's' in the abbreviation for density, mass and charge of the carriers. The meaning of the subscript will become clear during the discussion of the microscopic origin of superconductivity in Sect. 10.3. Unlike in ordinary metals, in superconductors it is not the current but its time derivative that is proportional to the electric field $\boldsymbol{\mathcal{E}}$. Equation (10.18) gives the impression that for long times the growth of the current could be unlimited. But because of the vanishing resistivity, the electrical field also vanishes in the stationary state. As a consequence, the temporal variation of the current also vanishes. The current stays constant and is exclusively determined by the current source.

To derive the second London equation we insert (10.18) in the Maxwell equation $\mathrm{curl}\,\boldsymbol{\mathcal{E}} = -\partial \boldsymbol{B}/\partial t$, and obtain

$$\frac{\partial}{\partial t}\left(\mathrm{curl}\,\boldsymbol{j}_s + \frac{n_s e_s^2}{m_s}\boldsymbol{B}\right) = 0 . \tag{10.19}$$

This equation reflects the fact that in ideal conductors the magnetic flux through an arbitrary loop inside the sample cannot vary with time. In superconductors, however, the magnetic flux is not only constant but is zero. Hence, the Meissner–Ochsenfeld effect requires that the expression in the bracket vanishes. This leads to the *second London equation* that relates supercurrent to magnetic induction:

$$\mathrm{curl}\,\boldsymbol{j}_s = -\frac{n_s e_s^2}{m_s}\boldsymbol{B} . \tag{10.20}$$

Penetration Depth

One of the most important applications of the second London equation is in describing the screening behavior of superconductors in the Meissner state. Until now we have argued that the magnetic field is completely expelled from a type I superconducting, as demonstrated by the Meissner–Ochsenfeld effect. If this were correct, infinitely high shielding currents would have to be present at the sample surface. In reality, the magnetic field penetrates the surface of the superconductor. To treat this phenomenon, we assume that a superconductor occupies the half-space $x > 0$, and the magnetic field B_0 is applied in the z-direction, as shown in Fig. 10.17. By inserting (10.20) in the Maxwell equation $\operatorname{curl} \boldsymbol{B} = \mu_0 \boldsymbol{j}$, we obtain the differential equation

$$\frac{\mathrm{d}^2 B_z(x)}{\mathrm{d}x^2} - \frac{\mu_0 n_\mathrm{s} e_\mathrm{s}^2}{m_\mathrm{s}} B_z(x) = 0 \tag{10.21}$$

for the given geometry. This equation can easily be solved and we obtain

$$B_z(x) = B_0 \,\mathrm{e}^{-x/\lambda_\mathrm{L}} , \tag{10.22}$$

where the *London penetration depth* λ_L is given by

$$\lambda_\mathrm{L} = \sqrt{\frac{m_\mathrm{s}}{\mu_0 n_\mathrm{s} e_\mathrm{s}^2}} . \tag{10.23}$$

Inserting the solution in the Maxwell equation $\operatorname{curl} \boldsymbol{B} = \mu_0 \boldsymbol{j}$, we find for the spatial variation of the screening current

$$j_{\mathrm{s},y}(x) = j_0 \,\mathrm{e}^{-x/\lambda_\mathrm{L}} . \tag{10.24}$$

Both magnetic field and current density decrease exponentially with the characteristic length λ_L. Furthermore, applied field and current are connected via the relation $B_0 = \mu_0 \lambda_\mathrm{L} j_0$.

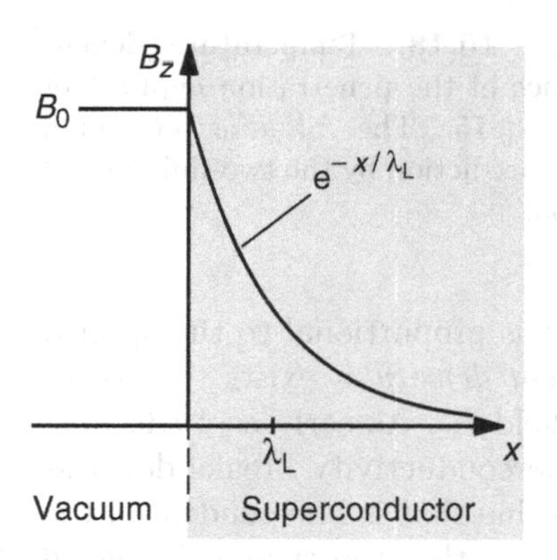

Fig. 10.17. Magnetic field inside a superconductor. The exponential decay is determined by the London penetration depth λ_L

The penetration depth λ_L depends on temperature since the density of the superconducting carriers varies with temperature. Without justification, we mention that the two-fluid model of superconductors predicts that the density of the superconducting carriers varies with temperature according to the relation $n_s \propto 1 - (T/T_c)^4$ [455]. Inserting this relation into (10.23), we find that the penetration depth is expected to vary with temperature as

$$\lambda_L(T) = \frac{\lambda_L(0)}{\sqrt{1-(T/T_c)^4}} . \tag{10.25}$$

As temperature approaches the transition temperature T_c, the penetration depth diverges since the carrier density goes to zero. Well below T_c, the penetration depth becomes temperature independent. Assuming that the density of superconducting carriers is of the order of one electron per atom, λ_L is expected to be of the order of 30 nm.

The validity of (10.25) can be examined by measuring the diamagnetic behavior of superconducting particles. In order to obtain a favorable ratio between the field-free volume and the penetration depth, these measurements are often carried out with very small spheres. As an example, we show in Fig. 10.18 the measurement made on thin lead cylinders where very good agreement with expectation was obtained.

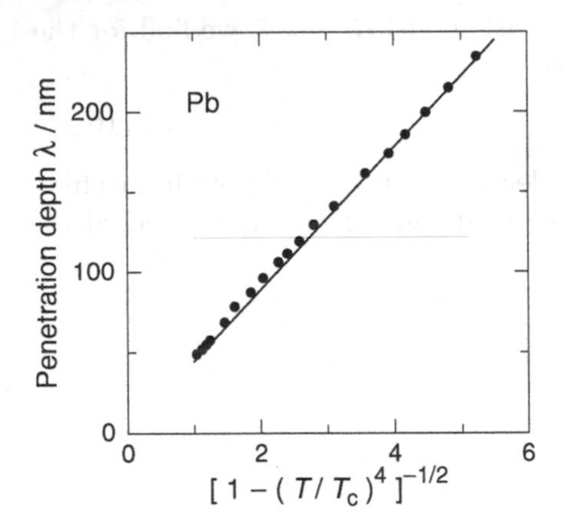

Fig. 10.18. Temperature dependence of the penetration depth λ of lead [475]. The *full line* represents the prediction by the two-fluid model [455]

As mentioned above, the shielding current is proportional to the applied magnetic field. Consequently, a *critical current density* j_c exists that is directly connected with the critical magnetic field B_c. As early as 1916, *Silsbee* [476] formulated the hypothesis that superconductivity breaks down as soon as the current density passes a critical value that is independent of the origin of the current, i.e., it does not matter whether the current is due to shielding effects or due to an external current source.

Superconducting Films

We use the second London equation to briefly investigate the shielding behavior of thin superconducting films. We will consider a film that is oriented parallel to the applied magnetic field in the z-direction, as indicated in Fig. 10.19. Starting from (10.22), we superimpose the contributions from the two sides of the film by writing $B_z(x) = \overline{B}\,[\exp(-x/\lambda_L) + \exp(x/\lambda_L)]$, where the value of the constant $\overline{B}$ follows from the boundary condition $B_z(d/2) = B_z(-d/2) = B_0$. With $\overline{B} = B_0/[2\cosh(d/2\lambda_L)]$, we obtain

$$B_z(x) = B_0 \frac{\cosh(x/\lambda_L)}{\cosh(d/2\lambda_L)} \tag{10.26}$$

for the magnetic field, and

$$j_{s,y}(x) = +\frac{B_0}{\mu_0\lambda_L}\frac{\sinh(x/\lambda_L)}{\cosh(d/2\lambda_L)} \tag{10.27}$$

for the current density. Setting $|x| = d/2$, we find for the current density at the film surface

$$|j_{s,y}(d/2)| = \frac{B_0}{\mu_0\lambda_L}\tanh\frac{d}{2\lambda_L}\,. \tag{10.28}$$

For thick films, i.e., for $d \gg \lambda_L$, we recover (10.24). However, with decreasing film thickness, the shielding current at the surface is reduced by the factor $\tanh(d/2\lambda_L) \approx d/2\lambda_L$. This means that the magnetic field is no longer completely screened, and the critical magnetic field is *enhanced* by the factor $2\lambda_L/d$.

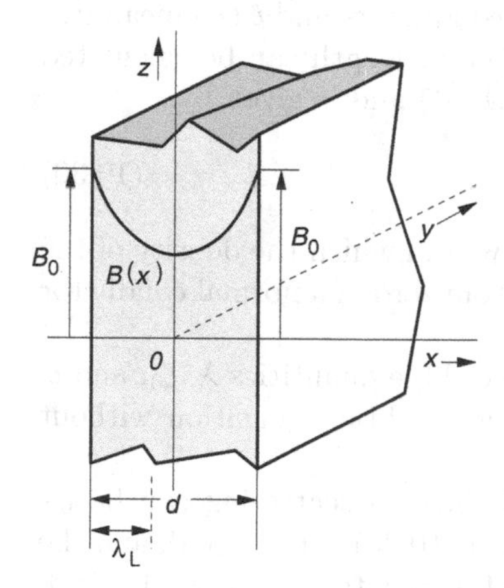

Fig. 10.19. Decay of the magnetic field $B_z(x)$ in a thin superconducting film. For this sketch, it was assumed that $d/\lambda_L \approx 3$

10.2.3 Pippard's Equation

As mentioned above, the penetration depth λ may be deduced from measurements of the magnetization of small particles. Around 1950, *Pippard* developed a different technique. He measured the frequency shift of superconducting resonators when driven normal by a magnetic field. This shift is caused by the change in the microwave surface impedance that, in turn, is determined by the penetration depth λ. He found that λ of pure tin exhibits an anisotropy, while λ of impure specimens is isotropic. From his studies, he concluded that the second London equation is inadequate and has to be replaced by a more complex, nonlocal expression [469]. The expression for the current density proposed by Pippard is analogous to the Reuter–Sondheimer equation that relates current density and electric-field strength in the case of the anomalous skin effect. Nonlocal electrodynamic equations are necessary when the electron mean free path ℓ is comparable to or greater than the skin depth. In this case, the current density $\boldsymbol{j}(\boldsymbol{r})$ is not simply governed by the local electric field strength $\boldsymbol{\mathcal{E}}(\boldsymbol{r})$ but rather by the electric field averaged over a volume determined by the mean free path. Similarly, Pippard assumed that in superconductors the current density is not governed by the local magnetic field but by an appropriate average of the field in the environment. Experimental observations and the predictions of Pippard's nonlocal theory were found to be in good agreement.

A characteristic parameter of Pippard's nonlocal theory is the *electromagnetic coherence length* ξ_{em}. It is defined by

$$\frac{1}{\xi_{\mathrm{em}}} = \frac{1}{\xi_0} + \frac{1}{\ell}\,, \tag{10.29}$$

where ξ_0 is the coherence length of pure superconductors and ℓ the mean free path of electrons in the normal state. The coherent length can be calculated within the framework of BCS theory (see (10.116)) and is given by

$$\xi_0 = 0.18\,\frac{\hbar v_{\mathrm{F}}}{k_{\mathrm{B}}T_{\mathrm{c}}}\,. \tag{10.30}$$

Roughly speaking, ξ_0 reflects the length scale within which the density of the superconducting carriers rises at an interface from zero in a normal conductor to the density n_{s} in a superconductor.

Depending on the relative magnitude of the three quantities λ, ξ_0, and ℓ, we may distinguish between various limits that we briefly mention without going into details:

For $\lambda \gg \xi_0$ and $\ell \gg \xi_0$, nonlocality and electron scattering can be ignored. In this so-called *London limit*, expression (10.23) is recovered, and the penetration depth is given by $\lambda = \lambda_{\mathrm{L}}$. It should be pointed out that 'classical' superconductors do not fall into this category. But here there is an interesting point: λ is temperature dependent, while ξ_{em} is not. Since $\lambda \to \infty$ for $T \to T_{\mathrm{c}}$, there is always a temperature interval near the transition temperature, even in pure superconductors, where the London limit holds.

The *Pippard limit* gives the correct description for $\lambda \ll \xi_0$, where it is still true that $\ell \gg \xi_0$. The analysis shows that in this case $\lambda \approx (0.28\,\lambda_L^2\,\xi_0)^{1/3}$, meaning that the measured penetration depth is considerably larger than λ_L. Clean aluminum, with $\lambda = 50\,\text{nm}$ and $\xi_0 = 1600\,\text{nm}$, and other 'classical' superconductors are in this limit.

In the *dirty limit* we have $\ell \ll \lambda$. In this case, the Pippard equation leads to the relation $\lambda = \lambda_L(1+\xi_0/\ell)^{1/2}$. This expression indicates that shortening the mean free path, for instance by alloying the superconductor, increases the penetration depth.

10.2.4 Ginzburg–Landau Theory

As we mentioned already, in 1950 *Ginzburg* and *Landau* succeeded in making an important step forward towards understanding superconductivity. Their description was based on Landau's theory of second-order phase transitions in which the free energy is expanded with respect to an order parameter Φ. A well-known example is the transition from the paramagnetic to the ferromagnetic state, where the spin orientation serves as the order parameter. If $n_\uparrow$ is the number of spins pointing upwards, and $n_\downarrow$ the number pointing downwards, the order parameter may be defined by $\Phi = (n_\uparrow - n_\downarrow)/(n_\uparrow + n_\downarrow)$. It varies between $\Phi = 1$ at $T = 0$, and $\Phi = 0$ at $T \geq T_c$. Since the free energy is expanded in a power series with respect to the order parameter, the theory is expected to give a good description only in the temperature range near T_c.

Before we consider special aspects of Ginzburg–Landau theory, let us have a brief look at predictions that classical Landau theory makes for superconductors. Expanding the density g_s of Gibbs free energy in a series, we write

$$g_s = g_n + \alpha(T)\Phi^2 + \frac{1}{2}\beta(T)\Phi^4\,, \tag{10.31}$$

where the temperature dependence is explicitly indicated. The equilibrium phase corresponds to a minimum in g_s, i.e., $\partial g_s/\partial\Phi = 0$. This leads to the condition $\alpha\Phi_0 + \beta\Phi_0^3 = 0$ for the order parameter Φ_0 in equilibrium. The solution $\Phi_0 = 0$ is associated with the disordered phase for $T > T_c$, while for the ordered phase, where $T < T_c$,

$$\Phi_0^2 = -\alpha/\beta \tag{10.32}$$

should hold. This condition can be achieved, if α/β is positive for $T > T_c$, and negative for $T < T_c$. β must be positive for all temperatures since otherwise g_s would decrease indefinitely for large values of Φ.

The simplest temperature dependence of the parameters α and β that is consistent with all the requirements, is the following:

$$\alpha(T) = A(T - T_c)\,, \tag{10.33}$$

$$\beta(T) = \beta(T_c) = \beta\,, \tag{10.34}$$

where A and β are positive constants.

In Sect. 10.2.1, we found that the difference in Gibbs free energy is related to the critical field via the relation $(g_\mathrm{n} - g_\mathrm{s}) = B_\mathrm{c}^2/2\mu_0$. From (10.31) and (10.32) we find for the same quantity the relation $(g_\mathrm{n} - g_\mathrm{s}) = \alpha^2/2\beta$. The critical field expressed in the parameters of Landau theory is thus given by

$$B_\mathrm{c}(T) = -\sqrt{\frac{\mu_0}{\beta}}\,\alpha(T)\,. \tag{10.35}$$

Close to the transition temperature, $B_\mathrm{c}(T)$ is expected to be proportional to $(T_\mathrm{c} - T)$ in agreement with (10.2).

In classical Landau theory, the order parameter is spatially invariant and a real quantity. In Ginzburg–Landau theory the macroscopic wave function of the superconducting state

$$\Psi(\boldsymbol{r}) = \Psi_0(\boldsymbol{r}) \exp[-\mathrm{i}\varphi(\boldsymbol{r})] \tag{10.36}$$

serves as the order parameter with the amplitude squared $|\Psi_0|^2 = n_\mathrm{s}$ being the density of the superconducting particles. We use here the macroscopic wave function that is a characteristic of the superfluid state of helium and of superconductivity. A more detailed consideration of $\Psi(\boldsymbol{r})$ will follow in Sect. 10.4. The use of the wave function $\Psi(\boldsymbol{r})$ as an order parameter introduces two complications: first, in the general case, the order parameter becomes a function of position; secondly, the magnetic-field energy and the coupling of the supercurrent to the magnetic field has to be included.

Let us first consider the situation without a magnetic field. In this case, we write for the density of Gibbs free energy

$$g_\mathrm{s} = g_\mathrm{n} + \alpha|\Psi(\boldsymbol{r})|^2 + \frac{1}{2}\beta|\Psi(\boldsymbol{r})|^4 + \frac{\hbar^2}{2m}|\nabla\Psi(\boldsymbol{r})|^2\,. \tag{10.37}$$

We use the modulus signs because $\Psi(\boldsymbol{r})$ is complex. With $\Psi(\boldsymbol{r})$ being a function of position, we expect a 'kinetic-energy term' proportional to $|\nabla\Psi|^2$ that does not exist in classical Landau theory. Following the usual conventions of Ginzburg–Landau theory, we write $\hbar^2/2m$ for the factor in front of this term, where m is the electronic mass. There is no physical content to this choice because it simply determines the normalization of $\Psi(\boldsymbol{r})$.

The inclusion of the magnetic field in the description of superconductivity requires two additional terms:

$$\begin{aligned} g_\mathrm{s} = g_\mathrm{n} + \alpha|\Psi(\boldsymbol{r})|^2 + \frac{1}{2}\beta|\Psi(\boldsymbol{r})|^4 &+ \frac{1}{2\mu_0}|B_\mathrm{a} - B_\mathrm{i}|^2 \\ &+ \frac{1}{2m}\left|(-\mathrm{i}\hbar\nabla + 2e\boldsymbol{A})\,\Psi(\boldsymbol{r})\right|^2\,. \end{aligned} \tag{10.38}$$

The first of these terms represents the expulsion energy necessary to change the field in the sample from B_a to B_i. In the last term, magnetic-field effects are included by making the usual replacement $\nabla \to \nabla - 2\mathrm{i}e\boldsymbol{A}/\hbar$. For the charge of the Cooper pairs here we have written $-2e$, although in the original formulation of the theory the electronic charge $-e$ was used.

The Gibbs free energy $G(\boldsymbol{B})$ of the whole specimen is obtained by integration over the volume. The next step is to minimize G with respect to Ψ and $\boldsymbol{A}$ in order to determine the equilibrium state. Using the variation method, the two *Ginzburg–Landau equations*

$$\frac{1}{2m}\left(-\mathrm{i}\hbar\nabla + 2e\boldsymbol{A}\right)^2 \Psi + \alpha\Psi + \beta|\Psi|^2\Psi = 0\,, \tag{10.39}$$

$$\boldsymbol{j}_\mathrm{s} = \frac{\mathrm{i}e\hbar}{m}\left(\Psi^*\nabla\Psi - \Psi\nabla\Psi^*\right) - \frac{4e^2}{m}|\Psi|^2\boldsymbol{A} \tag{10.40}$$

are found. In addition, the boundary condition has to be fulfilled that the current perpendicular to the sample surface vanishes, i.e.,

$$\boldsymbol{n}\cdot(-\mathrm{i}\hbar\nabla + 2e\boldsymbol{A})\Psi = 0\,, \tag{10.41}$$

where $\boldsymbol{n}$ is the normal to the surface.

The Ginzburg–Landau theory contains two *characteristic lengths*, namely the penetration depth λ and the *Ginzburg–Landau coherence length* ξ_GL. First, we will deduce the expression for the penetration depth. For this purpose, we consider a specimen with dimensions much greater than this quantity. In this case, the magnetic field vanishes inside the sample, and $\Psi = \text{const}$. Therefore, only the last two terms of (10.39) are relevant, leading to $|\Psi|^2 = -\alpha/\beta$. Of course, this result is identical to (10.32). Inserting this expression in (10.40), we obtain for the current density

$$\boldsymbol{j}_\mathrm{s} = \frac{4e^2}{m}\frac{|\alpha|}{\beta}\boldsymbol{A}\,. \tag{10.42}$$

Obviously, this expression for the supercurrent is identical with the second London equation.[4] Consequently, the Ginzburg–Landau expression for the penetration depth λ is given by

$$\lambda = \sqrt{\frac{m\beta}{4\mu_0 e^2|\alpha|}}\,. \tag{10.43}$$

The Ginzburg–Landau coherence length reflects the characteristic distance over which spatial changes in Ψ occur. To find the corresponding expression we consider a superconductor that takes up the half-space $x > 0$. Without a magnetic field, (10.39) reads:

$$-\frac{\hbar^2}{2m}\frac{\mathrm{d}^2\Psi}{\mathrm{d}x^2} + \alpha\Psi + \beta\Psi^3 = 0\,. \tag{10.44}$$

We introduce the normalized wave function $f(x) = \Psi(x)/\Psi_\infty$, where the index ∞ indicates that Ψ_∞ is the solution as $x \to \infty$, i.e., deep inside the specimen. We also introduce the Ginzburg–Landau coherence length via

[4] In this formulation of the Ginzburg–Landau equations we have taken into account that the supercurrent is due to Cooper pairs with the charge $-2e$, resulting in the factor of 4.

$$\xi_{\text{GL}} = \frac{\hbar}{\sqrt{2m|\alpha|}} \,. \tag{10.45}$$

Thus (10.44) takes the form

$$\xi_{\text{GL}}^2 \frac{\mathrm{d}^2 f(x)}{\mathrm{d}x^2} + f(x) - f^3(x) = 0 \,. \tag{10.46}$$

Using the boundary conditions

$$f(0) = 0, \qquad \lim_{x\to\infty} f(x) = 1, \qquad \lim_{x\to\infty} \frac{\mathrm{d}f(x)}{\mathrm{d}x} = 0 \,, \tag{10.47}$$

we finally obtain

$$f(x) = \tanh \frac{x}{\sqrt{2}\,\xi_{\text{GL}}} \,. \tag{10.48}$$

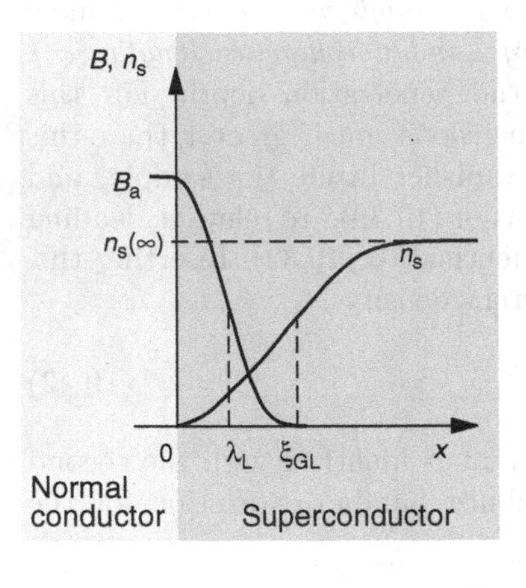

Fig. 10.20. Spatial variation of the magnetic field B and the Cooper pair density n_s at the boundary between a normal metal and a superconductor

This result is visualized in Fig. 10.20. As indicated, the density of the Cooper pairs $n_s(x)$ increases steadily from zero at the phase boundary to its bulk value $n_s(\infty) = |\Psi_\infty|^2$. This rise is characterized by the coherence length ξ_{GL}. In this figure, the decay of the magnetic field, determined by the penetration depth λ, is also drawn. The curve for the magnetic field does not exactly reflect (10.22) that was derived under the assumption that n_s is constant. Because of the vanishing density of Cooper pairs close to the phase boundary, the curve flattens in this region.[5]

Both characteristic length scales, λ and ξ_{GL}, have the same dependence on α, and hence on temperature. Their ratio $\kappa = \lambda/\xi_{\text{GL}}$ is the so-called *Ginzburg–Landau parameter* that allows a distinction between type I and

[5] It should be mentioned that we have neglected the so-called *proximity effect* that is due to the penetration of the wave function into the normal conducting metal, which leads to a finite density of Cooper pairs in the normal conductor very close to the surface.

type II superconductors to be made. Since κ is a function of β only, it is temperature independent and is given by

$$\kappa = \sqrt{\frac{m^2\beta}{2\mu_0\hbar^2e^2}}\,. \tag{10.49}$$

The penetration depth λ, the coherence length ξ_{GL}, and the critical fields are intimately connected. An 'in depth' treatment based on Ginzburg–Landau theory leads to the expressions listed below without derivation. In these expressions, the flux quantum $\Phi_0 = h/2e = 2.07\times10^{-15}\,\mathrm{T\,m^2}$ enters, the origin and significance of which we shall discuss in Sect. 10.4.

$$B_{\mathrm{c1}} \approx \frac{\Phi_0}{4\pi\lambda^2}\ln\kappa \qquad (\text{for } \kappa \gg 1)\,, \tag{10.50}$$

$$B_{\mathrm{c2}} = \frac{\Phi_0}{2\pi\xi^2}\,, \tag{10.51}$$

$$B_{\mathrm{c,th}} = \frac{\Phi_0}{\sqrt{8}\pi\lambda\xi}\,. \tag{10.52}$$

Rewriting these equations leads to

$$B_{\mathrm{c1}} \approx \frac{B_{\mathrm{c,th}}}{\sqrt{2}\kappa}\ln\kappa\,, \tag{10.53}$$

$$B_{\mathrm{c2}} = \sqrt{2}\kappa B_{\mathrm{c,th}}\,, \tag{10.54}$$

$$B_{\mathrm{c1}}\,B_{\mathrm{c2}} \approx B_{\mathrm{c,th}}^2\ln\kappa\,. \tag{10.55}$$

Phase-Boundary Energy

The behavior of type I superconductors, and in particular their intermediate state, can only be understood if a finite energy per area is required for the formation of an interface between superconducting and normal-conducting regions. If the superconducting specimen is homogenous and is at constant temperature, the magnetic field at a phase boundary will be the critical field $B_{\mathrm{c,th}}$. In the normal-conducting regions we have $B \geq B_{\mathrm{c,th}}$, but in the superconducting regions the magnetic field approaches zero within a distance determined by the penetration depth. In the normal state, the density of superconducting carriers $n_{\mathrm{s}} = 0$, while $n_{\mathrm{s}}(T) \neq 0$ in the superconducting phase.

Depending on the ratio between λ and ξ_{GL}, the boundaries contribute to the free energy of the system in a positive or negative sense. Two opposed effects are present. On the one hand, energy is required to expel the magnetic field, on the other hand, the specimen gains energy by its transition into the superconducting state, i.e., by forming Cooper pairs. This means that the field self-energy E_{B} and the condensation energy E_{C} are working against each other. At the critical field $B_{c,th}$, the two effects give the same contribution, but with opposite sign, i.e., $-E_B = E_C = VB_{c,th}^2/2\mu_0$. For a rough approximation we can replace the steady variation of B and n_s at boundaries by a step function at $x = \lambda$ and $x = \xi_{GL}$, respectively. With this simplification, we obtain for the energy change per area A the relations

$$\Delta E_B = A\,\lambda\,\frac{B_{c,th}^2}{2\mu_0}\,, \tag{10.56}$$

and

$$\Delta E_C = A\,\xi_{GL}\,\frac{B_{c,th}^2}{2\mu_0}\,. \tag{10.57}$$

The total energy change caused by the boundary is thus given by

$$\Delta E_C - \Delta E_B = (\xi_{GL} - \lambda)\;A\,\frac{B_{c,th}^2}{2\mu_0}\,. \tag{10.58}$$

The energy difference is positive, if $\xi_{GL} > \lambda$. In this case, the loss in condensation energy exceeds the gain in expulsion energy. This is the situation found in type I superconductors. In the intermediate state, the positive phase-boundary energy prevents the splitting up of the specimen into very tiny superconducting regions. In contrast, the energy difference is negative, if $\xi_{GL} < \lambda$, and the formation of boundaries is energetically favorable for the system. This is the situation found in type II superconductors.

Thus, the Ginzburg–Landau parameter $\kappa = \lambda/\xi_{GL}$ allows us to distinguish between the two types of superconductors. A more detailed analysis leads to the result

$$\kappa < \frac{1}{\sqrt{2}} \qquad \text{for type I superconductors}\,, \tag{10.59}$$

$$\kappa > \frac{1}{\sqrt{2}} \qquad \text{for type II superconductors}\,. \tag{10.60}$$

From the arguments given above, one could gain the impression that the magnetic field is only expected to penetrate into type II superconductors above the thermodynamic critical field $B_{c,th}$, but we have seen that flux already enters into type II superconductors in the field range $B_{c1} < B < B_{c,th}$. As discussed above, the formation of boundaries is energetically favorable as long as $(\xi_{GL}B_{c,th}^2 - \lambda B^2) < 0$. Taking into account the factor $\sqrt{2}$ in (10.58), this means that the magnetic field penetrates, if

$$B^2 > \frac{B_{c,th}^2}{\kappa\sqrt{2}}\,, \tag{10.61}$$

i.e., already at fields smaller than $B_{c,th}$.

In the previous discussion of type II superconductors, the magnetic field and shielding current associated with a flux line were shown in the schematic drawing of Fig. 10.14a. The variation of n_s, B, and j_s in and close to a flux line is depicted in more detail in Fig. 10.21. The Cooper pair density, or order parameter, is zero in the center of flux lines. The rise to the bulk value $n_s(\infty) = |\Psi_\infty|^2$ is determined by the coherence length ξ_{GL}. The magnetic field has its maximum in the center where the shielding current j_s is zero. The spatial extent of both quantities is governed by the penetration depth.

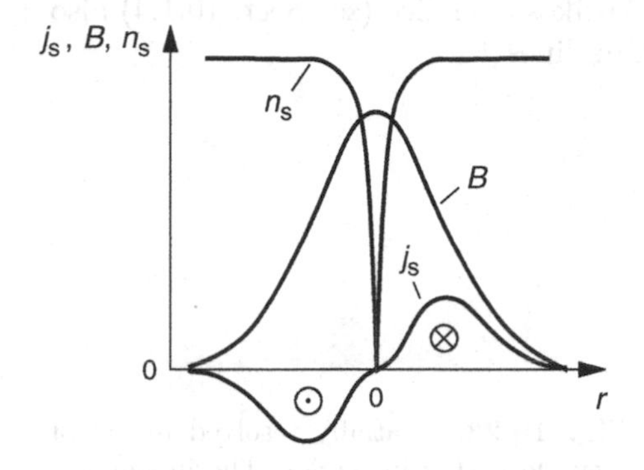

Fig. 10.21. Spatial variation of the Cooper pair density n_s, the magnetic field B, and supercurrent density j_s in and close to a flux line

Single flux lines can be studied with the help of scanning tunneling microscopy. The crucial quantity in measurements with this technique is the variation of the differential conductivity dI/dV with the voltage V applied between the tip of the microscope and the superconductor. As will be shown in Sect. 10.3.6, $dI/dV \propto D(E)$ and is thus reflecting the density of states of unpaired electrons. An example of this type, a measurement on $NbSe_2$, is shown in Fig. 10.22. The curve at the bottom represents the density of states without a magnetic field, i.e., in the Meissner phase. A gap in the density of states is found, followed by a maximum. This result is a characteristic of superconductors and will be discussed in detail within the framework of BCS theory (see Sect. 10.3). The curve at the top reflects the density of

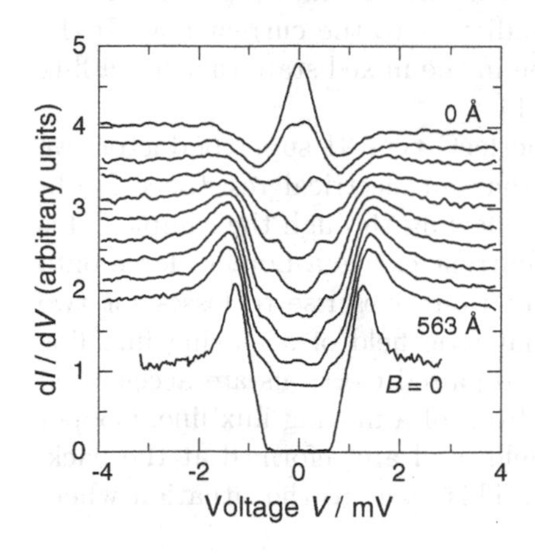

Fig. 10.22. Differential conductivity dI/dV at a flux line in $NbSe_2$ versus applied voltage. The curves are displaced vertically. The *curve at the top* was measured in the core of the vortex, the lower curves were obtained with increasing distance from the core up to 563 Å. The *lowest curve* was taken in the Meissner phase [477]

states in the core of the vortex where a maximum is found that is caused by unpaired electrons bound in the flux line. Away from the core, the density of states exhibits a more complicated energy dependence.

Spatially resolved measurements have shown that in $NbSe_2$ the vortices do not exhibit cylindrical symmetry as originally assumed (see Fig. 10.23). The shape reflects the hexagonal symmetry of the lattice that comes into play, for example, via the shape of the Fermi surface. In addition, it can be shown that the hexagonal symmetry of the Abrikosov lattice (see Sect. 10.1.4) also has an influence on the shape of the flux lines.

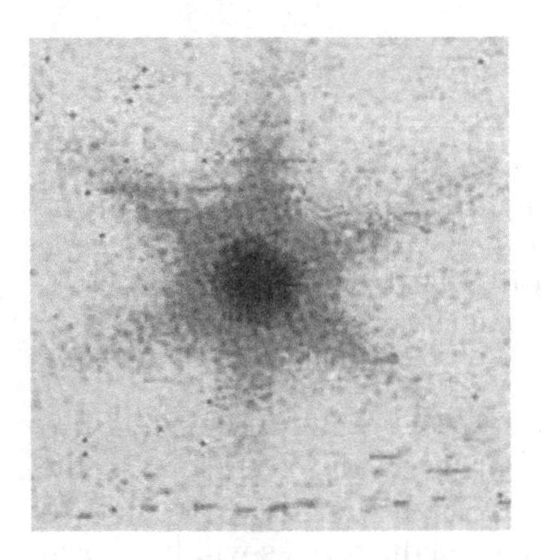

Fig. 10.23. Spatially resolved image of a single vortex in $NbSe_2$. The image was taken at a magnetic field of 50 mT and is 150×150 nm in size. The electronic density of states rises with *increasing blackening* [477]

Currents in the Mixed State

In type I superconductors, the electrical current flows through a wire without loss as long as the current does not exceed the critical current density (see Sect. 10.1.3). The same behavior is observed for type II superconductors in the Meissner state, but the behavior is completely different in the mixed state. For illustration, we discuss here the current flow in a superconducting slab with a magnetic field applied perpendicular to the current flow. If the field is sufficiently high, the sample will be in the mixed state that has a flux line pattern like that sketched in Fig. 10.14.

If we let a current flow through a 'perfect' type II superconductor, we would make the surprising observation that an electrical resistance is observed even if only a very small current is flowing through the sample. This phenomenon is due to the fact that the current exerts a Lorentz force onto the flux lines that start to move. This motion gives rise to losses for two reasons. First, the temporally varying magnetic field of a moving flux line causes an electric field. In this field, the unpaired electrons are accelerated and collide with the lattice. Secondly, in front of a moving flux line, Cooper pairs are broken by the high magnetic field, and are reformed at the back. However, time is needed for this process. This leads to the situation where the Cooper pairs are broken in a higher magnetic field than the field in which they are reformed. Since the condensation energy decreases with increasing magnetic field, this effect leads to dissipation of energy.

In a real type II superconductor, the flux lines are not freely movable. A force F_{p}, though very small, is always necessary to detach the flux lines from preferential places that are called *pinning centers*. As long as the Lorentz force F_{L} is smaller than F_{p}, the flux lines remain pinned, and the supercurrent can flow without resistance. Beyond the critical current, the flux lines become mobile and a voltage drop is observed.

This phenomenon is illustrated in Fig. 10.24 where the voltage–current characteristics of two slabs of $Nb_{0.5}Ta_{0.5}$ with different internal disorder are shown. The experiment was carried out in the mixed state at 3 K and a magnetic field of 0.2 T. Up to the critical current $I_{\mathrm{c}} = 1.2\,\mathrm{A}$, no voltage is detected in the sample with the higher disorder, while a voltage is measured above 0.2 A in the sample with less disorder.[6] In a 'perfect' sample made of the same material, the *dashed line* would be observed.

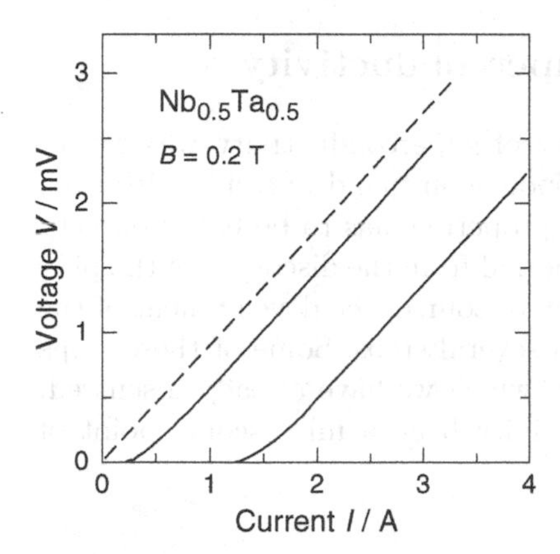

Fig. 10.24. Voltage–current characteristic in the mixed state. Data on $Nb_{0.5}Ta_{0.5}$ alloys with different defect content are shown. The expected curve for a defect-free sample is given by the *dashed line* [478]

In the linear part of the V–I characteristic, the voltage is caused by the movement of the flux lines. The differential resistance $\mathrm{d}V/\mathrm{d}I$ is obviously the same for both samples, i.e., the slope does not depend on the number of pinning centers. This observation can be understood in the following way: As soon as the flux lines are depinned, they move under the influence of the force $\boldsymbol{F}^* = \boldsymbol{F}_{\mathrm{L}} - \boldsymbol{F}_{\mathrm{p}}$. Once the flux lines are flowing, they reach a terminal velocity $v \propto F^* \propto (I - I_{\mathrm{c}})$, which is determined by the balance between the Lorentz force and the viscous drag force on a flux line. On the other hand, the voltage V is proportional to v, i.e., we obtain $V \propto v \propto (I - I_{\mathrm{c}})$. Thus, we find the relation $\mathrm{d}V/\mathrm{d}I = \mathrm{const.}$

[6] The statement 'no voltage' is an oversimplification. It is expected that very weak dissipation also occurs for $I < I_{\mathrm{c}}$, because thermally activated movement of flux lines ('flux creep') is also expected to occur at very small currents.

As mentioned above, the critical current depends on the internal disorder because most types of lattice defects produce pinning centers. Such centers may, for example, be produced by extensive cold working. The pinning is also very strong for small grains or fine precipitates of normal material. Without explanation, we would also like to mention that flux-pinning effects also have a strong influence on the magnetization curves of type II superconductors. The observed magnetization properties are far from an ideal reversible behavior and show highly hysteretic effects.

In general, we may state that as soon as the flux lines start to move across a superconductor, a voltage will develop along it. For instance, if a sufficiently heavy current is passed along a type II wire, flux lines in the form of rings will be nucleated at the surface and will be driven into the center of the wire where they will contract to a point and vanish. This steady flow will generate a voltage along the wire. The wire then has a resistance and is no longer perfectly superconducting. This state is often called the *resistive mixed state*.

10.3 Microscopic Theory of Superconductivity

The development of a microscopic theory of superconductivity was rather difficult, because a description of the phenomenon found in solids with such a wide variety of different structures and properties has to be based on very general premises. During the rather long period from the discovery of the phenomenon until its microscopic explanation, of course, the development of the theory of superconductivity took place in several steps. Some of these steps were based on phenomenological grounds, which we have already discussed. In this section, we consider superconductivity from a microscopic point of view.

10.3.1 Cooper Pairs

In 1956, it was shown by *Cooper* that the ground state of a Fermi gas becomes unstable if a small attractive interaction between a pair of electrons exists [479]. In other words: If we let two noninteracting electrons 'drop' onto the surface of the 'Fermi sea', i.e., we introduce two electrons with Fermi energy E_{F} into a metal, they will remain at the Fermi surface because the energetically lower lying (one-electron) states are already occupied. However, if an attraction between the two electrons exists, they will 'immerse into the Fermi sea' because of the lowering of the energy of the pair by the attractive interaction.

At first glance, it seems hardly conceivable that such an attractive interaction could exist because of the rather strong Coulomb repulsion between electrons. But in 1950 the so-called *isotope effect* was discovered that turned out to be a milestone in understanding superconductivity on a microscopic basis. It was found that the transition temperature of superconductors depends on the atomic mass, and thus on the lattice properties. An especially nice demonstration of this is the measurement of T_c of tin with different isotope concentrations. As shown in Fig. 10.25, the transition temperature is inversely proportional to the square root of the mass of the atoms, and hence proportional to the Debye frequency.

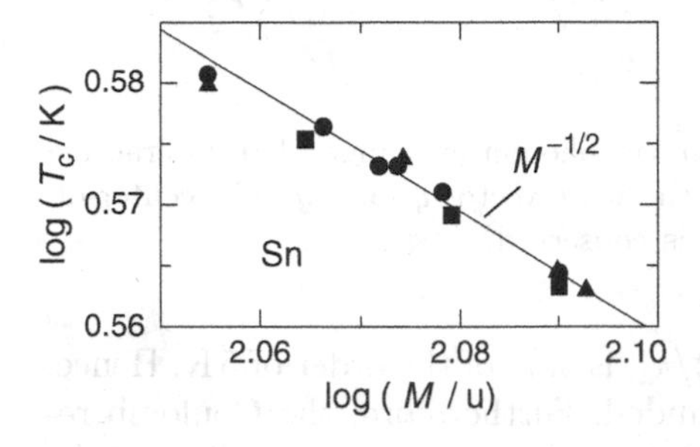

Fig. 10.25. Transition temperature of tin versus atomic mass M. The data are compiled from different publications and contain measurements on monoisotopic and mixed samples [480]

As early as 1950, *Fröhlich* [481] and *Bardeen* [482] independently proposed a mechanism leading to an attractive interaction. The underlying idea is as follows: An electron flying through a lattice of positive ion cores exerts an attractive force on the ions, and causes a positively charged cloud in its 'wake'. This surplus of positive charge ('overscreening') in turn attracts a second electron. The development of the ion cloud is retarded because it will take half of the period of the lattice vibration until the maximum charge density is reached. During the typical time $t \approx 10^{-13}$ s, the first electron has already covered a distance $s = v_F t \approx 10^8 \times 10^{-13}$ cm $= 1000$ Å. Because of this 'retardation effect', the interacting electrons are rather far apart, and the Coulomb repulsion between them is reduced.

As shown in Fig. 10.26, the attractive interaction can be pictured as an exchange of phonons that are virtual because at low temperatures electrons cannot undergo sufficient energy changes to create phonons of short wavelength. The energy change is due to the intermediate state in which one of the electrons emits a phonon with the wave vector $\boldsymbol{q}$ that is then absorbed by the other. In this process the center-of-mass momentum $\hbar\boldsymbol{K}$ is conserved, i.e., $\boldsymbol{k}_1 + \boldsymbol{k}_2 = \boldsymbol{k}_1' + \boldsymbol{k}_2' = \boldsymbol{K}$. In second-order perturbation theory the corresponding matrix element reads

$$\mathcal{V}(\boldsymbol{k}_1, \boldsymbol{k}_2, \boldsymbol{q}) = \frac{g^2 \hbar\omega_{\boldsymbol{q}}}{(\epsilon_{\boldsymbol{k}_1+\boldsymbol{q}} - \epsilon_{\boldsymbol{k}_1})^2 - (\hbar\omega_{\boldsymbol{q}})^2} . \tag{10.62}$$

Here, $\epsilon_{\boldsymbol{k}_1}$ is the energy of the electron in the state $\boldsymbol{k}_1$, $\hbar\omega_{\boldsymbol{q}}$ is the energy of the exchanged phonon, and g is the coupling constant for the electron–phonon interaction.

This interaction is only attractive if the matrix element $\mathcal{V}(\boldsymbol{k}_1, \boldsymbol{k}_2, \boldsymbol{q})$ is negative, i.e., if $|\epsilon_{\boldsymbol{k}_1+\boldsymbol{q}} - \epsilon_{\boldsymbol{k}_1}| < \hbar\omega_{\boldsymbol{q}}$. The average value of $\hbar\omega_{\boldsymbol{q}}/k_B$ is about 300 K,

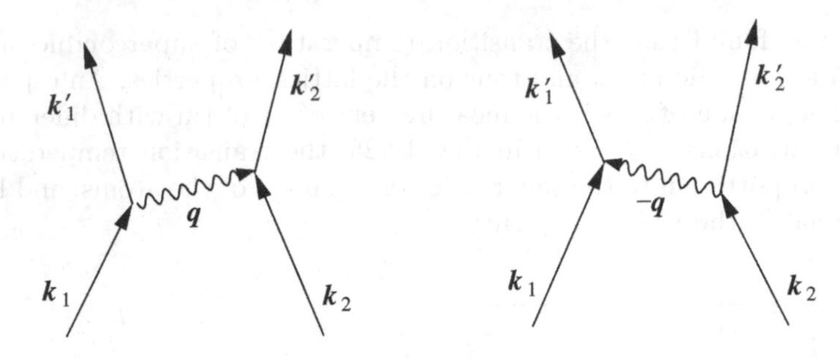

Fig. 10.26. Electron–electron interaction via phonon exchange. The interacting electrons exchange a virtual phonon with the wave vector $\boldsymbol{q}$, or $-\boldsymbol{q}$. The center-of-mass momentum $\boldsymbol{k}_1 + \boldsymbol{k}_2 = \boldsymbol{k}'_1 + \boldsymbol{k}'_2 = \boldsymbol{K}$ is conserved

while the average value of $(\epsilon_{\boldsymbol{k}_1+\boldsymbol{q}} - \epsilon_{\boldsymbol{k}_1})/k_\mathrm{B}$ is only of the order of 5 K. Hence, on the average, $\mathcal{V}$ is negative as demanded. Furthermore, the Coulomb repulsion between the electrons must be taken into account. Depending on the relative strengths of the two types of interaction, the resulting force can either be attractive or repulsive. Note that the attraction is proportional to g^2, therefore strong coupling between electrons and phonons is favorable for superconductivity. This is the explanation for the surprising observation that metals with high conductivity at room temperature like noble metals, are not superconducting. In these metals, the Coulomb repulsion exceeds the attraction caused by the rather weak phonon-exchange interaction.

In the following, we discuss the formation of *Cooper pairs* at $T = 0$, where all one-electron states below the Fermi level are occupied. Under this condition, only states $\epsilon(\boldsymbol{k})$ in the energy range $E_\mathrm{F} \leq \epsilon(\boldsymbol{k}) \leq E_\mathrm{F} + \hbar\omega_\mathrm{D}$ are involved in the phonon-exchange process, where ω_D is the Debye frequency. In $\boldsymbol{k}$ space, states with these energies are located on the Fermi surface within a thin spherical shell of thickness $\delta k = (m\omega_\mathrm{D}/\hbar k_\mathrm{F})$. Both electrons are subjected to this restriction, as schematically depicted in Fig. 10.27a. As indicated, the wave vectors $\boldsymbol{k}_1$ and $\boldsymbol{k}_2$ must start or end in the *dark tinted area of overlap*. Therefore, the effectiveness of the phonon exchange will have a sharp maximum at the center-of mass momentum $\hbar\boldsymbol{K} = 0$, because the whole shell will be accessible to the exchange processes. Thus, we come to the important conclusion that the center-of-mass momentum of Cooper pairs vanishes. The wave vectors of the electrons fulfill the condition $\boldsymbol{k}_1 = -\boldsymbol{k}_2 = \boldsymbol{k}$, meaning that the angular momentum of the Cooper pairs vanishes, i.e., $L = 0$. Using the nomenclature of atomic physics, such a pair is called an *s-state pair*. In the following, we symbolize these pair states by $(\boldsymbol{k}, -\boldsymbol{k})$.

To describe a Cooper pair, we need to construct an appropriate two-particle wave function $\psi(\boldsymbol{r}_1, \boldsymbol{r}_2)$. We use the superposition of two plane waves with the wave vectors $\boldsymbol{k}_1$ and $\boldsymbol{k}_2$, and write

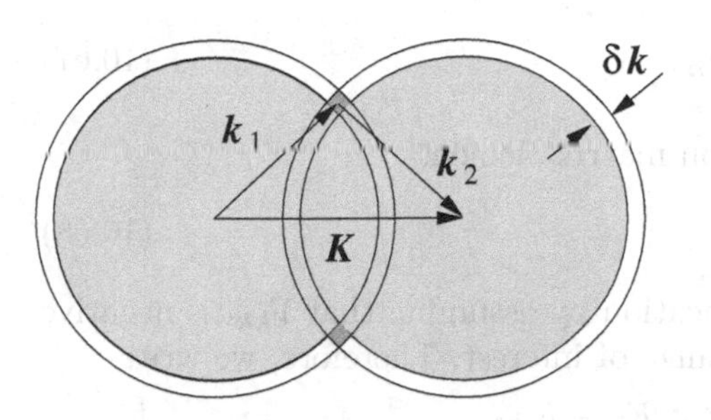

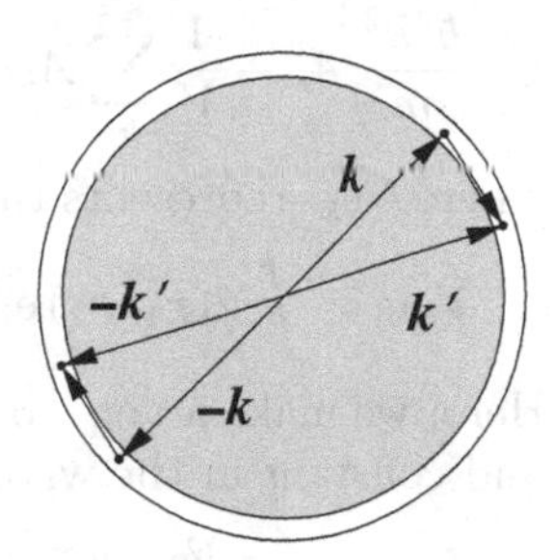

Fig. 10.27. (**a**) Construction to find the states that are accessible to an interacting electron pair with the center-of-mass momentum $\hbar \boldsymbol{K}$. (**b**) A typical transition occurring in a Cooper pair in which one pair of electrons interacts above a quiescent Fermi sea. The center-of-mass momentum of the pair is chosen to be zero

$$\psi(\boldsymbol{r}_1, \boldsymbol{r}_2) = \frac{1}{V}\, \mathrm{e}^{\mathrm{i}\boldsymbol{k}_1\cdot\boldsymbol{r}_1}\, \mathrm{e}^{\mathrm{i}\boldsymbol{k}_2\cdot\boldsymbol{r}_2} = \frac{1}{V}\, \mathrm{e}^{\boldsymbol{k}\cdot\boldsymbol{r}} = \Psi(\boldsymbol{r})\,, \tag{10.63}$$

where $\boldsymbol{r} = (\boldsymbol{r}_1 - \boldsymbol{r}_2)$ is the relative coordinate, and V stands for the volume of the sample. As depicted in Fig. 10.27b, the two electrons are permanently scattered into new states with different wave vectors $\boldsymbol{k}$. We take this fact into account by superimposing wave functions of the type (10.63), and write for the wave function of a Cooper pair

$$\Psi(\boldsymbol{r}) = \sum_{\boldsymbol{k}} A_{\boldsymbol{k}}\, \mathrm{e}^{\mathrm{i}\boldsymbol{k}\cdot\boldsymbol{r}}\,. \tag{10.64}$$

Here, $|A_{\boldsymbol{k}}|^2$ is a measure of the probability of finding a particular electron pair in the state $(\boldsymbol{k}, -\boldsymbol{k})$. As explained above, the only states that are accessible to electrons are in the thin shell at the Fermi surface, with a thickness defined by $\hbar\omega_\mathrm{D}$. Therefore, the expansion coefficient $A_{\boldsymbol{k}}$ is assumed to be nonzero only in the corresponding range of wave numbers, i.e.,

$$A_{\boldsymbol{k}} \begin{cases} \neq 0 & \text{for} \quad k_\mathrm{F} < k < \sqrt{2m(E_\mathrm{F} + \hbar\omega_\mathrm{D})/\hbar^2} \\ = 0 & \text{otherwise}\,. \end{cases} \tag{10.65}$$

To calculate the eigenvalue E of a Cooper pair, we start with the Schrödinger equation

$$\left[-\frac{\hbar^2}{2m}(\Delta_1 + \Delta_2) + \mathcal{V}(\boldsymbol{r}_1, \boldsymbol{r}_2)\right] \psi(\boldsymbol{r}_1, \boldsymbol{r}_2) = E\psi(\boldsymbol{r}_1, \boldsymbol{r}_2)\,. \tag{10.66}$$

The potential $\mathcal{V}(\boldsymbol{r}_1, \boldsymbol{r}_2)$ consists of two parts, the attractive part caused by phonon exchange and the repulsive part due to Coulomb repulsion. Its exact shape is unknown but for the time being, this is without importance.

The Schrödinger equation is solved as usual by inserting (10.64), multiplying from the left-hand side with $\exp(-\mathrm{i}\boldsymbol{k}' \cdot \boldsymbol{r})$, and integrating over the volume V. This procedure leads to

$$\frac{\hbar^2 k^2}{m} A_{\boldsymbol{k}} + \frac{1}{V} \sum_{\boldsymbol{k}'} A_{\boldsymbol{k}'} \mathcal{V}_{\boldsymbol{k}\boldsymbol{k}'} = E A_{\boldsymbol{k}}, \tag{10.67}$$

where $\mathcal{V}_{\boldsymbol{k}\boldsymbol{k}'}$ represents the interaction matrix element

$$\mathcal{V}_{\boldsymbol{k}\boldsymbol{k}'} = \int \mathcal{V}(\boldsymbol{r}_1, \boldsymbol{r}_2)\, \mathrm{e}^{\mathrm{i}(\boldsymbol{k}-\boldsymbol{k}')\cdot\boldsymbol{r}} \mathrm{d}^3 x. \tag{10.68}$$

Here, we make a very rough simplification by assuming that $\mathcal{V}_{\boldsymbol{k}\boldsymbol{k}'}$ is negative and constant in the whole energy range of interest. Therefore, we write

$$\mathcal{V}_{\boldsymbol{k}\boldsymbol{k}'} = \begin{cases} -\mathcal{V}_0 & \text{for } E_{\mathrm{F}} < \epsilon_{\boldsymbol{k}}, \epsilon_{\boldsymbol{k}'} < E_{\mathrm{F}} + \hbar\omega_{\mathrm{D}} \\ 0 & \text{otherwise}, \end{cases} \tag{10.69}$$

where $\mathcal{V}_0$ is a characteristic positive constant. With this simplification, (10.67) takes the form

$$\left(\frac{\hbar^2 k^2}{m} - E\right) A_{\boldsymbol{k}} = \frac{\mathcal{V}_0}{V} \sum_{\boldsymbol{k}'} A_{\boldsymbol{k}'}. \tag{10.70}$$

Introducing the abbreviation $z = \hbar^2 k^2/2m$, we obtain

$$A_{\boldsymbol{k}} = \frac{\mathcal{V}_0}{V} \frac{1}{2z - E} \sum_{\boldsymbol{k}'} A_{\boldsymbol{k}'}. \tag{10.71}$$

To simplify this equation further, we sum over all wave vectors $\boldsymbol{k}$. Since $\sum_{\boldsymbol{k}} A_{\boldsymbol{k}} = \sum_{\boldsymbol{k}'} A_{\boldsymbol{k}'}$, these sums may be cancelled, resulting in

$$1 = \frac{\mathcal{V}_0}{V} \sum_{\boldsymbol{k}} \frac{1}{2z - E}. \tag{10.72}$$

Furthermore, we may replace the sum in this equation by an integral. Since the density of states $D(E)$ in the vicinity of E_{F} is approximately constant, we put $D(E) \approx D(E_{\mathrm{F}})$, and write

$$1 = \mathcal{V}_0 \frac{D(E_{\mathrm{F}})}{2} \int_{E_{\mathrm{F}}}^{E_{\mathrm{F}}+\hbar\omega_{\mathrm{D}}} \frac{\mathrm{d}z}{2z - E}. \tag{10.73}$$

The factor 1/2 appears because we are considering pair states here.[7] Carrying out the integration, we obtain for the interaction energy the final result:

$$\delta E = E - 2E_{\mathrm{F}} = \frac{2\,\hbar\omega_{\mathrm{D}}}{1 - \exp[4/\mathcal{V}_0 D(E_{\mathrm{F}})]}. \tag{10.74}$$

As expected, the energy of the electron pair is smaller than $2E_{\mathrm{F}}$. The energy difference δE may be considered as the binding energy of a Cooper pair. In the so-called *weak coupling limit*, i.e., for $\mathcal{V}_0 D(E_{\mathrm{F}}) \ll 1$, we may approximate this quantity by

$$\delta E \approx -2\,\hbar\omega_{\mathrm{D}}\, \mathrm{e}^{-4/[\mathcal{V}_0 D(E_{\mathrm{F}})]}. \tag{10.75}$$

[7] In textbooks on superconductivity this factor 1/2 is often avoided by writing down the density of states for electrons with definite spin.

The wave function (10.64) that describes Cooper pairs tells us that it is not possible to relate well-defined wave vectors to the electrons of a Cooper pair. The wave function contains all wave vectors between the energy range E_F and $E_F + \hbar\omega_D$. From (10.71) it follows that the weighting factor $A_{\boldsymbol{k}}$ is largest for states with an energy $z = \hbar^2 k^2/2m$ comparable to E_F. Of course, the results obtained above are also valid for two electrons scattered from states below the Fermi surface to states above. Although their kinetic energy rises, the gain in potential energy predominates so that the electron pair is again in a bound state.

Until now we have not taken into account the fact that electrons are indistinguishable fermions with antisymmetric wave functions. Since (10.63) is symmetric in the position coordinates $(\boldsymbol{r}_1, \boldsymbol{r}_2)$, the spin part of the wave function has to be antisymmetric. This requires the opposite orientation of their spins, resulting in a *spin-singlet state*. Therefore, we symbolize a Cooper pair from now on by $(\boldsymbol{k}\uparrow, -\boldsymbol{k}\downarrow)$. The Cooper pairs with angular momentum $L = 0$ and total spin $S = 0$ behave like bosons and are able to condense into a common quantum-mechanical ground state. This possibility does not exist for single electrons, because their antisymmetric wave function does not allow multiple occupation of the same state.

If the exchange interaction is more complex than phonon exchange, the spins of the Cooper pair can be aligned, leading to so-called *spin-triplet pairing*. To fulfill the symmetry requirements, the orbital part of the wave function must then be antisymmetric. As we have already seen in our discussion of superfluid ^{3}He (see Chap. 4), pairs with spin $S = 1$ exhibit p-state symmetry. There is also the possibility that spin-singlet pairs possess a finite angular momentum, as in high-T_c superconductors, where $L = 2$. We will discuss such *unconventional superconductors* in Sect. 10.5.

10.3.2 BCS Ground State

In the previous section, it was shown that an attractive interaction between two quasifree electrons may lead to a reduction of their potential energy and thus to the formation of bound pairs. In a superconductor there are many pairs all residing in a common ground state. A theoretical description of this so-called *BCS ground state* is mathematically more involved than the treatment of a single pair because there is a subtle interplay between Cooper pairs and the remaining Fermi sea. Here, we only summarize the theoretical ideas, display the results, and try to make them plausible.

First, we consider the energy reduction due to the formation of the BCS ground state. The Hamiltonian $\mathcal{H}$ is taken as the kinetic energy of all the electrons, together with the interaction (10.69). The wave function is constructed so that if one member $(\boldsymbol{k}\uparrow)$ of a Cooper pair is present, so is $(-\boldsymbol{k}\downarrow)$. With $|1\rangle_{\boldsymbol{k}}$ we indicate that the pair $(\boldsymbol{k}\uparrow, -\boldsymbol{k}\downarrow)$ is occupied, and with $|0\rangle_{\boldsymbol{k}}$ that it is unoccupied. The general form of the wave function of a pair is the superposition of the two states $|1\rangle_{\boldsymbol{k}}$ and $|0\rangle_{\boldsymbol{k}}$, and has therefore the form

$$|\psi\rangle_{\boldsymbol{k}} = u_{\boldsymbol{k}}|0\rangle_{\boldsymbol{k}} + v_{\boldsymbol{k}}|1\rangle_{\boldsymbol{k}}\,, \tag{10.76}$$

with the real coefficients $u_{\boldsymbol{k}}$ and $v_{\boldsymbol{k}}$. The state $|1\rangle_{\boldsymbol{k}}$ is occupied with the probability $w_{\boldsymbol{k}} = v_{\boldsymbol{k}}^2$, while the probability for an unoccupied state is given by $u_{\boldsymbol{k}}^2 = 1 - w_{\boldsymbol{k}}$.

In BCS theory, the superconducting ground state $|\Psi\rangle$ is a common state of all Cooper pairs and is constructed by superimposing the wave functions of all pairs states, thus neglecting an interaction *between* the pairs. The wave function Ψ of the ground state is thus expressed by the product of the wave function of the individual Cooper pairs, i.e., by

$$|\Psi\rangle = \prod_{\boldsymbol{k}} |\psi\rangle_{\boldsymbol{k}} = \prod_{\boldsymbol{k}} \Big(u_{\boldsymbol{k}}|0\rangle_{\boldsymbol{k}} + v_{\boldsymbol{k}}|1\rangle_{\boldsymbol{k}} \Big)\,. \tag{10.77}$$

The coefficients $u_{\boldsymbol{k}}$ and $v_{\boldsymbol{k}}$ can be determined by minimizing the energy of the BCS ground state via the variation method.

Since the pair states are either occupied or unoccupied, we may use the analogy with two-state systems, and write

$$|1\rangle_{\boldsymbol{k}} = \begin{pmatrix} 1 \\ 0 \end{pmatrix}_{\boldsymbol{k}} \qquad \text{and} \qquad |0\rangle_{\boldsymbol{k}} = \begin{pmatrix} 0 \\ 1 \end{pmatrix}_{\boldsymbol{k}}\,. \tag{10.78}$$

The corresponding creation and destruction operators $\sigma_{\boldsymbol{k}}^+$ and $\sigma_{\boldsymbol{k}}^-$ can be expressed with the Pauli spin matrices σ^x and σ^y:

$$\begin{aligned} \sigma_{\boldsymbol{k}}^+ &= \frac{1}{2}(\sigma_{\boldsymbol{k}}^x + \mathrm{i}\sigma_{\boldsymbol{k}}^y) = \begin{pmatrix} 0 & 1 \\ 0 & 0 \end{pmatrix}_{\boldsymbol{k}}\,, \\ \sigma_{\boldsymbol{k}}^- &= \frac{1}{2}(\sigma_{\boldsymbol{k}}^x - \mathrm{i}\sigma_{\boldsymbol{k}}^y) = \begin{pmatrix} 0 & 0 \\ 1 & 0 \end{pmatrix}_{\boldsymbol{k}}\,. \end{aligned} \tag{10.79}$$

The effect of these operators can easily be demonstrated by inserting (10.79) into the definition of the pair states (10.78). As expected, we find

$$\begin{aligned} \sigma_{\boldsymbol{k}}^+|1\rangle_{\boldsymbol{k}} &= 0\,, & \sigma_{\boldsymbol{k}}^+|0\rangle_{\boldsymbol{k}} &= |1\rangle_{\boldsymbol{k}}\,, \\ \sigma_{\boldsymbol{k}}^-|1\rangle_{\boldsymbol{k}} &= |0\rangle_{\boldsymbol{k}}\,, & \sigma_{\boldsymbol{k}}^-|0\rangle_{\boldsymbol{k}} &= 0\,. \end{aligned} \tag{10.80}$$

With the help of some algebraic transformations, it can be shown that in the formalism of second quantization (occupation number formalism), the Hamiltonian $\mathcal{H}$ takes the form

$$\mathcal{H} = \sum_{\boldsymbol{k}} 2\eta_{\boldsymbol{k}}\, \sigma_{\boldsymbol{k}}^+\, \sigma_{\boldsymbol{k}}^- - \frac{\mathcal{V}_0}{V} \sum_{\boldsymbol{k},\boldsymbol{k}'} \sigma_{\boldsymbol{k}}^+\, \sigma_{\boldsymbol{k}'}^-\,. \tag{10.81}$$

The first term represents the kinetic energy of the Cooper pairs. As abbreviation we have introduced here $\eta_{\boldsymbol{k}} = \hbar^2 k^2/2m - E_\mathrm{F}$, representing the kinetic energy of a single electron with respect to the Fermi energy. The factor of 2 reflects the fact that a Cooper pair consists of two electrons with opposite spin. The second term describes the change of the potential energy due to the phonon-exchange interaction. With respect to the normal conducting state at $T = 0$, the kinetic energy of the Cooper pairs can be either positive or negative. The contribution of the pairs to the potential energy is, however, always negative, because of the assumption that the interaction is always attractive.

The expectation value $W_0 = \langle \Psi | \mathcal{H} | \Psi \rangle$ is easily calculated, and we obtain

$$W_0 = \sum_{\boldsymbol{k}} 2v_{\boldsymbol{k}}^2 \eta_{\boldsymbol{k}} - \frac{\mathcal{V}_0}{V} \sum_{\boldsymbol{k}',\boldsymbol{k}} v_{\boldsymbol{k}} u_{\boldsymbol{k}'} u_{\boldsymbol{k}} v_{\boldsymbol{k}'} \,. \tag{10.82}$$

Minimizing the energy W_0 with respect to $v_{\boldsymbol{k}}$ and $u_{\boldsymbol{k}}$, leads to the relation

$$2u_{\boldsymbol{k}} v_{\boldsymbol{k}} \eta_{\boldsymbol{k}} - \Delta_0 (u_{\boldsymbol{k}}^2 - v_{\boldsymbol{k}}^2) = 0 \,, \tag{10.83}$$

where we have used the abbreviation

$$\Delta_0 = \frac{\mathcal{V}_0}{V} \sum_{\boldsymbol{k}'} u_{\boldsymbol{k}'} v_{\boldsymbol{k}'} \,. \tag{10.84}$$

The quantity Δ_0 generally depends on $\boldsymbol{k}$, but this dependence does not appear here because of the above assumption $\mathcal{V}_{\boldsymbol{k}\boldsymbol{k}'} = -\mathcal{V}_0 = \text{const.}$ As we will see, this simplification is not applicable to unconventional superconductors that we consider in Sect. 10.5.

We express $u_{\boldsymbol{k}}$ and $v_{\boldsymbol{k}}$ in terms of a new variable $E_{\boldsymbol{k}}$ by writing

$$u_{\boldsymbol{k}}^2 = \frac{1}{2} \left(1 + \frac{\eta_{\boldsymbol{k}}}{E_{\boldsymbol{k}}} \right) , \tag{10.85}$$

$$v_{\boldsymbol{k}}^2 = \frac{1}{2} \left(1 - \frac{\eta_{\boldsymbol{k}}}{E_{\boldsymbol{k}}} \right) . \tag{10.86}$$

The significance of Δ_0 and $E_{\boldsymbol{k}}$ will become clear during the course of the following discussions. In terms of these two quantities, the minimum condition (10.83) takes the simple form

$$E_{\boldsymbol{k}}^2 = \eta_{\boldsymbol{k}}^2 + \Delta_0^2 \,. \tag{10.87}$$

With the new variables, the ground-state energy W_0 given by (10.82) reads

$$W_0 = \sum_{\boldsymbol{k}} \eta_{\boldsymbol{k}} \left(1 - \frac{\eta_{\boldsymbol{k}}}{E_{\boldsymbol{k}}} \right) - \frac{\Delta_0^2 \, V}{\mathcal{V}_0} \,. \tag{10.88}$$

Inserting relation (10.87) into (10.86), we obtain for the probability $w_{\boldsymbol{k}}$ that the pair state $(\boldsymbol{k} \uparrow, -\boldsymbol{k} \downarrow)$ is occupied:

$$w_{\boldsymbol{k}} = v_{\boldsymbol{k}}^2 = \frac{1}{2} \left(1 - \frac{\eta_{\boldsymbol{k}}}{E_{\boldsymbol{k}}} \right) = \frac{1}{2} \left(1 - \frac{\eta_{\boldsymbol{k}}}{\sqrt{\eta_{\boldsymbol{k}}^2 + \Delta_0^2}} \right) . \tag{10.89}$$

This expression is the counterpart of the Fermi function for one-electron states that is plotted in Fig. 10.28 for $T = T_c$, and compared with $w_{\boldsymbol{k}}$ at

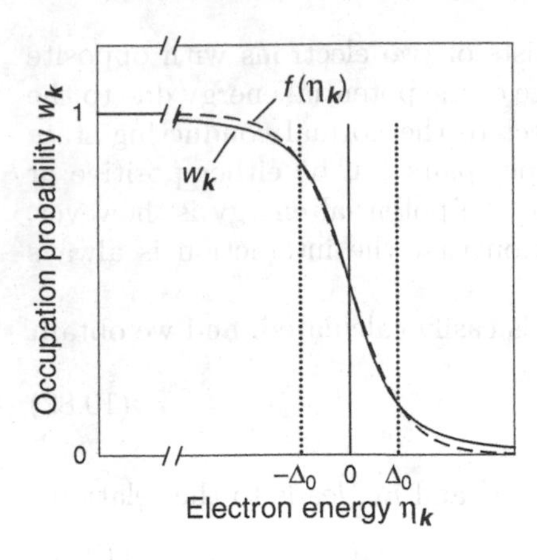

Fig. 10.28. Probability $w_{\boldsymbol{k}}$ for the occupation of pair states versus kinetic energy $\eta_{\boldsymbol{k}}$ at $T = 0$ (*full line*). For comparison the Fermi function (*dashed line*) at $T = T_c$ is also drawn

$T = 0$. Both curves drop steeply at E_F. The width is roughly determined by Δ_0 in both cases. The shape of the function (10.89) demonstrates that the interaction of the electrons with virtual phonons gives rise to an occupation of states above the Fermi energy, resulting in a kinetic energy of the system higher than in the normal state.

Even though the kinetic energy of the electrons is enhanced, energy is gained by the transition from the normal to the superconducting state due to the lowering of the potential energy. The difference between the (free) energy of the superconducting and the normal conducting state is the so-called *condensation energy*. It has already been introduced in Sect. 10.2.1, where it was shown that it is proportional to B_c^2. Its value at $T = 0$ can be found by subtracting the internal energy $W_0^n = 2\sum_{|k|<k_F} \eta_{\boldsymbol{k}}$ of the normal conductor from (10.88). After some algebraic manipulations, one obtains

$$\frac{W_{\text{con}}}{V} = \frac{W_0 - W_0^n}{V} = -\frac{1}{4} D(E_F)\, \Delta_0^2 \,. \tag{10.90}$$

This result indicates that the quantity Δ_0 is a measure of the condensation energy, and that $\Delta_0 \propto B_c(0)$.

At the end of this section we calculate the magnitude of Δ_0. For this purpose, we write down (10.84) once again, use (10.85) and (10.86), and obtain

$$\Delta_0 = \frac{\mathcal{V}_0}{V} \sum_{\boldsymbol{k}} u_{\boldsymbol{k}} v_{\boldsymbol{k}} = \frac{1}{2}\frac{\mathcal{V}_0}{V} \sum_{\boldsymbol{k}} \frac{\Delta_0}{E_{\boldsymbol{k}}} = \frac{1}{2}\frac{\mathcal{V}_0}{V} \sum_{\boldsymbol{k}} \frac{\Delta_0}{\sqrt{\eta_{\boldsymbol{k}}^2 + \Delta_0^2}} \,. \tag{10.91}$$

Replacing the sum by an integral and assuming, as in the derivation of (10.73), that the density of states at the Fermi energy is constant, we find

$$1 = \frac{\mathcal{V}_0}{2} \int_{-\hbar\omega_D}^{\hbar\omega_D} \frac{D(E_F)}{2} \frac{\mathrm{d}\eta}{\sqrt{\eta^2 + \Delta_0^2}} = \frac{\mathcal{V}_0\, D(E_F)}{2} \operatorname{arc\,sinh}\left(\frac{\hbar\omega_D}{\Delta_0}\right) . \tag{10.92}$$

Consequently, Δ_0 is given by

$$\Delta_0 = \frac{\hbar\omega_\mathrm{D}}{\sinh\left[\frac{2}{\mathcal{V}_0\, D(E_\mathrm{F})}\right]} \approx 2\,\hbar\omega_\mathrm{D}\,\mathrm{e}^{-2/\mathcal{V}_0\, D(E_\mathrm{F})}\,, \tag{10.93}$$

where the last relation holds for the weak coupling limit, i.e., for $\mathcal{V}_0\, D(E_\mathrm{F}) \ll 1$.

10.3.3 Excitation of the BCS Ground State

We now turn our attention to excited states of superconductors and give insight into the physical significance of Δ_0. As we shall see, Δ_0 is the *energy gap* for elementary excitations, meaning that excited states have an energy that is at least Δ_0 above the ground-state energy W_0. The simplest conceivable excited state is a broken pair state in which only one state of the pair $(\boldsymbol{k}'\uparrow, -\boldsymbol{k}'\downarrow)$ is occupied. With the help of some algebraic transformations, the ground-state energy (10.88) can be expressed in the form

$$W_0 = -2\sum_{\boldsymbol{k}} E_{\boldsymbol{k}} v_{\boldsymbol{k}}^4\,. \tag{10.94}$$

Correspondingly, the energy W_1 of the superconductor with one broken pair reads

$$W_1 = -2\sum_{\boldsymbol{k}\neq\boldsymbol{k}'} \mathrm{E}_{\boldsymbol{k}} v_{\boldsymbol{k}}^4\,. \tag{10.95}$$

Thus, the energy δE necessary to break up the pair is given by

$$\delta E = W_1 - W_0 = 2E_{\boldsymbol{k}'} = 2\sqrt{\eta_{\boldsymbol{k}'}^2 + \Delta_0^2}\,. \tag{10.96}$$

Since $\eta_{\boldsymbol{k}'} = \hbar^2 k'^2/2m - E_\mathrm{F}$ can be arbitrarily small, a minimum energy of $\delta E_\mathrm{min} = 2\Delta_0$ is required to excite the superconductor. The factor of 2 is due to the fact that two unpaired electrons are always created by breaking up one pair. Thus, an energy gap Δ_0 exists between the ground state and the excited states of a superconductor. This is an essential difference from the situation in a normal conductor. In a normal conductor, excitations can be achieved with arbitrarily small energies, but the energy δE_min has to be raised to excite a superconductor.

The excitation spectrum of superconductors given by (10.87) is sketched in Fig. 10.29. The excitations are part electron, part hole, and are generally called *quasiparticles* or *Bogoliubov quasiparticles.* For large positive values of $\eta_{\boldsymbol{k}}$, i.e., far from the Fermi surface, the dispersion relation (10.87) merges into that of free electrons. Quasiparticles with $\eta_{\boldsymbol{k}} < 0$ have hole-like character. This means that the quasiparticles change their character while passing through the Fermi level. At $\eta_{\boldsymbol{k}} \approx 0$, i.e., close to the Fermi energy, quasiparticles are a mixture of an electron with the wave vector $\boldsymbol{k}$ and a hole with the wave vector $-\boldsymbol{k}$. This becomes clearer in looking at the break up of a Cooper pair. Assume that an electron with wave vector $\boldsymbol{k}$ has been scattered out of the pair state into the state $\boldsymbol{k}'$. A hole at $\boldsymbol{k}$ stays behind that interacts

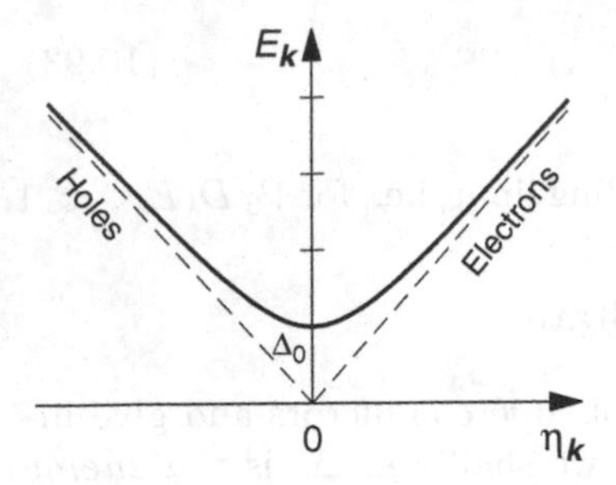

Fig. 10.29. Excitation energy of quasiparticles close to the Fermi energy. Hole-like states are *left* of the origin, electron-like states to the *right side*

with the second electron of the pair with the wave vector $-\boldsymbol{k}$. Similarly, the electron now being in state $\boldsymbol{k}'$, interacts with the hole at $-\boldsymbol{k}'$.

As mentioned above, the *common ground state* of the Cooper pairs is separated from the quasiparticle states by the energy gap Δ_0. The quasiparticle density of states $D_\mathrm{s}(E_{\boldsymbol{k}})$ follows directly from the density of the normal state since no state is lost in the superconducting transition, i.e., $D_\mathrm{s}(E_{\boldsymbol{k}})\,\mathrm{d}E_{\boldsymbol{k}} = D_\mathrm{n}(\eta_{\boldsymbol{k}})\,\mathrm{d}\eta_{\boldsymbol{k}}$, where $D_\mathrm{n}(\eta_{\boldsymbol{k}})$ represents the electronic density of states in the normal conductor. In the vicinity of the Fermi energy, we may put $D_\mathrm{n}(\eta_{\boldsymbol{k}}) \approx D_\mathrm{n}(E_\mathrm{F}) = \text{const.}$, and we obtain

$$D_\mathrm{s}(E_{\boldsymbol{k}}) = D_\mathrm{n}(\eta_{\boldsymbol{k}})\,\frac{\mathrm{d}\eta_{\boldsymbol{k}}}{\mathrm{d}E_{\boldsymbol{k}}} = \begin{cases} D_\mathrm{n}(E_\mathrm{F})\,\dfrac{E_{\boldsymbol{k}}}{\sqrt{E_{\boldsymbol{k}}^2 - \Delta_0^2}} & \text{for} \quad E_{\boldsymbol{k}} > \Delta_0 \\ 0 & \text{for} \quad E_{\boldsymbol{k}} < \Delta_0 \,. \end{cases} \tag{10.97}$$

In Fig. 10.30a, the predicted density of states $D_\mathrm{s}(E_{\boldsymbol{k}})$ of the quasiparticles is drawn. At $E_{\boldsymbol{k}} = \Delta_0$, the density of states is expected to diverge. For $\eta_{\boldsymbol{k}} \gg \Delta_0$, the quasiparticle density $D_\mathrm{s}(E_{\boldsymbol{k}})$ is expected to merge with $D_\mathrm{n}(\eta_{\boldsymbol{k}})$

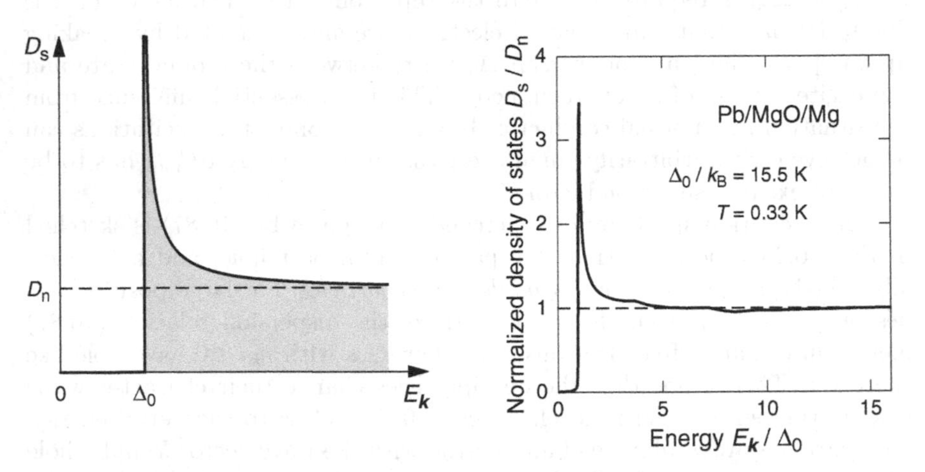

Fig. 10.30. (**a**) Quasiparticle density of states versus excitation energy. (**b**) Experimentally determined density of states of Pb versus normalized excitation energy. The measurement was carried out with a Pb/MgO/Mg-tunnel junction [483]

of the free-electron gas. The density of pair-excitation states corresponds to a δ-function at $E_k = 0$. Experimentally, the quasiparticle density of states can be measured by tunnelling spectroscopy (see Sect. 10.3.6). The result of an early measurement of this type on Pb is shown in Fig. 10.30b. Without discussion, we want to mention here that the features superimposed on the smooth curve at higher energies are due to the coupling between electrons and phonons.

At the end of this section we show how the density of states of the quasiparticles is represented in an 'one-electron representation' that stems from semiconductor physics. In this representation, hole-like states with positive energy are described as electron-like states with negative energy. The energy gap has the width $2\Delta_0$, in this representation Cooper pairs do not appear at all. As shown in Fig. 10.31a, at absolute zero all states up to the gap at $(E_F - \Delta_0)$ are occupied. As described in the next section, quasiparticles are thermally excited at finite temperatures resulting in a population of states above the gap. Consequently, empty states occur in the hitherto occupied part of the density of states, as depicted in Fig. 10.31b. In this figure it is also indicated that the energy gap $\Delta(T)$ depends on temperature. This aspect will be discussed in the next section, too. Since the properties of so-called tunnel junctions are more easily described in this representation, we shall use this approach again in Sect. 10.3.6.

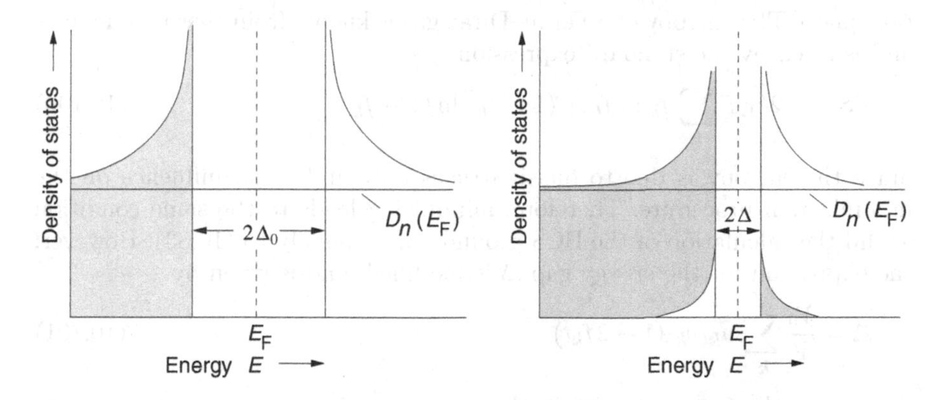

Fig. 10.31. One-particle representation of the density of states of a superconductor. Occupied states are represented by *tinted areas*. (**a**) At $T = 0$ all states below the gap are occupied. (**b**) At $0 < T < T_c$, quasiparticles are excited and empty states are present below the gap

10.3.4 BCS State at Finite Temperatures

At finite temperatures, thermally excited quasiparticles are also present in superconductors. They reduce the number of states accessible to electron pairs for phonon exchange. This leads to a reduction of the interaction energy and an increase of the energy of the system. The most important consequence is that the energy gap shrinks with increasing temperature, and vanishes at the transition temperature. In the following, we denote the energy gap by Δ, but we still abbreviate $\Delta(T=0)$ by Δ_0 to shorten the expressions.

The energy gap Δ at finite temperatures can be calculated in the same way as for $T=0$. We now have to consider the free energy $F = E_{\text{kin}} + E_{\text{pot}} - TS$ and must therefore also take into account the kinetic energy and entropy of the quasiparticles. We assume that quasiparticles obey Fermi statistics, i.e., we write $f_{\boldsymbol{k}} = [\exp(E_{\boldsymbol{k}}/k_{\text{B}}T)+1]^{-1}$ for the occupation number. The kinetic energy is thus given by

$$E_{\text{kin}} = 2\sum_{\boldsymbol{k}} \eta_{\boldsymbol{k}} f_{\boldsymbol{k}} + \eta_{\boldsymbol{k}}(1-2f_{\boldsymbol{k}})v_{\boldsymbol{k}}^2\,, \tag{10.98}$$

where the first term reflects the contribution from unpaired electrons. The factor $(1-2f_{\boldsymbol{k}})$ in the second term describes the probability for the pair state $|\Psi\rangle$ to be occupied. For the interaction energy we write

$$E_{\text{pot}} = -\frac{\mathcal{V}_0}{V}\sum_{\boldsymbol{k},\boldsymbol{k}'} u_{\boldsymbol{k}}v_{\boldsymbol{k}}u_{\boldsymbol{k}'}v_{\boldsymbol{k}'}\,(1-2f_{\boldsymbol{k}})\,(1-2f_{\boldsymbol{k}'})\,. \tag{10.99}$$

A comparison with (10.82) shows that the factor $(1-2f_{\boldsymbol{k}})(1-2f_{\boldsymbol{k}'})$ has been introduced because $|\psi\rangle_{\boldsymbol{k}}$ and $|\psi\rangle_{\boldsymbol{k}'}$ must be occupied for the interaction to take place. The entropy of a Fermi–Dirac gas is known from thermodynamics and is given by the standard expression

$$TS = -2k_{\text{B}}T\sum_{\boldsymbol{k}} f_{\boldsymbol{k}}\ln f_{\boldsymbol{k}} + (1-f_{\boldsymbol{k}})\ln(1-f_{\boldsymbol{k}})\,. \tag{10.100}$$

Since the entropy is due to quasiparticles only, it has no influence on the minimization procedure. Therefore, minimizing leads to the same condition as did the calculation of the BCS ground state, namely to (10.83). However, the expression for the energy gap Δ is modified, and is given by

$$\Delta = \frac{\mathcal{V}_0}{V}\sum_{\boldsymbol{k}'} u_{\boldsymbol{k}'}v_{\boldsymbol{k}'}(1-2f_{\boldsymbol{k}'})\,. \tag{10.101}$$

Because of the factor $(1-2f_{\boldsymbol{k}'})$, the energy gap is now temperature dependent. To calculate Δ, we proceed as in the previous section and find a result analogous to (10.92):

$$\begin{aligned} 1 &= \frac{\mathcal{V}_0\,D(E_{\text{F}})}{2}\int_{-\hbar\omega_{\text{D}}}^{\hbar\omega_{\text{D}}} \frac{\mathrm{d}\eta}{\sqrt{\eta^2+\Delta^2}}\,[1-2f(E)] \\ &= \frac{\mathcal{V}_0\,D(E_{\text{F}})}{2}\int_0^{\hbar\omega_{\text{D}}} \frac{\mathrm{d}\eta}{\sqrt{\eta^2+\Delta^2}}\tanh\left(\frac{\sqrt{\eta^2+\Delta^2}}{2k_{\text{B}}T}\right)\,. \end{aligned} \tag{10.102}$$

This equation has to be solved numerically. Before we consider the outcome of this calculation, we can use the equation to evaluate the transition temperature T_c. Knowing that the energy gap Δ vanishes at T_c, we insert $T = T_c$, and $\Delta = 0$, to obtain

$$\frac{2}{\mathcal{V}_0\, D(E_F)} = \int_0^{\hbar\omega_D} \frac{d\eta}{\eta} \tanh\left(\frac{\eta}{2k_B T_c}\right) . \tag{10.103}$$

In the weak coupling limit $\mathcal{V}_0\, D(E_F) \ll 1$, the numerical solution leads to

$$k_B T_c = 1.14\, \hbar\omega_D\, e^{-2/\mathcal{V}_0 D(E_F)} . \tag{10.104}$$

This relation explains the isotope effect mentioned at the beginning of this section, since $T_c \propto \omega_D \propto M^{-1/2}$, where M is the atomic mass. It also makes it plausible that deviations from this law exist. First, the derivation of (10.104) was based on the assumption that one-phonon exchange (10.62) gives a satisfactorily description of the attractive interaction between the electrons. Secondly, (10.69) drastically simplifies the corresponding expression, and finally, in all calculations a spherical Fermi surface was assumed. None of these simplifications are necessarily true for real metals.

Inserting (10.93) in (10.104), we find the important relation

$$\Delta_0 = 1.76\, k_B T_c , \tag{10.105}$$

connecting energy gap Δ_0 and transition temperature T_c. This simple equations is expected to hold for all superconductors as long as the simplifications mentioned above are fulfilled. In Table 10.3, experimental values for the ratio $\Delta_0/k_B T_c$ are given. In most cases fair agreement between theory and experiment is found, but in mercury and lead this ratio is considerably higher than expected. In these metals, the electrons are strongly coupled to phonons, thus creating pronounced charge clouds in the lattice. In this case, screening and retardation effects play an important role. These effects are correctly taken into account by the *Eliashberg theory* [484]. Good agreement with experimental results is found, but a treatment of this rather sophisticated theory would go far beyond the scope of this book.

Table 10.3. $\Delta_0/k_B T_c$ for some superconducting elements. After [485]

	Al	Cd	Hg	In	Nb	Pb	Zn
$\Delta_0/(k_B T_c)$	1.7	1.6	2.3	1.8	1.9	2.15	1.6

The numerical integration of (10.102) supplies us with the theoretical temperature dependence of the energy gap . In Fig. 10.32, the theoretical curve for the normalized energy gap $\Delta(T)/\Delta_0$ is plotted together with experimental data. Close to absolute zero, the energy gap is approximately constant. But

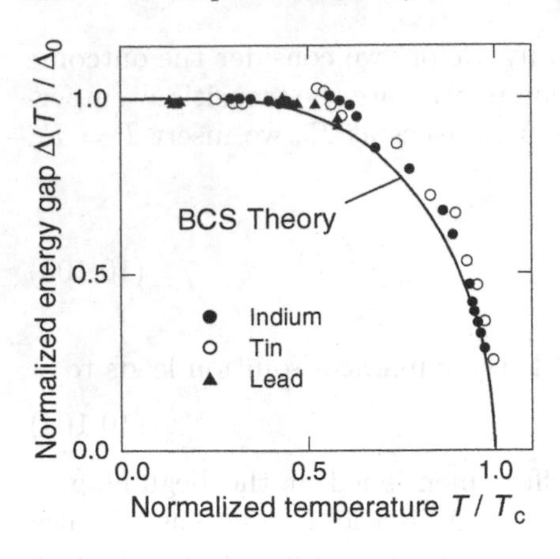

Fig. 10.32. Temperature dependence of the normalized energy gap $\Delta(T)/\Delta_0$ of In, Sn, and Pb [486]

with rising temperatures the number of Cooper pairs decreases faster and faster, resulting in a pronounced temperature dependence of $\Delta(T)$. Finally, at T_c, the gap approaches zero following the relation

$$\frac{\Delta(T)}{\Delta_0} = 1.74\sqrt{1 - \frac{T}{T_c}}\,. \tag{10.106}$$

The agreement is remarkably good bearing in mind that there is no free parameter. The occurrence of minor deviations is not surprising considering the crude simplifications made in the derivation of (10.103).

For a long time it was generally accepted that values of $\mathcal{V}_0 D(E_F)/2$ greater than about 0.4 were impossible, and in fact larger values could not be found in conventional superconductors. From (10.104) it follows that in this case transition temperatures greater than $0.1\,\hbar\omega_D/k_B$ cannot occur. Since the Debye temperature of metals generally does not exceed 300 K, transition temperatures much greater than 30 K are not expected for metals with superconductivity based on phonon exchange.

10.3.5 Measurement of the Energy Gap

Experiments with *tunnel junctions* are especially suited for determining the energy gap. Because of the great importance of this technique, we consider this method separately in the following section, but first we will describe several other experiments in which the gap value can be measured.

Infrared Measurements

The energy gap can be measured directly through infrared experiments. The mechanism for the absorption of infrared light depends on the ratio $\hbar\omega/2\Delta$, where $\omega/2\pi$ is the frequency of the infrared radiation. For $\hbar\omega < 2\Delta$, the electric field of the radiation accelerates thermally excited quasiparticles that subsequently interact with the lattice. For $\hbar\omega > 2\Delta$, this mechanism still exists, but infrared photons are now also able to break up Cooper pairs. This additional mechanism leads to a strong enhancement of the absorption In Fig. 10.33, we show the so-called *Mattis–Bardeen ratio* σ_1/σ_n of a thin indium film determined from the infrared transparency of the sample. Here, σ_1 is the real part of the high-frequency conductivity $\sigma_s(\omega) = \sigma_1(\omega) + i\sigma_2(\omega)$, and σ_n stands for the conductivity in the normal conducting state. Since σ_1 is proportional to the infrared absorption, the observed threshold indicates the onset of pair breaking, and allows us to directly determine the energy gap.

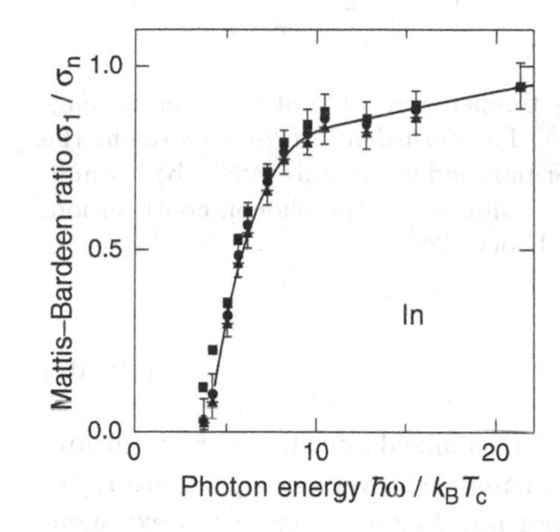

Fig. 10.33. Frequency dependence of the Mattis–Bardeen ratio σ_1/σ_n of a thin indium film measured at 1.4 K. This ratio is proportional to the film absorption [487]

Specific Heat

The existence of an energy gap also becomes evident from measurements of the specific heat. The low-temperature specific heat C of metals consists of contributions from electrons and phonons, i.e., $C = C_{el} + C_{ph}$. As shown in Fig. 10.34a for tin, in the normal conducting state the two contributions can easily be separated because of their different temperature dependence. Knowing the phonon contribution C_{ph}, the electronic part C_{el} can easily be deduced even in the superconducting state. Just as was the case in aluminum (Fig. 10.16), the specific heat of superconducting tin is characterized by a jump at T_c and a fast decrease upon cooling to values well below that of the normal conducting state.

Within the framework of BCS theory, the specific heat C_s of superconductors can easily be deduced from the free energy discussed in the previous section. At low temperatures, C_s increases exponentially, reflecting the energy gap Δ_0. An expression of general validity at low temperatures can be found by normalizing C_s to the electronic specific heat γT_c of the normal conductor at the transition temperature:

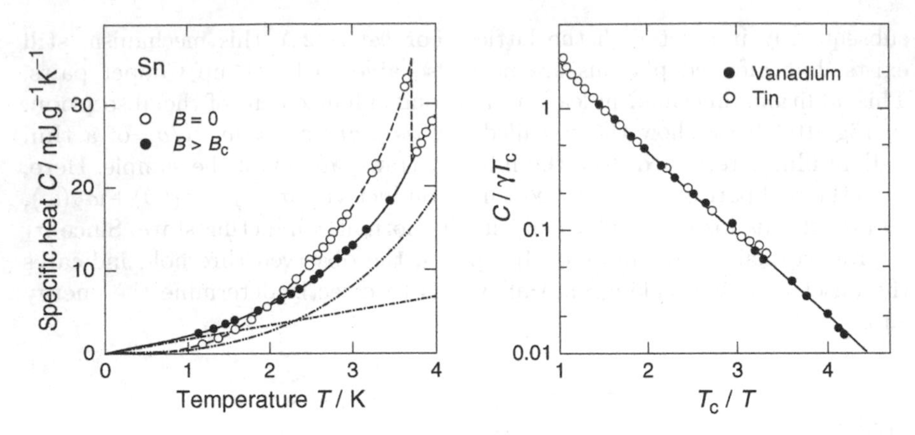

Fig. 10.34. (**a**) Specific heat of tin versus temperature without an external magnetic field (○), and with a field $B > B_c$ (•). The *dashed-dotted lines* represent the lattice and electronic contribution in the normal conducting state [488]. (**b**) Normalized specific heat of tin and vanadium after subtracting the phonon contribution. The *full line* is the prediction of the BCS theory [489]

$$\frac{C_s}{\gamma T_c} = 1.34 \left(\frac{\Delta_0}{k_B T} \right)^{3/2} e^{-\Delta_0/k_B T} . \tag{10.107}$$

In Fig. 10.34b, the logarithm of the normalized specific heat of tin and vanadium is plotted as a function of the normalized inverse temperature T_c/T. Both sets of data fall on top of each other and follow the expected exponential law. The small deviations, observed close to T_c, reflect the temperature dependence of $\Delta(T)$ and are in agreement with the prediction of BCS theory. For the jump of the specific heat at T_c, BCS theory predicts the universal relation $(C_s - C_n)/C_n = 1.43$. In Table 10.4, experimental values for some superconducting elements are given. The jump is much higher than predicted by BCS theory in strong-coupling superconductors like mercury or lead. These deviations can be understood within the framework of the Eliashberg theory mentioned above [484].

Table 10.4. Normalized jump $(C_s - C_n)/C_n$ of the specific heat at T_c for some superconductors. After [485]

	Al	Cd	Hg	In	Nb	Pb	Zn
$(C_s - C_n)/C_n$	1.4	1.4	2.4	1.7	1.9	2.7	1.3

Ultrasonic Absorption

In normal conductors, ultrasonic phonons collide with free electrons. The energy exchange in these collisions can be arbitrarily small. In superconductors at absolute zero such processes are impossible, since Cooper pairs cannot absorb energy because of the energy gap. Of course, pair breaking would be possible such as in infrared experiments, if phonons fulfilled the condition $\hbar\omega > 2\Delta(T)$. But in this case, frequencies of the order of 100 GHz would be required that are not accessible in classical experiments. Thus, ultrasonic phonons only interact with quasiparticles, and ultrasonic absorption is expected to decrease quickly below T_c because of the rapidly falling number of thermally excited quasiparticles. BCS theory predicts the simple relation

$$\frac{\alpha_s}{\alpha_n} = \frac{2}{e^{\Delta(T)/k_B T} + 1}, \tag{10.108}$$

where α_s and α_n represent the absorption of the superconducting and normal conducting states at T_c, respectively.

A measurement of the absorption of longitudinal ultrasonic waves in a single crystal of aluminum is shown in Fig. 10.35. Good agreement with the theoretical prediction was obtained, allowing precise measurements of Δ to be made. Studies of the ultrasonic absorption are also interesting for another reason. They allow the energy gap to be determined as a function of the crystal orientation. In fact, a relatively weak anisotropy of Δ is found in many superconductors that is mainly caused by the anisotropy of the Fermi surface.

Thermal Conductivity

The thermal conductivity Λ of lead in its superconducting and normal conducting state is shown in Fig. 10.36. While the conductivity rises in the normal conducting state, i.e., in magnetic fields $B > B_c$, the conductivity falls

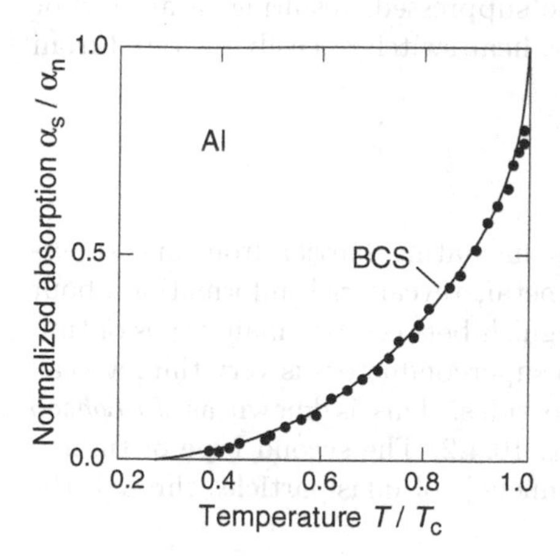

Fig. 10.35. Normalized ultrasonic absorption in aluminum as a function of T/T_c. The *full line* represents the prediction of BCS theory [490]

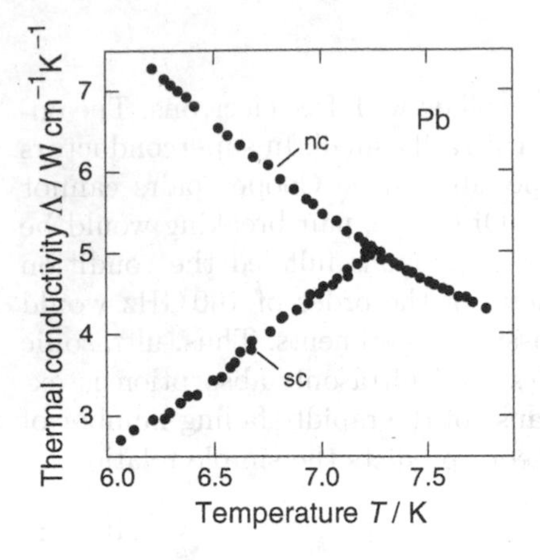

Fig. 10.36. Thermal conductivity of lead in the vicinity of T_c. The conductivity in the superconducting sample is smaller than in the normal conducting specimen (in a magnetic field $B > B_c$) [491]

in the superconducting state upon cooling. At first glance, it is surprising that the thermal conductivity in the superconducting phase is smaller than in the normal phase. A first prediction could have been that the friction-free motion of Cooper pairs facilitates the heat transport, however, Cooper pairs do not carry entropy. Consequently, they cannot contribute to heat transport. In normal conductors of high purity, the low-temperature thermal conductivity is limited by electron–phonon scattering. Since the number of phonons decreases upon cooling, the scattering probability decreases, leading to the observed rise of thermal conductivity. Below T_c, the conductivity of the superconductor decreases rapidly with decreasing temperature since the number of heat-carrying quasiparticles decreases very rapidly, more than compensating for their increasing mean free path.

As we will see in Sect. 11.7.4, the different thermal conductivity of normal and superconducting metals is used in low-temperature engineering for the production of 'heat switches'. At $T \ll T_c$, the thermal conductivity of superconductors is much smaller than that of normal conductors. With a magnetic field $B > B_c$, superconductivity can be suppressed, resulting in an abrupt rise of thermal conductivity. In this way, heat switches, easily operated from outside the cryostat, can be made.

10.3.6 Tunneling Experiments

The tunneling of electrons through an insulating barrier from one superconductor to another, or to a normal metal, reveals rich information about superconductivity. Here, we must distinguish between two main types of tunneling. First, if the barrier between two superconductors is very thin, a weak supercurrent can flow between the two sides. This is known as *Josephson tunneling* and will be discussed in Sect. 10.4.2. The second type of tunneling, discussed in this section, is the tunneling of quasiparticles through the barrier.

A typical experimental setup is depicted in Fig. 10.37. First, a thin metal strip is evaporated on a substrate. Then the surface of this film is oxidized to make the insulator, and finally the second strip is evaporated. The thickness of the oxide layer is typically of the order of 2 nm.

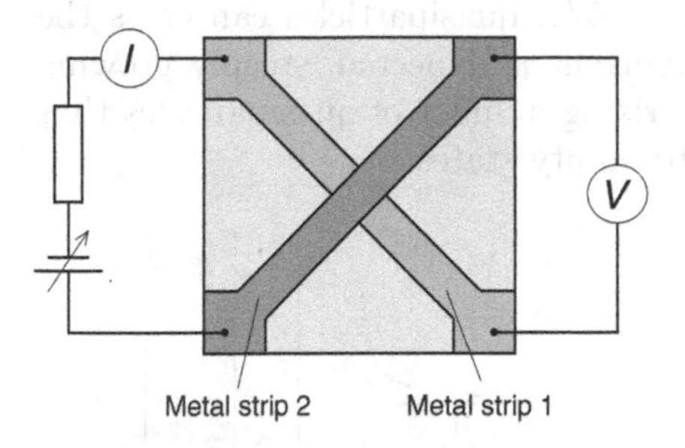

Fig. 10.37. Diagram of a tunnel junction. *Metal strip 1* is oxidized before evaporating *metal strip 2*. The oxide layer is typically 2 nm thick

We will first consider tunneling between two normal metals with an insulating layer in between. This configuration is often called an 'NIN junction'. In Fig. 10.38, the density of states close to the Fermi level of the two metals is drawn for $T = 0$, *dark areas* symbolize occupied states. Note that there are no states in the insulator between the two metals in the energy range considered here. Applying a voltage V across the tunnel junction causes a shift of the Fermi level by the amount eV. As indicated by the arrow, electrons are now able to tunnel from occupied states of the metal on the left to the empty states on the right. The resulting current I is proportional to V as in an ordinary ohmic resistance. Unfortunately, experiments such as this do not give much information about the electronic states.

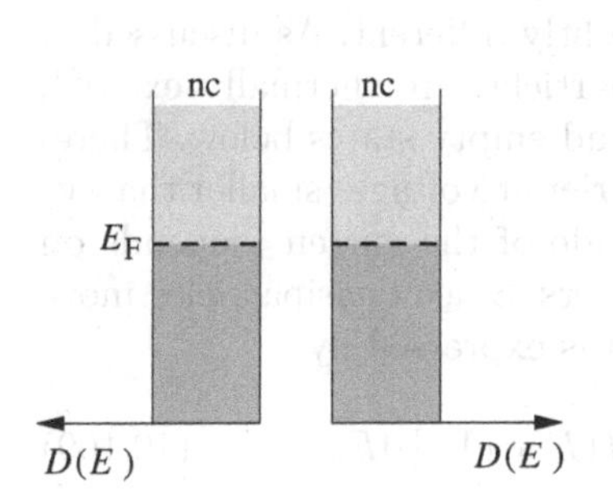

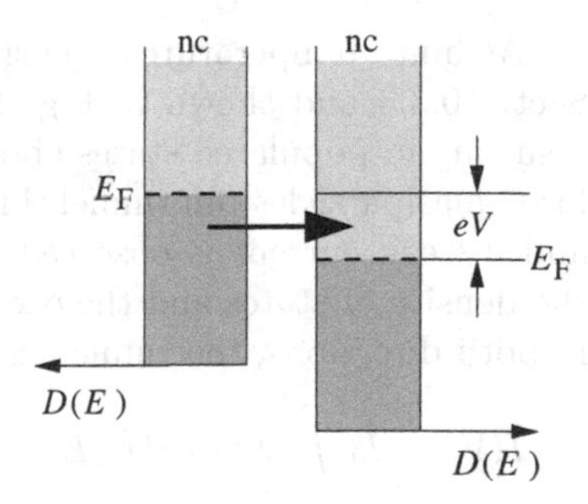

Fig. 10.38. Energy-level diagram for an NIN junction. The density of states is shown in the vicinity of E_F. The zero point of the energy is suppressed, occupied states are represented by *dark areas*. (**a**) $V = 0$, no current can flow. (**b**) $V \neq 0$, electrons from occupied states tunnel into empty states

More interesting results are obtained with a superconductor-insulator-normal metal junction, often called a 'SIN junction'. The energy-level diagram for $T = 0$ is depicted in Fig. 10.39 using the 'semiconductor representation' (see Sect. 10.3.3). The energy gap of the superconductor prevents the flow of quasiparticles through the barrier as long as $V < \Delta/e$. As soon as the applied voltage exceeds the critical voltage $V_c = \Delta/e$, quasiparticles can cross the barrier as indicated in Fig. 10.39b. A current is expected, steeply growing with the voltage because of the rapidly rising number of quasiparticles that are able to tunnel across the barrier into empty states.

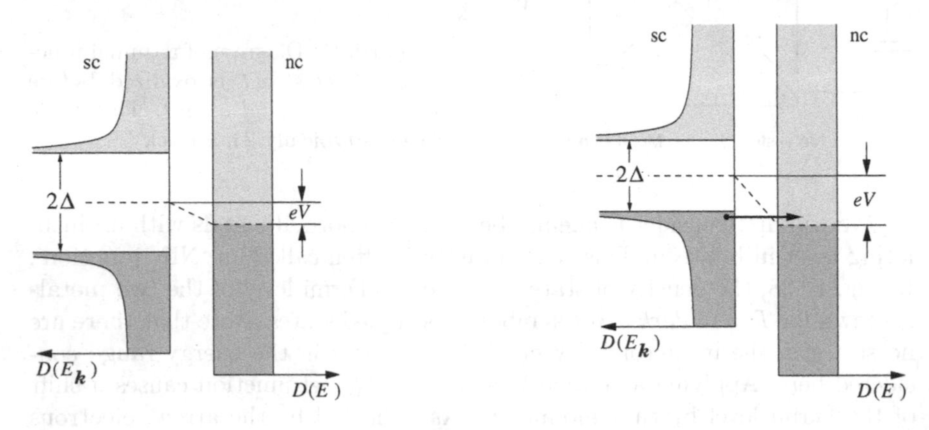

Fig. 10.39. Energy-level diagram for an SIN junction at absolute zero. (**a**) $V < V_c$, no free states are available for tunneling quasiparticles, (**b**) $V > V_c = eV$, quasiparticles tunnel from the superconductor to the normal conductor

At finite temperatures, the situation is slightly different. As discussed in Sect. 10.3.3 and shown in Fig. 10.40, quasiparticles are thermally excited, resulting in populated states above the gap, and empty states below. Therefore, quasiparticles can tunnel through the barrier at voltages smaller than V_c and a weak current is observed. The magnitude of the current depends on the density of states and the occupation numbers. Since quasiparticles move in both directions, the tunneling current $I(V)$ is expressed by

$$I(V) = I_0 \int D_s(E_{\boldsymbol{k}}) D_n(E + eV) \left[f(E) - f(E + eV) \right] \mathrm{d}E\,, \tag{10.109}$$

where I_0 is a constant depending on the geometry of the junction [492]. Of course, this formulation is also valid for other types of junctions if the appropriate densities of states are inserted. For SIN junctions, $D_n(E)$ can be replaced by $D_n(E_F)$, and $f(E)$ by a step function. Carrying out the integration and differentiating with respect to the voltage, we obtain the simple relation

$$\mathrm{d}I/\mathrm{d}V \propto D_s(E_{\boldsymbol{k}} = eV)\,. \tag{10.110}$$

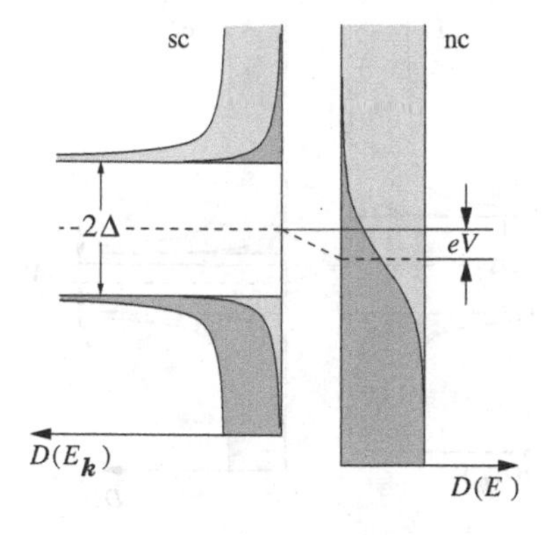

Fig. 10.40. Energy-level diagram for an SIN junction at finite temperature. Occupied states are represented by *dark tinted areas*. The applied voltage V is smaller than the critical voltage V_c

This result is remarkable. From the I–V characteristic of an SIN junction the energy gap as well as the quasiparticle density of states can be deduced.

The expected current–voltage characteristics of an SIN junction at $T = 0$ are depicted by the *full line* in Fig. 10.41. The current is zero for voltages smaller than V_c. It is followed by a steep rise at the critical voltage, reflecting the singularity of the density of states at $E_F \pm \Delta$. As mentioned above, at finite temperatures a weak current (*dashed-dotted line*) is already expected to flow at small applied voltages. At temperatures above T_c, the junction behaves like an NIN junction, and exhibits ohmic properties as shown by the *dashed line*. An example of the usefulness of this technique is the data that was presented in Fig. 10.30. The density of states of the quasiparticles in lead was determined with a Pb/MgO/Mg junction.

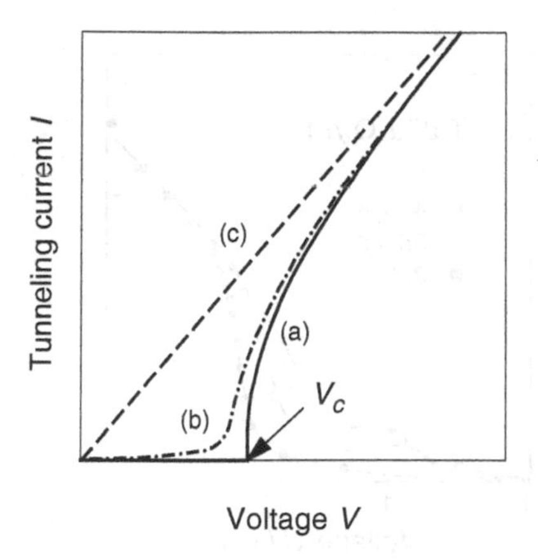

Fig. 10.41. Schematic I–V characteristics of an SIN junction with the critical voltage V_c. (**a**) $T = 0$, (**b**) $0 < T < T_c$, (**c**) $T > T_c$

As can directly be seen from Fig. 10.42a, the threshold voltage of SIS junctions is given by $eV_c = (\Delta_1 + \Delta_2)$. As in SIN junctions, at finite temperatures a current will flow because of the presence of excited quasiparticles. At the voltage $V = |\Delta_2 - \Delta_1|/e$, a maximum in the current is expected since the

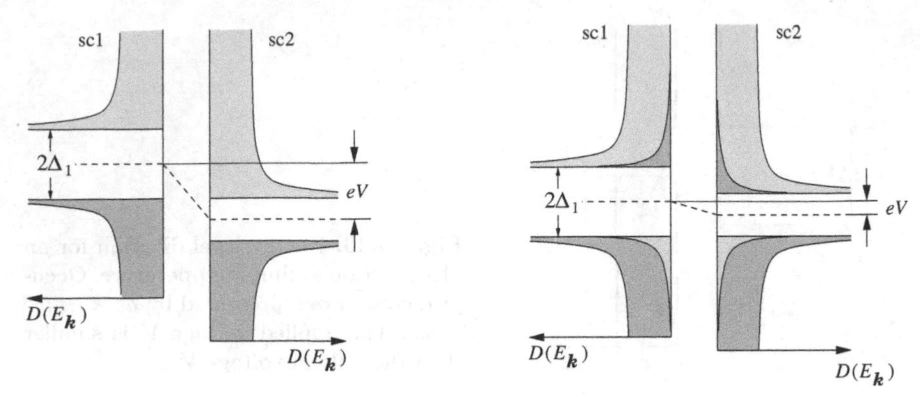

Fig. 10.42. Energy-level diagram for an SIS junction. (**a**) At $T = 0$ current starts to flow at $eV = (\Delta_1 + \Delta_2)$. (**b**) At $T \neq 0$, a current maximum is expected at $eV = (|\Delta_2 - \Delta_1|)$

poles of the density of states just face each other. This situation is depicted in Fig. 10.42b. The voltage dependence of the quasiparticle current for superconductors with the energy gaps Δ_1 and Δ_2 is schematically depicted in Fig. 10.43a. Experimental results obtained with a Ta/TaO/Pb junction are shown in Fig. 10.43b. As expected, a maximum of the tunneling current is found at the voltage $V = |\Delta_{\text{Pb}} - \Delta_{\text{Ta}}|/e$.

Finally, we add a short remark on *phonon generation*. As discussed, quasiparticles in the occupied band of superconductor 1 can tunnel in the empty

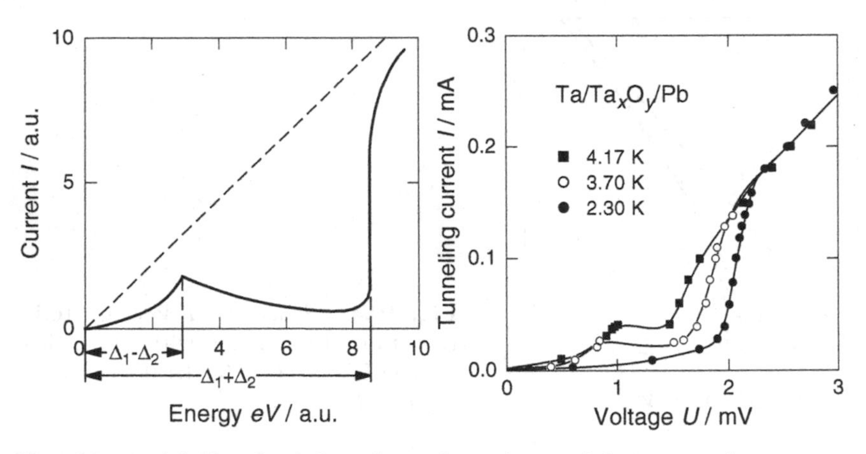

Fig. 10.43. (**a**) Sketch of the voltage dependence of the current between two superconductors at $0 < T < T_c$. The *dashed* line represents the ohmic variation at $T > T_c$. (**b**) I–V characteristics of a Ta/Ta$_x$O$_y$/Pb junction at different temperatures [493]

states of superconductor 2 if the applied voltage $V > (\Delta_1 + \Delta_2)/e$. After crossing the barrier these quasiparticles relax towards the band edge by emitting phonons. A continuous phonon-emission spectrum is generated, covering the energy range from zero to $eV - (\Delta_1 + \Delta_2)$. Because of the similarity to the spectrum in X-ray physics, these phonons are often called 'bremsstrahlung phonons'. The quasiparticles accumulated at the band edge, recombine with empty states in the 'valency band', and form new Cooper pairs. In this process, the energy $2\Delta_1$ is released in the form of 'recombination phonons'. Both mechanisms of phonon generation are exploited in phonon spectroscopy in the frequency range from 100 to 500 GHz (see, for example, [494]).

10.3.7 Critical Current and Energy Gap

In Sect. 10.2.1, it was found that a critical magnetic field exists for thermodynamic reasons. In Sect. 10.2.2, we discussed the shielding currents that prevent magnetic fields from penetrating into superconductors. In this context we have already pointed out that not only does a critical magnetic field exist but also a *critical current density*. Here, we establish the relation between the critical current and the energy gap Δ, the characteristic parameter of BCS theory.

The supercurrent flowing in a superconductor is due to the center-of-mass motion of the Cooper pairs. Their velocity $\boldsymbol{v} = \hbar\,\boldsymbol{\delta K}/m$ is directly connected with the change $\boldsymbol{\delta K}$ of the wave vector of the corresponding electrons. Therefore, the Cooper pairs in this state may be characterized by the abbreviation $(\boldsymbol{k} + \boldsymbol{\delta K}\uparrow, -\boldsymbol{k} + \boldsymbol{\delta K}\downarrow)$. For the sake of completeness, we mention that the center-of-mass motion changes neither $\mathcal{V}_{\boldsymbol{kk}'}$ nor Δ. This means that the energy gap moves with the Fermi sphere in $\boldsymbol{k}$-space.

It is plausible that superconductivity breaks down as soon as the kinetic energy of the Cooper pairs exceeds the condensation energy (10.90). Therefore, the relation

$$\frac{1}{2}\,n m v_{\mathrm{c}}^2 = \frac{1}{4} D(E_{\mathrm{F}})\Delta^2 \tag{10.111}$$

is expected to hold for the critical velocity v_{c} of the pair states. Inserting $D(E_{\mathrm{F}})$, the expression

$$v_{\mathrm{c}} = \sqrt{\frac{3}{2}}\,\frac{\Delta}{m v_{\mathrm{F}}} \tag{10.112}$$

is found for the critical velocity. Taking into account that the density of the supercurrent is given by the equation $\boldsymbol{j}_{\mathrm{s}} = -n_{\mathrm{s}} e \boldsymbol{v}$, we obtain for the critical current density

$$j_{\mathrm{c}} = \sqrt{\frac{3}{2}}\,\frac{e n_{\mathrm{s}} \Delta}{\hbar k_{\mathrm{F}}}\,. \tag{10.113}$$

Inserting typical numbers, the critical current density is estimated to be of the order of $j_{\mathrm{c}} \approx 10^7\ \mathrm{A\,cm^{-2}}$.

According to (10.5) the critical magnetic field B_c is linked to the critical current I_c via $B_c = \mu_0 I_c / 2\pi R$. Provided that the radius R of the wire is large in comparison with the penetration depth λ_L, we obtain, using (10.24), the relation $\int \boldsymbol{j} \cdot \mathrm{d}\boldsymbol{f} = 2\pi j_0 \lambda_L R$, where $\mathrm{d}\boldsymbol{f}$ represents the surface element of the cross section of the wire. Identifying j_0 in (10.24) with the critical current density j_c, and the field at the surface with the critical field, we finally obtain

$$B_c = \mu_0 \lambda_L j_c = \sqrt{\frac{3}{2} \frac{e n_s \mu_0 \lambda_L \Delta}{\hbar k_F}} \,. \tag{10.114}$$

The same relation is found by equating (10.11), with (10.90) reflecting the condensation energy deduced from thermodynamic and microscopic considerations, respectively.

Let us add some remarks on the vanishing resistance in the superconducting state. As discussed in Sect. 7.2, the resistance of normal conductors is caused by scattering of the conduction electrons by defects and phonons. In superconductors, the charge transport is brought about by the common motion of the Cooper pairs characterized by the superimposed wave vector $\delta \boldsymbol{K}$. Scattering of Cooper pairs is tantamount to leaving the common BCS state, resulting in breaking up of the pairs. Since in this case the binding energy has to be exceeded, elastic scattering processes of pairs as a whole are excluded from the beginning. Inelastic collisions with high-energy phonons are possible. In such a process the binding energy is supplied by the absorbed phonon, and the Cooper pair is destroyed. However, the inverse process also takes place where Cooper pairs are formed with the emission of a phonon. In thermal equilibrium, the two processes cancel each other because the newly formed Cooper pairs condense into those states that have been emptied before by the pair-breaking process. In addition, the question arises as to why thermally excited quasiparticles do not give rise to losses. The answer is simple: In the stationary state, electric fields are 'short-circuited' by the supercurrent. Therefore, quasiparticles are not accelerated and do not contribute to the current flow. This argument is not correct for ac voltages because in this case an electric field is present in the sample, as described by (10.18). The quasiparticles are accelerated, interact with the lattice, and give rise to losses. Hence the loss-free conduction is only observed for dc currents.

10.4 Flux Quantization – Josephson Effect

In the previous section, we learned that in the superconducting state the motion of electron pairs is highly correlated, thus forming the BCS state. This situation is not unlike the correlated motion of Cooper pairs in superfluid ^{3}He. Such a correlation is not surprising because Cooper pairs (in conventional superconductors) are not small, isolated units but penetrate each other mutually. This becomes obvious from a simple consideration of their concentration and 'size' that we can estimate with the help of the uncertainty principle. The upper limit for the energy uncertainty δE of the Cooper pairs is 2Δ. Consequently, the maximum uncertainty in momentum is given by

$$\delta E = \delta\left(\frac{p^2}{2m}\right) \approx \frac{p_\mathrm{F}}{m}\delta p \approx 2\Delta\,. \tag{10.115}$$

With this expression and relation (10.105), we obtain for the size

$$\delta x \approx \frac{\hbar}{\delta p} \approx \frac{\hbar^2 k_\mathrm{F}}{2m\Delta} \approx 0.28\frac{\hbar v_\mathrm{F}}{k_\mathrm{B}T_\mathrm{c}}\,. \tag{10.116}$$

Inserting typical numbers we find for δx values of the order of 1000 nm. Roughly speaking, the size δx determines the *coherence length* ξ_0 introduced by (10.30) in Sect. 10.2.3, i.e., $\delta x \approx \xi_0$. As we mentioned there, ξ_0 reflects the length scale within which the density of Cooper pairs at an interface rises from zero in a normal conductor to the bulk density in a superconductor. It is not surprising that δx and ξ_0 are comparable since the size of the Cooper pairs limits the number of pairs that can be present close to an interface.

The concentration of Cooper pairs is roughly given by Δ/E_F. Although this ratio is only of the order of 10^{-4}, there are 10^6 to 10^7 other pairs present within the volume defined by the size of a single Cooper pair. Therefore, it is intuitively plausible that the Cooper pairs do not move independently but in a highly correlated manner. In fact, the BCS ground state represents a coherent many-particle state within which *all* Cooper pairs occupy a *common quantum state*. This state may be described by the *macroscopic wave function* of the form

$$\Psi = \Psi_0\,\mathrm{e}^{\mathrm{i}\varphi(\boldsymbol{r})}\,, \tag{10.117}$$

where the magnitude of the wave function is given by $|\Psi_0|^2 = n_\mathrm{s}$. The existence of such a wave function with a well-defined phase in the whole sample has remarkable consequences for the behavior of superconductors and gives rise to peculiar effects in magnetic fields. We are first going to discuss flux quantization that follows from the fact that the macroscopic wave function is single-valued, and afterwards we consider the so-called *Josephson effect*.

10.4.1 Flux Quantization

The phase $\varphi(\boldsymbol{r})$ of the macroscopic wave function of superconductors is well defined. The phase difference $\Delta\varphi$ that exists between two points is obtained by carrying out the contour integral $\Delta\varphi = \int_1^2 \mathrm{grad}\,\varphi(\boldsymbol{r})\cdot \mathrm{d}\boldsymbol{s}$. Invoking the fact that the wave function is single-valued, the phase change has to be $2\pi p$ for a closed contour, where p is an integer.

To demonstrate the existence of flux quantization we consider a multiply connected superconductor, e.g., a specimen with a hole as shown in Fig. 10.44. Applying a magnetic field and cooling down below the transition temperature afterwards, the magnetic field inside the ring will not be expelled. The flux in

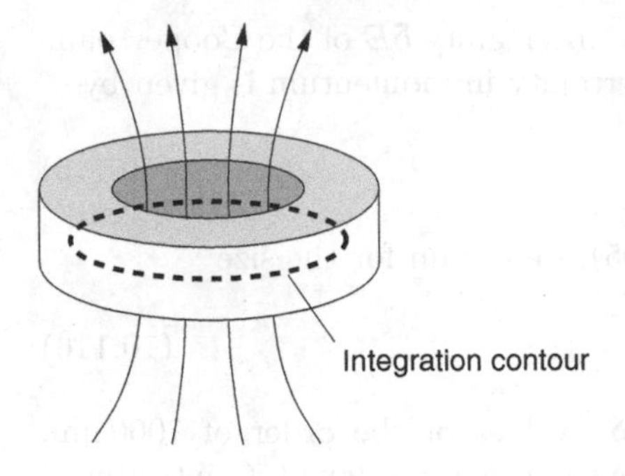

Fig. 10.44. Superconducting ring penetrated by a magnetic field. Shielding currents are only present at the surface of the specimen. The contour line L along which the integration of (10.120) is carried out, is indicated by a *dashed line* and runs inside the specimen

the interior of the annular specimen will remain trapped when the external field is switched off.

The existence of a macroscopic wave function has an important consequence for the trapped magnetic flux. To show this, we use the quantum-mechanical expression for the electrical current in a magnetic field

$$\boldsymbol{j} = \mathrm{i}\frac{\hbar q}{2M}\left(\Psi^\star \nabla \Psi - \Psi \nabla \Psi^\star\right) - \frac{q^2}{M}\boldsymbol{A}\Psi^\star\Psi\,, \tag{10.118}$$

and replace q by $-2e$, and M by $2m$, because we are dealing with Cooper pairs. Inserting wave function (10.117), we obtain

$$\mu_0\lambda_{\mathrm{L}}^2\boldsymbol{j} = \left(\frac{\hbar}{e}\nabla\varphi - 2\boldsymbol{A}\right), \tag{10.119}$$

where λ_{L} is the London penetration depth, given by (10.23). Next, we carry out an integration along a closed contour line L and obtain

$$\mu_0\lambda_{\mathrm{L}}^2 \oint\limits_L \boldsymbol{j}\cdot \mathrm{d}\boldsymbol{s} = \frac{\hbar}{e}\oint\limits_L \nabla\varphi\cdot \mathrm{d}\boldsymbol{s} - 2\oint\limits_L \boldsymbol{A}\cdot \mathrm{d}\boldsymbol{s}\,. \tag{10.120}$$

We take our integration contour deep inside the ring, as indicated in Fig. 10.44. Since in this case $\boldsymbol{j} = 0$, the integral on the left side vanishes. The first integral on the right side has already been discussed. Using Stokes' theorem we transform the line integral with the vector potential to a surface integral: $\oint_L \boldsymbol{A}\cdot \mathrm{d}\boldsymbol{s} = \int_\Sigma \boldsymbol{B}\cdot d\boldsymbol{f} = \Phi$, where the surface Σ spans the contour L. This integral is equal to the total magnetic flux Φ through the hole together with the region surrounding the hole where the field penetrates the superconductor. In this way, we obtain the final result

$$\Phi = p\,\frac{h}{2e} = p\,\Phi_0\,. \tag{10.121}$$

The magnetic flux is always a multiple of Φ_0, the *flux quantum*. Its value is rather small: $\Phi_0 = h/2e = 2.07\times 10^{-15}\,\mathrm{T\,m^2}$. For instance, a hollow cylinder with a diameter of only about 5 μm in the Earth's magnetic field is enough to encircle one flux quantum.

London pointed out as early as 1950 that magnetic flux ought to be quantized [11]. It is of great significance that the charge entering the expression for the flux quantization is $2e$, i.e., the charge of a Cooper pair. Therefore, the experimental determination of the flux quantum that was accomplished in 1961 by *Doll* and *Nähbauer* [495], and *Deaver* and *Fairbank* [496], was a direct demonstration of the existence of Cooper pairs. In these experiments, a thin superconducting hollow cylinder was cooled down in a very weak magnetic field. After switching off the external magnetic field, the magnetic dipole moment of the cylinders was measured for a number of different applied magnetic fields. The experimental results demonstrated that the trapped flux was quantized, and was in fact given by (10.121).

In Fig. 10.45 the result of a more recent measurement with a hollow cylinder of tin with a diameter of 56 μm is shown. Obviously, the trapped magnetic flux does not follow the steady variation of the cooling field but clearly exhibits the expected quantization. The rounding of the curve is caused by the fact that, under the given experimental conditions, in some cases a flux quantum is not trapped inside the cylinder along its whole length.

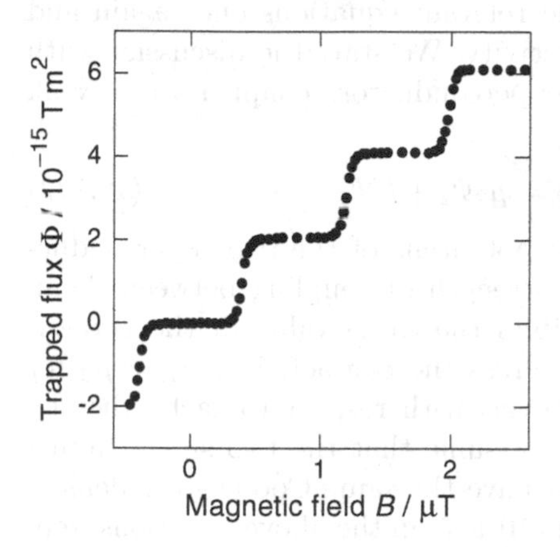

Fig. 10.45. Magnetic flux trapped by a thin hollow cylinder made of tin (length 24 mm, diameter 56 μm) as a function of the cooling field [497]

From the quantization of the magnetic flux it immediately follows that the current in a closed loop is quantized as well. A continuous variation of the current is not possible since the phase of the wave function can only be changed by a multiple of 2π. Thus, only phase jumps would be allowed, but such changes require a temporary destruction of the coherence of the wave function. In this case, the condensation energy of all the Cooper pairs would have to be raised. Therefore, a jump such as this does not occur, and no flux quantum can leave the superconducting loop, meaning that persistent currents are absolutely stable.

10.4.2 Pair Tunneling – Josephson Effect

In Sect. 10.3.6, we considered experiments with tunnel junctions and discussed tunneling of quasiparticles. Reducing the thickness of the insulating layer between the two superconductors to about 10 Å results in a noticeable overlap of the macroscopic wave functions and hence to a coupling between the superconducting states. In such a configuration, *Cooper pairs* are able to tunnel through the insulating layer. There are several possibilities to realize such a *weak link*. Besides the oxide barrier between evaporated thin films, so-called 'point contacts' and 'microbridges' are used. For instance, a niobium point contact can be made by grinding a Nb wire to a point, allowing the surface to oxidize, and pressing the tip against a piece of bulk Nb. A microbridge can be made by etching down a thin superconducting film in such a way that the film consists of two parts that are connected via a very narrow constriction.

The overlap of the wave functions gives rise to surprising effects that were predicted by *Josephson* in 1962 [95]. We have already considered this effect in some detail in Sect. 2.4.3, where the Josephson effect in helium II was discussed. Here, we write down the relevant equations once again and apply them to problems in superconductivity. We start the discussion with the Schrödinger equation for the two superconductors coupled via a weak link:

$$\mathrm{i}\hbar\dot{\Psi}_1 = \mu_1\Psi_1 + \mathcal{K}\Psi_2 \qquad \text{and} \qquad \mathrm{i}\hbar\dot{\Psi}_2 = \mu_2\Psi_2 + \mathcal{K}\Psi_1\,. \tag{10.122}$$

Here, μ_1 and μ_2 represent the chemical potentials of the two superconductors, while the constant $\mathcal{K}$ reflects the strength of coupling between them. Of course, the chemical potential exhibits the same value in the two superconductors if no voltage is present across the contact, i.e., $\mu_1 = \mu_2$ for $V = 0$. A voltage V shifts the Fermi levels with respect to each other by $\mu_2 - \mu_1 = -2eV$. For simplification, we assume that the two superconductors are made of the same material and have the same Cooper-pair density $n_{\mathrm{s}1} = n_{\mathrm{s}2} = n_{\mathrm{s}}$. We insert wave function (10.117) in the above equations, separate real and imaginary parts, and find

$$\dot{n}_{\mathrm{s}1} = \frac{2\mathcal{K}}{\hbar}\, n_{\mathrm{s}} \sin(\varphi_2 - \varphi_1) = -\dot{n}_{\mathrm{s}2}\,, \tag{10.123}$$

$$\hbar\,(\dot{\varphi}_2 - \dot{\varphi}_1) = -(\mu_2 - \mu_1) = 2eV\,. \tag{10.124}$$

First, we consider the voltage-free case where the phase difference $(\varphi_1 - \varphi_2)$ is time independent because of (10.124). In this case, $\dot{n}_{\mathrm{s}}$ is also constant and a dc current will flow through the junction. Since the current I_{s} through the contact is proportional to $\dot{n}_{\mathrm{s}1}$, we may write

$$I_{\mathrm{s}} = I_{\mathrm{c}} \sin(\varphi_2 - \varphi_1)\,. \tag{10.125}$$

The astonishing fact that a current crosses the insulating contact without a voltage drop is called the *dc Josephson effect*. The critical current I_{c} depends on the density of the Cooper pairs, the contact area A (typical 0.1 mm^2), and the coupling constant, i.e., $I_c \propto n_s \mathcal{K} A$.

In Fig. 10.46, the current–voltage characteristic of a lead junction is shown. As long as the current through the junction is smaller than the critical current I_c, no voltage is measured across the junction. The current, and consequently also the phase difference $(\varphi_2 - \varphi_1)$, is determined by the current source.

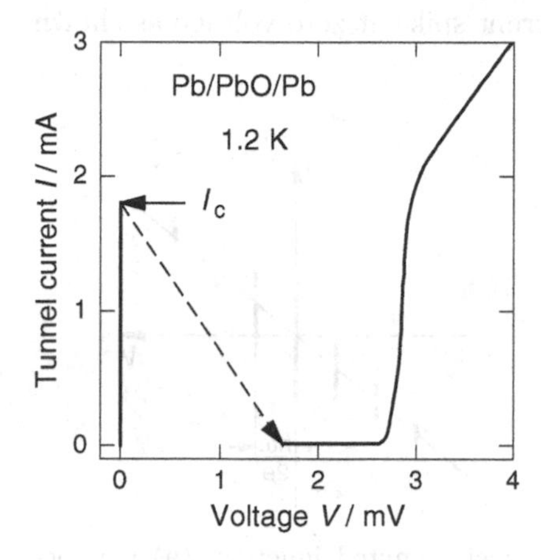

Fig. 10.46. Current–voltage characteristic of a lead junction displaying the dc Josephson current. When I_c is reached, the junction switches to the characteristic for quasiparticle tunneling [498]

Without justification, we want to add that for an ideal junction at $T = 0$ the relation $RI_c = \pi \Delta_0 / 2e$ should hold, where R is the junction resistance in its normal state. In a good niobium junction, for instance, typical values for the product RI_c are 2 mV, while I_c is found to be between 1 μA and 10 mA, depending on the geometry of the junction.

Increasing the current through the junction beyond the critical value I_c, the voltage across the insulating layer jumps to a finite value given by the quasiparticle characteristic of the junction. In this state, the weak link will carry a normal current in addition to the supercurrent given by (10.125). Integrating (10.124) we find

$$(\varphi_2 - \varphi_1) = \frac{2eU}{\hbar} t + \varphi_0 = \omega_J t + \varphi_0 \,, \tag{10.126}$$

meaning that the phase difference grows linearly with time. Inserting this result in (10.125), we obtain an entirely oscillatory supercurrent

$$I_s = I_c \sin(\omega_J t + \varphi_0) \,. \tag{10.127}$$

This is the *ac Josephson effect*. The frequency of the ac current is expected to be proportional to the voltage across the junction and should be given by $\omega_J = 2eV/\hbar$. The resulting frequency is relatively high, namely 48 GHz for a voltage of 100 μV. Of course, the current cannot be shown in Fig. 10.46 where the dc characteristic is depicted.

As mentioned above, the weak link will carry a normal current in addition to the supercurrent if the voltage across the junction is nonzero. The normal current will, in general, be a strongly nonlinear function of the voltage (see Sect. 10.3.6, and Fig. 10.46). To a first approximation the normal current can be treated as flowing in parallel to the supercurrent through a simple resistance R. This model provides a good description of SIS junctions, and is known as the 'resistively shunted junction'. In this simple model, the dc current is only flowing in the shunt resistance, resulting in a conventional ohmic I–V characteristic with a supercurrent spike at zero voltage as shown in Fig. 10.47a.

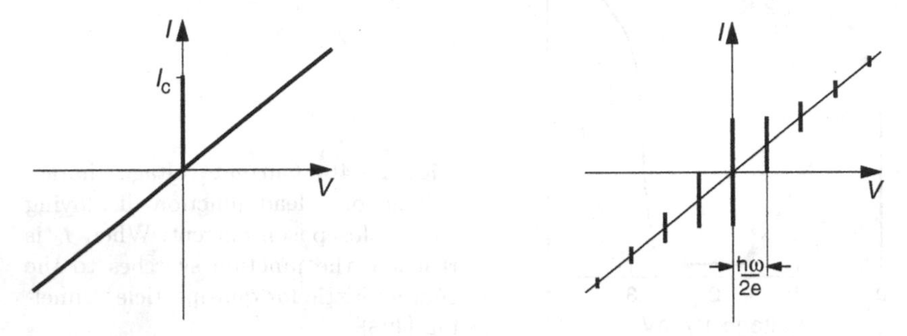

Fig. 10.47. dc I–V characteristic of a resistively shunted junction. (**a**) Characteristic showing the dc Josephson effect, (**b**) characteristic with radio-frequency excitation, showing the inverse ac effect

Direct proof of the ac Josephson current is rather difficult because the power generated by a single junction is of the order of RI_c^2, and therefore typically less than 1 μW. In addition, weak links represent a low-impedance source that is difficult to match. This problem can be solved by combining a dc and an ac voltage. With the voltage $V = V_0 + V_{rf}\cos(\omega_{rf}t)$ we find, upon integrating (10.124), that

$$(\varphi_2 - \varphi_1) = \omega_J t + (2eV_0/\hbar\omega_{rf})\sin(\omega_{rf}t) + \varphi_0 \,. \tag{10.128}$$

The second term reflects a phase modulation that not only causes an ac supercurrent at the Josephson frequency ω_J but also at the side frequencies $(\omega_J \pm n\omega_{rf})$. This result is illustrated in Fig. 10.47b. At most voltages, ac supercurrents have no effect on the dc I–V characteristic, but whenever the condition $2eV = n\omega_{rf}$ is fulfilled, one of the side frequencies will be zero. As shown in the figure, in this case a vertical spike on the dc I–V characteristic is expected to occur. These *Shapiro spikes* reflect a quantum condition: they occur when the energy $2eV$ that is accessible to Cooper pairs crossing the weak link is equal to a multiple of the photon energy $\hbar\omega_{rf}$. The appearance of the Shapiro spikes is the 'inverse ac Josephson effect'.

The position of the Shapiro spikes is extremely well defined. Therefore, a careful simultaneous measurement of their voltage position and the applied rf frequency provides an extremely precise value of the ratio e/h to 1 part in 10^8. Conversely, this effect is used by standards laboratories to realize volt references. In this case, the voltage is measured at which the Shapiro spike of an array of Josephson junctions is observed when the array is subjected to a microwave field of precisely known frequency.

Until now we have assumed that a voltage was applied to the junction. However, the impedance of most weak links is of the order of ohms, and consequently much smaller than the impedance of the external circuitry. Therefore, it is more realistic to treat Josephson junctions as fed from a current source. Unfortunately, this makes the analysis more complicated. We do not discuss this complication further but refer to [499] where this aspect is treated in detail. Here, we only mention two important consequences: First, the spike in the dc current–voltage characteristic at $V = 0$ is replaced by a current step of the same height. Secondly, the Shapiro spikes are transformed into Shapiro steps.

As an example, we show in Fig. 10.48 the voltage–current curves of a niobium Josephson junction exposed to a microwave source at 72 GHz. The step amplitude oscillates as a function of the microwave intensity, as expected from a detailed theory. In addition, we would like to mention that, in general, Josephson junctions exhibit an appreciable capacitance in parallel with the Josephson element and with the shunt resistance, resulting in a reduction of the supercurrent at finite voltage.

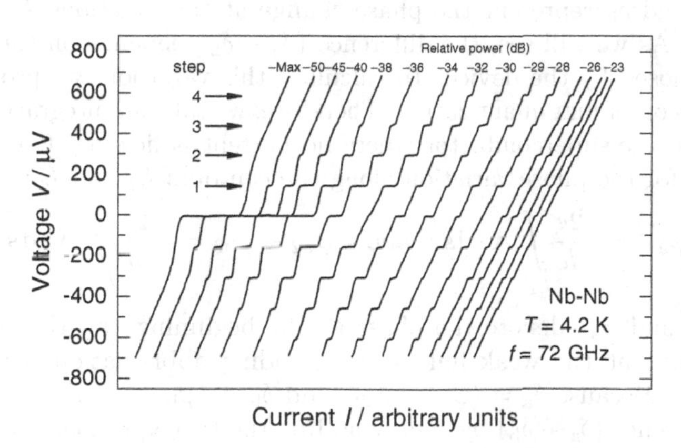

Fig. 10.48. Voltage–current curves of a niobium Josephson junction exposed to a microwave source at 72 GHz at various power levels [500]

10.4.3 Quantum Interference

In the following, we consider the influence of magnetic fields on the tunneling of Cooper pairs through a weak link. First, we consider two weak links connected in parallel with superconducting leads as drawn schematically in Fig. 10.49. Such a device shows the important phenomenon of *quantum interference*, i.e., it behaves like an interferometer.

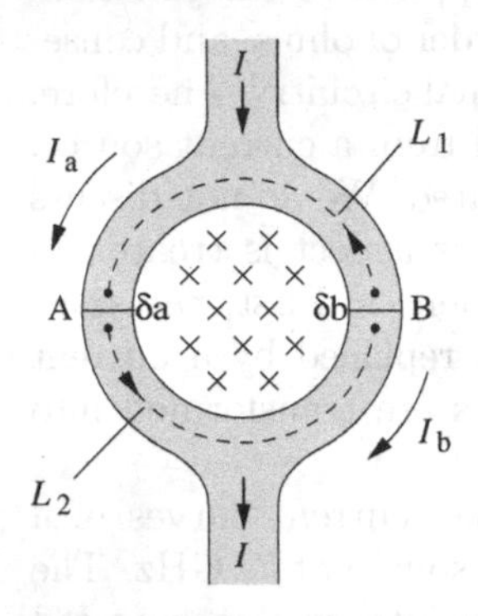

Fig. 10.49. Two Josephson junctions connected in parallel. The enclosed magnetic flux is indicated by *small crosses*. The phase changes are calculated along the contours L_1 and L_2 that are assumed to begin and end at the weak links

The total current I passing through the device is given by

$$I = I_c(\sin\delta_a + \sin\delta_b) = 2I_c \cos\left(\frac{\delta_a - \delta_b}{2}\right)\sin\left(\frac{\delta_a + \delta_b}{2}\right), \tag{10.129}$$

where δ_a and δ_b represent the phase change at the junctions A and B, respectively. As we will see, the difference $(\delta_a - \delta_b)$ depends on the magnetic flux Φ enclosed by the device. To calculate this variation, we proceed as in the discussion of flux quantization. Therefore, we take an integration contour deep inside the superconductor where no current is flowing. Using (10.120) we obtain for the phase variation along the contours L_1 and L_2:

$$\varphi_{b1} - \varphi_{a1} = -\frac{2e}{\hbar}\int_{L_1} \boldsymbol{A}\cdot d\boldsymbol{s} \quad \text{and} \quad \varphi_{a2} - \varphi_{b2} = -\frac{2e}{\hbar}\int_{L_2} \boldsymbol{A}\cdot d\boldsymbol{s}\,. \tag{10.130}$$

Here, φ_{b1} and φ_{a1} denote the phase at the beginning and the end of contour L_1, i.e., at the weak link. Corresponding abbreviations are used for contour L_2. Because $\delta_a = (\varphi_{a1} - \varphi_{a2})$ and $\delta_b = (\varphi_{b1} - \varphi_{b2})$, we obtain the phase difference $(\delta_a - \delta_b)$ by simply adding the two equations. We find

$$(\delta_a - \delta_b) = \frac{2e}{\hbar}\oint \boldsymbol{A}\cdot d\boldsymbol{s} = \frac{2e\Phi}{\hbar}\,. \tag{10.131}$$

Here, it should be mentioned that a contribution due to the magnetic field inside the junctions has been neglected. Inserting (10.131) in (10.129) we find

$$I = 2I_c \sin\left(\frac{\delta_a + \delta_b}{2}\right)\cos\left(\pi\frac{\Phi}{\Phi_0}\right) = I_{max}\cos\left(\pi\frac{\Phi}{\Phi_0}\right). \tag{10.132}$$

The phase angle $(\delta_a + \delta_b)$ varies to match the current I fed into the loop. The cosine term describes the oscillation of the current with respect to the magnetic field. An experimental verification of this prediction is shown in Fig. 10.50. As expected, small magnetic field changes cause a modulation of the current through the device that is in agreement with the theory.

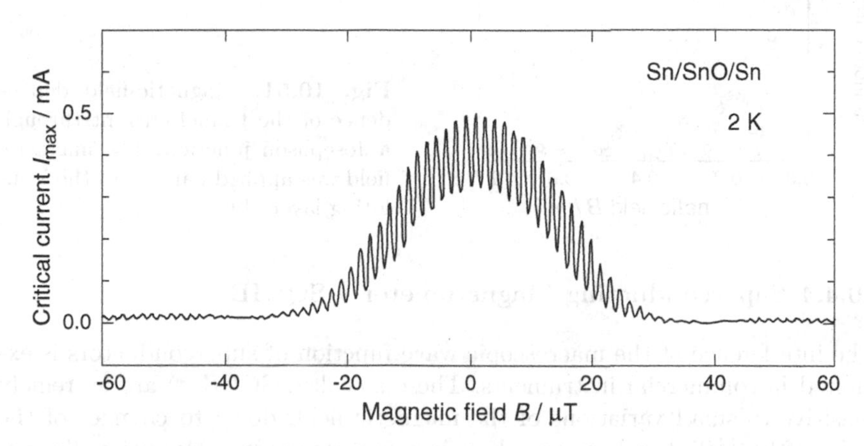

Fig. 10.50. Experimental trace of the current $I_{\max}$ in a Sn two-junction interferometer versus magnetic field. The period of oscillation is 16 μT [501]

A treatment of the current through a *single* Josephson junction requires a little more effort because the summation over the two junctions has to be replaced by an integration over the oxide layer. A calculation similar to the two-junction case yields

$$I = I_{\max} \left| \frac{\sin(\pi\Phi/\Phi_0)}{(\pi\Phi/\Phi_0)} \right| , \tag{10.133}$$

where Φ is the flux trapped in the junction. Figure 10.51 demonstrates the good agreement between theory and experiment.

Equation (10.133) has the form derived in optics for the Fraunhofer diffraction amplitude of a single slit. This result demonstrates the great similarity to optical diffraction. This is not so surprising because in both cases the common phenomenological basis is the interference of coherent waves. In optics, the envelope of the light intensity seen in a double-slit experiment is determined by the size of the individual slits. Similarly, the envelope of the pattern shown in Fig. 10.50 reflects the geometrical dimension of the individual junctions.

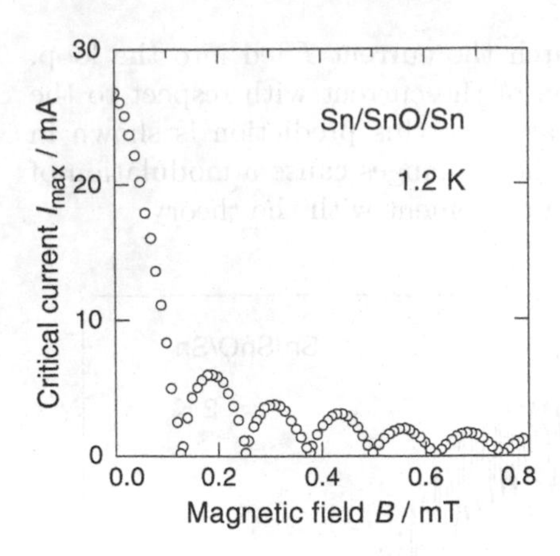

Fig. 10.51. Magnetic-field dependence of the tunnel current through a Josephson junction. The magnetic field was applied parallel to the insulating layer [498]

10.4.4 Superconducting Magnetometer – SQUID

The interference of the macroscopic wave function of superconductors is exploited in commercial instruments. These so-called SQUIDs[8] are extremely sensitive to small variations of the magnetic field, down to changes of the order of 10^{-14} T. Furthermore, these instruments serve as exceptionally sensitive ammeters and voltmeters.

In Sect. 10.4.1, we have shown that the magnetic flux enclosed by a superconducting ring is quantized and is given by $\Phi_i = p\Phi_0$, with p being an integer. The heart of a SQUID consists of a superconducting ring or cylinder with one or two weak links that allow magnetic fields to penetrate the junction. Before we consider the mode of operation of a SQUID we first discuss the properties of a superconducting ring with cross section A and a point contact that limits the supercurrent I_s to the critical value I_c. The flux generated by the current in the ring is given by LI_s, where L stands for the self-induction coefficient of the ring. Therefore, LI_c is the maximum flux that can be generated by the current in the ring. Henceforth, we assume that for the ring being considered, this quantity has the value $LI_c = 3\Phi_0/4$. Furthermore, $\Phi_a = B_a A$ is the flux due to the external field B_a, if no current is flowing.

In Fig. 10.52a, the field dependence of the magnetic flux in the ring is depicted. Here, we use normalized quantities, namely Φ_i/Φ_0 for the flux in the ring, and $B_a A/\Phi_0 = \Phi_a/\Phi_0$, for the external flux. Starting from $B_a = 0$, the shielding current I_s grows with increasing field until at $\Phi_a/\Phi_0 = 3/4$, the critical current I_c is reached (point A). A flux quantum penetrates the ring, and $\Phi_i = \Phi_0$. At point B the current is given by $I_s = I_c - \Phi_0/L = -\Phi_0/4L$, meaning that the current causes an *enhancement* of the flux inside the ring, i.e., the field inside exceeds the field outside. Increasing the external field further results in a reduction of I_s until at $\Phi_a/\Phi_0 = 1$ the supercurrent

[8] SQUID is the abbreviation for **S**uperconductive **QU**antum **I**nterference **D**evice.

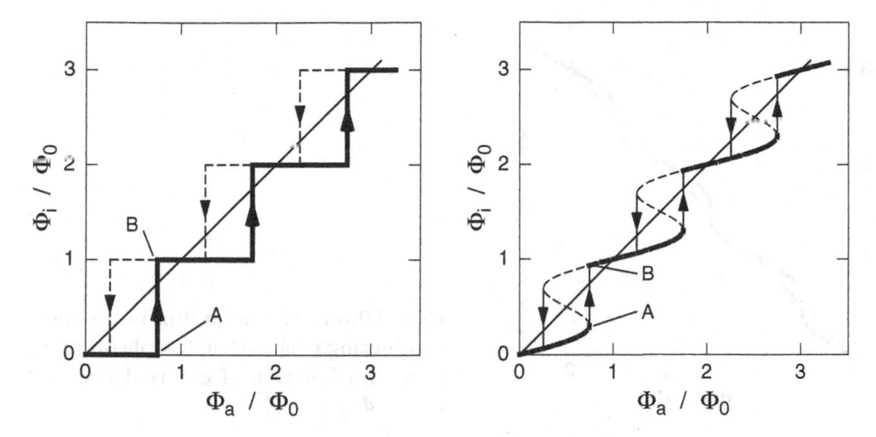

Fig. 10.52. Magnetic flux Φ_i/Φ_0 in a superconducting ring with the critical current $I_c = 3\Phi_0/4L$ versus external flux Φ_a/Φ_0, (**a**) with an ideal point contact, and (**b**) with an ideal Josephson junction. The *full lines* reflect the variation of Φ_i with rising field [502]

vanishes. With increasing field this phenomenon is repeated, as indicated in Fig. 10.52a. It can easily be shown that lowering the external field leads to the *dashed curve*. Clearly, the ring with the point contact exhibits a hysteretic behavior.

So far, we have considered an ideal point contact that keeps the flux $\Phi_i = n\Phi_0$ inside the ring stable up to the critical current $\pm I_c$. In fact, microbridges approximately show this type of behavior. However, superconducting rings with a Josephson junction behave differently in some respects. The flux encircled by the ring is not necessarily a multiple of Φ_0 since phase changes also occur in the junction, i.e., the relation $\Delta\varphi + 2\pi\Phi/\Phi_0 = 2\pi p$ has to be fulfilled, where $\Delta\varphi$ reflects the phase difference across the junction and p is an integer. As a consequence, the magnetic flux enters the ring even before the critical current is reached. In Fig. 10.52b, the relation between encircled flux and applied field is drawn for a ring with an ideal Josephson junction having the critical current $I_c = 3\Phi_0/4L$. At the point A the device becomes unstable and a sudden jump occurs to point B. With increasing field the *full line* is covered. Lowering the field obviously leads to hysteretic effects.

Of course, for technical applications hysteretic effects are undesirable. Luckily, the hysteretic properties can be influenced by external circuitry. The crucial quantity is the so-called *hysteresis parameter* defined by

$$\beta_c = \frac{2\pi R_s^2 I_c C}{\Phi_0}, \qquad (10.134)$$

where C is the intrinsic capacitance of the Josephson junction, and R_s the resistance connected in parallel. It can be shown that hysteretic effects vanish for $LI_c \leq \Phi_0/2$, corresponding to a hysteresis parameter $\beta \leq 0.7$. The relation

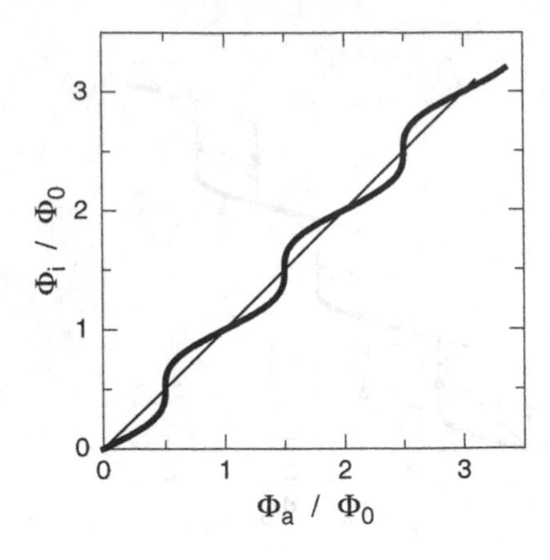

Fig. 10.53. Magnetic flux in a superconducting ring with a Josephson junction as a function of external field for $LI_c = \Phi_0/2$ [502]

between enclosed flux and applied field is depicted in Fig. 10.53 for the case where the relation $LI_c = \Phi_0/2$ holds.

The variation of the magnetic flux is connected with a variation of the current flowing in the ring. With appropriate circuitry, this variation can be detected as a voltage signal. Two different types of SQUIDs have been developed. In ac SQUIDs only one Josephson junction is used, while there are two junctions in dc SQUIDs. For a long time, ac SQUIDs were almost exclusively used in commercial applications. The main reason was that it was difficult to produce two well-defined Josephson junctions simultaneously. This situation has changed with the development of thin-film techniques. Nowadays, dc SQUIDs are routinely produced as planar SQUIDs and find applications in many fields.

As mentioned above, dc SQUIDs consist of two Josephson junctions connected in parallel (see Fig. 10.54a). A fixed current I is superimposed on the shielding current I_s that varies periodically with Φ_i/Φ_0, and therefore also with Φ_a/Φ_0. As a consequence, the voltage V across the junctions also reflects this characteristic periodicity. In Fig. 10.54b, the voltage output of a dc SQUID is shown. Clearly, the voltage varies periodically with the magnetic flux Φ_i through the SQUID loop, and hence with the external field. Without further external circuitry, the SQUID acts as a highly nonlinear current-to-voltage transformer. Therefore, an additional current is often coupled in via a feedback loop that compensates for any signals that are applied. Using feedback electronics the magnetic flux in the SQUID loop is kept constant even if the signal changes. In this way, the working point of the SQUID can be kept in the linear range of the characteristic curve. The current in the feedback loop is registered and serves as a measure for the variation of the magnetic field. Furthermore, SQUIDs are nowadays often used as voltage-to-current transformers by adding a low-resistance shunt across the SQUID. This shunt effectively keeps the voltage across the SQUID constant, and the current that flows is dependent on the magnetic flux.

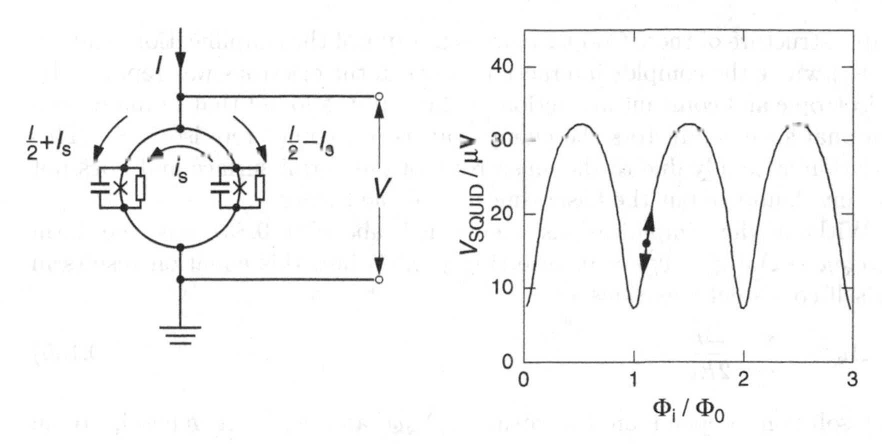

Fig. 10.54. (**a**) Schematic diagram for a dc SQUID. The two Josephson junctions are symbolized by *crosses*. The resistors are connected in parallel and together with the intrinsic capacitance of the junctions they allow a matching of the hysteresis parameter. (**b**) Voltage across a dc SQUID versus flux Φ_i in the SQUID loop. The *full circle* indicates the location of an appropriate working point of the feedback electronics

If the signal varies very quickly and the signal size is very large, the feedback electronics will not be able to compensate the flux variation properly. As a consequence, flux jumps with magnitude $p\Phi_0$ will occur in the SQUID resulting in a shift of the working point.

10.5 Superconductors with Unusual Properties

In Sect. 10.3, we have discussed the standard BCS theory that was originally developed for weak coupling between conduction electrons and the lattice. This theory led to the concept of Cooper pairs that occupy a common quantum state that can be described by a macroscopic wave function. The occurrence of Cooper pairs and the existence of a macroscopic wave function (also serving as an order parameter) is most likely a characteristic property of all superconductors. In addition, we have argued that the spin and the orbital angular momentum of Cooper pairs vanishes. It was a long-standing question as to whether so-called *unconventional superconductors* exist, exhibiting Cooper pairs with an internal structure. In fact, it was eventually found that superfluidity of ^{3}He is due to quasiparticle pairs with additional degrees of freedom.

In *conventional superconductors*, conduction electrons merge into pairs because of their attractive interaction via phonon exchange. The resulting Cooper pairs exhibit the orbital angular momentum $L = 0$. Analogous to atomic physics this state is called an *s-wave state*. From symmetry arguments it follows that the total spin $S = 0$, i.e., there is a *spin-singlet*. The simple structure of these Cooper pairs is a result of the simplification made in (10.69) where the complex interaction between the electrons was replaced by an isotropic and constant attraction. In fact, it was found that in many conventional superconductors the energy gap is not completely isotropic. This deviation is mainly due to the anisotropy of the Fermi surface and does not add any doubts about the basic concepts of the theory.

Without the simplifications mentioned above, (10.83) has the form $2u_{\boldsymbol{k}}v_{\boldsymbol{k}}\eta_{\boldsymbol{k}} - \Delta_{\boldsymbol{k}}(u_{\boldsymbol{k}}^2 - v_{\boldsymbol{k}}^2) = 0$. Inserting (10.87) into this equation results in the self-consistent equations

$$\Delta_{\boldsymbol{k}} = -\sum_{\boldsymbol{k}'} \frac{\Delta_{\boldsymbol{k}'}}{2E_{\boldsymbol{k}'}} \mathcal{V}_{\boldsymbol{k}\boldsymbol{k}'} \,. \tag{10.135}$$

The solutions depend on the form of $\mathcal{V}_{\boldsymbol{k}\boldsymbol{k}'}$ and lead, in general, to an anisotropic gap. It can be shown that the character of the Cooper pairs is unaltered as long as $\mathcal{V}_{\boldsymbol{k}\boldsymbol{k}'}$ is negative in all directions. But the anisotropy of the interaction may also lead to an energy gap with a symmetry lower than that of the Fermi surface. In these superconductors, the spin and orbital angular momentum of the Cooper pairs do not vanish, i.e., S or L, or both, are nonzero.

The orbital angular momentum of singlet pairs is zero in most cases because vanishing rotational energy leads to a lowering of the total energy of the pair. However, although in high-T_c superconductors Cooper pairs are singlet pairs, they have an orbital angular momentum $L = 2$. In this d-state, five different values for the z-component of the orbital angular momentum in principle exist. But because of the crystal field, these states are not degenerate as in free atoms, and thus only one of the states is found in high-T_c superconductors.

If the Cooper pairs carry the spin $S = 1$, then we will have *spin-triplets*, and the orbital angular momentum must be odd. We may classify them as p-, f-, etc. state pairs. Therefore, three z-components may occur for both the spin and the orbital angular momentum in the case of p-wave states, i.e., $3 \times 3 = 9$ combinations are possible. Again, crystal fields lead to a strong reduction of the relevant combinations. Furthermore, in heavy-fermion systems, the rather strong spin-orbit coupling must also be taken into account, meaning that in these systems the total angular momentum is the relevant quantity, and the pair states are classified according to their parity.

The symmetry of the gap function reflects the symmetry of the Cooper pairs. In s-wave superconductors a gap anisotropy may exist, but the gap does not vanish in any direction. In contrast, the energy gap of unconventional superconductors exhibits *nodes*. Depending on the symmetry, nodal points or nodal lines are expected to occur. The presence of nodes has a very important influence on the generation of quasiparticles. In s-wave superconductors, quasiparticles are generated via thermally activated processes, while in unconventional superconductors quasiparticles can be excited with arbitrarily small energies. As a result, the quasiparticle density n_{qp} is not given by $n_{\mathrm{qp}} \propto \exp(-\Delta/k_{\mathrm{B}}T)$ as in s-wave superconductors, but by $n_{\mathrm{qp}} \propto T^m$. The exponent m depends on the topological character of the nodes. It is $m = 2$ and $m = 3$ for nodal lines and nodal points, respectively.

Generally, there are two distinct classes of theories for unconventional pairing. The first approach is to replace phonons by another collective bosonic excitation of the solid. This approach successfully describes the physics of superfluid ^{3}He, where the intermediate bosons are failed ferromagnetic spin fluctuations (ferromagnetic paramagnons). This magnetically mediated interaction causes pairing in a state with orbital angular momentum $L = 1$. Magnetically mediated superconductivity has been proposed for various organic and heavy-fermion superconductors. A second approach to unconventional pairing is based upon the assumption that the superconducting condensation energy is not determined by the attractive interaction that is mediated by bosons, but rather by the energy gain due to feedback effects associated with pairing. Since so far no generally accepted pairing theory exists we will not pursue this important question any further.

Up to now we have considered 'classical' superconductors and have discussed their theoretical description in some detail. At the end of this chapter, we give a brief overview on some new developments that demonstrate the wealth of phenomena that have been observed during the last few decades in research on superconductivity. For this purpose, we select just a few interesting observations because it would be far beyond the scope of this book to try to give an overview of the entire field. First, we report on superconductors with 'unusual properties'. In particular, we consider organic superconductors and discuss the influence of magnetism on superconductivity. Finally, we will touch on the behavior of unconventional superconductors, in particular on *heavy-fermion systems* and *high-T_{c} superconductors* (see, for example, [503] and [504]).

10.5.1 Organic Superconductors

In 1964, *Little* hypothesized that high transition temperatures might be attainable in organic solids that consist of long conducting chains with conjugate double bonds and polarizable ligands [505]. The polarizable ligands should have a similar effect on electron movement as ions do in normal superconductors. Since this effect is based upon the movement of electrons and not of atoms, it should be noticeable at much higher temperatures. Little's proposal led to large efforts that did much to clarify ordering processes and conductivity in chain materials, but for many years superconductivity could not be found in organic systems.

Finally, superconductivity was discovered in 1980 in the organic compound $(\mathrm{TMTSF})_2\mathrm{PF}_6$ [506]. At a pressure of 12 kbar this compound is superconducting below 0.9 K. Later, superconductivity was also found in other members of the $(\mathrm{TMTSF})_2\mathrm{X}$ family and in a second class of organic compounds, namely in the $(BEDT\text{-}TTF)_2Y$ family.[9] The critical temperature of all these superconductors is relatively low and in most cases a pressure of up to several kbar has to be applied in order to avoid a metal–insulator transition. Although these systems have little in common with Little's original hypothesis, they do exhibit rather interesting low-temperature properties.

The compounds $(TMTSF)_2X$ are composed of the cation radical TMTSF and the inorganic anion X. The TMTSF molecules are arranged in stacks that are separated by the anions. The resulting crystals have a relatively high conductivity along the stacks, and a much smaller one perpendicular to them. The arrangement of the molecules in the BEDT-TTF salts is planar. They form conducting layers that are separated by quasi-insulating anion layers. Conductivity within the planes is much greater than between the planes, and therefore the BEDT-TTF salts display two-dimensional behavior.

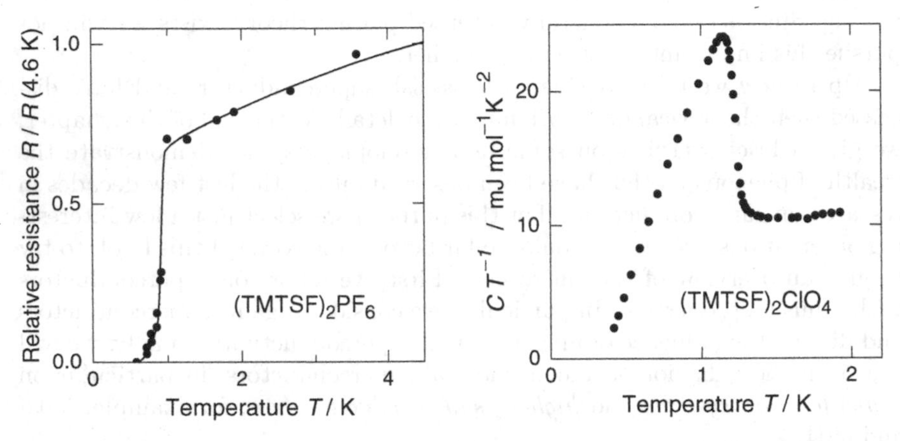

Fig. 10.55. (**a**) Temperature variation of the resistance of a $(TMTSF)_2PF_6$ sample at 12 kbar [507], (**b**) Specific heat C/T of $(TMTSF)_2ClO_4$ versus temperature [507]

As an example of a transition from the normal to the superconducting state of an organic solid, we show in Fig. 10.55a the temperature variation of the resistance of a $(TMTSF)_2PF_6$ sample. At high pressure, the resistance drops steeply at about 1 K, indicating that the sample becomes superconducting. The specific heat of $(TMTSF)_2ClO_4$ at ambient pressure is plotted in Fig. 10.55b. At the superconducting transition the typical anomalous form of the specific heat of superconductors is found. The anisotropic arrangement of the molecules has important consequences for the superconducting properties. For $(TMTSF)_2ClO_4$, the following critical fields were found at $T = 50\,\text{mK}$: $B_{c1} = 20\,\mu\text{T}$, $100\,\mu\text{T}$, and $1\,\text{mT}$ along the a-, b-, and c-axis, respectively. For

[9] The abbreviation TMTSF stands for tetramethyl-tetraseleniumafulvalene, and X for PF_6, TaF_6, ReO_4, ClO_4, etc., and BEDT-TTF for bisethylenedithiotetrathiofulvalene, and Y for ReO_4, I_3, AuI_2, etc.

the upper critical field the corresponding values are $B_{c2} = 2.8\,\mathrm{T}$, $2.1\,\mathrm{T}$, and $0.16\,\mathrm{T}$. For the penetration depth the values $\lambda = 0.5\,\mu\mathrm{m}$, $8\,\mu\mathrm{m}$, and $40\,\mu\mathrm{m}$ were measured [508].

Since the first discovery, a great number of other organic superconductors has been found. Among them are substances with higher transition temperature such as (BEDT-TTF)$_2$Cu[Ni(CN)$_2$]Br with $T_c = 11.5\,\mathrm{K}$. Although it is widely believed that superconductivity in these materials is caused by s-wave pairing, they exhibit many properties that are not predicted by BCS theory. Therefore, determining the nature of the pairing mechanism in organic superconductors is still a subject of great interest.

In 1985, the 'buckminster' fullerene C_{60} was discovered and has attracted much interest. In C_{60} crystals the molecules are weakly bound and form an fcc structure with the lattice parameter $a = 14.2\,\mathrm{Å}$. By doping with alkali or alkali-earth metals these crystals can be made metallic. Some of these so-called *fullerides* are even superconducting and exhibit remarkably high transition temperatures.[10] For example, for Na_2CsC_{60}, K_3C_{60}, and Cs_3C_{60} the transition temperatures 11.6 K, 19 K, and 40 K (at a pressure of 14.3 kbar) were measured. Apart from the high-T_c superconductors, Cs_3C_{60} is the solid with the highest transition temperature.

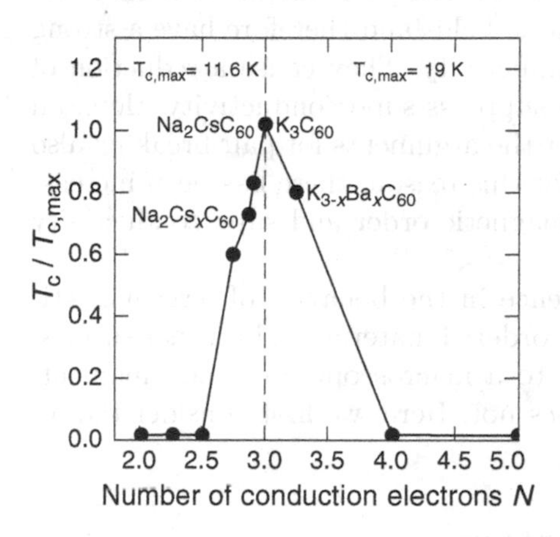

Fig. 10.56. Normalized transition temperature $T_c/T_c^{\max}$ of fulleride superconductors versus number N of conduction electrons per C_{60} molecule. For normalization, $T_c^{\max} = 11.6\,\mathrm{K}$ was chosen for $N \leq 3$, and $T_c^{\max} = 19\,\mathrm{K}$ for $N \geq 3$. The *full line* connecting the points is a guide for the eye [509]

In fullerides, the density of conduction electrons can be altered over a wide range by varying the dopant concentration. As shown in Fig. 10.56, the number $N = x + 2$ of conduction electrons per C_{60} molecule can be varied between $N = 2$ and $N = 5$ in the two systems $Na_2Cs_xC_{60}$ and $K_{3-x}Ba_xC_{60}$. Interestingly, the transition temperature depends strongly on this number. The quantity T_c has its highest value when the conduction band is just half filled, i.e., when three conduction electrons per C_{60} molecule are present.

[10] Although fullerenes are, in general, not considered to be organic molecules, we mention them in this section because many of their properties resemble those of organic superconductors.

The data shown in the figure are normalized to the critical temperature $T_c^{max} = 11.6\,K$ of Na_2CsC_{60} for the system $Na_2Cs_xC_{60}$, and to $T_c^{max} = 19\,K$ of K_3C_{60} for the system $K_{3-x}Ba_xC_{60}$. This interesting observation is still not completely understood.

Penetration depth and coherence length of fullerides are typically of the order of 500 nm and 3 nm, respectively, meaning that fullerides are type II superconductors just like all organic superconductors. Values around 10 mT and up to 50 T are observed for the lower and upper critical field. It is generally believed that the superconductivity in fullerides is 'conventional', i.e., Cooper pairs are in the *s*-wave state without resulting spin and orbital angular momentum. It seems that intramolecular vibrations give an important contribution to the pairing energy.

10.5.2 Magnetism and Superconductivity

Magnetic fields destroy singlet superconductivity by the orbital or paramagnetic effect. The orbital effect is a consequence of the opposite momenta of the electrons resulting in a Lorentz force acting in opposite directions, thus pulling the pair apart. The paramagnetic effect occurs since the applied field attempts to align the spins of both electrons along the magnetic field. Magnetic impurities give rise to local magnetic fields and therefore have a strong influence on the conventional superconductivity. They cause a reduction of the critical temperature and may even suppress superconductivity. Here, an interesting question arises as to whether the arguments for pair breaking also hold for magnetically ordered solids. For this reason, there has been increasing interest in the interplay between magnetic order and superconductivity in recent years.

There is a major qualitative difference in the behavior of ferromagnetically ordered and antiferromagnetically ordered materials. This is not surprising since ferromagnetic ordering leads to a macroscopic magnetic moment, whereas antiferromagnetic ordering does not. Here, we first consider the influence of antiferromagnetism.

Antiferromagnetically Ordered Systems

The coexistence of superconductivity and *long-range antiferromagnetic order* was most extensively studied in ternary rare-earth compounds including the molybdenum chalcogenides RMo_6S_8 and RMo_6Se_8 (Chevrel phases), and the rhodium borides RRh_4B_4, where R stands for the rare-earth atoms Gd, Tb, Dy, Er, etc. that carry magnetic moments. The occurrence of antiferromagnetic ordering in the superconducting state of RMo_6S_8 and RMo_6Se_8 compounds was inferred from a λ-type anomaly in the specific heat, a cusp in the magnetic susceptibility, and from features of the temperature dependence of the upper critical field B_{c2}. Later, magnetic order was investigated in detail by neutron-diffraction studies.

The growth of antiferromagnetic order in RMo_6S_8 with decreasing temperature is clearly demonstrated in Fig. 10.57a, where the effective magnetic moment per rare-earth ion is shown. In all substances the critical temperature T_c of superconductivity lies well above the Néel temperature T_N. As shown in Fig. 10.57b, an anomalous depression of B_{c2} in the vicinity of T_N is associated with the onset of antiferromagnetic order. The unusual variation of the critical field in the vicinity of T_N has been addressed by numerous theories that we do not pursue further here.

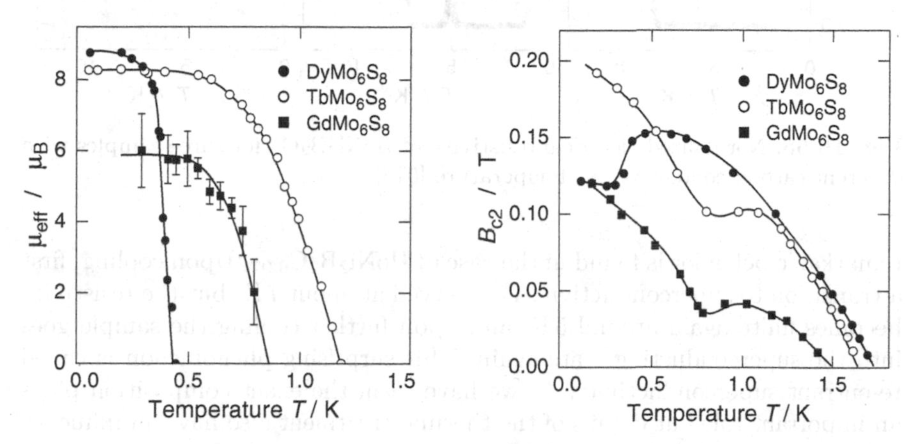

Fig. 10.57. (**a**) Temperature dependence of the effective magnetic moment per rare-earth ion of RMo_6S_8 as determined by neutron diffraction [510]. (**b**) Temperature variation of the critical field B_{c2} in RMo_6S_8 [511]

Quaternary boron carbides like RNi_2B_2C (with R = Tm, Er, Ho, etc.) are another class of substances with very interesting properties. These carbides exhibit a great variety of interesting physical phenomena due to the interplay between magnetic and superconducting properties. Let us first consider the so-called *re-entrant superconductivity*.[11] This phenomenon has been studied in great detail in compounds with the composition $HoNi_2B_2C_x$. In Fig. 10.58, the resistivity of three polycrystalline samples with different carbon content is shown. The compound $HoNi_2B_2C_{1.3}$ exhibits a well-defined transition to superconductivity at about 8 K and stays superconducting down to the lowest temperatures. In contrast, $HoNi_2B_2C_{0.88}$ remains normal conducting in the whole temperature range. Around 5 K, a change in the electrical resistivity is observed that can be attributed to antiferromagnetic ordering. The most

[11] Re-entrant superconductivity, i.e., the reappearance of a normal conducting phase due to the interaction between Cooper pairs and magnetic impurities was predicted by *Müller-Hartmann* and *Zitterartz* [512]. The first evidence for this phenomenon was found by *Winzer* in the system $La_{0.37}Ce_{0.67}Al_2$ [513].

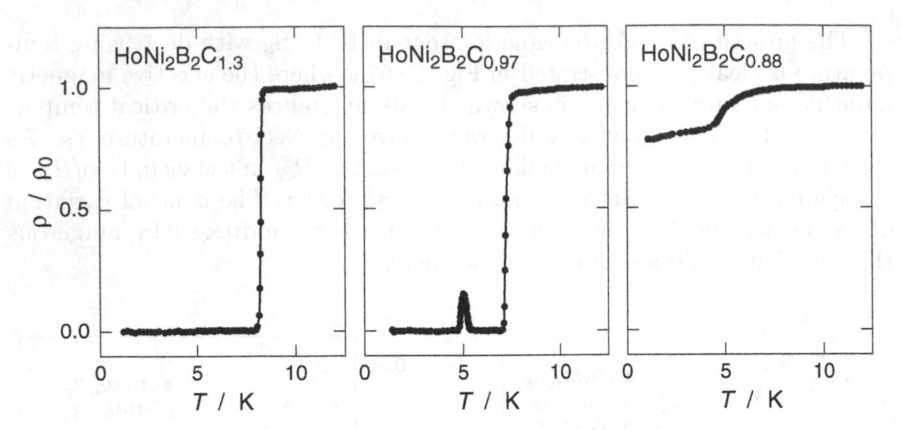

Fig. 10.58. Normalized electrical resistivity of $HoNi_2B_2C_x$ for three samples with different carbon content versus temperature [514]

remarkable behavior is found in the case of $HoNi_2B_2C_{0.97}$. Upon cooling, first a transition to superconductivity is observed at about 7 K, but the resistivity becomes finite again around 5 K, and upon further cooling the sample goes into the superconducting state again. This surprising phenomenon is called re-entrant superconductivity. As we have seen, the exact composition plays an important role but details of the thermal treatment also have an influence on the properties of these materials.

The so-called *near-re-entrant superconductivity* is closely related to the phenomenon just discussed. It is observed in $HoNi_2B_2C$, where three different types of antiferromagnetic order are found. The transitions from one state to another are easily detected in measurements of the specific heat (see Fig. 10.59a). Upon cooling, magnetic long-range order sets in at $T_M = 6$ K. An incommensurate 'spiral state' is formed where the magnetic moments spiral along the a-axis. Below $T^\star = 5.5$ K the spirals are oriented in the c-direction and exhibit a slightly different pitch. At still lower temperature, namely at $T_N = 5.2$ K, a first-order transition to the antiferromagnetic state occurs in which the spins order ferromagnetically in the ab-plane, and antiferromagnetically along the c-axis. The specific heat jump related to the superconducting transition at $T_c = 8$ K is hardly visible on the scale used in the figure.

As can be seen in Fig. 10.59b, the resistivity measured at zero magnetic field shows a sharp transition to superconductivity at $T_c = 8$ K, and magnetically ordered structures coexist with superconductivity. However, at relatively weak fields a finite resistivity is found around 5 K similar to the one reported for $HoNi_2B_2C_{0.97}$ in zero field. This means that in relatively weak fields the system re-enters the normal state. Simultaneously, an anomaly in the temperature variation of the upper critical field B_{c2} is observed that is similar to the Chevrel compounds RMo_6S_8. The re-entrant temperature depends little upon the strength of the magnetic field. Clearly, superconductivity is suppressed in

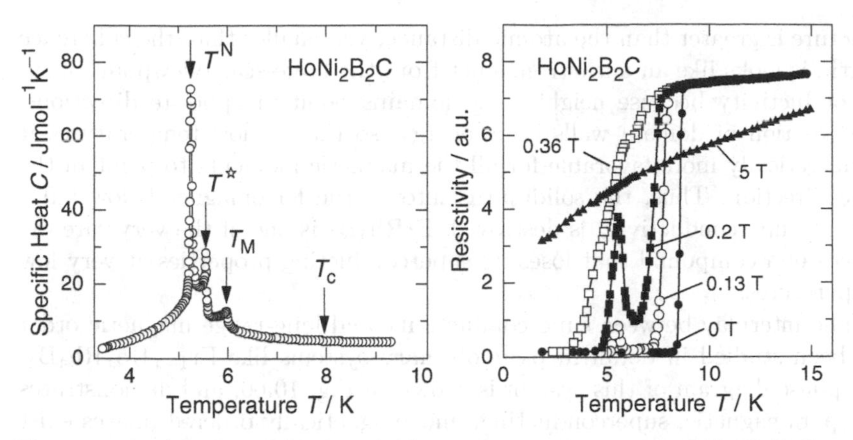

Fig. 10.59. **(a)** Zero-field specific heat of a single crystal of $HoNi_2B_2C$ versus temperature. At T_M and T^*, the magnetic moments arrange themselves in spirals along the a-axis and c-axis, respectively. The transition to antiferromagnetic order occurs at T_N [515]. **(b)** Resistivity versus temperature at different magnetic fields. Near-re-entrant behavior is observed around T_N [516]

the small temperature range where the two incommensurate magnetic structures are present, while in the antiferromagnetic state superconductivity is recovered.

Ferromagnetically Ordered Rare-Earth Systems

Re-entrant behavior due to the onset of *long-range ferromagnetic ordering* of the rare-earth ions was first discovered in $ErRh_4B_4$ and $HoMo_6S_8$ in 1977 [511, 517]. These two materials become superconducting at an upper critical temperature T_{c1} but lose this property again at a lower critical temperature $T_{c2} \approx T_{FM}$, where T_{FM} is the Curie temperature of the system. For $ErRh_4B_4$ the superconducting transition temperatures are $T_{c1} = 8.7\,K$ and $T_{c2} = 0.93\,K$. Thermal hysteresis in various physical properties and a spike-shaped feature in the heat capacity near T_{c2} indicate that a first-order transition from the superconducting to the ferromagnetic normal state occurs at that temperature.

Neutron-diffraction measurements confirmed that the ground states of $ErRh_4B_4$ and $HoMo_6S_8$ are ferromagnetic, but they also revealed an additional interesting feature. Just above T_{c2}, a 'modulated' magnetic structure appears in a narrow temperature interval where neighboring magnetic moments are aligned in the same direction, but the amplitude of the magnetization varies sinusoidally in space with a wavelength of about 10 nm [518]. Strictly speaking, the material is not ferromagnetic because it contains 'domain-like' structures with alternating magnetic moments. The coexistence of superconductivity and this sinusoidally modulated 'ferromagnetic' structure might be understood in the following way: If the period of the modulated structure is greater than the atomic distance, yet smaller than the coherence length, it looks like an antiferromagnet from the large-scale viewpoint of superconductivity because neighboring domains point in opposite directions. The creation of domain walls costs energy, so that at low temperatures it is energetically more favorable for all the magnetic moments to point in the same direction. Thus, the solid turns into a true ferromagnet below 0.9 K and the superconductivity is destroyed. $ErRh_4B_4$ is one of the very rare examples of a compound that loses its superconducting properties at very low temperatures.

The interplay between superconductivity and long-range magnetic order has been studied in detail in pseudoternary systems like $Er_{1-x}Ho_xRh_4B_4$. The phase diagram of this system is shown in Fig. 10.60, and demonstrates that paramagnetic, superconducting, and magnetically ordered phases exist and can be modified in a systematic way. The phase boundaries have been determined from ac magnetic susceptibility and neutron-diffraction studies. The temperature interval above T_{c2} is also indicated in this figure within which the sinusoidally modulated magnetic phase is observed.

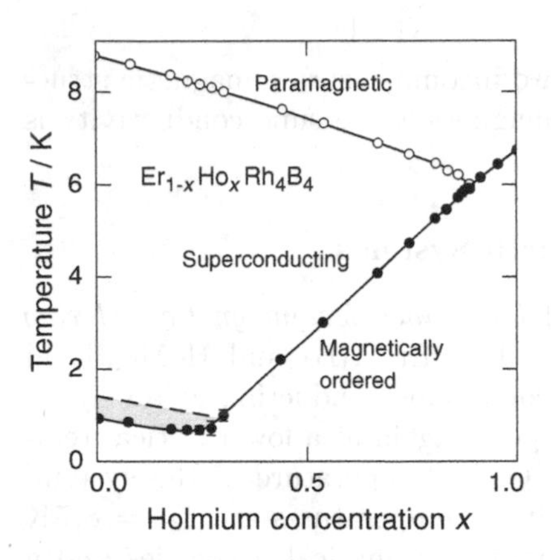

Fig. 10.60. Phase diagram of $(Er_{1-x}Ho_x)Rh_4B_4$ determined by different methods. In the *shaded area* superconductivity and the sinusoidally modulated ferromagnetically phase coexist. After [519]

Magnetically Mediated Superconductivity

One of the most interesting questions in low-temperature physics is the question of whether the electron–phonon interaction is necessary for the pairing of electrons in superconductors. In a classical description, the electrons move in a polarizable medium whose polarization leads via feedback to an energy gain for the electrons. In ordinary metals, the positively charged ions form this polarizable medium. In an alternative pairing mechanism either the electrons themselves must act as the polarizable medium or the ions possess an inner, polarizable structure. Since the only additional degree of freedom for electrons and ions is their magnetic moment, the question arises as to whether there is a scenario where the attractive interaction between the paired electrons is magnetically mediated.

The heavy-fermion system $CePd_2Si_2$ is one candidate for magnetically mediated superconductivity. As shown in the phase diagram (Fig. 10.61), this material exhibits superconductivity under high pressure. Under ambient pressure, $CePd_2Si_2$ is antiferromagnetic with a Néel temperature $T_N \approx 10\,K$. The ordering temperature decreases with increasing pressure and reaches absolute zero at about 26 kbar. It is conspicuous that superconductivity is observed in this pressure range, indicating that the formation of Cooper pairs is somehow related to this phase transition. The arguments in favor of this idea are as follows: Close to the critical pressure, namely at the so-called *quantum critical point*, fluctuations become very important that are not thermal but quantum mechanical in nature. Phase transitions of this kind are therefore called *quantum phase transitions*. They are ubiquitous in strongly correlated systems. In particular, many heavy-fermion systems can be tuned using pressure in between their antiferromagnetic and paramagnetic states. A fundamental question concerns the nature of the low-energy excitations close to a quantum critical point. It is generally assumed that the only important degree of freedom is the long-wavelength fluctuation of the order parameter, i.e., ordered regions of variable size are generated and disappear again. At a magnetic quantum critical point they are usually called *paramagnons*. Since electrons carry a magnetic moment, they interact with spin fluctuations. The range of interaction is determined by the typical size of the fluctuations, which grows when approaching the phase transition. In a series of theoretical considerations it has been shown that this interaction can give rise to an attraction between conduction electrons and hence to superconductivity. In particular, the magnetic interaction in nearly ferromagnetic metals should lead to Cooper pairs in spin-triplet state with an odd orbital quantum number, whereas in nearly antiferromagnetic metals the generation of spin-singlet states is expected. In addition, it has been shown that magnetic interactions tend to interfere with pairing via phonons, and are therefore expected to suppress conventional forms of superconductivity in nearly magnetic metals.

Let us have another look at the phase diagram of $CePd_2Si_2$ shown in Fig. 10.61. In the inset, the resistivity in the normal state is depicted. Near the critical pressure the resistivity varies proportional to $T^{1.2}$ over nearly two decades in temperature. This unusual temperature dependence is probably caused by the scattering of quasiparticles via magnetic interactions. As we have already mentioned, superconductivity is observed very close to the magnetically ordered state.[12] It seems that superconductivity occurs because

[12] A similar behavior is also observed for $CeIn_3$.

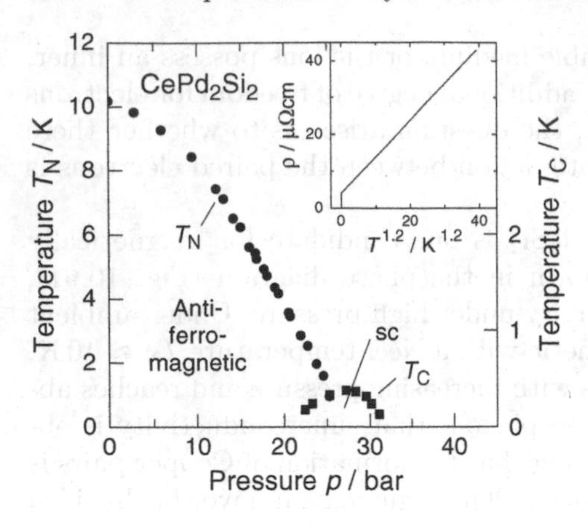

Fig. 10.61. Phase diagram of a high-purity single-crystal $CePd_2Si_2$. For clarity, the values of T_c have been scaled by a factor of three. In the inset, the resistivity ϱ in the a-direction is shown for a pressure of 28 kbar [520]

of magnetism rather than in spite of it, meaning that the Cooper pairs are probably held together by a 'magnetic glue'.

It should be pointed out that superconductivity is restricted to high-quality samples of $CePd_2Si_2$. The reason seems to be that superconductivity only occurs if there are attractive magnetic interactions that are strong enough to overcome competing interactions that may arise from impurities. This means that the Cooper pairs should effectively be small enough not to see the impurities. Therefore, an important condition for the occurrence of superconductivity is that the electron mean free path is greater than the coherence length. In fact, it was found that in samples showing superconductivity this condition was fulfilled. In contrast, superconductivity was not observed in samples with higher residual resistivity.

Superconducting Itinerant Ferromagnets

Until now we have discussed systems in which superconductivity and long-rang magnetic order coexist, but in all cases a precondition was that the period of the magnetically ordered structure is short compared with the coherence length of the superconductor. This argument holds for the antiferromagnetic systems considered above, and for the sinusoidally modulated ferromagnetic order in $ErRh_4B_4$ and $HoMo_6S_8$. In all the systems discussed so far, superconductivity and magnetism originate from different parts of the electron system. So the interesting question arises: Do superconductivity and long-rang magnetic order coexist in itinerant ferromagnets? Only recently has it been demonstrated that, in fact, such a coexistence occurs in UGe_2, $ZrZn_2$, and URhGe. In these compounds, the same band-like d- or f-electrons are responsible for both phenomena, for ferromagnetism and superconductivity.

We start with a brief look at the properties of URhGe, although its superconducting properties were the most recent to be discovered. Below the Curie temperature $T_{\mathrm{FM}} = 9.5\,\mathrm{K}$, this compound is a simple itinerant ferromagnet.

Its low-temperature specific heat $\gamma = C/T = 160\,\mathrm{mJ\,K^{-2}\,mol^{-1}}$ is relatively large, indicating an abundance of low-energy magnetic excitations in addition to conventional spin-wave contributions. In fact, superconductivity was found in high-purity samples with a residual resistivity of only $\varrho_0 \approx 2\,\mu\Omega\,\mathrm{cm}$. In Fig. 10.62, resistivity ρ, ac susceptibility χ_{ac}, and specific heat C/T are shown in the vicinity of $T_{\mathrm{c}} \approx 250\,\mathrm{mK}$. The relatively large width of the transition is, in part, due to an inhomogeneous magnetic field caused by the ferromagnetic domains of the sample and by material variations. The susceptibility χ_{ac} shows a clear diamagnetic response in the superconducting state, although the response is less than for an ideal Meissner phase. However, it has to be taken into account that each magnetic domain carries a magnetization of the order of $\mu_0 M_{\mathrm{s}} = 0.09\,\mathrm{T}$, leading to a nonzero local field. It is likely that the sample is in the mixed state, even without an external field, and therefore the susceptibility deviates from minus one.

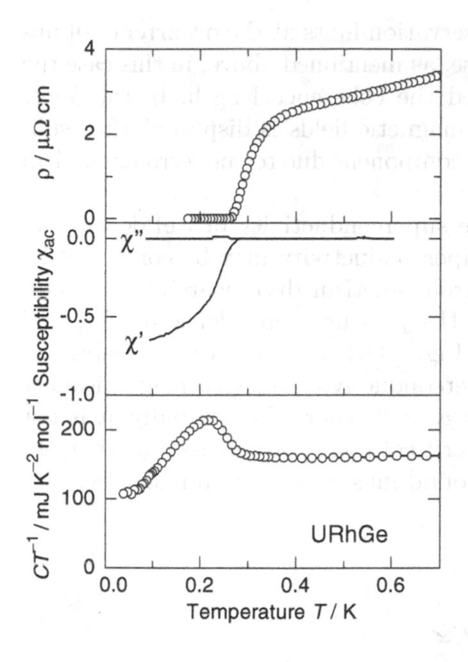

Fig. 10.62. Superconducting transition in polycrystalline URhGe. (**a**) Resistivity ρ, (**b**) ac susceptibility χ_{ac}, and (**c**) specific heat C/T in zero magnetic field [521]

$ZrZn_2$ is another interesting ferromagnetic compound, with Curie temperature $T_{\mathrm{FM}} = 28.5\,\mathrm{K}$, which also exhibits superconductivity. From its relatively large low-temperature heat capacity it can be concluded that low-energy magnetic excitations are present. As depicted in the upper panel of Fig. 10.63, there is a rapid drop of the resistivity below $T_{\mathrm{c}} \approx 0.29\,\mathrm{K}$, which is suppressed by a field of $0.2\,\mathrm{T}$, as expected for a superconducting transition. The purity of the sample is important. Superconductivity was only observed in the sample with the highest purity and a residual resistivity of only $0.62\,\mu\Omega$ cm, whereas the sample with a residual resistivity of $3.1\,\mu\Omega\,\mathrm{cm}$

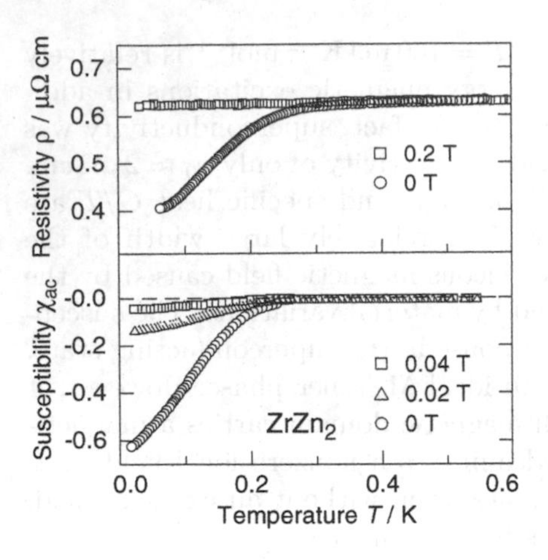

Fig. 10.63. Evidence for superconductivity in a $ZrZn_2$ single crystal of high purity. *Upper panel*: Resistivity versus temperature without field and at $B = 0.2\,\mathrm{T}$. *Lower panel*: ac susceptibility versus temperature at different magnetic fields after subtracting the background due to ferromagnetism [522]

remained normal conducting. This observation hints at the occurrence of unconventional superconductivity because, as mentioned above, in this case the electron mean free path has to exceed the coherence length. In the *lower panel* the ac susceptibility at various magnetic fields is displayed after subtracting the temperature-independent component due to the ferromagnetism of $ZrZn_2$.

As mentioned while discussing the superconductivity of $CePd_2Si_2$, theoretical considerations predict that superconductivity may be controlled by the quantum critical point, where ferromagnetism disappears at zero temperature. Therefore, investigations of the pressure dependence of T_{FM} and T_c are of great interest. As shown in Fig. 10.64, both critical temperatures decrease with pressure, and finally superconductivity and ferromagnetism are suppressed above the critical pressure $p_c = 21\,\mathrm{kbar}$. Unexpectedly, it is not sufficient to be close to the quantum critical point for superconductivity to occur in this material, since the compound must also be ferromagnetic.

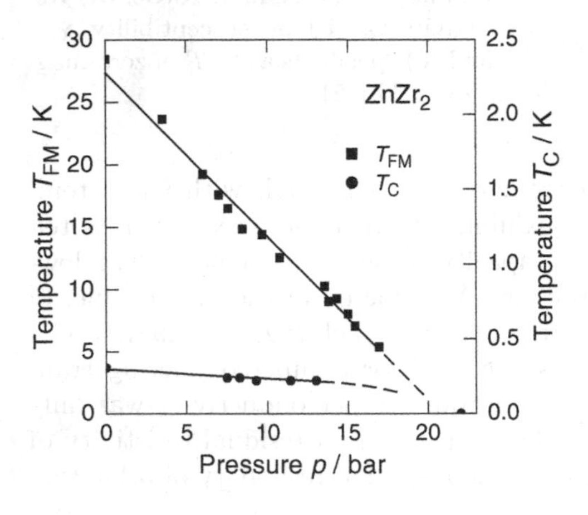

Fig. 10.64. Pressure dependence of the ferromagnetic ordering temperature T_{FM} and the superconducting transition temperature T_c. Note that for clarity T_c is magnified by a factor of ten [522]

The superconductivity of $ZrZn_2$ shows several incomprehensible properties. The flux expulsion in small magnetic fields is almost negligible, the electrical resistivity remains finite below its transition temperature. These observations seem to indicate that superconductivity in $ZrZn_2$ is inhomogeneous and only exists in clusters throughout the material. Furthermore, the specific heat does not exhibit a discontinuity at the transition temperature. From all these observations, it was concluded that superconductivity cannot be due to *s*-wave pairing but is rather caused by triplet states.

We will also briefly take a look at the properties of the ferromagnetic compound UGe_2 that becomes superconducting under pressure [523]. As shown in the phase diagram in Fig. 10.65, the ferromagnetic order is suppressed above the critical pressures $p_c = 15.8$ kbar. The resistivity and ac susceptibility show that UGe_2 is superconducting in the range between 9 kbar and p_c, i.e., well within the ferromagnetic state, where $T_c \ll T_{FM}$. The temperature dependence of the magnetization shows a sharp change at a certain pressure p_x. The corresponding temperature T_x is also depicted in the diagram. T_x approaches zero at about 12 kbar, where the superconducting transition temperature T_c goes through a maximum. Therefore, it was speculated that the transition at p_x would support the coexistence of itinerant ferromagnetism and superconductivity. In subsequent investigations it was shown that at p_x a transition between two ferromagnetic phases occurs, but this transition is not of second but of first order. Further studies have indicated that there is no quantum criticality associated with the suppression of this transition to zero temperature, although superconductivity in UGe_2 is intimately related to the proximity to a magnetic phase transition.

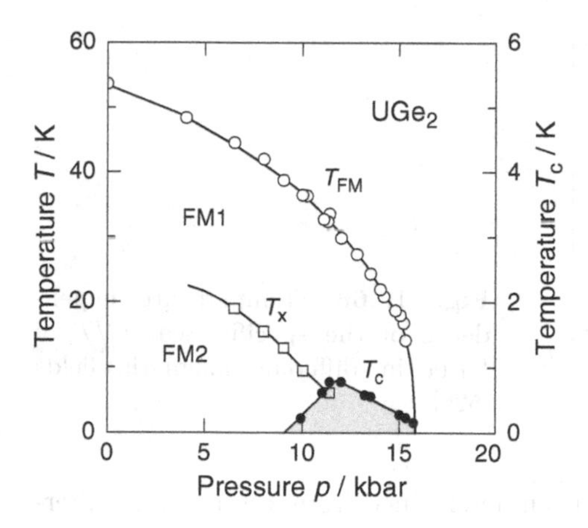

Fig. 10.65. Phase diagram of UGe_2. T_{FM} is the Curie temperature and T_c the superconducting transition temperature. FM1 and FM2 represent two different ferromagnetic phases above and below T_x, respectively. Note the magnified scale for T_c [524]

10.5.3 Heavy-Fermion Superconductors

In Sect. 7.5, we already considered heavy-fermion systems. Superconductivity in systems with 'heavy electrons' was first discovered in $CeCu_2Si_2$ in 1979 by *Steglich* [525]. Since then, many superconductors with heavy electrons have been found. They not only exhibit surprising thermal and magnetic properties but also have fascinating superconducting behavior. In particular, some of these compounds are, in fact, unconventional superconductors. As an example, we briefly consider here two members of the family of heavy-fermion systems, namely UPt_3 and UPd_2Al_3. A detailed discussion of the rather involved theoretical and experimental properties is far beyond the scope of this book (see, for example, [504]).

In the compound UPt_3, three phases with different superconducting properties are found. Since the s-wave state is unique, at least two of the three phases must have novel symmetry. Like superfluid ^3He, superconducting UPt_3 is thought to be in the spin-triplet p-wave state. Unfortunately, the pairing mechanism is still not really clear. However, it is almost certain that this superconductivity is not due to phonon exchange but has to do with an interaction mediated by spins.

The occurrence of different superconducting phases is clearly demonstrated in measurements of the specific heat shown in Fig. 10.66. Two well-defined steps are observed in zero field, indicating the surprising fact that transitions to different superconducting phases occur. The second step disappears at higher magnetic fields.

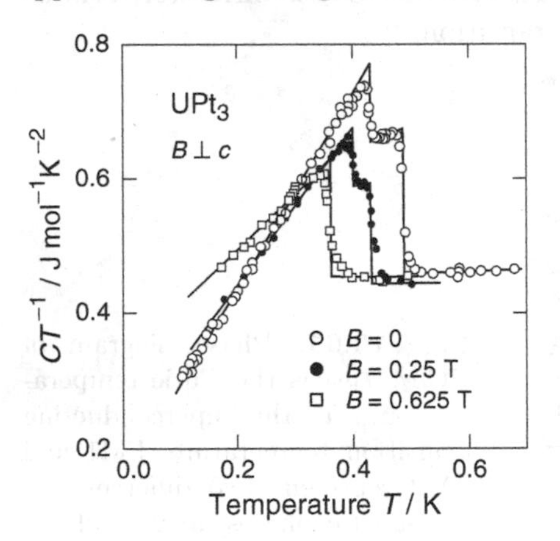

Fig. 10.66. Temperature dependence of the specific heat C/T of UPt_3 in different magnetic fields [526]

The phase diagram of UPt_3 has been determined from a number of different techniques such as measurements of specific heat, thermal dilatation and ultrasonic properties. The phase diagram shown in Fig. 10.67 was obtained by measurements of the sound velocity in an external magnetic field. By convention, the different superconducting phases are denoted by A, B and C. The transition temperatures to the A phase and to the B phase are generally

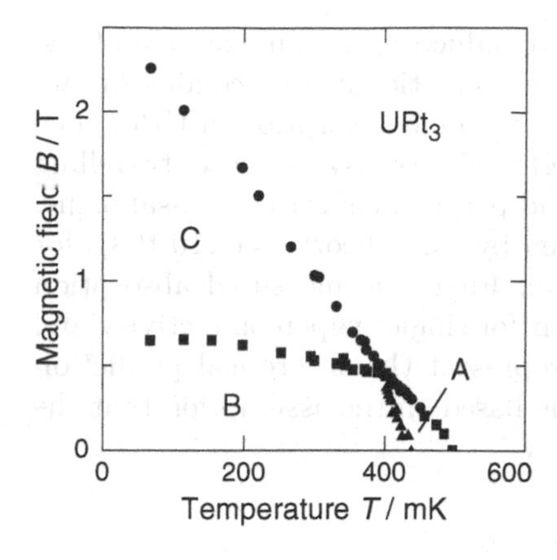

Fig. 10.67. Phase diagram of UPt_3 deduced from measurements of the sound velocity in a magnetic field. In the regions A, B, and C the sample exhibits superconductivity [527]

denoted by T_{c+} and T_{c-}, respectively. From this experiment and measurements by other techniques, it has been concluded that the phase transitions are of second order, and no structural or magnetic changes are involved.

Various different symmetries for the energy-gap function are conceivable, but it is most likely that the energy gap in UPt_3 has E_{2u} symmetry. The corresponding directional dependence of the surface of the energy gap in the three phases is visualized in Fig. 10.68. In the case of the B phase, an intersection is depicted where the magnitude of the gap is represented as the difference between gap surface and Fermi sphere. Note that this figure is not to scale, the magnitude of the energy gap is highly exaggerated. The frames define the hexagonal symmetry of the sample.

As already explained at the beginning of this section, the energy gap of unconventional superconductors vanishes in certain directions. As a consequence, the excitation of quasiparticles is much easier. Hence, below T_c, the drop of the specific heat and ultrasonic absorption with temperature is

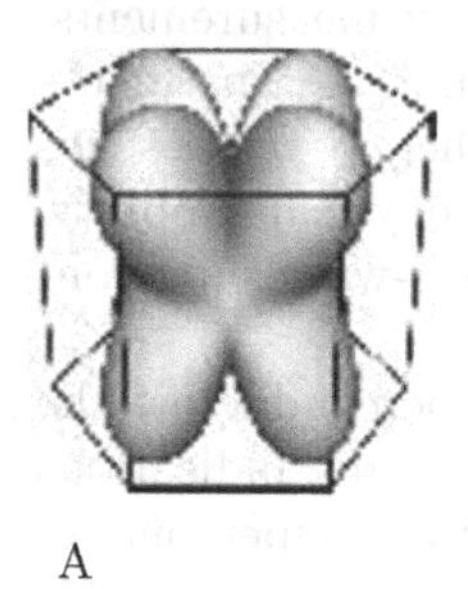

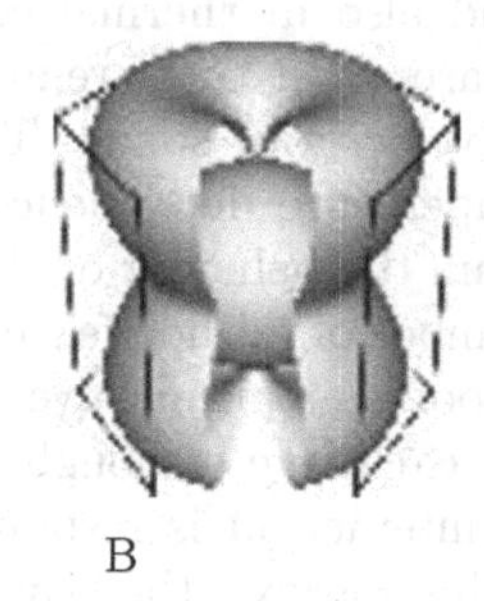

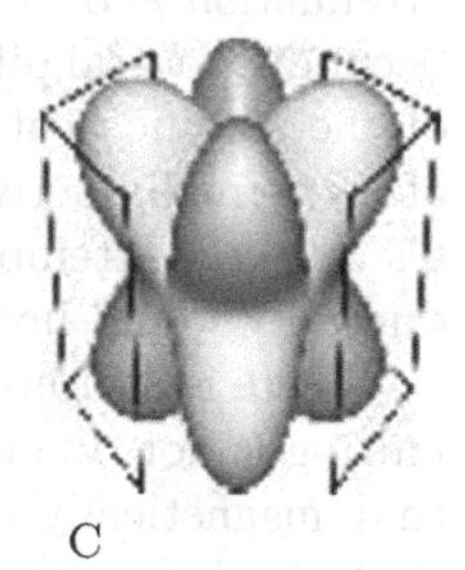

Fig. 10.68. Theoretical gap symmetries of the three phases for E_{2u} symmetry. For the B phase, an intersection is depicted where the magnitude of the gap is represented as the difference between the gap surface and the Fermi sphere. The frames define the hexagonal crystal symmetry

considerably weaker than in s-wave superconductors. As an example of the different properties of conventional and unconventional superconductors we show in Fig. 10.69 a measurement of the ultrasonic absorption in UPt_3. The experiment was carried out at about 200 MHz with transverse waves travelling in the a-direction and polarized parallel and perpendicular to the basal plane. For a comparison, the absorption predicted by BCS theory (see (10.108)) for s-wave pairing is shown by a *dashed line*. First, the measured absorption is considerably higher than the absorption for singlet superconductivity and rises much less steeply. The solid lines represent the theoretical prediction for the absorption of the two polarizations based on the assumption that the energy gap exhibits E_{2u} symmetry.

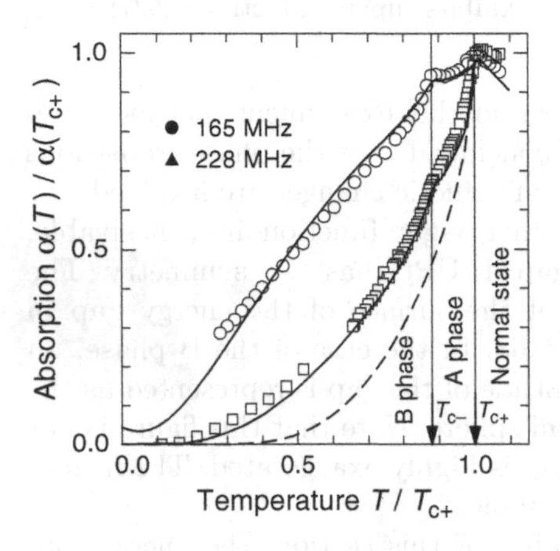

Fig. 10.69. Normalized ultrasonic absorption $\alpha(T)/\alpha(T_{c+})$ of UPt_3 versus reduced temperature T/T_{c+} of waves propagating in the a-direction, and polarized in the b-direction (∘) and c-direction (•). Theoretical predictions based on E_{2u} symmetry are shown by *full lines*, and by a *dashed line* for s-wave symmetry [528]

There is clear evidence that UPt_3 is an unconventional superconductor that we may summarize as follows: First, the transport properties are anisotropic in the superconducting state. This is obvious in the ultrasonic attenuation shown above and also in thermal conductivity measurements. Secondly, several different phases exist that are mainly seen in measurements of the specific heat and the velocity of sound. Thirdly, there is an absence of a thermally activated temperature dependence in any physical property. Clearly, any attempt to explain the behavior of UPt_3 using s-wave superconductivity would involve a number of ad hoc features.

There is still no direct proof that in unconventional superconductors the pairing is not due to phonon exchange although there are many indications that magnetically mediated interaction is responsible for the superconductivity in these materials. In this context, the heavy-fermion system UPd_2Al_3 plays an important role. It is an antiferromagnet with $T_N = 14.3$ K and transition temperature $T_c = 2$ K. The ratio $\Delta_0/k_B T_c \approx 6$ is remarkably high. The compound displays both pronounced local magnetic moments and heavy-mass itinerant quasiparticles due to the peculiar character of the $5f$-shell.

A combination of results from inelastic neutron-scattering measurements and tunneling spectroscopy yields an interesting indication of the participation of magnetic moments in the pairing process. First, we consider the differential conductivity measured at 0.3 K in a UPd_2Al_3-AlO_x-Pb tunnel junction (see Fig. 10.70a). The UPd_2Al_3 film was grown in situ epitaxially on a single-crystalline $LaAlO_3$ substrate. In the measurement displayed here, superconductivity in the lead film was suppressed by a magnetic field of 0.3 T. The *dashed line* represents a fit to the so-called Dynes formula [529] that describes the overall behavior rather well. A clear deviation is seen at 1.2 meV that is displayed in more detail in the inset. Because of the small energy involved it cannot be caused by a peak in the phonon spectrum.

In Fig. 10.70b, data is shown obtained by inelastic neutron scattering. In this plot, the intensity of scattered neutrons is depicted as a function of the energy transfer. A double peak is seen with maxima at 0.35 meV and 1.7 meV. The strong peak becomes weaker and broader with increasing temperature. It is attributed to a resonance associated with the itinerant heavy quasiparticles and is caused by the opening of the energy gap, but it is not a direct measure of 2Δ. The peak at 1.7 meV shows little intensity variation with temperature but considerable dispersion. It is assumed that it is associated with a collective mode of the localized magnetic moments due to an intersite interaction. The dispersion is caused by the strong interaction with the heavy superconducting electrons. This so-called 'magnetic exciton' is a localized excitation that moves through the lattice due to exchange forces between the magnetic moments. It is argued that these magnetic excitons may effectively introduce interactions between itinerant electrons, and so be responsible for the superconductivity.

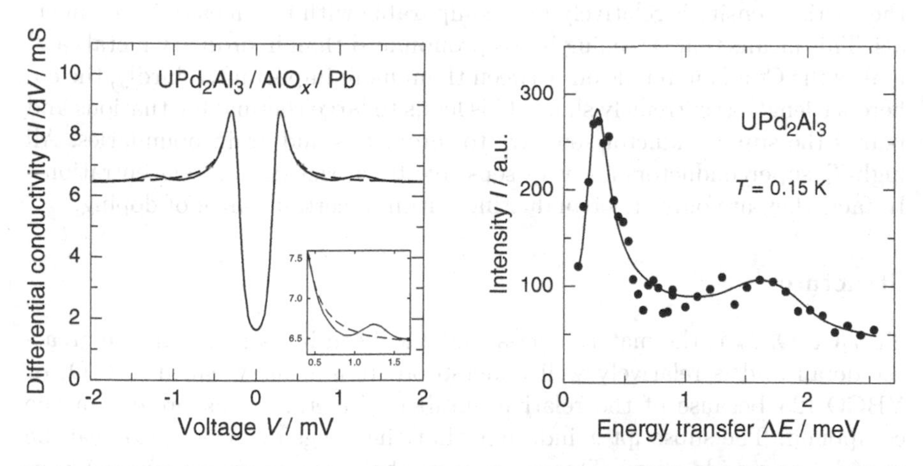

Fig. 10.70. **(a)** Differential conductivity of a UPd_2Al_3-AlO_x-Pb tunnel junction. The *dashed line* represents a fit to the Dynes formula [530]. **(b)** Low-temperature inelastic neutron scattering versus energy transfer measured at the antiferromagnetic wave vector (0 0 1/2), revealing a spin-wave mode at 1.5 meV [531]

10.5.4 High-T_c Superconductors

Ever since the discovery of superconductivity, large efforts have been made to develop or to discover superconductors with a high transition temperature. A transition temperature close to room temperature was always a pipe-dream, that would have enormous scientific and technical consequences. A less ambitious aim, but still one of outstanding importance, was the development of a superconductor with T_c above the boiling point of nitrogen of 77 K. In 1986 *Bednorz* and *Müller* made a crucial step forward in this direction when they studied the system Ba-La-Cu-O [459]. They found a transition temperature of about 30 K, i.e., a value considerably higher than the highest value known at that time. Later, transition temperatures up to 92 K, 127 K, and 135 K were reached in the systems Ba-Y-Cu-O, Tl-Ca-Ba-Cu-O, and Hg-Ba-Ca-Cu-O. Under pressure, a value for $T_c \approx 155\,\mathrm{K}$ has been reported for the system containing mercury.

As can be gathered from Table 10.1, the composition of high-T_c superconductors is rather complex. Hence, it is not easy to produce samples with exactly the desired stoichiometry, and the growth of single crystals is a very difficult task. For this reason, many of the early measurements on high-T_c superconductors were carried out on samples consisting of a mixture of different phases, making it rather difficult to interpret the data.

Apart from the high transition temperature, high-T_c superconductors have a number of features in common. First, they are all layered compounds with tetragonal (or close to tetragonal), or orthorhombic structure. They contain CuO planes lying perpendicular to the c-direction with the formula CuO_2. For this reason they are also called *cuprates*. These planes contain mobile carriers that are the source of superconductivity. Secondly, the carrier density is relatively low, comparable with the density in semimetals. This means that screening is less pronounced than in ordinary metals and makes the Coulomb repulsion between them more important. Thirdly, the coherence length is extremely short. This leads to large thermal fluctuations and makes the superconductor sensitive to impurities and grain boundaries. All high-T_c superconductors are very sensitive to carrier dopant concentrations. In fact, they are only superconducting within a certain range of doping.

Structure

$YBa_2Cu_3O_{6+x}$ is the material class that has been investigated in the greatest detail and is relatively well understood. It is usually called YBCO, or YBCO-123 because of the relative number of metal atoms present in the compound. The subscript x indicates that the oxygen concentration can be varied over a wide range. This also means that the compounds are not generally stoichiometric. Special cases are shown in Fig. 10.71, namely the case $x = 1$ and $x = 0$. The structures may be viewed as stacks of layers with different chemical composition. For $YBa_2Cu_3O_{6+x}$, the stacking sequence in the crystallographic unit cell is Y-CuO_2-BaO-CuO_x-BaO-CuO_2-Y. The metal atoms separate two identical blocks that contain two CuO_2 planes with Cu(2) sites, two BaO planes and a plane formed by Cu-O chains along the b-direction of the orthorhombic structure involving the Cu(1) sites. For $x = 1$, Cu-O chains are formed, as indicated in Fig. 10.71a. Upon depletion, oxygen vacancies are generated in these chains and they redistribute in a fashion such that, on average, oxygen atoms occupy sites along the a- and b-direction with equal probability. For $x = 0$ the former Cu-O chains are fully depleted of oxygen. A closer look at the chemistry reveals that upon doping, electrons are removed from the three layers with copper atoms. If the holes are distributed equally between the three copper layers, the number of holes per unit cell is given by $n_h = (2 + 2x)/3$.

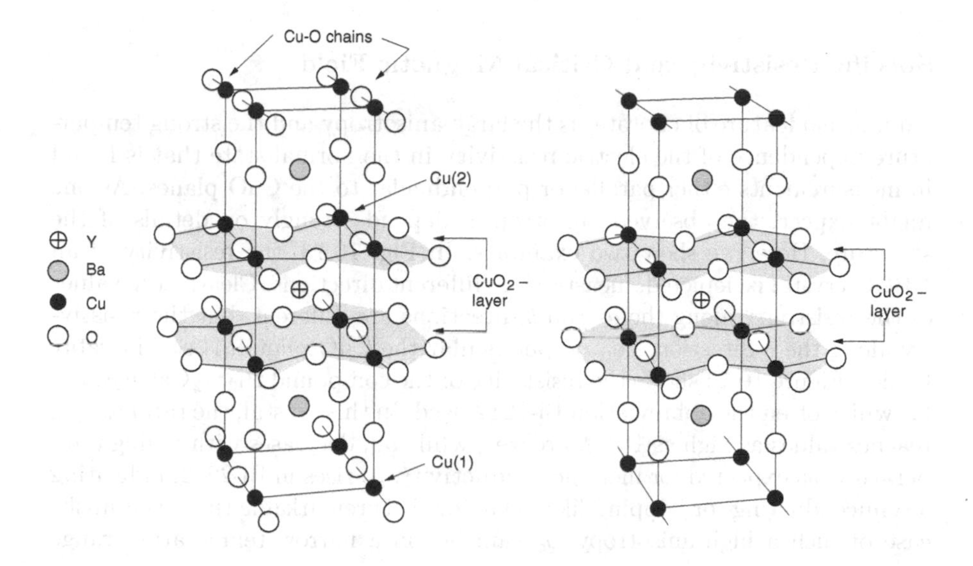

Fig. 10.71. Crystal structure of the cuprates **(a)** $YBa_2Cu_3O_7$ and **(b)** $YBa_2Cu_3O_6$

The oxygen content has an influence on the lattice structure and also has a pronounced effect on the electrical properties. The phase diagram of YBCO is shown in Fig. 10.72. $YBa_2Cu_3O_6$ is an insulating antiferromagnet with a tetragonal lattice. With increasing oxygen doping, the Néel temperature falls and finally the antiferromagnetism disappears, being replaced almost immediately by superconductivity. The structural transition from tetragonal to orthorhombic lattice in this dopant-concentration range has little effect on its superconductivity. Increasing x further causes T_c to rise and is at its maximum when *optimally doped* at about $x = 0.92$, where $T_c \approx 92\,K$. If the value of x is smaller, we speak of *underdoping*, and when it is larger of *overdoping*.

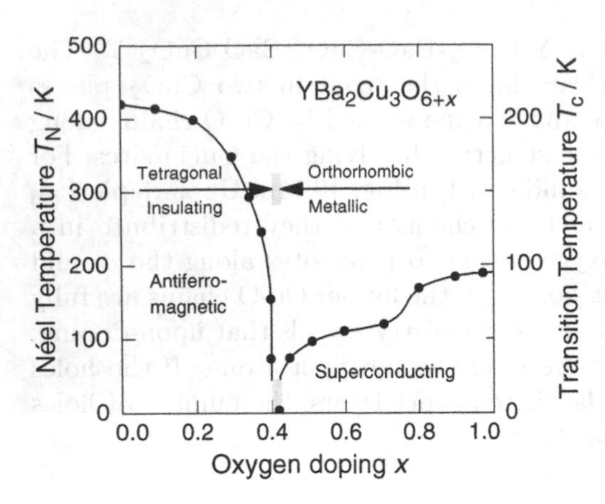

Fig. 10.72. Doping phase diagram for YBCO. With increasing oxygen content the hole concentration rises. The transition from the insulating to the conducting state occurs at $x \approx 0.4$. After [532]

Specific Resistivity and Critical Magnetic Field

An unusual feature of cuprates is the large anisotropy and the strong temperature dependence of the electric resistivity in the normal state that is found in measurements either parallel or perpendicular to the CuO planes. As one might expect, the observed anisotropies depend strongly on details of the structure. Here, we show two examples. In Fig. 10.73, the resistivity of an YBCO crystal is depicted, measured in different directions. Clearly, the values of the resistivity along the a- and b-directions are different, but the resistivity along the c-direction, i.e., perpendicular the CuO planes, is considerably higher. Figure 10.74 shows the resistivity of the compound $Bi_2Sr_2CaCu_2O_{8+x}$ for which often the abbreviation Bi-2212 used. In this crystal, the ratio ϱ_c/ϱ_{ab} reaches values as high as 10^5. Moreover, while ϱ_{ab} decreases with falling temperature, as expected for metallic conductivity, ϱ_c rises in Bi-2212, indicating a semiconducting- or hopping-like behavior. It is remarkable that even in the case of such a high anisotropy, ϱ_c vanishes in a narrow temperature range

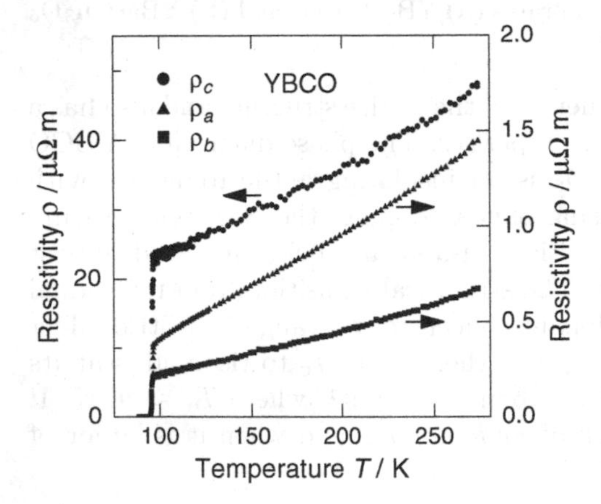

Fig. 10.73. Electrical resistivity of an YBCO single crystal, measured along the principal axes. Note the different scales for the resistivity in different directions [533]

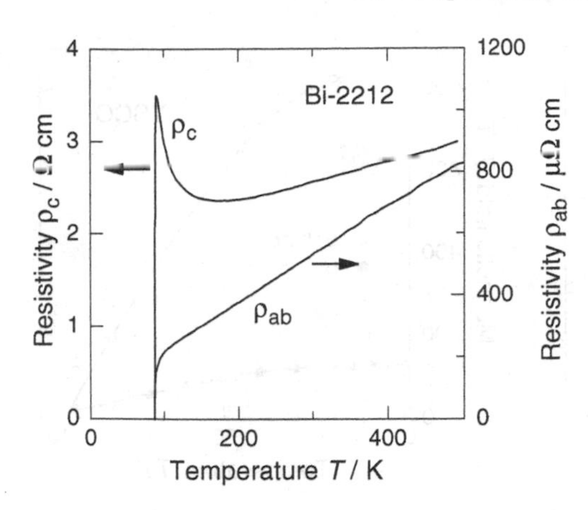

Fig. 10.74. Electrical resistivity of a Bi-2212 single crystal, measured in the basal plane, and in the *c*-direction. Note the different scales for the resistivity in different directions. After [534]

around the superconducting transition, confirming that superconductivity is a three-dimensional phenomenon.

Ginzburg–Landau theory is based on general principles, so it is not surprising that it applies well to high-T_c superconductors. However, the theory has to be modified in such a way that it takes into account the extreme anisotropy. For this purpose, different effective electron masses m_i are introduced. This concept leads to the penetration depth $\lambda_i \propto \sqrt{m_i}$, the coherence length $\xi_i \propto 1/\sqrt{m_i}$, and the Ginzburg–Landau constant $\kappa_i \propto m_i$. In cuprates, the penetration depths are relatively large. For example, the numerical values $\lambda_c = 890\,\text{nm}$, and $\lambda_{ab} = 135\,\text{nm}$ are found for YBCO. This means that according to (10.23) the density of Cooper pairs is relatively low since $\lambda^2 \propto n_s$.

The anisotropic structure of cuprates is also reflected in their superconducting properties, e.g., in the angular dependence of the critical magnetic fields. Measurements of critical fields are, however, experimentally difficult. For most materials, B_{c2} is outside the range of the magnets available, except close to T_c. Direct measurements of B_{c1} are also difficult because surface barriers hinder the flux entry into cuprate materials. Therefore, indirect methods have been used to determine the respective quantities. Via the relations (10.2) and (10.15) the thermodynamic critical field $B_{c,th}$ can be deduced from the jump ΔC in the specific heat at the transition temperature (see, e.g., the specific heat of YBCO in Fig. 10.76). If, in addition, the penetration depth λ is measured, the two quantities B_{c1} and B_{c2} can be calculated via (10.50) and (10.52).

The temperature variations of the critical fields B_{c1} and B_{c2} of YBCO are plotted in Fig. 10.75. As expected, the numerical values of the critical fields in the *c*-direction and in the *ab*-plane differ considerably. The low-temperature values of B_{c2} deduced from specific heat and penetration depth are extremely high and are not directly accessible to experiments. Using (10.51) the coherence length ξ can be calculated from the upper critical field, and the values $\xi_{ab} = 1.6\,\text{nm}$ and $\xi_c = 0.24\,\text{nm}$ are found. Since the size

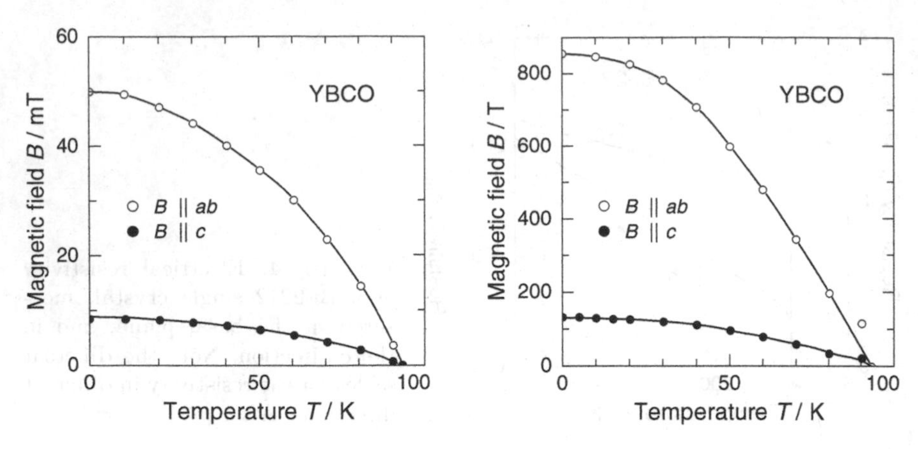

Fig. 10.75. Critical magnetic fields of YBCO deduced from specific heat and penetration-depth measurements. **(a)** Lower critical field B_{c1}, **(b)** upper critical field B_{c2}. The deviation of the data point near T_c is due to experimental difficulties caused by fluctuation effects [535]

of the Cooper pairs is comparable with ξ, this means that Cooper pairs in high-T_c superconductors are extremely small. In fact, their extension in the c-direction is of the order of the thickness of the CuO_2-layers, and is not much larger in the ab-plane. The data for λ and ξ tell us that the Ginzburg–Landau parameter κ of YBCO is, as in all cuprates, very large, meaning that high-T_c superconductors are type II superconductors.

The extremely small coherence length is responsible for many peculiarities of high-T_c superconductors. One important consequence is that fluctuation effects are extremely pronounced since the significance of fluctuations is inversely proportional to the number of Cooper pairs present within the coherence volume ξ^3. While in conventional superconductors this number is of the order of 10^6 to 10^7, it is about 10 in cuprates because of the small coherence length and the low density of Cooper pairs. Consequently, fluctuation effects are not negligible and distinct deviations are expected from the mean-field behavior of clean common superconductors. In particular, the temperature range within which fluctuations are noticeable is proportional to ξ^{-3} meaning that consequences of fluctuations are easily seen in cuprates over a relatively wide temperature range. Moreover, the two-dimensional character of the superconductivity in cuprates leads to a further enhancement of fluctuation effects.

It is obvious that strong fluctuation effects in the temperature dependence of measured quantities are annoying because they hinder exact evaluation of important parameters like the transition temperature. In the following figures, the importance of fluctuations is clearly demonstrated. In Fig. 10.76a, the specific heat of a YBCO single crystal is plotted. As can be seen from the scale of the ordinate, the anomaly is located on top of a large 'background' from the lattice because of the high transition temperature. Pronounced deviations from an ideal mean-field behavior occur close to T_c in both the normal and the superconducting state. To determine T_c or ΔC correctly, the background has to be taken into account. Similarly, fluctuation effects are also important for other properties like resistivity or magnetization. These fluctuation effects can be described on the basis of the so-called 3D XY model for phase transitions. A detailed discussion of the relevance of this model with respect to the phase transition of cuprates is given in [537]. In Fig. 10.76b, the resistivity $\varrho(T)$ of Bi-2212 is displayed. At temperatures that are substantially above T_c, a gradual decrease of the resistivity with decreasing temperature is observed, in contrast to the abrupt drop of $\varrho(T)$ observed in bulk s-wave superconductors where fluctuations are hardly observable.

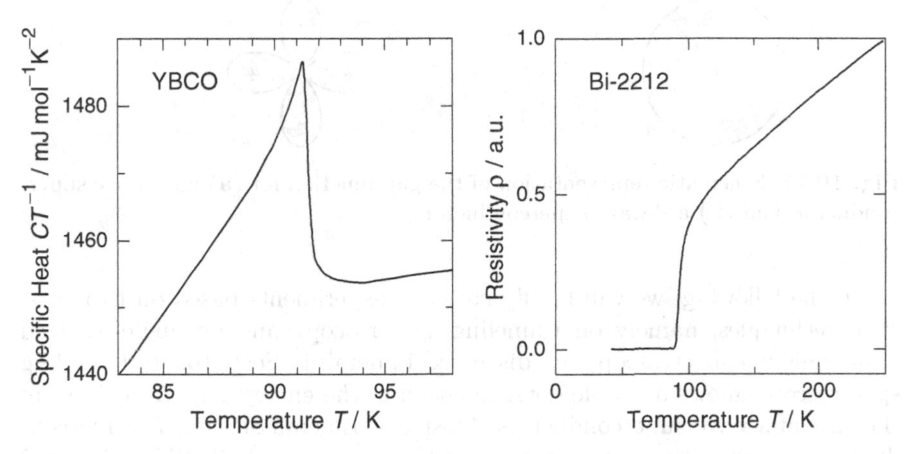

Fig. 10.76. Fluctuation effects. **(a)** Specific heat of YBCO at the transition temperature [536]. **(b)** Temperature dependence of the resistivity of Bi-2212 [534]

Energy Gap

Cuprates have many rather unusual properties. The unusually high transition temperature raises the question of whether superconductivity is caused by phonon exchange or by another mechanism in these materials. The strong anisotropy of the normal conducting and superconducting properties indicate that to a good approximation cuprates can be considered as two-dimensional systems. Therefore, it is unlikely that high-T_c superconductors exhibit s-wave superconductivity. As we already mentioned, the shape or symmetry of the gap function $\Delta_{\boldsymbol{k}}$ is tied to the symmetry of the pairing configuration and the corresponding order parameter of the superconducting state. Thus, experimental verification of the shape of $\Delta_{\boldsymbol{k}}$ is an important and extremely useful test of theoretical models. There is overwhelming evidence that superconductivity in cuprates is mainly determined by the special properties of the

CuO planes and also by the atomic arrangement of the copper and oxygen atoms in these planes. Taking these facts into account, the most likely configuration of the gap function has $d_{x^2-y^2}$ symmetry, with $L = 2$ and the gap function

$$\Delta_{\boldsymbol{k}} = \Delta_{\mathrm{m}} \cos 2\Phi \,, \tag{10.136}$$

where Δ_{m} is the maximum gap and Φ the angle between the a-axis and the wave vector $\boldsymbol{k}$. The occurrence of Cooper pairs with $L = 2$ is also compatible with the assumption that the pairing mechanism is not due to phonon exchange. The gap functions of s-wave and d-wave superconductors are compared in Fig. 10.77.

Fig. 10.77. Schematic representation of the gap function for **(a)** an s-wave superconductor, and **(b)** a d-wave superconductor

In the following, we will briefly consider experiments based on two powerful techniques, namely on tunneling spectroscopy and on angle-resolved photoemission spectroscopy. As discussed in detail in Sect. 10.3.6, tunneling spectroscopy allows a simple determination of the energy gap Δ, at least in the case of s-wave superconductors. Most experiments on high-T_{c} superconductors carried out in the past indicate that the ratio $\Delta_0/k_{\mathrm{B}}T$ is a factor 2 to 3 greater than the value predicted by (10.105). Obviously, electrons are strongly coupled in high-T_{c} superconductors.

The density of states of quasiparticles in d-wave superconductors differs considerably from that in conventional superconductors, mainly because of the existence of nodal lines. Figure 10.78a shows the density of states (in the 'one-electron representation') calculated for a d-wave superconductor with the maximum gap $\Delta_{\mathrm{m}} = 20\,\mathrm{meV}$. The characteristic features are a cusp at the Fermi energy, i.e., at zero bias, in the measurement with a tunnel junction, and a relatively small ratio of the peak height to the 'background' that is about two. These characteristic features are certainly distinctive and it might be expected that the presence of d-wave superconductivity could be demonstrated by tunneling experiments. However, it turned out that the spectrum of high-T_{c} superconductors is, in general, much richer than described here. Because of the strong coupling, additional features are observed, such as bound states. Since we do not want to discuss this in detail here, we show in Fig. 10.78b the tunneling conductance of $Tl_2Ba_2CuO_6$. This measurement was carried out with an SIN point contact, where a gold tip was used as the

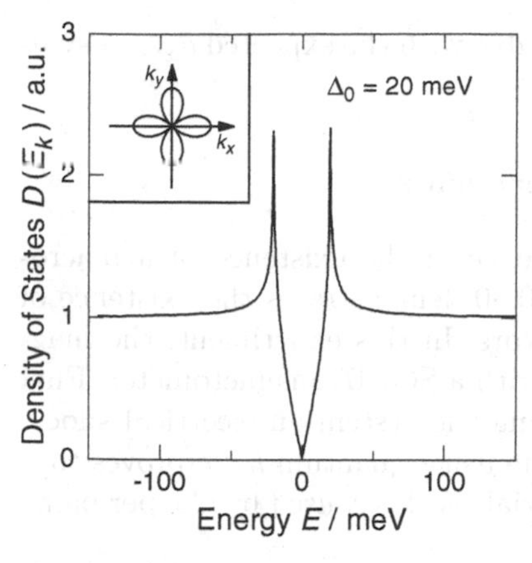

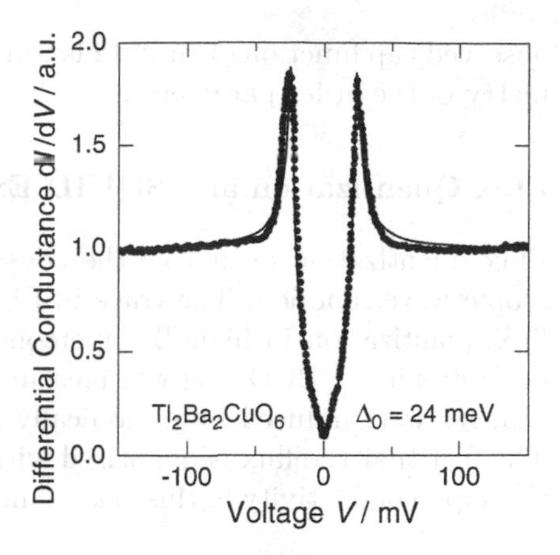

Fig. 10.78. Density of states of d-wave superconductors. **(a)** Theoretically expected spectrum for a d-wave superconductor with a gap maximum of 20 meV. **(b)** Comparison between the measured differential conductance (*dots*) and the theoretical prediction for $Tl_2Ba_2CuO_6$ (*solid line*) [538]

normal conducting material. The unit cell of $Tl_2Ba_2CuO_6$ contains only a single CuO layer and is therefore much simpler than those of common high-T_c superconductors. This seems to be why no additional features are found in the gap and a striking similarity exists between the measured and the expected spectrum demonstrating the d-wave character of this superconductor.

A more direct proof for the existence of nodal lines is possible with the help of angle-resolved photoemission spectroscopy. For superconductors with $d_{x^2-y^2}$ symmetry we expect the gap function to vanish in the $\langle 110 \rangle$ direction (see Fig. 10.77). Angle-resolved photoemission measurements are a suitable technique to test this prediction because they allows us to measure the $\boldsymbol{k}$-dependence of the energy gap. Figure 10.79 clearly demonstrates that the

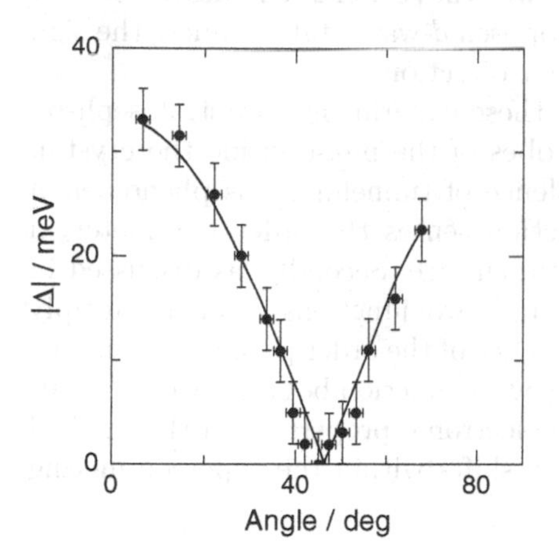

Fig. 10.79. Anglar dependence of the energy gap of Bi-2212 measured by photoemission spectroscopy. The *solid line* represents a fit to the data assuming a d-wave gap [539]

observed gap function of Bi-2212 is compatible with the expected $d_{x^2-y^2}$ symmetry of the order parameter.

Flux Quantization and SQUID Experiments

Flux quantization occurs as the consequence of the existence of a macroscopic wave function. The trace in Fig. 10.80 demonstrates the existence of flux quantization in high-T_c superconductors. In this experiment, the magnetic flux in a YBCO ring was measured with a SQUID magnetometer. Flux changes were induced by periodically giving the system an electrical shock. The fact that the flux is quantized with the usual quantum $h/2e$ proves that the superconductivity in this class of materials is also caused by Cooper pairs.

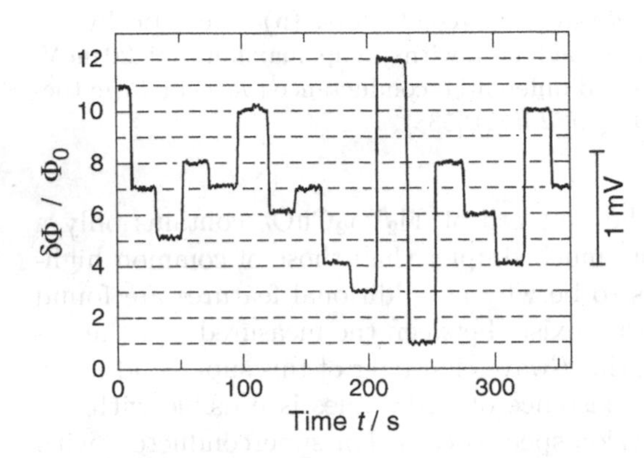

Fig. 10.80. Flux trapped in a sintered ring of YBCO. The flux was measured with a conventional SQUID. The distance between the *dashed lines* corresponds to the change by one flux quantum $h/2e$ [540]

An abundance of information on the nature of superconductivity is obtained by SQUID experiments. The key feature is the sensitivity to the anisotropy of the *phase* of the order parameter (or macroscopic wave function) rather than to its magnitude. As we will see, this enables us to test the most characteristic feature of the proposed *d*-wave state, namely the sign change of the order parameter in different directions.

There are two important aspects in these experiments. First, Josephson tunnel junctions serve as directional probes of the phase inside the crystal. Because of the strong thickness dependence of tunneling, this phenomenon is highly directional, so that each junction senses the order parameter in the direction perpendicular to the crystal surface. Secondly, as discussed in Sect. 10.4.3, the two superconductors and two junctions form a multiply connected loop around which phase coherence of the order parameter must be maintained in order that the macroscopic wave function be single valued. This leads to the periodic dependence of the electronic properties on the applied magnetic field and a sensitivity to phase shifts within the superconducting material.

We consider the 'corner SQUID interferometer' shown in Fig. 10.81. It consists of a YBCO single crystal with surfaces perpendicular to the a- and b-axis, and a loop made of a thin lead film. Between the crystal and the film there is a thin gold layer that serves as a tunneling barrier. Depending on the symmetry of the order parameter the phase constraint is different. According to (10.131) the relation $(\delta_a - \delta_b) = 2e\Phi/\hbar$ has to be fulfilled by s-wave superconductors. For d-wave superconductors, there is an additional phase jump by 180°, i.e., the corresponding relation reads $(\delta_a - \delta_b \pm \pi) = 2e\Phi/\hbar$. Consequently, (10.132) has to be replaced by

$$I = I_{\max} \cos\left(\frac{\pi\Phi}{\Phi_0} + \frac{\pi}{2}\right) . \tag{10.137}$$

The two principal cases are indicated in Fig. 10.81. If the YBCO crystal had s-wave symmetry, the phase of the order parameter would be the same at each junction inside the crystal, and the relation (10.132) would hold. In particular, the critical current for zero field would have a *maximum*. In contrast, for a superconductor with d-wave symmetry, one of the two junctions is a *π-junction* and the critical current is expected to vanish. If the finite inductance of the device is taken into account the current is expected to exhibit a *minimum* at zero field. The flux at the resistance minimum has been determined for a series of this type of corner SQUIDs. In agreement with expectation for d-wave superconductors the minimum was found at $\Phi = \Phi_0/2 \pm 0.1\Phi_0$ [541].

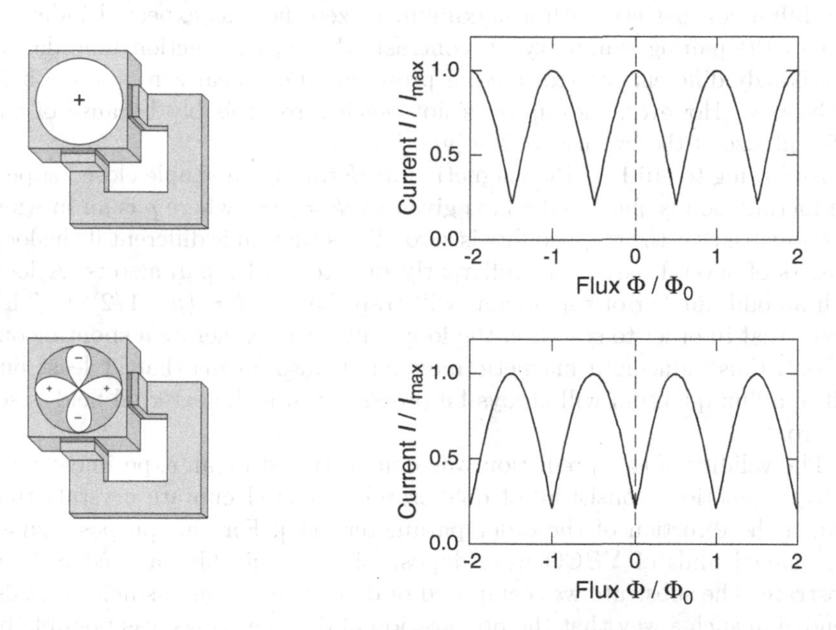

Fig. 10.81. Configuration of a corner SQUID interferometer experiment and the expected modulation of the critical current versus applied magnetic flux for **(a)** a superconductor with s-wave symmetry, and **(b)** with d-wave symmetry [542]

In Sect. 10.4.3, we also considered the current through a single junction. The symmetry of the pairing state of YBCO crystals can be tested by measuring the critical current of a junction fabricated at the corner of a crystal, as shown in Fig. 10.82b. In this geometry, part of the tunneling is into the *a-c* face and part into the *b-c* face. For an *s*-wave superconductor in a magnetic field in the *c*-direction, each face would see the same phase, and the critical current would have the usual Fraunhofer diffraction pattern (10.133). However, for a *d*-wave superconductor, the order parameter has the opposite sign in the *a*- and *b*-directions, modifying the single-junction diffraction pattern to

$$I = I_{\max} \left| \frac{\sin^2 (\pi \Phi / 2\Phi_0)}{(\pi \Phi / 2\Phi_0)} \right| . \qquad (10.138)$$

Without an applied field, the current through the two orthogonal faces cancels exactly and the critical current vanishes. It vanishes again when an integer number of flux quanta threads each half of the junction separately, giving a flux modulation period that is twice the value given by (10.133). The expected modulation of the critical current for the edge face and corner geometry, as well as for *s*- and *d*-wave order parameter, is schematically drawn in Fig. 10.82.

The experimentally observed magnetic-field modulation of single junctions on the edge face and on the corner of a YBCO crystal are plotted in Fig. 10.83. The critical current of the edge junction exhibits a Fraunhofer-like diffraction pattern with a maximum at zero field as expected independent of the pairing symmetry. In contrast, the corner-junction modulation is strikingly different. As expected, a pronounced dip near zero applied field is observed. However, the dip does not reach zero probably because of the different size of the two legs of the junction.

According to (10.121) the magnetic flux Φ through a simple closed superconducting loop is quantized and is given by $\Phi = p\Phi_0$, where p is an integer. It is possible that the trapped flux is zero. The situation is different if the loop consists of several parts with differently oriented order parameters. A loop with an odd number of π-junctions will trap the flux $\Phi = (p + 1/2)\,\Phi_0$. This means that in order to stabilize, the loop will always generate a spontaneous current, thus inducing a magnetic moment. It also means that at least one half of a flux quantum will always be present even if the external field is set to zero.

The validity of this prediction was demonstrated by an experiment with multiple junctions consisting of deliberately oriented cuprate crystals that defined the direction of the order parameter [544]. For this purpose, small ring-shaped films of YBCO were deposited on a suitably tailored $SrTiO_3$ substrate. The substrate was composed of differently oriented single crystals, mended in such a way that the preparation of different rings was possible by epitaxial deposition onto the substrates. The grain boundaries between the differently oriented YBCO single crystals caused weak links that act in this

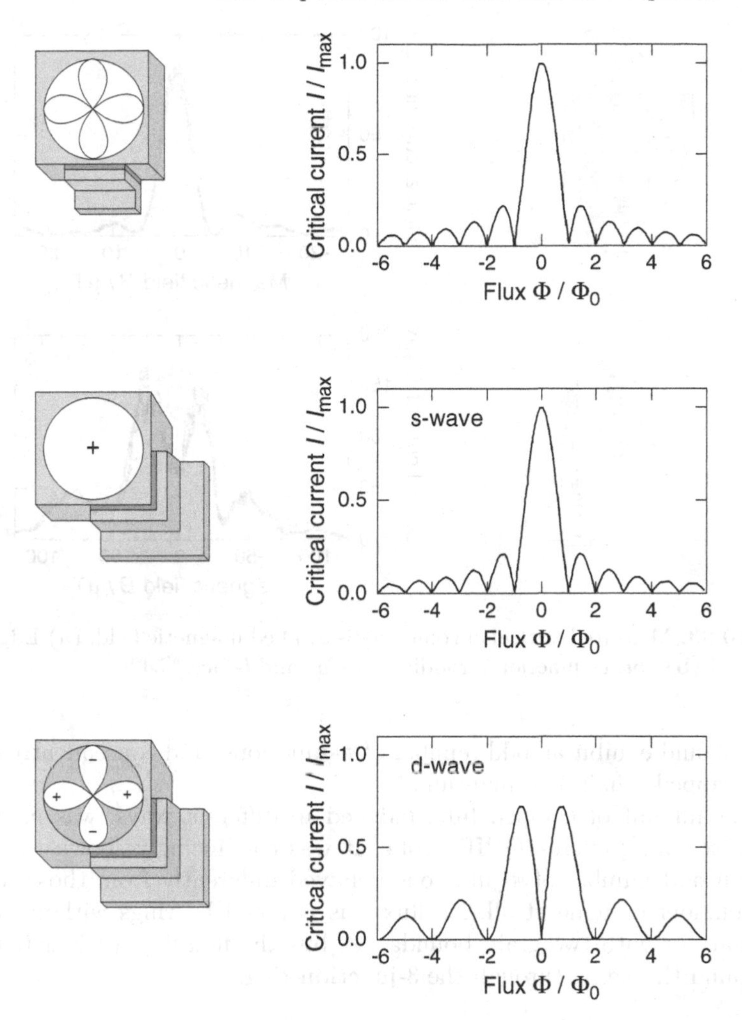

Fig. 10.82. Schematic representation of how to determine the symmetry from single junctions. **(a)** Fraunhofer diffraction pattern for the critical current modulation of a single Josephson junction in a magnetic field. **(b)** Expected modulation for a junction straddling the corner for an s-wave, and **(c)** for a d-wave symmetry [542]

particular case like π-junctions. The layout of the so-called 'tricrystal experiment' is shown schematically in Fig. 10.84. Given the relative orientation of the crystals, the expected relative orientations of the $d_{x^2-y^2}$ lobes of the order parameter in the corresponding segments of the YBCO films are shown in this drawing. The YBCO rings had an inner diameter of about 48 μm, were 10 μm wide, and about 120 nm thick. The loops contain either an even, including zero, or an odd number of π-junctions. For the sample shown in the figure, only the loop with its center at the location where all three crystals

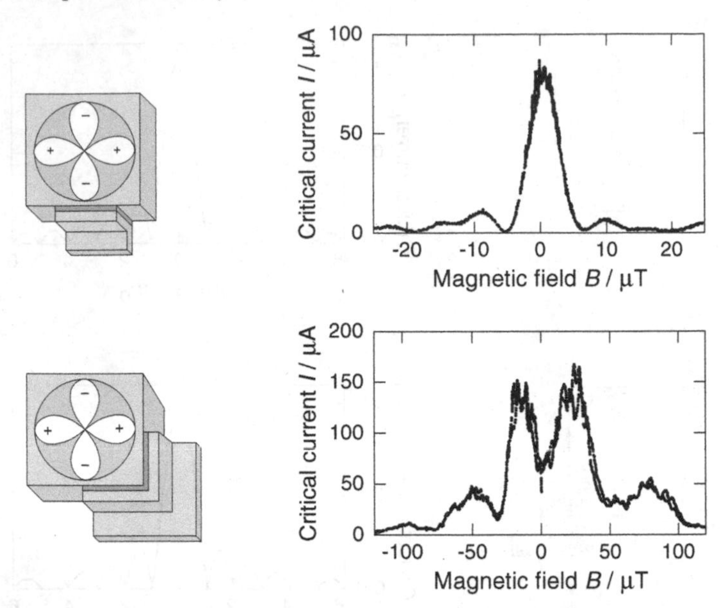

Fig. 10.83. Measured critical current versus applied magnetic field. **(a)** Edge junction, and **(b)** corner junction straddling the *a*- and *b*-faces [543]

meet, should exhibit an odd number of π-junctions, and consequently should have trapped a half-flux quantum.

The amount of trapped flux, induced in different ways, was monitored with a scanning micro-SQUID probe. It was convincingly shown that loops with an odd number of π-junctions behaved differently from those with an even number or none at all. No flux was trapped by rings without a grain boundary or with two grain boundaries, but the flux $\Phi = (0.49 \pm 0.015)\,\Phi_0$ was found threading through the 3-junction ring.

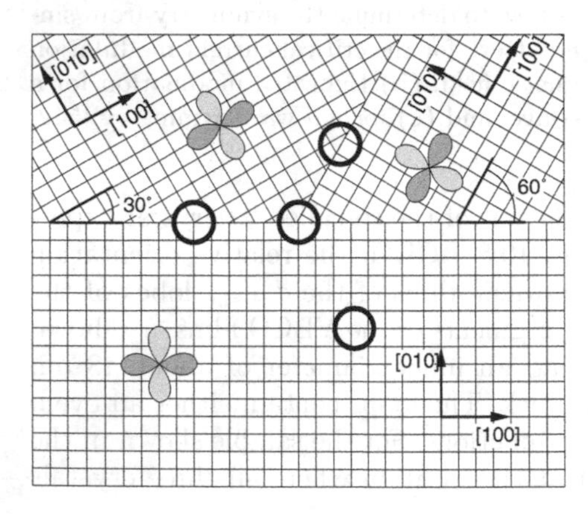

Fig. 10.84. Experimental configuration for the π-ring tricrystal experiment. The relative orientation of the *d*-wave order parameter contour in the ring-shaped films is also indicated. The central ring is expected to contain one π-junction. After [544]

Pseudogap

The temperature dependence of the magnetic susceptibility χ in the normal conducting state depends strongly on the dopant concentration. As can be seen from Fig. 10.85 for the compound $La_{2-x}Sr_xCuO_4$, $\partial\chi/\partial T$ is negative in the whole temperature range for specimens with high doping, but positive for low doping. A similar behavior in $\chi(T)$ with varying dopant concentration is also observed for the YBCO series, where $\chi(T)$ of underdoped specimens decreases strongly upon cooling. Obviously, a simple description invoking only a Pauli-like electron susceptibility determined by the density of states $D(E_F)$ at the Fermi energy, is not appropriate because in this case $\chi(T)$ is expected to be temperature independent. One possible explanation of the negative slope $\partial\chi/\partial T$ at low temperatures is that $D(E_F)$ decreases with temperature. In other words, a *pseudogap* in the quasiparticle excitation spectrum opens that is comparable in size to the superconducting gap.

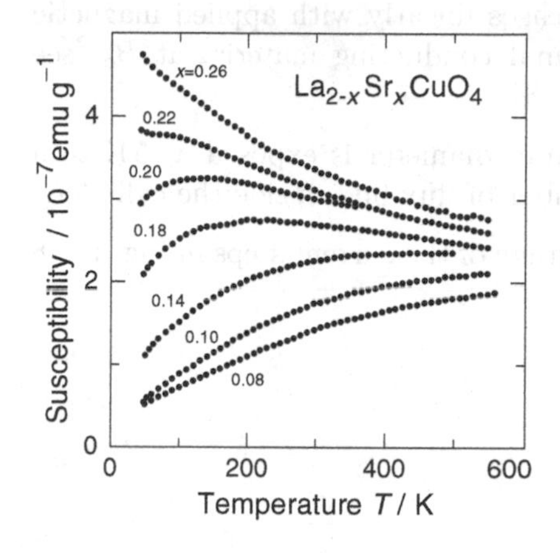

Fig. 10.85. Temperature dependence of the normal state susceptibility of $La_{2-x}Sr_xCuO_4$ [545]

The existence of such a pseudogap in the underdoped regime may also be concluded from measurements of the specific heat, NMR studies, ARPES investigations, or tunneling spectroscopy. Its seems that the gap starts to open at a doping-dependent characteristic temperature T^* that, in the underdoped region, is significantly larger than the superconducting transition temperature T_c. It has been argued that the pseudogap phenomenon and the anomalous normal-state properties of cuprates are due to a precursor to the formation of Cooper pairs in some incoherent fashion. But alternative scenarios have been proposed, and at present a final explanation does not exist.

Finally, we want to mention that it is possible to dope cuprates in such a way that the charge carriers are not holes, as in ordinary cuprates but electrons. An example is $M_{2-x}Ce_xCuO_{4-y}$ (M = Nd, Pr, or La), where x can vary between $0 < x < 1$. The properties of electron-doped cuprates differ from that of hole-doped cuprates in many respects. In particular, the symmetry of the superconducting gap is still under debate. It seems that in underdoped samples d-wave pairing occurs. However, near optimal doping the character of superconductivity seems to change and a transition from d-wave pairing to s-wave pairing is thought to occur. Furthermore, it seems that in underdoped samples a pseudogap also exists.

Exercises

10.1 Based on (10.2) and (10.12) the difference between the specific heat of normal conductors and superconductors can be calculated. In what temperature range does the specific heat of a superconducting lead sample exceed that of the normal conducting sample?

10.2 Show that above $B_\mathrm{a} = 2B_\mathrm{c}/3$, the magnetic flux through spherical samples of type I superconductors increases linearly with applied magnetic field and reaches the value of the normal conducting material at B_c (see Fig. 10.10).

10.3 A long cylinder of Nb_3Sn with 4 mm diameter is exposed at 5 K to a magnetic field of 1 T. Estimate the number of flux lines inside the cylinder.

10.4 Confirm that the spacing in the voltage of the current steps in Fig. 10.48 is given by $\omega = 2eV/\hbar$.

Part III

Principles of Refrigeration and Thermometry

11 Cooling Techniques

In the 19th century many gases were liquified for the first time by applying pressure at room temperature. However, for some gases, such as oxygen and hydrogen, this method was not successful in producing a liquid. Because of this, these gases were often referred to as *permanent gases*.

In 1877, *Cailletet* [546] and *Pictet* [547] independently succeeded in the liquefaction of oxygen, one of the permanent gases. To achieve this, Cailletet precooled the oxygen gas with liquid ethylene to about $-110°$C under a pressure of 200 bar. Subsequently, he opened a valve to produce a sudden expansion of the gas. As a result, he observed briefly oxygen droplets forming a wet fog. In contrast, Pictet used a cascade for precooling that consisted of two coolant loops, one with sulfur dioxide at $-20°$C and one with carbon dioxide at $-60°$C. Larger amounts of liquid oxygen and nitrogen were first produced in 1883 by *Wroblewski* and *Olczcwski* [548]. One year later, they also achieved the first liquefaction of hydrogen by precooling their apparatus with liquid oxygen. Whereas in these early experiments only a wet fog of hydrogen droplets was obtained, in 1898 *Dewar* succeeded in the production of quantities of liquid hydrogen that were sufficient to study its physical properties [549]. Precooling was provided by using a cascade with liquid ethylene and liquid air. Finally, in 1908 the last of the permanent gases, helium, was liquified by *Kamerlingh Onnes* [550].

As early as 1922, Kamerlingh Onnes reached temperatures below 1 K by reducing the vapor pressure above liquid helium to about 2×10^{-5} bar with a series of pumps [551]. Since ^{3}He was unavailable at that time, the only way to produce even lower temperatures was the adiabatic demagnetization of magnetic moments in solids. The basic idea of this cooling method was published independently by *Debye* [552] in 1926 and by *Giauque* [553] in 1927. Initially, only electronic spins in paramagnetic salts were demagnetized. Later, the technical prerequisites in terms of magnetic fields and precooling have allowed nuclear spins to also be used for refrigeration. With well-designed nuclear demagnetization cryostats temperatures below 10 μK can be produced.

With ^{3}He becoming available at the beginning of the 1950s additional cooling methods were made possible, ^{3}He evaporation cryostats, Pomeranchuk cells and dilution refrigerators. The latter are currently the workhorses of many low-temperature laboratories because they allow for continuous operation at temperatures down to about 5 mK.

In this chapter, we discuss some basic aspects of different cooling techniques. We focus on the physical principles on which these methods are based rather than on technical details.

11.1 Liquefaction of Gases

The standard technique for producing temperatures in the range between 300 K and 4.2 K involves the use of cryogenic liquids. One obtains these by the liquefaction of the corresponding gases, for example nitrogen or helium.

For many low-temperature experiments, a boiling cryogenic liquid at fixed pressure provides an ideal heat bath with constant temperature. For example, by varying the vapor pressure above a ^{3}He bath one can maintain temperatures from 3.3 K down to about 0.3 K. In general, in low-temperature experiments cryogenic liquids are used for precooling. In Table 11.1, some properties of various cryogenic liquids are listed. Despite the potential danger in handling liquid oxygen and liquid hydrogen these liquids were historically used frequently in low-temperatures laboratories. At present, liquid nitrogen and liquid helium are used for cooling almost exclusively.

Table 11.1. Thermodynamic data of some important cryogenic liquids. Here, T_b denotes the boiling temperature at atmospheric pressure, T_c the critical temperature, p_c the critical pressure, ΔH the latent heat of evaporation, and T_{inv} the inversion temperature. The values in brackets are calculated. After [12, 554]

Substance	T_b/K	T_c/K	p_c/bar	ΔH/kJ mol^{-1}	T_{inv}/K
CO_2	195	304	73.84	25.2	(2050)
CH_4	112	191	45.99	8.18	(1290)
O_2	90.2	154.6	50.43	6.82	762
N_2	77.4	126.0	33.99	5.57	625
H_2	20.3	32.9	12.84	0.90	203
^{4}He	4.21	5.19	2.29	0.082	43.2
^{3}He	3.19	3.32	1.16	0.025	(23)

The simplest method of gas liquefaction is based on an isothermal pressure rise. In fact, this procedure was employed exclusively in the 19th century. Based on the fundamental considerations of *Van der Waals* it is obvious today that this method of liquefaction can only work if the experiment is performed below the *critical temperature* T_c. Taking a look at Table 11.1 shows that this condition cannot be fulfilled for the most important cryogenic liquids. In order to liquify these gases one has to use different methods. Principally, one distinguishes between processes in which cooling of a gas is obtained by work against an external force at the expense of internal energy and methods in which work is performed against an internal force of the gas. Both techniques are discussed in the following sections.

11.1.1 Expansion Engines

A very important and frequently employed cooling technique involves performing work at the cost of internal energy by the expansion of a gas. An everyday example is a steam engine in which the steam does work in moving a piston. As a consequence of the work performed, the steam is cooled. Of course, in this case the power produced by the piston is the important final result, not the cooling of the steam.

Principle of Operation

In practical realizations of such cooling machines one tries to perform the expansion in such a way that the process runs adiabatically. In this limit, thermodynamics predicts that a pressure reduction from p_1 to p_2 is accompanied by a temperature reduction from T_1 to T_2, given by the expression

$$T_2 = T_1 \left(\frac{p_2}{p_1}\right)^{(\kappa-1)/\kappa} . \tag{11.1}$$

Here, the ratio $\kappa = C_p/C_V$ of the specific heats, which enters in the exponent, takes the values 5/3 and 7/5 for monatomic or diatomic gases. For monatomic helium gas, an expansion from 100 bar to atmospheric pressure is sufficient to cool the gas to 50 K starting from room temperature. With air as the working gas, one can reach about 80 K in the same expansion. Of course, in a real expansion irreversibilities are unavoidable. The process will not be fully adiabatic and thus (11.1) is only approximately valid.

The concept of an expansion engine for cooling is illustrated in Fig. 11.1. The gas is first compressed at room temperature, and the heat of compression is transferred via a cooler to the surroundings. Following this, the gas flows through a *counterflow heat exchanger* in which the high-pressure gas is precooled to the temperature T_1 by the cold returning gas. According to (11.1), the gas is cooled to the temperature T_2 in the expansion turbine. The gas can be used to provide cooling in the working volume. It might be used, for example, to liquify a gas with a higher condensation temperature. On the return to the compressor, the gas warms up in the counterflow heat exchanger as it cools the compressed gas.

The counterflow heat exchanger plays an important role in this cooling process. It should not impede the gas flow, but it should still have a very large surface area to allow an efficient exchange of heat. This can be realized, for

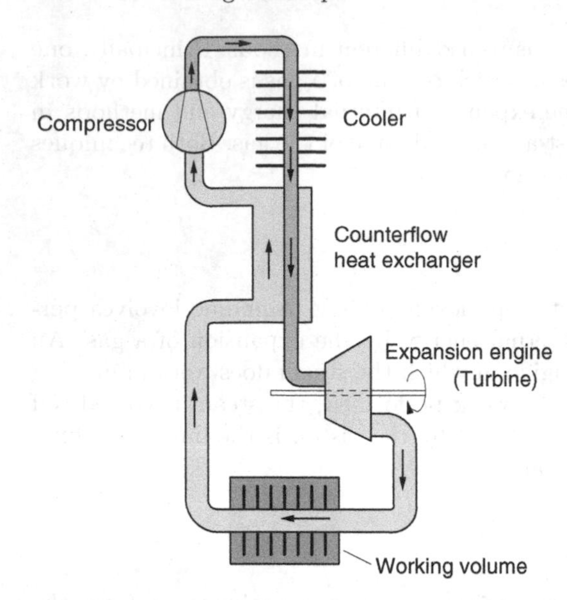

Fig. 11.1. Concept of a simple expansion cooling machine. The working gas is compressed, flows through a heat exchanger and is expanded in the expansion machine. After removing heat from the working volume the gas is directed back through the heat exchanger into the compressor

example, with concentric tubing or by winding the tube with the compressed gas around the tube carrying the expanded gas. Heat exchangers for use at very low temperatures will be discussed in Sect. 11.4.

Under ideal conditions, the enthalpy H is conserved in reversible heat-exchange processes. This means that the enthalpy is completely transferred from the warm to the cold gas. In this approximation the work W, which is removed from the gas by the expansion engine, corresponds exactly to the difference between the enthalpies of the gas at the start and at the end of the process. In this case, we can write for the work W

$$W = H_1 - H_2 = (U_1 + p_1 V_1) - (U_2 + p_2 V_2)\,. \tag{11.2}$$

For helium, which can be considered, to a good approximation, as an ideal gas, we can utilize the relations $U = \frac{3}{2} N k_\mathrm{B} T$ for the internal energy and $pV = N k_\mathrm{B} T$ to obtain $W = \frac{5}{2} N k_\mathrm{B}(T_1 - T_2)$ for the work performed by the gas.

In most gas liquefaction facilities, expansion engines are used to precool the gas to temperatures close to the temperature of liquefaction. A cooling to the liquefaction point is not the desired result, because the formation of liquid inside the expansion engine would cause severe mechanical problems. The last step in the liquefaction process is normally obtained via a Joule-Thomson stage, a technique that we discuss in Sect. 11.1.2. Helium liquefiers often contain several expansion engines in a multistage configuration, which are connected by a counterflow heat exchanger.

Technical Realization

Two different concepts exist for the technical realization of expansion engines for cooling. In one case, the expanding gas drives a piston that moves in a cylinder, as in a steam engine. In order to absorb energy the piston must be damped in a suitable way. Alternatively, in present designs, the expanding gas often performs work by driving a (slightly braked) turbine. In both cases, the mechanical problems are considerable and are principally associated with the difficulty of lubricating moving parts at low temperatures.

An example of the technical construction of a turbine cooler is shown in Fig. 11.2. The gas turbine, which turns at about 3000 revolutions per second, has air bearings because ordinary lubrications would freeze at low temperatures.

To illustrate the properties of a piston expansion engine we discuss a cooling process based on a Stirling cycle. Figure 11.3a shows a schematic diagram of such a cooling system. The machine consists of two cylinders, which are at the high temperature T_1 and the low temperature T_2, respectively. The two cylinders are connected via a regenerator, a device that we discuss below. A

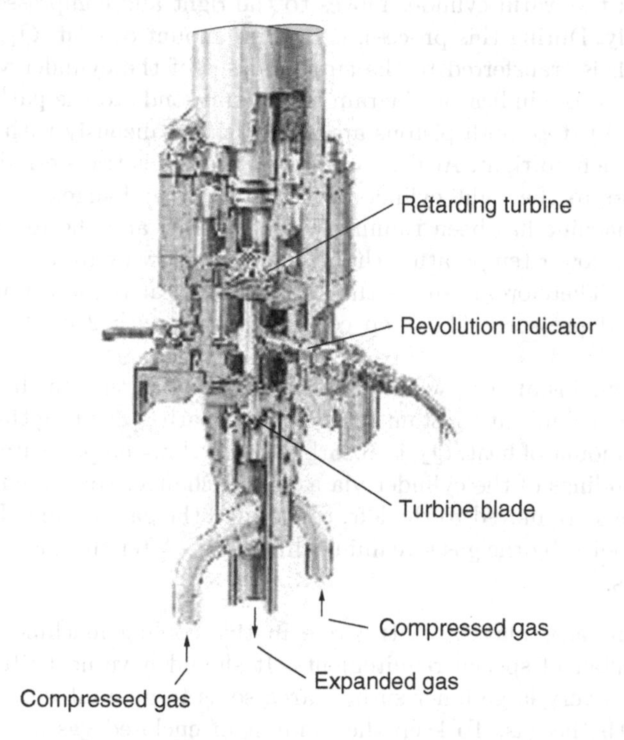

Fig. 11.2. Drawing of a turbine expansion engine. Turbines with air bearings are capable of 3000 revolutions per second. The expansion work performed by the gas is terminated by the braking turbine [555]

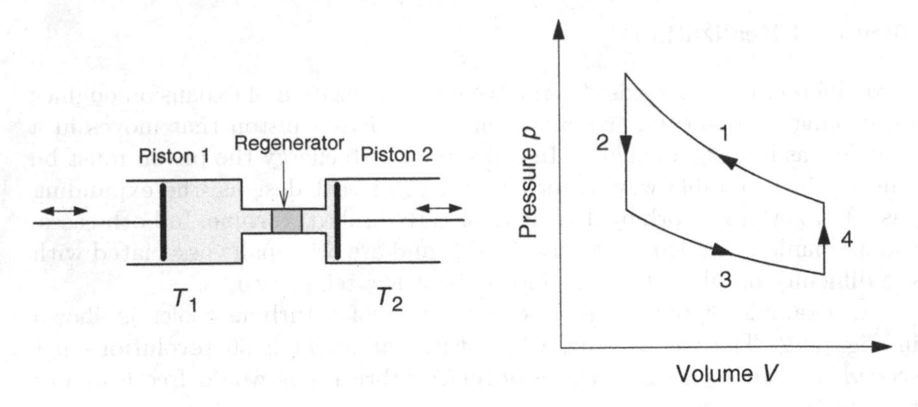

Fig. 11.3. (**a**) Schematic diagram of a Stirling cooling machine and (**b**) the corresponding indicator diagram of the four steps of the cooling cycle

working cycle has four steps in which the pistons move in different ways. The starting positions are shown in Fig. 11.3a.

- Piston 1 in the warm cylinder moves to the right and compresses the gas isothermally. During this process, a certain amount of heat, Q_1, is generated, which is transferred to the surroundings of the cylinder via a heat exchanger. In the indicator diagram this corresponds to the path 1.
- In the second step, both pistons are moved simultaneously with equal velocity from left to right. At the same time, the gas is transferred from the left cylinder to the right cylinder without volume change. If we assume that the machine has been running for some time and the regenerator is already at a lower temperature than T_1, the gas flowing from cylinder 1 to 2 is cooled. Therefore, it enters the right cylinder at reduced temperature and reduced pressure. This step corresponds to path 2 of the indicator diagram.
- Piston 1 remains at rest, while piston 2 is moved further to the right, increasing the volume at constant temperature (path 3). During this process, a certain amount of heat, Q_2, is absorbed on the low-temperature side from the surroundings of the cylinder via a heat exchanger connected to it.
- Both pistons are moved to the left, conserving the gas volume. The regenerator is cooled by the gas streaming through it. After this last step a new cycle starts.

The regenerator plays a central role in this cooling machine. It should satisfy a number of special requirements. It should have negligible flow resistance and a very large inner surface area so as to guarantee optimal heat exchange with the gas. To keep the amount of enclosed gas small, the regenerator should be as compact as possible. Furthermore, it should have a low thermal conductance in order to minimize the heat leak between the two cylinders. The biggest problem, however, is the demand for a very high heat

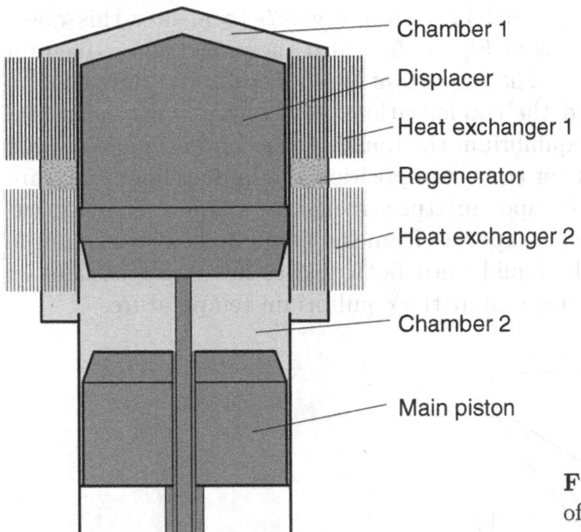

Fig. 11.4. Schematic diagram of the essential components of a Stirling cooler

capacity even at low temperatures. Usually, regenerators are made of a dense metal tangle, because this is a suitable compromise between the different requirements. However, it is still difficult to reach sufficiently low temperatures with these coolers to permit the liquefaction of helium.

Figure 11.4 shows the important elements of a Stirling cooler in an arrangement for efficient operation, as used in many practical designs. The two pistons move coaxially in a common cylinder. They are driven sinusoidally by a crank with a suitable phase relation between their motions. The indicator diagram for this design is smoothed as compared to the idealized one shown in Fig. 11.3b. However, the overall efficiency is changed only slightly. The main piston produces the compression, while the common motion of the main piston and the displacer moves the gas periodically between chambers 1 and 2. In order to move from one chamber to the other, the gas has to st-eam through the circular heat exchangers 1 and 2, which are mounted on a cylinder around the displacer. Heat exchanger 1 consists of thin tubes, which are surrounded by water to carry away the heat of compression. Heat exchanger 2 is an integral part of the cylinder head and has a large number of slits, through which the gas can flow. The regenerator is located between the heat exchangers and is made of fine copper wire.

Separation of Oxygen and Nitrogen

Because of obvious safety considerations, liquid air is no longer used in low-temperature experiments. Instead, nitrogen is used almost exclusively to cool to temperatures of about 80 K. Here, an interesting question arises, how can nitrogen and oxygen be separated in a simple way? To answer this question it is helpful to take a look at Fig. 11.5, which shows the phase diagram of nitrogen-oxygen mixtures. The important point here is the difference in the boiling point curve and the condensation point curve. This means for any given temperature, in equilibrium the liquid phase contains more oxygen than the gas phase, or that for equal composition the boiling liquid mixture is colder than the saturated vapor mixture. If one puts liquid air in an enclosed container, nitrogen will evaporate from the liquid and oxygen from the vapor will condense into the liquid until both phases have compositions in accordance with the phase diagram at the equilibrium temperature.

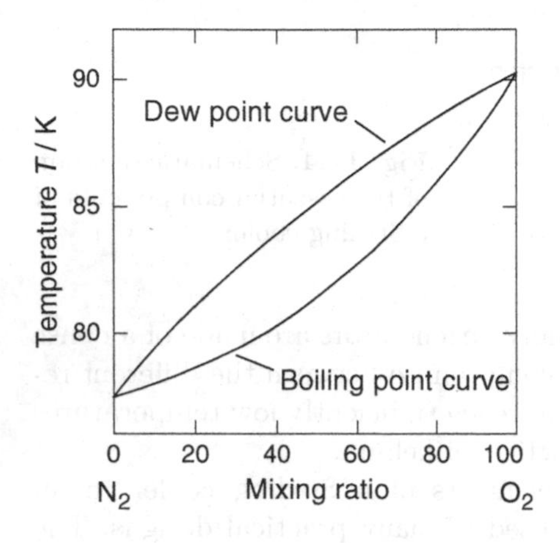

Fig. 11.5. Phase diagram of a nitrogen-oxygen mixture at atmospheric pressure [556]

This phenomenon is used in so-called *rectifiers* for the separation of nitrogen and oxygen. The construction of these devices is illustrated in Fig. 11.6. The rectifier consists of a condenser that is thermally connected to the expansion engine, the actual rectifying column, consisting of the two parts, A_1 and A_2, an evaporation unit and a heat exchanger. At the upper end of the rectifier the temperature is about 77 K, and at the lower end it is roughly 90 K. In the rectifying column, which mainly consists of a dense metal tangle, liquid and gas phase are flowing in opposite directions.

Liquid nitrogen (1) trickles from the condenser into the rectifying column. At the same time, precooled air, with a temperature of about 85 K (2), enters at the bottom end of the rectifying column A_1 and rises. The upstreaming air accumulates nitrogen, while oxygen condenses in the downstreaming liquid nitrogen. Almost pure nitrogen gas (3) arrives at the condenser (the purity is about 99.5%) and condenses. At this point, part of the liquid nitrogen is removed and the rest flows back in the rectifying column. The liquid drops (4) that enter the lower part of the rectifying column consist of roughly equal amounts of nitrogen and oxygen. The drops come in contact with the rising oxygen vapor (5) that originates from the evaporation unit, and the nitrogen further evaporates from the falling liquid. The drops leaving the rectifying

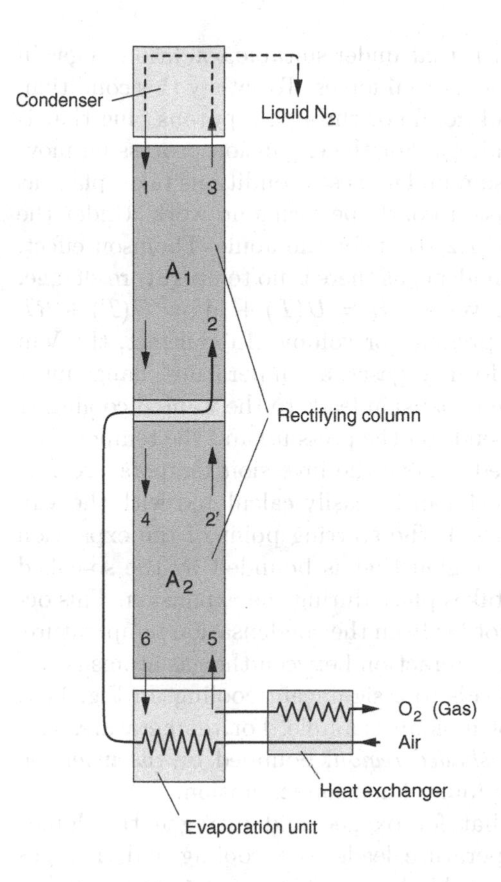

Fig. 11.6. Schematic sketch of a rectifier for the separation of nitrogen and oxygen

column as almost pure oxygen (6) are collected by the evaporation unit. The liquid oxygen is subsequently re-evaporated and part of the gas produced forms the rising oxygen (5). In removing the nitrogen from the falling liquid in A_2, the gas achieves a composition approximately that of air (2') by the time it reaches A_1. Part of the oxygen is removed and leaves the evaporation unit. At the same time, it cools the air that has been introduced into the heat exchanger. The required heat for the evaporation unit is provided by the incoming air, which is cooled further and simultaneously cleaned of water vapor and carbon dioxide.

11.1.2 Joule–Thomson Expansion

As we mentioned previously, liquefaction does not take place in expansion engines for technical reasons. In general, one uses a condenser, as in the above example, to liquify the gas. There is, however, another cooling process in which liquid is produced directly. This process uses the *Joule–Thomson effect* [557,558] to facilitate condensation, and is principally employed in the liquefaction of helium.

The basic idea behind this effect is that under suitable conditions a gas in expanding performs work against its internal forces. To satisfy this condition, the gas is expanded through a small nozzle or through a porous plug that is thermally isolated from its surroundings. For the expansion process no moving parts are necessary. The expansion under theses conditions takes place at constant enthalpy, since the expansion nozzle performs no work. Under the condition $W = 0$, it follows from (11.2) that for the Joule–Thomson effect, the relation $H_1 = H_2$ is valid. For an ideal gas there is no temperature change, since in this case the enthalpy is given by $H = U(T) + pV = U(T) + RT$ and therefore does not depend on pressure or volume. In contrast, the Van der Waals equation indicates that, for real gases, a temperature change must occur. Whether the Joule–Thomson expansion leads to the wanted cooling or instead to a warming, however, depends on the pressure and the temperature at which the expansion is performed. Above the inversion temperature T_{inv} and the inversion pressure p_{inv}, which can be easily calculated with the Van der Waals equation, warming occurs. If the starting point of the expansion lies in the p–V diagram within the region that is bounded by the so-called inversion curve, cooling of the gas takes place during the expansion. This occurs at low temperatures that are not far from the condensation temperature. Slightly above this temperature the interaction between the gas atoms is sufficiently strong so that expansion leads to a significant cooling. In Fig. 11.7, the p–T diagram of nitrogen is shown as an example. For temperatures and pressures that are within the *grey shaded region*, bounded by the inversion curve, cooling takes place during a Joule–Thomson expansion.

A look at Table 11.1 shows that for oxygen and nitrogen the Joule–Thomson expansion at room temperature leads to a cooling and thus gas liquefaction systems can be built on this basis. This is not true for hydrogen and helium, since for these gases the Joule–Thomson expansion at room temperature leads to a warming. In fact, the Joule–Thomson expansion is

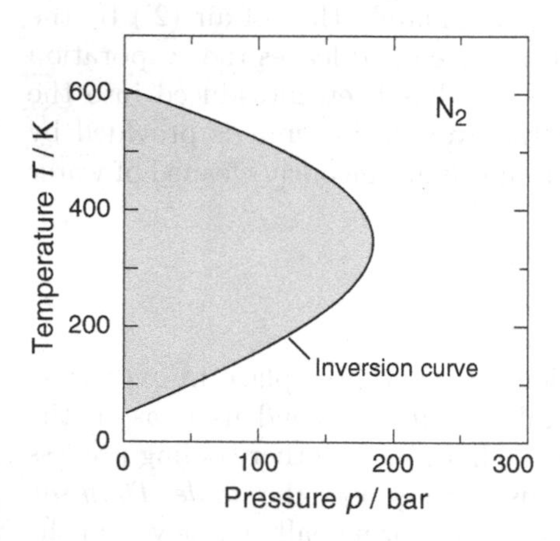

Fig. 11.7. p–T diagram of nitrogen [559]

usually only the last step in the liquefaction of helium, after the gas has been precooled using an expansion engine.

The first successful direct air liquefaction using the Joule–Thomson effect, which means without a moving piston or turbine, was obtained simultaneously but independently by *von Linde* [560] and *Hampson* [561] in 1895. Both developed compact and efficient air liquefiers, which were able to produce a few liters per hour of liquid air, large quantities in comparison to the typical air liquefiers at that time. They used the counterflow principle and the Joule–Thomson effect, as the schematic illustration of their liquefiers in Fig. 11.8 shows.

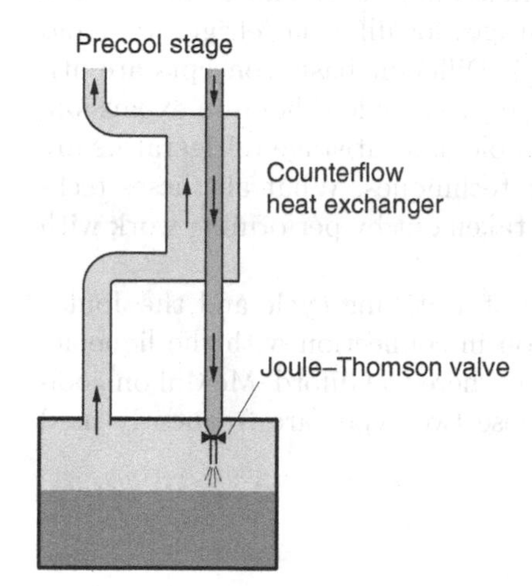

Fig. 11.8. Schematic illustration of the setup of a gas liquefier with Joule–Thomson expansion and counterflow heat exchanger

The arrangement with the heat exchanger and expansion valve connected in series leads to a process that takes place at constant enthalpy. It is often referred to as a *Linde liquefier*. If we denote the enthalpies of the gas streaming in, the liquid and the gas streaming back as H_{in}, H_ℓ and H_{back}, respectively, we can write the relation $H_{\text{in}} = \lambda H_\ell + (1-\lambda)H_{\text{back}}$, where λ represents the fraction of gas that has been liquefied. Since the three enthalpies that enter here are well known from various experiments, one can easily calculate λ using the equation

$$\lambda = \frac{H_{\text{back}} - H_{\text{in}}}{H_{\text{back}} - H_\ell} . \tag{11.3}$$

The magnitude of λ depends on the pressure difference as well as on the starting temperature of the expansion. Since, for a decent liquefaction $\lambda > 0.1$ is desirable, one usually needs to start with temperatures below $T_{\text{inv}}/3$. Because of this, it is sensible to use an expansion engine for precooling in the process of liquefaction of many gases. The combination of a Linde liquefier and an expansion engine is known as a *Claude cycle* [562].

To end this section, we would like to point out that modern large-scale air liquifiers for the industrial supply of cryogenic liquids are capable of producing 100 tons of oxygen and 1000 tons of nitrogen per day in one facility.

11.2 Closed-Cycle Refrigerators

Low temperatures can be obtained in modern-day applications using closed-cycle systems without liquefaction of helium. Such systems are particularly attractive for industrial applications, because no handling of cryogenic liquids is necessary. New cryogenic developments have also shown that closed-cycle systems can be used as precooling stages for dilution refrigerators and demagnetization cryostats (e.g., [563, 564]). Different basic concepts are utilized for closed-cycle refrigeration: Stirling cycles, Joule–Thomson expansion, Gifford–McMahon cycles and pulse-tube coolers. Multistage refrigerators are often based on a combination of different techniques. What all theses techniques have in common is that the heat is taken out by performing work with a gas, against interval or external forces.

Since we have discussed the principles of a Stirling cycle and the Joule–Thomson expansion in the previous section in connection with the liquefaction of gases, we shall restrict our discussion here to Gifford–McMahon coolers, and pulse-tube coolers. Currently, these two types are frequently used for closed-cycle refrigeration.

11.2.1 Gifford–McMahon Coolers

The Gifford–McMahon refrigeration cycle was introduced in 1960 [565]. Since then, many different forms of this cycle have been developed, and Gifford–McMahon coolers have become very popular because of their simplicity and rather high reliability. The idealized cycle is based upon two isothermal and two isobaric processes and can be viewed as a kind of heat pump in which the amount of heat transferred is simply given by

$$Q_{\text{ideal}} = (V_{\max} - V_{\min})\,(p_{\text{high}} - p_{\text{low}})\,. \tag{11.4}$$

Here, $V_{\max}$ and $V_{\min}$ denote the maximum and minimum volume, respectively. The quantities p_{high} and p_{low} represent the different pressures. The great advantage of the Gifford–McMahon principle is that the compressor unit and the cold head – or expansion unit – are separated. A schematic diagram of a single-stage Gifford–McMahon refrigerator is shown in Fig. 11.9. It consists of a compressor with aftercooler, a rotary valve, a regenerator, and an expansion unit with a free-floating displacer.[1] The closed volume created by the two coaxial cylinders of different size is divided into three variable chambers by

[1] We should point out that this realization is just one possibility of many and has been chosen for didactic reasons.

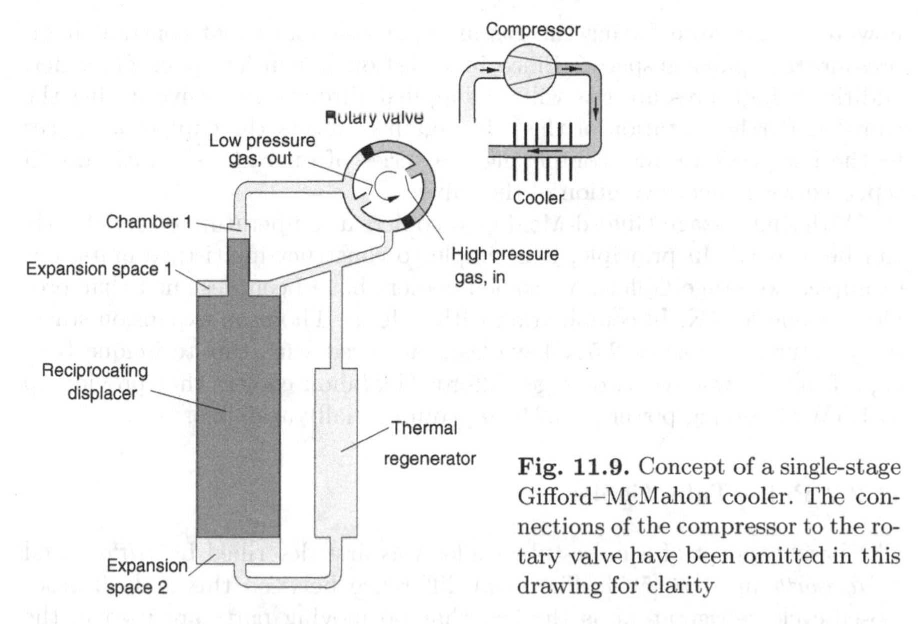

Fig. 11.9. Concept of a single-stage Gifford–McMahon cooler. The connections of the compressor to the rotary valve have been omitted in this drawing for clarity

the displacer. The expansion spaces 1 and 2 are interconnected by a thermal regenerator[2] of small pressure drop, so that both of these chambers are always maintained at approximately the same pressures. The regenerator consists of a porous solid-state matrix such as stainless steel screens, lead spheres, or materials with a magnetic ordering transition in the temperature range of interest. It should have a heat capacity as high as possible. A suitable seal separates chamber 1 from the expansion spaces, so that the pressure in chamber 1 can be different from that in the expansion spaces. A rotary valve is used to raise the pressure by delivering high-pressure gas from a compressor or to lower the pressure by exhausting to the low-pressure line from either chamber 1 or the two expansion spaces.

At the beginning of the cycle, the displacer is in its topmost position and the valve connects the low-pressure line to all chambers. This means that the volumes of chamber 1 and expansion space 1 are at their minimum volume, while the expansion space 2 is at its maximum volume. When the valve rotates counterclockwise, at some point high-pressure gas is connected to chamber 1, forcing the displacer to move to the bottom position. In this process, gas is transported from expansion space 2 to expansion space 1. If the gas in expansion space 2 is colder, some gas will be exhausted through the valve as a result of expansion back to the inlet temperature. When the value is rotated further, at some point high-pressure gas also flows into the two expansion spaces. Next, the valve is rotated into a position at which point the pressure decreases in chamber 1, and as a result the displacer moves

[2] The regenerator is drawn as a separate unit in Fig. 11.9, although in many realizations it is an integral part of the displacer.

upwards again, transferring the gas in expansion space 1 at constant high-pressure to expansion space 2. Since, in operation, expansion space 2 is colder, additional high-pressure gas will be supplied through the valve during the transfer. Further rotation of the valve again connects the expansion spaces to the low-pressure line, completing the series of operations, which are all repeated with each revolution of the valve.

With single-stage Gifford-McMahon coolers a temperature of about 15 K can be reached. In principle, it is simple to construct multistage units. For example, two-stage Gifford-McMahon coolers have been designed that provide cooling at 4 K. In combination with a Joule–Thomson expansion stage, temperatures as low as 2.5 K have been achieved with this technique (see, e.g., [566]). Nowadays, two-stage Gifford–McMahon coolers that provide up to 1.5 W of cooling power at 4.2 K are commercially available.

11.2.2 Pulse-Tube Coolers

The basic concept of a pulse-tube cooler was first described by *Gifford* and *Longsworth* in 1963 [567]. The main difference between this and all other closed-cycle refrigerators, is the fact that no moving parts are used in the low-temperature region of the cooler. This reduces the influence of vibrations on the low-temperature experiments and enhances the lifetime of the cooler. The original version of a pulse-tube refrigerator, hereafter referred to as the basic pulse tube, consisted of a compressor with rotary valve (as shown in Fig. 11.9) that is used to generate an oscillating pressure, a regenerator and a thin-walled tube with heat exchangers at both ends. The latter unit is the actual pulse tube, after which this type of cryocooler is named. Since the performance of the basic pulse tube was limited both in cooling power and temperature range, not much attention was paid to this concept, until *Mikulin*, *Tarasov* and *Shkrebyonock* introduced the orifice type of pulse-tube cooler in 1984 [568]. This was the crucial breakthrough for this technique. In 1990 it was followed by another important addition, the introduction of a second inlet by *Zhu*, *Wu* and *Chen* [569]. Figure 11.10 shows a schematic diagram of a double-inlet orifice-type pulse-tube cooler that is driven by a valveless compressor as used for Stirling coolers.[3]

The basic cooling effect relies on a periodic pressure variation and a displacement of the working gas in the pulse tube. In some ways, this process is similar to a Stirling cycle, which we have discussed in Sect. 11.1.1. The pressure variations are produced by the compressor. In response to the pressure changes, mass flows through the regenerator and exchanges heat with it. The regenerator serves as a heat reservoir. Up to this point, a pulse-tube

[3] This compressor version is chosen for didactic reasons. Alternatively, the pressure oscillation in a double-inlet pulse-tube cooler can be generated like that in a Gifford–McMahon cooler (Fig. 11.9). In particular, multistage pulse-tube coolers for cooling down to liquid-helium temperatures are operated in the latter way.

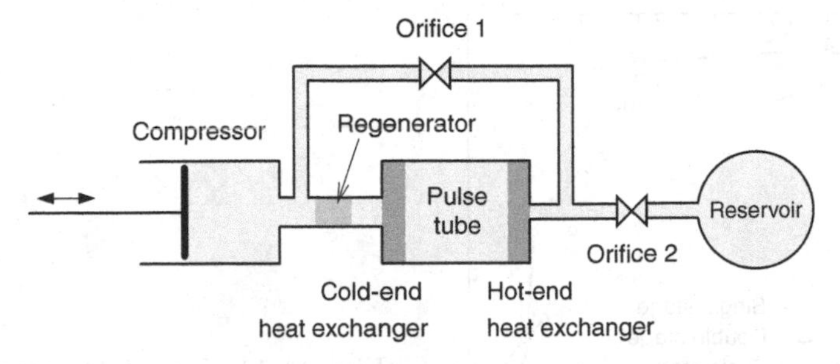

Fig. 11.10. Schematic diagram of a double-inlet orifice pulse-tube cooler

cooler is similar to a Stirling refrigerator. As we have seen in Sect. 11.1.1, a Stirling machine has a cold piston, which extracts heat from the gas by performing work. The maximum cooling occurs when the motion of the two pistons results in a pressure wave with the mass flow being in phase with it. This is the case if the pistons run 90° out of phase. A pulse-tube cooler does not have a piston to provide a phase shift. Instead, the gas within the pulse tube acts as a compressible displacer. The gas leaving the cold end of the regenerator does work on the gas inside the pulse tube, producing cooling. The heat of compression is taken out at the hot-end heat exchanger.

The motion of the gas displacer is determined by orifice 1 and the reservoir, acting together as a phase shifter. This is a passive method of producing a phase shift, as opposed to using an active piston. The process does not provide as much phase shift as a moving piston.[4] The use of a second inlet reduces the mass flow in the regenerator, decreasing the regenerator loss. To obtain this, a bypass line with a second orifice connects the hot-end heat exchanger to the entry of the regenerator.

Because the work of expansion is not recovered, the theoretical efficiency of a pulse-tube cooler is inherently lower than that of a Stirling engine. However, the importance of this loss decreases as the ratio of hot to cold temperature increases. For practical refrigerators, which span a large temperature ratio, the inherent loss is small compared to other losses and efficiencies similar to those of Stirling coolers can be achieved. The progress in developing pulse-tube coolers has been very rapid in the last decade. This is illustrated for single and multistage systems in Fig. 11.11. The lowest minimum temperature achieved up to this point is 1.27 K [571]. In this particular case, ^{3}He was used as the working gas. Presently, it is possible to buy commercial pulse-tube coolers able to remove 0.5 to 1 W at 4.2 K. It seems likely that at some

[4] A basic pulse-tube cooler has no orifice and no extra reservoir. In this case, the phase shift is produced by heat transfer between the gas and the walls of the pulse tube. The resulting phase shift is quite small and therefore the refrigeration power per unit mass flow is small as well.

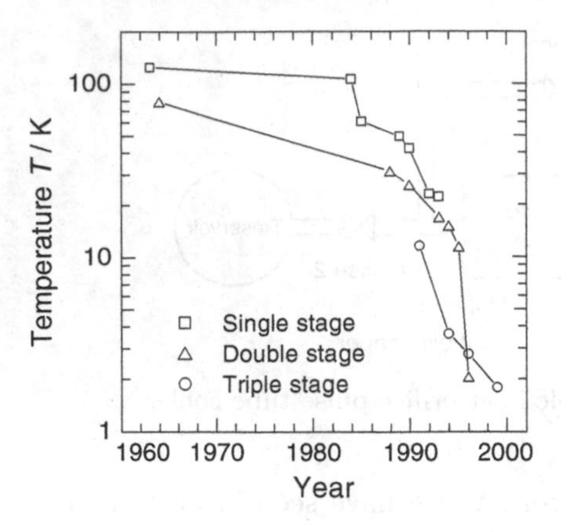

Fig. 11.11. Lowest temperature obtained with single-stage, double-stage, and three-stage pulse-tube coolers up to the year 2000. After [570]

point in the future such closed-cycle refrigerators will make the handling of cryogenic liquids in low-temperature laboratories unnecessary.

11.3 Simple Helium-Bath Cryostats

The vacuum-isolated glass vessels developed by *James Dewar* in 1893 and named after him are the precursors of all modern bath cryostats [572]. In this section, we will briefly discuss cryostats that provide low temperatures by using a ^{4}He bath at saturated vapor pressure and systems in which liquid ^{4}He or ^{3}He is pumped to obtain lower temperatures.

11.3.1 Bath Cryostats

A helium-bath cryostat that can be used to cool experiments down to 4.2 K is shown schematically in Fig. 11.12. In the simplest case, it consists of two dewar vessels that are arranged one inside the other. The outer dewar is filled with liquid nitrogen and serves to precool the system and for shielding from thermal radiation from warmer parts outside. The inner dewar contains liquid helium. The outer walls of glass dewars are usually mirror-coated to reduce the radiation entering the inner parts of the system. In most cases, this metalization covers all of the glass surface except for a narrow gap, which runs from the top to the bottom of the dewar. It is left uncoated so that it is possible to see the levels of the liquid nitrogen and liquid helium. The experiments are typically located in a vacuum vessel that is immersed in the helium. With these cryostats, experiments can be performed in the temperature range of 4.2 K to room temperature.

To shield radiation from the top of the cryostat and to thermally anchor the pumping tubes, a series of metal plates are attached to the tubes at different heights above the vacuum chamber. These radiation shields are cooled

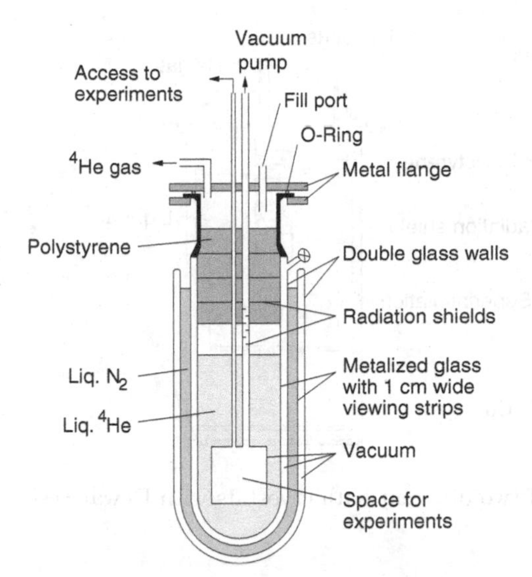

Fig. 11.12. Schematic sketch of a helium-bath cyrostat consisting of two glass dewars. After [573]

by the effluent gas and are used to force the cold helium gas to stream back along the wall of the dewar to cool it as well. To increase the efficiency of these gas baffles the space between the metal plates is often filled with polystyrene or a similar insulating material. The use of the cold gas for cooling is very important because it can significantly enhance the helium hold time of the dewar. The enthalpy change of helium gas between 4.2 K and 300 K is about 200 kJ/ℓ. In comparison, the latent heat of evaporation of helium is only 2.6 kJ/ℓ. Therefore, the use of the full enthalpy of ^{4}He to cool those parts of the inner cryostat making the transition from 300 K to 4 K is clearly very important for an efficient design.

A peculiarity that should be taken into account in operating a helium cryostat with glass dewars is the fact that helium gas can diffuse quite rapidly through glass walls at room temperature. This means that the pressure in the isolation vacuum of the dewar vessel increases over time. In order to maintain sufficient thermal insulation, one has to regularly pump the isolation vacuum of glass dewars. An alternative solution to this problem is the use of special types of glasses for which the diffusion problem is greatly reduced.

Presently, rather than using glass, dewar vessels are commonly constructed of a metal that has a very low thermal conductivity. In addition, there are an increasing number of dewars that are based on epoxy resin, structurally strengthened with glass fiber. The advantage of such systems lies in the fact that they are more robust and very flexible in their design. Also, the helium-diffusion problem can be avoided by using these materials.

In Fig. 11.13, two designs of bath cryostats with dewar vessels made of metal are shown schematically. The main difference lies in the arrangement of the nitrogen vessels. In the model on the left the helium vessel is completely

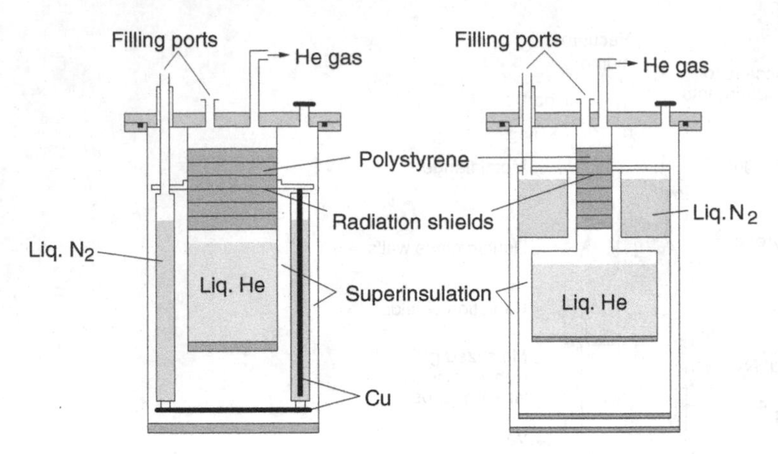

Fig. 11.13. Schematic illustration of two different bath cryostats with Dewar vessels made of metal. After [573]

surrounded by nitrogen, whereas in the one shown on the right the nitrogen vessel is located above the helium dewar. One advantage of the latter arrangement is the ability to perform optical experiments more simply since vacuum-tight optical windows are not required on the nitrogen reservoir. Most metallic dewars have a common isolation vacuum for nitrogen and helium vessels. The space of the isolation vacuum is often filled with a so-called *superinsulation*, which consists of multiple layers of metalized plastic foil that is used as additional radiation shielding. Usually, the superinsulation is made of mylar with an evaporated aluminum coating on one side. The heat transferred by radiation between two plates at temperatures T_{high} and T_{low} with N layers of superinsulation in between them is given by

$$\dot{Q} = \frac{\varepsilon \sigma A}{(N+1)} \left(T_{\text{high}}^4 - T_{\text{low}}^4 \right) , \tag{11.5}$$

where A represents the surface area of one side of a plate, σ the Stephan–Boltzmann constant and ε the emissivity. Highly reflecting aluminized mylar has a low emissivity of about 0.05.

For many years, an increasing number of helium dewars have been used that do not involve a nitrogen jacket. Such systems only have a thick superinsulation for radiation shielding. This technique is also used for helium-transport dewars. A schematic illustration of a typical transport vessel is shown in Fig. 11.14a. The helium space inside is completely surrounded by superinsulation. In the drawing, however, only a small portion of the shielding is indicated by *grey tinting* for clarity. Absorption and getter pumps are used to maintain and improve the vacuum in order to minimize the heat load by residual gas.

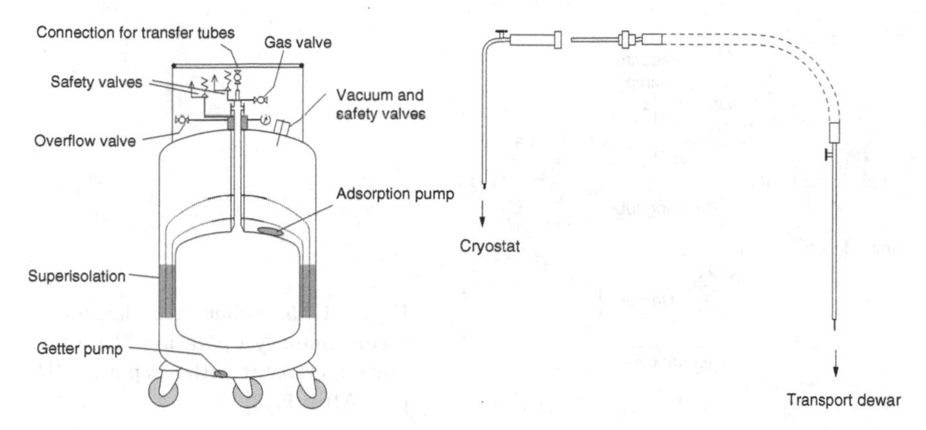

Fig. 11.14. Schematic illustration of (**a**) a transport dewar and of (**b**) a transfer tube for helium. The part of the transfer tube drawn with *dashed lines* is flexible. After [573]

Figure 11.14b shows a diagram of the typical tubing that is used to transfer liquid helium from the transport vessel to the cryostat. The transfer tube depicted consists of two parts that can be connected at room temperature outside of the dewar. Both parts of the transfer tube consist of double-walled concentric tubes with a vacuum between the tubes to provide the required thermal insolation.

11.3.2 Evaporation Cryostats

In order to reduce the temperature of liquid helium one often simply pumps the vapor above the bath. In this way, it is possible to reach temperatures down to about 1.3 K without much effort. However, this procedure results in enormous helium consumption, since cooling the liquid from 4.2 K to 1.3 K requires about 40% of the liquid helium to be evaporated. More elegant and efficient is the use of an additional vessel filled with ^{4}He or ^{3}He that is located inside a vacuum chamber (Fig. 11.15). In this case, the temperature is lowered only in a much smaller system, reducing the helium consumption needed to produce the low temperature. In addition, this arrangement requires a lower pumping capacity as compared to pumping the larger main bath. Continuous operation of such a cryostat can be achieved by refilling the separate small vessel via a small capillary from the main helium bath. To prevent the inflow of impurities, or small particles, which potentially could block the capillary, a filter made of sintered metal is usually installed in front of the capillary.

In the following, we discuss the cooling power of evaporation cryostats in which either liquid ^{3}He or ^{4}He are pumped. The evaporation process can be described by the *Clausius–Clapeyron equation*

$$\frac{\mathrm{d}p}{\mathrm{d}T} = \frac{L}{\Delta V\, T} . \tag{11.6}$$

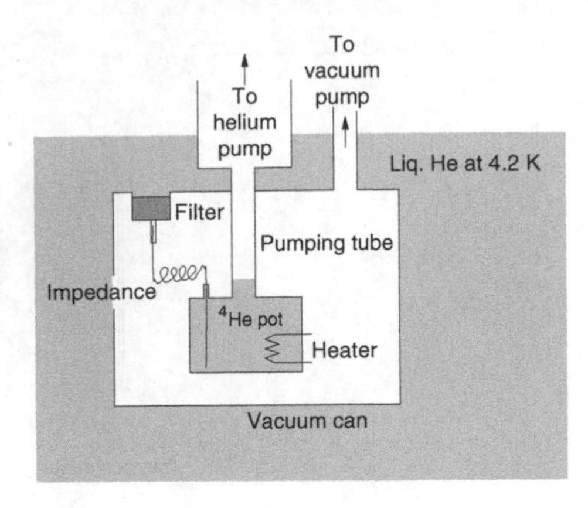

Fig. 11.15. Schematic diagram of a continuously operating ^{4}He evaporation cryostat with a separate ^{4}He pot. After [573]

The quantity L represents the latent heat (per mole) and ΔV denotes the volume change that occurs if one mole of the liquid is converted into gas. In comparison with the gas volume V_g the volume of the liquid V_ℓ is negligible and we can write, to a good approximation

$$\Delta V = V_g - V_\ell \approx V_g . \tag{11.7}$$

Using the ideal gas law $pV_g = RT$, we may express (11.6) as follows

$$\frac{dp}{dT} = \frac{L}{RT^2}\, p . \tag{11.8}$$

Solving this simple differential equation leads to the well-known exponential dependence of the vapor pressure on temperature

$$p(T) = p_0\, e^{-L/RT} . \tag{11.9}$$

As shown in Fig. 1.4, the latent heats of ^{3}He and ^{4}He differ considerably. At 1.5 K, the values $L = 40\,\mathrm{J\,mol^{-1}}$ and $L = 90\,\mathrm{J\,mol^{-1}}$ are found for ^{3}He and ^{4}He, respectively. Accordingly, the vapor pressure of ^{3}He at 1 K is about 70 times higher than that of ^{4}He, and at 0.5 K the ratio is 10^4:1. In Fig. 11.16, the vapor-pressure curves of ^{3}He and ^{4}He are shown in comparison with those of hydrogen and nitrogen.

The cooling power of evaporation cryostats is determined by the rate at which vapor can be pumped from the liquid, which is expressible as

$$\dot{Q} = \dot{n}_g\, L \propto p \propto e^{-L/RT} . \tag{11.10}$$

Thus, $\dot{Q}$ decreases exponentially with decreasing temperature. Here, $\dot{n}_g$ represents the rate at which atoms are evaporated. The lowest temperatures that can be obtained with evaporation cryostats depend on the *heat leak*, which is the heat input to the system. This heat input can be due to different sources. Examples include heat conduction through the tubes, radiation from warmer parts of the apparatus, and the heat generated by the experiment. At a certain temperature, depending on the heat leak, increasing the pumping capacity does not result in a significantly lower temperature because of

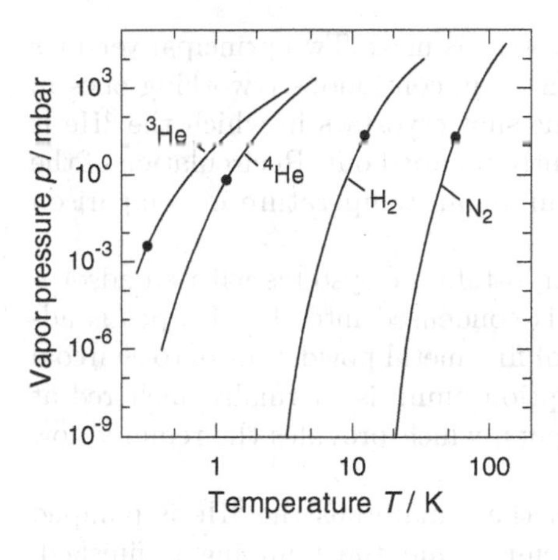

Fig. 11.16. Vapor pressure of different cryogenic liquids as a function of temperature (after [574]). The *points* (•) indicate the lowest temperatures that can be achieved in practice with moderate efforts

the exponential relation between vapor pressure and temperature. Therefore, the evaporation rate, and in turn the cooling power, can hardly be increased under these conditions.

The use of ^{3}He has two essential advantages over ^{4}He. First, the high vapor pressure allows temperatures as low as 0.3 K to be reached with modest pumping systems. Secondly, ^{3}He does not possess the problem created by the creeping superfluid film, which leads to an unwanted heat transport from warm to cold parts in ^{4}He evaporation cryostats. However, the cost for ^{3}He is much higher than for ^{4}He. The first ^{3}He evaporation cryostat was build by Roberts and Sydoriak in 1954 to measure the vapor pressure and the specific heat of ^{3}He [575].

Since then, many different types of ^{3}He evaporation cryostats have been designed, two of which are shown schematically in Fig. 11.17. To precool the

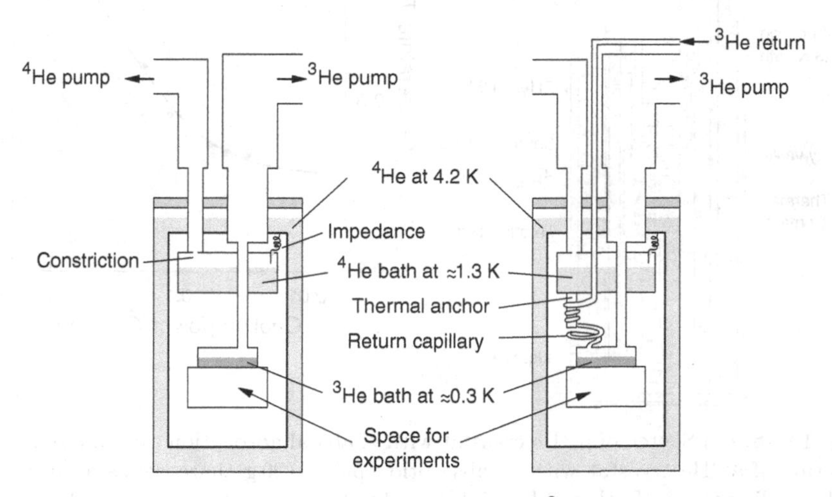

Fig. 11.17. Schematic illustration of two types of ^{3}He evaporation cryostats for (**a**) single shot and (**b**) continuous operation (After [576])

system, normally a ^{4}He evaporation cryostat is used. Two principal versions of ^{3}He cryostats can be distinguished, namely continuously working ones in which the ^{3}He is recirculated, and single-shot cryostats in which the ^{3}He is first condensed into a pot and subsequently pumped out. Recirculation of the ^{3}He typically results in a 10% higher minimum temperature in comparison to single-shot operation.

A special class of ^{3}He evaporation cryostats are systems with an adsorption pump. In this case, the ^{3}He that is condensed into the ^{3}He pot is adsorbed onto the cold surface of a sinter of fine metal powder, or onto charcoal powder. While in operation, this adsorption pump is thermally anchored at a pumped ^{4}He pot (often called a 1 K pot), which provides the required low temperatures.

Because of the large surface area of these materials the ^{3}He is pumped very efficiently. After the ^{3}He pot is empty and the pumping is finished, the adsorption pump is pulled up to a warmer region in the cryostat. Upon warming, the ^{3}He is desorbed from the powder and then condenses into the ^{3}He pot. The cycle can be started again by cooling the adsorption pump back

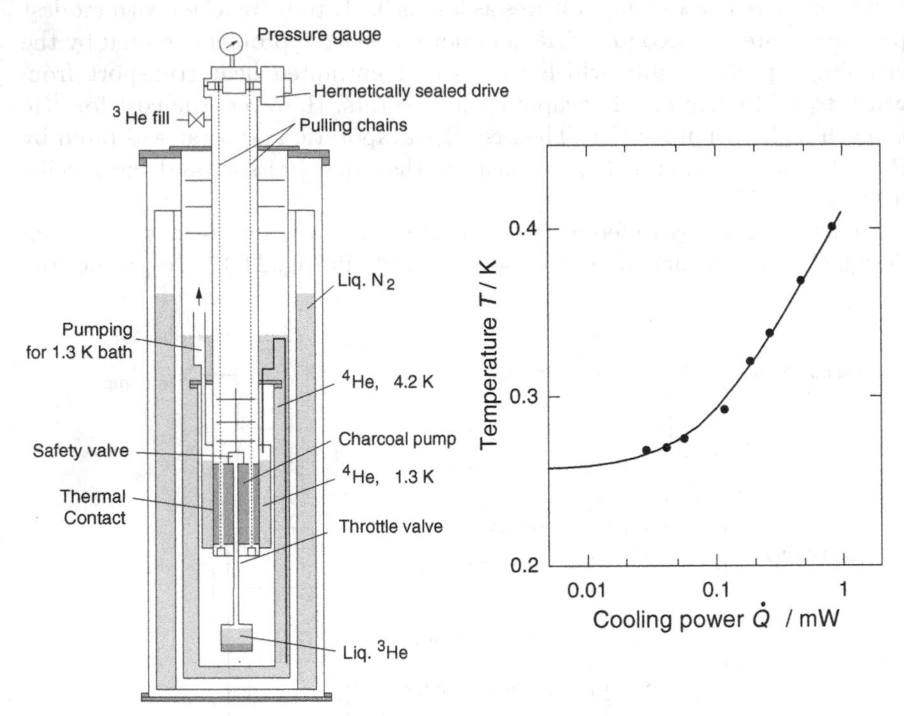

Fig. 11.18. (**a**) Sketch of a ^{3}He cryostat with charcoal absorption pump. (**b**) Temperature of a ^{3}He cryostat with an absorption pump (20 g charcoal) as a function of the cooling power. In thermal equilibrium, the latter quantity corresponds to the heat leak [577]

to the temperature of the 1 K pot. A design of such a cryostat is shown in Fig. 11.18a. In this case, the charcoal pump is manipulated via chains using a spindle valve. Because of their enormous pumping power, ^{3}He cryostats with absorption pumps have high cooling powers compared to those with mechanical pumps. The temperature of a ^{3}He cryostat with a 20 g charcoal absorption pump is shown in Fig. 11.18b as a function of the cooling power. The cooling power has been determined in this experiment with the help of a heater at the ^{3}He pot. The result of the measurement shows that it is possible to reach temperatures of about 0.26 K with such a cryostat and that the cooling power at about 0.3 K can still be of the order of 0.1 mW.

11.4 Dilution Refrigerators

As we discussed in the previous section, a liquid can be cooled by pumping the vapor above it. Using ^{3}He, the lowest temperature that has been achieved in this way is about 0.24 K [578]. In order to produce lower temperatures one has to employ other techniques. Currently, temperatures in the mK range are obtained using a *dilution refrigerator*. The principle of operation relies on the specific properties of dilute ^{3}He/^{4}He mixtures, which we considered in Chap. 5. The basic idea of this method was suggested in 1951 by *H. London* [579]. About ten years later, in 1962, *London*, *Clarke* and *Mendoza* published a detailed concept for realizing such a cooling system [580]. The first successful experimental demonstration of a dilution refrigerator was achieved by *Das, De Bruyn Ouboter* and *Taconis* in 1965 [581]. They reached a lowest temperature of about 0.22 K.

11.4.1 Principle of Operation

The cooling mechanism of a dilution refrigerator is, in some ways, similar to the evaporative cooling discussed in the previous section. However, in the case of a dilution refrigerator, the heat of solution plays the role of the latent heat. The fact that in ^{3}He/^{4}He mixtures a phase separation at very low temperatures occurs is crucial. As we have seen in Chap. 5, ^{3}He atoms are more strongly bound to ^{4}He than among each other, because the ^{4}He atoms have a lower zero-point energy due to their larger mass. The ^{3}He atoms obey the Fermi statistics and thus their kinetic energy increases with their number density. Because of this, the effective binding energy is reduced. At a ^{3}He concentration of 6.5% in ^{4}He, the effective binding energy vanishes for $T \to 0$, and no further ^{3}He atoms can be dissolved in the ^{4}He. Energetically, two phases are more favorable when the average ^{3}He concentration is greater than 6.5%. A light, ^{3}He-rich phase and a heavy phase mainly consisting of ^{4}He atoms are formed. The maximum solubility of the ^{3}He atoms in the heavy phase depends on temperature and pressure. Figure 11.19 shows the temperature and concentration dependence of different ^{3}He/^{4}He mixtures.

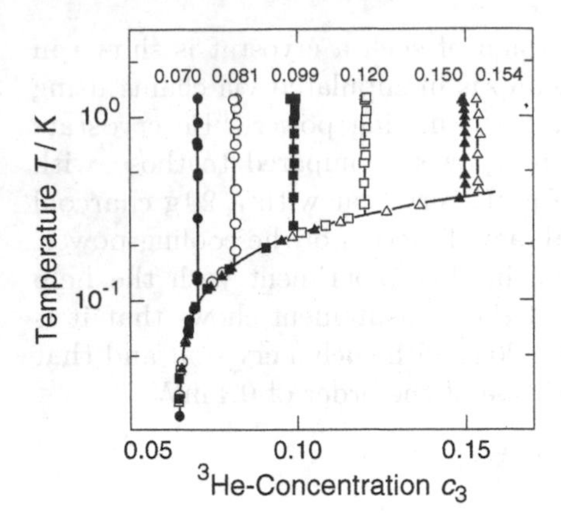

Fig. 11.19. Concentration and temperature dependence of ^{3}He dissolved in ^{4}He. The starting concentration of the ^{3}He is indicated at the *top* of each curve [582]

At high temperatures, the ^{3}He concentration is constant and leads into a common curve at very low temperatures. This curve reflects the maximal solubility of ^{3}He in ^{4}He under normal pressure as a function of temperature. Below this curve, a miscibility gap opens up and two separate phases with different ^{3}He concentrations coexist at the same temperature.

The ^{3}He atoms in the ^{3}He-rich phase have a lower entropy than the ^{3}He atoms in the ^{4}He-rich phase. The cooling process of a dilution refrigerator takes place in the so-called mixing chamber and consists of the transfer of ^{3}He atoms from the ^{3}He-rich phase into the dilute phase. In analogy to the evaporation of gases, one sometimes refers to this process as an evaporation into the quasivacuum of the superfluid ^{4}He. The heat of solution that occurs in this process for one mole of ^{3}He is given by

$$\Delta Q = T\Delta S = aT^2 \,, \tag{11.11}$$

with $a = -84\,\mathrm{J\,K^{-2}}$. In order to use this cooling mechanism for continuous operation one has to remove ^{3}He from the dilute phase in the mixing chamber and feed it back into the ^{3}He-rich phase. How this can be achieved technically will be discussed in the next section.

11.4.2 Principles of a Dilution Refrigerator

The centerpiece of a dilution refrigerator is the cold part of the ^{3}He/^{4}He circuit that is shown schematically in Fig. 11.20a. This part of the apparatus is located in a vacuum chamber that is immersed into a ^{4}He bath. It essentially consists of the *mixing chamber*, the *still* and a *counterflow heat exchanger*. How these components are integrated in a dilution refrigerator is illustrated in Fig. 11.20b in the case of a commercial apparatus.

The relatively complex ^{3}He/^{4}He circuit shown in Fig. 11.20a is necessary to realize a sufficiently high circulation of ^{3}He, while maintaining a low heat load at the mixing chamber. The circulation of the ^{3}He/^{4}He mixture

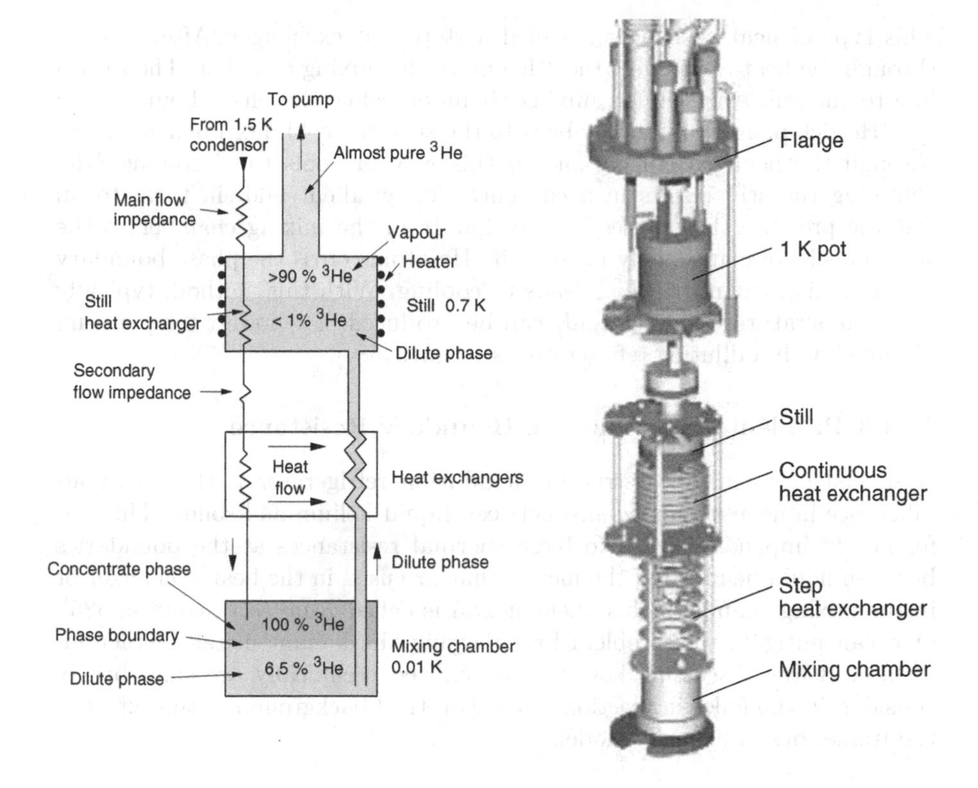

Fig. 11.20. (**a**) Scheme of the inner ^{3}He/^{4}He circuit of a dilution refrigerator After [573]. (**b**) Design of the dilution unit of a commercial dilution refrigerator [583]

is driven by pumping the still. The still is heated to about 0.7 K to increase the efficiency of the pumping. Because of its higher vapor pressure, ^{3}He is predominantly evaporated from the liquid, although the ^{3}He concentration in the liquid in the still is only about 1%. Once it has been pumped, the ^{3}He is cleaned outside the cryostat in a nitrogen trap before being returned to the cryostat. Further cleaning often takes place in a helium trap in the helium bath. Following this step, the ^{3}He enters the vacuum chamber in a capillary and is precooled at the 1 K pot.

The pressure of the ^{3}He is maintained sufficiently high by using a flow impedance before the still so that it condenses. After the still, the ^{3}He is led into the counterflow heat exchanger that consists, in most systems, of two different types of heat exchangers (see Sect. 11.4.3). The first one is called a continuous heat exchanger and is normally made of two tubes that are arranged with one inside the other in a rather complicated manner so that the interface between the two is as large as possible. The second heat exchanger consists of several chambers each of which has a dividing wall with sintered silver attached to it in order to increase the thermal contact area.

This type of heat exchanger is called a step heat exchanger. After passing through the heat exchangers the ^{3}He enters the mixing chamber. The return line to the still starts in the mixing chamber below the phase boundary in the ^{4}He-rich phase. On the way back to the still, the cold mixture again flows through the heat exchangers and in this way precools the incoming ^{3}He. Pumping the still results in a concentration gradient and, in turn, to an osmotic pressure that causes ^{3}He to flow from the mixing chamber to the still. This is, of course, only possible if ^{3}He atoms cross the phase boundary in the mixing chamber, which leads to cooling. With this method, typically base temperatures of about 5 mK can be produced. The lowest temperature obtained with a dilution refrigerator is 1.5 mK [584].

11.4.3 Problem of the Thermal Boundary Resistance

A special problem in constructing a dilution refrigerator is the enormous difference in acoustic impedance between liquid helium and solids. This difference in impedance leads to large thermal resistances at the boundaries between liquid helium and the metals that are used in the heat exchanger or in the mixing chamber. This phenomenon is called *Kapitza resistance* [585]. One can mitigate this problem by enlarging the helium–metal contact interface. Before discussing how this is realized technically, we shall briefly consider, in the following section, the theoretical background of this effect in the framework of a simple model.

Kapitza Resistance

In the so-called acoustic mismatch model, which was first introduced by *Khalatnikov* in 1952 [586], the fraction of phonons that can pass from one material to the other despite the refraction taking place at the contact interface due to the different acoustic impedances is calculated. A sketch of the situation at the boundary is shown in Fig. 11.21.

A crucial quantity in these considerations is the critical angle of total reflection. Using Snell's law of refraction

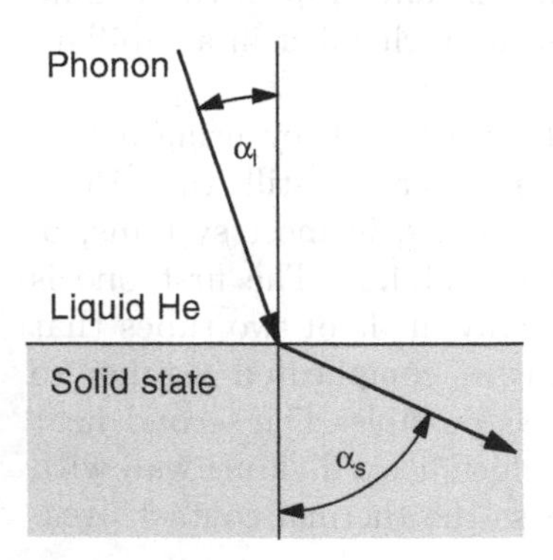

Fig. 11.21. Schematic illustration of the transition of a phonon from liquid helium into a solid

$$\frac{\sin \alpha_\ell}{\sin \alpha_s} = \frac{v_\ell}{v_s} \tag{11.12}$$

one can calculate the critical angle α_ℓ^c from the speeds of sound v_ℓ in the liquid and v_s in the solid:

$$\alpha_\ell^c = \arcsin\left(\frac{v_\ell}{v_s}\right) . \tag{11.13}$$

We obtain a critical angle of only $\alpha_\ell^c \approx 4°$ using the sound velocities of ^{4}He $v_\ell = 238\,\mathrm{m\,s^{-1}}$ and a typical value $v_s = 3\times 10^3\,\mathrm{m\,s^{-1}}$ for the solid. Calculating the fraction f of phonons entering the interface within this angle one finds

$$f = \frac{1}{2}\sin^2 \alpha_\ell^c = \frac{1}{2}\left(\frac{v_\ell}{v_s}\right)^2 < 10^{-2} . \tag{11.14}$$

Actually, because of the acoustic impedance jump at the interface not even this fraction of the phonons is completely transmitted. Since the critical angle is very small, we can calculate the transmission coefficient t to a good approximation under the assumption that the transmitted phonons incident perpendicular to the interface. In this case, we find for the transmission coefficient

$$t = \frac{4Z_\ell Z_s}{(Z_\ell + Z_s)^2} \approx \frac{4Z_\ell}{Z_s} = \frac{4\varrho_\ell v_\ell}{\varrho_s v_s} , \tag{11.15}$$

where $Z_\ell = \varrho_\ell v_\ell$ and $Z_s = \varrho_s v_s$ represent the acoustic impedances for liquid and solid, respectively. For phonons incident on a helium–copper interface within the critical angle one finds a transmission coefficient of only $t \approx 10^{-3}$.

Considering (11.14) and (11.15) the fraction of phonons that actually cross the interface is given by

$$ft = \frac{2\varrho_\ell v_\ell^3}{\varrho_s v_s^3} . \tag{11.16}$$

For a helium–copper interface this fraction is smaller than 10^{-5}. The heat flow from the liquid to the solid can be expressed by

$$\dot{Q} = \frac{1}{2} f t u v_\ell A = \frac{\pi^2 k_B^4 \varrho_\ell v_\ell}{30\hbar^3 \varrho_s v_s^3} A T^4 . \tag{11.17}$$

Here, $u = U/V = \pi^2 k_B^4 T^4/(30\hbar^3 v_\ell^3)$ denotes the energy density of longitudinal phonons in the liquid and A the size of the contact interface. In thermal equilibrium, the identical heat flow takes place in the opposite direction and therefore the net flow is zero. If, however, the temperature of the liquid is higher than that of the solid, a resulting net heat flow occurs. Under the assumption that the temperature difference ΔT is much smaller than the temperature T_ℓ of the liquid, one finds for the resulting heat flow across the interface

$$\dot{Q} = \frac{\mathrm{d}\dot{Q}}{\mathrm{d}T}\Delta T = \frac{2\pi^2 k_B^4 \varrho_\ell v_\ell}{15\hbar^3 \varrho_s v_s^3} A T^3 \Delta T . \tag{11.18}$$

Finally, we obtain for the thermal boundary resistance, or Kapitza resistance, the relation

$$R_{\mathrm{K}} = \frac{A\Delta T}{\dot{Q}} = \frac{15\hbar^3 \varrho_{\mathrm{s}} v_{\mathrm{s}}^3}{2\pi^2 k_{\mathrm{B}}{}^4 \varrho_\ell v_\ell} \frac{1}{T^3} . \tag{11.19}$$

Whereas the heat flow described by (11.18) is proportional to the interface area A, the Kapitza resistance R_{K} itself is independent of the area. As shown in Fig. 11.22, the experimental data obtained in the temperature range between $20\,\mathrm{mK} < T < 100\,\mathrm{mK}$ are well described by a T^{-3} dependence. Below 10 mK, however, the simple model presented here is not sufficient. In this temperature range, several additional effects have to be taken into account. For example, the interaction between the phonons in the metal with the zero-sound modes in ^{3}He or second-sound modes in ^{4}He have to be considered. In addition, the discussion given above of the acoustic mismatch model is certainly too simple to describe the Kapitza resistance in layers of silver sinter with very small grain sizes, because the wavelengths of the dominant phonons at low temperatures become larger than the typical dimensions within the grains. Experimentally, one finds a T^{-1} dependence for the thermal boundary resistance between metal sinters and pure ^{3}He, and a T^{-2} dependence for ^{3}He/^{4}He mixtures and metal sinters.

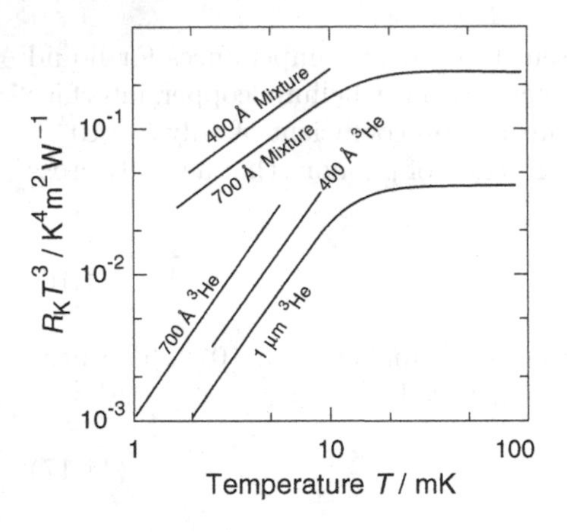

Fig. 11.22. Kapitza resistance between sintered silver powders of different grain sizes (indicated on each curve) and pure ^{3}He or ^{3}He/^{4}He mixtures as a function of temperature. Note that the product $R_{\mathrm{K}}T^3$ is plotted [587]

At higher temperatures, which are not shown in Fig. 11.22, the Kapitza resistance R_{K} is significantly lower than expected from the impedance-mismatch model. This phenomenon is often referred to as *anomalous Kapitza resistance*, the origin of which is not fully understood. In this connection, we remark that the cleanliness of the metal surfaces is important for the absolute magnitude of the thermal resistance. In general, one finds a reduction of the Kapitza resistance due to the influence of impurities at the contact surface.

Construction of Heat Exchangers

As we have already discussed, the incoming ^{3}He in a dilution refrigerator is precooled in a counterflow heat exchanger before it reaches the mixing chamber. In this section, we shall briefly consider the construction of such counterflow heat exchangers. Figure 11.23 shows a so-called continuous heat exchanger. It consists of an outer brass tube and within it an inner tube that is divided into two parts. The first part, located at the warm end of the heat exchanger, is a copper-nickel capillary that serves as a secondary flow impedance for the incoming concentrated phase of condensed ^{3}He and prevents re-evaporation after the main impedance. The second part of the inner tubing consists of a brass tube that is rolled into a tight spiral. The concentrated ^{3}He flows through the capillary, while the ^{3}He in the dilute phase flows in the opposite direction in the outer tube from the mixing chamber to the still driven by osmotic pressure.

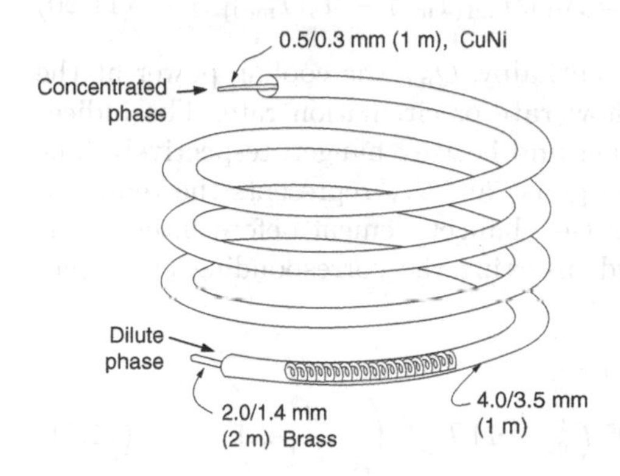

Fig. 11.23. Sketch of a continuous heat exchanger. After [573]

Small and simple dilution refrigerators often only have a continuous heat exchanger and can reach temperatures of about 20 mK at the mixing chamber. More powerful units normally include a system of step heat exchangers. The principle construction of this type of heat exchanger is shown in Fig. 11.24. Note that this version of a step heat exchanger is only one of

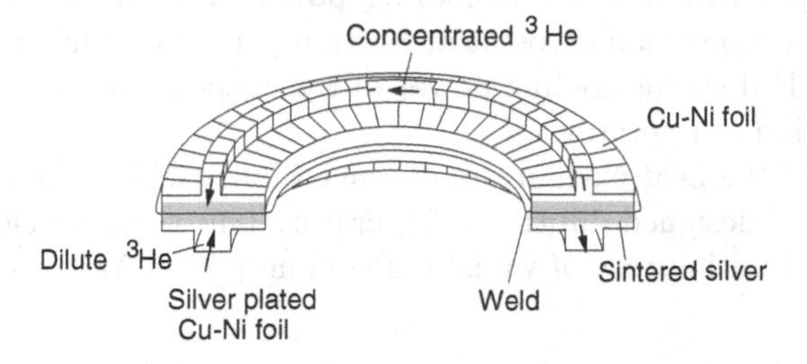

Fig. 11.24. Sketch of a cut through the cross section of one element of a step heat exchanger. After [587]

many different designs for making such a device. The crucial element is the layer of silver sinter that is attached to a copper-nickel foil, which divides the heat exchanger into two chambers, one for the ^{3}He-rich phase and one for the dilute phase. The silver sinter provides an enormous increase of the contact surface and, in this way, reduces the thermal resistance. Normally, such a heat exchanger consists of five steps with contact surfaces ranging from $10\,\mathrm{m}^2$ to $100\,\mathrm{m}^2$. With well-designed exchangers, temperatures as low as 4 mK can be reached.

11.4.4 Cooling Power

Neglecting external heat leaks such as radiation, residual gas or vibrations and assuming that the heat input into the mixing chamber is entirely due to the incoming ^{3}He the following balance is valid:

$$\dot{Q}_{\mathrm{mc}} + \dot{N}_3 \left[H_3(T_{\mathrm{ex}}) - H_3(T_{\mathrm{mc}})\right] = \dot{N}_3 \left[H_{3,\mathrm{d}}(T_{\mathrm{mc}}) - H_3(T_{\mathrm{mc}})\right] . \tag{11.20}$$

Here, $H = U + PV$ represents the enthalpy, $\dot{Q}_{\mathrm{mc}}$ the cooling power at the mixing chamber and $\dot{N}_3$ the ^{3}He flow rate or circulation rate. The indices mc and ex denote the mixing chamber and heat-exchanger, respectively. The index d denotes the dilute ^{3}He/^{4}He phase and T_{ex} represents the temperature of the mixture after the last heat-exchanger element before entering the mixing chamber. Using (11.20) and inserting the corresponding enthalpies one finds for the cooling power:

$$\begin{aligned}\dot{Q}_{\mathrm{mc}} &= \dot{N}_3 \left[H_{3,\mathrm{d}}(T_{\mathrm{mc}}) - H_3(T_{\mathrm{ex}})\right] \\ &= \dot{N}_3 \left(95\, T_{\mathrm{mc}}^2 - 11\, T_{\mathrm{ex}}^2\right) \left(\frac{\mathrm{J}}{\mathrm{mol\ K}^2}\right) .\end{aligned} \tag{11.21}$$

Figure 11.25 shows the cooling power of a dilution refrigerator in comparison with that of a ^{3}He evaporation cryostat. In this plot, it has been assumed that the pumps used can handle $5\,\ell/\mathrm{s}$ at all relevant pressures. For the case of the dilution refrigerator this corresponds to a ^{3}He circulation rate of $30\,\mu\mathrm{mol\,s}^{-1}$ in the entire temperature range. For the ^{3}He evaporation cryostat, this rate is reached at 0.5 K. The plot demonstrates the advantage of the weakly temperature-dependent cooling power of a dilution refrigerator relative to a ^{3}He evaporation cryostat at low temperatures. Only at temperatures above 0.35 K does the cooling power of the evaporation cryostat exceed that of the dilution refrigerator.

The quality of the heat exchangers is crucial for reaching the lowest temperatures in a well-designed dilution refrigerator. This becomes clear looking at (11.21) in the limiting case of vanishing cooling power. With $\dot{Q}_{\mathrm{mc}} = 0$ one finds

$$\frac{T_{\mathrm{ex}}}{T_{\mathrm{mc}}} = 2.8 . \tag{11.22}$$

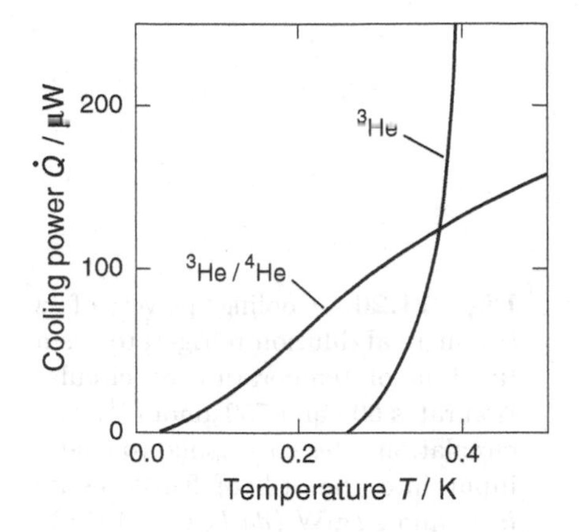

Fig. 11.25. Cooling power of a dilution refrigerator in comparison to a ^{3}He evaporation cryostat as a function of temperature [588]

This means that for a given temperature of the ^{3}He leaving the last heat exchanger, the mixing chamber can, at most, reach a temperature that is 2.8 times colder. Therefore, for powerful dilution refrigerators one needs very efficient heat exchangers as we have discussed before.

According to (11.21), the cooling power is proportional to the ^{3}He circulation rate $\dot{N}_3$, which is determined by the pumping speed and the temperature of the still. Varying the heat input into the still enables one to adjust the cooling power within certain limits. Typical operational temperatures of the still are in the range between 0.6 K and 0.7 K. Increasing the temperature much above 0.7 K results in an enhanced fraction of ^{4}He in the gas phase and therefore ^{4}He has to be circulated as well. The important point is that the ^{4}He adds to the heat input into the mixing chamber, but does not contribute to the cooling process itself.

Figure 11.26 shows the cooling power of a commercial dilution refrigerator operated at two different circulation rates. At temperatures above 20 mK, the cooling power varies proportional to T^2 and is higher at the higher circulation rate, as expected according to (11.21). The minimum temperature is lower for the smaller circulation rate because in this case the heat input into the mixing chamber is less.

In the balance of heat input and cooling power (11.20) we have so far neglected all heat leaks, except for the incoming warm ^{3}He. Surprisingly, for a properly designed dilution refrigerator the limiting heat leak is the viscous friction of the circulated ^{3}He in the heat exchanger. This process determines the temperature T_{ex} of the ^{3}He leaving the last heat exchanger, and thus the mixing chamber temperature. For laminar mass flow through the heat exchanger the volume flow rate $\dot{V}$ of the mixture is determined by $\Delta p = G\eta\dot{V}$. Here, Δp denotes the pressure difference along the heat exchanger and η represents the viscosity of the mixture. The quantity G depends on the geometry of the heat exchanger. According to the Hagen–Poiseuille law for a

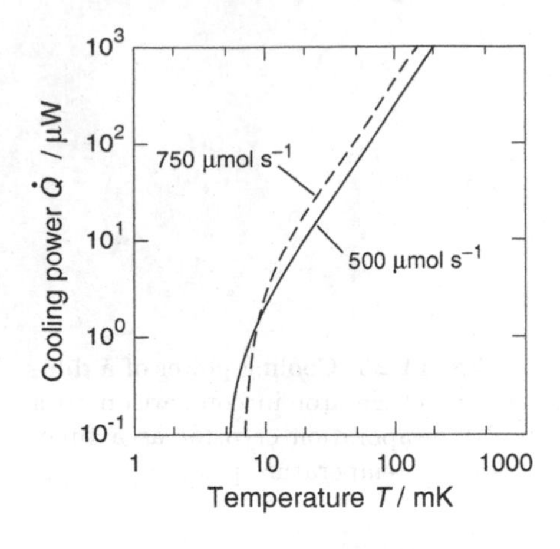

Fig. 11.26. Cooling power of a commercial dilution refrigerator as a function of temperature at circulation rates 500 and 750 µmol s^{-1}. The circulation rates correspond to a heat input into the still of 8 mW (*solid line*) and 22 mW (*dashed line*) [573]

tube with radius r and length L one has $G = 8L/(\pi r^4)$. The heat input due to viscous friction is given by

$$\dot{Q}_{\text{visc}} = \dot{V}\Delta p \;=\; G\eta\dot{V}^2 \,. \tag{11.23}$$

The relevance of this heat input comes from the enormous viscosity of ^{3}He at low temperatures. As we have seen in Sect. 3.1, the viscosity is proportional to T^{-2}, increasing with decreasing temperature until the transition to the superfluid phase occurs. Close to the transition, the absolute value is comparable to that of honey. It is noteworthy that the temperature rise caused by the viscous friction is higher for the dilute phase, because it has a much lower heat capacity than the concentrated phase. At a ^{3}He fraction of 6.4%, the temperature rise is eight times higher than for pure ^{3}He. For compensation, the dilute phase flows through a heat exchanger that has a correspondingly larger cross section (see Figs. 11.23 and 11.24).

11.5 Pomeranchuk Cooling

In 1950, *Pomeranchuk* suggested that very low temperatures could be obtained by solidification of ^{3}He [589]. At that time ^{3}He was available only in very small quantities and had just been liquified for the first time. For many years very little attention was paid to Pomeranchuk's idea and it took 15 years until the first successful experimental demonstration by *Anufriev* [590].

Well-designed Pomeranchuk cells work at temperatures below 300 mK and reach a minimum temperature of about 1 mK. This cooling technique was important mainly between 1965 and 1975, because the development of dilution refrigerators had not yet evolved sufficiently. Today, Pomeranchuk cells are very rarely used and are not available commercially.

11.5.1 Cooling by Solidification of ^{3}He

The cooling mechanism of this method is based upon the unusual fact that below 320 mK the entropy of solid ^{3}He is larger than that of liquid ^{3}He. We have already seen in Chap. 1 that the entropy of ^{3}He at low temperatures is essentially given by the contribution of the spins. In liquid ^{3}He, the spin entropy varies linearly with temperature as expected for a Fermi liquid. In solid ^{3}He, the situation is quite different because the spins are strongly localized and their behavior is not governed by Fermi statistics. Above 0.9 mK, the spins in the solid are in the paramagnetic state, while below that temperature they form an antiferromagnet.

Since the entropy of the liquid is less than that of the solid below 320 mK, the temperature of ^{3}He can be reduced by applying adiabatically a pressure and converting liquid to solid as illustrated by the arrow (I $\rightarrow$ F) in Fig. 11.27. It is not necessary to solidify the ^{3}He completely to obtain the lowest temperatures. The minimum temperature obtainable depends mainly on the starting entropy and thus on the starting temperature. Starting from $T_\text{i} = 25$ mK, the temperature can be reduced under ideal conditions by solidification of 20% of the ^{3}He to about $T_\text{f} = 3$ mK.

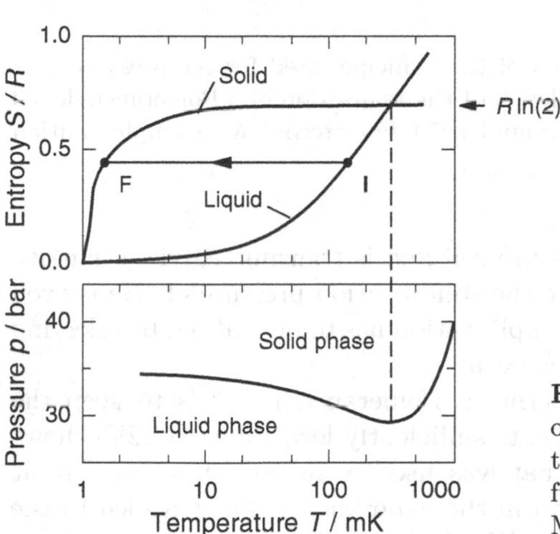

Fig. 11.27. *Upper part*: Comparison of the reduced entropies $S/(Nk_\text{B})$ of the solid and liquid phase of ^{3}He as a function of temperature. *Lower part*: Melting curve of ^{3}He

11.5.2 Technical Realization

Unfortunately, it is not possible to realize the necessary pressure increase in the cell for this cooling process by introducing additional ^{3}He via a capillary, because the ^{3}He would solidify in the capillary at the temperature of the minimum pressure of the melting curve and block any further inflow of ^{3}He. How this problem can be solved is shown schematically in Fig. 11.28a. The

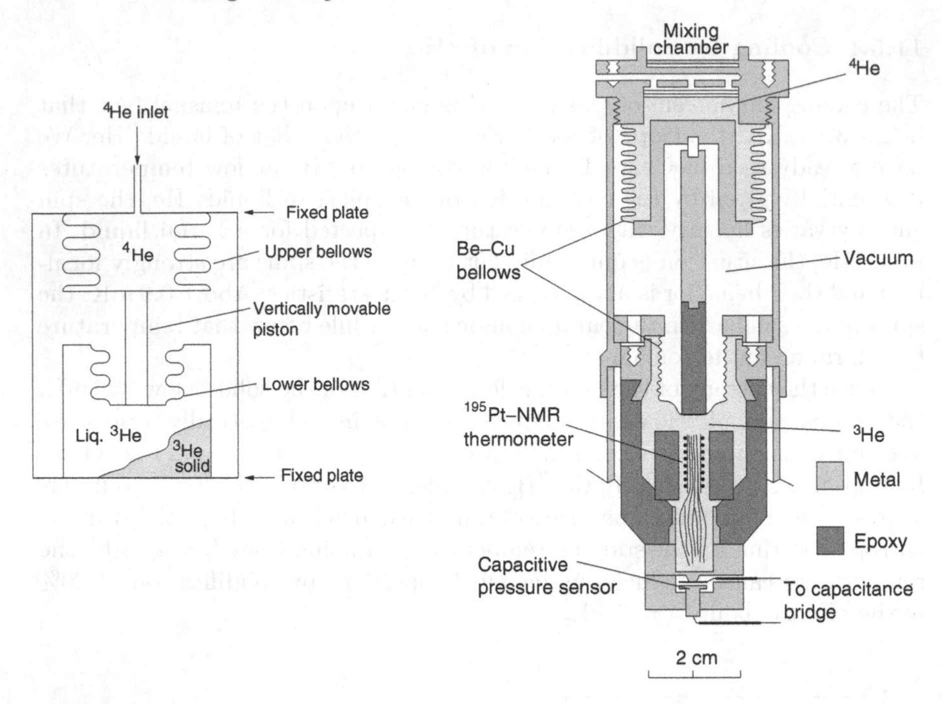

Fig. 11.28. (**a**) Schematic illustration of the principle used for the pressure production in a Pomeranchuk cell. (**b**) Sketch of the realization of a Pomeranchuk cell that was used at Cornell University around 1970. For precooling, a simple dilution refrigerator was used

pressure is produced in a ^{4}He pressure cell and is transmitted via a massive rod to the Pomeranchuk cell. Since the solidification pressure of ^{4}He is even lower than that of ^{3}He, a pressure amplification has to be realized by selecting the appropriate diameters of the two cells.

A difficult problem in constructing a Pomeranchuk cell is to keep the friction within mechanical components sufficiently low. Figure 11.28b shows a sketch of a Pomeranchuk cell that was used in investigations of ^{3}He at Cornell University around 1970 and in the experiments that have led to the discovery of the superfluid phases of ^{3}He. In this cell, the ^{3}He is located in a volume that can be reduced by the expansion of a beryllium-cooper bellows. The actual pressure can be determined in a capacitance measurement using a membrane that is part of the bottom of the cell. Inside the cell, a platinum NMR thermometer is mounted (see Sect. 12.2.5).

11.5.3 Cooling Power

Using $S_\ell = \pi^2 RT/(2T_\mathrm{F})$ for the entropy of the liquid, and assuming an isentropic change of state by increasing the pressure in the Pomeranchuk cell at temperatures above 10 mK where the entropy of the solid phase is approximately constant, we find for the cooling power

$$\dot{Q} = \dot{n}_3 T \left(S_\mathrm{s} - S_\ell\right) \approx TR\left(\ln 2 - \frac{\pi^2}{2}\frac{T}{T_\mathrm{F}}\right) . \qquad (11.24)$$

Here, $\dot{n}_3$ denotes the ^{3}He solidification rate. Since the entropy of the liquid can be neglected at very low temperatures, the cooling power varies approximately linearly with temperature. This is favorable in comparison with most other cooling techniques because their cooling power vanishes more rapidly at low temperatures. However, one should keep in mind that this is only a theoretical prediction for an ideal system. In practice, friction in the cell produces heat and therefore the effective cooling power is reduced.

Figure 11.29 shows a comparison of the cooling powers of a Pomeranchuk cell and a dilution refrigerator. The cooling power divided by the rate $\dot{n}_3$ is plotted versus temperature. The rate $\dot{n}_3$ corresponds to the solidification rate for Pomeranchuk cooling and the circulation rate for the dilution refrigerator. Below 50 mK, the cooling power of the Pomeranchuk cell is higher and above that temperature the dilution refrigerator provides more cooling power.

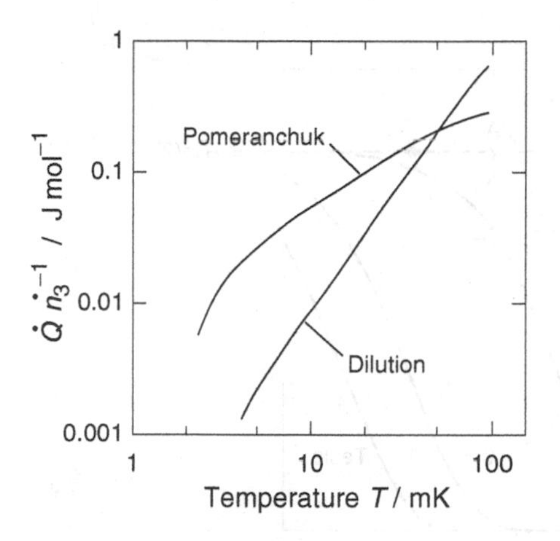

Fig. 11.29. Comparison of the normalized cooling powers $\dot{Q}/\dot{n}_3$ of a Pomeranchuk cell and a dilution refrigerator as a function of the temperature. The solidification rate and the circulation rate are 10 µmol s^{-1}. After [588]

11.6 Adiabatic Demagnetization

Enormous progress has been made in the last few years in the production of ultralow temperatures ($T < 1$ mK). Cooling via adiabatic demagnetization of nuclear spins is the only process that makes temperatures in the µK range accessible. This process is similar to the demagnetization of atomic magnetic moments. The possibility of cooling by the demagnetization of atomic magnetic moments was pointed out independently in 1926 by *Debye* [552] and in 1927 by *Giauque* [553]. Only a few years later, in 1933, two groups exploited this process almost simultaneously for the first time reaching temperatures well below 1 K [591, 592].

As mentioned above there are two variants of cooling via adiabatic demagnetization, namely electron spin and nuclear spin demagnetization. Initially, only the electron spins in paramagnetic salts were used, because of limitations on precooling and the unavailability of sufficiently high magnetic fields. Now, dilution refrigerators and superconducting magnets enable us to reach starting conditions that make the cooling via adiabatic demagnetization of nuclear spins possible. Although the cooling mechanism is in principle the same, the technical realization of the two processes differs significantly. In the following section, we will first discuss the cooling process and the demagnetization of paramagnetic salts. Following this, the peculiarities of the nuclear spin demagnetization will be discussed.

11.6.1 Cooling Mechanism

Cooling via adiabatic demagnetization relies on the magnetic-field dependence of the spin entropy. The temperature dependence of the entropy for a system with spin 1/2 in different magnetic fields is shown in Fig. 11.30.

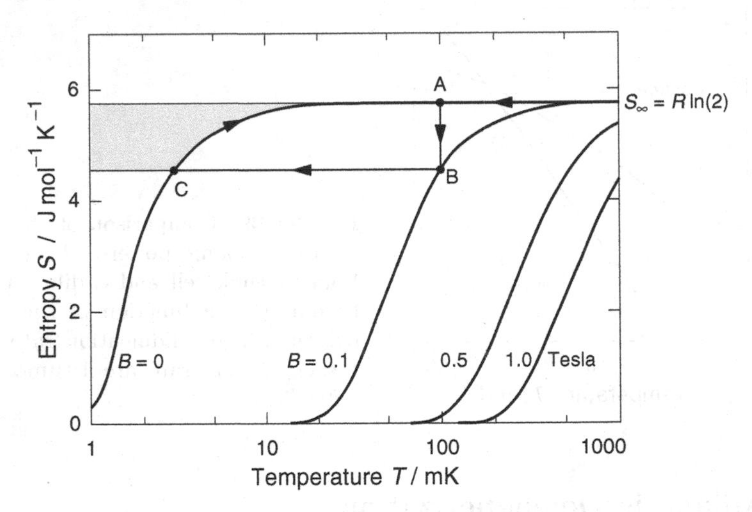

Fig. 11.30. Entropy of a CMN single crystal in different magnetic fields as a function of temperature, calculated for fields along the crystallographic *a*-axis

Without an external magnetic field the temperature dependence of the entropy is determined entirely by the interaction between the spins. At high temperatures, the spins are randomly oriented and carry the entropy $S = R\ln 2$. To calculate the curves shown in Fig. 11.30, the parameters of the paramagnetic salt cerium magnesium nitrate $2Ce(NO_3)_3 \cdot 3Mg(NO_3)_2 \cdot 24H_2O$

have been used. In this system, a transition into a ferromagnetic order occurs at a temperature of $T_c = 1.9\,\mathrm{mK}$. In finite external magnetic fields this transition is shifted towards higher temperatures, as depicted in Fig. 11.30. The entropy of such a spin system is enormous in comparison to the entropies of other systems at this temperature, such as phonons or conduction electrons in metallic cooling media. We will therefore only consider the entropy of the spins and will neglect the contribution of all other degrees of freedom.

Figure 11.31 shows the essential components of a demagnetization apparatus, consisting of a precooling stage, a heat switch, a cooling medium and a magnet. Initially, the cooling medium, e.g., a paramagnetic salt, is precooled to a temperature T_i (point A in Fig. 11.30). During this process, the heat switch to the precooling stage is closed to provide thermal contact. Without external magnetic field, the spins in the cooling medium are randomly oriented at this temperature.

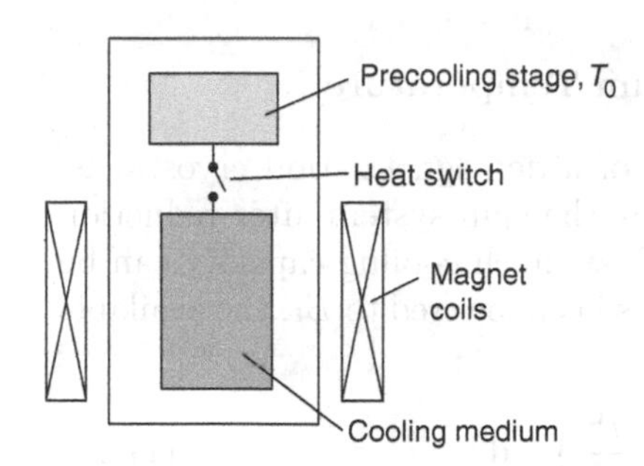

Fig. 11.31. Schematic illustration of the essential components of a demagnetization cryostat

The application of a strong magnetic field leads to a partial alignment of the magnetic moments. During this isothermal magnetization, the amount of heat

$$\Delta Q_{\mathrm{mag}} = -T_i[\,S(B_i, T_i) - S(0, T_i)\,] \tag{11.25}$$

is transferred to the precooling stage. The corresponding change of state is indicated in the entropy–temperature diagram in Fig. 11.30 by an arrow (A → B). After the isothermal magnetization, the cooling medium is thermally isolated by opening the heat switch. For systems with a starting temperature between 1 K and 4 K, mechanical heat switches are mostly used. In the millikelvin temperature range, superconductors are often employed as heat switches. The thermal conductivity of a superconductor can be greatly different in the normal and superconducting states and can be changed from one to the other using a magnetic field (see Sect. 10.3.5). The technical design of this crucial component of a demagnetization facility will be discussed in Sect. 11.7.4.

The next step in the cooling process is the adiabatic demagnetization itself, corresponding to the step (B → C) in Fig. 11.30. In this step, the magnetic field is slowly reduced. When the relaxation rate of the spins is longer or at least equal to the duration of the demagnetization, a large proportion of the spins keep their alignment during this process. After the magnetic field has been reduced, the spins exhibit a low occupation temperature (see Sect. 8.4). At this point, it is important that the spin system comes into thermal equilibrium with the other degrees of freedom, namely phonons and electrons. Scattering processes between spins and phonons or electrons lead to a gradual reduction of the alignment of the spins. Since the total system is thermally isolated, phonons and electrons lose energy in this process. In practice, the demagnetization is conducted in such a way that during the decrease of the magnetic field a thermalization between the spin system and the phonons and electrons is reached, because the typical relaxation times at the lowest temperatures are too long. After reaching the minimal equilibrium temperature, internal and external heat leaks – such as the experiment itself – cause a slow warming of the cooling medium along the entropy curve (C → A).

11.6.2 Cooling Capacity and Minimum Temperature

An important question regarding the use of a demagnetization cryostat is: how much heat ΔQ_{spin} can be absorbed by the spin system after reduction of the magnetic field? Or in other words, how much cooling capacity can be used for experiments. Assuming the field has been lowered to B_{f}, the available cooling capacity is given by

$$\Delta Q_{\mathrm{spin}}(B_{\mathrm{f}}) = \int_{T_{\mathrm{f}}}^{T_{\mathrm{i}}} C_{\mathrm{spin}}\,\mathrm{d}T = \int_{T_{\mathrm{f}}}^{T_{\mathrm{i}}} T \left(\frac{\partial S}{\partial T}\right)_{B_{\mathrm{f}}} \mathrm{d}T\,. \tag{11.26}$$

This expression corresponds to the *grey tinted region* in Fig. 11.30 for $B_f = 0$. In order to increase the cooling capacity (i.e., the heat capacity of the spins) it is advantageous not to lower the magnetic field B completely, although the minimum spin temperature will be higher in this case. The final temperature T_{f} that can be reached depends mainly on the material used, or more precisely, on the magnitude of the internal fields that lead to a finite transition temperature T_{c} into a magnetically ordered state. In addition, T_{f} is determined by the start temperature and the magnetic field at the beginning of the demagnetization. As we have seen, at high temperatures and without external magnetic field the spins are oriented randomly and carry the entropy $S = Nk_{\mathrm{B}}\ln(2J+1)$. Here, J represents the quantum number of the total angular momentum. In calculating the spin entropy, one has to include the interaction between the magnetic moments that can be described in terms of an internal field B_{int}. In the limit $\mu_{\mathrm{B}} gB \ll k_{\mathrm{B}}T$, the entropy is given by

$$S = Nk_{\mathrm{B}}\left\{\ln(2J+1) - \frac{g^2 J(J+1)\mu_{\mathrm{B}}^2}{6\,{k_{\mathrm{B}}}^2}\,\frac{B^2 + B_{\mathrm{int}}^2}{T^2}\right\}. \tag{11.27}$$

Under ideal conditions ($S_{\mathrm{i}} = S_{\mathrm{f}}$) for an adiabatic demagnetization the relation between the start and minimum temperature is given by

$$T_{\mathrm{f}} = T_{\mathrm{i}} \sqrt{\left(\frac{B_{\mathrm{f}}^2 + B_{\mathrm{int}}^2}{B_{\mathrm{i}}^2 + B_{\mathrm{int}}^2} \right)} . \tag{11.28}$$

Hence, the internal field B_{int} limits the lowest obtainable temperature.

11.6.3 Electron Spins – Paramagnetic Salts

The first demonstration of this cooling method was made by *Giauque* and *McDougal* in 1933 [591]. After precooling to 3.4 K they reached a minimum temperature of 0.54 K by the demagnetization of a 61-g gadolinium sulfate crystal $Gd_2(SO_4)_3{\cdot}8H_2O$ from 0.8 T. In a further experiment shortly afterwards, they obtained a temperature of 0.25 K by precooling to 1.5 K. Almost simultaneously, *De Haas*, *Wiersma* and *Kramers* demagnetized CeF_3 starting from 2.76 T and 1.3 K and reached 0.27 K [592].

For a long time, the adiabatic demagnetization of electron spins, or more general magnetic moments of atoms, was the only technique available to produce temperatures well below 1 K, and was therefore widely used. After the development of powerful dilution refrigerators, however, the importance of this cooling technique has diminished and such cryostats are not commonly used in low-temperature laboratories. The main reason for this is that a dilution refrigerator can be operated continuously. In the last few years, however, the interest in cooling by demagnetization of electron spins has revived, because for certain applications, such as the cooling of special satellite-based instruments, this technique has advantages. In addition, new cryogenic developments have generated the expectation that in the near future demagnetization cryostats, reaching 10 mK can be operated without liquid nitrogen or helium. The basic idea is to use closed-cycle systems (see Sect. 11.2) for precooling that can reach temperatures of below 4 K. Further cooling can then be obtained in a multistage demagnetization process. Today, it is already possible to reach 60 mK in a two-stage process starting from 4 K. In particular, for low-temperature applications in industry, cryostats without liquid coolants are highly desirable.[5]

In Fig. 11.32, the reduced entropy S/R of four different paramagnetic salts is plotted. As mentioned before, for applications for which a very low minimum temperature is not so important, but the availability of a large cooling capacity is needed, the magnetic field will not be reduced completely during the demagnetization. In addition, high cooling capacities result from the use of materials with high spin entropy. For such experiments the paramagnetic salts MAS and FAA in particular are suitable.[6] However, as shown

[5] One should add here that there are also developments under way with the aim of a closed-cycle dilution refrigerator system, that also could be handled without cryogenic liquids

[6] The following abbreviations are used: MAS for $MnSO_4{\cdot}(NH_4)_2SO_4{\cdot}6H_2O$, FAA for $Fe_2(SO_4)_3 \cdot (NH_4)_2SO_4{\cdot}24H_2O$, CPA for $Cr_2(SO_4)_3{\cdot}K_2SO_4{\cdot}24H_2O$, and CMN for $2Ce(NO_3)_3 \cdot 3Mg(NO_3)_2 \cdot 24H_2O$

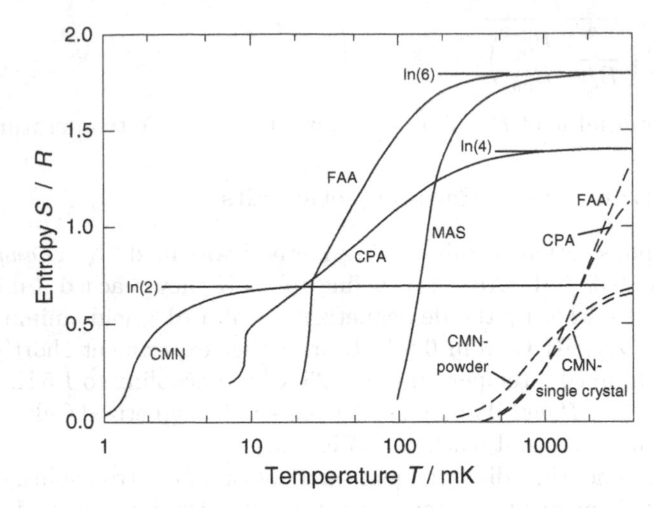

Fig. 11.32. Reduced entropy S/R of ferric ammonium alum (FAA), manganous ammonium alum (MAS), chromic potassium alum (CPA) und cerium magnesium nitrate (CMN) as a function of temperature in zero magnetic field (*solid lines*) and for a field of 2 T (*dashed lines*) [593–596]

in Fig. 11.32, these systems not only have high spin entropies, they also have high transition temperatures to ordered states.

If the priority is to obtain as low a temperature as possible, CMN is very often the choice as cooling medium. As mentioned before, CMN has a ordering temperature of only $T_c = 1.9$ mK, which corresponds to an internal field of $B_{\text{int}} = 4$ mT. Starting from temperatures well below 0.5 K, a minimum temperature of 2 mK has been achieved by several groups using CMN. By diluting the cerium concentration, the internal field can even be reduced, thus, in principle, allowing lower minimum temperatures to be obtained. For example, it should be possible to reach temperatures as low as 0.7 mK by substituting 90% of the cerium ions by lanthanum ions, which are nonmagnetic. Unfortunately, at the same time the entropy of the salt and thus the cooling power are reduced correspondingly. Therefore, it is perhaps not surprising that to date sub-mK temperatures have not been obtained by the demagnetization of electron spin systems.

A serious problem in connection with the demagnetization of paramagnetic salts is the difficulty of establishing sufficient thermal contact between the cooling medium and the experiment. In order to reduce the thermal resistance, the salt crystals are typically grown directly onto wire bundles consisting of a large number of thin copper or gold wires. In some cases, as many as 50 000 copper wires have been used to provide good thermal contact between the salt pill and the experimental stage.

11.7 Nuclear Spin Demagnetization

The possibility of using nuclear magnetic moments for cooling was pointed out in 1934 by *Gorter* [597] and independently in 1935 by *Kurti* and *Simon* [594]. The first successful experimental demonstration was achieved in 1956 by Kurti and his coworkers at Oxford. Using a 3 T magnet for the demagnetization they were able to reduce the spin temperature of copper nuclei from 12 mK to about 1 μK [598]. However, the temperature of the other degrees of freedom in the sample, phonons and conduction electrons, remained unchanged, since the precooling system was in close thermal contact with the cooling medium during the demagnetization. No heat switch was used in the first experiments. Therefore, the spin temperature increased to the starting temperature within minutes.

This indicates a central problem of the cooling via nuclear spin demagnetization, namely the transfer of energy to the ultracold spins from phonons and conduction electrons. In order to achieve this transfer, the system to be cooled must be thermally isolated for a sufficiently long time to allow the nuclear spins and the other degrees of freedom to come into equilibrium. This is difficult to achieve because of the extremely long thermal relaxation times of nuclear spins at very low temperatures. For adiabatic nuclear spin demagnetization only metals are a suitable host medium because the relaxation times of nuclei in insulators are of the order of weeks at 1 mK. The long relaxation times require a large reduction of the heat input into the cooling system. To obtain temperatures below 100 μK the heat leak should typically be smaller than 10^{-9} W. This figure is difficult to achieve in a complex apparatus such as a nuclear demagnetization cryostat. For comparison, we note that the power dissipation of a typical quartz wrist watch is about 2×10^{-6} W and is therefore more than 2000 times higher than the tolerable heat leak into a nuclear demagnetization cryostat.

11.7.1 Coupling of Nuclear Spins and Conduction Electrons

In order to describe the temporal development of the nuclear magnetization M after a perturbation, we may use the relaxation approximation introduced in Sect. 7.2. Therefore, we write

$$\frac{\mathrm{d}M}{\mathrm{d}t} = -\frac{(M - M_0)}{\tau}, \tag{11.29}$$

where M_0 denotes the equilibrium magnetization and τ the relaxation time that is determined by the hyperfine interaction between the conduction electrons at the Fermi surface and the nuclear spins. For this process, the expression

$$\tau = \frac{2\kappa k_\mathrm{B}}{g_\mathrm{n}\mu_\mathrm{n}B} \tanh\left(\frac{g_\mathrm{n}\mu_\mathrm{n}B}{2k_\mathrm{B}T_\mathrm{e}}\right) \tag{11.30}$$

is found, where g_n and μ_n represent the nuclear g-factor and the magnetic moment of the nuclei. In Sect. 8.4 we have already introduced the Korringa constant κ and the temperature T_e of the conduction electrons. The well-known *Korringa relation* (8.42), namely $\tau = \kappa / T_e$, follows from (11.30) in the limiting case $k_B T_e \gg g_n \mu_n B$. Using the Curie law $M \propto 1/T$, we can transform (11.29) into an equivalent expression for T_e and T_n, the temperatures of the conduction electrons and the nuclear spin system. We obtain

$$\frac{dT_n^{-1}}{dt} = -\frac{(T_n^{-1} - T_e^{-1})}{\tau_1} . \tag{11.31}$$

Using the Korringa relation (8.42), this equation reads

$$\dot{T}_n = \frac{(T_e - T_n)\, T_n}{\kappa} . \tag{11.32}$$

The thermalization of the conduction electrons with the nuclear spins leads to the heat flow

$$\dot{Q}_n = C_e \dot{T}_e = -C_n \dot{T}_n , \tag{11.33}$$

where C_e and C_n represent the heat capacities of the conduction electrons and the nuclear spins, respectively. The magnitude of the latter is determined by the applied magnetic field. Inserting (11.32) into (11.33) one obtains, for the temporal development of the electron temperature, the relation

$$\dot{T}_e = -\Delta T \frac{T_n C_n}{\kappa C_e} . \tag{11.34}$$

Here, $\Delta T = (T_e - T_n)$ denotes the difference between the temperature of the conduction electrons and the nuclear spin system. The rate $\Delta \dot{T}$ at which T_n and T_e equilibrate can be calculated using (11.32) and (11.34). For an ideal adiabatic demagnetization process (no external heat input) one finds:

$$\Delta \dot{T} = -\frac{T_n}{\kappa} \left(1 + \frac{C_n}{C_e} \right) \Delta T . \tag{11.35}$$

This leads to the interesting result that the temperature difference between conduction electrons and nuclear spins equalizes with the effective time constant

$$\tau_{\text{eff}} = \tau \frac{C_e}{C_n + C_e} . \tag{11.36}$$

Since, in general, the heat capacity of the spins is much larger than that of the conduction electrons, the effective relaxation time τ_{eff} is considerably shorter than τ. Therefore, the temperature of the nuclei stays nearly unchanged, whereas the temperature of the conduction electrons varies rapidly towards the nuclear spin temperature.

11.7.2 Influence of Heat Leaks

So far, we have neglected the influence of heat leaks in our discussion. In the calculation of the optimal value to which the magnetic field should be reduced during the demagnetization, one should include any heat leaks to the cooling medium. Incoming heat is first absorbed by the electron system and therefore increases the temperature difference ΔT between T_e and T_n. In order to calculate ΔT we consider the power $\dot{Q}_\mathrm{n} = C_\mathrm{n}\dot{T}_\mathrm{n}$ of the nuclear spins available for cooling the conduction electrons. Using the high-temperature approximation for the specific heat of the nuclear spins $C_\mathrm{n} = nI(I+1)\mu_\mathrm{n}^2 g_\mathrm{n}^2 B_\mathrm{f}^2/(3k_\mathrm{B}T^2)$ in the final field B_f together with (11.32), one finds

$$\dot{Q}_\mathrm{n} = \frac{n g_\mathrm{n}^2 I(I+1)\mu_\mathrm{n}^2 B_\mathrm{f}^2}{3k_\mathrm{B}\kappa T_\mathrm{n}}\,\Delta T\,. \tag{11.37}$$

Assuming that the heat input $\dot{Q}$ is equivalent to the cooling power $\dot{Q}_\mathrm{n}$ provided by the nuclear spins, (11.32) and (11.37) lead to the ratio of the temperature of conduction electrons and nuclear spins given by

$$\frac{T_\mathrm{e}}{T_\mathrm{n}} = 1 + \frac{3k_\mathrm{B}\kappa\dot{Q}}{n g_\mathrm{n}^2 I(I+1)\mu_\mathrm{n}^2 B_\mathrm{f}^2}\,. \tag{11.38}$$

The optimum magnetic field $B_\mathrm{f,opt}$, at which the demagnetization process should be stopped to obtain the minimum electron temperature can be calculated from the condition $\mathrm{d}T_\mathrm{e}/\mathrm{d}B_\mathrm{f} = 0$ using (11.28) and (11.38), and leads to

$$B_\mathrm{f,opt} = \sqrt{\frac{3k_\mathrm{B}\kappa\dot{Q}}{n g_\mathrm{n}^2 I(I+1)\mu_\mathrm{n}^2}}\,. \tag{11.39}$$

Here, we have assumed $B_\mathrm{f,opt} \gg B_\mathrm{int}$, a condition that is usually fulfilled in the demagnetization of nuclear spins. Inserting this result into (11.38) one finally obtains a minimal electron temperature $T_\mathrm{e,min}$ after the demagnetization that is twice the minimum temperature $T_\mathrm{n,min}$ of the nuclear spins:

$$T_\mathrm{e,min} = 2B_\mathrm{f,opt}\frac{T_\mathrm{i}}{B_\mathrm{i}} = 2T_\mathrm{n,min}\,. \tag{11.40}$$

11.7.3 Sources of Heat Leaks

The presence of heat leaks determines not only the minimum attainable temperature, but also the time in which the nuclear spin system warms up, or in other words, the time in which experiments at ultralow temperatures can be performed. The heat leaks of the best nuclear demagnetization cryostats are of the order of a few 10^{-10} W. At this level, there are many sources for such unwanted heat leaks. In the following, we will discuss a few important examples.

Heat leaks arise from the radiation of warm parts in the cryostat. By using a radiation shield around the nuclear stage that is thermally coupled to the mixing chamber, this contribution can be reduced significantly. The radiation shield also has the advantage that any residual gas in the vacuum chamber, evaporated from warm sections, condenses preferentially not at the nuclear stages but onto the radiation shield. A further important source of heat is high-frequency electromagnetic radiation that is picked up by leads outside the cryostat and enters the cold part of the system inside via leads that are connected to thermometers and heaters or some other components of the experiment. The situation can be improved by operating the cryostat in a shielded room or by filtering each line going into the cryostat. In addition, vibrations can generate significant amounts of heat in nuclear stages. Because of this, nuclear spin demagnetization cryostats are often mounted on heavy plates that have a support system with vibration isolation and damping, such as air cushions. In addition, the pumping tubes should have mechanical filters that isolate the cryostat from the vibrations generated by the pumps. In practice, solutions to the vibration problem often require great experimental efforts. Furthermore, the heat input by cosmic rays (approximately $10^{-11}\,\mathrm{W\,kg^{-1}}$) cannot be neglected completely. Additional heat leaks may arise from the construction of the cryostat itself, such as the heat flow through an open heat switch.

In the following, we discuss in somewhat greater detail several important additional heat leaks beyond the ones mentioned above.

Eddy-Current Heating

A heat input that always occurs in demagnetization cryostats is produced by eddy currents. Such a contribution can arise, for example, from mechanical vibrations of the nuclear stages in the presence of a constant magnetic field, or from changes of the applied magnetic field during the demagnetization process. If the field is uniformly reduced at rate $\dot{B}$, the heat produced by eddy-current heating in a conductor with volume V and resistivity ϱ is given by

$$\dot{Q}_{\mathrm{eddy}} = f\frac{V\dot{B}^2}{\varrho} \,. \tag{11.41}$$

Here, f denotes a geometry-dependent factor. For a cylinder with radius r oriented parallel to a magnetic field, this factor is $f = r^2/8$. For example, a copper cylinder with radius $r = 1\,\mathrm{mm}$ and resistivity $\varrho = 2\,\mathrm{n\Omega\,cm}$, corresponding to a resistance ratio of $\varrho(300\,\mathrm{K})/\varrho(4\,\mathrm{K}) \approx 1000$, one finds with $\dot{B} = 0.5\,\mathrm{T\,h^{-1}}$, a heat input of $\dot{Q}_{\mathrm{eddy}}/V \approx 1\,\mathrm{nW\,mol^{-1}}$. To minimize the eddy-current heating, the nuclear stage is slitted, and the reduction of the magnetic field is done very slowly. However, one has to take into account that too slow a demagnetization would also not be advantageous because of other heat leaks. In practice, rates of $\dot{B} \approx 1\,\mathrm{T\,h^{-1}}$ have led to satisfactory results.

Time-Dependent Heat Leaks

If all heat leaks previously discussed were sufficiently reduced, time-varying heat leaks can remain a limiting factor. These originate in the material of the nuclear stage itself or in pieces that are attached to the nuclear stage. Figure 11.33 shows the heat input into a nuclear stage as a function of time after cooling from 4 K. In the same figure, the minimum temperature reached by this cryostat is also plotted as a function of time. One clearly sees that the heat input and thus the minimum temperature continuously decreases over 40 days.

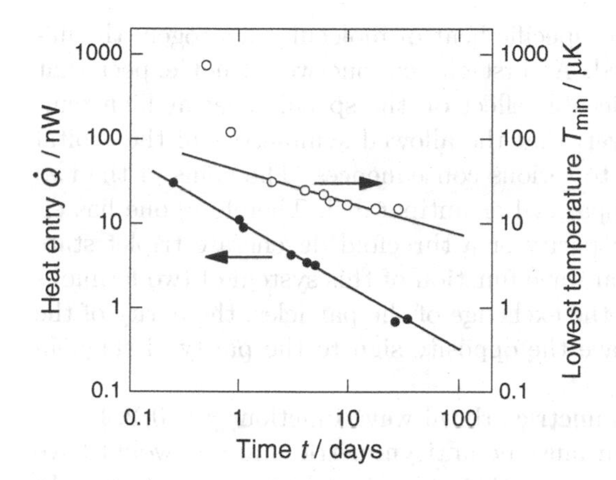

Fig. 11.33. Heat input into a nuclear stage of a demagnetization cryostat and minimum temperature as a function of time after cooling down from 4 K [599]

Such time-varying heat leaks are caused by systems that are not in thermal equilibrium after cooling. They slowly relax, and in this process they generate heat. Systems that can cause such an effect are inclusions of molecular hydrogen, tunneling defects, and radioactive impurities. In the following, we will discuss the first two of these systems in more detail, because of their interesting physical background.

Ortho–Para Conversion of H_2

Due to the production process, hydrogen is found in many metals with concentrations typically of the order of 10 to 100 ppm. In copper, for example, only a very small amount of this hydrogen is incorporated in the lattice as atomic hydrogen. Almost all the hydrogen is present in molecular form in small bubbles with a typical diameter of 100 nm. To understand the origin of the time-dependent heat leak caused by hydrogen we first discuss the specific heat of gaseous hydrogen.

Besides translational degrees of freedom, molecular gases also have rotational degrees of freedom. In a classical picture, we can view such a molecule as a dumbbell with the moment of inertia $\mathcal{I}$ rotating about its center of mass. Quantum mechanically the solution of the Schrödinger equation for this problem provides eigenfunctions for the angular momentum with the eigenvalues

$$E = \frac{j(j+1)\hbar^2}{2\mathcal{I}} . \tag{11.42}$$

The eigenfunctions for even values j are even and for odd values of j are odd. Taking into account the degeneracy $(2j+1)$, the partition function becomes

$$Z = \sum_{j=0}^{\infty} (2j+1)\, \mathrm{e}^{-E/k_\mathrm{B}T} . \tag{11.43}$$

For a full description of the specific heat of molecular hydrogen, the nuclear spin has to be considered. At first glance, one would not expect that the nuclear spin causes a noticeable effect on the specific heat at high temperatures. It turns out, however, that the allowed symmetries of the orbital and spin wave functions lead to serious consequences. The spins of the two protons can either be aligned parallel or antiparallel. Therefore, one has either a singlet state with even parity or a threefold degenerate triplet state with odd parity. Since the total wave function of this system of two fermions must be antisymmetric under the exchange of the particles, the parity of the orbital wave function must have the opposite sign to the parity of the spin wave function.

Therefore, if we have a symmetric orbital wave function ($j = 0, 2, 4, \ldots$), the nuclear spin wave function must be antisymmetric and one would have a nondegenerate state. Hydrogen in this state is called *para-hydrogen*. In contrast, if the orbital wave function is antisymmetric ($j = 1, 3, 5, \ldots$), the nuclear spin wave function has to be symmetric. This threefold degenerate state of hydrogen is called *ortho-hydrogen*. The energetically lowest state of ortho-hydrogen is that of $j = 1$. In contrast, para-hydrogen can occupy the ground state with $j = 0$. The energy difference between the lowest states of ortho- and para-hydrogen is $\Delta E/k_\mathrm{B} = 172\,\mathrm{K}$.

The specific heat contribution of the rotational states taking into account the nuclear spins was first derived by *Hund* in 1927 [600]. The result was

$$C_V = \frac{\mathrm{d}}{\mathrm{d}T}\left[RT^2 \frac{\mathrm{d}}{\mathrm{d}T} \ln(Z_\mathrm{e} + 3Z_\mathrm{u}) \right] . \tag{11.44}$$

The partition function is split into Z_e and Z_u that correspond to the sums over terms with even and odd j, respectively. At high temperature ($k_\mathrm{B}T \gg \Delta E$) one finds three times more molecules with an antisymmetric orbital wave function than with a symmetric orbital function, in accordance with the degeneracy of the two configurations.

Surprisingly, at low temperatures the experimental results differ from the behavior predicted by (11.44)! The reason for this discrepancy lies in the fact that between these different states hardly any transitions occur. Collisions between the hydrogen molecules have very little influence on the concentration of ortho- and para-hydrogen because the conversion demands a change of the nuclear spin state. However, this process is forbidden in first order and requires an interaction of the magnetic moments with each other. Because of the smallness of the magnetic moments this interaction is rather weak and leads to a slow self-conversion rate of only 1.9% per hour. Therefore, the system can be considered as a metastable mixture of two different substances.

Since, at room temperature, one has $Z_e \approx Z_u$, the concentration ratio of ortho-to para-hydrogen is roughly 3:1. For the total specific heat of the 'two independent' substances we would therefore expect

$$C_V = \frac{1}{4}\frac{\mathrm{d}}{\mathrm{d}T}\left(RT^2\frac{\mathrm{d}}{\mathrm{d}T}\ln Z_e\right) + \frac{3}{4}\frac{\mathrm{d}}{\mathrm{d}T}\left(RT^2\frac{\mathrm{d}}{\mathrm{d}T}\ln Z_u\right). \quad (11.45)$$

This equation is in very good agreement with the measured values. The curve, which reflects the temperature dependence of the specific heat for such a mixture, is labelled 'normal mixture' in Fig. 11.34. In thermal equilibrium at 20 K, the hydrogen is 99% para-hydrogen. The maximum of the curve for the equilibrium mixture results from the temperature-dependent change of the ortho-para ratio, which means from the conversion of one into the other type of hydrogen.

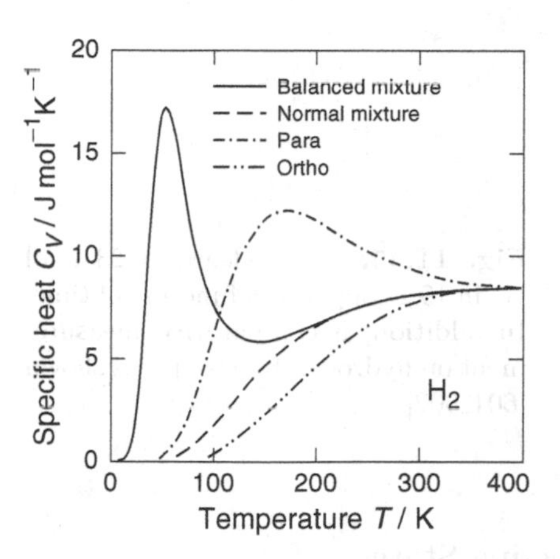

Fig. 11.34. Specific heat of pure ortho- and pure para-hydrogen and of mixtures of ortho- and para-hydrogen corresponding to the composition at room temperature (*dashed line*) and corresponding to thermal equilibrium (*solid line*) as a function of temperature. The different concentrations can be obtained using catalytic agents such as charcoal or certain paramagnetic salts. After [258]

The molecular hydrogen in the material of a nuclear stage is solid at low temperatures. In this case, the rotational degrees of freedom of the gas correspond to librational degrees of freedom in the solid. In the process of cooling, the ortho-para ratio remains nearly unchanged. If the system stays cold for a sufficiently long time, the nonequilibrium concentration of ortho-hydrogen slowly transforms to para-hydrogen. This is an exothermic reaction during which the relatively large amount of heat $Q_R \approx 1.06\,\mathrm{kJ\,mol^{-1}}$ is released. The self-conversion by pairwise collisions of ortho-hydrogen is described by the rate equation $\mathrm{d}c/\mathrm{d}t = -kc^2$, where c is the concentration of the ortho-molecules and k the conversion rate of 1.9% per hour. Thus, the time dependence of the ortho-hydrogen concentrations is given by

$$c(t) = \frac{c_0}{1 + c_0 kt}, \tag{11.46}$$

with c_0 representing the starting concentration of ortho-hydrogen. Finally, the heat release per mole resulting from the conversion is

$$\dot{Q} = -Q_\mathrm{R}\frac{\mathrm{d}c}{\mathrm{d}t} = Q_\mathrm{R}\frac{kc_0}{(1 - c_0 kt)^2}. \tag{11.47}$$

Figure 11.35 shows the heat release of 23 μmol H_2 in 19 g Cu at $T < 100$ mK as a function of time after cooling below 4 K. The heat release continuously decreases within 130 h from 50 nW to 10 nW. The *solid line* corresponds to the expected behavior (11.47). For comparison, the result of a measurement of the heat release of a copper sample without hydrogen is also shown in Fig. 11.35. Here, a time-independent background heat input of 0.1 nW is observed.

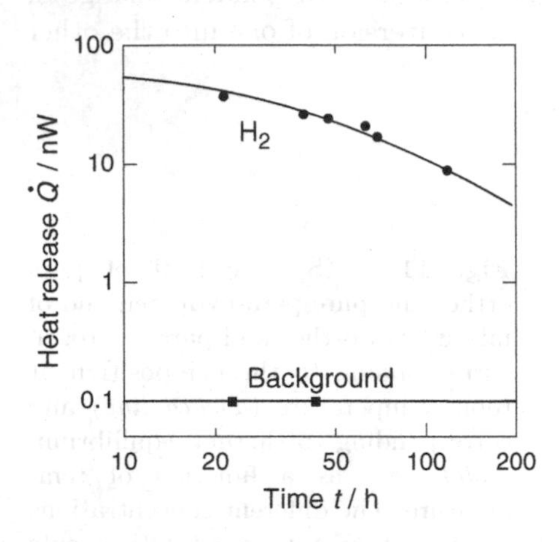

Fig. 11.35. Heat release of 23 μmol H_2 in 19 g copper as a function of time. In addition, a comparative measurement on hydrogen-free copper is shown [601, 602]

Heat Release due to Atomic Tunneling States

As we have seen in Sect. 9.5, tunneling systems in disordered materials have a broad distribution of relaxation times. For our discussion here, it is important that in this distribution, systems exist that have very long relaxation times, and may reach thermal equilibrium on a timescale of many hours or even days and weeks. The heat release of such systems in amorphous materials can be observed experimentally over days.

Using the tunneling model (see Sect. 9.5) the heat release can be calculated and (9.69) is found. For a sample that has been cooled from an initial temperature T_1 to a final temperature T_0, the heat release is given by

$$\dot{Q} = \frac{\pi^2 k_B^2}{24} P_0 \left(T_1^2 - T_0^2\right) \frac{1}{t}, \tag{11.48}$$

where the constant P_0 was defined in (9.59). The time dependence $\dot{Q} \propto 1/t$ is in good agreement with the experimental results on vitreous silica (Suprasil W) (see Fig. 9.34) and for PMMA (Plexiglass). For other materials shown in Fig. 11.36, one finds more or less pronounced deviations from the predicted behavior, the detailed origin of which is unknown to date.

These results imply that in the design of a demagnetization cryostat, amorphous materials should not be used, at least directly attached to the nuclear stage. However, this kind of heat leak has also been observed in annealed metals, such as those used for the construction of the nuclear stage itself. For example, the heat release shown in Fig. 11.33 is due to relaxation processes within the material of the nuclear stage. Although the origin of these kinds of heat release is not fully understood, it seems clear that it is related to structural relaxation processes associated with grain boundaries or other lattice defects of these polycrystalline materials.

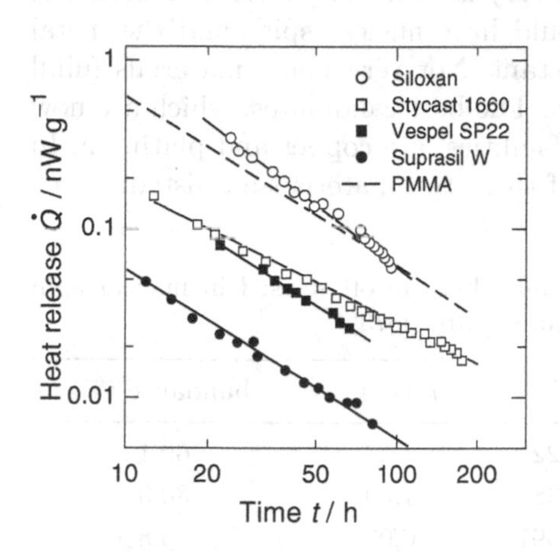

Fig. 11.36. Heat release of various disordered materials after cooling them down below 1 K [603]

11.7.4 Technical Features

In the design of a nuclear spin demagnetization facility, one has to take into account that the magnetic moments of nuclei are typically a factor of 1000 smaller than those of the localized electron spins in paramagnetic salts. This means that in order to obtain the same reduction of entropy as is achievable with a paramagnetic salt, a roughly 1000 times higher value of the ratio B/T is required. Using superconducting magnets it is relatively simple to obtain fields of the order of a few Tesla. For copper at 10 mK, a field of 6 T would lead to a 5% reduction of the entropy. However, an isothermal magnetization at 10 mK is very difficult to realize, because the heat of magnetization would be too high for the cooling power of a dilution refrigerator. Because of this, the field is applied at a temperature between 0.1 and 1 K and the nuclear stage is then cooled down in this field to 10 mK, the starting temperature of the demagnetization process. In the following, we briefly discuss the question of which materials are suitable for nuclear spin demagnetization and will make a few remarks about the design of heat switches. Finally, we take a look at the actual cooling process and the minimum temperatures achieved with this technique.

Cooling Media

Selecting a material that is suitable for constructing the nuclear stage of a demagnetization refrigerator requires the consideration of many characteristics. It should be a normal conducting metal with a low Korringa constant, high thermal conductivity, low ordering temperature for the nuclear spins, and no electronic magnetic moments. In addition, the material should be machineable and it should be obtainable with a very low impurity level. Furthermore, a large proportion of the isotopes should have nuclear spins and the metal should have a high nuclear Curie constant. Not very many materials fulfill all these criteria to the required extent. The best candidates, which are now used in nuclear spin demagnetization facilities, are copper and platinum. In Table 11.2, a few relevant properties of these two materials are listed.

Table 11.2. Various properties of materials that are often used in nuclear spin demagnetization cryostats as cooling medium. After [573]

	Structure	I	μ/μ_N	κ (K s)	Abundance (%)
^{63}Cu	fcc	3/2	2.22	1.27	69.1
^{65}Cu	fcc	3/2	2.38	1.09	30.9
^{195}Pt	fcc	1/2	0.597	0.03	33.8
$^{141}PrNi_5$	fcc	5/2	4.28	<0.001	100

In addition, the parameters of the system $PrNi_5$ are listed in Table 11.2, because it is an interesting choice for work in the temperature range between 0.5 mK and 5 mK. The intermetallic compound $PrNi_5$ belongs to the class of so-called *Van-Vleck paramagnets*, and has a rather large cooling capacity. The distinctive feature of these materials is the fact that in applied magnetic fields, an induced atomic magnetic moment enhances the hyperfine field. At low temperatures, the Pr^{3+} ions in the hexagonal $PrNi_5$ have a temperature-independent magnetic susceptibility, since the $4f$-electrons in this material are in a singlet ground state, which is nonmagnetic. In magnetic fields, the configuration of the f-electrons is changed by the admixture of exited states.

In this way, an atomic magnetic moment is induced that causes an internal hyperfine field B_{hyp} that is *stronger* than the external field B_{ext}. The average enhancement factor is $B_{hyp}/B_{ext} \approx 11.2$. With the help of this effect, the nuclear spin entropy can be reduced significantly even at relatively high temperatures or modest magnetic fields. As an example, we note that at 25 mK and 6 T the reduction of the entropy for $PrNi_5$ is $\Delta S/S \approx 0.7$. Of course, the internal field in such systems is quite high, so that the minimum temperature obtained with $PrNi_5$ is 'only' 0.19 mK, which is roughly two orders of magnitude higher than the lowest temperatures that have been reached in experiments with a copper/platinum double-stage system [604]. A further disadvantage of $PrNi_5$ is the fact that this material is very brittle and difficult to make thermal contact with. Because of these reasons, $PrNi_5$ is not widely used and most nuclear spin demagnetization facilities consist of nuclear stages made of copper. As we shall see, in some cases a first copper stage is used for precooling, with a second stage made of platinum or copper.

Superconducting Heat Switch

As mentioned above, thermal isolation between the nuclear stage and the precool system is needed after magnetization. This requires a heat switch. In most nuclear spin demagnetization facilities, aluminum switches are used that can be switched on and off with a modest magnetic field. The thermal conductivity of aluminum in the two states is shown in Fig. 11.37. In this figure, the thermal conductivity of copper and of the glue 'Epibond 121', which is often used in low-temperatures laboratories, are also plotted for comparison.

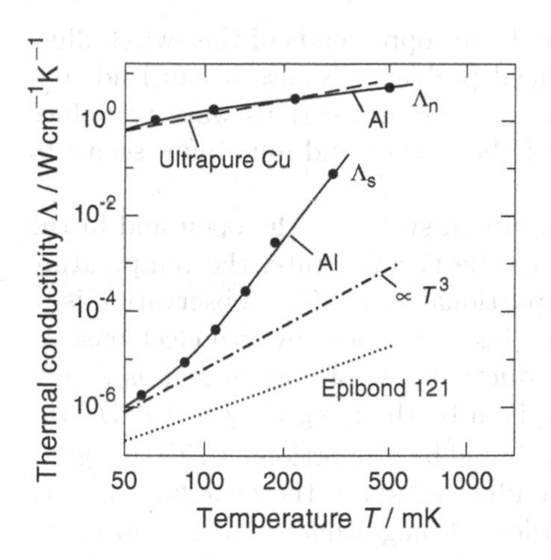

Fig. 11.37. Comparison of the thermal conductivity of aluminum in the normal (Λ_n) and in the superconducting state (Λ_s) as a function of temperature. In addition, the conductivity of pure copper (*dashed line*) and of the Epibond 121 (*dotted line*) are plotted. The *dashed-dotted line* reflects the T^3 dependence [605, 606]

The construction of a superconducting heat switch is schematically shown in Fig. 11.38a. The coil for generating the magnetic field is mounted at the mixing chamber. The aluminum is arranged in such a way that at least part

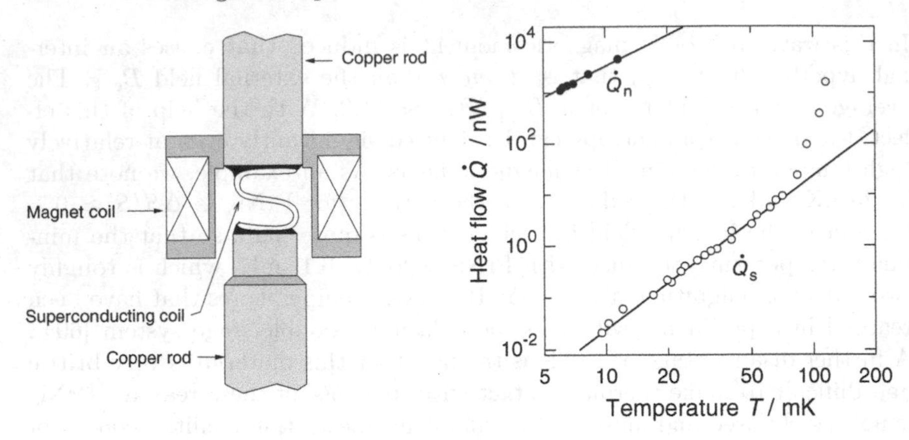

Fig. 11.38. (**a**) Schematic sketch of the design of a superconducting heat switch. After [573]. (**b**) Heat flow through an aluminum heat switch in the normal and in the superconducting states as a function of temperature [313]

of it is parallel to the magnetic field. This is done to ensure that magnetic flux that is trapped in the superconductor after switching the field off, does not form a continuous 'normal' path through the aluminum and thus lead to a short in the thermal heat switch. In addition, the geometry is chosen in such a way that eddy-current heating during the field change is minimized. In general, thin foils are used that have the additional advantage that the thermal conductivity associated with the lattice is also reduced, since at low temperatures the mean free path of the phonons is limited by the dimensions of the sample (see Sect. 6.2).

The joint between the aluminum and the copper ends of the switch illustrated in Fig. 11.38a presents a technical problem because it can lead to a large thermal resistance at the interface. There are various ways to reduce the contact resistance. Cold welding of the copper and aluminum seems to give the best results.

The heat flow through such an aluminum switch in the open and in the closed state is shown in Fig. 11.38b. In the normal state, the temperature dependence of the heat flow $\dot{Q}_\mathrm{n}$ is proportional to T^2. This observation is in agreement with the expectation for the heat transport by free electrons. As discussed in Sect. 7.3, the thermal conductivity Λ_n of normal metals is proportional to T. Since the heat flow is given by the integral $\dot{Q} \propto \int \Lambda \,\mathrm{d}T$, the heat flow $\dot{Q}_\mathrm{n}$ through the closed switch should be proportional to T^2, in agreement with observation. In the superconducting state, the heat flow at very low temperatures $T \ll T_\mathrm{c}$ is several orders of magnitude smaller than in the normal state, since in this case the heat is only transported by phonons. From measurements of the thermal conductivity of perfect dielectric crystals in the Casimir regime (see Sect. 6.2), it is known that in this case $\Lambda \propto T^3$ leading to $\dot{Q}_\mathrm{s} \propto T^4$. Surprisingly, for the heat flow across the switch $\dot{Q}_\mathrm{s} \propto T^3$ is found.

Obviously, the phonons are not predominantly scattered at the surface of the aluminum foil and other scattering mechanisms must be responsible for the heat resistance. It is interesting to note that a similar temperature dependence has also been observed for massive aluminum samples. Those studies indicate that the scattering of the phonons by dislocations is the dominant process in this temperature range. Therefore, it seems that phonon scattering by dislocations limits the heat flow through the open switch.

Cooling Process

In Fig. 11.39, a typical example of the temperature evolution in a nuclear spin demagnetization experiment is shown. In this experiment, a double-stage system was used, consisting of a first stage, made of $PrNi_5$ and a second stage made of copper. With the help of the first stage, the second stage was precooled in a magnetic field of 8 T to a starting temperature of 3.8 mK. Subsequently, the heat switch between the two stages was opened and the field was reduced exponentially with time from 8 T to 0.01 T over 10 h. Four days after the demagnetization of the second stage, the minimum temperature of 41 μK was reached. Subsequently, the nuclear stages started to slowly warm up. Overall, the copper stage (nuclei and conduction electrons) remained below 50 μK for more than ten days. After 13 days, the system was heated for a few hours with a power of 1 nW.

The construction of the low-temperature part of a double-stage nuclear spin demagnetization apparatus of the low-temperature group at the University of Bayreuth is shown in Fig. 11.40. The first stage consists of 17 kg of copper. First, of a massive copper block with the desired form was cast in a graphite crucible. Subsequently, 36 slits with a width of 0.4 mm were cut into the block between the two end flanges in order to reduce eddy-current heating during demagnetization. Before installing the copper block, it was carefully annealed for several weeks. During this process, large crystallites

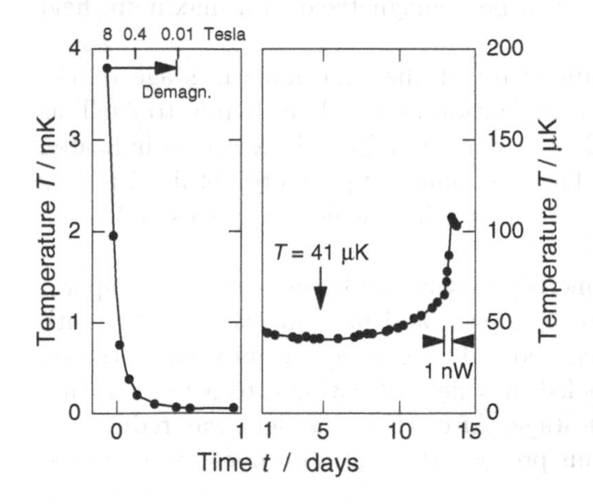

Fig. 11.39. Temperature of a copper nuclear stage during and after a demagnetization cycle. The actual magnetic field at the nuclear stage during the demagnetization process is indicated by the scale on top of the *left part* of the figure [607]

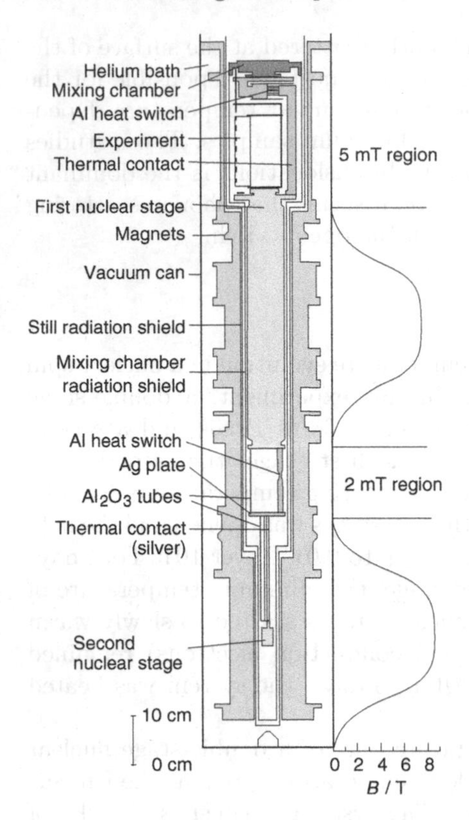

Fig. 11.40. (**a**) Sketch of the cross section of the low-temperature part of a nuclear spin demagnetization cryostat of the University of Bayreuth. In addition, the field distribution is shown on the *right side*. After [599]. (**b**) Photo of the first nuclear stage [599]

with diameters of up to 2 cm were formed as can be seen in Fig. 11.40b. The total length of this stage is 52.5 cm. According to the field distribution, 6.6 kg of copper in the center of the stage are effectively magnetized in a field of 8 T. Above and below this stage there are regions in which the magnetic field is compensated and in which experiments can be installed. The second stage consists of 130 g of copper, which can be demagnetized in a maximum field of 9 T.

Figure 11.41 shows the temperature of the first nuclear stage of the Bayreuth cryostat during a demagnetization from 8 T at 10 mK to 4 mT as a function of the magnetic field. The corresponding timescale is indicated on the *upper axis* of this figure. The minimum temperature obtained in this demagnetization run was 15 μK, measured in the field-compensated 5 mT region (see Fig. 11.40).

The lowest temperature, or more precisely the lowest electron temperature, ever obtained was produced by *Pobell* and his coworkers at Bayreuth in 1996 with the facility described above during experiments with platinum. The platinum sample was precooled in a field of 0.37 mT to a temperature of 100 μK using the first nuclear stage. After that, the field was reduced to a value just below 0.05 mT. In this process, the platinum nuclei were cooled

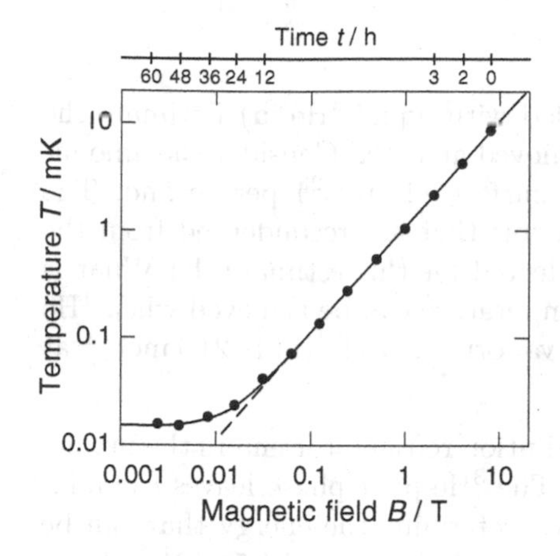

Fig. 11.41. Temperature of the first copper stage of the Bayreuth cryostat during a demagnetization from 8 T to 4 mT [599]

to 0.3 μK. In this very low field, the electrons were significantly warmer than the nuclei. Figure 11.42 shows the ac susceptibility of the platinum sample as a function of temperature.[7] At temperatures of 2 mK a maximum is observed that is caused by iron impurities (see Sect. 8.3). The rise below 10 μK with decreasing temperature is due to the contribution of the nuclear spins.

In a further experiment, the demagnetization was stopped at 2.5 mT. At this field, the nuclei had a temperature of 0.8 μK. The temperature of the electron system was estimated to be at most 2 μK, using the known heat leak of only 12 pW and the measured Korringa constant of 7 mK s.

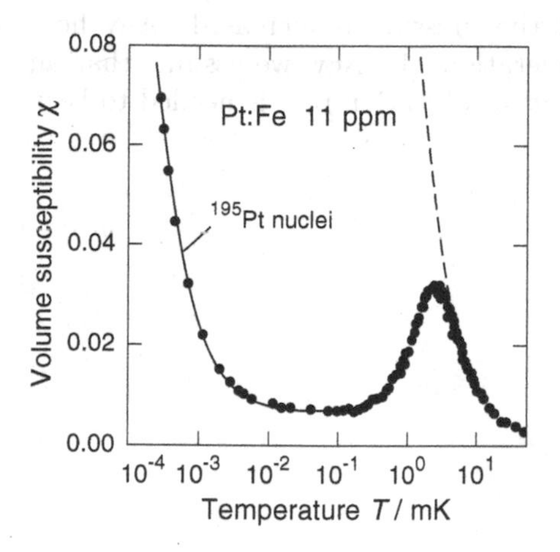

Fig. 11.42. Ac susceptibility of a platinum sample with 10 ppm iron impurities as a function of temperature. The *solid line* corresponds to the expected Curie dependence of the nuclear susceptibility, the *dashed line* reflects the Curie dependence of the electronic susceptibility [604]

[7] At very low temperatures, i.e., at $T < 10\,\mu$K, the temperature scale was determined by the temperature of the nuclear spin system.

Exercises

11.1 An evaporation cryostat is half filled with liquid ^{4}He. **a)** Estimate the maximum cooling power that can be achieved at 1.2 K. Consider the amount of heat that is carried away from the surface (100 cm^2) per second. The latent heat of ^{4}He is 80 J mol^{-1}. The atoms that are recondensed from the gas phase into the liquid should be neglected for this estimate. **b)** What is the cooling power at 0.8 K? **c)** What temperature can be achieved when ^{4}He is replaced by ^{3}He? The latent heat of vaporization of ^{3}He is 21 J mol^{-1} at that temperature.

11.2 The ^{3}He-rich phase of a ^{3}He/^{4}He-dilution refrigerator enters the mixing chamber with a temperature of 40 mK. The ^{3}He-poor phase leaves the mixing chamber at a temperature of 25 mK. Determine the energy that can be absorbed by the mixing chamber at a ^{3}He circulation rate of 1.5×10^4 mol s^{-1}.

11.3 Estimate the Kapitza resistance at the interface of a metal with sound velocity 1000 m s^{-1} and density 5 g cm^{-3} and liquid ^{4}He at 4.2 K and at 1 K.

11.4 At a temperature T an ensemble of N noninteracting magnetic dipoles with magnetic moment μ is located in a magnetic field B. Reducing the magnetic field to the value B' results in a partial demagnetization of the ensemble, which can be used to cool the environment by the amount of heat ΔQ. Express ΔQ as a function of μ, B, B', T, and N.

11.5 A Pomeranchuk cell is operated in the temperature range where the slope of the melting curve is negative. It is half filled with liquid and half with solid ^{3}He. (**a**) What happens if the pressure is increased? Any heat leak should be neglected in this consideration. (**b**) Now we assume that an external heat leak of 1 µW exists. What solidification rate is needed to keep the temperature at 10 mK?

12 Thermometry

Temperature is a thermodynamic property of state that can be defined, for example, via a reversible cycle. The equation $\oint T^{-1}\mathrm{d}Q = 0$ holds for a reversible process, which means that no change of the entropy occurs. In considering a *Carnot cycle*, *Lord Kelvin* showed in 1854 that for systems in thermal equilibrium there exists a lowest temperature, i.e., an absolute zero of temperature [608, 609]. The temperature scale named after Lord Kelvin starts at $T_0 = 0\,\mathrm{K}$ and has a value of 273.16 K at the triple point of water. It is related to the temperatures scale proposed by *Celsius* in 1742 [610] by the relation $T_0 = -273.15°\mathrm{C}$.

Although the definition of temperature via a reversible cycle is theoretically satisfying, an experimental determination by this process is rather difficult. In practice, one uses reproducible fixed points, in order to calibrate other thermometers. Between these fixed points, the temperature scale can be described via polynomial functions or tables obtained in precise measurements. After many years of work, performed in different countries, in 1989 the International Temperatures Scale ITS-90 was accepted by the *Comité International des Poids et Messures* as the official temperature scale down to 0.65 K. At low temperatures, this scale is based mainly on the vapor pressures of ^{3}He and ^{4}He, and on phase transitions in pure materials that occur in this temperature range. An extension of this scale using the ^{3}He melting curve has been worked out recently. The so-called PLTS-2000 (Provisional Low Temperature Scale 2000) was accepted in October 2000 by the international committee and extends the temperature scale down to 0.9 mK [611].

There are a surprisingly large number of possible ways to measure low temperatures. However, in practice, the determination of the temperature is often as involved as the process of obtaining the low temperatures. We shall discuss in this chapter a number of possibilities to determine the temperature. We shall distinguish between primary and secondary thermometers, although there is no sharp definition for the distinction and this classification is somewhat arbitrary.

12.1 Primary Thermometers

Primary thermometers are generally defined as devices that can be used to measure the temperature without any prior calibration. This is possible if the temperature dependence of the relevant property is described by a fundamental law of physics, or if a reproducible measurement of the corresponding quantity is possible independent of the special setup used in the measurement. For example, this is the case for the melting curve of ^{3}He. In contrast, the temperature curve of *secondary thermometers* is obtained by an individual calibration to other thermometers. In this case, the calibration is only valid for each individual thermometer.

12.1.1 Gas Thermometers

For a gas thermometer, the fixed relation of the thermodynamic variables of state, pressure p, volume V and temperature T of gases are used for the determination of the temperature [612]. For an ideal gas, in which there are no interactions between the molecules, and the molecules occupy a negligible volume compared to the volume of the container, the ideal-gas law

$$pV = \nu RT \,, \tag{12.1}$$

applies, with ν denoting the number of moles of the gas and R representing the universal gas constant. In the limiting case of sufficiently high dilution or sufficiently high temperature, all gases approach the ideal-gas state. The relation (12.1) is valid independent of the nature of the gas. The use of the material-independent ideal-gas law provides a fundamental method for obtaining the absolute temperature by measuring pressure and volume. The practical realization, however, always involves real gases at finite pressures, which means that for high-precision measurements the deviations from the ideal-gas law have to be taken into account. In practice, one has to consider the viral coefficients of the gas.

A gas thermometer can be operated either at constant pressure or at constant volume. The precision of both variants can be spoiled by a large number of systematic errors. Dead volumes in the apparatus, thermal or elastic changes of the volume and absorption or desorption of gas molecules at the walls of the container are examples of the many effects that can cause systematic errors. High-precision thermometry with gases is therefore involved and is usually only used in metrology laboratories for calibration or for improving existing temperature scales.

12.1.2 Vapor-Pressure Thermometers

With a vapor-pressure thermometer the temperature is deduced from the equilibrium pressure of the vapor above a (cold) liquid [612, 613]. The choice of the liquid depends on the temperature range in which the thermometer is to be operated. For measuring low temperatures, ^{3}He, ^{4}He, and H_2 are suitable in different ranges.

Although the relation between temperature and pressure for real gases cannot be calculated theoretically with accuracy, this type of thermometer is classified as a primary thermometer, because the vapor pressure can be measured in a reproducible way and the temperature can be assigned without prior calibration using published tables. This means that it is unnecessary to calibrate each individual vapor-pressure thermometer separately. However, it is essential to measure the absolute pressure with a precise pressure gauge.

Possible experimental arrangements for vapor-pressure thermometers are shown in Fig. 12.1. A general problem connected with this kind of measurement is the occurrence of unwanted temperature gradients in the liquid, i.e., the temperature at the surface of the liquid can be different from the temperature at the position of the experiment. Because of this, the setup shown to the *left* in Fig. 12.1 is, in general, not a good choice. A better solution is the arrangement shown to the *right* that avoids, to a large extent, the problems arising from temperature gradients in the liquid. In this case, the vapor pressure is measured in a small container, a *vapor-pressure cell*, in which pure gas is condensed and that is in direct thermal contact with the actual experiment. A tube, thermally isolated from the liquid since it passes through a vacuum tube, connects the vapor-pressure cell to the pressure gauge outside the cryostat. However, problems can also occur in this configuration because of the so-called *thermomolecular pressure difference* resulting from thermal creep if the vapor pressures are small or if the tube connecting the pressure gauge to the cell is too narrow [614]. Such problems are eliminated if the pressure gauge can be operated at low temperatures. In this case, the pressure in the cell can be measured directly. Special capacitive pressure gauges have been developed that can be operated at low temperatures [615, 616].

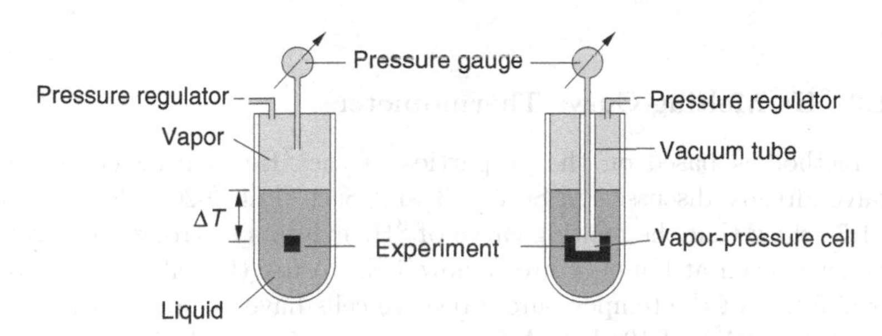

Fig. 12.1. Schematic illustration of two different realizations of a vapor-pressure thermometer

As mentioned above, the temperature that corresponds to a certain vapor pressure can be taken from tables or can be calculated using a polynomial expression of the form

$$T = \sum_{i=0}^{9} A_i \left[\frac{\ln(p) - B}{C} \right]^i , \qquad (12.2)$$

for which the coefficients A_i and the constants B and C have been obtained in previous calibrations. In Table 12.1, the official values of the coefficients and constants for ^{3}He and ^{4}He are listed according to the International Temperature Scale ITS-90. Inserting in (12.2) the vapor pressure in units of 10^{-5} bar, one obtains the temperature in K.

Table 12.1. The coefficients A_i and the constants B and C of (12.2) for the vapor pressure curves of ^{3}He and ^{4}He in different temperature ranges according to the ITS-90 [617–619]

	^{3}He (0.65–3.2 K)	^{4}He (1.25–2.1768 K)	^{4}He (2.1768–5.0 K)
A_0	1.053477	1.392408	3.146631
A_1	0.980106	0.527153	1.357655
A_2	0.676380	0.166756	0.413923
A_3	0.372692	0.050988	0.091159
A_4	0.151656	0.026514	0.016349
A_5	−0.002263	0.001975	0.001826
A_6	0.006595	−0.017976	−0.004325
A_7	0.088966	0.005409	−0.004973
A_8	−0.006596	0.013259	0
A_9	−0.054943	0	0
B	7.3	5.6	10.3
C	4.3	2.9	1.9

12.1.3 ^{3}He Melting-Curve Thermometer

This method is based on the properties of the ^{3}He melting curve, which we have already discussed in Sect. 1.3 and Sect. 11.5 [620–622]. A look at Fig. 1.7a show that the melting curve of ^{3}He exhibits a strong temperatures dependence even at temperatures below 1 K. To use this effect for a precise determination of the temperature, pressure cells have been constructed providing a resolution of 10 μbar. An example of a high-resolution pressure cell is shown schematically in Fig. 12.2. In this case, the pressure is determined capacitively. An elastic membrane, forming the top to the inner part of the pressure cell is connected with the electrode of a parallel-plate capacitor. The separation of the capacitor plates depends on the displacement of the diaphragm and hence on pressure.

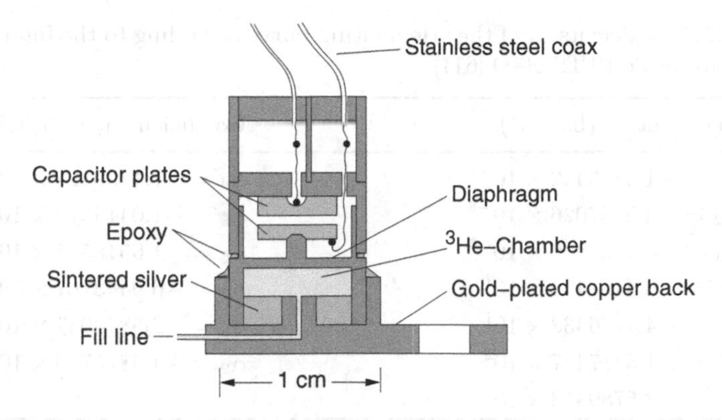

Fig. 12.2. Sketch of a pressure cell used as a ^{3}He melting-curve thermometer

Similar to the vapor-pressure thermometer, the ^{3}He melting-curve thermometer also belongs to the primary thermometer category. No precise analytical description exists for the pressure–temperature relationship of the ^{3}He melting curve. But the measurement is reproducible, independent of the specific experimental setup , and therefore a calibration obtained in one particular pressure cell can also be used for other pressure cells.

Because of its very high reproducibility and accuracy, the ^{3}He melting-curve thermometer is an essential element of the new International Temperature Scale PLTS-2000. However, its high sensitivity to magnetic fields and the very large heat capacity of ^{3}He is a disadvantage. During the last fifteen years several laboratories world-wide have worked on an as precise as possible determination of the ^{3}He melting curve. Despite some considerable discrepancies between the calibration curves obtained at ultralow temperatures ($T < 10$ mK) by the different groups, in 2000 an agreement was finally reached for the new Provisional Low Temperature Scale PLTS-2000 for the range between 0.9 mK and 1 K. In this range, the ^{3}He melting curve is represented by a polynomial of the form

$$p = \sum_{i=-3}^{9} \alpha_i T^i \,. \tag{12.3}$$

The coefficients α_i corresponding to the PLTS-2000 scale are listed in Table 12.2. In the range between 1 K and 0.5 K, the absolute accuracy of this scale is roughly ± 0.5 mK. Between 0.5 K and 100 mK, the absolute accuracy increases linearly to about ± 0.2 mK. At the lowest temperature of 900 μK, the absolute error is ± 18 μK, which means that the relative accuracy at this temperature is about 2%.

In addition to the polynomial function (12.3) four fixed points can be used for thermometry with a ^{3}He melting-curve thermometer. These are the minimum of the melting curve, the transitions into the superfluid phases A and B and the transition into an antiferromagnetic order of the nuclear spins

Table 12.2. Coefficients α_i of the ^{3}He melting curve according to the International Temperature Scale PLTS-2000 [611]

Coefficient α_i (bar/K^i)	Coefficient α_i (bar/K^i)
$\alpha_{-3} = -1.3855442 \times 10^{-11}$	$\alpha_4 = 7.1499125 \times 10^2$
$\alpha_{-2} = 4.5557026 \times 10^{-8}$	$\alpha_5 = -1.0414379 \times 10^3$
$\alpha_{-1} = -6.4430869 \times 10^{-5}$	$\alpha_6 = 1.0518538 \times 10^3$
$\alpha_0 = 3.4467434 \times 10^1$	$\alpha_7 = -6.9443767 \times 10^2$
$\alpha_1 = -4.4176438 \times 10^1$	$\alpha_8 = 2.6833087 \times 10^2$
$\alpha_2 = 1.5417437 \times 10^2$	$\alpha_9 = -4.5875709 \times 10^1$
$\alpha_3 = -3.5789853 \times 10^2$	

Table 12.3. Fixed points of the Provisional Low Temperature Scale PLTS-2000 [611]

Fixed Points	Pressure p (bar)	Temperature T (mK)
T_{min}	29.3113	315.24
T_{A}	34.3407	2.444
T_{B}	34.3609	1.896
T_{N}	34.3934	0.902

in solid ^{3}He. In Table 12.3, the temperature and pressure values of these fixed points are listed.

12.1.4 Noise Thermometers

From a theoretical point of view, this type of thermometer is particularly attractive, since only fundamental thermodynamic relations are necessary to calculate the magnitude of the thermodynamic fluctuations. Noise thermometers are based on the statistical thermal motion (Brownian motion) of the conduction electrons in resistors. Independently, in 1927 *Johnson* and *Nyquist* investigated experimentally and theoretically the voltage fluctuations across a resistor [623, 624].

According to the *Nyquist theorem,* the mean square of the voltage fluctuations is given by

$$\langle V_{\mathrm{N}}^2 \rangle = 4 k_{\mathrm{B}} T R \, \Delta f \, . \tag{12.4}$$

Here, R denotes the resistance and Δf the bandwidth of the measurement. This expression is an approximation and is only valid if the upper limit of the bandwidth in the measurement fulfills the condition $\Delta f_{\mathrm{max}} \ll k_{\mathrm{B}} T / h$.

In practice, this is not a serious limitation, since in most cases the bandwidths that can be realized are much smaller. For example, at 1 K one finds $k_B T/h \approx 20\,\text{GHz}$.

Figure 12.3 shows the mean square voltage for different conductors as a function of the resistance. As expected from (12.4) a linear relation between the resistance and the mean square voltage is found, independent of the type of the conductor.

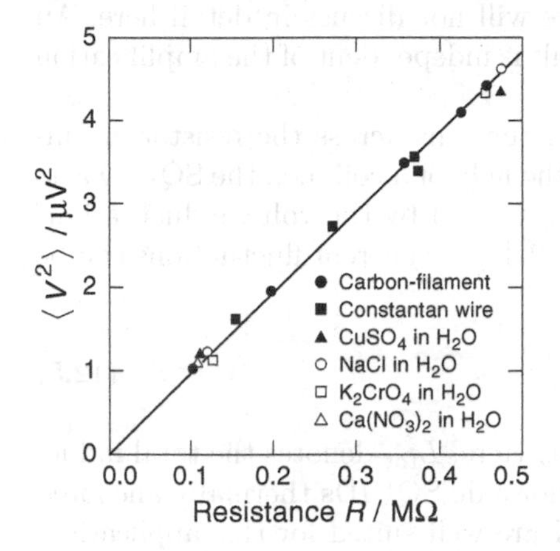

Fig. 12.3. Mean square voltage of different conductors as a function of the resistance [623]

To determine the temperature, the quantities $\langle V_N^2 \rangle$, Δf, and R must be known [620,622]. The measurement of each of these quantities has its specific demands and attendant problems. For example, the voltage fluctuations V_N are extremely small and must be amplified. In doing this, the amplifier noise is superimposed on the thermodynamic fluctuations. In addition, the amplification factor must be known precisely. A further problem is the accurate determination of the bandwidth, because it strongly depends on the specific filter characteristic of the setup.[1] Therefore, absolute measurements have a typical error margin of a few per cent. Because of this, a noise thermometer is usually calibrated against a second thermometer, at least at one temperature, to obtain the amplification factor and the bandwidth.

To get a feeling for the difficulty of measuring voltage fluctuations we consider a specific example: Assuming a resistance of 10 kΩ, a bandwidth of $\Delta f = 10^5\,\text{Hz}$, and a temperature of 1 K, the voltage fluctuations are of the order of $\langle V_N^2 \rangle^{1/2} \approx 2 \times 10^{-7}\,\text{V}$. The corresponding power is only $P \approx 10^{-18}\,\text{W}$. Precise measurements of these very small signals require extremely good amplifiers. In many cases, SQUIDs (see Sect. 10.4) are used for noise thermometry, since conventional semiconductor amplifiers do not fulfill the required criteria.

[1] With the use of modern analogue-to-digital converters and FFT this problem has become much less significant.

There are several types of noise thermometers based on SQUID amplifiers. One method is to connect the resistor in parallel with a Josephson contact, that serves as a precise voltage–frequency converter making use of the ac Josephson effect (see, for example, [625]). In this case, the transfer function contains only fundamental constants. A voltage V across the Josephson contact results in a frequency of $f_{\mathrm{J}} = (2e/h)V$. Therefore, for the determination of the voltage fluctuations, only a frequency counter is necessary. However, the amplitude of the oscillations is extremely small and special techniques are required for the measurement, which we will not discuss in detail here. An advantage of this method is that the result is independent of the amplification factor.

In another technique, the voltage fluctuations across the resistor are inductively coupled into the SQUID with the help of a coil, i.e., the SQUID acts as a low-noise preamplifier of the current caused by the voltage fluctuations (see, for example, [626]). Analogous to (12.4), the current fluctuations can be expressed by

$$\langle I_{\mathrm{N}}^2 \rangle = \frac{4k_{\mathrm{B}}T}{R} \left[\frac{1}{1+(2\pi f \tau)^2} \right] \Delta f \,, \tag{12.5}$$

with $\tau = L_{\mathrm{tot}}/R$ being the time constant. Here, L_{tot} denotes the total inductance of the input circuit. Modern low-noise dc SQUIDs thermally anchored at a fixed temperature of 4.2 K or 1.2 K are well suited for this application. A sketch of a recent realization of such a current-sensing noise thermometer is shown in Fig. 12.4a.

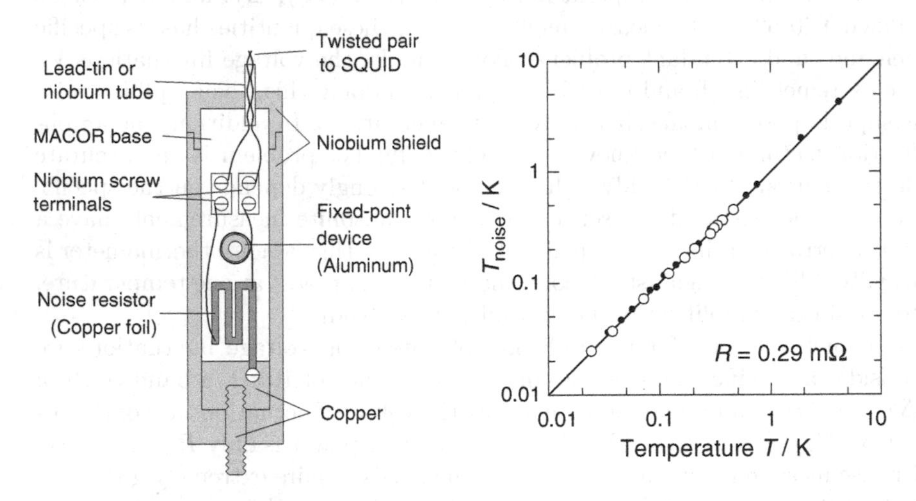

Fig. 12.4. (**a**)Schematic diagram of a current-sensing noise thermometer (after [626]). A superconducting aluminum fixed-point thermometer is incorporated for a one-point calibration. (**b**) Temperature derived from the noise of a 29-mΩ resistor plotted versus the temperature measured with a ^{3}He melting-curve thermometer (*open circles*) and with a resistive thermometer (*closed circles*) [626]

At very low temperatures, a potential problem arises from the occurrence of hot electrons because of the weak electron–phonon coupling. To avoid this phenomenon, the noise resistor is grounded at the base of the copper holder. In this way, the electrons in the noise resistor are in direct thermal contact with the free electrons of the copper support. The temperatures derived from the noise of a 29-mΩ resistor are plotted in Fig. 12.4b versus the temperature measured by other thermometers. Obviously, very good agreement is found in the whole temperature range between 24 mK and 4.2 K.

Very recently, a novel noncontact noise thermometer has been developed [627]. In this thermometer, the magnetic Johnson noise from a solid gold post was measured inductively using a low-noise dc SQUID operated at 1.2 K. The main advantage of this technique is that no leads are in contact with the conductor that is used for noise thermometry. A schematic of the setup is shown in Fig. 12.5a. The pickup loop was arranged as a first-order gradiometer to suppress possible fluctuations of external magnetic fields that couple simultaneously into both loops.

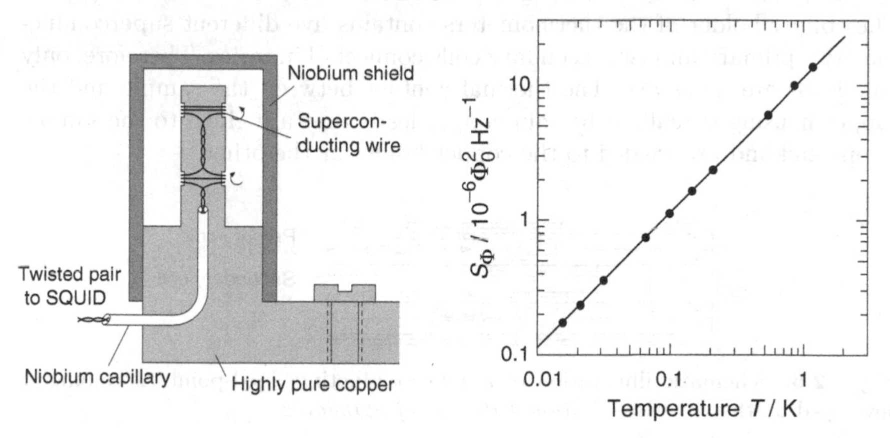

Fig. 12.5. (**a**) Sketch of an inductively coupled noise thermometer. (**b**) Spectral density of the flux noise caused by the magnetic Johnson noise of a gold post plotted against the temperatures of a superconducting fixed-point device SDR 1000 [627]

At sufficiently low frequencies $\omega \ll \omega_{\mathrm{ro}}$, the power density of the flux noise through the cross section of a cylindrical sample with radius r and conductivity σ is given by [627]

$$S_\Phi = 4k_{\mathrm{B}}T\mu_0^2\sigma G r^3, \tag{12.6}$$

where G is a geometry-dependent numerical constant, which is of the order of 0.1 for a single turn pickup loop tightly wound around the conductor. Depending on the geometry of the setup, the so-called roll-off frequency ω_{ro} can be determined either by the radius of the sample and its conductivity via the skin depth, or by the total inductance of the superconducting flux transformer, which forms an RL low-pass filter with the real part of the complex impedance of the pickup coil. As shown in Fig. 12.5b, successful operation of this device was demonstrated by investigating the flux noise at temperatures predestined by the calibrated superconducting fixed-point thermometer SDR1000 (see following section). As expected, the power density of the flux noise varies proportional to temperature.

12.1.5 Superconducting Fixed-Point Thermometers

This type of thermometer is based on the precise measurement of the transition temperature of superconductors [573, 622]. To avoid contact problems and mechanical tension due to electrical leads, the transition temperature is deduced from measurements of the magnetic susceptibility making use of the Meissner effect (see Sect. 10.1). To determine the transition temperature magnetically the mutual inductance of a coil containing the sample is measured.

The general construction of fixed-point thermometers is shown in Fig. 12.6. The copper holder of the thermometers contains five different superconductors, two primary and two secondary coils connected in series. Therefore, only four leads are necessary. The thermal contact between the sample and the copper housing is realized by thin copper leads that are glued to the sample at one end and are welded to the copper holder at the other.

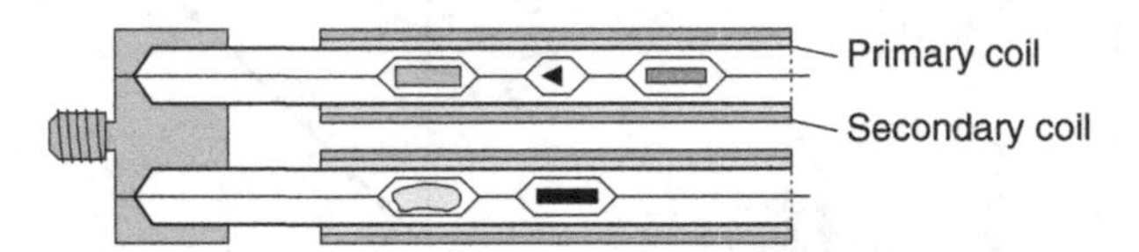

Fig. 12.6. Schematic illustration of a superconducting fixed-point thermometer developed by the American *National Bureau of Standards*

Two different fixed-point thermometers were offered by the the American *National Bureau of Standards* (NBS, today NIST) in the mid-1980s [628,629]. They are still in use in various low-temperature laboratories. One thermometer, named SRM 767, covers the temperature range from 0.5 K to 7.2 K. The other, SRM 768, is suitable for lower temperatures. It has five fixed points between 15 mK and 207 mK. The superconducting materials and the transition temperatures of both NBS devices are listed in Table 12.4.

The realization and the use of such thermometers is not trivial, because certain external influences can significantly effect the temperature at which the transition occurs. For example, the influence of magnetic fields on the superconducting transition is rather strong (see Sect. 10.1). Figure 12.7 shows the result of a measurement of the change of the magnetization of beryllium in the vicinity of the superconducting transition in different magnetic fields. In general, the transition is shifted towards lower temperatures in magnetic

Table 12.4. Materials and transition temperatures used for the SRM 767 and SRM 768 fixed-point thermometers. The transition temperatures quoted for SRM 768 are mean values [630]

SRM 767:	Substance	T_c (K)	SRM 768:	Substance	T_c (mK)
	Cd	0.519		W	15.6
	Zn	0.851		Be	22.7
	Al	1.1796		$Ir_{0.8}Ru_{0.2}$	99.2
	In	3.4145		$AuAl_2$	159.8
	Pb	7.1999		$AuIn_2$	204.0

fields. In finite fields, it also depends on whether the measurement is performed on cooling or warming. The hysteresis effect grows with increasing magnetic field. Although the magnetic fields used in the measurements depicted in Fig. 12.7 were lower than the Earth's field, the influence on the transition temperature is quite strong.

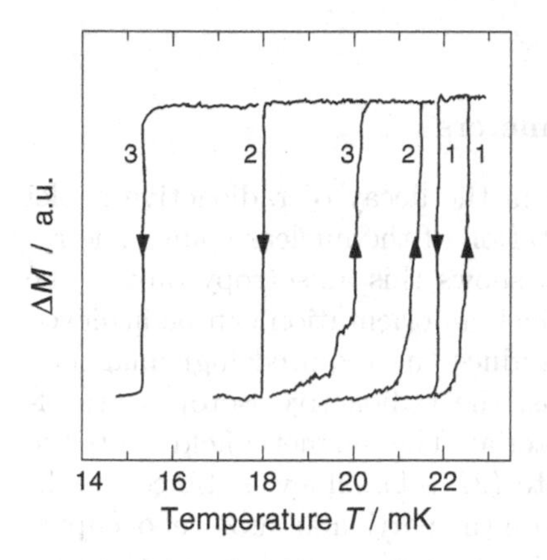

Fig. 12.7. Influence of magnetic fields on the superconducting transition of beryllium. The *arrows* indicate the direction in which the temperature was changed in this experiment. The magnetic fields were: *curve (1)* 0.5 μT, *curve (2)* 9.5 μT and *curve (3)* 19 μT [628, 629]

This means that superconducting fixed-point thermometers have to be very well shielded from magnetic fields or that the field at the sample has to be accurately compensated. However, for compensating the magnetic field one needs three pairs of Helmholtz coils, because all three spatial components have to be compensated individually. The NBS recommended a reduction of the magnetic fields to below 1 μT at the sample.

Recently, a collaboration of Dutch physicists has developed a new superconducting fixed-point thermometer with ten fixed points in the range between 15 mK and 1.1 K [631, 632]. Prototypes of these new devices have been made available to several low-temperature laboratories for testing. This new superconducting fixed-point thermometer is commercially obtainable under the trade name SRD 1000. Figure 12.8 shows the voltage output of the mutual inductance bridge used to measure the temperature dependence of the magnetic susceptibility of the SDR 1000. The ten steps associated with the superconducting transitions are clearly visible.

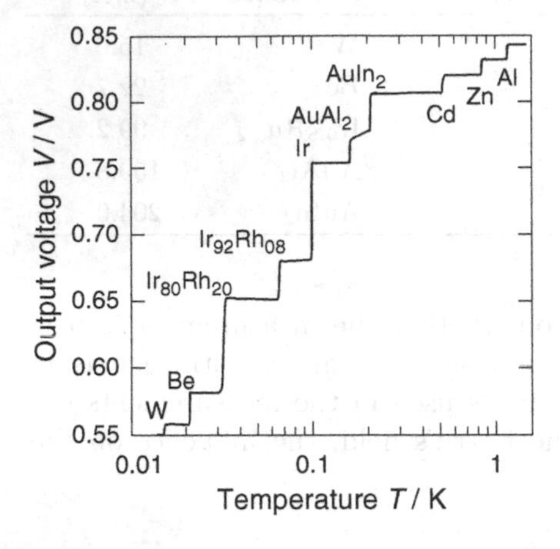

Fig. 12.8. Output signal of a fixed-point thermometer SRD 1000 as a function of temperature [633]

12.1.6 Nuclear-Orientation Thermometers

In many cases, the emission of γ-quanta in the decay of radioactive nuclei is not isotropic but depends on the orientation of the nuclear spins. The radiation emitted by an ensemble of nuclei shows this anisotropy only if the nuclei exhibit a preferential orientation. Such an orientation can be achieved by strong magnetic fields. In order to produce the required high magnetic field for the polarization of nuclear spins, the radioactive isotopes are often imbedded in a ferromagnetic host material. The magnetic field splits the degenerate levels of nuclei with spin I into $(2I+1)$ sublevels. Since the direction of an emitted γ-quantum depends on the particular sublevel occupied by the spin, it is possible to determine the thermal occupation of the sublevels, and thus the temperature via a measurement of the anisotropy of the γ-emission [634, 635].

The temperature and angular dependence of the total emission intensity is described by the function

$$W(T,\Theta) = 1 + \sum_{k=2,4,\ldots}^{k_{\max}} G_k\, U_k\, F_k\, B_k(T)\, P_k(\cos\Theta)\,, \tag{12.7}$$

where Θ denotes the angle between the preferred nuclear orientation and the direction of observation. The sum only runs over even values of k, since γ-quanta carry an angular momentum $\hbar$ and the transition conserves parity. The upper limit $k_{\max}$ is given by $2I$ or $2L$, whichever is smaller, where L represents the multipole order of the emitted radiation. The factors G_k reflect the geometry of the experiment and can be close to unity. They are determined by the properties of the detector, the size of the radioactive source and the finite angle of the measurement. The coefficients U_k depend on the characteristics of the decay chain prior to the transition of interest. For all systems that are considered for thermometry, these coefficients are known and can be looked up in published tables. The quantities F_k denote the angular momentum coupling coefficients for the actual nuclear transition. These coefficients are known and can be obtained from the literature. The functions $P_k(\cos\Theta)$ represent the Legendre polynomial expansion. The temperature dependence of W is reflected by factors B_k, given by

$$B_k = \frac{(2k)!}{(k!)^2} I^k \sqrt{\frac{(2I+1)(2k+1)(2I-k)!}{(2I+k+1)!}}\, f_k(I)\,, \tag{12.8}$$

which describe the thermal occupation of the levels. The first two terms of the coefficients f_k are given by:

$$f_2 = -\frac{I+1}{3I} + \frac{1}{I^2} \sum_{m=-I}^{I} m^2 \mathcal{P}(m)\,,$$

$$f_4 = -\frac{(6I^2+6I-5)}{7\,I^2} f_2 - \frac{(3I^2+3I-1)(I+1)}{15\,I^3} + \frac{1}{I^4} \sum_{m=-I}^{I} m^4 \mathcal{P}(m)\,,$$

where $\mathcal{P}(m) = Z^{-1}\exp(-E_m/k_\mathrm{B}T)$ represents the occupation of the m-th level, with Z being the partition function. The temperature dependence of $\mathcal{P}(m)$ is the basis for the determination of the absolute temperature.

Of course, not all radioactive nuclei are equally suitable for nuclear orientation thermometry. In most cases, the radioactive nuclei ^{60}Co or ^{54}Mn are used. These two isotopes are particularly attractive since all relevant parameters are well known and the lifetimes of the intermediate states are short enough not to be influenced by spin-flip processes. In both cases, the γ-quanta originate from a pure electric quadropole transition (E2) of the nuclei. This means that the multipole order is $L = 2$, and k_max is 4 for both isotopes, although the spins of ^{54}Mn and ^{60}Co nuclei have the values $I = 3$ and $I = 5$, respectively.

In the following, we consider ^{60}Co in a single crystal of ^{59}Co as an example of nuclear orientation thermometry. Figure 12.9 shows the slightly simplified decay scheme of ^{60}Co. The cobalt isotope ^{60}Co has a half-life of 5.26 years and decays via a β^- process to ^{60}Ni. The initially excited Ni nucleus goes into the ground state by emitting two γ-quanta of energies 1.17 MeV and 1.33 MeV. The magnetic hyperfine interaction in the cobalt crystal results in eleven equidistant sublevels with an energy splitting of $\Delta E/k_\mathrm{B} \approx 6$ mK.

As already mentioned, all parameters for the determination of the temperature and angular dependence of the γ-emission of ^{60}Co are known, so that one can calculate the theoretical curves quantitatively. Figure 12.10a shows the expected angular distribution of the γ-emission for different values

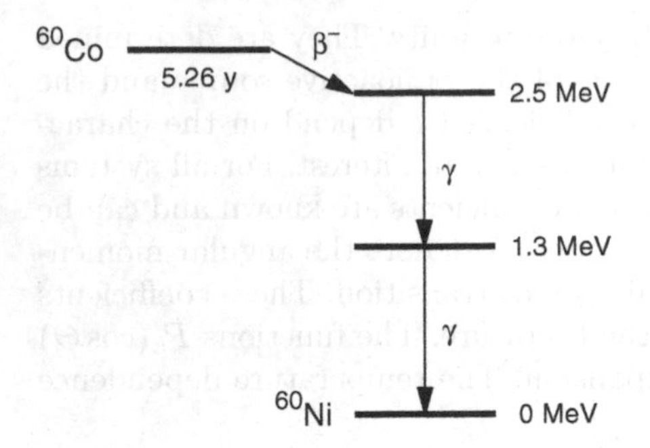

Fig. 12.9. Simplified decay scheme of ^{60}Co

of $k_{\mathrm{B}}T/\Delta E$, where ΔE represents the splitting of the magnetic sublevels. At high temperatures ($k_{\mathrm{B}}T \gg \Delta E$) the radiation is isotropic. With decreasing temperature it becomes more anisotropic. As $T \to 0$, one finds the ideal angular dependence of a pure electric quadropole transition.

Figure 12.10b shows the temperature dependence of $W(T, \Theta)$ for ^{60}Co in a cobalt single crystal. The largest total change is observed for the γ-emission in the direction of the nuclear spin orientation. Whereas for this direction the signal decreases with decreasing temperature, it increases for $\Theta = 90°$. The maximum variation with temperature for $\Theta = 90°$ is at roughly 6 mK.

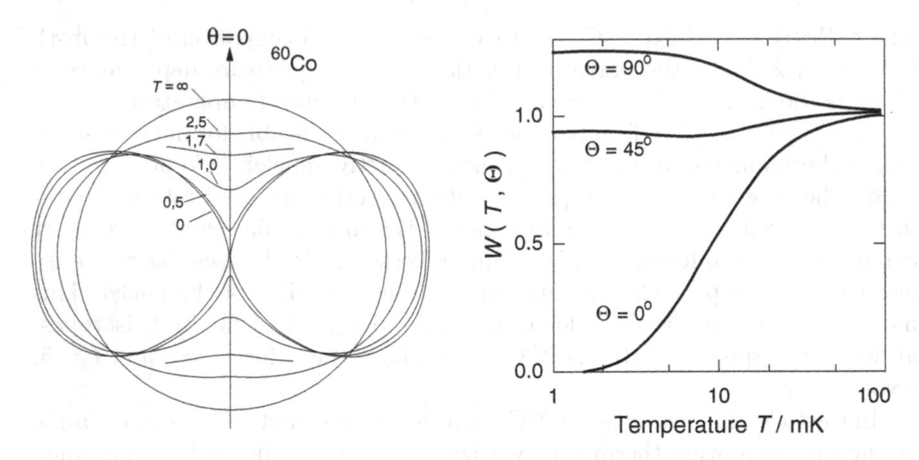

Fig. 12.10. (**a**) Temperature dependence of the angular distribution for the γ-emission of ^{60}Co in a cobalt single crystal for different values of $k_{\mathrm{B}}T/\Delta E$. (**b**) Emission probability $W(T, \Theta)$ as a function of temperature for three different angles Θ. The curves have been calculated for ^{60}Co nuclei in a cobalt single crystal

In practice, one often uses cobalt single crystals that are doped with small amounts of ^{60}Co. Using a suitable shape, for example a needle-like shape, oriented parallel to the c-axis of the crystal, one can omit the use of an exter-

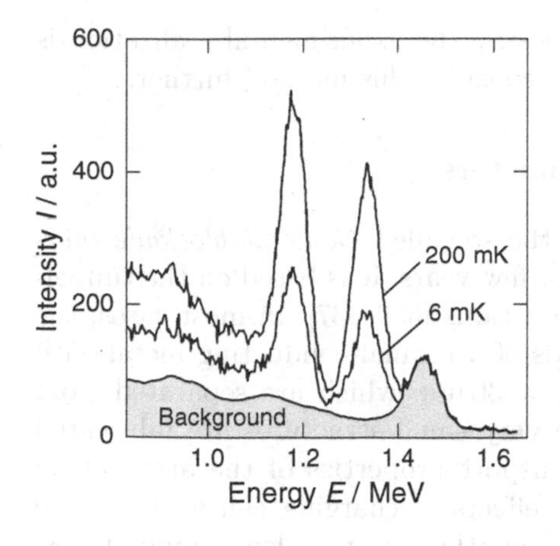

Fig. 12.11. ^{60}Co spectra measured with a NaI detector under $\Theta = 0$ at two different temperatures. The background is shown by the *tinted area*. The intensities are normalized with respect to the ^{40}K line [622]

nal magnetic field to orient the ferromagnetic domains. Figure 12.11 shows two ^{60}Co spectra obtained with a NaI scintillation counter at two different temperatures. The recording time was ten minutes for each spectrum. The reduction in intensity of the two ^{60}Co lines at low temperatures is clearly visible. For comparison, a long-time background counting rate (16 h) without the cobalt source was taken in the same geometry. The result is also plotted in Fig. 12.11. The ^{40}K line appears in this spectrum at 1.45 MeV, originating from the natural radioactivity of the surroundings. The intensities of all the spectra have been normalized with respect to the counting time.

A ^{60}Co nuclear orientation thermometer can be used for temperatures between 2 mK and 50 mK, since in this range the temperature dependence is sufficiently large. These thermometers are not suitable for use in strong magnetic fields. To obtain spectra with good statistics one needs sources with high activity. However, the amount of ^{60}Co in the cobalt host crystal cannot be increased beyond a certain limit, since the source represents a heat leak for the cryostat that is proportional to the activity. In practice, activities between 1 μCi and 10 μCi have been considered as a suitable compromise. NaI scintillation counters are often used as the detector because they are relatively inexpensive and easy to handle. For more precise measurements Ge ionization detectors are more suitable, since they provide a much better energy resolution compared to NaI counters. A particular advantage of nuclear orientation thermometers is the fact that no leads into the cryostat are needed. The disadvantages are the limited temperature range and the radioactivity.

12.1.7 Mössbauer-Effect Thermometers

A further possibility for determining the temperature between 2 mK and 20 mK, which also relies on the thermal occupation of the hyperfine levels of nuclei, is the Mössbauer effect, i.e., the recoilless emission and absorption of γ-quanta from nuclei embedded in suitable solids [620]. However, this type of thermometer is hardly been used, because the experimental realization is rather involved. Therefore, we shall not consider this method further.

12.1.8 Coulomb-Blockade Thermometers

A new type of primary thermometer, the so-called *Coulomb-blockade thermometer*, has been developed in the last few years. It is based on the tunneling of electrons in an array of tunnel junctions [636, 637]. In most cases, the array consists of a chain of small islands of a normal-conducting metal with typical dimensions of 1 mm × 100 nm × 30 nm, which are separated from each other by tunnel junctions. These very small structures are fabricated by electron-beam lithography. The transport properties of the array along the chain are strongly affected by the effects of charging caused by single electrons that tunnel from one island to another. At very low temperatures, when the Coulomb energy $E_C = e^2/2C_{eff}$ is much larger than the thermal energy $k_B T$, the so-called Coulomb blockade is observed and the conduction along the chain vanishes for small voltages [638]. Here, C_{eff} denotes the effective capacitance of the array. In this regime, the effect is nearly temperature independent. In the opposite limit $E_C \ll k_B T$, however, it is found that the electrical conductance G of a tunnel junction chain has a minimum when the voltage across the chain is zero. The width of the dip in the conduction strongly depends on temperature. In the high-temperature limit, the differential conduction can be described by

$$G = G_T \left[1 - \left(\frac{E_C}{k_B T} \right) \right] g(x) , \tag{12.9}$$

where G_T denotes the asymptotic conductance at high bias voltage and $x = eU/(N k_B T)$. The quantity N represents the number of junctions. The function $g(x)$ is given by

$$g(x) = \frac{x \sinh(x) - 4 \sinh^2(x/2)}{8 \sinh^4(x/2)} . \tag{12.10}$$

Figure 12.12 shows the conduction of a one-dimensional array of 40 tunnel junctions as a function of the voltage across the array at a temperature of 0.5 K. An important point regarding the use of Coulomb-blockade thermometers as primary thermometers is that the width $U_{1/2}$ of the dip in the conduction at half-minimum divided by the temperature is a universal constant for all arrays with the same number of junctions, provided that the variation in the junction parameters is not too large. The full width at half-minimum of the conductance dip described by (12.9) is given by

$$U_{1/2} = 5.439 \, N \frac{k_B T}{e} . \tag{12.11}$$

The numerical factor 5.439 originates from the shape of the dip given by (12.10). The absolute temperature can be determined directly by measur-

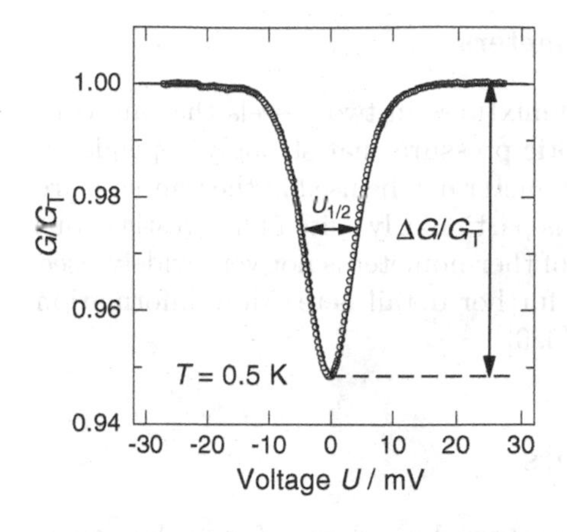

Fig. 12.12. Normalized conduction G/G_{T} of a Coulomb-blockade thermometer with $N = 40$ as a function of the bias voltage U at a temperature of 0.5 K [639]

ing $U_{1/2}$, because only universal constants and the number of tunnel junctions enter the relation (12.11).

Another interesting aspect of Coulomb-blockade thermometers is the fact that one can also use a second feature for thermometry, namely the change of the conduction $\Delta G/G_{\mathrm{T}}$ with temperature. This quantity is inversely proportional to temperature. In the high-temperature limit one finds $\Delta G/G_{\mathrm{T}} = E_{\mathrm{c}}/6k_{\mathrm{B}}T$. Of course, this phenomenon cannot be used for primary thermometry, because the Coulomb energy depends on the individual capacitance of each device. An example of the temperature dependence of $\Delta G/G_{\mathrm{T}}$ is shown in Fig. 12.13.

The temperature range in which a Coulomb-blockade thermometer can be used as a primary thermometer can be tailored by fabrication. There are two devices commercially available covering the temperature ranges from 20 mK to 1 K, and 1 K to 30 K, respectively. A very remarkable feature of Coulomb-blockade thermometers is that they are practically independent of magnetic field. This has been established experimentally for fields up to 23 T [640].

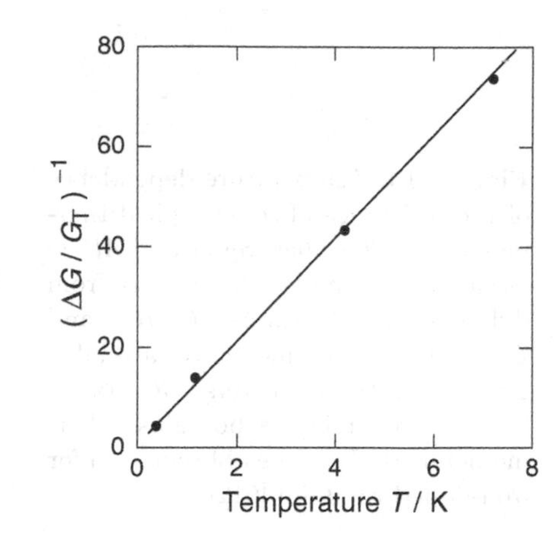

Fig. 12.13. Inverse relative change $(\Delta G/G_{\mathrm{T}})^{-1}$ of the conduction of a Coulomb-blockade thermometer as a function of temperature [636]

12.1.9 Osmotic-Pressure Thermometers

A concentration difference of ^{3}He/^{4}He mixtures in two vessels that are connected via a superleak causes an osmotic pressure that strongly depends on temperature (see Sect. 5.2). This thermometer can be used in the temperature range between 10 mK and 700 mK and is particularly suited for investigations of dilute ^{3}He/^{4}He mixtures. This type of thermometer is not very widely used and will therefore not be discussed in further detail here. More information about this technique can be found in [620].

12.2 Secondary Thermometers

As the name already implies, the temperature dependence of secondary thermometers is obtained via calibration against other thermometers, which have been calibrated previously. The calibration of all secondary thermometers must originate from a primary thermometer. After calibration, secondary thermometers are often easier to use than the primary thermometers.

12.2.1 Resistance Thermometers

The most widely used type of a thermometer in low-temperature laboratories is based on the temperature dependence of the electrical resistance of certain conducting materials [573, 620]. Figure 12.14 shows the temperature dependence for various types of resistors. The physical origin of the temperature dependence of the different types of systems differs considerably and will therefore be discussed in separate sections.

The resistance can be measured, for example, using a modified Wheatstone bridge, that is adapted for use at low temperatures. The principle of operation of a Wheatstone bridge is based on the comparison of the resistor

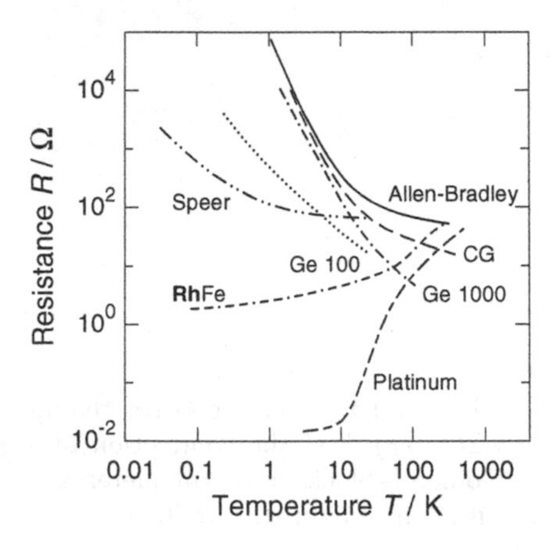

Fig. 12.14. Temperature dependence of the resistance of some typical thermometers. The labels *Speer* and *Allen-Bradley* refer to carbon resistors from different manufacturers. *Ge 100* and *Ge 1000* are commercially available germanium thermometers. *CG* stands for a commercial carbon-glass thermometer. **Rh**Fe is the abbreviation for iron-doped rhodium [612]

under consideration to a known reference resistor. In general, the so-called four-wire method is applied (Fig. 12.15), in order to separate current and voltage leads. In this way, the resistance of the leads is not measured along with the resistance of the thermometer. To enhance the sensitivity and to avoid contact potentials, the bridges are usually operated with an alternating current. A lockin amplifier is used as a null-balancing instrument. To balance the capacitive differences in the two measuring branches, an adjustable capacitor is also built in.

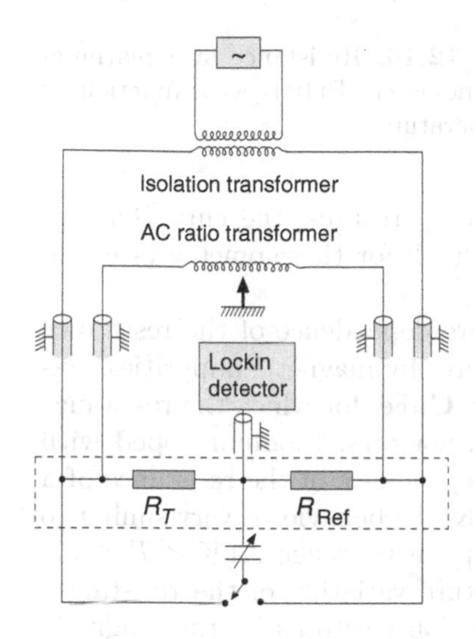

Fig. 12.15. Schematic sketch of a resistance bridge used for thermometry. The *dashed frame* indicates the components of the circuit, which are located at low temperatures

Some modern resistance bridges for thermometry do not use a reference resistor, but compensate the voltage drop across the resistor using feedback electronics. In this way, it is possible to realize fully automatic bridges, that can be controlled and read out by a computer.

Metals

As we have seen in Sect. 7.2, the electrical resistance of simple metals at low temperatures is determined by the scattering of electrons by defects and is temperature independent. At higher temperatures, the resistance increases, because the electrons are scattered more and more efficiently by the increasing number of phonons. Among the metals, platinum is most frequently used for thermometry, since it is chemically very stable and can be produced with high purity. Because of the highly reproducible properties, platinum thermometers are even used for the interpolation between the fixed points of the ITS-90.

Figure 12.16 shows the temperature dependence of the resistance of a commercial platinum thermometer, which is named Pt100, because its resistance at 0°C is precisely 100 Ω. Above 50 K, one finds an almost perfectly

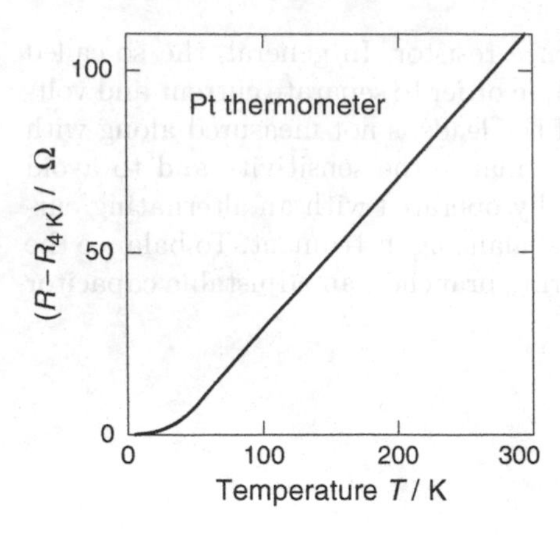

Fig. 12.16. Resistance of a platinum thermometer (Pt100) as a function of temperature

linear temperature dependence. At lower temperatures, the curve flattens and below 20 K the variation becomes too small for thermometry (see also Fig. 12.14).

In Sect. 7.4, we saw that the temperature dependence of the resistance of metals can be enhanced at low temperatures by magnetic impurities. Examples are Kondo systems, such as **Au**Fe or **Cu**Fe, for which the resistance rises with decreasing temperature. For thermometry, rhodium doped with 0.5% iron is often used. The temperature dependence of the resistance of a **Rh**Fe alloy is shown in Fig. 12.14. Above 30 K, the behavior is very similar to that of pure platinum. However, in the temperature range $0.1\,\mathrm{K} < T < 1\,\mathrm{K}$ one finds an approximately linear temperature variation of the resistance, which is caused by scattering of the conduction electrons by the magnetic impurities.

Doped Semiconductors

Intrinsic or lightly doped semiconductors are usually not used for thermometry since their electrical conductivity becomes extremely small at low temperatures. In contrast, certain types of highly doped semiconductors with dopant concentrations close to those giving rise to a insulator–metal transition, are rather popular thermometers for low-temperature work. Because of elastic and electric interactions between the dopant atoms, the energy levels of the donors and acceptors are widely distributed in highly doped semiconductors. With increasing dopant concentration, the situation becomes similar to that of amorphous semiconductors. In the latter system, one also finds a broad distribution of defect states near the Fermi level. Because of this similarity we will briefly discuss the well-understood low-temperature conduction mechanism of amorphous semiconductors.

The conductivity of amorphous semiconductors at temperatures below room temperature is based on thermally activated hopping of carriers (electrons or holes) from one defect state to another. At low temperatures, the so-called *variable-range hopping* is the dominant process [641]. The explanation of this conduction mechanism is based upon the idea that the mean energy difference ΔE of the states of nearest neighbors becomes large compared to the thermal energy. This means that the hopping probability for this process decreases rapidly with decreasing temperature. In this regime, the motion of the carriers is due to thermally activated hopping, in combination with tunneling. With decreasing temperature, the average hopping distance increases, because it becomes more favorable for the carriers to tunnel further to remote potential minima, which require lower activation energies than nearby sites. One could call this process a 'high-long' jump. The corresponding situation for a defect band is schematically illustrated in Fig. 12.17.

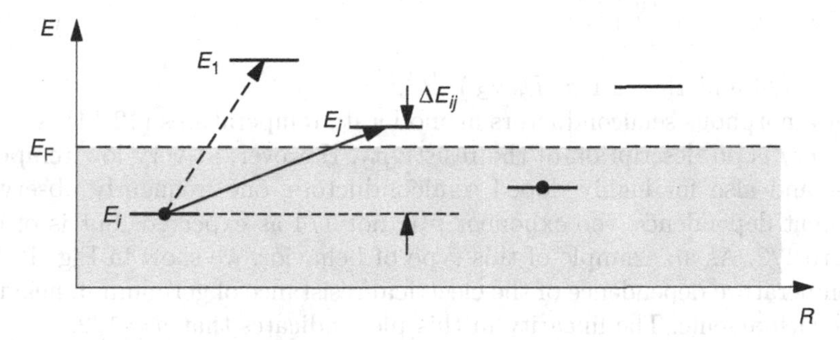

Fig. 12.17. Schematic illustration of the variable-range hopping in the defect band of a highly doped semiconductor. At low temperatures, it is more favorable for the carrier in the state with the energy E_i to move to state E_j by a combined thermally activated hopping and tunneling process than to jump thermally activated to the neighboring state with energy E_1

The hopping probability W_{ij} for such a combined process is given to a good approximation by the expression

$$W_{ij} = \nu_0 \, \mathrm{e}^{-(\Delta E_{ij}/k_\mathrm{B}T + 2R_{ij}/a_0)} \,. \tag{12.12}$$

Here, R_{ij} denotes the distance between the two defect states that are separated by a potential barrier. The first term in the exponent reflects the thermally activated process and the second one the tunneling process. The prefactor ν_0 represents the attempt frequency, which describes how often the carrier approaches the potential barrier. The attempt frequency is determined by the strength of the electron–phonon coupling. The quantity ΔE_{ij} is the activation energy for this process, i.e., the energy difference between the initial and the final state. a_0 denotes the so-called *localization length* that describes the decay of the wave function of the bound carriers. To a first approximation, the wave function is assumed to be exponential. Thus, the factor e^{-2R/a_0} reflects the overlap of the wave function of the initial and the final state.

For a jump to occur, there must be at least one state within a certain predetermined energetic and spatial distance. This condition can be expressed by

$$\frac{4\pi}{3} D_0 \, \Delta E \, R^3 = 1 \, , \tag{12.13}$$

where D_0 denotes the density of states at the Fermi level that is assumed to be approximately constant. Inserting (12.13) into the exponent of (12.12) and maximizing W_{ij} with respect to distance (i.e., using $\partial W_{ij}/\partial R_{ij} = 0$), one finds for the optimal hopping distance $\overline{R} = [9a_0/(8\pi \, D_0 \, k_B T)]^{1/4}$. Since the resistance is proportional to the inverse hopping rate, the temperature dependence of the resistivity is finally obtained as

$$\varrho(T) = \varrho_0 \exp\left(\frac{T_0}{T}\right)^p , \tag{12.14}$$

with $p = 1/4$ and $T_0 = 2.1(a_0^3 \, D_0 \, k_B)^{-1/4}$.

For amorphous semiconductors at moderate temperatures (12.14) is usually a very good description of the resistivity. However, at very low temperatures, and also for highly doped semiconductors, one frequently observes a different dependence: the exponent p is not 1/4 as expected, but is often closer to 1/2. As an example of this type of behavior, we show in Fig. 12.18 the temperature dependence of the electrical resistance of germanium heavily doped with arsenic. The linearity in this plot indicates that $p = 1/2$.

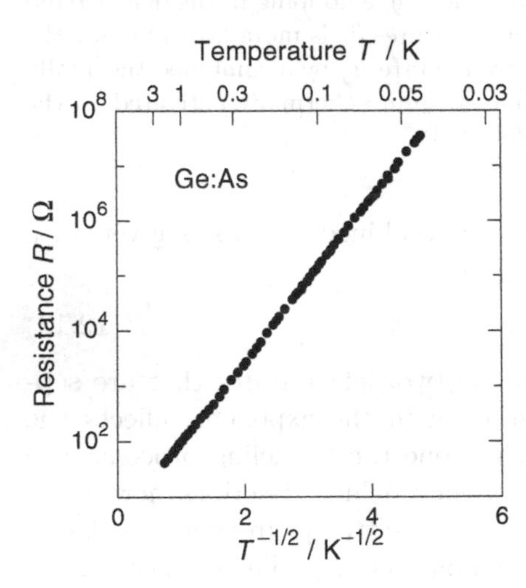

Fig. 12.18. Electrical resistance of germanium doped with arsenic as a function of $T^{-1/2}$ [642]

The reason for the systematic discrepancy is a modification of the density of states at low temperatures caused by the Coulomb interaction resulting in the so-called *Coulomb gap*. This originates from the fact that the electrostatic equilibrium existing prior to the jump of a carrier is disturbed by the hopping process [643]. In thermal equilibrium, the Coulomb energy is minimal. A carrier jumping to a neighboring defect has to raise not only the difference in energy between the two relevant sites, but also the electrostatic energy arising from a nonoptimal charge distribution after the jump, since the jump of the carrier disturbs the energetically optimized charge distribution. At high temperatures this effect is irrelevant, since structural rearrangements occur rapidly and the system can adapt quasi-instantly. At very low temperatures, however, this relaxation becomes too slow. In this case, the additional electrostatic energy has to be accounted for and therefore no states in the direct vicinity of the Fermi level are accessible. Therefore the effective density of states exhibits a gap at the Fermi energy that can be described in the simplest case by a quadratic energy variation. A derivation of the temperature dependence of the resistance of such systems, accounting for the Coulomb gap, leads to the experimentally observed exponent $p = 1/2$.

For thermometers, usually highly doped germanium resistors are used with doping levels between 10^{15} and 10^{19} dopant atoms per cm^3. Although it can be shown that the conduction mechanism of these systems is based upon variable-range hopping, the temperature dependence often cannot be described by (12.14). The main reason for this is that usually an inhomogeneous distribution of the dopant atoms is present in highly doped semiconductors. This was not taken into account in the theoretical considerations discussed above.

Figure 12.19a shows the resistance of different germanium thermometers as a function of temperature. Clearly, no well-defined exponent is observed. The calibration of such germanium thermometers can be expressed by a polynomial function of the form

$$\ln R = \sum_{n=0}^{m} \alpha_n (\ln T)^n \,. \tag{12.15}$$

The stability and reproducibility of commercially available calibrated germanium thermometers is very high. To achieve this, the germanium crystals are fixed with thin gold wires to avoid any mechanical tension. A schematic illustration of the construction of such a thermometer is shown in Fig. 12.19b.

The main problems that arise from using germanium thermometers at very low temperatures result from the relatively poor thermal contact via the thin gold wires and the poor thermal conductivity of the material itself. Because of this, the resistor is easily overheated at very low temperatures by using measuring currents that are too large. In Fig. 12.19a, the measuring currents, which should be used in order to avoid self-heating, are indicated on the right scale. In addition, low-pass filters have to be used in the electrical leads to reduce high-frequency disturbances entering the cryostat from external sources, which would also cause a heating of the resistors. A further problem with using highly doped semiconductors for thermometry is their high sensitivity to magnetic fields. The magnetoresistance makes these thermometers unusable for work in high magnetic fields.

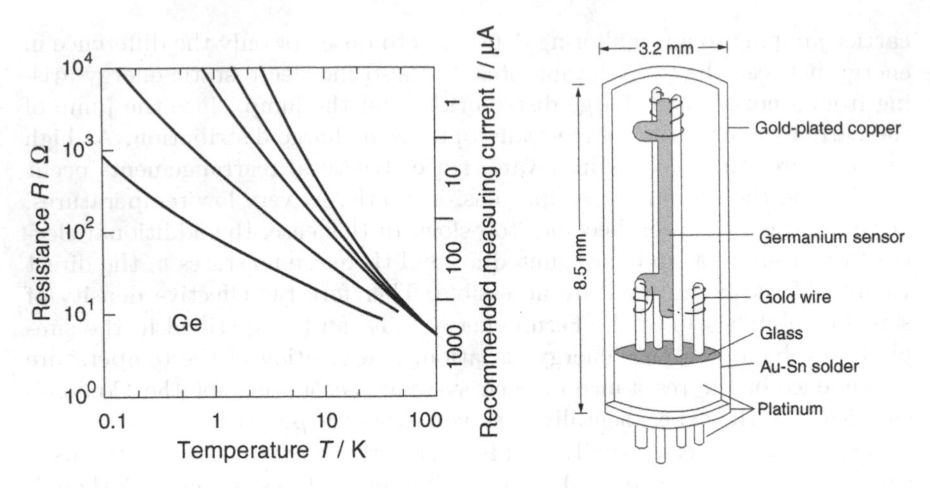

Fig. 12.19. (**a**) Resistance of various commercially available germanium thermometers as a function of temperature. The scale on the right indicates the recommended measuring current. (**b**) Schematic sketch of the construction of a germanium thermometer

Finally, we remark that very homogenous distributions of dopant atoms have been realized in semiconductors with high doping concentrations using a special technique called *neutron transmutation doping* (NTD) [644]. In this case, the semiconductor crystals are irradiated with thermal neutrons from a nuclear reactor. With some low probability, the neutrons are captured by nuclei, which transform into isotopes of neighboring elements. By setting the duration of the radiation and by choosing the isotopic composition of the starting material, one can control the doping process very precisely. As an example, we note the reaction for the production of p-doped germanium

$$^{70}\mathrm{Ge} + \mathrm{n} \longrightarrow {}^{71}\mathrm{Ge} \longrightarrow {}^{71}\mathrm{Ga} + \mathrm{e}^- . \tag{12.16}$$

Using this technique, NTD-Ge resistors with very uniform doping concentration have been produced [645, 646]. Thermometers based on this technique can be used at very low temperatures down to about 20 mK. The temperature dependence usually agrees with that expected by (12.14) with $p = 1/2$. However, such thermometers are not yet available commercially.

Carbon Resistors

Carbon resistors have been used very often as thermometer in low-temperature experiments. These resistors are very cheap because they are produced in large quantities for use in electronics. Although pure carbon is not a semiconductor, carbon resistors exhibit a similar temperature dependence of their electrical resistance as semiconductors. The resistance characteristic is essentially determined by the production process, in which small carbon grains are pressed and sintered. The resistance is mainly given by the contact resistance between the grains. Since the carbon resistors have properties that are specific to a certain production company, the name of the company is often used to label the curves, resistance versus temperature, which are obtained with the thermometers. In fact, only resistors from particular producers are suitable for thermometry. In general, sheet resistors, where the carbon is deposited as a thin film on a small ceramic tube, are not a good choice for low-temperature thermometry.

As an example, we show in Fig. 12.20 the temperature dependence of the electrical resistance of six different carbon thermometers based on resistors of the company *Speer*. One outstanding feature of these resistors is that their temperature dependence is not too steep, making possible the use in a wide temperature range and down to relatively low temperatures.

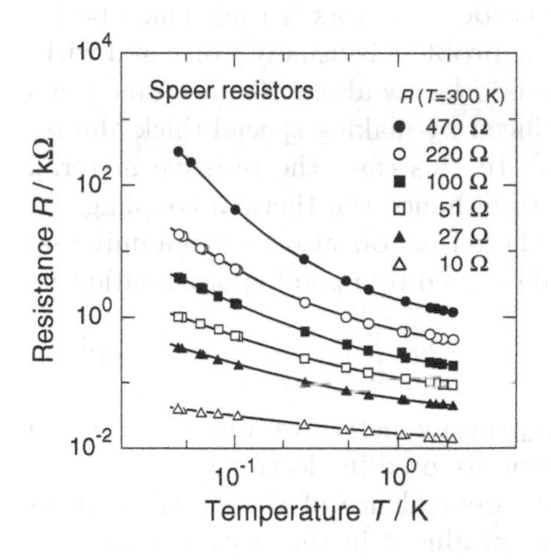

Fig. 12.20. Resistance of carbon thermometers made of Speer resistors as a function of temperature [647]

Before such resistors can be used as a thermometer at low temperatures ($T < 1\,\mathrm{K}$) they are often ground down and glued to a copper holder. This is done to obtain better thermal contact. Despite this method, carbon thermometers are very sensitive to heating effects by the measuring current or due to external noise sources. Since the resistance of a carbon thermometer rises roughly exponentially with decreasing temperature, this problem becomes significantly worse, and finally limits the use of carbon thermometers at low temperatures. In special cases, however, carbon thermometers have successfully been prepared and operated down to $5\,\mathrm{mK}$ [648] without self-heating problems. In general, the use of carbon resistors with suitable resistance characteristics is not recommended at temperatures below $15\,\mathrm{mK}$, even if they are well prepared.

A further problem connected with the use of carbon resistors is their relatively poor reproducibility. Mechanical stress, occurring for example during the cool-down and on the warm-up may influence both the absolute value of the resistance and its temperature dependence. In order to reduce this problem, carbon thermometers are normally thermally cycled a few times after being produced, until the characteristics becomes stable and a meaningful calibration can be obtained. Nevertheless, the calibration must be re-examined from time to time because ageing effects might still occur.

RuO_2 Thick-Film Resistors

In recent years, thick-film resistors, which are commercially available and inexpensive, have turned out to be an attractive alternative to carbon resistors. They consist of a mixture of the two conductive compounds RuO_2 and Bi_2RuO_2 together with lead-silica glass (PbO-B_2O_3-SiO_2). This mixture is deposited on a Al_2O_3-ceramic substrate. Just as for carbon resistors, the characteristics of these thick-film resistors depends on the producer. The advantage of these thermometers over carbon resistors is their much better reproducibility. However, the self-heating problem is usually worse and without special precautions they cannot be used below about 20 mK. Some years ago, this problem was substantially reduced by making special thick-film resistors for thermometry purposes [649]. In this case, the resistive material was deposited on sapphire substrates to enhance the thermal coupling. As well as having a better thermalization, these resistors also showed a universal temperature dependence, that can be described to a good approximation by

$$R(T) = R_0 \exp \left(\frac{T_0}{T} \right)^{-0.345} . \tag{12.17}$$

However, these specially produced RuO_2 thermometers are very expensive in comparison to the thick-film resistors usually used in electronics.

Figure 12.21 shows the temperature dependence of the electrical resistance of RuO_2 thermometers that were produced in this way. Just as with

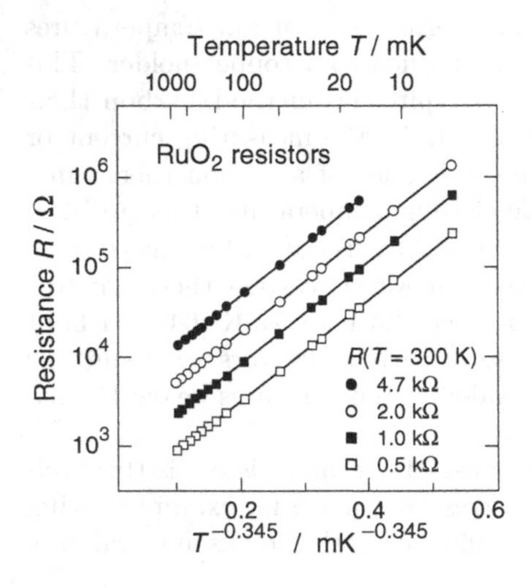

Fig. 12.21. Resistance of RuO_2 thermometers as a function of temperature [649]

semiconductor thermometers and carbon thermometers, the thick-film resistors show a magnetic-field dependence. However, the magnitude of the magnetoresistance is somewhat smaller.

12.2.2 Thermoelectric Elements

If a thermal gradient is generated in a metallic conductor, a voltage between the warm and cold end occurs, termed the *Seebeck effect*. This thermoelectric voltage is given by

$$\Delta U = \mathcal{S}\Delta T\,, \tag{12.18}$$

where the constant of proportionality $\mathcal{S}$ is called the thermal power. A precise measurement of the thermoelectric voltage using a single piece of conducting material is very difficult because the contacts to a voltmeter through the electrical leads are at different temperatures and therefore can influence the measurement. To avoid these complications, the difference between the thermoelectric voltage of two different conductors is usually measured. In a typical setup, the joint of the two conductors is at a temperature T_1 and the other ends are thermally anchored at a fixed reference temperature $T_2 \neq T_1$. The voltage between these two ends is measured using a voltmeter. The classic arrangement of a thermoelectric element with the conductors A and B is shown in Fig. 12.22.

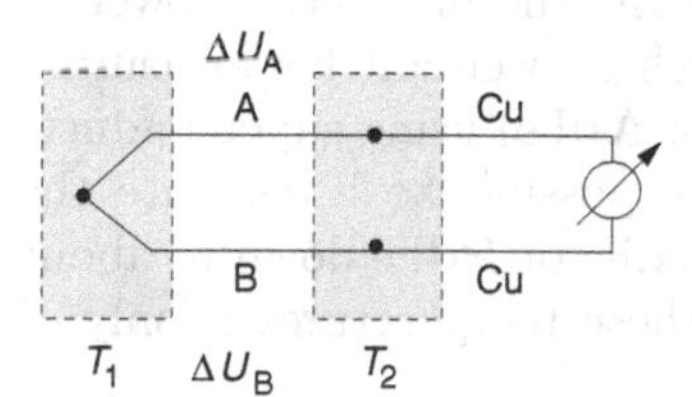

Fig. 12.22. Schematic illustration of the classical setup to determine the temperature with a thermoelectric element

Thermoelectric elements have been used in thermometry for a long time, because this type of temperature determination is inexpensive and very simple [573,613]. Since the thermoelectric voltages are low (μV to mV), one must take care that no contact voltages between leads, switches and plugs disturb the measurement. For this reason, one should not use a third metal to connect thermoelectric elements. This means that in practice the wires should be welded together rather then soldered. A welded joint has the additional advantage of ensuring good reproducibility, even after many thermal cycles.

Thermoelectric elements have the advantage that they are small and thus have a relatively low heat capacity. A disadvantage is, however, that the thermoelectric power vanishes for $T \to 0$. At high temperatures, copper-constantan thermocouples are often used, but the thermoelectric power of this system is generally too small for thermometry below about 10 K. Some special materials, certain metals with magnetic impurities, show thermoelectric powers that are sufficiently large at low temperatures. They can be used for thermometry down to about 1 K. An example of this kind of system is Au containing 300 ppm Fe connected to Ag. Figure 12.23 shows the thermoelectric power of different commercially available thermocouples as a function of temperature.

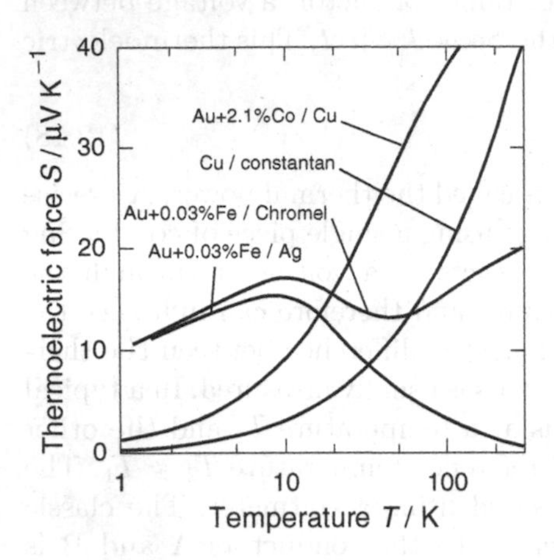

Fig. 12.23. Thermoelectric power S of different thermoelectric elements as a function of temperature [612]

Since superconductors do not exhibit the Seebeck effect they are well suited as connecting leads for measurements of the thermoelectric power at low temperatures. Figure 12.24 shows the sketch of a circuit for measuring the thermoelectric voltage of a thermowire (here **Au**Fe) using superconducting leads and a SQUID. With this setup, it is possible to investigate the thermoelectric power of certain materials like **Au**Fe or **Pd**Fe down to about 10 mK, although the thermoelectric power at these temperatures is only of the order of 10 nV K^{-1} [650, 651].

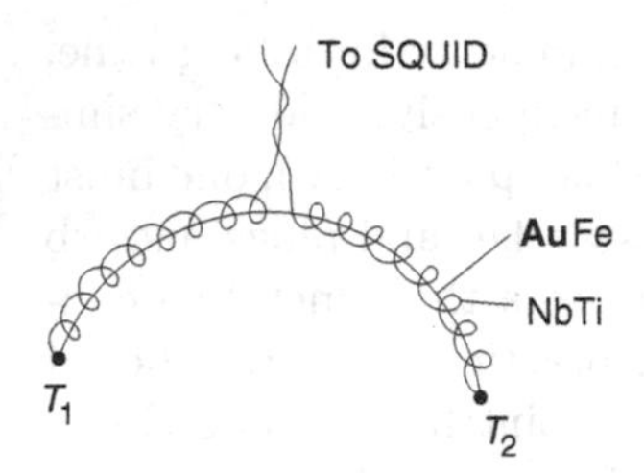

Fig. 12.24. Sketch of a setup for measuring the thermoelectric power at very low temperatures

12.2.3 Capacitive Thermometers

The temperature dependence of the dielectric properties can also be used for thermometry [573]. As we have seen in Sect. 9.5, the dielectric susceptibility of amorphous solids is governed by the contribution of atomic tunneling systems. According to the tunneling model, one expects for $\hbar\omega \ll k_B T$ at low temperatures a logarithmic decrease of the dielectric constant independent of the measuring frequency. In the last twenty-fivc years, capacitive thermometers have been developed that make use of this temperature dependence. The main interest for this type of thermometry arises from the fact that the dielectric properties of glasses are, to a large extent, independent of magnetic fields and therefore these thermometers are suitable for measurements in a magnetic field. Recent investigations of the dielectric properties of glasses in magnetic fields have shown, however, that in some cases strong magnetic field effects in the dielectric properties of insulating glasses are present [652]. It has been suggested that tunneling systems with nuclear spin $I > 1/2$, i.e., with electric quadrupole moments, are responsible for the occurrence of these strong magnetic field effects (see Sect. 9.7) [452].

Figure 12.25 shows low-temperature data of a glass capacitance thermometer based on vitreous silica containing 1200 ppm OH^-. One finds, as expected, a logarithmic decrease of the capacitance with temperature, which turns into an increase at about 100 mK, because of the contributions from relaxation processes. There is virtually no difference between the data with and without magnetic field. One unattractive feature of these thermometers is the fact that the result of the measurement usually depends on the excitation voltage. Since this is a nonlinear effect this can be avoided by using very small amplitudes, but of course, the resolution of the measurement is reduced accordingly.

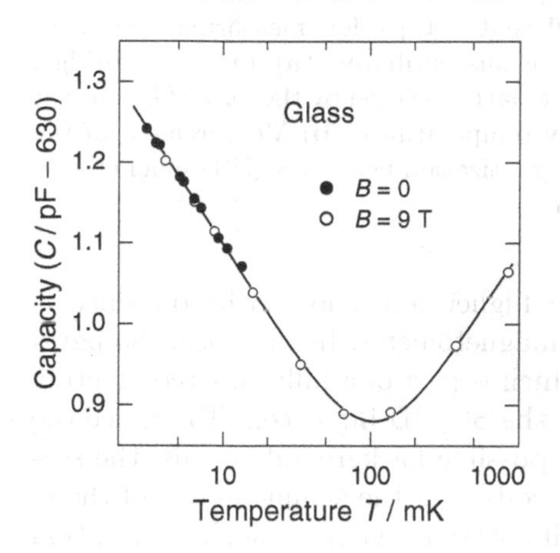

Fig. 12.25. Capacitance C of a glass capacitance thermometer as a function of temperature. The measurement was performed without magnetic field (•) and in a magnetic field of 9 T (○). From the measured capacitance 630 pF have been subtracted [653]

12.2.4 Magnetic Thermometers

Classifying this type of thermometer as a secondary thermometer seems unjustified at first glance, because the temperature dependence of the magnetic susceptibility of paramagnets is described by the fundamental *Curie–Weiss law*

$$\chi = \frac{C}{T - T_c}, \tag{12.19}$$

where T_c represents the Curie temperature. However, in general, one has to fix the two parameters C and T_c for each thermometer individually by calibrating it against other thermometers.

There are many possible ways to measure the magnetic susceptibility of paramagnetic substances at low temperatures [573, 620]. Two different possibilities are depicted schematically in Fig. 12.26. The classical method is to use a mutual inductance bridge. In this method, the paramagnet of interest is located inside a coil, the inductance of which is compared to a known reference inductance.

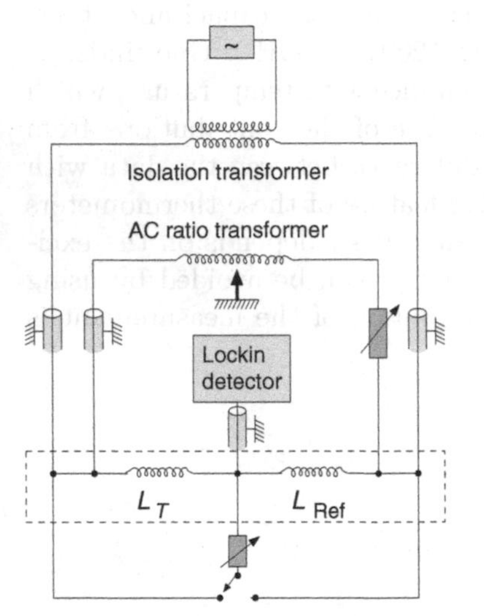

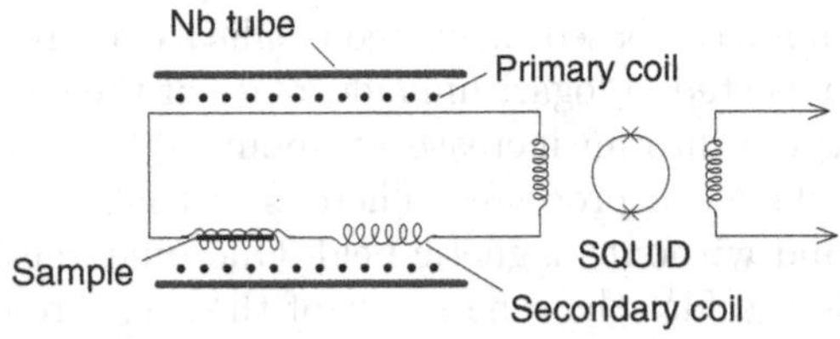

Fig. 12.26. Schematic illustration of two different setups for measuring the magnetic susceptibility: (**a**) Inductive bridge. The parts enclosed by the *dashed line* are at low temperatures. (**b**) Measurement of the magnetization using a SQUID magnetometer

A second method allowing a much higher resolution to be obtained is based on the deployment of a SQUID magnetometer. In this case, the paramagnetic sample is located in a coil, which is part of a fully superconducting loop formed by the secondary coil and the SQUID input coil. To reduce the influence of external disturbances and possible background signals, the secondary coil consists of a pair of astatic coils with the sample in one of them. The secondary coil is located in a small constant magnetic field. A variation of the magnetization of the paramagnetic sample leads to a change of the current in the superconducting loop and therefore to a change of the magnetic flux coupled into the SQUID. The feedback electronics of the SQUID controller provides an additional flux at the SQUID, which is regulated in such a way that the total flux at the SQUID is kept constant. In this case, the signal is given by the current that is used to compensate the flux. With this experimental setup, one can measure the relative change of the magnetic susceptibility very precisely. Susceptibility thermometers designed for a special purpose have provided the best ever relative temperature resolution at 1 K of $\Delta T/T \approx 10^{-10}$ [654]. However, the absolute value of the temperature cannot be obtained with high accuracy. An interesting feature of susceptibility thermometers is that the relative accuracy increases with decreasing temperature, because of the $1/T$ dependence of the susceptibility.

Electron Magnetic Moments

For a long time, paramagnetic salts were exclusively used for magnetization thermometers. For the operation range of a dilution refrigerator, the salt cerium magnesium nitrate (CMN) is particularly suitable. As we have seen in Sect. 11.6, CMN has a very low ordering temperature, $T_c \approx 2\,\text{mK}$, into an antiferromagnetic state. At the same time, CMN has a relatively large Curie constant so that the temperature variation of the magnetic susceptibility at low temperature is rather high in comparison to other systems. On the other hand, this substance and other paramagnetic salts also have serious disadvantages. The response time at low temperatures can be long, since thermal coupling to the salt crystals is difficult to establish. Because of this, CMN thermometers have often been used inside the mixing chamber where they are surrounded by liquid ^{3}He. In addition, CMN thermometers cannot be operated in vacuum without precautions, since the water of hydration would be pumped out and the magnetic susceptibility would change.

These disadvantages can be avoided by using nonmagnetic metals doped with small amounts of magnetic impurities. One such system, which has been successfully used in the past is palladium containing a few ppm of iron. The response time of this type of thermometer at 10 mK is less than one second and therefore at least 100 times shorter than that of a typical CMN thermometer. Figure 12.27a shows a sketch of the construction of a **Pd**Fe thermometer. A measurement of the magnetic susceptibility of Pd containing 15 ppm Fe is shown in Fig. 12.27b.

A similar system that is in some ways even more favorable for thermometry purposes is gold containing small amounts of erbium. The interaction between the erbium ions in gold is much weaker than the corresponding interaction between the Fe ions in palladium. This means that the transition into a spin-glass state (see Sect. 8.3) takes place at much lower temperatures. Figure 12.28 shows the magnetic susceptibility of gold containing 60 ppm erbium as a function of temperature. The *solid line* in Fig. 12.28 corresponds to the Curie law. A hint of the existence of a transition into the spin-glass state is visible at about 250 μK.

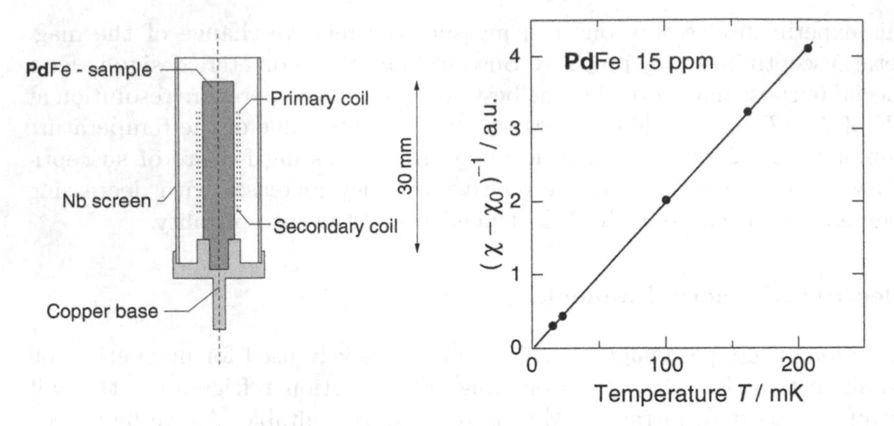

Fig. 12.27. (a) Schematic illustration of the construction of a **Pd**Fe susceptibility thermometer. The primary and secondary coils are wound directly onto the **Pd**Fe sample. A niobium tube is used to screen magnetic fields. (b) Inverse magnetic susceptibility $(\chi - \chi_0)^{-1}$ of Pd doped with 15 ppm Fe as a function of temperature. A SRM 768 fixed-point thermometer was used to determine the temperature. The occurrence of the constant χ_0 can be explained by the presence of a temperature-independent background. The *solid line* corresponds to the Curie behavior [655]

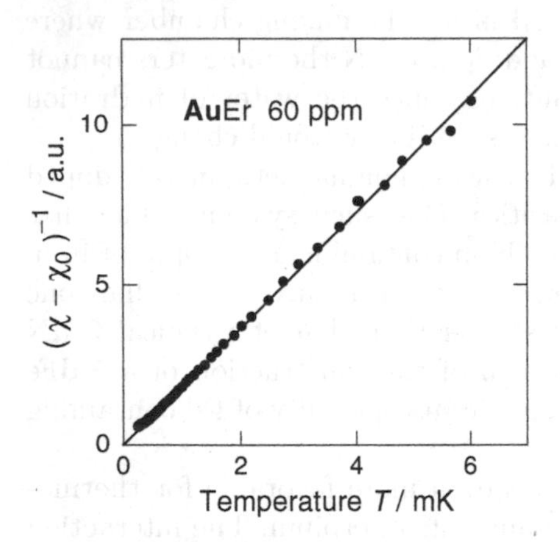

Fig. 12.28. Inverse magnetic susceptibility of Au doped with 60 ppm of Er [656]

Nuclear Magnetic Moments

Although the magnetic moments of nuclei are roughly a factor of 1000 smaller than that of electrons, they can still be used for magnetic thermometry. In particular, for temperatures below 1 mK the measurement of the magnetization of nuclear spins is an important means of determining the temperature.

Using a high-resolution SQUID magnetometer it is also possible to use the magnetic susceptibility of nuclei at higher temperatures as a measure for the temperature. As an example, we show in Fig. 12.29 the measurement of the magnetization of a very pure copper sample (6N) in a magnetic

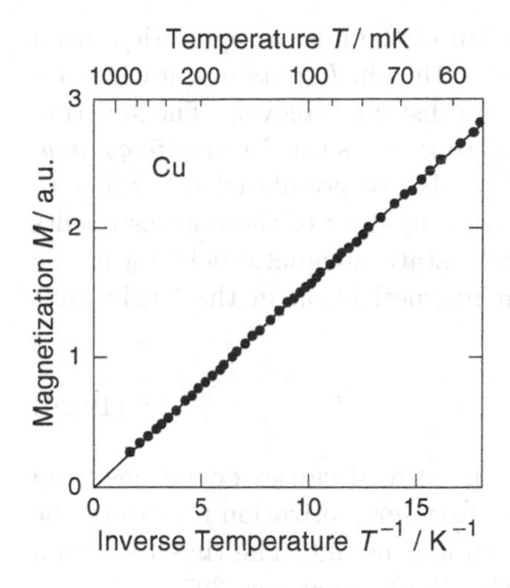

Fig. 12.29. Magnetization of pure copper as a function of $1/T$ as determined with a SQUID [657]

field of 0.25 mT. To determine the temperature, a ^{3}He melting-curve thermometer was used. The resolution of this setup was sufficiently high to measure the magnetization even at 900 mK without difficulty. The data shown in Fig. 12.29 perfectly follows the Curie law. Within the experimental accuracy, the observed Curie constant agrees well with the calculated value for copper.

Nevertheless, one has to be careful in measurements of the static magnetization of nuclear spins, because even very small amounts of impurities with electronic magnetic moments can cause a significant temperature-dependent contribution. At low fields, for example, the presence of just 1 ppm of iron can give the same contribution to the signal as the copper nuclei themselves. To suppress this effect, one should measure the magnetization in large fields, because under these conditions the contribution of the electronic spins saturates. A second method to reduce this problem is to perform a resonant measurement of the magnetization, as we shall see in the following section.

12.2.5 Nuclear Spin Resonance Thermometers

The classical method of nuclear magnetic resonance (NMR) is based upon the alignment of nuclear spins by a static magnetic field and the changes of the resulting magnetization caused by an rf field applied perpendicular to the static field. Compared with the measurement of the static magnetization discussed in the previous section, this method has the advantage of being more sensitive and at the same time selective in exciting a certain ensemble of resonant nuclear spins [573,620]. Because of the lack of alternatives, NMR thermometry is very important at temperatures below 1 mK. ^{195}Pt is used for this application almost exclusively because its properties are most suitable. Principally, there are two techniques, namely stationary and pulsed NMR. In practice, pulsed measurements have turned out to be more favorable and are used today almost exclusively.

As we have already seen in our discussion of the nuclear spin orientation thermometry, the energy levels of a nucleus with spin I splits in a static magnetic field $\boldsymbol{B}_0$ into $(2I+1)$ energetically equidistant sublevels. The splitting is given by $\Delta E = g_{\mathrm{n}}\mu_{\mathrm{n}}B_0 = \hbar\omega_{\mathrm{L}}$, where ω_{L} represents the *Larmor frequency*. In thermal equilibrium, the different sublevels are populated according to the Boltzmann distribution. A nonuniform occupation of these levels results in a polarization of the nuclear spins. For a static magnetic field B_0 in the z-direction, one finds for the equilibrium magnetization in the Curie limit $\Delta E \ll k_{\mathrm{B}}T$ the relation

$$M_z^0 = \frac{N\gamma^2\hbar^2 I(I+1)}{3k_{\mathrm{B}}T} B_0 \,. \tag{12.20}$$

Here, γ denotes the gyromagnetic ratio. The rf field causes deviations from the equilibrium magnetization. The momentary magnetization relaxes to the equilibrium magnetization with a certain time constant. The time evolution of the magnetization $\boldsymbol{M}$ is described by the *Bloch equations* [397]

$$\frac{\mathrm{d}M_x}{\mathrm{d}t} = \gamma\,(\boldsymbol{M}\times\boldsymbol{B})_x - \frac{M_x}{\tau_2}\,, \tag{12.21}$$

$$\frac{\mathrm{d}M_y}{\mathrm{d}t} = \gamma\,(\boldsymbol{M}\times\boldsymbol{B})_y - \frac{M_y}{\tau_2}\,, \tag{12.22}$$

$$\frac{\mathrm{d}M_z}{\mathrm{d}t} = \gamma\,(\boldsymbol{M}\times\boldsymbol{B})_z - \frac{M_z - M_z^0}{\tau_1}\,. \tag{12.23}$$

The so-called longitudinal relaxation time τ_1 is the characteristic time in which the energy from the nuclear spin system is transferred to the conduction electrons. As we have seen already, this time is given by the Korringa relation (8.42). The so-called transverse relaxation time τ_2 measures how long the individual nuclear magnetic moments that contribute to M_x and M_y, are in phase. Since the destruction of the phase coherence is due to the interaction between the nuclear spins, τ_2 is often referred to as the spin-spin relaxation time.

Stationary Method

In this method, a high-frequency field $B_y = B_1\cos(\omega t)$ is applied perpendicular to the static magnetic field, which is oriented along the z-direction. The field induces transitions between the levels of the nuclear spins. In the course of the experiment the static magnetic field – or the frequency of the rf field – is swept slowly until the resonance condition $\omega = \omega_{\mathrm{L}}$ is met and the maximum absorption is reached. The frequency dependence of the absorption can be calculated using the the Bloch equations. In the stationary case, i.e., for $\mathrm{d}\boldsymbol{M}/\mathrm{d}t = 0$, (12.21) to (12.23) reduce to algebraic equations. Under these conditions, the amplitude of the transverse magnetization M_y is given by

$$M_y = \frac{1}{2} M_0(T) \frac{\gamma B_1 \tau_2}{1 + (\omega_L - \omega)^2 \tau_2^2 + \gamma^2 B_1^2 \tau_1 \tau_2} . \qquad (12.24)$$

At the resonance ($\omega = \omega_L$) one therefore finds

$$M_y^{max} = \frac{1}{2} M_0(T) \frac{\gamma B_1 \tau_2}{1 + \gamma^2 B_1^2 \tau_2 \tau_1} . \qquad (12.25)$$

For small amplitudes $\gamma^2 B_1^2 \tau_1 \tau_2 \ll 1$, i.e., without saturation effects, this expression can be simplified further to $M_y^{max} = M_0(T) \gamma B_1 \tau_2$. Assuming that the linewidth does not depend on temperature, we finally obtain the relation $M_y^{max} \propto M_0(T) \propto 1/T_n$. Of course, using this method only the occupation temperature T_n of the nuclear spins can be determined. Any statement about the temperature of the conduction electrons requires the two systems to be in equilibrium.

The main problem in applying stationary NMR for thermometry lies in the demand of very small amplitudes of the rf field ($B_1 \approx 1\,\mu T$), because saturation effects and heating due to eddy-current heating would be unavoidable. Of course, the resolution obtainable with very small rf amplitudes is poor, which is one reason why the stationary method is infrequently used for thermometry. A variant of this method is a direct measurement of the change of magnetization caused by the high-frequency field. Since these measurements are usually performed with a SQUID magnetometer to obtain the required resolution, one refers to this method as SQUID-NMR. This technique combines the advantages of the direct magnetization measurement and the selectivity and resolution of nuclear spin resonance experiments. An example of this kind of measurement on copper is shown in Fig. 12.30. Clearly, a dip in the total magnetization is seen in the regions where the resonance conditions for the two copper isotopes ^{63}Cu and ^{65}Cu are met.

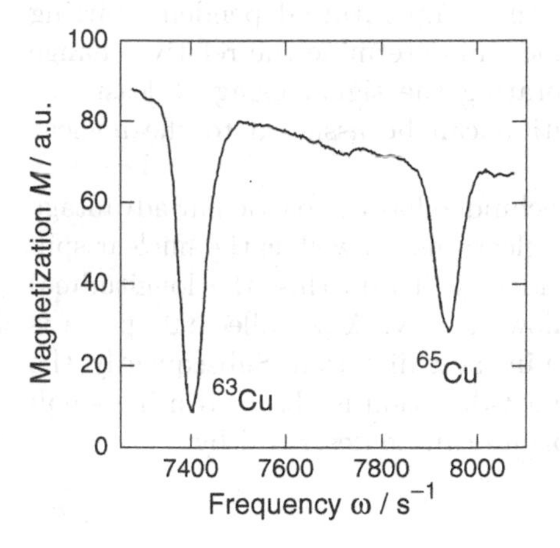

Fig. 12.30. Magnetization of pure copper as function of frequency ω. The measurement was performed in a static field of $B_0 = 104\,\mathrm{mT}$ at 0.65 K [658]

Pulsed Nuclear Resonance

In the case of the pulsed nuclear resonance, short rf pulses are applied perpendicular to the static magnetic field B_0. The amplitudes of these pulses are usually considerably larger than those used in the stationary method. As a result, the magnetization vector is turned away from the z-direction, where the angle Θ between the z-axis and the magnetization is determined by the amplitude and the duration t_p of the pulses of the rf field[2]

$$\Theta = \gamma B_1 t_p \, . \tag{12.26}$$

After application of the pulse, the magnetization precesses around the z-axis. The component $M_0 \sin\Theta$ of the magnetization rotates in the xy-plane and is monitored by a pickup coil, which in many cases is identical to the transmitter coil.

Selecting the pulse length and the amplitude so that the angle is $\Theta = 90°$, the magnetization in the z-direction becomes zero. The magnetization observed in the y-direction oscillates with the frequency $\omega/2\pi$ and its amplitude is given by

$$M_y(t) = \frac{1}{2} M_0 \cos(\Omega t)\, \mathrm{e}^{-t/\tau_2^*} \, . \tag{12.27}$$

The amplitude oscillates with the frequency difference $\Omega = (\omega - \omega_L)$ and the decay is determined by the relaxation time τ_2^*. The phenomenon is called *free induction decay*. The time constant τ_2^* is not necessarily identical to the spin-spin relaxation time, since inhomogeneities of the magnetic field, causing a distribution of the Larmor frequencies with width $\delta\omega_L$, also contribute to the destruction of the phase coherence. Thus, the effective time constant is given by $1/\tau_2^* = 1/\tau_2 + \delta\omega_L$.

Figure 12.31a shows the signal induced in the pickup coil during a free induction decay of ^{195}Pt nuclei. Using the temperature-dependent starting amplitude (or the integrated signal), one can determine the relative change of magnetization $\delta M_0/M_0$. After calibrating the signal using at least one temperature, the measured magnetization can be assigned to the nuclear spin temperature T_n.

The pulsed nuclear spin resonance technique has the particular advantage that the temperature of the conduction electrons, as well as the nuclear spin temperature, can be measured independently. To do this, the longitudinal relaxation time is measured in the following way: A so-called 90° pulse is applied that destroys the magnetization in the z-direction. Subsequently, the recovery of the magnetization M_z in the z-direction is observed using small test pulses. The recovery to the equilibrium can be described by

$$M_z(t) = \frac{1}{2} M_0 \left[1 - \mathrm{e}^{-t/\tau_1}\right] . \tag{12.28}$$

[2] Formally, there is a great similarity with the dielectric echo phenomena discussed in Sect. 9.7.

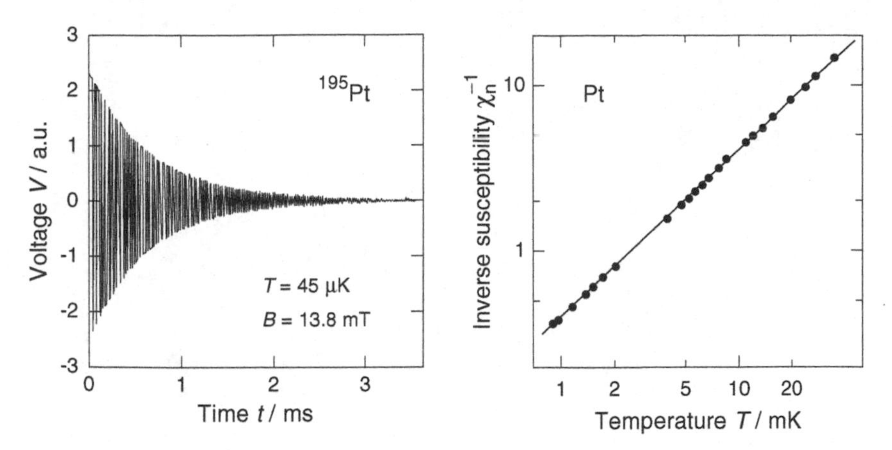

Fig. 12.31. (**a**) Free induction decay of ^{195}Pt nuclear spins at a temperature of 45 μK [573] (**b**) Inverse magnetic susceptibility of platinum as a function of the electron temperature [659, 660]

From the measurement of the relaxation time τ_1, the temperature of the electrons can be calculated via the Korringa relation $T_e = \kappa/\tau_1$. This method can be used, for example, to investigate whether electrons and nuclear spins are in thermal equilibrium. Figure 12.31b shows the inverse magnetic susceptibility χ_n^{-1} of ^{195}Pt nuclei as a function of the electron temperature that has been determined independently. From the fact that a linear relation between χ_n^{-1} and T_e exists and that the relation $\chi_n^{-1} \propto T_n$ holds, it can be concluded that electrons and nuclear spins were in thermal equilibrium in this experiment.

Exercises

12.1 Calculate the power resulting from thermal voltage fluctuations of a resistor with 100 Ω operated at 100 mK with a bandwidth of 10 kHz.

12.2 At low temperatures, the electrical current in highly doped semiconductors is transported by variable-range hopping. Estimate the optimal hopping distance for an electron in an n-type germanium resistor doped with 10^{19} As atoms per cm^3 assuming a localization length of 10 Å and a temperature of 100 mK.

12.3 Gold containing a small amount of erbium is used as a susceptibility thermometer at very low temperatures. (**a**) Calculate the Er concentration at which the magnetic susceptibility of the Er spins is equivalent to the nuclear magnetic susceptibility of the host. Note that the effective g factor of erbium in gold is $g = 6.8$ and the nuclear magnetic moment of gold is $\mu_{\mathrm{Au}} = 0.143\,\mu_{\mathrm{n}}$. (**b**) Compare the heat capacity of the erbium spins at 2 mT and 20 mK with the heat capacity of the conduction electrons assuming the erbium concentration calculated in (a).

[illegible] the temperature [illegible] a function of the [illegible] electron temperature [illegible]

[illegible] of the relaxation [illegible] temperature of the [illegible] can be calculated via the Korringa relation [illegible]. This method can be used, for example, to [illegible] whether electrons and nuclear spins are in thermal equilibrium [illegible] the inverse magnetic susceptibility [illegible] temperature that has been determined [illegible] a linear relation between [illegible] and that the relation [illegible] can be concluded that electrons and nuclear spins [illegible] in this experiment.

Exercises

12.1 [illegible] of a [illegible] with 100 [illegible] with a [illegible] 10 [illegible]

12.2 At low temperatures [illegible] highly doped semiconductors [illegible] optimal hopping distance [illegible] doped with [illegible] per [illegible] and a temperature [illegible]

[illegible] equivalent to the [illegible] factor [illegible]

References

Chapter 1

1. P.J.C. Janssen, C.R. Acad. Sci. Paris **67**, 494 (1868)
2. J.N. Lockyer, C.R. Acad. Sci. Paris **67**, 836 (1868)
3. W. Ramsay, Nature (London) **51**, 512 (1895) and **52**, 7 (1895)
4. P.T. Cleve, C.R. Acad. Sci. Paris **120**, 843 (1995)
5. H. Kayser, Chem. Ztg. **19**, 1549 (1895)
6. H. Kamerlingh Onnes, Leiden Commun. **105**, Proc. Roy. Acad. Sci. Amsterdam **11**, 168 (1908)
7. M.L.E. Oliphant, B.B. Kinsey, E. Rutherford, Proc. Roy. Soc. London **A141**, 722 (1933)
8. L.W. Alvarez, R. Cornog, Phys. Rev. **56**, 379 and 613 (1939)
9. S.G. Sydoriak, E.R. Grilly, E.F. Hammel, Phys. Rev. **75**, 303 (1949)
10. J.E. Lennard-Jones, Proc. Phys. Soc. **43**, 461 (1931)
11. F. London, *Superfluids* Vol. 1 (Wiley, New York 1950) and Vol. 2 (Wiley, New York 1954)
12. R.C. Reid, J.M. Prausnitz, T.K. Sherwood, *The Properties of Gases and Liquids* (Mc Graw-Hill, New York 1977)
13. *Handbook of Cryogenic Engineering*, (J.G. Weisend II ed.) (Taylor & Francis, London 1998)
14. H. Kamerlingh Onnes, Leiden Commun. **119**, Proc. Roy. Acad. Sci. Amsterdam **13**, 1093 (1911)
15. H. Kamerlingh Onnes, J.D.A. Boks, Leiden Commun. **170b**, 18 (1924)
16. E.R. Grilly, E.F. Hammel, S.G. Sydoriak, Phys. Rev. **75**, 1103 (1949)
17. E.C. Kerr, Phys. Rev. **96**, 551 (1954); J. Chem. Phys. **26**, 511 (1957)
18. L.I. Dana, H. Kamerlingh Onnes, Leiden Commun. **190b**, Proc. Roy. Acad. Sci. Amsterdam **29**, 1061 (1926)
19. W.H. Keesom, K. Clusius, Leiden Commun. **219e**, Proc. Roy. Acad. Sci. Amsterdam **35**, 307 (1932)
20. K.R. Atkins, *Liquid Helium* (Cambridge Univ. Press, Cambridge 1959)
21. T.A. Alvesalo, T. Haavasoju, M.T. Manninen, J. Low Temp. Phys. **45**, 373 (1981)
22. D.D. Osheroff, R.C. Richardson, D.M. Lee, Phys. Rev. Lett. **28**, 885 (1972)
23. R. Berman, J. Poulter, Philos. Mag. **42**, 1047 (1952)
24. H. van Dijk, M. Durieux, in *Progress in Low Temperature Physics*, Vol. II, (C.J. Gorter ed.), (North-Holland, Amsterdam 1955), p. 431
25. C.A. Swenson, Phys. Rev. **79**, 626 (1950)

26. W.P. Halperin, F.B. Rasmussen, C.N. Archie, R.C. Richardson, J. Low Temp. Phys. **31**, 617 (1978)
27. E.R. Grilly, J. Low Temp. Phys. **4**, 615 (1971) and (erratum, J. Low Temp. Phys. **11**, 243 (1973)
28. D.S. Greywall, P.A. Busch, J. Low Temp. Phys. **46**, 451 (1982)
29. D.S. Greywall, J. Low Temp. Phys. **31**, 2675 (1985)
30. D.S. Greywall, Phys. Rev. B **27**, 2747 (1983)

Chapter 2

31. P. Kapitza, Nature **141**, 74 (1938)
32. J.F. Allen, A.D. Misener, Nature **141**, 75 (1938)
33. F. London, Nature **141**, 643 (1938) and Phys. Rev. **54**, 947 (1938)
34. L. Tisza, Nature **141**, 913 (1938)
35. L.D. Landau, J. Phys. USSR **5**, 71 (1941); **8**, 1 (1944), and **11**, 91 (1947), English translation: Z.M. Galasiewicz, *Helium-4* (Pergamon Press, Oxford 1971)
36. R.P. Feynman, Phys. Rev. **90**, 1116 (1953)
37. J.D. Reppy, Phys. Rev. Lett. **14**, 733 (1965)
38. J.B. Mehl, W. Zimmermann Jr., Phys. Rev. Lett. **14**, 815 (1965); Phys. Rev. **167**, 214 (1968)
39. J.F. Allen, A.D. Misener, Proc. R. Soc. **A172**, 467 (1939)
40. K.R. Atkins, Philos. Mag. Supp. **1**, 169 (1952)
41. W.J. Heikkila, A.C. Hollis Hallett, Can. J. Phys. **33**, 420 (1955)
42. A.D.B. Woods, A.C. Hollis Hallett, Can. J. Phys. **41**, 596 (1963)
43. A. De Troyer, A. van Itterbeek, G.J. van den Berg, Physica **17**, 50 (1951)
44. J.G. Daunt, K. Mendelssohn, Proc. Roy. Soc. A **170**, 423 and 439 (1939)
45. K.R. Atkins, Proc. Roy. Soc. A **203**, 119 (1950)
46. J.F. Allen, H. Jones, Nature **141**, 243 (1938)
47. J.F. Allen, photographed 1971, unpublished
48. N. Kurti, B.V. Rollin, F. Simon, Physica **3**, 269 (1936)
49. W.H. Keesom, A.P. Keesom, Physica **3**, 359 (1940)
50. W.H. Keesom, B.F. Saris, L. Meyer, Physica **7**, 817 (1940)
51. P. Winkel, A. Broese von Groenou, C.J. Gorter, Physica **21**, 345 (1955)
52. D.F. Brewer, D.O. Edwards, Philos. Mag. **6**, 775 (1961)
53. V.P. Peshkov, J. Phys. (Moscow) **10**, 389 (1946); Soviet Phys. JETP **18**, 950 (1948); English translation also by: Z.M. Galasiewicz, *Helium-4* (Pergamon Press, Oxford 1971)
54. R.B. Dingle, Proc. Phys. Soc. A **62**, 648 (1949)
55. E.L. Andronikashvili, Zh. Eksperim. i. Teor. Fiz. **18**, 424 (1948)
56. E.M. Lifshitz, E.L. Andronikashvili, *A Supplement to Helium*, (Consultants Bureau Inc. 1959)
57. J.H. Scholtz, E.O. McLean, I. Rudnick, Phys. Rev. Lett. **32**, 147 (1974)
58. H. London, Nature **142**, 612 (1938); Proc. Roy. Soc. **A171**, 484 (1939)
59. P.L. Kapitza, Soviet Phys. JETP **11**, 581 (1941)
60. P.L. Kapitza, Zh. Eksperim. i. Teor. Fiz. **11**, 1 (1941); J. Phys. Moscow **4**, 181 (1941)
61. E.H. Hall, Proc. Phys. Soc. A **67**, 485 (1954)
62. K.R. Atkins, C.E. Chase, Proc. Phys. Soc. A **64**, 826 (1951)

63. W.M. Whitney, C.E. Chase, Phys. Rev. Lett. **9**, 243 (1962)
64. V.P. Peshkov, Soviet Phys. JETP **11**, 580 (1960)
65. I.M. Khalatnikov, in *The Physics of Liquid Helium and Solid Helium*, Part I (K.H. Bennemann, J.B. Ketterson eds.) (Wiley-Interscience, New York 1976), p. 1
66. B. Ratnam, J. Mochel, Phys. Rev. Lett. **25**, 711 (1970)
67. K.R. Atkins, Phys. Rev. **113**, 962 (1959)
68. K.R. Atkins, I. Rudnick, in *Progress in Low Temperature Physics*, Vol. VI, (C.J. Gorter ed.) (North-Holland, Amsterdam 1970), p. 37
69. C.W.F. Everitt, K.R. Atkins, A. Denenstein, Phys. Rev. Lett. **8**, 161 (1962)
70. G.A. Williams, R. Rosenbaum, I. Rudnick, Phys. Rev. Lett. **42**, 1282 (1979)
71. G.J. Jelatis, J.A. Roth, J.D. Maynard, Phys. Rev. Lett. **42**, 1285 (1979)
72. R.S. Shapiro, I. Rudnick, Phys. Rev. **137**, A1383 (1965)
73. A. Einstein, Akademie der Wissenschaften, Berlin, Sitzungsberichte, 261 (1924), 3 (1925)
74. M.J. Buckingham, W.M. Fairbank, in *Progress in Low Temperature Physics*, Vol. III, (C.J. Gorter ed.), (North-Holland, Amsterdam 1961), p. 80
75. A. Robert, O. Sirjean, A. Browaeys, J. Poupard, S. Nowak, D. Boiron, C.I. Westbrook, A. Aspect, Science **292**, 461 (2001)
76. N.N. Bogoliubov, J. Phys. (Moscow) **11**, 23 (1947), English translation: Z.M. Galasiewicz, *Helium-4*, (Pergamon Press, Oxford 1971)
77. H.N. Robkoff, D.A. Ewen, R.B. Hallock, Phys. Rev. Lett. **43**, 2006 (1979)
78. V.F. Sears, E.C. Svensson, Phys. Rev. Lett. **43**, 2009 (1979)
79. V.F. Sears, E.C. Svensson, P. Martel, A.D.B. Woods, Phys. Rev. Lett. **49**, 279 (1982)
80. H.A. Mook, Phys. Rev. Lett. **51**, 1454 (1983)
81. L.J. Campbell, Phys. Rev. B **27**, 1913 (1983)
82. N.J. van Druten, C.G. Townsend, M.R. Andrews, D.S. Durfee, D.M. Kurn, M.O. Mewes, W. Ketterle, Czech. J. Phys. **46**, 3077 (1996)
83. W. Ketterle, D.S. Durfee, D.M. Stamper-Kurn, in *Proc. Intern. School of Physics 'Enrico Fermi'*, Course CXL (M. Inguscio, S. Stringari and C.E. Wieman eds.)(IOS Press, Amsterdam, 1999), p. 67
84. E.A. Cornell, J.R. Ensher, C.E. Wieman, in *Proc. Intern. School of Physics 'Enrico Fermi'*, Course CXL (M. Inguscio, S. Stringari and C.E. Wieman eds.) (IOS Press, Amsterdam, 1999), p. 135
85. L. Onsager, Nuovo Cimento, Suppl. **2** 249 (1949)
86. R.P. Feynman, in *Progress in Low Temperature Physics*, Vol. I, (C.J. Gorter ed.), (North-Holland, Amsterdam 1955), p.17
87. W.F. Vinen, Proc. Roy. Soc. **A260**, 218 (1961)
88. D.V. Osborne, Proc. Roy. Soc. **A63**, 909 (1950)
89. R. Meservey, Phys. Rev. **133**, 1471 (1961)
90. P.W. Karn, D.R. Starks, W. Zimmermann Jr., Phys. Rev. B **21**, 1797 (1980)
91. D.R. Tilley, J. Tilley, *Superfluidity and Superconductivity* (IOP Publishing, Bristol 1990)
92. R.E. Packard, T.M. Sanders Jr., Phys. Rev. A **6**, 799 (1972)
93. E.J. Yarmchuk, M.J.V. Gordon, R.E. Packard, Phys. Rev. Lett. **43**, 214 (1979)
94. E.J. Yarmchuk, R.E. Packard, J. Low Temp. Phys. **46**, 479 (1982)
95. B.D. Josephson, Phys. Rev. Lett. **1**, 251 (1962)
96. K. Sukhatme, Y. Mukharsky, T. Chui, D. Pearson, Nature **411**, 280 (2001)

97. D. de Klerk, R.P. Hudson, J.R. Pellam, Phys. Rev. **89**, 326 and 662 (1953)
98. R.J. Donnelly, J.A. Donnelly, R.H. Hills, J. Low Temp. Phys. **44**, 471 (1981)
99. W.G. Stirling, in *75th Jubilee Conference on Helium-4*, (J.G. Armitage ed.), (World Scientific, Singapore 1983), p. 109
100. D. Ruger, J.S. Foster, Phys. Rev. B **30**, 2595 (1984)
101. H.J. Maris, Rev. Mod. Phys. **49**, 341 (1977)
102. D.S. Greywall, Phys. Rev. B **18**, 2127 (1978) and **21**, 1329 (1979)
103. D.S. Greywall, Phys. Rev. B **23**, 2152 (1981)
104. T. Ellis, P.V.E. McClintock, Philos. Trans. R. Soc. A **315**, 259 (1985)
105. G.W. Rayfield, F. Reif, Phys. Rev. Lett. **11**, 305 (1963)
106. G.W. Rayfield, F. Reif, Phys. Rev. **136**, 1194 (1965)
107. R.D. Bruyn Ouboter, K.W. Taconis, W.M. van Alphen, in *Progress in Low Temperature Physics*, Vol. V, (C.J. Gorter ed.), (North-Holland, Amsterdam 1964), p. 44
108. R.J. Donnelly, *Quantized Vortices in Helium II* (Cambridge Univ. Press, Cambridge 1991)
109. J.D. Van der Waals, Ph.D. Thesis (Univ. Leiden, 1873)
110. P. Weiss, J. Phys. Radium **6**, 667 (1907)
111. L.D. Landau, Phys. Z. Sowjetunion **11**, 26 and 545 (1937); Zh. Eksperim. i. Teor. Fiz. **7**, 19 and 627 (1937)
112. K.G. Wilson, Phys. Rev. **B4**, 3174 (1971)
113. J.C.L. Guillou, J. Zinn-Justin, Phys. Rev. Lett. **39**, 95 (1977)
114. J.A. Lipa, T.C.P. Chui, Phys. Rev. Lett. **51**, 2291 (1983)
115. T.C. P.Chui, J.A. Lipa, in *75th Jubilee Conference on Helium-4*, (J.G. Armitage ed.), (World Scientific, Singapore 1983), p. 206
116. J.A. Lipa, D.R. Swanson, J.A. Nissen, T.C.P. Chui, Physica **B197**, 239 (1994)
117. D.S. Greywall, G. Ahlers, Phys. Rev. Lett. **28**, 1251 (1972)
118. R.P. Henkel, E.N. Smith, J.D. Reppy, Phys. Rev. Lett. **23**, 1276 (1969)

Chapter 3

119. L.D. Landau, Zh. Eksperim. i. Teor. Fiz. **30**, 1058 (1956); Soviet Phys. JETP **3**, 920 (1957)
120. L.D. Landau, Zh. Eksperim. i. Teor. Fiz. **32**, 59 (1957); Soviet Phys. JETP **5**, 101 (1957)
121. D.S. Greywall, Phys. Rev. B **27**, 2747 (1983)
122. A.L. Thomas, H. Meyer, E.D. Adams, Phys. Rev. **128**, 509 (1962)
123. W.R. Abel, A.C. Anderson, W.C. Black, J.C. Wheatley, Physics **1**, 337 (1965)
124. D.S. Betts, D.W. Osborne, B. Weber, J. Wilks, Philos. Mag. **8**, 977 (1963)
125. D.S. Betts, B.E. Keen, J. Wilks, Proc. R. Soc. **A289**, 34 (1965)
126. J.M. Parpia, D.J. Sandiford, J.E. Berthold, J.D. Reppy, Phys. Rev. Lett. **40**, 565 (1978)
127. H.R. Hart, J.C. Wheatley, Phys. Rev. **120**, 1111 (1960)
128. A.C. Anderson, W. Reese, R.J. Sarwinski, J.C. Wheatley, Phys. Rev. Lett. **7**, 220 (1961)
129. A.C. Anderson, W. Reese, J.C. Wheatley, Phys. Rev. **127**, 671 (1962)
130. D.S. Greywall, Phys. Rev. B **29**, 4933 (1984)
131. B.E. Keen, P.W. Matthews, J. Wilks, Phys. Lett **5**, 5 (1963); Proc. Roy. Soc. A **284**, 125 (1965)

132. A.A. Abrikosov, I.M. Khalatnikov, Rep. Prog. Phys., Phys. Soc. London **22**, 329 (1959)
133. D. Vollhardt, P. Wölfle, *The Superfluid Phases of Helium 3*, (Taylor & Francis, 1990)
134. W.R. Abel, A.C. Anderson, J.C. Wheatley, Phys. Rev. Lett. **17**, 74 (1966)
135. P.R. Roach, J.B. Ketterson, Phys. Rev. Lett. **36**, 736 (1976)
136. V. Silin, Zh. Eksperim. i. Teor. Fiz. **33**, 1227 (1957); Soviet Phys. JETP **6**, 945 (1958)
137. M . Masuhara, D. Candela, D.O. Edwards, R.F. Hoyt, H.N. Scholz, D.S. Sherrill, R. Combescot, Phys. Rev. Lett. **53**, 1168 (1984)
138. K. Sköld, C.A. Pelizzani, R. Kleb, G.E. Ostrowski, Phys. Rev. Lett. **37**, 842 (1978)
139. K. Sköld, C.A. Pelizzani, J. Phys. C **11**, L 589 (1978)
140. C.H. Aldrich III. D. Pines, J. Low Temp. Phys. **32**, 689 (1978)

Chapter 4

141. D.D. Osheroff, R.C. Richardson, D.M. Lee, Phys. Rev. Lett. **28**, 885 (1972)
142. D.D. Osheroff, W.J. Gully, R.C. Richardson, D.M. Lee, Phys. Rev. Lett. **29**, 920 (1972)
143. E.R. Dobbs, *Helium Three*, (Oxford Univ. Press, Oxford 2000)
144. G.E. Volovik, *The Universe in a Helium Droplet*, (Oxford Univ. Press, Oxford 2003)
145. W.P. Halperin, F.B. Rasmussen, C.N. Archie, R.C. Richardson, J. Low Temp. Phys. **31**, 617 (1978)
146. N.D. Paulson, H. Kojima, J.C. Wheatley, Phys. Rev. Lett. **32**, 1098 (1974)
147. W.P. Halperin, C.N. Archie, F.B. Rasmussen, Phys. Rev. B **13**, 2124 (1976)
148. J.C. Wheatley, Rev. Mod. Phys. **47**, 415 (1975)
149. P.L. Gammel, H.E. Hall, J.D. Reppy, Phys. Rev. Lett. **52**, 121 (1984)
150. J.P. Pekola, J.T. Simola, K.K. Nummila, O.V. Lounasmaa, R.E. Packard, Phys. Rev. Lett. **53**, 70 (1984)
151. P.L. Gammel, T.L. Ho, J.D. Reppy, Phys. Rev. Lett. **55**, 2708 (1985)
152. R.E. Packard, S. Vitale, Phys. Rev. B **46**, 3540 (1992)
153. J.P. Pekola, J.T. Simola, J. Low Temp. Phys. **58**, 555 (1985)
154. B.C. Crooker, Ph. D. Thesis, Cornell University (1984)
155. A.J. Leggett, Phys. Rev. Lett. **29**, 1227 (1972); J. Phys. **C6**, 3187 (1973)
156. D. Einzel, in *Low Temperature Physics*, Lecture Notes in Physics **394** (M.J.R. Hock, R.H. Lemmer eds.) (Springer, Heidelberg 1991), p. 275
157. M.T. Manninen, J.P. Pekola, Phys. Rev. Lett. **48**, 812 (1982)
158. J.M. Parpia, D.G. Wildes, J. Saunders, E.K. Zeise, J.D. Reppy, R.C. Richardson, J. Low Temp. Phys. **61**, 337 (1985)
159. Y.A. Ono, J. Hara, K. Nagai, J. Low Temp. Phys. **48**, 167 (1982)
160. C.N. Archie, T.A. Alvesalo, J.D. Reppy, R.C. Richardson, J. Low Temp. Phys. **42**, 295 (1981)
161. D.C. Carless, H.E. Hall, J.R. Hook, J. Low Temp. Phys. **50**, 605 (1983)
162. D. Einzel, P.K. Wölfle, H. Højgaard Jensen, H. Smith, Phys. Rev. Lett. **52**, 1705 (1984)
163. D. Einzel, J.M. Parpia, Phys. Rev. Lett. **58**, 1937 (1987)

164. D. Einzel, J.M. Parpia, J. Low Temp. Phys. **109**, 1 (1997)
165. Lord Rayleigh, Philos. Mag. **14**, 186 (1982)
166. J. Pellam, W. Hanson, Phys. Rev. **85**, 216 (1952)
167. P.W. Anderson, P. Morel, Phys. Rev. **123**, 1911 (1961)
168. P.W. Anderson, W.F. Brinkman, Phys. Rev. Lett. **30**, 1108 (1973)
169. R. Balian, N.R. Werthamer, Phys. Rev. **131**, 1553 (1963)
170. M. Liu, Physica B + C, **109 & 110**, 1615 (1982)
171. G.E. Volovik, *Exotic Properties of Superfluid* ^{3}*He* (World Scientific, Singapore 1992)
172. G.E. Volovik, Czech. J. Phys. **46**, 3048 (1996)
173. P.R. Roach, B.M. Abraham, M. Kuchnir, B.B. Ketterson, Phys. Rev. Lett. **34**, 711 (1975)
174. P.G. de Gennes, *The Physics of Liquid Crystals* (Clarendon Press, Oxford 1974)
175. V. Ambegaokar, P.G. de Gennes, D. Rainer, Phys. Rev. A **9**, 2676 (1974)
176. D.M. Lee, R.C. Richardson, in *The Physics of Liquid and Solid Helium*, (K.H. Bennemann and J.B. Ketterson eds.) Part II (Wiley-Interscience 1978), p. 287
177. J.E. Berthold, R.W. Giannetta, E.N. Smith, J.D. Reppy, Phys. Rev. Lett. **37**, 1138 (1976)
178. D.D. Osheroff, S. Engelsberg, W.F. Brinkman, L.R. Corruccini, Phys. Rev. Lett. **34**, 190 (1975)
179. A.J. Leggett, Ann. Phys. **85**, 11 (1974)
180. D.D. Osheroff, L.R. Corruccini, Phys. Lett. A **51**, 447 (1975)
181. D.D. Osheroff, W.F. Brinkman, Phys. Rev. Lett. **32**, 584 (1974)
182. D.D. Osheroff, P.W. Anderson, Phys. Rev. Lett. **33**, 686 (1974)
183. P. Wölfe, Phys. Lett. A **47**, 224 (1974)
184. J.C. Davis, J.D. Close, R. Zieve, R.E. Packard, Phys. Rev. Lett. **66**, 329 (1991)
185. P.J. Hakonen, O.T. Ikkala, S.T. Islander, O.V. Lounasmaa, G.E. Volovik, J. Low Temp. Phys. **53**, 425 (1983)
186. M.M. Salomaa, G.E. Volovik, Rev. Mod. Phys. **59**, 533 (1987)
187. P. Hakonen, O. Lounasmaa, J. Simola, Physica B **160**, 1(1989)
188. G.E. Volovik, V.P. Mineev, Zh. Eksperim. i. Teor. Fiz. **24**, 605 (1976); Soviet Phys. JETP Lett. **24**, 561 (1976)
189. M.C. Cross, W.F. Brinkman, J. Low Temp. Phys. **27**, 683 (1977)
190. V.P. Mineev, M.M. Salomaa, O.V. Lounasmaa, Nature **324**, 333 (1986)
191. P.J. Hakonen, M. Krusius, M.M Salomaa, J.T. Simola, Y.M Bunkov, V.P. Mineev, G.E. Volovik, Phys. Rev. Lett **51**, 1362 (1983)
192. S.V. Pereverzev, A. Loshak, S. Backhaus, J.C. Davis, R.E. Packard, Nature **388**, 449 (1997)
193. S. Backhaus, S.V. Pereverzev, A. Loshak, J.C. Davis, R.E. Packard, Science **278**, 449 (1997)
194. R.W. Simmonds, A. Marchenkov, E. Hoskinson, J.C. Davis, R.E. Packard, Nature **412**, 55 (1997)
195. P. Wölfle, Rep. Prog. Phys. **42**, 269 (1979)
196. A.F. Andreev, Zh. Eksperim. i. Teor. Fiz. **46**, 1823 (1964); Soviet Phys. JETP **19**, 1228 (1964)
197. D.S. Fisher, A.M. Guénault, C.J. Kennedy, G.R. Pickett, Phys. Rev. Lett. **63**, 2566 (1989)
198. S.T. Lu, H. Kojima, Phys. Rev. Lett. **55**, 1677 (1985)

199. M .Liu, Phys. Rev. Lett. **43**, 1740 (1979)
200. L.R. Corruccini, D.D. Osheroff, Phys. Rev. Lett. **45**, 2029 (1980)
201. A.M.R. Schechter, R.W. Simmonds, R.E. Packard, J.C. Davis, Nature, **396**, 554 (1998)
202. H. Kojima, D.N. Paulson, J.C. Wheatley, Phys. Rev. Lett. **32**, 141 (1974)
203. A.W. Yanof, J.D. Reppy, Phys. Rev. Lett. **33**, 631 (1974)
204. T. Chainer, Y. Morii, H. Kojima, Phys. Rev. B **21**, 3941 (1980)
205. L. Tewordt, N. Schopohl, J. Low Temp. Phys. **34**, 489 (1979)
206. P. Wölfe, V.E. Koch, J. Low Temp. Phys. **30**, 61 (1978)
207. R. Ling, W. Wojtanowski, J. Saunders, E.R. Dobbs, J. Low Temp. Phys. **78**, 187 (1990)
208. M.E. Daniels, E.R. Dobbs, J. Saunders, P.L. Ward, Phys. Rev. B **27**, 6988 (1983)
209. R.W. Gianetta, A. Ahonen, E. Polturak, J. Saunders, R.K. Zeise, R.C. Richardson, D.M. Lee, Phys. Rev. Lett. **45**, 262 (1980)
210. O. Avenel, M.E. Varoquaux, H. Ebisawa, Phys. Rev. Lett. **45**, 1952 (1980)

Chapter 5

211. H.A. Fairbank, C.T. Lane, L.T. Aldrich, A.O. Nier, Phys. Rev. **71**, 911 (1947)
212. J.G. Daunt, R.E. Probst, H.L. Johnston, L.T. Aldrich, A.O. Nier, Phys. Rev. **72**, 502 (1947); J. Chem. Phys. **15**, 759 (1947)
213. C.T. Lane, H.A. Fairbank, L.T. Aldrich, A.O. Nier, Phys. Rev. **73**, 911 (1948)
214. J. Franck, Phys. Rev. **70**, 561 (1946)
215. E.A. Lynton, H.A. Fairbank, Phys. Rev. **79**, 735 (1950); **80**, 1043 (1950)
216. Z. Dokoupil, G. Van Soest, D.H.N. Wansink, D.G. Kapadnis, Physica **20**, 1181 (1954)
217. G.K. Walters, W.M. Fairbanks, Phys. Rev. **103**, 262 (1956)
218. Z. Dokoupil, D.G. Kapadnis, K. Sreeramamurty, K.W. Taconis, Physica **25**, 1369 (1959)
219. D.O. Edwards, D.F. Brewer, P. Seligman, M. Skertic, M. Yaqub, Phys. Rev. Lett. **15**, 773 (1965)
220. T.A. Alvesalo, P.M. Berglund, S.T. Islander, G.R. Pickett, W. Zimmermann Jr., Phys. Rev. A **4**, 2354 (1971)
221. J.P. Laheurte, J.R.G. Keyston, Cryogenics **11**, 485 (1971)
222. C. Ebner, D.O. Edwards, Phys. Rep. **2C**, 77 (1970)
223. R. de Bruyn Ouboter, K.W. Taconis, C. le Pair, J.J.M. Beenakker, Physica **26**, 853 (1960)
224. A.C. Anderson, D.O. Edwards, W.R. Roach, R.E. Sarwinski, J.C. Wheatley, Phys. Rev. Lett. **17**, 367 (1966)
225. T.R. Roberts, R.H. Sherman, S.G. Sydoriak, J. Res. Nat. Bur. Std. **68A**, 567 (1964)
226. A.E. Watson, J.D. Reppy, R.C. Richardson, Phys. Rev. **188**, 384 (1969)
227. J. Landau, J.T. Tough, N.R. Brubaker, D.O. Edwards, Phys. Rev. Lett. **23**, 283 (1969)
228. B.M. Abraham, O.G. Brandt, Y. Eckstein, Proc. 12th Int. Conf. Low Temp. Physics, (K. Kanda ed.), 161, (Academic Press of Japan, Tokyo 1971), p. 161
229. C. Ebner, D.O. Edwards, Phys. Rep. C **1**, 77 (1971)

230. J. Bardeen, G. Baym, D. Pines, Phys. Rev. Lett. **17**, 372 (1966); Phys. Rev. **156**, 207 (1967)
231. C. Ebner, Phys. Rev. **156**, 222 (1967)
232. L.D. Landau, I. Pomeranchuk, Dokl. Akad. Nauk. SSSR **59**, 669 (1948)
233. J.R. Pellam, Phys. Rev. **99**, 1327 (1955)
234. J. Landau, J.T. Tough, N.R. Brubaker, D.O. Edwards, Phys. Rev. A **2**, 2472 (1970)
235. I.M. Khalatnikov, *An Introduction to the Theory of Superfluidity*, (Benjamin, New York 1965)
236. B.M. Abraham, Y. Eckstein, J.B. Ketterson, M. Kuchnir, Phys. Rev. Lett. **20**, 251 (1968)
237. D.A. Rockwell, R.F. Benjamin, T.J. Greytak, J. Low Temp. Phys. **18**, 389 (1975)
238. S.A.J. Wiegers, R. Jochemsen, C.C. Kranenburg, G. Frossati, J. Low Temp. Phys. **71**, 69 (1988)
239. I. Pomeranchuk, Zh. Eksperim. i. Teor. Fiz. **34**, 33 (1949),
240. D.S. Greywall, Phys. Rev. B **20**, 2643 (1979)
241. W.R. Abel, R.T. Johnson, J.C. Wheatley, Phys. Rev. Lett. **18**, 737 (1967)
242. F.A. Staas, K.W. Taconis, K. Fokkens, Physica **26**, 669 (1960)
243. D.A. Ritchie, J. Saunders, D.F. Brewer, Phys. Rev. Lett. **59**, 737 (1987)
244. P.G. van de Haar, G. Frossati, K.S. Bedell, J. Low. Temp. Phys. **77**, 35 (1989)

Chapter 6

245. F.H. Weber, Ann. d. Phys. (2) **147**, 311 (1872)
246. U. Behn, Ann. d. Phys. (3) **66**, 237 (1898)
247. J. Dewar, Proc. Roy. Soc. A **76**, 325 (1905)
248. P.L. Dulong, A.T. Petit, Ann. Chim. Phys. **10**, 395 (1818)
249. A. Einstein, Ann. d. Phys. **22**, 180 (1906), and erratum **22**, 800 (1907)
250. A. Einstein, Ann. d. Phys. **35**, 679 (1911)
251. M. Born, T. von Kármán, Z. Physik **13**, 297 (1912)
252. P. Debye, Ann. d. Phys. (4) **39**, 789 (1912)
253. L. Finegold, N.E. Phillips, Phys. Rev. **177**, 1383 (1964)
254. G. Dolling, R.A. Cowley, Proc. Roy. Soc. London **88**, 463 (1966)
255. W.T. Berg, J.A. Morrison, Proc. Roy. Soc. A **242**, 467 (1957)
256. M. Blackman, Proc. Roy. Soc. A **22**, 365 (1935), and **159**, 416 (1937)
257. G.R. Stewart, Rev. Sci. Instrum. **54**, 1 (1983)
258. E.S.R. Gopal, *Specific Heats at Low Temperatures*, (Plenum Press, New York 1966)
259. J.T. Lewis, A. Lehoczky, C.V. Briscoe, Phys. Rev. **161**, 877 (1967)
260. G.A. Alers, in *Physical Acoustics* Vol. III - B (W.P. Mason, ed.), (Academic Press, New York 1965), p. 1
261. M. Blackman, Proc. Roy. Soc. A **181**, 58 (1942)
262. F. Lindemann, Phys. Z. **11** 609 (1910)
263. S.V. Hering, S.W. Van Sciver, O.E. Vilches, J. Low Temp. Phys. **25**, 793 (1976)
264. H.P. Baltes, E.R. Hilf, Solid State Commun. **12**, 369 (1973)
265. T.F. Nonnenmacher, Phys. Lett. **51A**, 213 (1975)

266. G. Goll, H.v. Löhneysen, Nanostructured Mater. **6**, 559 (1995)
267. O. Vergara, D. Heitkamp, H.v.Löhneysen, J. Phys. Chem. Solids **45**, 251 (1984)
268. R. Krüger, M. Meissner, J. Mimkes, A. Tausend, phys. stat. sol. (a) **17**, 471 (1973)
269. H.E. Jackson, C.T. Walker, T.F. McNelly, Phys. Rev. Lett. **25**, 26 (1970)
270. See for example: H.J. Maris in *Physical Acoustics*, Vol. VII, (W.P. Mason, R.N. Thurston, eds.), (Academic Press, New York 1971), p. 279
271. H.B.G. Casimir, Physica **5**, 495 (1938)
272. P.D. Thacher, Phys. Rev. **156**, 975 (1967)
273. V. Röhring, private communication (1992)
274. R. Berman, J.C.F. Brock, Proc. Roy. Soc. A **289**, 46 (1965)
275. P.G. Klemens, in *Solid State Physics* **7**, (D. Turnbull, F. Seitz, eds.), (Academic Press, New York 1958), p.1
276. M. Sato, K. Sumino, J. Phys. Soc. Jap. **36**, 1075 (1974)
277. A.V. Granato, K. Lücke, J. Appl. Phys. **27**, 583 (1956)
278. A. Seeger, Philos. Mag. **1**, 651 (1956)
279. W. Wasserbäch, S. Abens, S. Sahling, phys. sat. sol. (b) **222**, 425 (2000)
280. R. Berman, Proc. Phys. Soc. A **65**, 1029 (1952)
281. L.P. Mezhov-Deglin, Zh. Eksperim. i. Teor. Fiz. **49**, 66 (1965), Soviet Phys. JETP **22**, 47 (1966)
282. C.C. Ackerman, B. Bertman, H.A. Fairbank, R.A. Guyer, Phys. Rev. Lett. **16**, 789 (1966)
283. W.C. Thomlinson, Phys. Rev. Lett. **23**, 1333 (1969)
284. A. Smontara, J.C. Lasjaunias, R. Maynard, Phys. Rev. Lett. **77**, 5397 (1996)
285. E.M. Hogan, R.A. Guyer, H.A. Fairbank, Phys. Rev. **185**, 356 (1969)
286. A.A. Levchenko, L.P. Mezhov-Deglin, Sov. Phys. JETP **55**, 166 (1982)
287. C.C. Ackerman, R.A. Guyer, Ann. Phys. **50**, 128 (1968)
288. C.C. Ackerman, W.C. Overton, Phys. Rev. Lett. **22**, 764 (1969)
289. T.F. McNelly, S.J. Rogers, D.J. Channin, R.J. Rollefson, W.M. Goubau, G.E. Schmidt, J.A. Krumhansl, R.O. Pohl, Phys. Rev. Lett. **24**, 100 (1970)
290. V. Narayanamurti, R.C. Dynes, Phys. Rev. Lett. **28**, 1461 (1972)
291. T.F. McNelly, Ph.D. Thesis (Cornell Univ. 1974)
292. J.P. Maneval, A. Zylbersztejn, D. Huet, Phys. Rev. Lett. **27**, 1375 (1971)
293. V. Narayanamurti, M.A. Chin, R..A. Logan, Appl. Phys. Lett. **33**, 481 (1978)
294. J.P. Wolfe, Physics Today, **9**, 34 (1995)
295. D.C. Hurley, M.T. Ramsbey, J.P. Wolfe, in *Solid State Sciences* **68**, (A.C. Anderson, J.P. Wolfe eds.), (Springer, Heidelberg 1986), p. 299
296. W. Eisenmenger, in *Phonon Scattering in Condensed Matter*, (H.J. Maris ed.), (Plenum Press, New York 1980), p. 303
297. K. Schwab, E.A. Henriksen, J.M. Worlock, M.L. Roukes, Nature **404**, 974 (2000)
298. L.G.C. Rego, G. Kirczenow, Phys. Rev. Lett. **81**, 232 (1998)

Chapter 7

299. P. Drude, Ann. Phys. **1**, 566, and **3**, 369 (1900)
300. A. Sommerfeld, Naturwiss. **15**, 825 (1927); Z. Phys. **47**, 1, and 43 (1928)
301. J.A. Rayne, Austral. J. Phys. **9**, 189 (1956)
302. N.W. Ashcroft, N.D. Mermin, *Solid State Physics* (Holt, Rinehart and Winston, New York 1975)
303. J. Callaway, C.S. Wang, Phys. Rev. B **7**, 1096 (1973)
304. J.O. Willis, R.H. Aiken, Z. Fisk, E. Zirngiebl, J.D. Thompson, H.R. Ott, B. Batlogg, Proc. Int. Conf. Valence Fluctuations, Bangalore (1987)
305. G.R. Stewart, Z. Fisk, J.O. Willis, Phys. Rev. B **28**, 172 (1983)
306. G.E. Brodale, R.A. Fisher, N.E. Phillips, J. Flouquet, Phys. Rev. Lett. **56**, 390 (1986)
307. J.M. Ziman, *Electrons and Phonons* (Oxford Univ. Press, London 1960)
308. J.O. Linde, Ann. Phys. **15**, 219 (1932)
309. L. Nordheim, Naturw. **16**, 1042 (1928); Ann. Phys. **9**, 641 (1931)
310. C.H. Johansson, J.O. Linde, Ann. Phys. **25**, 1 (1936)
311. W. Meissner, B. Voigt, Ann. Phys. **7**, 761 and 892 (1930)
312. N.H. Anderson, H. Smith, J. Phys. **39**, C6-824 (1978)
313. K. Gloos, C. Mitschka, F. Pobell, P. Smeibidl, Cryogenics **30**, 14 (1990)
314. R. Berman, D.K.C. Mac Donald, Proc. Roy. Soc. A **211**, 122 (1952)
315. W.B. Pearson, Philos. Mag. **46**, 911 (1955) and **46**, 920 (1955)
316. J. Kondo, Prog. Theoret. Phys. (Kyoto) **42**, 37 (1966)
317. P.W. Anderson, Phys. Rev. **124**, 41 (1961)
318. H.P. Myers, L. Wallden, A. Karlsson, Philos. Mag. **18**, 725 (1968)
319. M.D. Daybell, W.A. Steyert, Rev. Mod. Phys. **40**, 380 (1968)
320. J. Friedel, Advan. Phys. **3**, 446 (1954)
321. M.A. Ruderman, C. Kittel, Phys. Rev. **96**, 99 (1954); T. Kasuya, Prog. Theoret. Phys. (Kyoto) **16**, 45 (1956); K. Yosida, Phys. Rev. **106**, 893 (1957)
322. J.P. Franck, F.D. Manchester, D.L. Martin, Proc. Roy. Soc. London A **263**, 494 (1961)
323. J. Loram, T.E. Whall, P.J. Ford, Phys. Rev. B **2**, 857 (1970)
324. J.E. Van Dam, P.C.M. Gubbens, G.J. van den Berg, Physica **62**, 389 (1972)
325. M.D. Daybell, W.A. Steyert, Phys. Rev. Lett. **18**, 398 (1967)
326. A.J. Heeger, Solid State Phys. **23**, 283 (1969)
327. N. Andrei, K. Furuya, J.H. Lowenstein, Rev. Mod. Phys. **55**, 331 (1983)
328. A.C. Hewson, *The Kondo Problem to Heavy Fermions*, (Cambridge Univ. Press, Cambridge 1993)
329. G.R. Stewart, Rev. Mod. Phys. **56**, 755, (1984)
330. P. Fulde, J. Keller, G. Zwicknagl, Solid State Phys. **41**, 1 (1988)
331. N. Grewe, F. Steglich, in *Handbook of Physics and Chemistry of Rare Earths* Vol. **14** (K.A. Geschneidner, Jr. and L. Eyring, eds.), (Elsevier Sciencies Publishers, Amsterdam 1991), p. 343
332. A. Amato, Rev. Mod. Phys. **69**, 1119, (1997)
333. H. Aoki, S. Uji, A.K. Albessard, Y. Onuki, Phys. Rev. Lett. **71**, 2110 (1993)
334. K.H. Marder, W.M. Swift, J. Phys. Chem. Solids **29**, 1759 (1968)
335. K. Andres, J.E. Graebner, H.R. Ott, Phys. Rev. Lett. **35**, 1779 (1975)
336. Z. Fisk, H.R. Ott, G. Aeppli, Jpn. J. Appl. Phys. **26**, Suppl. 26-3, 1882 (1987)
337. H.R. Ott, O. Marti, F. Hullinger, Solid State Commun. **49**, 1129 (1984)

338. P. Nozieres, A. Blandin, J. de Phys. **41**, 193 (1980)
339. K. Kadowaki, S.B. Woods, Solid State Commun. **58**, 507 (1986)
340. E.A. Schuberth, J. Schupp, R. Freese, K. Andres Phys. Rev. B **51**, 12892 (1995); L. Pollack, M.J.R. Hoch, C. Jin, E.N. Smith, J.M. Parpia, D.L. Hawthorne, D.A. Geller, D.M. Lee, R.C. Richardson, Phys. Rev. B **52**, R15707 (1995)
341. T. Pietrus, B. Bogenberger, S. Mock, M. Sieck, H.v. Löhneysen, Physica **206 & 207**, 317 (1995)
342. H.v. Löhneysen, M. Sieck, O. Stockert, M. Waffenschmidt, Physica **223 & 224**, 471 (1996)
343. H.v. Löhneysen, T. Pietrus, G. Portisch, H.G. Schlager, A. Schröder, M. Sieck, T. Trappmann, Phys. Rev. Lett. **72**, 3262 (1994)
344. B. Bogenberger, H. v. Löhneysen, Phys. Rev. Lett. **74**, 1016 (1995)
345. G.R. Stewart, Rev. Mod. Phys. **73**, 797 (2001)

Chapter 8

346. G.E. Uhlenbeck, S. Goudsmit, Naturwiss. **13**, 953 (1925)
347. P.A.M. Dirac, Proc. Roy. Soc. **117**, 610 and **118**, 35 (1928)
348. J.H. Van Vleck, *The Theory of Electric and Magnetic Susceptibilities*, (Oxford Univ. Press, London 1952), p. 243
349. W.E. Henry, Phys. Rev. **88**, 559 (1952)
350. J.M. Parpia, W.P. Kirk, P.S. Kobiela, Z. Olejniczak, J. Low Temp. Phys. **60**, 57 (1985)
351. J.W. Stout, W.B. Hadley, J. Chem. Phys. **40**, 55 (1964)
352. B. Bleaney, R.S. Trenam, Proc. Roy. Soc. (London), Ser. A **223**, 1 (1954)
353. A.H. Cooke, H. Meyer, W.P. Wolf, Proc. Roy. Soc. (London), Ser. A **237**, 404 (1956)
354. B. Bleaney, J. Appl. Phys. **34**, 1024 (1963)
355. O.V. Lounasmaa, Phys. Rev. **128**, 1136 (1962)
356. H.v. Kempen, A.R. Miedema, W.J. Huiskamp, Physica **30**, 229 (1964)
357. W. Heisenberg, Z. Phys. **49**, 619 (1928)
358. F. Bloch, Z. Phys. **61**, 206 (1930)
359. R.N. Sinclair, B.N. Brockhouse, Phys. Rev. **120**, 1638 (1960)
360. B.E. Argyle, S.H. Charap, E.W. Pugh, Phys. Rev. **132**, 2051 (1963)
361. D.T. Edmonds, R.G. Petersen, Phys. Rev. Lett. **2**, 499 (1959)
362. C.G. Windsor, R.W.H. Stevenson, Proc. Phys. Soc. **87**, 501 (1960)
363. A.S. Borovik-Romanov, Sov. Phys. JETP **9**, 539 (1959)
364. A.S. Borovik-Romanov, I.N. Kalinkina, Sov. Phys. JETP **14**, 1205 (1962)
365. E. Ising, Z. Phys. **31**, 253 (1925)
366. S.F. Edwards, P.W. Anderson, J. Phys. F, **5**, 965 (1975)
367. D. Sherrington, S. Kirkpatrik, Phys. Rev. Lett. **35**, 1792 (1975)
368. A.J. Bray, M.A. Moore, in *Heidelberg Colloquium on Glassy Dynamics*, Lecture Notes in Physics **275** (J.L. van Hemmen, I. Morgenstern eds.) (Springer, Heidelberg 1987), p. 121
369. G. Parisi, Phys. Rev. Lett. **43**, 1754 (1979)
370. K.H. Fischer, J.A. Hertz, *Spin Glasses*, (Cambridge University Press 1991)
371. *Spin Glasses and Random Fields*, (A.P. Young ed.) (World Scientific, Singapore 1997)

372. H. Maletta, J. Appl. Phys. **53**, 2185 (1982)
373. S. Nagata, P.H. Keesom, H.R. Harrison, Phys. Rev. **B19**, 1633 (1979)
374. P. Nordblad, P. Svendlindh, in *Spin Glasses and Random Fields*, (A.P. Young ed.) (World Scientific, Singapore 1998), p.1
375. P. Granberg, L. Sandlund, P. Nordblad, P. Svedlindh, L. Lundgren, Phys. Rev. B **38**, 7097 (1988)
376. F. Lefloch, J. Hammann, M. Ocio, E. Vincent, Europhys. Lett, **18**, 647 (1992)
377. M. Chapellier, M. Goldmann, V.H. Chau, A. Abragam, C.R. Acad. Sci. **268**, 1530 (1969); J. Appl. Phys. **41**, 849 (1970)
378. W.P. Halperin, C.N. Archie, F.B. Rasmussen, R.A. Buhrmann, R.C. Richardson, Phys. Rev. Lett. **32**, 927 (1974)
379. J. Babcock, J. Kiely, T. Manley, W. Weyhmann, Phys. Rev. Lett. **43**, 380 (1979)
380. J. Korringa, Physica **16**, 601 (1950)
381. T. Herrmannsdörfer, F. Pobell, Physica B **194–196**, 339 (1994); J. Low Temp. Phys. **100**, 253 (1995)
382. S. Rehmann, T. Herrmannsdörfer, F. Pobell, Phys. Rev. Lett. **78**, 1122 (1997)
383. M.T. Huiku, M.T. Loponen, Phys. Rev. Lett. **49**, 1288 (1982)
384. M.T. Huiku, T.A. Jyrkkiö, J M. Kyynäräinen, M.T. Loponen, O.V. Lounasmaa, A.S. Oja, J. Low Temp. Phys. **62**, 433 (1986)
385. M.T. Huiku, T.A. Jyrkkiö, J.M. Kyynäräinen, A.S. Oja, O.V. Lounasmaa, Phys. Rev. Lett. **53**, 1692 (1984)
386. A.J. Annila, K.N. Clausen, P.-A. Lindgård, O.V. Lounasmaa, A.S. Oja, K. Siemensmeyer, M. Steiner, J.T. Tuoriniemi, H.W. Weinfurter, Phys. Rev. Lett. **64**, 1421 (1990)
387. H.B.G. Casimir, F.K. Du Pré, Physica **5**, 507 (1938)
388. A.S. Oja, O.V. Lounasmaa, Rev. Mod. Phys. **69**, 1 (1997)
389. E.M. Purcell, R.V. Pound, Phys. Rev. **81**, 279 (1951)
390. P.J. Hakonen, S. Yin, O.V. Lounasmaa, Phys. Rev. Lett. **64**, 2707 (1990)
391. R.T. Vuorienen, Diploma Thesis, (Helsinki Univ. Technology 1992)

Chapter 9

392. F. Hund, Z. Physik **43**, 805 (1927)
393. C.E. Cleeton, H.N. Williams, Phys. Rev. **45**, 234 (1934)
394. L. Pauling, Phys. Rev. **36**, 430 (1930)
395. W. Känzig, J. Phys. Chem. Solids **32**, 479 (1962)
396. G. Wentzel, Z. Physik **38**, 518 (1926); H.A. Kramers, Z. Phys. **39**, 828 (1926); L. Brillouin, Compt. Rend. **183**, 24 (1926)
397. F. Bloch, Phys. Rev. **70**, 460 (1946)
398. R.L. Kronig, J. Opt. Soc. Am. **12**, 547 (1926); H.A. Kramers, Atti. Congr. Intern. Fis. **2**, 545 (1927)
399. M. Gomez, S.P. Bowen, J.A. Krumhansl, Phys. Rev. **153**, 1009 (1967)
400. J.P. Sethna, Phys. Rev. B **12**, 1546 (1982)
401. J.P. Harrison, P.P. Peressini, R.O. Pohl, Phys. Rev. **171**, 1037 (1968)
402. P.P. Peressini, J.P. Harrison, R.O. Pohl, Phys. Rev. **182**, 939 (1969)
403. T.F. McNelly, Ph.D. thesis (Cornell University 1974)
404. P.P. Peressini, J.P. Harrison, R.O. Pohl, Phys. Rev. **180**, 926 (1969)

405. D. Walton, Phys. Rev. Lett. **19**, 305 (1967), and *Localized Excitations in Solids*, (R.F. Wallis ed.), (Plenum Press, New York 1968), p. 395
406. F. Holuj, F. Bridges, Phys. Rev. B **27**, 5286 (1967)
407. X. Wang, F. Bridges, Phys. Rev. B **46**, 5122 (1992)
408. C. Enss, M. Gaukler, S. Hunklinger, M. Tornow, R. Weis, A. Würger, Phys. Rev. B **53**, 12094 (1996)
409. N.E. Byer, H.S. Sack, J. Phys. Chem. Solids **29**, 677 (1968)
410. G. Weiss, M. Hübner, C. Enss, Physica B **263 & 364**, 388 (1999)
411. M.E. Baur, W.R. Salzmann, Phys. Rev. **178**, 1440 (1969)
412. M.W. Klein, Phys. Rev. B **35**, 1397 (1987); Phys. Rev. B **40**, 1918 (1989); Phys. Rev. Lett **65**, 3017 (1990)
413. A.Würger, Z. Phys. B **94**, 173 (1994) and **98**, 561 (1995)
414. H. Mori, Progr. Thoer. Phys. **33**, 423 (1965) and **34**, 399 (1965)
415. R. Zwanzig, J. Chem. Phys. **33**, 1338 (1960); Physica **30**, 1109 (1964)
416. K. Kawasaki, Phys. Rev. **150**, 291 (1966)
417. W. Götze, Solid State Commun. **27**, 3728 (1978); J. Phys. C **12**, 1279 (1979); Philos. Mag. B **43**, 219 (1981)
418. A. Würger, R. Weis, M. Gaukler, C. Enss, Europhys. Lett. **33**, 533 (1996)
419. H. Wipf, K. Neumeier, Phys. Rev. Lett. **52**, 1308 (1984)
420. S.K. Watson, Phys. Rev. Lett. **75**, 1965 (1995)
421. R.C. Zeller, R.O. Pohl, Phys. Rev. B **4**, 2029 (1971)
422. W.A. Phillips, J. Low Temp. Phys. **7**, 351 (1972)
423. P.W. Anderson, B.I. Halperin, C.M. Varma, Philos. Mag. **25**, 1 (1972)
424. S. Hunklinger, Adv. Solid State Phys. **17**, 1 (1977)
425. A. Heuer, in *Tunneling Systems in Amorphous and Crystalline Solids*, (P. Esquinazi ed.), (Springer, Heidelberg 1998), p. 459
426. K. Trachenko, M.T. Dove, M. Harris, V. Heine, J. Phys.: Condens. Matter **12**, 8041 (2000)
427. J.C. Lasjaunias, A. Raver, M. Vandorpe, S. Hunklinger, Solid State Commun. **17**, 1045 (1975)
428. A. Nittke, S. Sahling, P. Esquinazi, in *Tunneling Systems in Amorphous and Crystalline Solids*, (P. Esquinazi ed.), (Springer, Heidelberg 1998), p. 9
429. M. Schwark, F. Pobell, M. Kubota, R.M. Mueller J. Low Temp. Phys. **58**, 171 (1985)
430. V.G. Karpov, M.I. Klinger, and F.N. Ignat'ev, Sov. Phys. JETP **57**, 439 (1983)
431. D.A. Parshin, Phys. Rev. B **49**, 9400 (1994)
432. M.A. Ramos, U. Buchenau in *Tunneling Systems in Amorphous and Crystalline Solids*, (P. Esquinazi ed.), (Springer, Heidelberg 1998), p. 527
433. P. Jund, R. Jullien, Phys. Rev. B **59**, 13707 (1999)
434. J. Classen, C. Enss, C. Bechinger, G. Weiss, S. Hunklinger, Ann. Phys. **3**, 315 (1994)
435. S. Ludwig, C. Enss, pivate communication
436. S. Hunklinger, W. Arnold, in *Physical Acoustics* Vol. XII, (W.P. Mason, R.N. Thurston eds.), (Academic Press, New York 1976), p. 155
437. B. Golding, J.E. Graebner, R.J. Schutz, Phys. Rev. B **14**, 1660 (1976)
438. S. Hunklinger, H. Sussner, K. Dransfeld, Adv. Solid State Phys. **16**, 267 (1976)
439. M.v. Schickfus, S. Hunklinger, Phys. Lett. **64A**, 144 (1977)
440. J. Classen, T. Burkert, C. Enss, S. Hunklinger, Phys. Rev. Lett. **84**, 2176 (2000)

441. C. Enss, C. Bechinger, M.v. Schichfus, *Phonons 89*, (S. Hunklinger, W. Ludwig, G. Weiss eds.), (World Scientific, Singapore 1990), p. 474
442. D.A. Parshin, Z. Phys. B **91**, 367 (1993)
443. J.T. Stockburger, M. Grifoni, M. Sassetti, Phys. Rev. B **51**, 2835 (1995)
444. D. Salvino, S. Rogge, B. Tigner, D.D. Osheroff, Phys. Rev. Lett. **73**, 268 (1994)
445. S. Rogge, D. Natelson, D.D. Osheroff, Phys. Rev. Lett. **76**, 3136 (1996)
446. D. Natelson, D. Rosenberg, D.D. Osheroff, Phys. Rev. Lett. **80**, 4689 (1998)
447. A.L. Burin, J. Low Temp. Phys. **100**, 309 (1995)
448. P. Doussineau, A. Levelut, G. Bellessa, O. Béthoux, J. Phys. Lett. **38**, L-483 (1977)
449. A.K. Raychaudhuri, S. Hunklinger, Z. Phys. B **57**, 113 (1984)
450. R.P. Feynman, F.L. Veron, R.W. Hellwarth, J. Appl. Phys. **28**, 49 (1957)
451. C. Enss, H. Schwoerer, D. Arndt, M. v. Schickfus, Phys. Rev. B **51**, 811 (1995)
452. A. Würger, A. Fleischmann, C. Enss, Phys. Rev. Lett. **89**, 237601 (2002)
453. P. Nagel, A. Fleischmann, S. Hunklinger, C. Enss, Phys. Rev. Lett. **92**, 245511 (2004)

Chapter 10

454. H.K. Onnes, Commun. Leiden, **120b**, (1911)
455. C.J. Gorter, H.B.J. Casimir, Phys. Z. **35**, 963 (1934)
456. F. London, H. London, Z. Phys. **96**, 359 (1935)
457. V.L. Ginzburg, L.D. Landau, Zh. Eksperim. i. Teor. Fiz. **20**, 1064 (1950)
458. J. Bardeen, L.N. Cooper, J.R. Schrieffer, Phys. Rev. **108**, 1175 (1957)
459. J.G. Bednorz, K.A. Müller, Z. Phys. B **64**, 189 (1986)
460. D.J. Quinn, W.B. Ittner, J. Appl. Phys. **33**, 748 (1962)
461. B.T. Matthias, Phys. Rev. **97**, 74 (1955); *Prog. Low Temp. Phys.* Vol. II, (C.J. Gorter ed.), (North Holland, Amsterdam 1957), p. 138
462. J.K. Hulm, R.D. Blaugher, Phys. Rev. **123**, 1569 (1961); V.B. Compton, E. Corenzwit, E. Maita, B.T. Matthias, F.J. Morin, Phys. Rev. **123**, 1567 (1961); W. Gey, Z. Phys. **229**, 85 (1969)
463. W. Meißner, R. Ochsenfeld, Naturwissenschaften **21**, 787 (1933)
464. F. Haenssler, L. Rinderer, Helv. Phys. Acta **40**, 659 (1967)
465. G. Otto, E. Saur, H. Witzgall, J. Low Temp. Phys. **1**, 19 (1969); W. DeSorbo, Phys. Rev. **104**, A914 (1965); R. Odermatt, Ø. Fischer, Proc. 14th Int. Conf. Low Temp. Physics, Vol.5, (North-Holland, Amsterdam 1975), p. 172
466. T. Kinsel, E.A. Lynton, B. Serin, Rev. Mod. Phys. **36**, 105 (1964)
467. U. Eßmann, H. Träuble, Phys. Lett. **24 A**, 526 (1967)
468. H.F. Hess, R.B. Robinson, R.C. Dynes, J.M. Valles Jr., J.V. Waszczak, Phys. Rev. Lett. **62**, 214 (1989)
469. A.B. Pippard, Proc. Roy. Soc. A **216**, 547 (1953)
470. A.A. Abrikosov, Sov. Phys. JETP **5**, 1174 (1957)
471. L.P. Gorkov, Sov. Phys. JETP **9**, 1364 (1960)
472. N.E. Phillips, Phys. Rev. **114**, 676 (1959)
473. D.E. Mapother, IBM Journal **6**, 77 (1962)
474. D. Shoenberg, *Superconductivity*, (Cambridge Univ. Press, Cambridge 1952)
475. R.F. Gasparovic, W.L. McLean, Phys. Rev. B **2**, 2519 (1970)
476. F.B. Silsbee, Washington Academy of Sciences Journal **6**, 597 (1916)

477. H.F. Hess, R.B. Robinson, J.V. Waszczak, Phys. Rev. Lett. **64**, 2711 (1990)
478. A.R. Strnad, C.F. Hempstead, Y.B. Kim, Phys. Rev. Lett. **13**, 794 (1964)
479. L.N. Cooper, Phys. Rev. **104**, 1189 (1956)
480. E. Maxwell, Phys. Rev. **86**, 235 (1952), J.E. Lock, A.B. Pippard, D. Shoenberg, Proc. Cambridge Philos. Soc. **47**, 811 (1951), B. Serin, C.A. Reynolds, C. Lohmann, Phys. Rev. **86**, 162 (1952)
481. H. Fröhlich, Proc. Roy. Soc. A **63**, 778 (1950) and Phys. Rev. **79**, 845 (1950)
482. J. Bardeen, Phys. Rev. **80**, 567 (1950)
483. I. Giaever, H. Hart, K. Megerle, Phys. Rev. **126**, 941 (1962)
484. G.M. Eliashberg, Sov. Phys. JETP **11**, 696 (1960)
485. R. Mersevey, B.B. Schwartz, in *Superconductivity*, (R.D. Parks ed.) (Dekker, New York 1969), p.141
486. I. Giaever, K. Megerle, Phys. Rev. **122**, 1101 (1961)
487. D.M. Ginsberg, M. Tinkham, Phys. Rev. **118**, 990 (1960)
488. W.H. Keesom, P.H. van Laer, Physica **5**, 193 (1938)
489. M.A. Biondi, A.T. Forester, M.P. Garfunkel, C.B. Satterthwaite, Rev. Mod. Phys. **30**, 1109 (1958)
490. R. David, N.J. Ponlis, Proc. 8th Inter. Conf. Low Temp. Phys. (1962), p. 193
491. J.H.P. Watson, G.M. Graham, Can. J. Phys. **41**, 1738 (1963)
492. M.H. Cohen, L.M. Falicov, J.C. Phillips, Phys. Rev. Lett. **8**, 316 (1962)
493. I. Dietrich, Proc. 8th Int. Conf. Low Temp. Phys. (1962), p. 173
494. W. Eisenmenger, in *Physical Acoustics*, Vol. XII, (R.N. Thurston, W.P. Mason eds.), (Academic Press, New York 1976), p. 79
495. R. Doll, M. Nähbauer, Phys. Rev. Lett. **7**, 51 (1961)
496. B.S. Deaver Jr., W.M. Fairbank, Phys. Rev. Lett. **7**, 43 (1961)
497. W.L. Goodman, W.D. Willis, D.A. Vincent, B.S. Deaver, Phys. Rev. B **4**, 1530 (1971)
498. D.N. Langenberg, D.J. Scalapino, B.N. Taylor, Proc. IEEE **54**, 560 (1966)
499. J.R. Waldram, *Superconductivity of Metals and Cuprates*, (IOP Publishing, London 1996)
500. C.C. Grimes, S. Shapiro, Phys. Rev. **169**, 397 (1968)
501. R.C. Jaklevic, J. Lambe, J.E. Mercereau, A.H. Silver, Phys. Rev. **140**, A 1628 (1965)
502. A.H. Silver, J.E. Zimmermann, Phys. Rev. **157**, 317 (1967)
503. *The Physics of Superconductors*, Vol. I, (K.H. Bennemann, J.B. Ketterson eds.) (Springer, Heidelberg 2004)
504. *The Physics of Superconductors*, Vol. II, (K.H. Bennemann, J.B. Ketterson eds.) (Springer, Heidelberg 2004)
505. W.A. Little, Phys. Rev. A **134**, 1416 (1964)
506. D. Jérome, A. Mazand, M. Ribault, K. Bechgaard, J. Phys. Lett. **41**, L95 (1980)
507. J. Friedel, D. Jérome, Contemp. Phys. **23**, 583 (1982)
508. G. Saito, H. Yamochi, T. Nakamura, T. Komatsu, M. Nakashima, H. Mori, K. Oshima, Physica B **169**, 372 (1991)
509. T. Yildirim, L. Barbedette, J.E. Fischer, C.L. Lin, J. Robert, P. Petit, T.T.M. Palstra, Phys. Rev. Lett. **77**, 167 (1996)
510. C.F. Majkrzak, G. Shirane, W. Thomlinson, M. Ishikawa, Ø. Fischer, D.E. Moncton, Solid State Commun. **31**, 773 (1979)
511. M. Ishikawa, Ø. Fischer, Solid State Commun. **24**, 747 (1977)

512. E. Müller-Hartmann, J. Zittartz, Phys. Rev. Lett. **26**, 428 (1970)
513. K. Winzer, Z. Phys. **265**, 139 (1973)
514. H. Schmidt, H.F. Braun, in *Studies of High Temp. Supercond.* **26**, (A.V. Narlikar ed.), (Nova Science Publishers, New York 1998), p. 47
515. T. Park, M.B. Salamon, E.M. Choi, H.J. Kim, S. Lee, Phys. Rev. Lett. **90**, 177001 (2003)
516. K-H. Müller, V.N. Narozhnyi, Rep. Prog. Phys. **64**, 943 (2001)
517. W.A. Fertig, D.C. Johnston, L.E. DeLong, R.W. McCallum, M.B. Marple, B.T. Matthias, Phys. Rev. Lett. **38**, 987 (1977)
518. S.K. Sinha, G.W. Crabtree, D.G. Hinks, H.A. Mook, Phys. Rev. Lett. **48**, 950 (1982)
519. M.B. Maple, Physica B **215**, 110 (1995)
520. N.D. Mathur, F.M. Grosche, S.R. Julian, I.R. Walker, D.M. Freye, R.K.W. Haselwimmer, G.G. Lonzarich, Nature **394**, 39 (1998)
521. D. Aoki, A. Huxley, E. Ressouche, D. Braithwaite, J. Flouquet, J-P. Brison, E. Lhotel, C. Paulsen, Nature **413**, 613 (2001)
522. C. Pfleiderer, M. Uhlarz, S.M. Hayden, R. Vollmer, H.v. Löhneysen, N.R. Bernhoeft, G.G. Lonzarich, Nature **412**, 58 (2001)
523. S.S. Saxena, P. Agarwal, K. Ahilan, F.M. Grosche, R.K.W. Haselwimmer, M.J. Steiner, E. Pugh, I.R. Walker, S.R. Julian, P. Monthoux, G.G. Lonzarich, A. Huxley, I. Sheikin, D. Braithwaite, J. Flouquet, Nature **406**, 587 (2000)
524. C. Pfleiderer, A.D. Huxley, Phys. Rev. Lett. **89**, 147005 (2002)
525. F. Steglich, J. Aarts, C.D. Bredl, W. Lieke, D. Meschede, W. Franz, H. Schäfer, Phys. Rev. Lett. **43**, 1892 (1979)
526. K. Hasselbach, L. Taillefer, J. Flouquet, Phys. Rev. Lett. **63**, 93 (1989)
527. S. Adenwalla, S.W. Lin, Q.Z. Ran, Z.Zhao, J.B. Ketterson, J.A. Sauls, L. Taillefer, D.G. Hinks, M. Levy, B.K. Sarma, Phys. Rev. Lett. **65**, 2298 (1990)
528. M.J. Graf, S.-K. Yip, J.A. Sauls, Phys. Rev. B **62**, 14393 (2000)
529. R.C Dynes, V. Narayanamurti, J. Garno, Phys. Rev. Lett. **41**, 1509 (1978)
530. M. Jourdan, M. Huth, H. Adrian, Nature **398**, 47 (1999)
531. N.K. Sato, N. Aso, K. Miyake, R. Shiina, P. Thalmeier, G. Varelogiannis, C. Gibel, F. Steglich, P. Fulde, T. Komatsubara, Nature **410**, 340 (2001)
532. J. Rossat-Mignod, L.P. Regnault, C. Vettier, P. Bourges, P. Burlet, J. Bossy, J.Y. Henry, G. Lapertot, Physica B, **169**, 58 (1991)
533. T.A. Friedmann, M.V. Rabin, J. Giapintzakis, J.P. Rice, D.M. Ginzberg, Phys. Rev. **42**, 6217 (1990)
534. H.R. Ott, in [503]
535. D.N. Zheng, A.M. Campell, J.D. Johnson, J.R. Cooper, F.J. Blunt, A. Porch, P.A. Freeman, Phys. Rev. B **49**, 1417 (1994)
536. A. Schilling, O. Jeandupeux, Phys. Rev. B **52**, 9714 (1995)
537. T. Schneider, H. Keller, Physica C **207**, 336 (1993)
538. L. Ozyuzer, Z. Yusof, J.F. Zasadzinski, T-W. Li, D.G. Hinks, G.E. Gray, Physica C **320**, 9 (1999)
539. H. Ding, M.R. Norman, J.C. Campuzano, M. Randeria, A.F. Bellman, T. Yokoya, T. Takahashi, T. Mochiku, K. Kadowaki, Phys. Rev. B **54**, R9678 (1996)
540. C.E. Gough, M.S. Colclough, E.M. Forgan, R.G. Jordan, M. Keene, C.M. Muirhead, A.I.M. Rae, N. Thomas, J.S. Abel, S. Sutton, Nature **326**, 855 (1987)

541. D.A. Wollman, D.J. Van Harlingen, W.C. Lee, D.M. Ginsberg, A.J. Leggett, Phys. Rev. Lett. **71**, 2134 (1993)
542. D.J. Van Harlingen, Rev. Mod. Phys. **67**, 515 (1995)
543. D.A. Wollman, D.J. Van Harlingen, J. Giapintzakis, D.M. Ginsberg, Phys. Rev. Lett. **74**, 797 (1995)
544. C.C. Tsuei, J.R. Kirtley, C.C. Chi, Lock See Yu-Jahnes, A. Gupta, T. Shaw, J.Z. Sun, M.B. Ketchen, Phys. Rev. Lett. **73**, 593 (1994)
545. T. Nakano, M. Oda, C. Manabe, N. Miura, M. Ido, Phys. Rev. B,**49**, 16000 (1994)

Chapter 11

546. L. Cailletet, C.R. Acad. Sci. Paris **85**, 1213 (1877)
547. R. Pictet, C.R. Acad. Sci. Paris **85**, 1214 and 1220 (1877)
548. K. Olszewski, Ann. Phys. Chem. **31**, 58 (1887); Philos. Mag. **39**, 188 (1895)
549. J. Dewar, Proc. R. Inst. Gt. Br. **15**, 815 (1898)
550. H. Kamerlingh Onnes, Leiden Commun. **105**, Proc. Roy. Acad. Sci. Amsterdam **11**, 168 (1908)
551. H. Kamerlingh Onnes, Leiden Commun. **159**; Trans. Faraday Soc. **18** (1922)
552. P. Debye, Ann. Phys. **81**, 1154 (1926)
553. W.F. Giauque, J. Am. Chem. Soc. **49**, 1864 (1927)
554. *Handbook of Cryogenic Engineering*, (J.G. Weisend II ed.) (Taylor & Francis, London 1998)
555. by courtesy of Linde AG
556. F.E. Hoare, L.C. Jackson, N. Kurti, *Experimental Cryophysics* (Butterworths, London 1961)
557. J.P. Joule, W. Thomson (later Lord Kelvin), Philos. Mag. **4**, 481 (1852)
558. J.P. Joule, Sci. Pap. **2**, 216 (1852)
559. A. Kent, *Experimental Low-Temperature Physics* (AIP, New York 1993)
560. C.v. Linde, german patent 88824, (1895); Z. ges. Kälteind. **4**, 23 (1897)
561. W. Hampson, english patent 10165, (1895)
562. G. Claude, C.R. Acad. Sci. Paris **134**, 1568 (1902)
563. Y. Koike, Y. Morii, T. Igarashi, M. Kubota, Y. Hiresaki, K. Tanida, Cryogenics **39**, 579 (1999)
564. K. Uhlig, Cryogenics **42**, 73 (2002)
565. W.E. Gifford, H.O. McMahon, Adv. Cryog. Eng. **5**, 354 (1960) and 368 (1960)
566. M. Britcliffe, T. Hanson, J. Fernandez, IPN Progress Report **42–147**, 1 (2001)
567. W.E. Gifford, R.C. Longsworth, Adv. Cryo. Eng. **3b**, 69 (1963)
568. E.I. Mikulin, A.A. Tarasov, M.P. Shkrebyonock, Adv. Cryo. Eng. **31**, 629 (1984)
569. S. Zhu, P. Wu, Z. Chen, Cryogenics **30**, 257 (1990)
570. A.T.A.M. de Waele, Physica B **280**, 479 (2000)
571. N. Jiang, U. Lindemann, F. Giebeler, G. Thummes, Cryogenics **44**, 809 (2004)
572. J. Dewar, Proc. R. Inst. Gt. Br. **14**, 1 (1893) and **15**, 815 (1898)
573. F. Pobell, *Matter and Methods at Low Temperatures*, (Springer, Heidelberg 1996)
574. Landolt-Börnstein, *Zahlenwerte und Funktionen aus Physik, Chemie, Astronomie, Geophysik*, Vol. IV, (Springer, Heidelberg 1967)

575. T.R. Roberts, S.G. Sydoriak, Phys. Rev. **98**, 1672 (1955)
576. D.S. Betts, *Refrigeration and Thermometry Below One Kelvin*, (Sussex Univ. Press, Brighton 1976)
577. W. Wiedemann, E. Smolic, Proc. 2nd International Cryogenics Eng. Conf., Brighton, (1968), p. 559
578. G. Batey, V. Mikheev, J. Low Temp. Phys. **113**, 933 (1998)
579. H. London, Proc. 2nd Int. Conf. on Low Temp. Phys., (Oxford Univ. Press, London 1951), p. 157
580. H. London, G.R. Clarke, E. Mendoza, Phys. Rev. **128**, 1992 (1962)
581. P. Das, R. De Bruyn Ouboter, K.W. Taconis, Proc. 9th Int. Conf. on Low Temp. Phys., (Plenum Press, London 1965), p. 1253
582. D.O. Edwards, E.M. Ifft, R.E. Sarwinski, Phys. Rev. **177**, 380 (1969)
583. by courtesy of Oxford Instruments
584. D.J. Cousins, S.N. Fisher, A.M. Guénault, R.P. Haley, I.E. Miller, G.R. Pickett, G.N. Plenderleith, P. Skyba, P.Y.A. Thibault, M.G. Ward, J. Low Temp. Phys. **114**, 547 (1999)
585. P.L. Kapitza, Zh. Eksperim. i. Teor. Fiz. **11**, 1 (1941), english translation: J. Phys. USSR **4**, 181 (1941)
586. I.M. Khalatnikov, Zh. Eksperim. i. Teor. Fiz. **22**, 687 (1952)
587. G. Frossati, J. Phys. **39** (C6), 1578 (1978), J. Low Temp. Phys. **87**, 595 (1992)
588. O.V. Lounasmaa, *Experimental Principles and Methods Below 1 K*, (Academic Press London 1974)
589. I. Pomeranchuk, Zh. Eksperim. i. Teor. Fiz. **20**, 919 (1950)
590. Y.D. Anufrier Sov. Phys. JETP Lett. **1**, 155 (1965)
591. W.F. Giauque, D.P. Mac Dougall, Phys. Rev. **43**, 768 (1933)
592. W.J. De Haas, E.C. Wiersma, H.A. Kramers, Physica **1**, 1 (1933)
593. A.H. Cooke, Proc. Phys. Soc. London, Sect. A **62**, 269 (1949)
594. N. Kurti, F.E. Simon, Proc. Roy. Soc. Ser. A **149**, 152 (1935)
595. O.E. Vilches, J.C. Wheatley, Rev. Sci. Instrum. **37**, 819 (1966)
596. O.E. Vilches, J.C. Wheatley, Phys. Rev. **148**, 509 (1966)
597. C.J. Gorter, Z. Phys. **35**, 928 (1934)
598. N. Kurti, F.N. Robinson, F. Simon, D.A. Spohr, Nature **178**, 450 (1956)
599. K. Gloos, P. Smeibidl, C. Kennedy, A. Singsaas, P. Sekowski, R.M. Mueller, F. Pobell, J. Low Temp. Phys. **73**, 101 (1988)
600. F. Hund, Z. Phys. **42**, 93 (1927)
601. K. Motizuki, T. Nagamiya, J. Phys. Soc. Jpn. **12**, 163 (1957)
602. M. Schwark, F. Pobell, W.P. Halperin, C. Buchal, J. Hanssen, M. Kubota, R.M. Mueller, J. Low Temp. Phys. **53**, 685 (1983)
603. M. Schwark, M. Kubota, R.M. Mueller, F. Pobell, J. Low Temp. Phys. **58**, 171 (1985)
604. W. Wendler, T. Herrmannsdörfer, S. Rehmann, F. Pobell, Euro. Phys. Lett. **38**, 619 (1997)
605. R.M. Mueller, C. Buchal, T. Oversluizen, F. Pobell, Rev. Sci. Instrum. **49**, 515 (1978)
606. R.M. Mueller, C. Buchal, H.R. Folle, M. Kubota, F. Pobell, Cryogenics **20**, 395 (1980)
607. F. Pobell, Physica B, C **109 & 110**, 1485 (1982)

Chapter 12

608. W. Thomson (later Lord Kelvin), Philos. Mag. **33**, 313 (1848)
609. J.P. Joule, W. Thomson (later Lord Kelvin), Philos. Trans. **144**, 321 (1854)
610. A. Celsius, Kgl. Svenska Vetenskaps akad. Handl. **4**, 197 (1742)
611. R.L. Rusby, M. Durieux, A.L. Reesink, R.P. Hudson, G. Schuster, M. Kühne, W.E. Fogle, R.J. Soulen, E.D. Adams, J. Low Temp. Phys. **126**, 633 (2002).
612. G.K. White, P.J. Meeson, *Experimental Techniques in Low-Temperature Physics*, (Oxford Univ. Press, Oxford 2002)
613. H. Neumann, K. Stecker, *Temperaturmessung* (Akademie-Verlag, Berlin 1983)
614. O. Reynolds, Philos. Trans. R. Soc. London, B**170**, 727 (1880)
615. D.S. Greywall, P.A. Busch, Rev. Sci. Instrum. **51**, 509 (1980)
616. V. Steinberg, G. Ahlers, J. Low. Temp. Phys. **53**, 255 (1983)
617. H. Preston-Thomas, Metrologia **27**, 3 (1990)
618. B.W. Magnum, J. Res. Nat. Inst. Stand. Technol. **95**, 69 (1990)
619. B.W. Magnum, G.T. Furukawa, Nat. Inst. Stand. Technol., Technical Note 1265 (1990)
620. R.P. Hudson, H. Marshak, R.J. Soulen Jr., D.B. Utton, J. Low Temp. Phys. **20**, 1 (1975)
621. D.S. Greywall, Phys. Rev. **33**, 7520 (1986)
622. G. Schuster, D. Hechtfischer, B. Fellmuth, Rep. Prog. Phys. **57**, 187 (1994)
623. J.B. Johnson, Nature **119**, 50 (1927); Phys. Rev. **29**, 367 (1927) and **32**, 97 (1928)
624. H. Nyquist, Phys. Rev. **29** 614 (1927) and **32**, 110 (1928)
625. S. Menkel, D. Drung, Y.S. Greensberg, T. Schurig, J. Low Temp. Phys. **120**, 382 (2000)
626. C.P. Lusher, J. Li, V.A. Maidanov, M.E. Digby, H. Dyball, A. Casey, J. Ny'e ki, V.V. Dmitriev, B.P. Cowan, J. Saunders, Meas. Sci. Technol. **12**, 1 (2001)
627. A. Netsch, E. Hassinger, A. Fleischmann, C. Enss, to be published
628. R.J. Soulen Jr., R.B. Dove, Standard Reference Materials, SRM 768, Temperature Reference Standard for use Below 0.5 K, National Bureau of Standards, US Department of Commerce, Special Publication, (1979), p. 260
629. J.H. Colwell, W.E. Fogle, R.J. Soulen Jr., Proc. 17th International Conf. on Low Temp. Phys. (E. Eckern, A. Schmid, W. Weber, H. Wühl eds.) (North-Holland, Amsterdam 1984), p. 395
630. J.F. Schooley, R.J. Soulen Jr., *Temperature, its Measurement and Control in Science and Industry* **6**, (J.F. Schooley ed.), (AIP, New York 1992), p. 251
631. A.J. Storm, W.A. Bosch, M.J. deGroot, R. Jochemsen, F. Mathu, G.J. Nieuwenhuys, Physica B **284–288**, 2008 (2000)
632. W.A. Bosch, J. Flokstra, G.E. de Groot, M.J. de Groot, R. Jochemsen, F. Mathu, P. Peruzzi, D. Veldhuis, in *Temperature: Its Measurement and Control in Sience and Industry* **7** (D.C. Ripple ed.), (AIP, New York 2003), p. 155
633. S. Schöttl, R. Rusby, H. Godfrin, M. Meschke, V. Goudon, S. Trqueneaux, A. Peruzzi, M.J. de Groot, R. Jochemsen, W.A. Bosch, Y. Hermier, L. Pitre, C. Rives, B. Fellmuth, J. Engert, Proc. Quantum Fluids Solids 2004
634. H. Marshak, J. Res. NBS **88**, 175 (1983)
635. H. Marshak, *Low-Temperature Nuclear Orientation*, (N.J. Stone, H. Postma, eds.), (North-Holland, Amsterdam 1986), p. 769

636. J.P. Pekola, K.P. Hirvi, J.P. Kauppinen, M.A. Paalanen, Phys. Rev. Lett. **73**, 2903 (1994)
637. J.P. Kauppinen, K.T. Loberg, A.J. Manninen, J.P. Pekola, R.A. Voutilaninen, Rev. Sci. Instrum. **69**, 4166 (1998)
638. *Single Charge Tunneling, Coulomb Blockade Phenomena in Nanostructures*, (H. Grabert, M.H. Devoret eds.), (Plenum Press, New York, 1992)
639. K.P. Hirvi, J.P. Kauppinen, A.N. Korotkov, M.A. Paalanen, J.P. Pekola, Appl. Phys. Lett. **67**, 2096 (1995)
640. J.P. Pekola, J.J. Toppari, J.P. Kauppinen, K.M. Kinnunen, A.J. Manninen, A.G.M. Jansen, J. Appl. Phys. **83**, 5582 (1998)
641. N.F. Mott, Philos. Mag. **19**, 835 (1969)
642. W. Schoepe, Z. Phys. B **71**, 455 (1988)
643. B.I. Shklovskii, A.L. Efros, *Electronic Properties of Doped Semiconductors*, (Springer, Heidelberg 1984)
644. H. Fritsche, M. Cuevas, Phys. Rev. **119**, 1238 (1960)
645. E.E. Haller, Infrared Phys. **25**, 263 (1985)
646. E.E. Haller, J. Appl. Phys. **77**, 2857 (1995)
647. W.C. Black, W.R. Roach, J.C. Wheatley, Rev. Sci. Instrum. **35**, 587 (1964)
648. K. Neumeier, G. Eska, Cryogenics **23**, 84 (1983)
649. R.W. Willekers, F. Mathu, H.C. Meijer, H. Postma, Cryogenics **30**, 351 (1990)
650. H. Armbrüster, W.P. Kirk, Physica B **107**, 335 (1981)
651. D.J. Bradley, A.M. Guénault, V. Keith, G.R. Pickett, W.P. Pratt Jr., J. Low Temp. Phys. **45**, 357 (1981)
652. P. Strehlow, M. Wohlfahrt, A.G.M. Jansen, R. Haueisen, G. Weiss, C. Enss, S. Hunklinger, Phys. Rev. Lett. **84**, 1938 (2000)
653. S.A.J. Wiegers, R. Jochemsen, C.C. Kranenburg, G. Frossati, Rev. Sci. Instrum. **57**, 1413 (1986)
654. T.C.P. Chui, D.R. Swanson, M.J. Adriaans, J.A. Nissen, J.A. Lipa, Phys. Rev. Lett. **69**, 3005 (1992)
655. M. Jutzler, B. Schröder, K. Gloos, F. Pobell, Z. Phys. B **64**, 115 (1986)
656. R. König, T. Herrmannsdörfer, C. Enss, private communication
657. R.A. Buhrmann, W.P. Halperin, S.W. Schwenterly, J. Reppy, R.C. Richardson, W.W. Webb, Proc. 12th International Conf. Low Temp. Phys., (E. Kanda ed.) (Academic Press Japan, Tokyo 1971), p. 831
658. D.J. Meredith, G.R. Pickett, O.G. Symko, J. Low Temp. Phys. **13**, 607 (1973)
659. A.I. Ahonen, M. Krusius, M.A. Paalanen, J. Low Temp. Phys. **25**, 421 (1976)
660. A.I. Ahonen, P.M. Berglund, M.T. Haikala, M. Krusius, O.V. Lounasmaa, M.A. Paalanen, Cryogenics **16**, 521 (1976)

Index